Wetlands

Fifth Edition

William J. Mitsch

James G. Gosselink

WILEY

Dedication

This fifth edition of Mitsch and Gosselink is dedicated to my long-time coauthor and friend, Professor James G. Gosselink (1931–2015). He was a gentleman and a scholar whom I will greatly miss.

—WJM

Cover image: Bernard Master
Cover design: C. Wallace

This book is printed on acid-free paper. ∞

Copyright © 2015 by John Wiley & Sons, Inc. All rights reserved.

Published by John Wiley & Sons, Inc., Hoboken, New Jersey.
Published simultaneously in Canada.

Acknowledgement: Figure 1.6 features four images that are ™ and © DC Comics

No part of this publication may be reproduced, stored in a retrieval system, or transmitted in any form or by any means, electronic, mechanical, photocopying, recording, scanning, or otherwise, except as permitted under Section 107 or 108 of the 1976 United States Copyright Act, without either the prior written permission of the Publisher, or authorization through payment of the appropriate per-copy fee to the Copyright Clearance Center, 222 Rosewood Drive, Danvers, MA 01923, (978) 750-8400, fax (978) 646-8600, or on the web at www.copyright.com. Requests to the Publisher for permission should be addressed to the Permissions Department, John Wiley & Sons, Inc., 111 River Street, Hoboken, NJ 07030, (201) 748-6011, fax (201) 748-6008, or online at www.wiley.com/go/permissions.

Limit of Liability/Disclaimer of Warranty: While the publisher and author have used their best efforts in preparing this book, they make no representations or warranties with the respect to the accuracy or completeness of the contents of this book and specifically disclaim any implied warranties of merchantability or fitness for a particular purpose. No warranty may be created or extended by sales representatives or written sales materials. The advice and strategies contained herein may not be suitable for your situation. You should consult with a professional where appropriate. Neither the publisher nor the author shall be liable for damages arising herefrom.

For general information about our other products and services, please contact our Customer Care Department within the United States at (800) 762-2974, outside the United States at (317) 572-3993 or fax (317) 572-4002.

Wiley publishes in a variety of print and electronic formats and by print-on-demand. Some material included with standard print versions of this book may not be included in e-books or in print-on-demand. If this book refers to media such as a CD or DVD that is not included in the version you purchased, you may download this material at http://booksupport.wiley.com. For more information about Wiley products, visit www.wiley.com.

Library of Congress Cataloging-in-Publication Data:

Mitsch, William J.
　Wetlands / William J. Mitsch, James G. Gosselink. — Fifth edition.
　　　pages cm
　Includes bibliographical references and index.
　　ISBN 978-1-118-67682-0 (cloth); 978-1-119-01978-7 (ebk.); 978-1-119-01979-4 (ebk.);
　1. Wetland ecology—United States. 2. Wetlands—United States. 3. Wetland management–United States. I. Gosselink, James G. II. Title.
　QH104.M57 2015
　577.68—dc23

2014040786

978-1-118-67682-0

Printed in the United States of America

10 9 8 7 6 5 4 3 2 1

Contents

Preface v

Part I — Introduction

1. Wetlands: Human Use and Science 3
2. Wetland Definitions 27
3. Wetlands of the World 45

Part II — The Wetland Environment

4. Wetland Hydrology 111
5. Wetland Soils 161
6. Wetland Biogeochemistry 179
7. Wetland Vegetation and Succession 215

Part III — Wetland Ecosystems

8. Tidal Marshes 259
9. Mangrove Swamps 311
10. Freshwater Marshes 341
11. Freshwater Swamps and Riparian Ecosystems 373
12. Peatlands 413

Part IV — Traditional Wetland Management

13 Wetland Classification   455
14 Human Impacts and Management of Wetlands   477
15 Wetland Laws and Protection   503

Part V — Ecosystem Services

16 Wetland Ecosystem Services   527
17 Wetlands and Climate Change   563
18 Wetland Creation and Restoration   591
19 Wetlands and Water Quality   647
    Appendix A. Wetland Losses by State in the United States, 1780s–1980s   701
    Appendix B. Useful Wetland Web Pages   703
    Appendix C. Useful Conversion Factors   705
    Glossary   708
    Index   721

Preface

This is the fifth edition of *Wetlands*—we updated the book every seven years from 1993 to 2007—since Van Nostrand Reinhold published the first edition in 1986. This fifth edition (referred to here as *Wetlands 5*) is eight years after the *Wetlands 4* but the additional one-year wait is well worth it, especially because so much new has happened the last year in the world of wetlands.

Because of requests by many instructors using this textbook, we reincorporated updated versions of our "ecosystem chapters" that were popular parts of the first three editions of *Wetlands*. Theses ecosystem chapters—now in Part III: Wetland Ecosystems (chapters 8 through 12)—bring back the ecosystem view of tidal marshes, mangroves, freshwater marshes and swamps, and northern peatlands. We had spit the 2000 edition of *Wetlands* into essentially two books—*Wetlands 4* (2007) and *Wetland Ecosystems* (2009), partially because students were asking for a shorter textbook. Most if not all of the pertinent information in those two books, all updated, is now included in one book. Yet *Wetlands 5* is 744 pages long, 20 percent shorter than the 920-page *Wetlands 3*. Instructors now have the choice of including or not including these ecosystem chapters, which were always among our favorites because of their "systems" view, in their syllabi. The chapters, by definition, integrate the otherwise separate fields of hydrology, biogeochemistry, microbiology, vegetation, consumers, and ecosystem function for the main types of wetlands found in the world in single chapters.

There is much new in *Wetlands 5* in addition to the five reinserted and updated ecosystem chapters in Part III. We provide a newly published trend of wetland publications in the world, a summary and list of publications from the every-four-year INTECOL international wetland conferences and the addition of *MegaPython vs. Gatoroid* campy science fiction wetland movie playbill to replace the long-reigning *Swamp Thing* movie playbill in Part I: Introduction. Updates of many of the great wetlands of the world are also provided in this section of the book, including new photos and descriptions of several wetlands in China.

Part II: The Wetland Environment (chapters 4 through 7), is significantly different from previous editions. There are now separate new chapters, "Wetland Soils" (Chapter 5) and "Wetland Vegetation and Succession" (Chapter 7), to complement the updated "Wetland Hydrology" (Chapter 4) and "Wetland Biogeochemistry" (Chapter 6) chapters. This fits better with wetland science as it is now practiced but also fits better the way in which we manage wetlands. The book is now more compatible with hydrology, soils, and vegetation, the three-legged stool of wetland definitions in many countries including the United States.

The management section of the book is now divided into two parts: Part IV: Traditional Wetland Management (chapters 13–15) and Part V: Ecosystem Services (chapters 16–19). Chapter 13, "Wetland Classification," now has an update on the U.S. National Wetland Inventory that was just completed for the lower 48 states after a 35-year effort on May 1, 2014. A web connection is also provided where readers can obtain wetland maps from almost anywhere in the United States. A description of methods that are being used in the United States to rate wetlands is also provided in that chapter, emphasizing systems developed in the states of Washington, Ohio, and Florida. New peat production rates for countries in the world are provided in Chapter 14, "Human Impacts and Management of Wetlands," and compared to rates from 14 years prior. The new regional wetland delineation manuals in the United States are described in Chapter 15, "Wetland Laws and Protection," as is a new U.S. Supreme Court decision on wetland mitigation that occurred in the summer of 2013. That makes three Supreme Court decision of wetlands in the United States since the new century began. The status of the international Ramsar Convention on Wetlands, which is growing in international importance by leaps and bounds, is also brought up to date in Chapter 15.

Chapter 16, "Wetland Ecosystem Services," now provides a new description of the ecosystem services that wetlands provide to society, newly categorized into the system developed by the Millennium Ecosystem Assessment of 2005 and also updates new economic value of different types of wetlands as published in mid-2014. Chapter 17, "Wetlands and Climate Change," updates the trends in greenhouse gases in the atmosphere and sea level rise, both of which affect and are affected by wetlands. A newly published model that provides a way of balancing the fluxes of methane with carbon sequestration in the same wetlands is also presented in that chapter as are more references on these two wetland processes. We updated the "permitted" versus "mitigated" data from the U.S. Army Corps of Engineers on trends in the USA on mitigating wetlands loss in Chapter 18, "Wetland Creation and Restoration." We now have seven wetland restoration case studies thoroughly updated in this chapter: the Florida Everglades, the Mesopotamia Marshlands in Iraq, the Bois-des-Bel experimental peatlands in Quebec, Canada, the Delaware Bay and Hackensack Meadowlands salt marsh restorations in the eastern United States, mangrove around the Indian Ocean, and the Skjern River channel and floodplain system in western Denmark.

Chapter 19, "Wetlands and Water Quality," provides updates on long-term studies that have investigated improving water quality by wetlands at the Houghton Lake treatment peatlands in Michigan and the freshwater marshes at the Olentangy River

Wetlands in Ohio, and the wetlands filtering agricultural runoff in south Florida called stormwater treatment areas (STAs). A new promising design of a stormwater treatment wetland, located at Freedom Park in Naples Florida is presented with preliminary data. New estimates of the costs of creating wetlands to improve water quality are also provided in this last chapter.

We continued the tradition of "boxes" or sidebars in *Wetlands 5*. There are now 41 such sidebars or case studies, especially in Chapter 16, "Wetland Ecosystem Services," and Chapter 18, "Wetland Creation and Restoration." New citations were added in this edition, with over 120 from 2010 or later, to augment some classics from the past. Many older citations, particularly those that would be hard to find, were eliminated.

On a personal note, I am pleased to write this edition of *Wetlands* from my new venue as director and professor at the Everglades Wetland Research Park of Florida Gulf Coast University, located at the Naples Botanical Garden in Naples Florida.

We could not have completed this edition without help from many friends and colleagues. Anne Mischo provided dozens more of new illustrations for *Wetlands 5* to supplement her beautiful work carried over from previous editions. We are honored to have the cover photo of a mangrove swamp from southwest Florida, on the fringe of the Florida Everglades, taken by a long-time friend and world-class birder Bernie Master. Ruthmarie Mitsch provided assistance in editing some parts of this edition. Li Zhang and Chris Anderson need to be especially thanked for the updates that they provided in *Wetland Ecosystems* that were used in this book. Li Zhang also helped on many technical details related to publishing the book. We also appreciate the input, illustrations, or insight provided by the following, listed in alphabetical order: Jim Aber, Andy Baldwin, Jim Bays, Jenny Davis, Frank Day, Max Finlayson, Brij Gopal, Glenn Guntenspergen, Wenshan He, Wolfgang Junk, David Latt, Pierrick Marion, Mike Rochford, Line Rochefort, Clayton Rubec, Kenneth Strait, and Ralph Tiner. We also appreciate the professional effort on the part of editors and assistants at Wiley & Sons, Inc. It has always been a pleasure to work with the Wiley brand.

Finally, you will note that there is only one author of this preface. My coauthor Jim Gosselink, Professor Emeritus of Louisiana State University, had been ill for several years and was unable to participate in this new edition. Jim died on January 18, 2015, at the age of 83. But his spirit and incredible knowledge of wetlands are embedded in this book from his contributions to the previous editions, so there was no question that his name should remain on the front of this book. I have also taken the liberty to dedicate this book to my long-time friend and coauthor Jim Gosselink.

William J. Mitsch, Ph.D.
Naples, Florida

February 2015

Part I

Introduction

Chapter 1

Wetlands: Human Use and Science

Wetlands are found in almost all parts of the world. They are sometimes referred to as "kidneys of the landscape" and "nature's supermarkets" to bring attention to the important ecosystem services and habitat values that they provide. Although many cultures have lived among and even depended on wetlands for centuries, the modern history of wetlands until the 1970s is fraught with misunderstanding and fear, as described in much of our Western literature. Wetlands have been destroyed at alarming rates throughout the developed and developing worlds. Now, as their many benefits are being recognized, wetland conservation has become the norm. In many parts of the world, wetlands are now revered, protected, and restored; in other parts, they are still being drained for human development.

Because wetlands have properties that are not adequately covered by current terrestrial and aquatic ecology paradigms, a case can be made for wetland science as a unique discipline encompassing many fields, including terrestrial and aquatic ecology, chemistry, hydrology, and engineering. Wetland management, as the applied side of wetland science, requires an understanding of the scientific aspects of wetlands balanced with legal, institutional, and economic realities. As awareness of the ecosystem services of wetlands has grown, so too have public interest for wetland protection, wetland science programs in universities, and publications about wetlands in scientific journals.

Wetlands are among the most important ecosystems on Earth. In the great scheme of things, the swampy environment of the Carboniferous period produced and preserved many of the fossil fuels on which our society now depends. In more recent biological and human time periods, wetlands have been valuable as sources, sinks, and transformers of a multitude of chemical, biological, and genetic materials. Although the value of wetlands for fish and wildlife protection has been known for a century, some of the other benefits have been identified more recently.

Wetlands are sometimes described as kidneys of the landscape because they function as the downstream receivers of water and waste from both natural and human sources. They stabilize water supplies, thus mitigating both floods and drought. They have been found to cleanse polluted waters, protect shorelines, and recharge groundwater aquifers.

Wetlands also have been called nature's supermarkets because of the extensive food chain and rich biodiversity that they support. They play major roles in the landscape by providing unique habitats for a wide variety of flora and fauna. Now that we have become concerned about the health of our entire planet, wetlands are being described by some as important carbon sinks and climate stabilizers on a global scale.

These values of wetlands are now recognized worldwide and have led to wetland conservation, protection laws, regulations, and management plans. But our history before current times with wetlands had been to drain, ditch, and fill them, never as quickly or as effectively as was undertaken in countries such as the United States beginning in the mid-1800s. In some regions of the world that destruction of wetlands continues.

Wetlands have become the cause célèbre for conservation-minded people and organizations throughout the world, in part because they have become symptoms of our systematic dismantling of our water resources and in part because their disappearance represents an easily recognizable loss of natural areas to economic "progress." Scientists, engineers, lawyers, and regulators are now finding it both useful and necessary to become specialists in wetland ecology and wetland management in order to understand, preserve, and even reconstruct these fragile ecosystems. This book is for these aspiring wetland specialists as well as for those who would like to know more about the structure and function of these unique ecosystems. It is a book about wetlands—how they work and how we manage them.

Human History and Wetlands

There is no way to estimate the impact humans have had on the global extent of wetlands except to observe that, in developed and heavily populated regions of the world, the impact has ranged from significant to total. The importance of wetland environments to the development and sustenance of cultures throughout human history, however, is unmistakable. Since early civilization, many cultures have learned to live in harmony with wetlands and have benefited economically from surrounding wetlands, whereas other cultures quickly drained the landscape. The ancient Babylonians, Egyptians, and the Aztec in what is now Mexico developed specialized systems of water delivery involving wetlands. Major cities of the world, such as Chicago and Washington, DC, in the United States, Christchurch, New Zealand, and Paris, France, stand on sites that were once part wetlands. Many of the large airports (in Boston, New Orleans, and J. F. Kennedy in New York, to name a few) are situated on former wetlands.

While global generalizations are sometimes misleading, there was and is a propensity in Eastern cultures not to drain valuable wetlands entirely, as has been done in the West, but to work within the aquatic landscape, albeit in a heavily managed way. Dugan (1993) makes the interesting comparison between *hydraulic civilizations* (European in origin) that controlled water flow through the use of dikes, dams, pumps, and drainage tile, in part because water was only seasonally plentiful, and *aquatic civilizations* (Asian in origin) that better adapted to their surroundings of water-abundant floodplains and deltas and took advantage of nature's pulses, such as flooding. It is because the former approach of controlling nature rather than working with it is so dominant today that we find such high losses of wetlands worldwide.

Wetlands have been and continue to be part of many human cultures in the world. Coles and Coles (1989) referred to the people who live in proximity to wetlands and whose culture is linked to them as *wetlanders*.

Sustainable Cultures in Wetlands

Some of the original wetlander cultures are described here. The Marsh Arabs of southern Iraq (Fig. 1.1) and the Camarguais of southern France's Rhone River Delta (Fig. 1.2) are two examples of ancient cultures that have lived in harmony and sustainably with their wetland environments for centuries. In North America, the Cajuns of Louisiana and several Native Americans tribes have lived in harmony with wetlands for hundreds of years. The Louisiana Cajuns, descendants of the French colonists of Acadia (present-day Nova Scotia, Canada), were forced out of Nova Scotia by the English and moved to the Louisiana delta in the last half of the

Figure 1.1 The Marsh Arabs of present-day southern Iraq lived for centuries on artificial islands in marshes at the confluence of the Tigris and Euphrates rivers in Mesopotamia. The marshes were mostly drained by Saddam Hussein in the 1990s and are now being restored.

Figure 1.2 The Camargue region of southern France in the Rhone River Delta is a historically important wetland region in Europe where Camarguais have lived since the Middle Ages. (Photo by Tom Nebbia, reprinted with permission)

Figure 1.3 A Cajun lumberjack camp in the Atchafalaya Swamp of coastal Louisiana. (Photo courtesy of the Louisiana Collection, Tulane University Library, reprinted with permission)

eighteenth century. Their society and culture flourished within the bayou wetlands (Fig. 1.3). The Chippewa in Wisconsin and Minnesota have harvested and reseeded wild rice (*Zizania aquatica*) along the littoral zone of lakes and streams for centuries (Fig. 1.4). They have a saying: "Wild rice is like money in the bank."

Likewise, several Native American tribes lived and even thrived in large-scale wetlands, such as the Florida Everglades. These include the ancient Calusa, a culture that based its economy on estuarine fisheries rather than agriculture. The Calusa disappeared primarily as a result of imported European disease. In the nineteenth century, the Seminoles and especially one of its tribes, the Miccosukee, moved south to the Everglades while being pursued by the U.S. Army during the Seminole Indian wars. They never surrendered. The Miccosukee adapted to living in hammock-style camps

Figure 1.4 "Ricer" poling and "knocking" wild rice (*Zizania aquatica*) into canoes as Anishinaabe (Chippewa, Ojibwe) tribes and others have done for hundreds of years on Rice Lake in Crow Wing County, Minnesota. (Photo by John Overland, reprinted with permission)

spread throughout the Everglades and relied on fishing, hunting, and harvesting of native fruits from the hammocks (Fig. 1.5). A recent quote in a Florida newspaper by Miccosukee tribal member Michael Frank is poignant yet hopeful about living sustainably in the Florida Everglades:

> We were taught to never, ever leave the Everglades. If you leave the Everglades you will lose your culture, you lose your language, you lose your way of life.
> —Michael Frank, as quoted by William E. Gibson, "Pollution Is Killing Everglades, Miccosukee Warn," *South Florida Sun Sentinel*, August 10, 2013

Literary References to Wetlands

With all of these important cultures vitally depending on wetlands, not to mention the aesthetics of a landscape in which water and land often provide a striking panorama, one might expect wetlands to be more respected by humanity; this has certainly not always been the case. Wetlands have been depicted as sinister and forbidding and as having little economic value throughout most of Western literature and history. For example, in the *Divine Comedy*, Dante describes a marsh of the Styx in Upper Hell as the final resting place for the wrathful:

> Thus we pursued our path round a wide arc of that ghast pool,
> Between the soggy marsh and arid shore,
> Still eyeing those who gulp the marsh [marsh] foul.

8 Chapter 1 Wetlands: Human Use and Science

Figure 1.5 The Miccosukee Native Americans adapted to life in the Florida Everglades in hammock-style camps. They relied on fishing, hunting, and harvesting of native fruits from the hammocks. (Photo by W. J. Mitsch of panorama at Miccosukee Indian Village, Florida Everglades)

Centuries later, Carl Linnaeus, crossing the Lapland peatlands in 1732, compared that region to that same Styx of Hell:

> Shortly afterwards began the muskegs, which mostly stood under water; these we had to cross for miles; think with what misery, every step up to our knees. The whole of this land of the Lapps was mostly muskeg, hinc vocavi Styx. Never can the priest so describe hell, because it is no worse. Never have poets been able to picture Styx so foul, since that is no fouler.

In the eighteenth century, an Englishman who surveyed the Great Dismal Swamp on the Virginia–North Carolina border and is credited with naming it described the wetland as

> [a] horrible desert, the foul damps ascend without ceasing, corrupt the air and render it unfit for respiration.... Never was Rum, that cordial of Life, found more necessary than in this Dirty Place.
> —**Colonel William Byrd III,** "Historie of the Dividing Line Betwixt Virginia and North Carolina," in *The Westover Manuscripts,* written 1728–1736 (Petersburg, VA: E. and J. C. Ruffin, printers, 1841)

Even those who study and have been associated with wetlands have been belittled in literature:

> Hardy went down to botanise in the swamp, while Meredith climbed towards the sun. Meredith became, at his best, a sort of daintily dressed Walt Whitman: Hardy

became a sort of village atheist brooding and blaspheming over the village idiot.
—G. K. Chesterton, Chapter 12 in *The Victorian Age in Literature*
(New York, NY: Henry Holt and Company, 1913)

The English language is filled with words that suggest negative images of wetlands. We get *bogged down* in detail; we are *swamped* with work. Even the mythical *bogeyman*, the character featured in stories that frighten children in many countries, may be associated with European bogs. Grendel, the mythical monster in *Beowulf*, one of the oldest surviving pieces of Old English literature and Germanic epic, comes from the peatlands of present-day northern Europe:

> Grendel, the famous stalker through waste places, who held the rolling marshes in his sway, his fen and his stronghold. A man cut off from joy, he had ruled the domain of his huge misshapen kind a long time, since God had condemned him in condemning the race of Cain.
> —*Beowulf*, translated by William Alfred, *Medieval Epics*
> (New York, NY: The Modern Library, 1993)

Hollywood has continued the depiction of the sinister and foreboding nature of wetlands and their inhabitants, in the tradition of Grendel, with movies such as the classic *Creature from the Black Lagoon* (1954), a comic-book-turned-cult-movie *Swamp Thing* (1982), and its sequel *Return of the Swamp Thing* (1989). Even Swamp Thing, the man/monster depicted in Figure 1.6, evolved in the 1980s from a feared creature to a protector of wetlands, biodiversity, and the environment. A more modern approach to scaring and entertaining the public with megafauna from the swamps is a science fiction movie *Mega Python vs. Gatoroid* (2011) that is set in the Florida Everglades (Fig. 1.7). The movie exaggerates much of the current dynamics about the Florida Everglades including conservation, invasive species, genetically altered organisms, fund-raising by conservationists, and conflicts among hunters, conservation agencies, and environmentalists. In some respects, current life in the Everglades imitates art. Big snakes and alligators from wetlands continue to strike fear.

As long as wetlands remain more difficult to stroll through than a forest and more difficult to cross by boat than a lake, they will remain misunderstood by the general public unless a continued effort of education takes place.

Food from Wetlands

Domestic wetlands such as rice paddies feed an estimated half of the world's population (Fig. 1.8). Countless other plant and animal products are harvested from wetlands throughout the world. Many aquatic plants besides rice, such as Manchurian wild rice (*Zizania latifolia*), are harvested as vegetables in China. Cranberries are harvested from bogs, and the industry continues to thrive today in North America (Fig. 1.9). Coastal marshes in northern Europe, the British Isles, and New England were used for centuries and are still used today for grazing of animals and production of salt hay. Salt marsh coastlines of Europe are still used for the production of salt.

10 Chapter 1 Wetlands: Human Use and Science

Figure 1.6 The sinister image of wetlands, especially swamps, has often been promoted in popular media such as Hollywood movies and comic books. Shown here are four examples: (a) Swamp Thing movie poster; (b) Swamp Thing: Dark Genesis cover; (c) Saga of Swamp Thing #26; and (d) Swamp Thing #9. All ™ and © DC Comics.

Wetlands can be important sources of protein. The production of fish in shallow ponds or rice paddies developed several thousands of years ago in China and Southeast Asia, and crayfish harvesting is still practiced in the wetlands of Louisiana and the Philippines. Shallow lakes and wetlands are an important provider of protein in many parts of sub-Saharan Africa (Fig. 1.10).

Peat and Building Materials

Russians, Finns, Estonians, and Irish, among other cultures, have mined their peatlands for centuries, using peat as a source of energy in small-scale production

Figure 1.7 The playbill for the *Mega Python vs. Gatoroid* science fiction movie published by The Asylum in 2011 (http://www.theasylum.cc). (Permission from David Latt, President, The Asylum, Burbank, CA)

(Fig. 1.11) and in large-scale extraction processes (Fig. 1.12). *Sphagnum* peat is now harvested for horticultural purposes throughout the world. In southwestern New Zealand, for example, surface sphagnum has been harvested since the 1970s for export as a potting medium. Reeds and even the mud from coastal and inland marshes have been used for thatching for roofs in Europe, Iraq, Japan, and China as well as in wall construction, as fence material, and for lamps and other household goods (Fig. 1.13). Coastal mangroves are harvested for timber, food, and tannin in many countries throughout Indo-Malaysia, East Africa, and Central and South America.

Figure 1.8 Rice production occurs in "managed" wetlands throughout Asia and other parts of the world. Half of the world's population is fed by rice paddy systems. (Photo by W. J. Mitsch)

Figure 1.9 Cranberry wet harvesting is accomplished by flooding bogs in several regions of North America. The cranberry plant (*Vaccinium macrocarpon*) is native to the bogs and marshes of North America and was first cultivated in Massachusetts. It is now also an important fruit crop in Wisconsin, New Jersey, Washington, Oregon, and parts of Canada. (Photo courtesy of Ocean Spray Cranberries, Inc.)

Figure 1.10 Humans use the wetlands of sub-Saharan Africa for sustenance, as with this man fishing for lungfish (*Proptopterus aethiopicus*) in Lake Kanyaboli, western Kenya. (Photo by M. K. Mavuti, reprinted with permission)

Figure 1.11 Harvesting of peat, or "turf," as a fuel has been a tradition in several parts of the world, as shown by this scene of turf carts in Ireland.

Wetlands and Ecotourism

Ecotourism is a modern version of wetland use. Wetlands have been the focus of attempts by several countries to increase tourist flow into their countries. The Okavango Delta in Botswana is one of the natural resource jewels of Africa, and protection of this wetland for tourists and hunters has been a priority in that country since the

Figure 1.12 Large-scale peat mining in Estonia. (Photo by W. J. Mitsch)

Figure 1.13 A wetland house in the Ebro River Delta region on the Mediterranean Sea, Spain. The walls are made from wetland mud, and the roof is thatched with reed grass and other wetland vegetation. (Photo by W. J. Mitsch)

1960s. Local tribes provide manpower for boat tours (in dugout canoes called *mokoros*) through the basin and assist with wildlife tours on the uplands as well (Fig. 1.14). In Senegal, West Africa, there is keen interest in attracting European birder tourists to the mangrove swamps along the Atlantic coastline. For many people, ecotourism in the wetlands is all about the wildlife and especially the birds (Fig. 1.15). It has been reported that bird-watching, or "birding," is a $32 billion per year industry in the United States alone.

Figure 1.14 The vast seasonally flooded Okavango Delta of northern Botswana in southern Africa is a mecca for ecotourism. The wetlands attract tourists, as shown in this illustration, and also wildlife hunting. In addition, the wetlands provides basic sustenance to these communities. (Photo by W. J. Mitsch)

(a)   (b)

Figure 1.15 Intense ecotourism interest in the wetlands in Asia is shown by (a) crowds that surround Lake Biwa in Shiga Prefecture, Japan, at a winter 2006 international wetlands forum, and (b) press coverage at the Ramsar Convention held in Changwon, Korea, in 2008. (Photos by W. J. Mitsch)

The advantage of ecotourism as a management strategy is obvious—it provides income to the country where the wetland is found without requiring or even allowing resource harvest from the area. The potential disadvantage is that if the site becomes too popular, human pressures will begin to deteriorate the landscape and the very ecosystem that initially drew the tourism.

Wetland Conservation

Prior to the mid-1970s, drainage and destruction of wetlands were accepted practices around the world and were even encouraged by specific government policies. Wetlands were replaced by agricultural fields and by commercial and residential development. Had those trends continued, wetlands would have been in danger of extinction in some parts of the world decades ago. Some countries and states, such as New Zealand and California and Ohio in the United States, reported 90 percent loss of their wetlands. Only through the combined activities of hunters and anglers, scientists and engineers, and lawyers and conservationists has the case been made for wetlands as a valuable resource whose destruction has serious economic as well as ecological and aesthetic consequences for the nations of the world. This increased level of respect was reflected in activities such as the sale of federal "duck stamps" to waterfowl hunters that began in 1934 in the United States (Fig. 1.16); other countries, such as New

Figure 1.16 Federal Migratory Bird Hunting and Conservation Stamps are more commonly known as duck stamps. They are produced by the U.S. Postal Service for the U.S. Fish & Wildlife Service and are not valid for postage. Originally created in 1934 as the federal licenses required for hunting migratory waterfowl, today income derived from their sale is used to purchase or lease wetlands. *Top*; First duck stamp from 1934 (mallards); *Bottom*; 2013 duck stamp (wood duck).

Zealand, have followed suit. Approximately 2.4 million hectares (ha) of wetlands have been purchased or leased as waterfowl habitat by the U.S. duck stamp program alone since 1934.

The U.S. government now supports a variety of other wetland protection programs through at least a dozen federal agencies; individual states have also enacted wetland protection laws or have used existing statutes to preserve these valuable resources. On an international scale, the Convention of Wetlands of International Importance, or the Ramsar Convention, a multinational agreement for the conservation of wetlands, has formally registered as "Wetlands of International Importance" 210 million ha of wetlands in 168 contracting parties. The Ramsar Convention is the only global international treaty specific to the conservation and wise management of a specific ecosystem.

Wetland Science and Wetland Scientists

A specialization in the study of wetlands is often termed *wetland science* or *wetland ecology*, and those who carry out such investigations are called *wetland scientists* or *wetland ecologists*. The term *mire ecologist* has also been used. Some have suggested that the study of all wetlands be called *telmatology* (*telma* being Greek for "bog"), a term originally coined to mean "bog science" (Zobel and Masing, 1987). No matter what the field is called, it is apparent that there are four good reasons for treating wetland ecology as a distinct field of ecological study:

1. Wetlands have unique properties that are not adequately covered by present ecological paradigms and by fields such as limnology, estuarine ecology, and terrestrial ecology.
2. Wetland studies have begun to identify some common properties of seemingly disparate wetland types.
3. Wetland investigations require a multidisciplinary approach or training in several fields not routinely combined in university academic programs.
4. There is a great deal of interest in formulating sound policy for the regulation and management of wetlands. These regulations and management approaches need a strong scientific underpinning integrated as wetland ecology.

A growing body of evidence suggests that the unique characteristics of wetlands—standing water or waterlogged soils, anoxic conditions, and plant and animal adaptations—may provide some common ground for study that is neither terrestrial ecology nor aquatic ecology. Wetlands provide opportunities for testing "universal" ecological theories and principles involving succession and energy flow, theories that were developed for aquatic or terrestrial ecosystems. For example, wetlands provided the setting for the establishment of the current system used for lake trophic status (e.g., oligotrophic, eutrophic [Weber, 1907]), the successional theories of Clements (1916), and the energy flow approaches of Lindeman (1942).

They also provide an excellent laboratory for the study of principles related to transition zones, ecological interfaces, and ecotones.

Our knowledge of different wetland types such as those discussed in this book is often isolated in distinctive literatures and scientific circles. One set of literature deals with coastal wetlands, another with forested wetlands and freshwater marshes, and still another with peatlands. Very few investigators have analyzed the properties and functions common to all wetlands. This is probably one of the most exciting areas for wetland research because there is so much to be learned. Comparisons of wetland types have shown, for example, the importance of hydrologic flow-through for the maintenance and productivity of these ecosystems. The anoxic biochemical processes that are common to all wetlands provide another area for comparative research and pose many questions: What are the roles of different wetland types in local and global biochemical cycles? How do the activities of humans influence these cycles in various wetlands? What are the synergistic effects of hydrology, chemical inputs, and climatic conditions on wetland biological productivity? How can plant and animal adaptations to anoxic stress be compared in various wetland types?

The true wetland ecologist must be an ecological generalist because of the number of sciences that bear on those ecosystems. Knowledge of wetland flora and fauna, which are often uniquely adapted to a substrate that may vary from submerged to dry, is necessary. Emergent wetland plant species support both aquatic animals and terrestrial insects. Because hydrologic conditions are so important in determining the structure and function of the wetland ecosystems, a wetland scientist should be well versed in surface and groundwater hydrology. The shallow-water environment means that chemistry—particularly for water, sediments, soils, and water–sediment interactions—is an important science. Similarly, questions about wetlands as sources, sinks, or transformers of chemicals require investigators to be versed in many biological and chemical techniques. While the identification of wetland vegetation and animals requires botanical and zoological skills, backgrounds in microbial biochemistry and soil science contribute significantly to the understanding of the anoxic environment. Understanding adaptations of wetland biota to the flooded environment requires both biochemistry and physiology. If wetland scientists are to become more involved in the management of wetlands, some engineering techniques, particularly for wetland hydrologic control or wetland creation, need to be learned.

Wetlands are seldom, if ever, isolated systems. Rather, they interact strongly with adjacent terrestrial and aquatic ecosystems. Hence, a holistic view of these complex landscapes can be achieved only through an understanding of the principles of ecology, especially those that are part of ecosystem and landscape ecology and systems analysis. Finally, if wetland management involves the implementation of wetland policy, then training in the legal and policy-making aspects of wetlands is warranted.

Thousands of scientists and engineers are now studying and managing wetlands. Only a relatively few pioneers, however, investigated these systems in any detail prior to the 1960s. Most of the early scientific studies dealt with classical botanical surveys or investigations of peat structure. Several early scientific studies of peatland hydrology were produced, particularly in Europe and Russia. Later, investigators such

Table 1.1 Pioneer researchers in wetland ecology and representative citations for their work

Wetland Type and Researcher	Country	Representative Citations
Coastal Marshes/Mangroves		
Valentine J. Chapman	New Zealand	Chapman (1938, 1940)
John Henry Davis	USA	Davis (1940, 1943)
John M. Teal	USA	Teal (1958, 1962); Teal and Teal (1969)
Howard T. Odum	USA	H. T. Odum et al. (1974)
D. S. Ranwell	UK	D. S. Ranwell (1972)
Peatlands/Freshwater Wetlands		
C. A. Weber	Germany	Weber (1907)
Herman Kurz	USA	Kurz (1928)
A. P. Dachnowski-Stokes	USA	Dachnowski-Stokes (1935)
R. L. Lindeman	USA	Lindeman (1941, 1942)
Eville Gorham	UK/USA	Gorham (1956, 1961)
Hugo Sjörs	Sweden	Sjörs (1948, 1950)
G. Einar Du Rietz	Sweden	Du Rietz (1949, 1954)
P. D. Moore/D. J. Bellamy	UK	Moore and Bellamy (1974)
S. Kulczynski	Poland	Kulczynski (1949)
Paul R. Errington	USA	Errington (1957)
R. S. Clymo	UK	Clymo (1963, 1965)
Milton Weller	USA	Weller (1981)
William H. Patrick	USA	Patrick and Delaune (1972)

as Chapman, Teal, Sjörs, Gorham, Eugene and H. T. Odum, Weller, Patrick, and their colleagues and students began to use modern ecosystem and biogeochemical approaches in wetland studies (Table 1.1). Currently active research centers devoted to the study of wetlands include the School of Coast and Environment at Louisiana State University; the H. T. Odum Center for Wetlands at the University of Florida; the Duke Wetland Center at Duke University; Florida Gulf Coast University's Everglades Wetland Research Park in Naples, Florida; the Harry Oppenheimer Okavango Research Centre (HOORC) in Botswana, Africa: and the Institute for Land, Water, and Society at Charles Stuart University in Australia.

In addition, a professional society now exists, the Society of Wetland Scientists, which has among its goals to provide a forum for the exchange of ideas within wetland science and to develop wetland science as a distinct discipline. The Association of State Wetland Managers (ASWM) is an organization based primarily in the USA as a place for state, federal, and local managers and consultants to meet and discuss wetland management issues. They currently sponsor popular webinars on subjects related to wetlands. The International Association of Ecology (INTECOL) has sponsored a major international wetland conference every four years somewhere in the world since 1980. Table 1.2 lists the locations around the world where the INTECOL Wetland conference has been held and each meeting's theme, attendance, and resulting publications.

Table 1.2 INTECOL wetland conferences, 1980 to 2012, indicating year, location, theme, approximate attendance, chair/organizer, and resulting publications

Year	Location	Theme	Attendance	Organizer	Key Publication(s)
1980	New Delhi, India		90	B. Gopal	Gopal et al., 1982a,b
1984	Trebon, Czechoslovakia		210	J. Kvet/J. Pokorny	Pokorny et al., 1987; Mitsch et al., 1988; Bernard, 1988; Whigham et al., 1990, 1993
1988	Rennes, France	Conservation and Development: The Sustainable Use of Wetland Resources	400	J. C. Lefeuvre	Lefeuvre, 1989, 1990; Maltby et al., 1992
1992	Columbus, USA	Global Wetlands: Old World and New	905	W. J. Mitsch	Mitsch, 1993, 1994; Wetzel et al., 1994; Finlayson and van der Valk, 1995; Gopal and Mitsch, 1995; Jørgensen, 1995
1996	Perth, Australia	Wetlands for the Future	550	A. J. McComb; J. A. Davis	McComb and Davis, 1998; Tanner et al., 1999; Zedler and Rhea, 1998
2000	Quebec City, Canada	Quebec 2000: Wetlands at the Millennium*	2160	C. Rubec, B. Belangér, and G. Hood	11 books/special reports; 6 special journal issues; 8 International Peat Society Proceedings
2004	Utrecht, Netherlands		787	J. T. A. Verhoeven	Vymazal, 2005; Bobbink et al., 2006; Junk 2006; van Diggelen et al., 2006; Verhoeven et al., 2006; Davidson and Finlayson, 2007; Whitehouse and Bunting, 2008
2008	Cuiaba, Brazil	Big Wetlands, Big Concerns	700	P. Teixeira de Sousa Jr.; C. Nunes da Cunha	Vymazal, 2011; Junk, 2013
2012	Orlando, USA	Wetlands in a Complex World**	1240	R. Best/ K.R. Reddy	

*INTECOL met with three additional societies in 2000: International Peat Society; International Mire Conservation Group; Society of Wetland Scientists.
**INTECOL met with Society of Wetland Scientists in 2012.

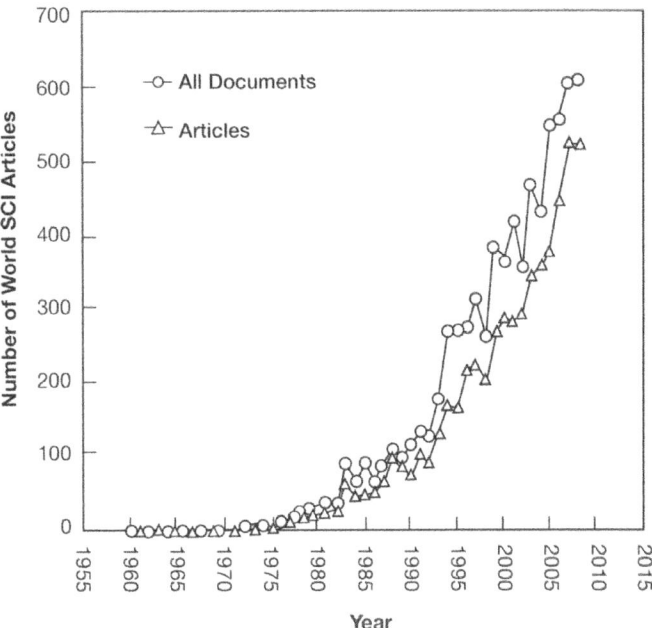

Figure 1.17 Science Citation Index (SCI) listed scientific articles that included "wetland" in their title or in keywords during the period 1960 to 2010 (From Zhang et al., 2010).

The increasing interest and emphasis on wetland science and management has been demonstrated by a veritable flood of books, reports, scientific journal articles, and conference proceedings, most in the last two decades of the twentieth century and the first decade of the twenty-first century. From 1991 to 2008, the annual number of wetland research journal articles published and the number of wetland articles cited increased six- and nine-fold, respectively (Fig. 1.17). The journal citations in this book are only the tip of the iceberg of the literature on wetlands. Two journals specific to wetlands—*Wetlands* and *Wetlands Ecology and Management*—are now published to disseminate scientific and management papers on wetlands, and several other scholarly journals frequently publish papers on the topic. Dozens of wetland meeting proceedings and journal special issues have been published from conferences on wetlands held throughout the world.

Wetland Managers and Wetland Management

Just as there are wetland scientists who are uncovering the processes that determine wetland functions and values, so too there are those who are involved, by choice or by vocation, in some of the many aspects of wetland management. These individuals, whom we call *wetland managers*, are engaged in activities that range from waterfowl production to wastewater treatment. They must be able to balance the scientific aspects of wetlands with myriad legal, institutional, and economic constraints to provide optimum wetland management. The management of wetlands has become increasingly important in many countries because government policy and wetland regulation seek to reverse historic wetland losses in the face of continuing draining or encroachment by agricultural enterprises and urban expansion. The simple act of being able to identify the boundaries of wetlands has become an important skill for a new type of wetland technician in the United States called a *wetland delineator*.

Private organizations, such as Ducks Unlimited, Inc. and The Nature Conservancy have protected wetlands by purchasing thousands of hectares of wetlands throughout North America. Through the Ramsar Convention and an agreement jointly signed by the United States and Canada in 1986 called the North American Waterfowl Management Plan, wetlands are now being protected primarily for their waterfowl value on an international scale. In 1988, a federally sponsored National Wetlands Policy Forum (1988) in the United States raised public and political awareness of wetland loss and recommended a policy of "no net loss" of wetlands. This recommendation has stimulated widespread interest in wetland restoration and creation to replace lost wetlands, and "no net loss" has remained the policy of wetland protection in the United States since the late 1980s.

Subsequently, a National Research Council (NRC) report in the United States (NRC, 1992) called for the fulfillment of an ambitious goal of gaining 4 million ha of wetlands by the year 2010, largely through the reconversion of crop- and pastureland. That goal was not met. Wetland creation for specific functions remains an exciting new area of wetland management that needs trained specialists and may eventually stem the tide of loss and lead to an increase in this important resource. Another NRC report (1995) reviewed the scientific basis for wetland delineation and classification, particularly as it related to the regulation of wetlands in the United States at that time, and yet another NRC (2001) study investigated the effectiveness of the national policy of mitigation of wetland loss in the United States.

Wetland management organizations, such as the Association of State Wetland Managers (ASWM) and the Society of Wetland Scientists (SWS), focus on disseminating information on wetlands, particularly in North America. The International Union for the Conservation of Nature and Natural Resources (IUCN) and the Ramsar Convention, both based in Switzerland, have developed a series of publications on wetlands of the world. Wetlands International (www.wetlands.org) is the world's leading nonprofit organization concerned with the conservation of wetlands and wetland species. It comprises a global network of governmental and nongovernmental experts working on wetlands. Activities are undertaken in more than 120 countries worldwide. The head office is located in Wageningen, Netherlands.

Recommended Readings

Beautifully illustrated popular books and articles, many with color photographs, were developed on wetlands by many authors in years past. Here are some of our timeless favorites.

Dugan, P. 1993. *Wetlands in Danger*. London: Oxford University Press.
Finlayson, M., and M. Moser, eds. 1991. *Wetlands*. Oxford, UK: Facts On File.

Kusler, J., W. J. Mitsch, and J. S. Larson. 1994. Wetlands. *Scientific American* 270(1): 64–70.

Littlehales, B., and W. A. Niering. 1991. *Wetlands of North America*. Charlottesville, VA: Thomasson-Grant.

Lockwood, C. C., and R. Gary. 2005. *Marsh Mission: Capturing the Vanishing Wetlands*. Baton Rouge: Louisiana State University Press.

McComb, A. J., and P. S. Lake. 1990. *Australian Wetlands*. London: Angus and Robertson.

Mendelsohn, J., and S. el Obeid. 2004. *Okavango River: The Flow of a Lifeline*. Cape Town, South Africa: Struik.

Mitchell, J. G., R. Gehman, and J. Richardson. 1992. Our Disappearing Wetlands. *National Geographic* 182(4): 3–45.

Niering, W. A. 1985. *Wetlands*. New York: Knopf.

Rezendes, P., and P. Roy. 1996. *Wetlands: The Web of Life*. San Francisco: Sierra Club Books.

References

Bernard, J. M., ed. 1998, *Carex*. Special Issue of *Aquatic Botany* 30: 1–168.

Bobbink R., B. Beltman, J. T. A. Verhoeven, and D. F. Whigham, eds. 2006. Wetlands: Functioning, Biodiversity, Conservation and Restoration. *Ecological Studies* 191, Springer, Berlin, 315 pp.

Chapman, V. J. 1938. Studies in salt marsh ecology. I-III. *Journal of Ecology* 26: 144–221.

Chapman, V. J. 1940. Studies in salt marsh ecology. VI-VII. *Journal of Ecology* 28: 118–179.

Clements, F. E. 1916. *Plant Succession*. Publication 242. Carnegie Institution of Washington. 512 pp.

Clymo, R. S. 1963. Ion exchange in *Sphagnum* and its relation to bog ecology. *Annals of Botany (London) New Series* 27: 309–324.

Clymo, R. S. 1965. Experiments on breakdown of *Sphagnum* in two bogs. *Journal of Ecology* 53: 747–758.

Coles, B., and J. Coles. 1989. *People of the Wetlands, Bogs, Bodies and Lake-Dwellers*. Thames & Hudson, New York. 215 pp.

Dachnowski-Stokes, A. P. 1935. Peat land as a conserver of rainfall and water supplies. *Ecology* 16: 173–177.

Davidson, N. C., and M. Finlayson, eds. 2007. Satellite-based radar - Developing tools for wetlands management. *Aquatic Conservation: Marine and Freshwater Ecosystems* 17(3): 219–329.

Davis, J. H. 1940. The ecology and geologic role of mangroves in Florida. Publication 517. Carnegie Institution of Washington, pp. 303–412.

Davis, J. H. 1943. The natural features of southern Florida, especially the vegetation and the Everglades. *Florida Geological Survey Bulletin* 25. 311 pp.

Dugan, P. 1993. *Wetlands in Danger*. Michael Beasley, Reed International Books, London. 192 pp.

Du Rietz, G. E. 1949. Huvudenheter och huvudgränser i Svensk myrvegetation. *Svensk Botanisk Tidkrift* 43: 274–309.

Du Rietz, G. E. 1954. Die Mineralbodenwasserzeigergrenze als Grundlage Einer Natürlichen Zweigleiderung der Nord-und Mitteleuropäischen Moore. *Vegetatio* 5– 6: 571–585.

Errington, P. L. 1957. *Of Men and Marshes*. The Iowa State University Press, Ames, IA.

Finlayson, C. M. and A. G. van der Valk. 1995. Classification and inventory of the world's wetlands. *Special Issue Vegetatio* 118: 1–192.

Gopal, B., R. E. Turner, R. G. Wetzel, and D. F. Whigham, eds. 1982a. Wetlands: Ecology and Management. International Scientific Publications, Jaipur, India. Vol. 1, 514 pp.

Gopal, B., R. E. Turner, R. G. Wetzel, and D. F. Whigham, eds. 1982b. Wetlands: Ecology and Management. International Scientific Publications, Jaipur, India. Vol. 2, 156 pp.

Gopal, B., and W. J. Mitsch, eds. 1995. The role of vegetation in created and restored wetlands. *Special Issue Ecological Engineering* 5:1–121.

Gorham, E. 1956. The ionic composition of some bogs and fen waters in the English lake district. *Journal of Ecology* 44: 142–152.

Gorham, E. 1961. Factors influencing supply of major ions to inland waters, with special references to the atmosphere. *Geological Society of America Bulletin* 72: 795–840.

Jørgensen, S. E., ed. 1995. Wetlands: Interactions with watersheds, lakes, and riparian zones. *Special issue Wetlands Ecology and Management* 3:79–137.

Junk, W., ed. 2006. The comparative biodiversity of seven globally important wetlands. *Special Issue of Aquatic Sciences* 68(3): 239–414.

Junk, W., ed. 2013. The world's wetlands and their future under global climate change. *Special Issue of Aquatic Sciences* 75(1): 1–167.

Kulczynski, S. 1949. Peat bogs of Polesie. *Acad. Pol. Sci. Mem.*, Ser. B, No.15. 356 pp.

Kurz, H. 1928. Influence of Sphagnum and other mosses on bog reactions. *Ecology* 9: 56–69.

Lefeuvre J. C., ed .1989. Conservation et développement : gestion intégrée des zones humides. Troisième conférence internationale sur les zones humides, Rennes, 19-23 Septembre 1988. Ed. Muséum National d'Histoire Naturelle, Laboratoire d'Evolution des Systèmes Naturels et Modifiés, Paris. 371 pp.

Lefeuvre J. C. 1990. INTECOL's Third International Wetlands Conference. Rennes, 1988. *Bull. Ecol.*, 21(3), 80 pp.

Lindeman, R. L. 1941. The developmental history of Cedar Creek Lake, Minnesota. *American Midland Naturalist* 25: 101–112.

Lindeman, R. L. 1942. The trophic-dynamic aspect of ecology. *Ecology* 23: 399–418.

Maltby E., P. J. Dugan, and J. C. Lefeuvre, eds. 1992. Conservation and Development: The Sustainable Use of Wetland Resources. Proceedings of the Third International Wetlands Conference. IUCN, Gland, Switzerland. 219 pp.

McComb, A. J. and J. A. Davis, eds. 1998. Wetlands for the Future - Contributions from INTECOL's V International Wetlands Conference. Gleneagles Press, Adelaide, Australia, 750 pp.

Mitsch, W. J. 1993. INTECOL's IV International Wetlands Conference: A report. *International Journal of Ecology and Environmental Sciences* 19:129–134.

Mitsch, W. J., ed. 1994. *Global Wetlands: Old World and New*. Elsevier, Amsterdam. 967+ xxiv pp.

Mitsch, W. J., and J. G. Gosselink. 1986. *Wetlands*, Van Nostrand Reinhold, New York. 539 pp.

Mitsch, W. J., M. Straskraba, and S.E. Jørgensen, eds. 1988. Wetland Modelling. Elsevier, Amsterdam, 227 pp.

Mitsch, W. J., and J. G. Gosselink. 1993. *Wetlands*, 2nd ed. Van Nostrand Reinhold and John Wiley & Sons, New York. 722 pp.

Mitsch, W. J., and J. G. Gosselink. 2000. *Wetlands*, 3rd ed. John Wiley & Sons, New York. 920 pp.

Mitsch, W. J., and J. G. Gosselink. 2007. *Wetlands*, 4th ed. John Wiley & Sons, Hoboken, NJ. 582 pp.

Moore, P. D., and D. J. Bellamy. 1974. *Peatlands*. Springer-Verlag, New York. 221 pp.

National Research Council (NRC). 1992. *Restoration of Aquatic Ecosystems*. National Academy Press, Washington, DC. 552 pp.

National Research Council (NRC). 1995. *Wetlands: Characteristics and Boundaries*. National Academy Press, Washington, DC. 306 pp.

National Research Council (NRC). 2001. *Compensating for Wetland Losses under the Clean Water Act*. National Academy Press, Washington, DC, 158 pp.

National Wetlands Policy Forum. 1988. *Protecting America's Wetlands: An Action Agenda*. Conservation Foundation, Washington, DC. 69 pp.

Odum, H. T., B. J. Copeland, and E. A. McMahan, eds. 1974. *Coastal Ecological Systems of the United States*. Conservation Foundation, Washington, DC. 4 vols.

Patrick, W. H., Jr., and R. D. Delaune. 1972. Characterization of the oxidized and reduced zones in flooded soil. *Proceedings of the Soil Science Society of America* 36: 573–576.

Pokorny, J., O. Lhotsky, P. Denny, and R. E. Turner, eds. 1987. Waterplants and wetland processes. Special issue of *Archiv Fur Hydrobiologie* 27: I-VIII and 1–265.

Ranwell, D. S. 1972. *Ecology of Salt Marshes and Sand Dunes*. Chapman & Hall, London. 258 pp.

Sjörs, H. 1948. Myrvegetation i bergslagen. *Acta Phytogeographica Suecica* 21: 1–299.

Sjörs, H. 1950. On the relationship between vegetation and electrolytes in North Swedish mire waters. *Oikos* 2: 239–258.

Tanner, C. C., G. Raisin, G. Ho, and W. J. Mitsch, eds. 1999. Constructed and Natural Wetlands for Pollution Control. *Special Issue of Ecological Engineering* 12: 1–170.

Teal, J. M. 1958. Distribution of fiddler crabs in Georgia salt marshes. *Ecology* 39: 18–19.

Teal, J. M. 1962. Energy flow in the salt marsh ecosystem of Georgia. *Ecology* 43: 614–624.

Teal, J. M., and M. Teal. 1969. *Life and Death of the Salt Marsh*. Little, Brown, Boston. 278 pp.

van Diggelen, R., Middleton, B., Bakker, J.P., Grootjans, A.P., Wassen, M.J. (eds.) 2006. Fens and floodplains of the temperate zone: Present status, threats, conservation and restoration. *Special Issue of Applied Vegetation Science* 9(2): 157–316.

Verhoeven J. T. A., B. Beltman, R. Bobbink, and D. F. Whigham, eds. 2006. Wetlands and Natural Resource Management. Ecological Studies 190, Springer, Berlin, 347 pp.

Vymazal, J., ed. 2005. Constructed wetlands for wastewater treatment. *Special Issue of Ecological Engineering* 25: 475–621.

Vymazal, J., ed. 2011. Enhancing ecosystem services on the landscape with created, constructed and restored wetlands. *Special Issue of Ecological Engineering* 37(1): 1–98.

Weber, C. A. 1907. Autbau und Vegetation der Moore Norddutschlands. *Beibl. Bot. Jahrb.* 90: 19–34.

Weller, M. W. 1981. *Freshwater Marshes*. University of Minnesota Press, Minneapolis. 146 pp.

Wetzel, R. G., A. van der Valk, R. E. Turner, W. J. Mitsch and B. Gopal, eds. 1994. Recent studies on ecology and management of wetlands. *Special Issue of International Journal of Ecology and Environmental Sciences* 20(1-2): 1–254.

Whigham, D. F., R. E. Good, and J. Květ, eds. 1990. Wetland Ecology and Management: Case Studies, Tasks for Vegetation Science 23, Kluwer Academic Publishers, Dordrecht, 180 pp.

Whigham, D. F., D. Dykyjová, and S. Hejný, eds. 1993. Wetlands of the World: Inventory, Ecology and Management. Volume 1. Africa, Australia, Canada and Greenland, Mediterranean, Mexico, Papua New Guinea, South Asia, Tropical South America, United States. Kluwer Academic Publishers, Dordrecht, 768 pp.

Whitehouse, N. J., and M.J. Bunting, eds. 2008. Palaeoecology and long-term wetland function dynamics: A tool for wetland conservation and management. *Special Issue of Biodiversity and Conservation* 17(9): 2051–2304.

Zedler, J., and N. Rhea, eds. 1998. Ecology and management of wetland plant invasions. *Special issue of Wetlands Ecology and Management* 5(3): 159–242.

Zhang. L., M. H. Wang, J. Hu, and Y.-S. Ho. 2010. A review of published wetland research, 1991–2008: Ecological engineering and ecosystem restoration. *Ecological Engineering* 36: 973–980.

Zobel, M., and V. Masing. 1987. Bog changing in time and space. *Archiv für Hydrobiologie, Beiheft: Ergebnisse der Limnologie* 27: 41–55.

Chapter **2**

Wetland Definitions

Wetlands have many distinguishing features, the most notable of which are the presence of standing water for some period during the growing season, unique soil conditions, and organisms, especially vegetation, adapted to or tolerant of saturated soils. Wetlands are unique because of their hydrologic conditions and their role as ecotones between terrestrial and aquatic systems. Terms such as swamp, marsh, fen, *and* bog *have been used in common speech for centuries to define wetlands and are frequently used and misused today. Formal definitions have been developed by scientists and federal agencies in the United States and Canada and through an international treaty known as the Ramsar Convention. These definitions are used for both scientific and management purposes. Wetlands are not easily defined, however, especially for legal purposes, because they have a considerable range of hydrologic conditions, because they are found along a gradient at the margins of well-defined uplands and deepwater systems, and because of their great variation in size, location, and human influence. No absolute answer to "What is a wetland?" should be expected, but legal definitions involving wetland protection have become the norm.*

The most common questions that the uninitiated ask about wetlands are "What exactly is a wetland?" or "Is that the same as a swamp?" These are surprisingly good questions, and it is not altogether clear that they have been answered completely by wetland scientists and managers. Wetland definitions and terms are many and are often confusing or even contradictory. Nevertheless, definitions are important both for the scientific understanding of these systems and for their proper management.

In the nineteenth century, when the drainage of wetlands was the norm, a wetland definition was unimportant because it was considered desirable to produce uplands from wetlands by draining them. In fact, the word *wetland* did not come into common use until the mid-twentieth century. One of the first references to the word was in the publication *Wetlands of the United States* (Shaw and Fredine, 1956). Before that time, wetlands were referred to by the many common terms that developed in the nineteenth century and before, such as *swamp, marsh, bog, fen, mire*, and *moor*. Even as the value of wetlands was being recognized in the early 1970s, there was little interest in precise definitions until it was realized that a better accounting of the remaining wetland resources was needed, and definitions were necessary to achieve that inventory.

When national and international laws and regulations pertaining to wetland preservation began to be written in the late 1970s, the need for precision became even greater as individuals recognized that definitions were having an impact on what they could or could not do with their land. The definition of a wetland, and by implication its boundaries (referred to as "delineation" in the United States), became important when society began to recognize the value of these systems and began to translate that recognition into laws to protect itself from further wetland loss. However, just as an estimate of the boundary of a forest, desert, or grassland is based on scientifically defensible criteria, so too should the definition of wetlands be based on scientific measures to as great a degree as possible. What society chooses to do with wetlands, once the definition has been chosen, remains a political decision.

Wetlands in the Landscape

Even after the ecological and economic benefits of wetlands were determined and became widely appreciated, wetlands have remained an enigma to scientists. They are difficult to define precisely, not only because of their great geographical extent but also because of the wide variety of hydrologic conditions in which they are found. Wetlands are usually found at the interface of terrestrial ecosystems, such as upland forests and grasslands, and aquatic systems, such as deep lakes and oceans (Fig. 2.1a), making them different from each yet highly dependent on both. They are also found in seemingly isolated situations, where the nearby aquatic system is often a groundwater aquifer (Fig. 2.1b). Sometimes these wetlands are referred to as *isolated wetlands*, a somewhat misleading term because they are usually connected hydrologically to groundwater and biologically through the movement of many mobile organisms. And, of course, all wetland ecosystems are open to solar radiation and precipitation.

Because wetlands combine attributes of both aquatic and terrestrial ecosystems but are neither, they have fallen between the cracks of the scientific disciplines of terrestrial and aquatic ecology. They serve as sources, sinks, and transformers of nutrients; deepwater aquatic systems (at least lakes and oceans) are almost always sinks, and terrestrial systems are usually sources. Wetlands are also among the most productive ecosystems on the planet when compared to adjacent terrestrial and deepwater aquatic systems, but it is not correct to say that all wetlands are highly productive. Peatlands and cypress swamps are examples of low-productivity wetlands.

Distinguishing Features of Wetlands

We can easily identify a coastal salt marsh, with its great uniformity of grasses and its maze of tidal creeks, as a wetland. A cypress swamp, with majestic trees festooned with Spanish moss and standing in knee-deep water, provides an unmistakable image of a wetland. A northern *Sphagnum* bog, surrounded by tamarack trees that quake as people trudge by, is another easily recognized wetland. All of those sites have several features in common: (1) all have shallow water or saturated soil; (2) all accumulate organic plant material that decomposes slowly; and (3) all support a variety of plants and animals adapted to the saturated conditions. Wetland definitions, then, often include three main components:

1. Wetlands are distinguished by the presence of water, either at the surface or within the root zone.
2. Wetlands often have unique soil conditions that differ from adjacent uplands.
3. Wetlands support biota such as vegetation adapted to the wet conditions (*hydrophytes*) and, conversely, are characterized by an absence of flooding-intolerant biota.

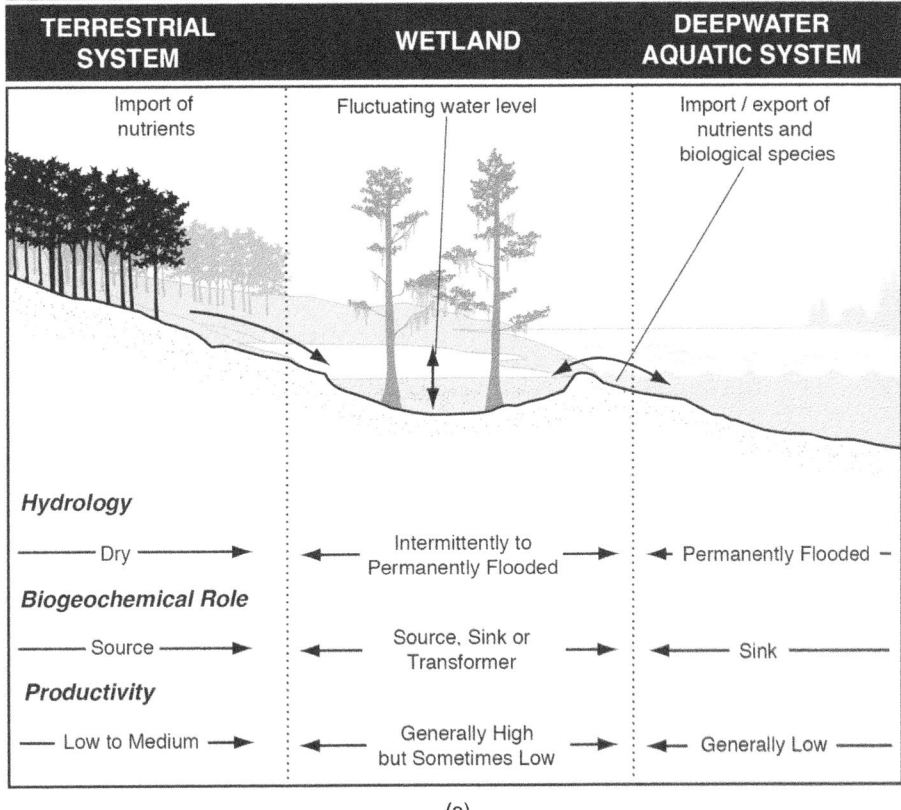

(a)

Figure 2.1 Wetlands are often located (a) between dry terrestrial systems and permanently flooded deepwater aquatic systems such as rivers, lakes, estuaries, or oceans or (b) as isolated basins with little outflow and no adjacent deepwater system.

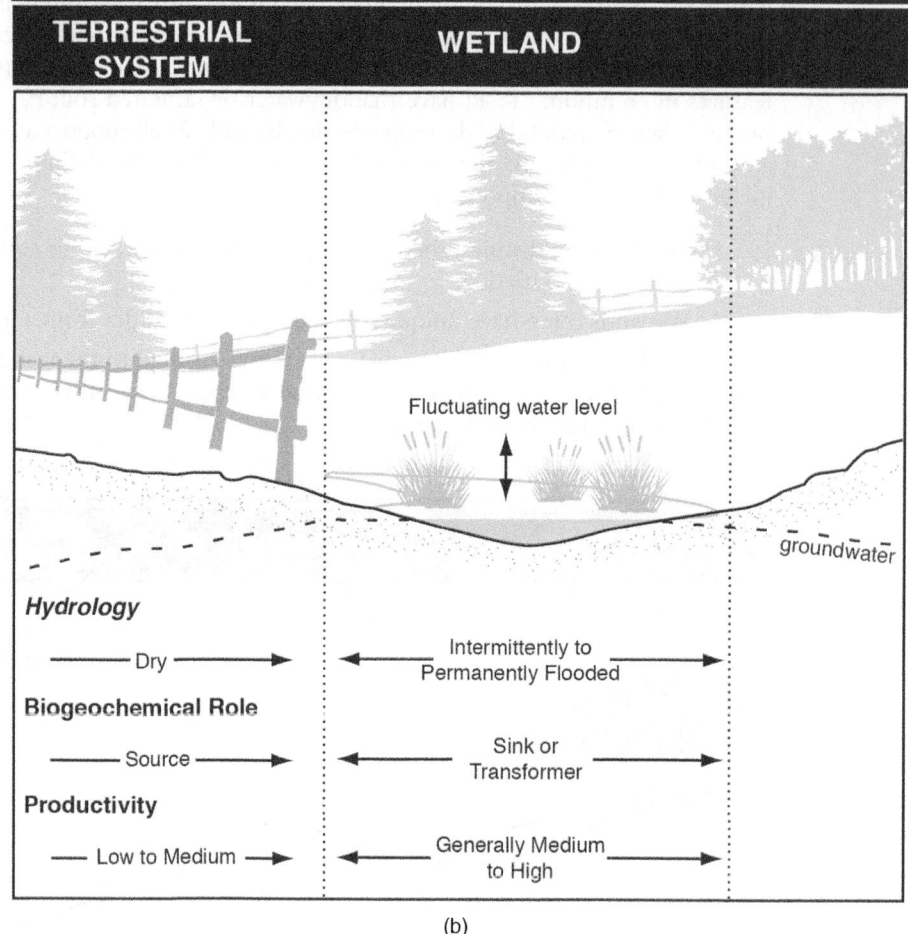

Figure 2.1 (Continued)

This three-level approach to the definition of wetlands is illustrated in Figure 2.2. Climate and geomorphology define the degree to which wetlands can exist, but the starting point is the *hydrology*, which, in turn, affects the *physiochemical environment*, including the soils, which, in turn, determines with the hydrology what and how much *biota*, including vegetation, is found in the wetland. This model is reintroduced and discussed in more detail in Chapter 4: "Wetland Hydrology."

Difficulty of Defining Wetlands

Although the concepts of shallow water or saturated conditions, unique wetland soils, and vegetation adapted to wet conditions are fairly straightforward, combining these

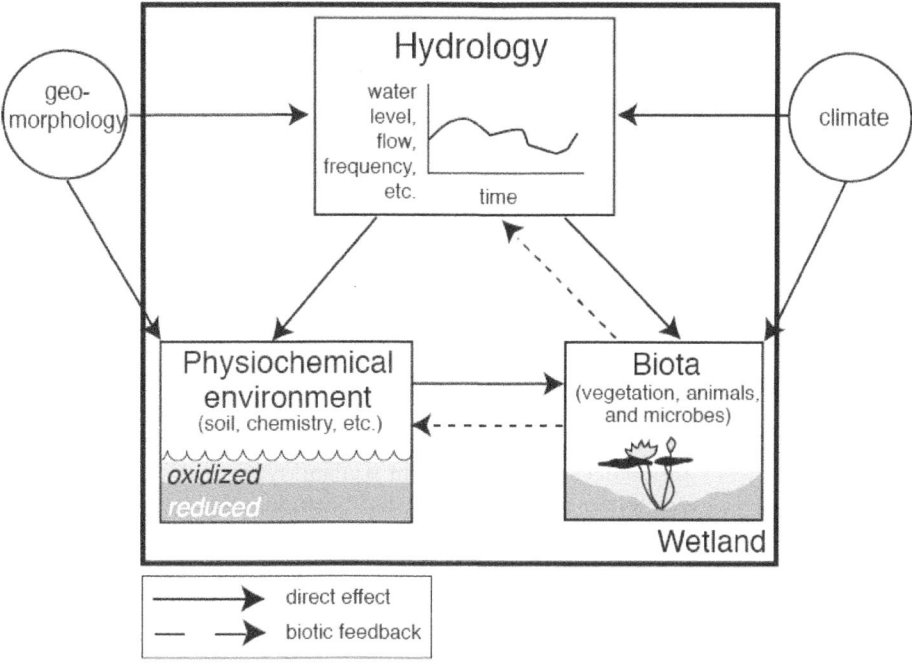

Figure 2.2 The three-component basis of a wetland definition: hydrology, physiochemical environment, and biota. From these components, the current approach to defining jurisdictional wetlands in the United States is based on three indicators: hydrology, soils, and vegetation. Note that these three components are not independent and that there is significant feedback from the biota to the physics, chemistry, and hydrology.

three factors to obtain a precise definition is difficult because of six characteristics that distinguish wetlands from other ecosystems yet make them less easy to define:

1. *Although water is present for at least part of the time, the depth and duration of flooding vary considerably from wetland to wetland and from year to year.* Some wetlands are continually flooded, whereas others are flooded only briefly at the surface or even just below the surface. Similarly, because fluctuating water levels can vary from season to season and year to year in the same wetland type, the boundaries of wetlands cannot always be determined by the presence of water at any one time.

2. *Wetlands are often located at the margins between deep water and terrestrial uplands and are influenced by both systems.* This ecotone position has been suggested by some as evidence that wetlands are mere extensions of either the terrestrial or the aquatic ecosystems or both and have no separate identity. There are, however, emergent properties in wetlands not contained in either upland or deepwater systems.

3. *Wetland species (plants, animals, and microbes) range from those that have adapted to live in either wet or dry conditions (facultative), which makes difficult their use as wetland indicators, to those adapted to only a wet environment (obligate).*
4. *Wetlands vary widely in size, ranging from small prairie potholes of a few hectares in size to large expanses of wetlands several hundreds of square kilometers in area.* Although this range in scale is not unique to wetlands, the question of scale is important for their conservation. Wetlands can be lost in large parcels or, more commonly, one small piece at a time in a process called *cumulative loss*. Are wetlands better defined functionally on a large scale or in small parcels?
5. *Wetland location can vary greatly, from inland to coastal wetlands and from rural to urban regions.* Whereas most ecosystem types—for example, forests or lakes—have similar ecosystem structure and function, there are great differences among different wetland types such as coastal salt marshes, inland pothole marshes, and forested bottomland hardwoods.
6. *Wetland condition, or the degree to which a wetland has been modified by humans, varies greatly from region to region and from wetland to wetland.* In rural areas, wetlands are likely to be associated with farmlands, whereas wetlands in urban areas are often subjected to the impact of extreme pollution and altered hydrology associated with housing, feeding, and transporting a large population. Many wetlands can easily be drained and turned into dry lands by human intervention; similarly, altered hydrology or increased runoff can cause wetlands to develop where they were not found before.

Wetlands have been described as a halfway world between terrestrial and aquatic ecosystems, exhibiting some of the characteristics of each system. They form part of a continuous gradient between uplands and open water. As a result, the exact upper and the lower limits of wetlands are arbitrary boundaries in any definition. Consequently, few definitions adequately describe all wetlands.

The problem of definition arises at the edges of wetlands, toward either wetter or drier conditions. How far upland and how infrequently should the land flood before we can declare that it is not a wetland? At the other edge, how far can we venture into a lake, pond, estuary, or ocean before we leave a wetland? Does a floating mat of vegetation define a wetland? What about a submerged bed of rooted vascular vegetation?

The frequency of flooding is another variable that has made the definition of wetlands particularly controversial. Some classifications include seasonally flooded bottomland hardwood forests, whereas others exclude them because they are dry for most of the year. Because wetland characteristics grade continuously from aquatic to terrestrial, there is no single, universally recognized definition of a wetland. This lack has caused confusion and inconsistencies in the management, classification, and inventorying of wetland systems, but considering the diversity of types, sizes, locations, and conditions of wetlands in this country, inconsistencies should be no surprise.

Wetland Common Terms

A number of common terms have been used over the years to describe different types of wetlands (Table 2.1). The number of common wetland words has risen from the 15 listed in the first edition of *Wetlands* (Mitsch and Gosselink, 1986) to 40 terms in this edition as we continue to discover terms in use. The history of the use and misuse of these words has often revealed a decidedly regional or at least continental origin. Although the lack of standardization of terms is confusing, many of the old terms are rich in meaning to those familiar with them. They often bring to mind vivid images of specific kinds of ecosystems that have distinct vegetation, animals, and other characteristics. Each of the terms has a specific meaning to some people, and many are still widely used by both scientists and laypersons alike. A *marsh* is known by most as an herbaceous plant wetland. A *swamp*, however, has woody vegetation, either shrubs or trees. There are subtle differences among marshes. A marsh with significant (>30 cm) standing water throughout much of the year is often called a *deepwater marsh*. A shallow marsh with waterlogged soil or shallow standing water is sometimes referred to as a *sedge meadow* or a *wet meadow*. Intermediate between a marsh and a meadow is a *wet prairie*. Several terms are used to denote peat-accumulating systems. The most general term is *peatland*, which is generally synonymous with *moor* and *muskeg*. There are many types of peatlands, however, the most general being *fens* and *bogs*.

Within the international scientific community, these common terms do not always convey the same meaning relative to a specific type of wetland. In fact, some languages have no direct equivalents for certain kinds of wetlands. The word *swamp* has no direct equivalent in Russian because the forested wetlands there are simply a variety of peatlands or bogs. *Bog*, however, can easily be translated because bogs are a common feature of the Russian landscape. The word *swamp* in North America clearly refers to a wetland dominated by woody plants—shrubs or trees. In Europe, *reedswamps* are dominated by reed grass (*Phragmites*), a dense-growing but nonwoody plant. In Africa, what would be called a *marsh* in the United States is referred to as a *swamp*. A cutoff meander of a river is called a *billabong* in Australia (Shiel, 1994) and an *oxbow* in North America.

Even common and scientific names of plants and animals can become confusing on a global scale. *Typha* spp., a cosmopolitan wetland plant, is called *cattail* in the United States, *reedmace* in the United Kingdom, *bulrush* in Africa, *cumbungi* in Australia, and *raupo* or *bulrush* in New Zealand. True bulrush is still called *Scirpus* spp. by some in North America and *Schoenoplectus* spp. in much of the rest of the world. *Scirpus fluviatilis* (river bulrush) is *Bolboschoenus fluviatilis* in much of the rest of the world. The great egret in North America is *Casmerodius albus*, whereas the great egret in Australia is *Ardea alba*. To further complicate matters, the Australian version of the great egret is called *Egretta alba* in New Zealand and is not called an egret at all but a white heron.

Confusion in terminology occurs because of different regional or continental uses of terms for similar types of wetlands. In North America, nonforested inland wetlands are often casually classified either as peat-forming, low-nutrient acid bogs or as

Table 2.1 Common terms used for various wetland types in the world

Billabong—Australian term for a riparian wetland that is periodically flooded by the adjacent stream or river.

Bog—A peat-accumulating wetland that has no significant inflows or outflows and supports acidophilic mosses, particularly *Sphagnum*.

Bottomland—Lowland along streams and rivers, usually on alluvial floodplains, that is periodically flooded. When forested, it is called a bottomland hardwood forest in the southeastern and eastern United States.

Carr—Term used in Europe for forested wetlands characterized by alders (*Alnus*) and willows (*Salix*).

Cumbungi swamp—Cattail (*Typha*) marsh in Australia.

Dambo—A seasonally waterlogged and grass-covered linear depression in headwater zone of rivers with no marked stream channel or woodland vegetation. The term is from the ChiChewa (Central Africa) dialect meaning "meadow grazing."

Delta—A wetland-river-upland complex located where a river forms distributaries as it merges with the sea; there are also examples of inland deltas, such as the Peace-Athabasca Delta in Canada and the Okavango Delta in Botswana (see Chapter 3: "Wetlands of the World").

Fen—A peat-accumulating wetland that receives some drainage from surrounding mineral soil and usually supports marshlike vegetation.

Lagoon—Term frequently used in Europe to denote a deepwater enclosed or partially opened aquatic system, especially in coastal delta regions.

Mangal—Same as mangrove.

Mangrove—Subtropical and tropical coastal ecosystem dominated by halophytic trees, shrubs, and other plants growing in brackish to saline tidal waters. The word *mangrove* also refers to the dozens of tree and shrub species that dominate mangrove wetlands.

Marsh—A frequently or continually inundated wetland characterized by emergent herbaceous vegetation adapted to saturated soil conditions. In European terminology, a marsh has a mineral soil substrate and does not accumulate peat. See also *tidal freshwater marsh* and *salt marsh*.

Mire—Synonymous with any peat-accumulating wetland (European definition); from the Norse word *myrr*. The Danish and Swedish word for peatland is now *mose*.

Moor—Synonymous with peatland (European definition). A highmoor is a raised bog; a lowmoor is a peatland in a basin or depression that is not elevated above its perimeter. The primitive sense of the Old Norse root is "dead" or barren land.

Muskeg—Large expanse of peatlands or bogs; particularly used in Canada and Alaska.

Oxbow—Abandoned river channel, often developing into a swamp or marsh.

Pakihi—Peatland in southwestern New Zealand dominated by sedges, rushes, ferns, and scattered shrubs. Most pakihi form on terraces or plains of glacial or fluvial outwash origin and are acid and exceedingly infertile.

Peatland—A generic term of any wetland that accumulates partially decayed plant matter (peat).

Playa—An arid- to semiarid-region wetland that has distinct wet and dry seasons. Term is used for shallow depressional recharge wetlands occurring in the Great Plains region of North America "that are formed through a combination of wind, wave, and dissolution processes" (Smith, 2003).

Pocosin—Peat-accumulating, nonriparian freshwater wetland, generally dominated by evergreen shrubs and trees and found on the southeastern coastal plain of the United States. The term comes from the Algonquin for "swamp on a hill."

Pokelogan—Northeastern U.S. marshy or stagnant water that has branched off from a stream or lake.

Pothole—Shallow marshlike pond, particularly as found in the Dakotas and central Canadian provinces, the so-called prairie pothole region.

Raupo swamp—Cattail (*Typha*) marsh in New Zealand.

Reedmace swamp—Cattail (*Typha*) marsh in the United Kingdom.

Reedswamp—Marsh dominated by *Phragmites* (common reed); term used particularly in Europe.

Table 2.1 (Continued)

Riparian ecosystem—Ecosystem with a high water table because of proximity to an aquatic ecosystem, usually a stream or river. Also called bottomland hardwood forest, floodplain forest, bosque, riparian buffer, and streamside vegetation strip.

Salt marsh—A halophytic grassland on alluvial sediments bordering saline water bodies where water level fluctuates either tidally or nontidally.

Sedge meadow—Very shallow wetland dominated by several species of sedges (e.g., *Carex, Scirpus, Cyperus*).

Shrub-Scrub Swamp—A freshwater wetland transitional between a forested swamp and a wet meadow or marsh, dominated by shrubs, with trees having less than 20 percent cover and less that 10 m height.

Slough—An elongated swamp or shallow lake system, often adjacent to a river or stream. A slowly flowing shallow swamp or marsh in the southeastern United States (e.g., cypress slough). From the Old English word *sloh*, meaning a watercourse running in a hollow.

Strand—Similar to a slough; a slow-flowing riverine/wetland system, often forested, found especially in south Florida, where gradients are low.

Swamp—Wetland dominated by trees or shrubs (U.S. definition). In Europe, forested fens and wetlands dominated by reed grass (*Phragmites*) are also called swamps (see *reedswamp*).

Tidal freshwater marsh—Marsh along rivers and estuaries close enough to the coastline to experience significant tides by nonsaline water. Vegetation is often similar to nontidal freshwater marshes.

Turlough—Areas seasonally flooded by karst groundwater with sufficient frequency and duration to produce wetland characteristics. They generally flood in winter and are dry in summer and fill and empty through underground passages. Term is specific for these types of wetlands found mostly in western Ireland.

Várzea—A seasonally flooded forest in the Amazon River Basin. It usually refers to forests flooded by whitewater (sediment-laden) river water.

Vernal pool—Shallow, intermittently flooded wet meadow, generally typical of Mediterranean climate with dry season for most of the summer and fall. Term is now used to indicate wetlands temporarily flooded in the spring throughout the United States.

Vleis—Seasonal wetland similar to a dambo; term used in southern Africa.

Wad (pl. wadden)—Unvegetated tidal flat originally referring to the northern Netherlands and northwestern German coastline. Now used throughout the world for coastal areas.

Wet meadow—Grassland with waterlogged soil near the surface but without standing water for most of the year.

Wet prairie—Similar to a marsh, but with water levels usually intermediate between a marsh and a wet meadow.

marshes. European terminology, which is much older, is also much richer and distinguishes at least four different kinds of freshwater wetlands—from mineral-rich reed beds, called reedswamps, to wet grassland marshes, to fens, and, finally, to bogs or moors. To some, all of these wetland types are considered *mires*. According to others, mires are limited to peat-building wetlands. The European classification is based on the amount of surface water and nutrient inflow (rheotrophy), type of vegetation, pH, and peat-building characteristics.

Two points can be made about the use of common terms in classifying wetland types: First, the physical and biotic characteristics grade continuously from one of these wetland types to the next; hence, any classification based on common terms is, to an extent, arbitrary. Second, the same term may refer to different systems in different regions. The common terms continue to be used, even in the scientific literature; we simply suggest that they be used with caution and with an appreciation for an international audience.

Formal Wetland Definitions

Precise wetland definitions are needed for two distinct interest groups: (1) wetland scientists and (2) wetland managers and regulators. The wetland scientist is interested in a flexible yet rigorous definition that facilitates classification, inventory, and research. The wetland manager is concerned with laws or regulations designed to prevent or control wetland modification and, thus, needs clear, legally binding definitions. Because of these differing needs, different definitions have evolved for the two groups. The discrepancy between the regulatory definition of *jurisdictional wetlands* and other definitions in the United States has meant, for example, that maps developed for wetland inventory purposes cannot be used for regulating wetland development. This is a source of considerable confusion to regulators and landowners.

Definitions that are more scientific in nature are presented in this section. Definitions that are more used in a legal sense are presented in the next section.

Early U.S. Definition: Circular 39 Definition

One of the earliest definitions of the term *wetlands* was presented by the U.S. Fish and Wildlife Service in 1956 in a publication that is frequently referred to as Circular 39 (Shaw and Fredine, 1956):

> The term "wetlands" ... refers to lowlands covered with shallow and sometimes temporary or intermittent waters. They are referred to by such names as marshes, swamps, bogs, wet meadows, potholes, sloughs, and river-overflow lands. Shallow lakes and ponds, usually with emergent vegetation as a conspicuous feature, are included in the definition, but the permanent waters of streams, reservoirs, and deep lakes are not included. Neither are water areas that are so temporary as to have little or no effect on the development of moist-soil vegetation.

The Circular 39 definition (1) emphasized wetlands that were important as waterfowl habitats and (2) included 20 types of wetlands that served as the basis for the main wetland classification used in the United States until the 1970s (see Chapter 13). It thus served the limited needs of both wetland managers and wetland scientists.

U.S. Fish and Wildlife Service Definition

Perhaps the most comprehensive definition of wetlands was adopted by wetland scientists in the U.S. Fish and Wildlife Service in 1979, after several years of review. The definition was presented in a report entitled *Classification of Wetlands and Deepwater Habitats of the United States* (Cowardin et al., 1979):

> Wetlands are lands transitional between terrestrial and aquatic systems where the water table is usually at or near the surface or the land is covered by shallow water.... Wetlands must have one or more of the following three attributes: (1) at least periodically, the land supports predominantly hydrophytes; (2) the substrate is predominantly undrained hydric soil; and (3) the substrate is nonsoil and is saturated with water or covered by shallow water at some time during the growing season of each year.

This definition was significant for its introduction of several important concepts in wetland ecology. It was one of the first definitions to introduce the concepts of *hydric soils* and *hydrophytes*, and it served as the impetus for scientists and managers to define these terms more accurately (National Research Council, 1995). Designed for scientists as well as managers, it is broad, flexible, and comprehensive, and includes descriptions of vegetation, hydrology, and soil. It has its main utility in scientific studies and inventories and generally has been more difficult to apply to the management and regulation of wetlands. It is still frequently accepted and employed in the United States today and was, at one time, accepted as the official definition of wetlands by India. Like the Circular 39 definition, this definition serves as the basis for a detailed wetland classification and an updated and comprehensive inventory of wetlands in the United States. The classification and inventory are described in more detail in Chapter 13.

Canadian Wetland Definitions

Canadians, who deal with vast areas of inland northern peatlands, have developed a specific national definition of wetlands. Two definitions were formally published in the book *Wetlands of Canada* by the National Wetlands Working Group (1988). First, Zoltai (1988) defined a wetland as:

> Land that has the water table at, near, or above the land surface or which is saturated for a long enough period to promote wetland or aquatic processes as indicated by hydric soils, hydrophytic vegetation, and various kinds of biological activity which are adapted to the wet environment.

Zoltai (1988) also noted that "wetlands include waterlogged soils where in some cases the production of plant materials exceeds the rate of decomposition." He describes the wet and dry extremes of wetlands as:

- Shallow open waters, generally less than 2 m; and
- Periodically inundated areas only if waterlogged conditions dominate throughout the development of the ecosystem.

Tarnocai et al. (1988) offered a slightly reworded definition in that same publication as the basis of the Canadian wetland classification system. That definition, repeated by Zoltai and Vitt (1995) and Warner and Rubec (1997) in later years, remains the official definition of wetlands in Canada:

> Land that is saturated with water long enough to promote wetland or aquatic processes as indicated by poorly drained soils, hydrophytic vegetation and various kinds of biological activity which are adapted to a wet environment.

These definitions emphasize wet soils, hydrophytic vegetation, and "various kinds" of other biological activity. The distinction between "hydric soils" in the Zoltai definition and "poorly drained soils" in the current, more accepted definition may be a reflection of the reluctance by some to use hydric soils exclusively to define wetlands. Hydric soils are discussed in more detail in Chapter 5: "Wetland Soils."

U.S. National Academy of Sciences Definition

In the early 1990s, amid renewed regulatory controversy in the United States as to what constitutes a wetland, the U.S. Congress asked the private, nonprofit National Academy of Sciences to appoint a committee through its principal operating agency, the National Research Council (NRC), to undertake a review of the scientific aspects of wetland characterization. The committee was charged with considering: (1) the adequacy of the existing definition of wetlands; (2) the adequacy of science for evaluating the hydrologic, biological, and other ways that wetlands function; and (3) regional variation in wetland definitions. The report produced by that committee two years later was entitled *Wetlands: Characteristics and Boundaries* (NRC, 1995) and included yet another scientific definition, referred to as a "reference definition" in that it was meant to stand "outside the context of any particular agency, policy or regulation":

> A wetland is an ecosystem that depends on constant or recurrent, shallow inundation or saturation at or near the surface of the substrate. The minimum essential characteristics of a wetland are recurrent, sustained inundation or saturation at or near the surface and the presence of physical, chemical, and biological features reflective of recurrent, sustained inundation or saturation. Common diagnostic features of wetlands are hydric soils and hydrophytic vegetation. These features will be present except where specific physiochemical, biotic, or anthropogenic factors have removed them or prevented their development.

Although little formal use has been made of this definition, it remains the most comprehensively developed scientific wetland definition. It uses the terms *hydric soils* and *hydrophytic vegetation*, as did the early U.S. Fish and Wildlife Service definition, but indicates that they are "common diagnostic features" rather than absolute necessities in designating a wetland.

An International Definition

The International Union for the Conservation of Nature and Natural Resources (IUCN) at the Convention on Wetlands of International Importance Especially as Waterfowl Habitat, better known as the Ramsar Convention, adopted the following definition of wetlands in Article 1.1 of the Convention of Wetlands (Finlayson and Moser, 1991):

> For the purposes of this Convention wetlands are areas of marsh, fen, peatland or water, whether natural or artificial, permanent or temporary, with water that is static or flowing, fresh, brackish, or salt including areas of marine water, the depth of which at low tide does not exceed six meters.

The Ramsar definition further provides in Article 1.2 of the Convention that wetlands

> may incorporate riparian and coastal zones adjacent to the wetlands, and islands or bodies of marine water deeper than six metres at low tide lying within the wetlands.

This definition, which was adopted at the first meeting of the convention in Ramsar, Iran, in 1971, does not include vegetation or soil and extends wetlands to water depths of 6 meters or more, well beyond the depth usually considered wetlands in the United States and Canada. The rationale for such a broad definition of wetlands "stemmed from a desire to embrace all the wetland habitats of migratory water birds" (Scott and Jones, 1995).

Legal Definitions

When protection of wetlands began in earnest in the mid-1970s in the United States, there arose an almost immediate need for precise definitions that were based as much on closing legal loopholes as on science. Two such definitions have developed in U.S. agencies—one for the U.S. Army Corps of Engineers to enforce its legal responsibilities with a "dredge-and-fill" permit program in the Clean Water Act and the other for the U.S. Natural Resources Conservation Service to administer wetland protection under the so-called swampbuster provision of the Food Security Act. Both agencies were parties to an agreement in 1993 to work together on administering a unified policy of wetland protection in the United States, yet the two separate definitions remain.

U.S. Army Corps of Engineers Definition

A U.S. government regulatory definition of wetlands is found in the regulations used by the U.S. Army Corps of Engineers for the implementation of a dredge-and-fill permit system required by Section 404 of the 1977 Clean Water Act amendments. That definition has now survived several decades in the legal world and is given as follows:

> The term "wetlands" means those areas that are inundated or saturated by surface or ground water at a frequency and duration sufficient to support, and that under normal circumstances do support, a prevalence of vegetation typically adapted for life in saturated soil conditions. Wetlands generally include swamps, marshes, bogs, and similar areas. (33 CFR 328.3(b); 1984)

This definition replaced a 1975 definition that stated "those areas that normally are characterized by the prevalence of vegetation that *requires* saturated soil conditions for growth and reproduction" (42 *Fed. Reg.* 3712X, July 19, 1977; italics added), because the Corps of Engineers found that the old definition excluded "many forms of truly aquatic vegetation that are prevalent in an inundated or saturated area, but that do not require saturated soil from a biological standpoint for their growth and reproduction." The words *normally* in the old definition and *that under normal circumstances do support* in the new definition were intended "to respond to situations in which an individual would attempt to eliminate the permit review requirements of Section 404 by destroying the aquatic vegetation" (quotes from 42 *Fed. Reg.* 37128, July 19, 1977). The need to revise the 1975 definition illustrates how difficult it has been to develop a legally useful definition that also accurately reflects the ecological reality of a wetland site.

This legal definition of wetlands has been debated in the courts in several cases, some of which have become landmark cases. In one of the first court tests of wetland protection, the Fifth Circuit of the U.S. Court of Appeals ruled in 1972, in *Zabel v. Tabb*, that the U.S. Army Corps of Engineers has the right to refuse a permit for filling of a mangrove wetland in Florida. In 1975, in *Natural Resources Defense Council v. Callaway*, wetlands were included in the category "waters of the United States," as described by the Clean Water Act. Prior to that time, the Corps of Engineers regulated dredge-and-fill activities (Section 404 of the Clean Water Act) for navigable waterways only; since that decision, wetlands have been legally included in the definition of waters of the United States.

In 1985, the question of regulation of wetlands reached the U.S. Supreme Court for the first time. The court upheld the broad definition of wetlands to include groundwater-fed wetlands in *United States v. Riverside Bayview Homes, Inc.* In that case, the Supreme Court affirmed that the U.S. Army Corps of Engineers had jurisdiction over wetlands that were adjacent to navigable waters, but it left open the question as to whether it had jurisdiction over nonadjacent wetlands (NRC, 1995). The legal definition of wetlands has been involved in the U.S. Supreme Court three times more, in 2001, 2006, and 2013; these cases are discussed in more detail in Chapter 15 "Wetland Laws and Protection."

Food Security Act Definition

In December 1985, the U.S. Department of Agriculture, through its Soil Conservation Service [now known as the Natural Resources Conservation Service (NRCS)], was brought into the arena of wetland definitions and wetland protection by means of a provision known as swampbuster in the 1985 Food Security Act. On agricultural land in the United States that, prior to December 1985, had been exempt from regulation, wetlands were now protected. As a result of this swampbuster provision, a definition, known as the NRCS or Food Security Act definition, was included in the Act (16 CFR 801(a)(16); 1985):

The term "wetland" except when such term is part of the term "converted wetland" means land that—

(A) has a predominance of hydric soils;
(B) is inundated or saturated by surface or ground water at a frequency and duration sufficient to support a prevalence of hydrophytic vegetation typically adapted for life in saturated soil conditions; and
(C) under normal circumstances does support a prevalence of such vegetation.

For purposes of this Act and any other Act, this term shall not include lands in Alaska identified as having high potential for agricultural development which have a predominance of permafrost soils.

The emphasis on this agriculture-based definition is on hydric soils. The omission of wetlands that do not have hydric soils, while not invalidating this definition, makes it less comprehensive than some others—for example, the NRC (1995) definition. A curious feature of this definition is its wholesale exclusion of the largest state in the United States from the definition of wetlands. The exclusion of Alaskan wetlands that have a high potential for agriculture makes this definition even less of a scientific and more of a regulatory or even political definition. There is no scientific distinction between the characteristics of Alaskan wetlands and wetlands in the rest of the United States except for climatic differences and the presence of permafrost under many but certainly not all Alaskan wetlands (NRC, 1995).

Jurisdictional Wetlands

Since 1989, the term *jurisdictional wetland* has been used for legally defined wetlands in the United States to delineate those areas that are under the jurisdiction of Section 404 of the Clean Water Act or the swampbuster provision of the Food Security Act. The U.S. Army Corps of Engineers' definition cited previously emphasizes only one indicator, vegetative cover, to determine the presence or absence of a wetland. It is difficult to include soil information and water conditions in a wetland definition when its main purpose is to determine jurisdiction for regulatory purposes and there is little time to examine the site in detail. The Food Security Act definition, however, includes hydric soils as the principal determinant of wetlands.

It is likely that most of the wetlands that are considered jurisdictional wetlands by the preceding two legal definitions fit the scientific definition of wetlands. It is also just as likely that some types of wetlands, particularly those that have less chance of developing hydric soil characteristics or hydrophytic vegetation (e.g., riparian wetlands), would not be identified as jurisdictional wetlands with the legal definitions. And of course, excluding Alaskan wetlands "having high potential for agricultural development" from the Food Security Act definition has no scientific basis at all but is a political decision.

Those who delineate wetlands are interested in a definition that allows the rapid identification of a wetland and the degree to which it has been or could be altered. They are interested in the delineation of wetland boundaries, and establishing boundaries is facilitated by defining the wetland simply, according to the presence or absence of certain species of vegetation or aquatic life or the presence of simple indicators, such as hydric soils. Several U.S. federal manuals spelling out specific methodologies for identifying jurisdictional wetlands were written or proposed in the 1980s and early 1990s. The manuals differed, however, in the prescribed ways these three criteria are proved in the field. The first of these manuals (U.S. Army Corps of Engineers, 1987), is now the accepted version and is widely used to field-identify wetlands in the United States. All three manuals indicated that the three criteria for wetlands—namely, wetland hydrology, wetland soils, and hydrophytic vegetation—must be present. As illustrated in Figure 2.2, these three variables are not independent; strong evidence of long-term wetland hydrology, for example, should almost ensure that the other two variables are present. Furthermore, potentially other indicators of the physiochemistry and biota beyond hydric soils and hydrophytic vegetation may one day serve as useful indicators of wetlands.

Choice of a Definition

A wetland definition that will prove satisfactory to all users has not yet been developed because the definition of wetlands depends on the objectives and the field of interest of the user. Different definitions can be formulated by the geologist, soil scientist, hydrologist, biologist, ecologist, sociologist, economist, political scientist, public health scientist, and lawyer. This variance is a natural result of the differences in emphasis in the definer's training and of the different ways in which individual disciplines deal with wetlands. For ecological studies and inventories, the U.S. Fish and Wildlife Service definition has been and should continue to be applied to wetlands in the United States. Although somewhat generous in defining wetlands on the wet edge, the Ramsar definition is firmly entrenched in international circles. When wetland management, particularly regulation, is necessary, the U.S. Army Corps of Engineers' definition, as modified, is probably most appropriate.

Just as important as the precision of the definition of a wetland, however, is the consistency with which it is used. That is the difficulty we face when science and legal issues meet, as they often do, in resource management questions such as wetland conservation versus wetland drainage. Applying a comprehensive definition in a uniform and fair way requires a generation of well-trained wetland scientists and managers armed with a fundamental understanding of the processes that are important and unique to wetlands.

Recommended Reading

National Research Council. 1995. *Wetlands: Characteristics and Boundaries*. Washington, DC: National Academies Press.

References

Cowardin, L. M., V. Carter, F. C. Golet, and E. T. LaRoe. 1979. *Classification of Wetlands and Deepwater Habitats of the United States.* FWS/OBS-79/31. U.S. Fish and Wildlife Service, Washington, DC. 103 pp.

Finlayson, M., and M. Moser, eds. 1991. *Wetlands.* Facts on File, Oxford, UK. 224 pp.

Mitsch, W. J., and J. G. Gosselink. 1986. *Wetlands.* Van Nostrand Reinhold, New York. 539 pp.

National Research Council (NRC). 1995. *Wetlands: Characteristics and Boundaries.* National Academy Press, Washington, DC. 306 pp.

National Wetlands Working Group. 1988. *Wetlands of Canada.* Ecological and Classification Series 24, Environment Canada, Ottawa, Ontario, and Polyscience Publications, Montreal, Quebec. 452 pp.

Scott, D. A., and T. A. Jones. 1995. Classification and inventory of wetlands: A global overview. *Vegetatio* 118: 3–16.

Shaw, S. P., and C. G. Fredine. 1956. *Wetlands of the United States, Their Extent, and Their Value for Waterfowl and Other Wildlife.* Circular 39, U.S. Fish and Wildlife Service, U.S. Department of Interior, Washington, DC. 67 pp.

Shiel, R. J. 1994. Death and life of the billabong. In X. Collier, ed. *Restoration of Aquatic Habitats.* Selected Papers from New Zealand Limnological Society 1993 Annual Conference, Department of Conservation, pp. 19–37.

Smith, L. M. 2003. *Playas of the Great Plains.* University of Texas Press, Austin, Texas, 257 pp.

Tarnocai, C., G. D. Adams, V. Glooschenko, W. A. Glooschenko, P. Grondin, H. E. Hirvonen, P. Lynch-Stewart, G. F. Mills, E. T. Oswald, F. C. Pollett, C. D. A. Rubec, E. D. Wells, and S. C. Zoltai. 1988. The Canadian wetland classification system. In National Wetlands Working Group, ed. Wetlands of Canada. Ecological Land Classification Series 24, Environment Canada, Ottawa, Ontario, and Polyscience Publications, Montreal, Quebec, pp. 413–427.

U.S. Army Corps of Engineers. 1987. *Corps of Engineers Wetlands Delineation Manual.* Technical Report Y-87-1. U.S. Army Corps of Engineers Waterways Experiment Station, Vicksburg, MS. 100 pp. and appendices.

Warner, B. G., and C. D. A. Rubec, eds. 1997. *The Canadian Wetland Classification System.* National Wetlands Working Group, Wetlands Research Centre, University of Waterloo, Ontario.

Zoltai, S. C. 1988. Wetland environments and classification. In National Wetlands Working Group, ed. *Wetlands of Canada.* Ecological Land Classification Series 24, Environment Canada, Ottawa, Ontario, and Polyscience Publications, Montreal, Quebec, pp. 1–26.

Zoltai, S. C., and D. H. Vitt. 1995. Canadian wetlands: Environmental gradients and classification. *Vegetatio* 118: 131–137.

Chapter 3

Wetlands of the World

The extent of the world's wetlands is now thought to be from 7 to 10 million km^2, or about 5 to 8 percent of the land surface of Earth. The loss of wetlands in the world is difficult to determine, but recent estimates suggest that we have lost more than half of the world's wetlands, with much of that occurring in the twentieth century. The United States had a 50 percent loss rate for the lower 48 states from the 1770s to 1970s. There are also high rates of wetland loss in Europe and parts of Australia, Canada, and Asia and lower rates in less developed areas like Africa, South America, and northern boreal regions. Estimated areas of wetlands in North America are 44 million hectares (ha) in the lower 48 states, 71 million ha in Alaska, and 127 million ha in Canada, representing in total about 30 percent of the world's wetlands. In this chapter we also describe a number of important wetlands from around the world, including the Florida Everglades and the Louisiana Delta in the United States, the Pantanal and Amazon in South America, the Okavango Delta and the Congolian Swamp in Africa, the Mesopotamian Marshlands in the Middle East, Australian billabongs, and wetlands in natural areas and parks throughout China. All of these wetlands are impacted by human activities to some degree, yet most remain functional ecosystems.

The Global Extent of Wetlands

Wetlands include the swamps, bogs, marshes, mires, fens, and other wet ecosystems found throughout the world. They are found on every continent except Antarctica and in every clime, from the tropics to the tundra (Fig. 3.1a). Any estimate of the extent of wetlands in the world is difficult and depends on the definition used as described in Chapter 2: "Wetland Definitions"; there also is the pragmatic difficulty of quantifying

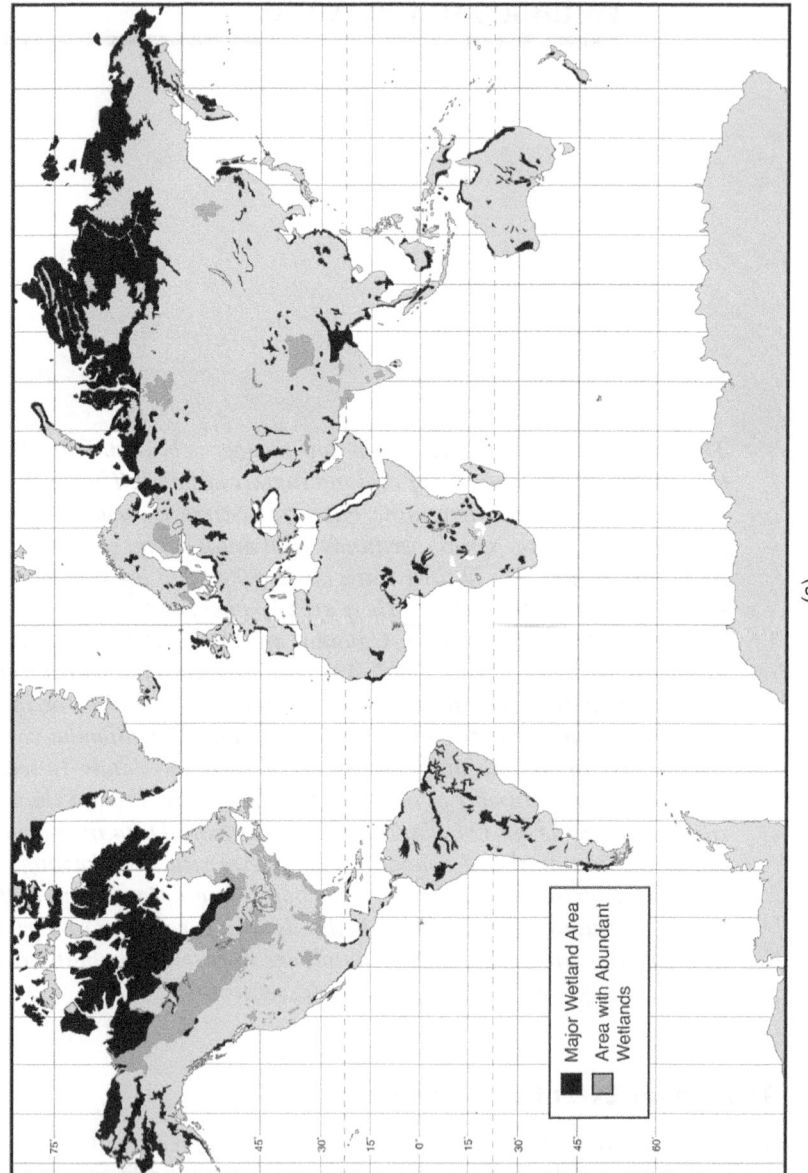

Figure 3.1 Wetlands of the world: (a) general extent determined a composite from a number of separate sources, and (b) distribution of wetlands with latitude based on data from Matthews and Fung (1987) and Lehner and Döll (2004).

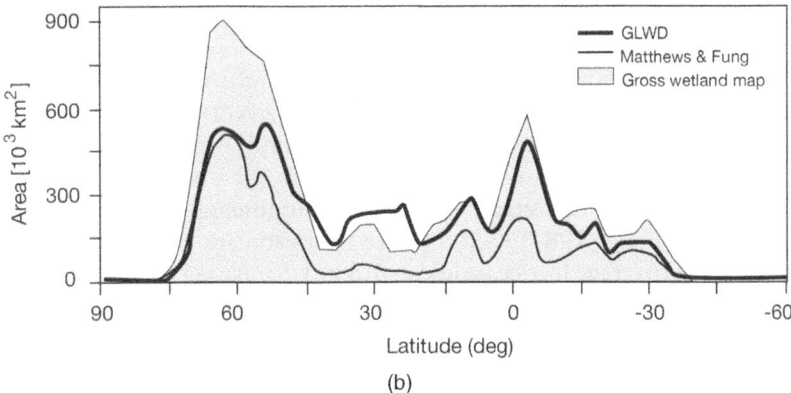

Figure 3.1 *(Continued)*

wetlands in aerial and satellite images that are now the most common sources of data. It is now fairly well established that most of the world's wetlands are found in both boreal and tropical regions of the world and the least amount of wetlands are found in temperate zones (Fig. 3.1b).

Based on several studies (Table 3.1), we now estimate that the extent of the world's wetlands is 7 to 10 million km^2, or about 5 to 8 percent of the land surface of Earth. We estimate that number by deleting the highest (Finlayson and Davidson, 1999) and lowest (Matthews and Fung, 1987) estimates in Table 3.1 and using the remaining numbers to provide a range. We believe that the estimate provided by Lehner and Döll (2004) of 8 to 10 million km^2 is the most detailed study on global wetland inventory and may be the most accurate.

Earlier wetland estimates provided a narrower range. Maltby and Turner (1983), based on the work of Russian geographers, estimated that more than 6.4 percent of the land surface of the world, or 8.6 million km^2, is wetland. Almost 56 percent of this estimated total wetland area is found in tropical (2.6 million km^2) and subtropical (2.1 million km^2) regions. Using global digital databases (1 degree resolution),

Table 3.1 Comparison of estimates of extent of wetlands in the world by climatic zone

Zone[1]	Wetland Area (× 10^6 km^2)						
	Maltby and Turner (1983)[2]	Matthews and Fung (1987)	Aselmann and Crutzen (1989)	Gorham (1991)	Finlayson and Davidson (1999)	Ramsar Convention Secretariat (2004)	Lehner and Döll (2004)
Polar/boreal	2.8	2.7	2.4	3.5	–	–	–
Temperate	1.0	0.7	1.1	–	–	–	–
Subtropical/tropical	4.8	1.9	2.1	–	–	–	–
Rice paddies	–	1.5	1.3	–	–	1.3	–
Total wetland area	8.6	6.8	6.9	–	12.8	7.2	8.2–10.1

[1] Definitions of polar, boreal, temperate, and tropical vary among studies.
[2] Based on Bazilevich et al. (1971).

Matthews and Fung (1987) estimated that there were 5.3 million km^2 of wetlands in the world, with a higher percentage of wetlands being boreal and a far lower percentage of wetlands being subtropical and tropical than those estimated by Maltby and Turner (1983). Aselmann and Crutzen (1989) estimated that there were 5.6 million km^2 of natural wetlands in the world, with a higher amount and percentage of wetlands in the temperate region than given in either of the earlier estimates. They used regional wetland surveys and monographs rather than maps, which Matthews and Fung (1987) used to make their estimate. These two research groups estimated the coverage by rice paddies—1.3 to 1.5 million km^2—but did not include this in their total wetland area. By including rice fields, their estimates of the extent of the world's wetlands are 6.8 and 6.9 million km^2, respectively. Bogs and fens accounted for about 60 percent of the world's wetlands (3.35 million km^2) in the Matthews and Fung (1987) study, an estimate that is very close to Gorham's (1991) 3.46 million km^2 estimate for northern boreal and subarctic peatlands. Aselmann and Crutzen (1989) described bogs and fens as also occurring in both temperate (40°–50° N) and tropical latitudes. Both Matthews and Fung (1987) and Aselmann and Crutzen (1989) showed a much lower extent of wetlands in tropical and subtropical regions than did Maltby and Turner (1983), although definitions of zones differ.

Finlayson and Davidson (1999) estimated that there were 12.8 million km^2 of wetlands using the international Ramsar definition described in Chapter 2. This estimate, which is 30 percent or more higher than the other estimates reported in the literature, was repeated in a Millennium Ecosystem Assessment (2005) report (coauthored by Finlayson and Davidson) on wetlands and water. This estimate includes all freshwater lakes, reservoirs, and rivers and near-shore marine ecosystems up to 6 m depth in the world, aquatic ecosystems that are not included in all wetland definitions. Ironically the Millennium Ecosystem Assessment (2005) report, in describing this high estimate, suggests that "it is well established that this estimate is an underestimate."

Lehner and Döll (2004) provide one of the most comprehensive and recent examinations of the global extent of wetlands. Their geographic information system (GIS)–based Global Lakes and Wetlands Database (GLWD) system focused on three coordinated levels: (1) large lakes and reservoirs, (2) smaller water bodies, and (3) wetlands. With the first two categories excluded, 8.3 to 10.2 million km^2 of wetlands in the world was estimated. As with several of the other studies summarized above and in Table 3.1, the greatest proportion of wetlands were found in the northern boreal regions (peaking at 60° N latitude) with another peak of tropical wetlands exactly at the equator (Fig. 3.1b).

Worldwide Wetland Losses

The rate at which wetlands are being lost on a global scale is only now becoming clear, in part with the use of new technologies associated with satellite imagery. But there are still many vast areas of wetlands where accurate records have not been kept, and many wetlands in the world were drained centuries ago. These impacts are discussed in more detail in Chapter 14 "Human Impacts and Management of Wetlands." It is probably safe to assume that (1) we are still losing wetlands at a fairly rapid rate globally, particularly in developing countries; and (2) we have lost half or more of the world's original wetlands. A study published by The Economics of Ecosystems & Biodiversity (TEEB) (Russi et al. 2013) reported that the world actually lost half of its wetlands in the twentieth century alone, with the expanse being reduced from 25 million km^2 to the current 12.8 million km^2. Davidson (2014), in an analysis of 63 reports and other publications, determined that the world lost 53:5 percent of its wetlands "long-term" (i.e., multi-century) with higher loss rates in inland vs. coastal wetlands (60.8 versus 46.4 percent, respectively). An extrapolation of data in a different calculation gives another statistic—that the world lost 87 percent of its wetlands since 1700. He also found out that the wetland rate of loss in the twentieth to early twenty-first centuries was 3.7 times faster than the long-term loss rate.

Prigent et al. (2012) found a 6 percent decrease in land-surface water on the world from 1993 to 2007 alone, presumably mostly due to wetland drainage and increased water withdrawals. This represents a net reduction of 0.33 million km^2 of wetlands in 15 years. Fifty-seven percent of the decrease occurred in tropical and subtropical regions.

There are some areas where the loss rate has been documented (Table 3.2). The estimate of about 53 percent loss of wetlands since European settlement in the lower 48 United States is fairly accurate. By 1985, 56 to 65 percent of wetlands in North America and Europe, 27 percent in Asia, 6 percent in South America, and 2 percent in Africa had been drained for intensive agriculture (Ramsar Convention Secretariat, 2004). Several regions of the world have lost considerable wetlands. A 90 percent loss of wetlands in New Zealand is documented. An early loss rate of 60 percent from China is based on the estimate of 250,000 km^2 of natural wetlands in the country out of a total of 620,000 km^2, including artificial wetlands such as rice paddies (Lu, 1995). More recent studies suggest that China may have lost 33 percent of its wetlands but historically had much higher rates on coastal areas and in the Tibetan Plateau. Europe has lost an estimated 60 to 80 percent of its wetlands, most due to agricultural conversion. Spain has lost more than 60 percent of its inland wetlands, and Lithuania, 70 percent of its total wetlands since 1970; Sweden drained 67 percent of its wetlands and ponds since the 1950s (Revenga et al., 2000).

North American Wetland Changes

The best and most recent estimate is that there are 44.6 million ha of wetlands in the lower 48 (conterminous) states of the United States (Table 3.3). In addition, there are an estimated 71 million ha of wetlands in Alaska. The inclusion of Alaska in wetland surveys of the United States increases the wetland inventory in the country by 160 percent. Combining these numbers with estimates of wetland areas from Canada and Mexico (described below), North America has about 2.5 million km^2 of wetlands, or an estimated 30 percent of the world's wetlands.

Overall, 53 percent of the wetlands in the conterminous United States were estimated to have been lost from the 1780s to the 1980s (Table 3.4). Estimates of the area of wetlands in the United States, while they vary widely, are becoming quite accurate

Table 3.2 Loss of wetlands in various locations in the world

Location	Percentage Loss (%)	Reference
United States (1780s–1980s[1])	53	Dahl (1990)
Canada		National Wetlands Working Group (1988)
Atlantic tidal and salt marshes	65	
Lower Great Lakes–St. Lawrence River	71	
Prairie potholes and sloughs	71	
Pacific coastal estuarine Wetlands	80	
Australia	>50	Australian Nature Conservation Agency (1996)
Swan Coastal Plain	75	
Coastal New South Wales	75	
Victoria	33	
River Murray Basin	35	
New Zealand	>90	Dugan (1993)
Philippines (mangroves)	67	Dugan (1993)
China	60	Lu (1995)
Coastal wetlands, 1950–2010	57	Qiu (2011)
Mangroves, 1950–2010	73	
All China, 1978–2008	33	Niu et al. (2011)
Tibetan Plateau, 1978–1990	66	
Tibetan Plateau, 2000–2008	6	
Europe		
Loss due to agriculture	80	Rovonga et al. (2000)
Overall estimated loss	80	Verhoeven (2014)

[1]Lower 48 states only.

(Table 3.3), and most studies indicate a rapid rate of wetland loss in the United States prior to the mid-1970s, a steady but significant reduction in the loss rate to about the mid-1980s, and almost no loss in wetland area over the most recent 12 years of record of 1997 to 2009 (Table 3.4).

The early numbers of wetland area vary widely for four reasons:

1. *The purposes of the inventories varied from study to study.* Early wetland censuses—for example, Wright (1907) and Gray et al. (1924)—were undertaken to identify lands suitable for drainage for agriculture. Later inventories of wetlands (Shaw and Fredine, 1956) were concerned with only those wetlands important for waterfowl protection. Only within the last three decades have wetland inventories considered all of the wetland ecosystem services.

2. *The definition and classification of wetlands varied with each study*, ranging from simple terms to complex hierarchical classifications.

Table 3.3 Estimates of wetland area in the United States at different times

Period or Year of Estimate	Wetland Area ($\times 10^6$ ha)[1]	Reference
Presettlement	87	Roe and Ayres (1954)
	86.2	USDA estimate, in Dahl (1990)
	89.5	Dahl (1990)
1906	32[2]	Wright (1907)
1922	37 (total)	Gray et al. (1924)
	3 (tidal)	
	34 (inland)	
1940	39.4[3]	Whooten and Purcell (1949)
1954	30.1[4] (total)	Shaw and Fredine (1956)
	3.8 (coastal)	
	26.3 (inland)	
1954	43.8 (total)	Frayer et al. (1983)
	2.3 (estuarine)	
	41.5 (inland)	
1974	40.1 (total)	Frayer et al. (1983); Tiner (1984)
	2.1 (estuarine)	
	38.0 (inland)	
mid-1970s	42.8[5] (total)	Dahl and Johnson (1991)
	2.2 (estuarine)	
	40.6 (inland)	
mid-1980s	41.8[5]	
	2.2 (estuarine)	
	39.3 (inland)	
1997	42.7	Dahl (2000)
	2.14 (estuarine)	
	40.56 (inland)	
2004	43.6	Dahl (2006)
	2.15 (estuarine)	
	41.45 (inland)	
2009	44.56	Dahl (2011)
	2.34 (estuarine)	
	42.22 (inland)	

[1] For 48 conterminous states unless otherwise noted.
[2] Does not include tidal wetlands or eight public land states in West.
[3] Outside of organized drainage enterprises.
[4] Only included wetlands important for waterfowl.
[5] Based on estimates of National Wetland Inventory (NWI) classes for vegetated estuarine and palustrine wetlands.

3. *The methods available for estimating wetlands changed over the years or varied in accuracy.* Remote sensing from aircraft and satellites is one example of a technique for wetland studies that was not generally available or used before the 1970s. Early estimates, in contrast, were often based on fragmentary records.

4. *In a number of instances, the borders of geographical or political units changed between censuses,* leading to gaps or overlaps in data.

Table 3.4 Estimates of wetland changes in the conterminous United States. (All changes were losses until the most recent measurements, which indicated wetland gains.)

Period	Wetland Change			Reference
	million ha	ha/yr	Percentage (%)	
Presettlement–1980s	−47.3	−236,500	−53	Dahl (1990)
1950s–1970s	−3.7	−185,000	−8.5	Frayer et al. (1983)
1970s–1980s	−1.06	−105,700	−2.5	Dahl and Johnson (1991)
1986–1997	−0.26	−23,700	−0.6	Dahl (2000)
1997–2004	+0.19	+12,900	+0.44	Dahl (2006)
2004–2009	−0.25	−5,590	−0.1	Dahl (2011)

Several states in the midwestern United States (Illinois, Indiana, Iowa, Kentucky, Missouri, and Ohio) plus California all have had wetland losses of more than 80 percent, principally for agricultural production; these seven states collectively show a loss of 14.1 million ha of wetlands during the past 200 years, or 30 percent of the wetland loss of the entire conterminous United States. States with high densities of wetlands—Minnesota, Illinois, Louisiana, and Florida—had among the highest losses of total area of wetlands—2.6, 2.8, 3.0, and 3.8 million ha, respectively.

Estimates of wetland loss in the last 30 years suggest a substantial decrease in the wetland loss rate in the lower 48 states. Frayer et al. (1983) estimated a net loss from the 1950s to the 1970s of more than 3.7 million ha (8.5 percent loss), or an average annual loss of 185,000 ha. This loss represents a wetland area equivalent to the combined size of Massachusetts, Connecticut, and Rhode Island. Freshwater marshes and forested wetlands were hardest hit. Wetland losses continued into the 1980s and 1990s, but the enactment of strong wetland protection laws in the mid-1980s, combined with interest in wetland restoration and stormwater pond creation, has had a dramatic effect. Wetland losses decreased from about 105,700 ha for the 1970s to 1980s (2.5 percent loss) to 23,700 ha from the mid-1980s to mid-1990s (0.6 percent loss). The loss changed to a gain of 12,900 ha of wetlands (0.44 percent gain) from 1998 to 2004, albeit mostly as gains in open-water ponds. The comparison of wetland area between 2004 and 2009 showed no statistical difference in wetland coverage between the two years. While it has been difficult to document, wetland losses have been at least partially offset in area by creation and restoration of wetlands and the creation of rural and suburban ponds during this period. The question remains as to whether these ponds and other additions to the wetland ledger are functioning wetlands.

Wetland Conversions—What Wetlands Are We Really Losing (and Gaining)?

By themselves, estimates of net wetland losses or gains provide an incomplete picture of the dynamics of change. A more complete picture would show that human activities converted millions of hectares of wetlands from one class to another. Through these conversions, some wetland classes increased in area at the expense of other types. Considering the period from the mid-1970s to the mid-1980s, for example, swamps and forested riparian wetlands in the United States suffered the greatest loss, 1.4 million ha (Fig. 3.2). Although 800,000 ha were converted to agricultural and other

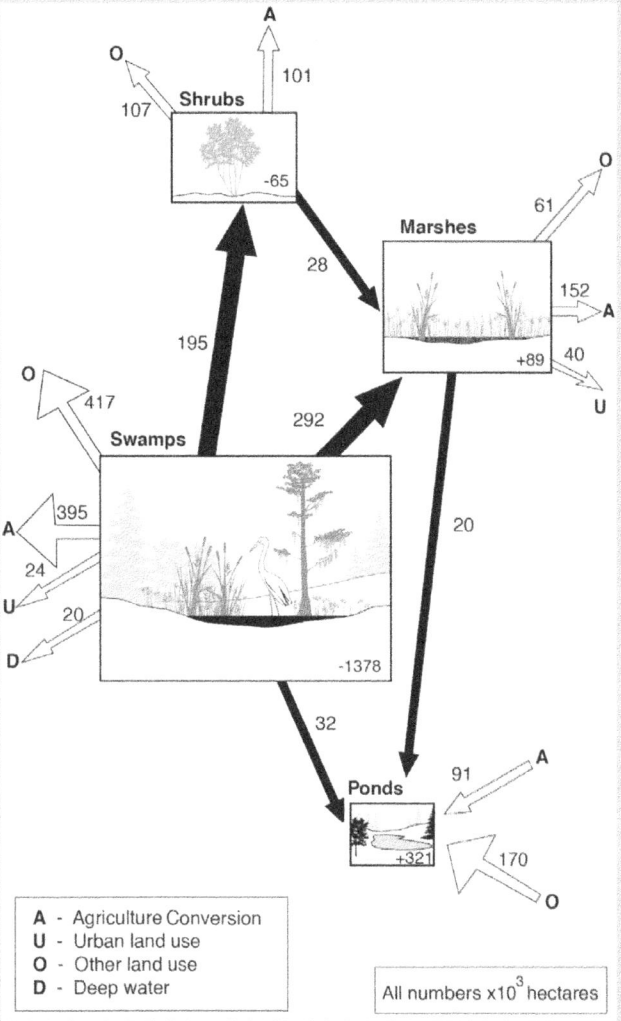

Figure 3.2 Wetland conversion in the conterminous United States, mid-1970s to mid-1980s. The figure shows how misleading the net change figures are. For example, although there was a net gain in freshwater marshes (89,000 ha), it occurred along with a loss of about 1,378,000 ha of swamps, some of which were converted to freshwater marshes. (After Dahl and Johnson, 1991)

land uses, large areas were converted to other wetland types: 292,000 ha to marshes, 195,000 ha to scrub and shrubs, and 32,000 ha to nonvegetated wetlands. Although shrub wetlands lost 208,000 ha to agriculture and other nonwetland uses, this was almost offset by the conversion of forested wetlands to shrubs, leaving a net loss of 65,000 ha. A net gain of 89,000 ha of marshes occurred despite a loss of 213,000 ha to agriculture and other land uses, because 320,000 ha of swamps and shrub wetlands changed to marshes. In this example, most of the scrub–shrub wetlands are probably areas recently cut over for their timber.

For 1998 to 2004, wetlands were shown to actually increase in the United States by 12,900 ha per year (ha/yr). The excitement of actually seeing an increase in wetlands for the first time in 200 years in the United States was dampened by the fact that this increase was a result of an increase of 46,900 ha/yr of freshwater ponds (13 percent increase). Furthermore, there was a net gain of 37,000 ha/yr of forested wetlands (1.1 percent increase), but these gains were balanced by losses of 60,800 ha/yr of scrub wetlands (4.9 percent decrease), 9,600 ha/yr of freshwater emergent marshes (0.5 percent decrease), and 2,240 ha/yr of estuarine emergent marshes (0.7 percent decrease). In essence, there were large gains in unvegetated ponds in human developments (farms, suburban developments, and even golf course ponds) and forested wetlands that were countered, respectively, by losses in marshes and shrub wetlands (many of which became forested wetlands). Describing wetland losses and gains is not a simple exercise.

Canada has about three times the area of wetlands found in the lower 48 states of the United States, or about 127 million ha of wetlands (about 14 percent of the country). Most of that area (111.3 million ha) is defined as peatland. The greatest concentration of Canadian wetlands can be found in the provinces of Manitoba and Ontario. The National Wetlands Working Group (1988), which provided a particularly comprehensive description of major regional wetlands in Canada, estimated that there were 22.5 million ha and 29.2 million ha, respectively, of wetlands in these two provinces, or about 41 percent of the total wetlands of Canada. Much of this total is boreal forested peatlands as bogs and fens, but there are also many shoreline marshes and floodplain swamps in the region.

Because of the vastness of Canada and its wetlands, and because the low-population regions have had less impact on wetland loss than the coastal and southern regions of Canada, there have been few attempts to summarize the loss of wetlands in Canada to one number, as has been the case for the conterminous United States. Locally, there are many regions of southern and coastal Canada where high rates of wetland loss have been experienced, and some detailed estimates do exist for the more populated regions of Canada. There has been a 65 to 80 percent loss of coastal marshes in the Atlantic and Pacific regions, respectively, a 71 percent loss of all wetlands in the lower Great Lakes, and a 71 percent loss of wetlands in the prairie pothole region (Table 3.2). Even higher loss rates have occurred in the major urban areas of Canada. The most extensive wetland loss has occurred in southern Ontario, Canada's most populated region, particularly from Windsor on the west toward and past Toronto on the east, where 80 to more than 90 percent wetland loss is common. Farther north to Quebec City, Quebec, and farther west to Thunder Bay, Ontario, loss rates are lower. Few data are available on the conversion of wetlands to other uses in rural areas, even in eastern Canada. However, studies have suggested that 32 percent of the tidal marshes along the St. Lawrence Estuary were converted to agricultural use and that, on the St. Lawrence River between Cornwall and Quebec, there was a 7 percent loss in wetland area from 1950 to 1978 alone (National Wetlands Working Group, 1988).

Regional Wetlands of the World

The remainder of this chapter describes some of the regionally important wetlands found around the world (Fig. 3.3). We cannot possibly include every major wetland in the world in this section, but we chose to present a wide diversity of international wetlands and attempted to give a broad range of wetland ecosystems. Each of these regional wetland areas or specific wetlands has or had a significant influence on the culture and development of its region. Some areas, such as the Florida Everglades, have had the luxury of major investigations by wetland scientists or books written for both academicians and the public. These studies and books have taught us much about wetlands and have identified much of their intrinsic values.

North America

Many regions in the United States and Canada support, or once supported, large contiguous wetlands or many smaller and more numerous wetlands. Some are often large, heterogeneous wetland areas, such as the Okefenokee Swamp in Georgia and Florida, that defy categorization as one type of wetland ecosystem. Others can also be large regions containing a single class of small wetlands, such as the prairie pothole region of Manitoba, Saskatchewan, and Alberta in Canada and the Dakotas and Minnesota in the United States. Some regional wetlands, such as the Great Dismal Swamp on the Virginia–North Carolina border, have been drastically altered since presettlement times, and others, such as the Great Kankakee Marsh of northern Indiana and Illinois and the Great Black Swamp of northwestern Ohio, have virtually disappeared as a result of extensive drainage programs.

The Florida Everglades

The southern tip of Florida, from Lake Okeechobee southward to the Florida Bay, harbors one of the unique regional wetlands in the world. The region encompasses three major types of wetlands in its 34,000-km^2 area: the Everglades, the Big Cypress

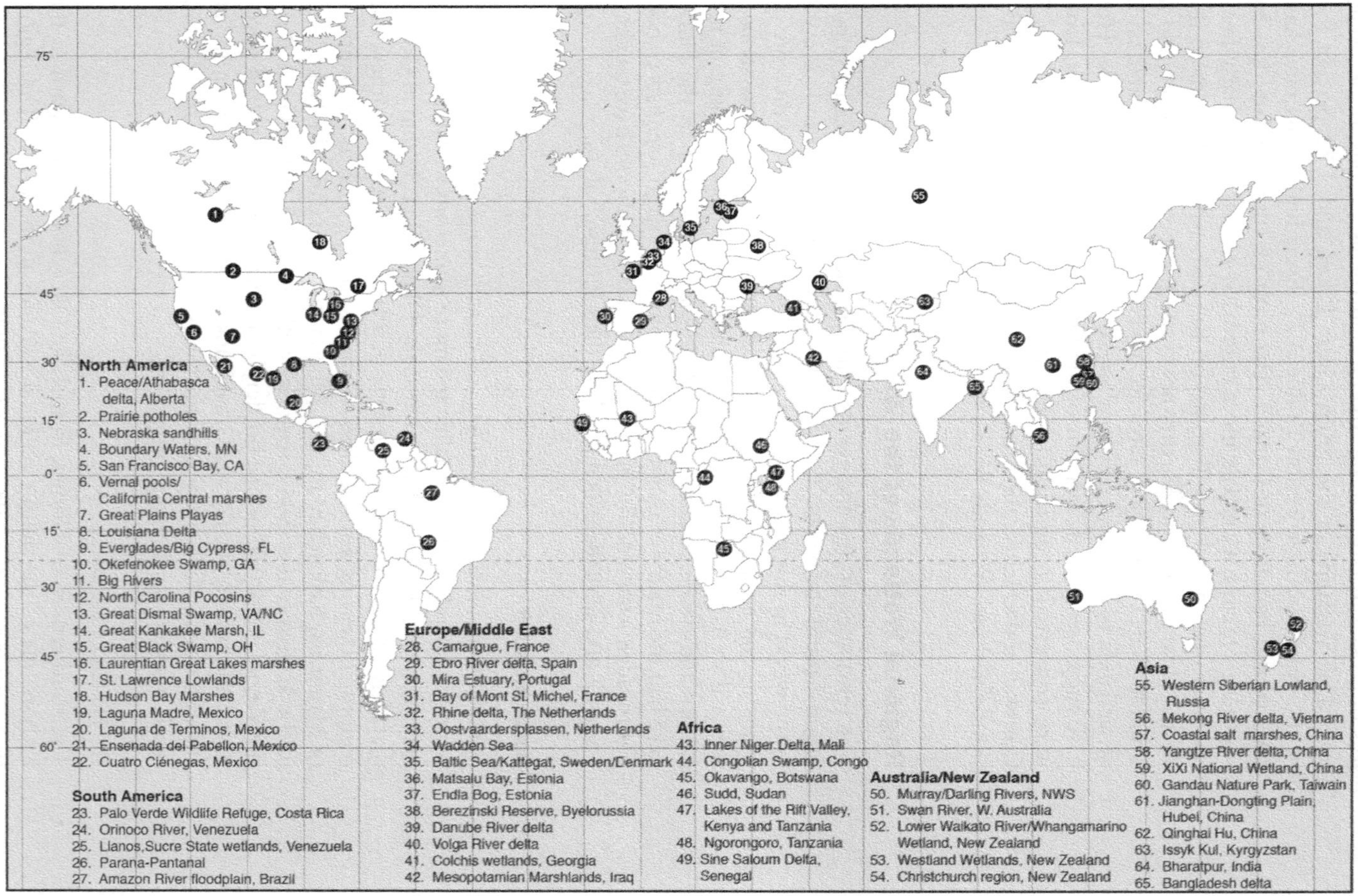

Figure 3.3 Major international wetlands discussed in this chapter.

North America
1. Peace/Athabasca delta, Alberta
2. Prairie potholes
3. Nebraska sandhills
4. Boundary Waters, MN
5. San Francisco Bay, CA
6. Vernal pools/ California Central marshes
7. Great Plains Playas
8. Louisiana Delta
9. Everglades/Big Cypress, FL
10. Okefenokee Swamp, GA
11. Big Rivers
12. North Carolina Pocosins
13. Great Dismal Swamp, VA/NC
14. Great Kankakee Marsh, IL
15. Great Black Swamp, OH
16. Laurentian Great Lakes marshes
17. St. Lawrence Lowlands
18. Hudson Bay Marshes
19. Laguna Madre, Mexico
20. Laguna de Terminos, Mexico
21. Ensenada del Pabellon, Mexico
22. Cuatro Ciénegas, Mexico

South America
23. Palo Verde Wildlife Refuge, Costa Rica
24. Orinoco River, Venezuela
25. Llanos,Sucre State wetlands, Venezuela
26. Parana-Pantanal
27. Amazon River floodplain, Brazil

Europe/Middle East
28. Camargue, France
29. Ebro River delta, Spain
30. Mira Estuary, Portugal
31. Bay of Mont St. Michel, France
32. Rhine delta, The Netherlands
33. Oostvaardersplassen, Netherlands
34. Wadden Sea
35. Baltic Sea/Kattegat, Sweden/Denmark
36. Matsalu Bay, Estonia
37. Endla Bog, Estonia
38. Berezinski Reserve, Byelorussia
39. Danube River delta
40. Volga River delta
41. Colchis wetlands, Georgia
42. Mesopotamian Marshlands, Iraq

Africa
43. Inner Niger Delta, Mali
44. Congolian Swamp, Congo
45. Okavango, Botswana
46. Sudd, Sudan
47. Lakes of the Rift Valley, Kenya and Tanzania
48. Ngorongoro, Tanzania
49. Sine Saloum Delta, Senegal

Australia/New Zealand
50. Murray/Darling Rivers, NWS
51. Swan River, W. Australia
52. Lower Waikato River/Whangamarino Wetland, New Zealand
53. Westland Wetlands, New Zealand
54. Christchurch region, New Zealand

Asia
55. Western Siberian Lowland, Russia
56. Mekong River delta, Vietnam
57. Coastal salt marshes, China
58. Yangtze River delta, China
59. XiXi National Wetland, China
60. Gandau Natura Park, Taiwan
61. Jianghan-Dongting Plain, Hubei, China
62. Qinghai Hu, China
63. Issyk Kul, Kyrgyzstan
64. Bharatpur, India
65. Bangladesh delta

Figure 3.4 The Florida Everglades including (a) its "river of grass," (b) coastal area on the south where freshwater plants give way to mangroves, and (c) extensive forested wetland swamps, such as Audubon's Corkscrew Sanctuary in the Big Cypress Swamp. (Photos by W. J. Mitsch)

Swamp, and the coastal mangroves and Florida Bay (Fig. 3.4). The water that passes through the Everglades on its journey from Lake Okeechobee is often referred to conceptually as a "river of grass" that is often only centimeters in depth and 80 km wide. The Everglades is dominated by sawgrass (*Cladium jamaicense*), which is actually a sedge, not a grass. The expanses of sawgrass, which can be flooded by up to 1 meter of water in the wet season (summer) and burned in a fire in the dry season (winter/spring), are interspersed with deeper water sloughs and tree islands, or *hammocks*, that support a vast diversity of tropical and subtropical plants, including hardwood trees, palms, orchids, and other air plants. To the west of the sawgrass Everglades is the Big Cypress Swamp, called big because of its great expanse, not because of the size of the trees. The swamp is dominated by cypress (*Taxodium* spp.) interspersed with pine flatwoods and wet prairie. The cypress swamps receive about 125 cm of rainfall per year but do not receive major amounts of overland flow as the Everglades river of grass does. The third major wetland type, mangrove swamps, form impenetrable thickets where the sawgrass and cypress swamps give way to saline waters on the coastline.

Since about half of the original Everglades has been lost to agriculture in the north and to urban development in the east and west, concern for the remaining wetlands has been extended to the quality and quantity of water delivered to the Everglades through a series of canals and water conservation areas. The Everglades is currently the site of one of the largest wetland restoration efforts in the United States. The project includes the expertise of all major federal and state environmental agencies and universities in the region as well as a commitment of $20 billion by the federal government and the state of Florida (See details in Chapter 18: "Wetland Creation and Restoration.") The comprehensive restoration blueprint includes plans for improving the water quality as it leaves the agricultural areas and for modifying the hydrology to conserve and restore habitat for declining populations of wading birds, such as the wood stork and the white ibis, and mammals such as the Florida panther (*Puma concolor coryi*). North of the Everglades, there is a renewed effort to restore the ecological functions of the Kissimmee River, including many of its backswamp areas. This river feeds Lake Okeechobee, which, in turn, originally spilled over to the Everglades.

Numerous popular books and articles, including the classic *The Everglades: River of Grass* by Marjory Stoneman Douglas (1947), have been written about the Everglades and its natural and human history. A textbook specific to the Florida Everglades is now in its third edition (Lodge, 2010), and a wonderful historical account called *The Swamp* describes the many attempts to manage, drain, and restore the Everglades (Grunwald, 2006). The wetlands and south Florida have been through a history of several drainage attempts, a land-grab boom, a hurricane in 1926 that killed 400 people, a massive water management system developed by the U.S. Army Corps of Engineers, and today's attempt to restore the hydrology and part of the Everglades to something resembling what it was before.

Okefenokee Swamp

The Okefenokee Swamp on the Atlantic Coastal Plain of southeastern Georgia and northeastern Florida is a 1,750-km^2 mosaic of several different types of wetland communities. It is believed to have been formed during the Pleistocene or later when ocean water was impounded and isolated from the receding sea by a sand ridge (now referred to as the Trail Ridge) that kept water from flowing directly toward the Atlantic. The swamp forms the headwaters of two river systems: the Suwannee River, which flows southwest through Florida to the Gulf of Mexico, and the St. Mary's River, which flows southward and then eastward to the Atlantic Ocean.

Much of the swamp is now part of the Okefenokee National Wildlife Refuge, established in 1937 by Congress. The Okefenokee is named for an Indian word meaning "land of trembling earth" because of the numerous vegetated floating islands that dot the wet prairies. Six major wetland communities comprise the Okefenokee Swamp:

1. Pond cypress forest
2. Emergent and aquatic bed prairie
3. Broad-leaved evergreen forest
4. Broad-leaved shrub wetland
5. Mixed cypress forest
6. Black gum forest

Pond cypress (*Taxodium distichum* var. *imbricarium*), black gum (*Nyssa sylvatica* var. *biflora*), and various evergreen bays (e.g., *Magnolia virginiana*) are found in slightly elevated areas where water and peat deposits are shallow. Open areas, called prairies, include lakes, emergent marshes of *Panicum* and *Carex*, floating-leaved marshes of water lilies (e.g., *Nuphar* and *Nymphaea*), and bladderwort (*Utricularia*). Fires that actually burn peat layers are an important part of this ecosystem and have recurred in a 20- to 30-year cycle when water levels became very low. Many people believe that the open prairies represent early successional stages, maintained by burning and logging, of what would otherwise be a swamp forest.

The Pocosins of the Carolinas

Pocosins are evergreen shrub bogs found on the Atlantic Coastal Plain from Virginia to northern Florida. These wetlands are particularly dominant in North Carolina, where an estimated 3,700 km^2 remained undisturbed or only slightly altered in 1980, whereas 8,300 km^2 were drained for other land uses between 1962 and 1979 alone (Richardson et al., 1981). The word *pocosin* comes from the Algonquin phrase for "swamp on a hill." In successional progression and in nutrient-poor acid conditions, pocosins resemble bogs typical of much colder climes and, in fact, were classified as bogs in an early wetland survey (Shaw and Fredine, 1956). A typical pocosin ecosystem in North Carolina is dominated by evergreen shrubs and pine (*Pinus serotina*). Draining and ditching for agriculture and forestry have affected pocosins in North Carolina.

Great Dismal Swamp

The Great Dismal Swamp is one of the northernmost "southern" swamps on the Atlantic Coastal Plain and one of the most studied and romanticized wetlands in the United States. The swamp covers approximately 850 km^2 in southeastern Virginia and northeastern North Carolina near the urban sprawl of the Norfolk–Newport News–Virginia Beach metropolitan area. It once extended over 2,000 km^2. The swamp has been severely affected by human activity during the past 200 years. Draining, ditching, logging, and fire played a role in diminishing its size and altering its ecological communities. The Great Dismal Swamp was once primarily a magnificent bald cypress–gum swamp that contained extensive stands of Atlantic white cedar (*Chamaecyparis thyoides*). Although remnants of those communities still exist today, much of the swamp is dominated by red maple (*Acer rubrum*), and mixed hardwoods are found in drier ridges. In the center of the swamp lies Lake Drummond, a shallow, tea-colored, acidic body of water. The source of water for the swamp is thought to be underground along its western edge as well as surface runoff and precipitation. Drainage occurred in the Great Dismal Swamp as early as 1763 when a corporation called the Dismal Swamp Land Company, which was owned in part by George Washington, built a canal from the western edge of the swamp to Lake Drummond to establish farms in the basin (Fig. 3.5). That effort, like several others in the ensuing years, failed, and Mr. Washington went on to help found a new country. Timber companies, however, found economic reward in the swamp by harvesting the cypress and cedar for shipbuilding and other uses. One of the last timber companies that owned the swamp, the Union Camp Corporation, gave almost 250 km^2 of the swamp

Figure 3.5 Washington's Ditch in the Great Dismal Swamp in eastern Virginia. This ditch was part of an unsuccessful effort began by George Washington to drain the swamp for commercial reasons in the mid-eighteenth century. (Photo by Frank Day, reprinted with permission)

to the federal government to be maintained as a national wildlife refuge. At least one book, *The Great Dismal Swamp* (Kirk, 1979), describes the ecological and historical aspects of this important wetland. The extent and management of Atlantic white cedar, a dominant species in the Great Dismal Swamp, are presented by Sheffield et al. (1998).

Swamp Rivers of the South Atlantic Coast

The Atlantic Coastal Plain, extending from North Carolina to the Savannah River in Georgia, is a land dominated by forested wetlands and marshes and cut by large rivers that drain the Piedmont and cross the Coastal Plain in a northwest–southeast direction to the ocean. These rivers include the Roanoke, Chowan, Little Pee Dee, Great Pee Dee, Lynches, Black, Santee, Congaree, Altamaha, Cooper, Edisto, Combahee, Coosawhatchie, and Savannah, as well as a host of smaller tributaries. Extensive bottomland hardwood forests and cypress swamps line these rivers and spread into the lowlands between them. Interspersed among these forests are hundreds of Carolina bays, small elliptical lakes of uncertain origin surrounded by or overgrown with marshes and forested wetlands (Lide et al., 1995). The origin of these lake–wetland complexes, of which there are more than 500,000 along the eastern Coastal Plain, has been suggested to be meteor showers, wind, or groundwater flow (D. C. Johnson, 1942; H. T. Odum, 1951; Prouty, 1952; Savage, 1983). Along the coast, freshwater tides on the lower rivers formerly overflowed extensive forests, but many of these were cleared in the early 1800s to establish rice plantations. Most of the rice plantations have since been abandoned, and the former fields are now extensive freshwater marshes that have become a paradise for ducks and geese. The estuaries at the mouths of the rivers support the most extensive salt marshes on the Southeast Coast.

In 1825, Robert Mills wrote of Richland County, South Carolina: "What clouds of miasma, invisible to sight, almost continually rise from these sinks of corruption, and who can calculate the extent of its pestilential influence?" (quoted in Dennis, 1988). At that time, only 10,000 ha of the 163,000-ha county were being cultivated. Almost all the rest was a vast, untouched swamp. Our appreciation of these swamps has changed dramatically since that time, and parts of this swamp are now the Congaree Swamp National Monument and the Francis Beidler Forest; the latter includes the world's largest virgin cypress–tupelo (*Taxodium–Nyssa*) swamp and is now an Audubon sanctuary. Both preserves contain extensive stands of cypress more than 500 years old that escaped the logger's ax in the late 1800s.

Prairie Potholes

A significant number of small wetlands, primarily freshwater marshes, are found in a 780,000-km^2 region in the states of North Dakota, South Dakota, and Minnesota and in the Canadian provinces of Manitoba, Saskatchewan, and Alberta (Fig. 3.6). It has been estimated that there are only about 10 percent of the original wetlands remaining from presettlement times. These wetlands, called *prairie potholes*, were formed by glacial action during the Pleistocene. This region is considered one of the most

Figure 3.6 Oblique aerial view of prairie pothole wetlands, showing many small ponds surrounded by wetland vegetation, in the middle of large agricultural fields. (File photograph, U.S. Fish and Wildlife Service, Jamestown, North Dakota)

important wetland regions in the world because of its numerous shallow lakes and marshes, its rich soils, and its warm summers, which are optimum for waterfowl. Wet-and-dry cycles are a natural part of the ecology of these prairie wetlands. In fact, many of the prairie potholes might not exist if there were no periodic dry periods. In some cases, dry periods of 1 to 2 years every 5 to 10 years are required to maintain emergent marshes. Another feature of this wetland region is the occasional presence of saline wetlands and lakes caused by high evapotranspiration/precipitation ratios. Salinities as high as 370 parts per thousand (ppt) have been recorded for some hyper-saline lakes in Saskatchewan. It is estimated that 50 to 75 percent of all the waterfowl originating in North America in any given year comes from this region.

More than half of the original wetlands in the prairie pothole region have been drained or altered, primarily for agriculture. An estimated 500 km^2 of prairie pothole wetlands in North Dakota, South Dakota, and Minnesota were lost between 1964 and 1968 alone. More recently it was estimated that there was a net loss of 300 km^2, or 1.1 percent of the 26,000 ha of the U.S. prairie pothole region wetlands, over the 12-year period of 1997 to 2009 (Dahl, 2014). Most of that loss was as emergent and farmed marshes. However, major efforts to protect the remaining prairie potholes are progressing. There was an estimated 355 km^2 of marshes restored in the region from 1997 to 2009. But this was overshadowed by the loss of 510 km^2 of emergent wetlands converted to agriculture. Thousands of square kilometers of wetlands have been purchased under the U.S. Fish and Wildlife Service Waterfowl Production Area program in North Dakota alone since the early 1960s. The Nature Conservancy and other private foundations have also purchased many wetlands in the region for conservation.

The Nebraska Sandhills and Great Plains Playas

South of the prairie pothole region is an irregular-shaped region of 52,000 km^2 in northern Nebraska described as "the largest stabilized dune field in the Western Hemisphere" (Novacek, 1989). These Nebraska sandhills, which constitute one-fourth of the state, represent an interesting and sensitive coexistence of wetlands, agriculture, and a very important aquifer-recharge area. The area was originally mixed-grass prairie composed of thousands of small wetlands in the interdunal valleys. Much of the region is now used for farming and rangeland agriculture, and many of the wetlands in the region have been preserved, even though the vegetation is often harvested for hay or grazed by cattle. The Ogallala Aquifer is an important source of water for the region and is recharged to a significant degree through overlying dune sands and to some extent through the wetlands. It has been estimated that there are 558,000 ha of wetlands in the Nebraska sandhills, many of which are interconnected wet meadows or shallow lakes that contain water levels determined by both runoff and regional water table levels. The wetlands in the region have been threatened by agricultural development, especially pivot irrigation systems that cause a lowering of the local water tables despite increased wetland flooding in the vicinity of the irrigation systems. Like the prairie potholes to the north, the Nebraska sandhill wetlands are important breeding grounds for numerous waterfowl, including about 2 percent of the Mallard breeding population in the north-central flyway.

Smith (2003) has argued that many of the wetlands that occur in Nebraska, particularly in southwestern Nebraska, could be defined as *playas* (see definition in Table 2.1) because of their seasonal flooding patterns in a semiarid environment in the Great Plains Region. Most of the playas in North America are found in the Southern Great Plains that includes western Texas, southern New Mexico, southeastern Colorado, and southwestern Kansas. These temporarily or seasonally flooded wetlands are characterized as being depressional (i.e., isolated) and recharge (i.e., they recharge groundwater; see Chapter 4: "Wetland Hydrology") wetland basins. It is estimated that there are over 25,000 playas in the United States Great Plains (Sabin and Holliday, 1995) and that they cover 1,800 km^2 in what is otherwise a semiarid to arid agricultural landscape (Smith, 2003).

Great Kankakee Marsh

For all practical purposes, this wetland no longer exists, although until about 100 years ago it was one of the largest marsh-swamp basins in the interior United States. Located primarily in northwestern Indiana and northeastern Illinois, the Kankakee River basin is 13,700 km^2 in size, including 8,100 km^2 in Indiana, where most of the original Kankakee Marsh was located. From the river's source to the Illinois line, a direct distance of only 120 km, the river originally meandered through 2,000 bends along 390 km, with a nearly level fall of only 8 cm per km. Numerous wetlands, primarily wet prairies and marshes, remained virtually undisturbed until the 1830s, when settlers began to enter the region. The naturalist Charles Bartlett (1904) described the wetland as follows:

> More than a million acres of swaying reeds, fluttering flags, clumps of wild rice, thick-crowding lily pads, soft beds of cool green mosses, shimmering ponds and black mire and trembling bogs—such is Kankakee Land. These wonderful fens, or marshes, together with their wide-reaching lateral extensions, spread themselves over an area far greater than that of the Dismal Swamp of Virginia and North Carolina.

The Kankakee region was considered a prime hunting area until the wholesale draining of the land for crops and pasture began in the 1850s. The Kankakee River and almost all of its tributaries in Indiana were channelized into a straight ditch in the late nineteenth century and early twentieth century. In 1938, the Kankakee River in Indiana was reported to be one of the largest drainage ditches in the United States; the Great Kankakee Marsh was essentially gone by then. Early accounts of the region were given by Bartlett (1904) and Meyer (1935). More recently, there has been some effort to restore parts of the Great Kankakee Marsh in northwestern Indiana.

Black Swamp

Another vast wetland of the Midwest that has ceased to exist is the Black Swamp in what is now northwestern Ohio. The Black Swamp (Fig. 3.7) was once a combination of marshland and forested swamps that extended about 160 km long and 40 km wide in a northeasterly direction from Indiana toward the lake and covered an estimated 4,000 km^2. The bottom of an ancient extension of Lake Erie, the Black Swamp was

64 Chapter 3 Wetlands of the World

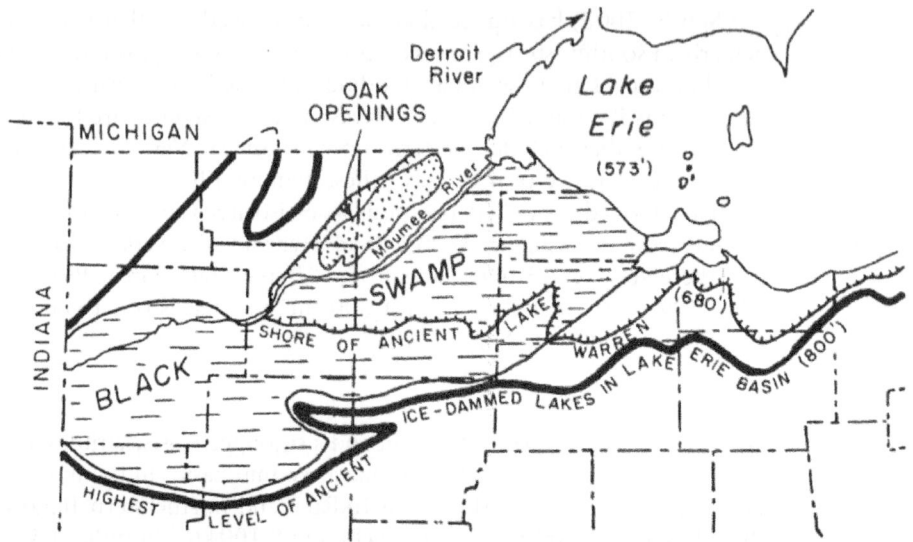

Figure 3.7 The Black Swamp as it probably existed 200 years ago in northwestern Ohio. Essentially none of this 4,000-km² wetland remains. (From Forsyth, 1960)

named for the rich, black muck that developed in areas where drainage was poor as a result of several ridges that existed perpendicular to the direction of the flow to the lake. There are numerous accounts of the difficulty that early settlers and armies (especially during the War of 1812) had in negotiating this region, and few towns of significant size have developed in the location of the original swamp. One account of travel through the region in the late 1700s suggested that "man and horse had to travel mid-leg deep in mud" for three days just to cover a distance of only 50 km (Kaatz, 1955). As with many other wetlands in the Midwest, state and federal drainage acts led to the rapid drainage of this wetland, until little of it was left by the beginning of the twentieth century. Only one small example of an interior forested wetland and several coastal marshes (about 150 km²) remain of the original western Lake Erie wetlands. The Maumee River, which now drains a mostly agricultural watershed, is identified as the major source of phosphorus pollution to Lake Erie (Scavia et al., 2014). Lake Erie is now experiencing frequent harmful algal blooms in its western basin (Michalak et al., 2013). Discussions have begun on the restoration of the Black Swamp to help mitigate this pollution.*

The Louisiana Delta

As the Mississippi River reaches the last phase of its journey to the Gulf of Mexico in southeastern Louisiana, it enters one of the most wetland-rich regions of the world. The total area of marshes, swamps, and shallow coastal lakes covers more than 36,000 km². As the Mississippi River distributaries reach the sea, forested wetlands

---

*See "Restoring the Black Swamp to Save Lake Erie" at www.wef.org/blogs/blog.aspx?id=12884904840&blogid=17296.

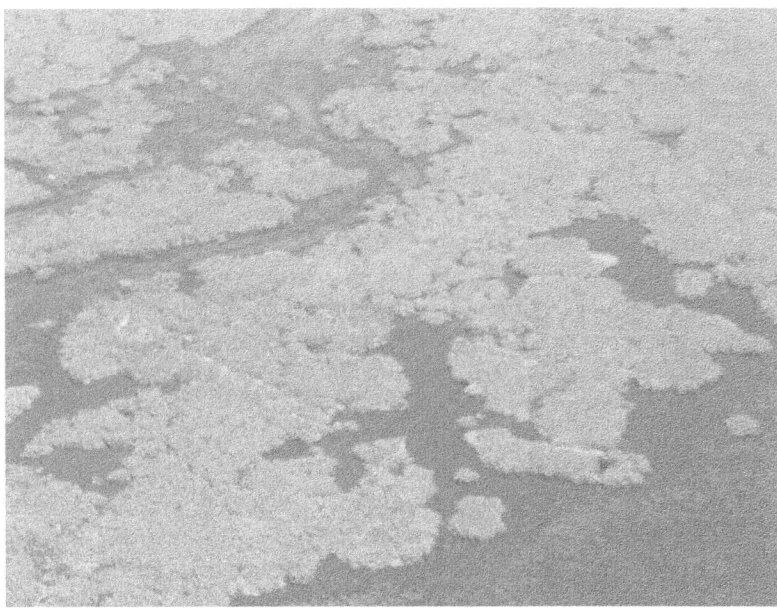

Figure 3.8 Coastal marshlands of the Mississippi River Delta in southern Louisiana; some breakup of marshes caused by land subsidence and lack of sediment inputs from the river is evident. (Photo by W. J. Mitsch)

give way to freshwater marshes and then to salt marshes. The salt marshes are some of the most extensive and productive in the United States (Fig. 3.8) and depend on the influx of fresh water, nutrients, sediments, and organic matter from upstream swamps. Freshwater and saltwater wetlands has been decreasing at a rapid rate in coastal Louisiana, amounting to a total wetland loss of 4,800 km^2 since the 1930s and annual loss rates between 60 and 100 km^2 yr^{-1} (Day et al., 2005, 2007). These losses have been attributed to both natural and artificial causes although the main cause has been the isolation of the river from the delta (Day et al., 2007).

Characteristic of the riverine portion of the delta, the Atchafalaya River, a distributary of the Mississippi River, serves as both a flood-relief valve for the Mississippi River and a potential captor of its main flow. The Atchafalaya Basin by itself is the third-largest continuous wetland area in the United States and contains 30 percent of all the remaining bottomland forests in the entire lower Mississippi alluvial valley. The river passes through this narrow 4,700-km^2 basin for 190 km, supplying water for 1,700 km^2 of bottomland forests and cypress–tupelo swamps and another 260 km^2 of permanent bodies of water. The Atchafalaya Basin, contained within a system of artificial and natural levees, has had a controversial history of human intervention. Its flow is controlled by structures located where it diverges from the Mississippi River main channel, and it has been dredged for navigation and to prevent further infilling of the basin by Mississippi River silt. It has been channelized for oil and gas production. The old-growth forests were logged at the beginning of the twentieth century, and the higher lands are now in agricultural production.

Another frequently studied wetland area in the delta is the Barataria Bay estuary in Louisiana, an interdistributary basin of the Mississippi River that is now isolated from the river by a series of flood-control levees. This basin, 6,500 km^2 in size, contains 700 km^2 of wetlands, including cypress–tupelo swamps, bottomland hardwood forests, marshes, and shallow lakes.

The U.S. Army Corps of Engineers, in cooperation with other federal and state agencies, began designing a comprehensive strategy for conservation and restoration of the delta two decades or more ago. Then in late August 2005, Hurricanes Katrina and Rita battered the Louisiana coastline and destroyed much of the city of New Orleans (see Costanza et al., 2006; Day et al., 2007), prompting a redirection of some funds from wetland restoration to levee construction. The 2010 Gulf of Mexico oil spill (Mitsch, 2010) continued that redirection of funds. The delta restoration plan is described in more detail in Chapter 18: "Wetland Creation and Restoration."

San Francisco Bay

One of the most altered and most urbanized wetland areas in the United States is San Francisco Bay in northern California. The marshes surrounding the bay covered more than 2,200 km^2 when the first European settlers arrived. Almost 95 percent of these marshes have since been destroyed. The ecological systems that make up San Francisco Bay range from deep, open water to salt and brackish marshes. The salt marshes are dominated by Pacific cordgrass (*Spartina foliosa*) and pickleweed (*Salicornia virginica*), and the brackish marshes support bulrushes (*Scirpus* spp.) and cattails (*Typha* spp.). Soon after the beginning of the Gold Rush in 1849, the demise of the bay's wetlands began. Industries such as agriculture and salt production first used the wetlands, clearing the native vegetation and diking and draining the marsh. At the same time, other marshes were developing in the bay as a result of rapid sedimentation. The sedimentation was caused primarily by upstream hydraulic mining. Sedimentation and erosion continue to be the greatest problems encountered in the remaining tidal wetlands.

Great Lakes Wetlands/St. Lawrence Lowlands

The Canadian marshes of the Great Lakes Wetlands/St. Lawrence Lowlands region, especially those along the Great Lakes in Ontario and in the St. Lawrence lowlands of Ontario and Quebec (Fig. 3.9), are important habitats for migratory waterfowl. Several of the notable wetlands in the region include Long Point and Point Pelee on northern Lake Erie, the St. Clair National Wildlife Area on Lake St. Clair along the Great Lakes in southern Ontario, and many wetlands along the St. Lawrence River in eastern Ontario and southwestern Quebec.

The St. Lawrence River wetlands, generally defined as being from Cornwall, Ontario, on the upstream edge to Trois-Pistoles in the lower estuary near the gulf, supports 34,000 ha of marshes and swamps along its corridor. Cap Tourmente, a 2,400-ha tidal freshwater marsh complex located about 50 km northeast of Quebec City, was the first wetland in Canada designated as a Ramsar site of international importance (Fig. 3.10). It consists of both intertidal mud flats and freshwater

Figure 3.9 Wetland scientist botanizing in a *Scirpus americanus* marsh adjacent to the St. Lawrence River near Quebec City, Canada. (Photo by W. J. Mitsch)

Figure 3.10 Snow geese at Cap Tourmente National Wildlife Area, Quebec, Canada. (Photo by Robbie Sproule, provided by Creative Commons license)

marshes as well as nontidal marshes, swamps, shrub swamps, and peatlands. The Cap Tourmente freshwater tidal marshes are subjected to heavy tidal flooding, with tidal amplitudes of 4.1 m at mean tides and 5.8 m during spring tides. The Cap Tourmente National Wildlife Area has a wide range of communities, including 400 ha of tidal marsh, 100 ha of coastal meadow, 700 ha of agricultural land, and 1,200 ha of forest. *Scirpus americanus* (American bulrush) marshes of the St. Lawrence, such as those found at Cap Tourmente, are restricted to the freshwater tidal portion of the river, with only 4,000 ha remaining in the entire region. Although increasing numbers of greater snow geese have led to a depletion of *Scirpus* rhizomes, which may eventually cause a deterioration of the marshes at Cap Tourmente, the snow geese remain one of the notable features of this wetland during the migratory season. Tens of thousands of the geese migrate in both the spring and fall and feed on the bulrushes. Environment Canada estimated that wetland cover along the St. Lawrence actually increased by 3 percent from the early 1990s to the early 2000s.

The marshes along the Great Lakes are generally diked and heavily managed to buffer them from the year-to-year fluctuations in lake levels as they are in much of the United States. This temperate region in Canada also has a considerable number of hardwood forested swamps dominated by red and silver maples (*Acer rubrum* and *A. saccharinum*) and ash (*Fraxinus* spp.). Without human intervention, these swamps are quite stable; however, logging has been frequent, including clear-cutting. A clear-cut swamp is often replaced by a marsh, and the successional pattern starts all over again.

Canada's Central and Eastern Province Peatlands

The peatlands of northern Ontario and Manitoba are extensive regions that are used less by waterfowl and more by a wide variety of mammals, including moose, wolf, beaver, and muskrat. Wild rice (*Zizania palustris*), a common plant in littoral zones of boreal lakes, is often harvested for human consumption. Some of the boreal wetlands are mined for peat that is used for horticultural purposes or fuel. Fens of the region can be quite stable and are fairly common; bogs are also stable in this region but are less common. Radiocarbon dating of the bottom peat layers in Quebec bogs suggests that they began as fens between 9,000 and 5,500 years ago. Once formed, open bogs are quite stable, and forested bogs are even more stable, although they can revert to open bogs if fire occurs.

Hudson–James Bay Lowlands

A large wetland complex is found in northern Ontario and Manitoba and the eastern Northwest Territories, wrapping around the southern shore of the Hudson Bay (Fig. 3.11) and its southern extension, James Bay. These Hudson–James Bay lowlands are part of the vast subarctic wetland region of Canada, which stretches from the Hudson Bay northwestward to the northwestern corner of Canada and into Alaska and covers 760,000 km^2 of Canada (Zoltai et al., 1988). This region has been described as the region with the highest density and percentage cover of wetlands in North America (76 to 100 percent) (Abraham and Keddy, 2005). One of the largest and best-described wetland sites in this region is the 24,000-km^2 Polar Bear Provincial Park in northern

Figure 3.11 Extensive peatlands and marshes of Hudson Bay lowlands. (Photo by C. Rubec, reprinted with permission)

Ontario. Two additional sanctuaries of note are located in the southern James Bay: the Hannah Bay Bird Sanctuary and the Moose River Bird Sanctuary, which total 250 km^2. The region is dominated by extensive areas of mud flats, intertidal marshes, and supertidal meadow marshes, which grade into peatlands, interspersed with small lakes, thicket swamps, forested bogs and fens, and open bogs, fens, and marshes away from the shorelines. The southern shore of the bay is dominated by sedges (*Carex* spp.), cotton grasses (*Eriophorum* spp.), and clumps of birches (*Betula* spp.). The more southerly low subarctic wetland region is made up of low, open bogs, sedge–shrub fens, moist sedge-covered depressions, and open pools and small lakes separated by ridges of peat, lichen–peat-capped hummocks, raised bogs, and beach ridges. Even though the tidal range from the Hudson Bay is small, the gradual slope of land allows much tidal inundation of flats that vary from 1 to 5 km in width. Low-energy coasts with wide coastal marshes occur in the southern James Bay; high-energy coasts with sand flats and sand beaches are found along the Hudson Bay shoreline itself. Isostatic rebound following glacial retreat has resulted in the emergence of land from the bay at a rate of 1.2 m per century for the past 1,000 years, the greatest rate of glacial rebound in North America.

The coastal marshes, intertidal sand flats, and river mouths of the Hudson–James Bay lowlands serve as breeding and staging grounds for a large number of migratory waterfowl, including the lesser snow goose, which was once in danger of disappearing but is now flourishing; Canada goose; black duck; pintail; green-winged teal; mallard; American wigeon; shoveler; and blue-winged teal. The western and southwestern coasts of the Hudson and James bays form a major migration pathway for many shorebird species as well, including red knot, short-billed dowitcher, dunlin, greater yellowlegs, lesser yellowlegs, ruddy turnstone, and black-bellied plover. The wetlands of Polar Bear Provincial Park provide nesting habitat for red-throated, Arctic, and common loons; American bittern; common and red-breasted merganser; yellow rail; sora; sandhill crane; and several gulls and terns.

Peace–Athabasca Delta

The Peace–Athabasca Delta in Alberta, Canada (Fig. 3.12), is the largest freshwater inland boreal delta in the world and is relatively undisturbed by humans. It actually comprises three deltas: the Athabasca River delta ($1,970 \text{ km}^2$), the Peace River delta ($1,684 \text{ km}^2$), and the Birch River delta (168 km^2). It is one of the most important waterfowl nesting and staging areas in North America and is the staging area for breeding ducks and geese on their way to the MacKenzie River lowlands, Arctic river deltas, and Arctic islands. The major lakes of the delta are very shallow (0.6–3.0 m) and have a thick growth of submerged and emergent vegetation during the growing season. The delta consists of very large flat areas of deposited sediments with some outcropping islands of the granitic Canadian Shield. The site has the following types of wetlands: emergent marshes, mud flats, fens, sedge meadows, grass meadows,

Figure 3.12 Athabasca River in the Peace–Athabasca Delta in Jaspar National Park. (Photo by AudeVivere; courtesy of Wikimedia Commons)

shrub–scrub wetlands, deciduous forests of balsam (*Populus balsamifera*) and birch (*Betula* spp.), and coniferous forests dominated by white and black spruce (*Picea glauca* and *P. mariana*). Owing to the shallow water, high fertility, and relatively long growing season for that latitude, the area is an abundant food source of particular importance during drought years on the prairie potholes to the south. All four major North American flyways cross the delta, with the most important being the Mississippi and central flyways.

At least 215 species of birds, 44 species of mammals, 18 species of fish, and thousands of species of insects and invertebrates are found in the delta. Up to 400,000 birds use this wetland in the spring and more than 1 million birds in autumn. Waterfowl species recorded in the delta area include lesser snow goose, white-fronted goose, Canada goose, tundra swan, all four species of the loon, all seven species of North American grebe, and 25 species of duck. The world's entire population of the endangered whooping crane nests in the northern part of the delta area. The site also contains the largest undisturbed grass and sedge meadows in North America, which support an estimated 10,000 wood and plains buffalo.

Wetlands of Mexico

Mexico has about 8 million hectares of wetlands (Mitsch and Hernandez, 2013). Because of extensive arid regions in its interior, Mexico was initially underrepresented in the number and area of Ramsar "Wetlands of International Importance," with only seven Ramsar sites designated in 2001 (Pérez-Arteaga et al., 2002). That situation has changed dramatically since then, with Mexico having 142 Ramsar sites covering 8.8 million ha as of late 2014. Many of the priority wetland sites in Mexico are associated with or near the Gulf of Mexico and the Pacific Ocean coastlines (Fig. 3.13). Mexico has an estimated 1.6 million ha of wetlands adjacent to or on its coasts, with 75,000 ha on the Pacific and 675,000 ha on the Gulf of Mexico. Coastal wetlands in Mexico include about 118 major wetlands complexes and at least another 538 smaller systems representing a wide variety of types (Contreras-Espinosa and Warner, 2004). Freshwater coastal wetlands include swamps dominated by *Annona* and *Pachira* trees, marshes dominated by *Typha*, mixed broadleaved communities with *Pontederia* and *Sagittaria*, and freshwater open water lagoons with submerged and floating macrophytes (Mitsch and Hernandez, 2013).

One of the largest coastal wetlands in Mexico is the 700,000-ha Laguna de Términos in Campeche on the Gulf of Mexico. This area includes mangrove swamps on its coastline as well as coastal dune vegetation, freshwater swamps, flooded vegetation, lowland forest, palms, spiny scrubs, forests, secondary forests, and sea grass beds. The Ensenada del Pabellon on the Gulf of California on the Pacific Coast was estimated to account for almost 10 percent of the birds wintering in Mexico (Pérez-Arteaga et al., 2002). Laguna Madre on the Gulf of Mexico just south of the Texas coastline is another important coastal wetland in Mexico, with 200,000 ha of shallow water and mudflats and 42,000 ha of sea grass beds (dominated by *Halodule wrightii*). Other important Mexican wetlands are in the arid north region of the country in the Sonoran and Chihuhuan deserts.

Figure 3.13 Freshwater marsh at the natural reserve of the Coastal Research Center La Mancha (CICOLMA) in Veracruz, Mexico, near the Gulf of Mexico. This reserve is part of the International Ramsar site "La Mancha y El Llano." (Photo by W.J. Mitsch)

Central and South America

There are extensive and relatively understudied tropical and subtropical wetlands throughout Central and South America. Some of the more significant ones are located on the South American map in Figure 3.14 and are discussed here.

Central American Wetlands

Although poorly mapped, there are an estimated 40,000 km^2 of wetlands in Central America (Ellison, 2004). Mangrove swamps occur on both coastlines and cover 6,500 to 12,000 km^2 in Central America. Forested freshwater wetlands, the most common type of wetland in Central America, cover an estimated 15,000 km^2 of land. One type of forested wetlands—palm swamps dominated by *Raphia taedigera*—account for 1.2 percent of the land cover of Costa Rica alone, particularly in the Atlantic lowlands. There are also some freshwater marshes (1,000–2,000 km^2) in Central America, often dominated by floating aquatic plants (*Azolla, Salvinia, Pistia, Eichhornia crassipes*) rather than emergent plants. These same floating aquatic plants often dominate wastewater treatment wetlands in Central America (Nahlik and Mitsch, 2006).

Rivers on the Pacific Coast side of Central America are shorter and more seasonal than their counterparts on the Caribbean side of the isthmus. As a result of this and the prevailing climate, which causes more even monthly distribution of precipitation on the Caribbean Sea (Atlantic) side than on the Pacific side, wetlands near the Pacific

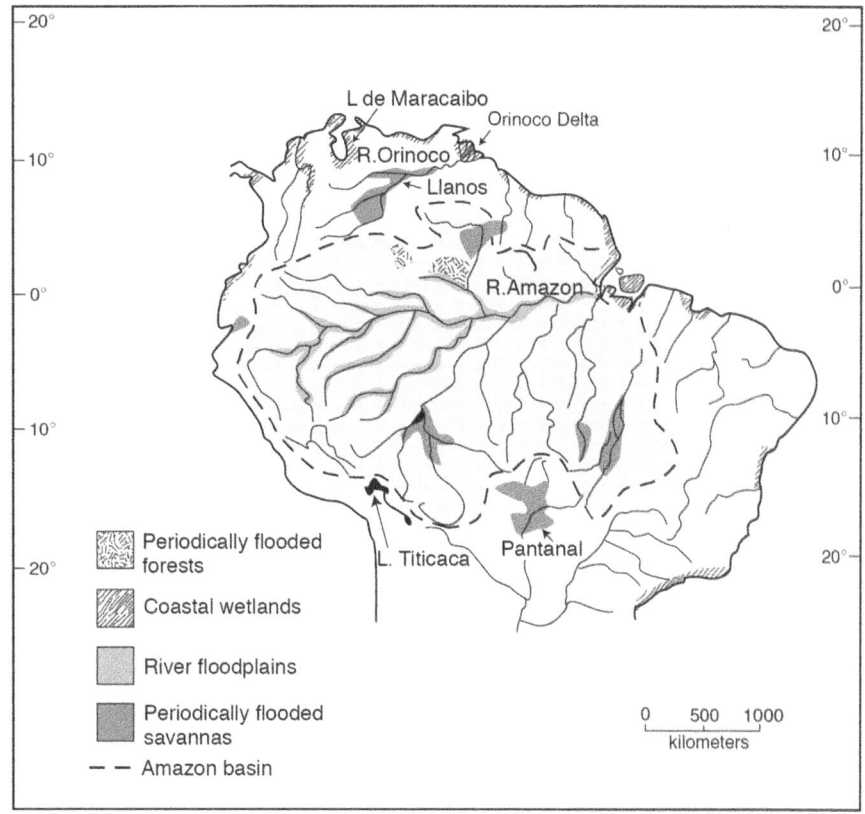

Figure 3.14 Major wetland areas of tropical South America.

Coast tend to be very seasonal with wet summers and dry winters. One of the most important wetlands in this Central American setting is a seasonal, freshwater marsh at the Palo Verde National Park in Costa Rica (Fig. 3.15). The 500-ha tidal freshwater marsh receives rainwater, agricultural runoff, and overflow water from the Tempisque River during the wet season, which, in turn, discharges to the Gulf of Nicoya about 20 km downstream of the wetland on Costa Rica's Pacific Coast. The marsh dries out almost completely by March during the dry season. It was habitat for about 60 resident and migratory birds, and thousands of migrating black-bellied whistling ducks and blue-winged teal and hundreds of northern shoveler, American wigeon, and ring-necked ducks visited the wetland during the dry season. More recently, after cattle were removed because of the marsh's designation as a wildlife refuge, the marsh was completely taken over by cattail (*Typha domingensis*), which covered 95 percent of the marsh by the late 1980s. This is a common problem in wetlands throughout the world, where clonal dominants such as *Typha* tend to choke off any other vegetation and make a poor habitat for many waterfowl and other birds. Curiously, the diversity

Figure 3.15 Palo Verde National Park in western Costa Rica: (a) seasonally flooded freshwater marsh. (b) Northern jacana (*Jacana spinosa*), a bird capable of walking on floating vegetation. (Photos by W. J. Mitsch)

of birds was partially maintained because of cattle grazing, which was permitted until 1980. Site managers tried to reintroduce cattle grazing, burning, disking, below-water mowing, and mechanical crushing to control the *Typha*. The only method that was consistently successful was crushing the cattails (Trama et al., 2009).

Orinoco River Delta

The Orinoco River delta of Venezuela was explored by Columbus during one of his early voyages. It covers 36,000 km² and is dominated along its brackish shoreline by

Figure 3.16 Mangroves of the Orinoco River delta in Venezuela. (Reprinted from Mitsch et al., 1994, with permission from Elsevier)

magnificent mangrove forests (Fig. 3.16). The Orinoco Delta economy is based on cattle ranching, with the cattle being shipped out during the high-water season, as well as on cacao production and palm heart canning. The delta's indigenous population practices subsistence farming and fishing, and exports salted fish to the population centers bordering the region (Dugan, 1993). Although some regions are protected and conservation efforts have been made by government and industry, grazing and illegal hunting have been detrimental to the area's flora and fauna.

Llanos

The western part of the Orinoco River basin in western Venezuela and northern Colombia (Fig. 3.14) is a very large (450,000 km^2) sedimentary basin called the Llanos. This region represents one of the largest inland wetland areas of South America. The Llanos has a winter wet season coupled with a summer dry season, which causes it to be a wetland dominated by savanna grasslands and scattered palms rather than floodplain forests typical of the Orinoco Delta (Junk, 1993). The region is an important wading-bird habitat and is rich with such animals as the caiman (*Caiman crocodilus*), the giant green anaconda (*Eunectes murinus*), and the red piranha (*Serrasalmus nattereri*). It supports about 470 bird species, although only one species is considered endemic. Dominant mammals include the giant anteater (*Myrmecophaga tridactyla*) and the abundant capybara (*Hydrochaeris hydrochaeris*).

Pantanal

One of the largest regional wetlands in the world is the Gran Pantanal of the Paraguay–Paraná River basin and Mato Grosso and Mato Grosso do Sul, Brazil (Por, 1995; da Silva and Girard, 2004; Harris et al., 2005; Junk and Nunes de Cunha, 2005; Ioris, 2012), located almost exactly in the geographic center of South America (Fig. 3.14). The wetland complex is 160,000 km², four times the size of the Florida Everglades, with about 130,000 km² of that area flooded annually. The annual period of flooding (called the *cheia*) from March through May supports luxurious aquatic plant and animal life and is followed by a dry season (called the *seca*) from September through November, when the Pantanal reverts to vegetation typical of dry savannas. There are also specific terms for the period of rising waters (*enchente*) from December through February and the period of declining waters (*vazante*) from June through August. There is also an asynchronous pattern to flooding in the Pantanal: While maximum rainfall and upstream flows occur in January, water stage does not peak until May in downstream reaches.

Just as in the Florida Everglades cycle of wet and dry seasons, the biota spread across the landscape during the wet season and concentrate in fewer wet areas in a food-chain frenzy during the dry season. Even though the Pantanal is one of the

Figure 3.17 The seasonally flooded Pantanal of South America is a haven to abundant wildlife including over 450 species of birds including egrets, herons, and the jabiru (*Jabiru mycteria*), intermixed with jacaré, or caiman (*Caiman yacare*) and, during the dry period, cattle. (Photo by W. J. Mitsch)

least-known regions of the globe, it is legendary for its bird life (Fig. 3.17). The Pantanal has been described as the "bird richest wetland in the world" with 463 species of birds recorded there (Harris et al., 2005). There are 13 species of herons and egrets, 3 stork species, 6 ibis and spoonbill species, 6 duck species, 11 rail species, and 5 kingfisher species. Wetland birds also include the Anhinga and the magnificent symbol of the Pantanal, the jabiru, the largest flying bird of the Western Hemisphere. In addition, the wetland supports abundant populations of the jacaré, or caiman, a relative of the North American crocodile, and the large rodent capybara (*Hydrochoerus hydrochaeris*).

There are many threats to the Pantanal as the Upper Paraguay River watershed continues to develop including accelerated cattle ranching and agricultural land use, deforestation, water pollution from point and nonpoint sources, mining activity for diamonds and gold, excessive burning, exotic species introduction, and plans for the Paraguay-Parana Waterway and potentially up to 135 hydroelectric power dams and/or reservoirs on upstream tributaries (Calheiros et al., 2012). The hydroelectric power dams are part of a Brazilian national goal for increased domestic energy production.

The threats to the Pantanal are many, but until recently, there was a semibalance between human use of the Pantanal region, particularly for cattle ranching during the dry season, and the ecological functions of the region. The ecological health of the Pantanal, however, is in a state of developmental uneasiness. Some of the rivers are polluted with metals, particularly mercury, from gold-mining activity and by agrochemicals from farms. Although the Pantanal provides tourist revenues, it is also the site of illegal wildlife trafficking and cocaine smuggling. In such a vast and remote wetland, law enforcement is physically difficult and prohibitively expensive.

The Amazon

Vast wetlands are found along many of the world's rivers, well before they reach the sea, especially in tropical regions. The Amazon River in South America is one of the best examples; wetlands cover about 20 to 25 percent of the 7-million-km^2 Amazon basin (Junk and Piedade, 2004, 2005). The Amazon is considered one of the world's major rivers, with a flow that results in about one-sixth to one-fifth of all the fresh water in the world. Many Amazonian streams and rivers are characterized as being either "black water" or "white water," with the former dominated by dissolved humic materials and low dissolved materials and the latter dominated by suspended sediments derived from the eroding Andes Mountains. Floodplains on the white-water or high-sediment rivers (called várzea) are nutrient rich while floodplains on the black-water streams (called igapó) are nutrient poor (Junk and Piedade, 2005). Deforestation from development threatens many Amazon aquatic ecosystems and has great social ramifications for people displaced in the process. Some of the floodplain-forested wetlands of the Amazon, which are estimated to cover about 300,000 km^2, undergo flooding with flood levels reaching 5 to 15 m or more (see Chapter 4). During the flood season, it is possible to boat around the canopy of trees (Fig. 3.18).

Figure 3.18 When the Amazon River is flooded annually, it is possible to boat around the treetops of the riparian forests. (Photo by W. Junk, reprinted with permission)

Europe

Mediterranean Sea Deltas

The saline deltaic marshes of the mostly tideless Mediterranean Sea are among the most biologically rich in Europe. The Rhone River delta created France's most important wetland, the Camargue (Fig. 3.19; see also Chapter 1: "Wetlands: Human Use and Science"), an expanse of wetlands centered around the 9,000-ha Étang du Vaccarès. This land is home to the free-roaming horses celebrated in literature and film; here, too, is a species of bull that inhabited Gaul several thousand years ago before being driven south by encroaching human settlements. The Camargue is also home to one of the world's 25 major flamingo nesting sites and France's only such site. The sense of mystery and the feeling for space and freedom pervading the Camargue are linked with the Gypsies, who have gathered at Les Saintes-Maries-de-la-Mer since the fifteenth century, as well as with the Camarguais cowboys, the *gardians*, who ride their herds over the lands (see Fig.1.2).

Aquatic plants and plant communities differ distinctly from those of northern Europe or tropical Africa, as the landscape transitions from dune to lagoon, to marshland, to grassland, and then to forest. Set-aside agricultural policies in Europe called for restoration of some of the rice fields in the Camargue, and some restoration of former wetlands along rivers in the region has already taken place (Mauchamp et al., 2002).

A principal delta on the Spanish Mediterranean coast is the Ebro Delta, located halfway between Barcelona and Valencia and fed by the Ebro River, which flows hundreds of kilometers through arid landscape to the sea. The delta itself, covered with

Figure 3.19 The Camargue of the Rhone River delta in southern France is highly affected by a Mediterranean climate of hot, dry summers and cool, wet winters. (Photo by W. J. Mitsch)

extensive and ancient rice paddies, also has salt marshes dominated by several species of *Salicornia* and other halophytes. Lagoons are populated with a wide variety of avian species. Some restoration of rice paddies to *Phragmites* marshes has been attempted in the delta (Comin et al., 1997).

Rhine River Delta

The Rhine River is a highly managed river and a major transportation artery in Europe. The Netherlands, the name of which comes from *Nederland*, meaning "low country," is essentially the Rhine River delta, and although the Dutch language did not even have a word for *wetlands*, the English word was adopted in the 1970s. The Netherlands is one of the most hydraulically controlled locations on Earth (Fig. 3.20). It is estimated that 16 percent of the Netherlands is wetland; the Dutch have warmed to the idea of the importance of wetlands and have registered 7 percent of the country as internationally important wetlands with the Ramsar Convention on Wetlands of International Importance. Today, several governmental initiatives are designed to encourage some water to enter, or at least remain, on the lands, in great contrast to earlier Dutch traditions of controlling water in this close-to-sea-level environment.

Earlier in the twentieth century, thousands of hectares were reclaimed from the Zuiderzee; today, some of these areas are reverting back to wetlands. For example, beginning in 1968, the Oostvaardersplassen in the Flevoland Polder, originally created as a site for industrial development, was artificially flooded in order to create a wildlife sanctuary. The 5,600-ha site is now a habitat for birds such as herons, cormorants, and spoonbills (250 bird species have been recorded there, 90 of which have bred there) as well as for Konik horses, descended from the original Tarpan wild horses of

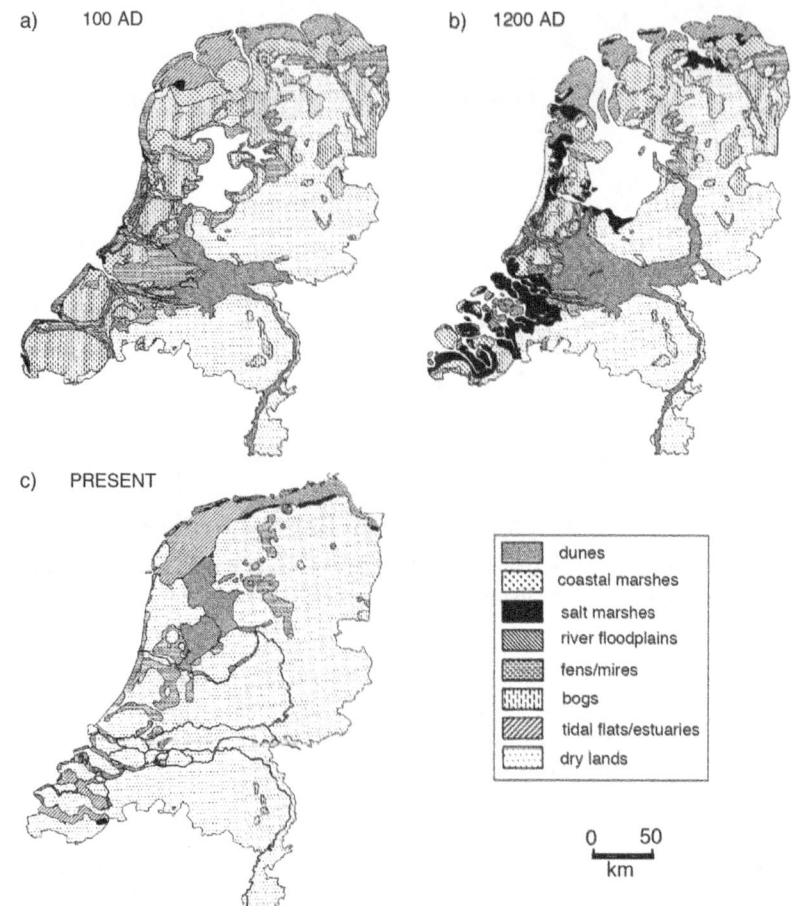

Figure 3.20 Estimated extent of wetlands in the present-day Netherlands and Rhine River delta in: (a) A.D. 100, (b) A.D. 1200, and (c) present day. (Wolff, 1993, reprinted with permission, Springer)

Western Europe (Fig. 3.21). Cattle have been crossbred from Scottish, Hungarian, and Camarguais breeds in an effort to re-create the original oxen of Europe. The Oostvaardersplassen is now one of the most popular places in the Netherlands for bird watching, and this wetland has become a national treasure.

Coastal Marshes, Mud Flats, and Bays of Northern Europe

Extensive salt marshes and mud flats are found along the Atlantic Ocean and the North Sea coastlines of Europe from the Mira Estuary in Portugal to the Wadden Sea of the Netherlands, Germany, and Denmark. These marshes contrast with the extensive salt marshes of North America, which stretch from the Bay of Fundy in Canada to southern Florida and the Gulf of Mexico, in dominant vegetation, tidal inundation,

Figure 3.21 Konik horses (descended from the Tarpan wild horses of Western Europe) are among the unusual features of the Oostvaardersplassen, one of the largest and best-known created wetlands in the Netherlands. It was originally designed for industrial development and is now one of the Netherlands' best birding locations. (Photo by W. J. Mitsch)

and sediment transport. One of the better-known coastal wetland areas in France is at the Normandy–Brittany border near the world-famous abbey of Mont St. Michel, perched atop a promontory in a bay of the English Channel and accessible to pilgrims and tourists by day only, until the tides turn it into an island. Some of the most extensive salt marshes of Europe are found surrounding the abbey. There has been 60 percent drainage of coastal marshes since the beginning of the twentieth century in this region, but now coastal wetlands are better protected, even though sheep grazing is still commonly practiced on these marshes. At nearby Le Vivier-sur-Mer, mussels are grown on *bouchots* (mussel beds created by sinking poles into the mud flats) in the shelter of a 30-km dike built in the eleventh century.

The Wadden Sea, making up over 8,000 km^2 of shallow water, extensive tidal mud flats, marsh, and sand, is considered by some to be Western Europe's most important coastal wetlands. Over the past five centuries or more, drainage of the coastal land by local residents created hundreds of square kilometers of arable land. The wetlands extend for more than 500 km along the coasts of Denmark (10 percent), Germany (60 percent), and the Netherlands (30 percent), supporting a productive North Sea fisheries.

Numerous bays surround the Baltic Sea and adjacent seas in northern Europe, many with extensive wetlands, although most of the rivers that feed these brackish seas are relatively small. Matsalu Bay, a water meadow and reed marsh in northwestern Estonia, has been known for years as a very important bird habitat. The wetland covers about 500 km^2, with much of that designated the Matsalu State Nature Preserve.

As many as 300,000 to 350,000 birds, including swans, mallards, pintails, coots, geese, and cranes, stay in the Matsalu wetland during migration in the spring.

Southeastern Europe Inland Deltas

Many important wetlands around the world form not as coastal deltas but as inland deltas or coastal marshes along large bodies of brackish and freshwater systems. There are several significant inland deltas in southeastern Europe. The 6,000-km^2 Danube River delta, one of the largest and most natural European wetlands, has been degraded by drainage and by activities related to agricultural development, gravel extraction, and dumping. The delta occurs where the Danube River spills into the Black Sea, spreading its sediments over 4,000 km^2. Plans to dike the delta and grow rice and corn ended when the communist regime of Nicolae Ceausescu fell in 1990 (Schmidt, 2001). Now there is significant international research in the delta, and plans continue for its restoration. Much of the restoration has been simple: Restore the natural hydrology by breeching dams and reconnecting waterways. The Danube Delta supports 320 species of birds and is the home of white water lilies, oak-ash forests, and floating marshes of *Phragmites australis*.

On the edge of the Caspian Sea, the Volga River forms one of the world's largest inland deltas (19,000 km^2), a highly "braided" delta over 120 km in length and spreading over 200 km at the sea's edge. The most extensive wetland area occurs in the delta of the Caspian Sea as the sea declined in water level, creating extensive *Phragmites* marshes and water lotus (*Nelumbo nucifera*) beds (Fig. 3.22). A large percentage of the world's sturgeon comes from the Caspian Sea, and the delta is a wintering site in mild winters for water birds and a major staging area for a broad variety of water bird, raptor, and passerine species. A series of dams destroyed the river's natural hydrology, and heavy industrial and agricultural pollution, as well as sea-level decline in the Caspian Sea, are making an impact.

Yet another lowland inland delta is the Colchis wetlands of eastern Georgia, a 13,000-km^2 region of subtropical alder (*Alnus glutinosa, A. barbata*) swamps and sedge–rush–reed marshes created by tectonic settling plus backwaters from the rivers discharging into the eastern Black Sea. This wetland is found in an area of great mythological interest because it is supposedly where Jason and the Argonauts (the Greek story of Argonautica as told by Apollonius) "hid their ship in a bed of reeds" (Grant, 1962) as they attempted to claim the Golden Fleece from the King of Colchis.

European Peatlands

A good portion of the world's peatlands are found in the Old World, where peatlands spread across a significant portion of Ireland, Scandinavia, Finland, northern Russia, and many of the former Soviet republics. There are about 960,000 km^2 of peatlands in Europe, or about 20 percent of Europe. About 60 percent of those peatlands have been altered for agriculture, forestry, and peat extraction (Vasander et al., 2003), and about 25 percent of the peatlands are in the Baltic Sea Basin. The Endla Bog in Estonia (Fig. 3.23) and the Berezinski Bog in Byelorussia are but two examples of many peatlands that have been protected as nature preserves and are in seminatural states in this

Figure 3.22 Lotus bed in the Volga Delta, Russia. (From C. M. Finlayson, reprinted with permission)

Figure 3.23 The Endla Bog in central Estonia. (Photo by W. J. Mitsch)

region of Europe. The 76,000-ha Berezinski reservation in northeastern Byelorussia is over half peatland and predominantly forested peatland dominated by pine (*Pinus*), birch (*Betula*), and black alder (*Alnus*).

Africa

An abundance of wetlands are found in sub-Saharan Africa (Fig. 3.24). Some of these major wetlands are far larger than those found in the Western world; examples include

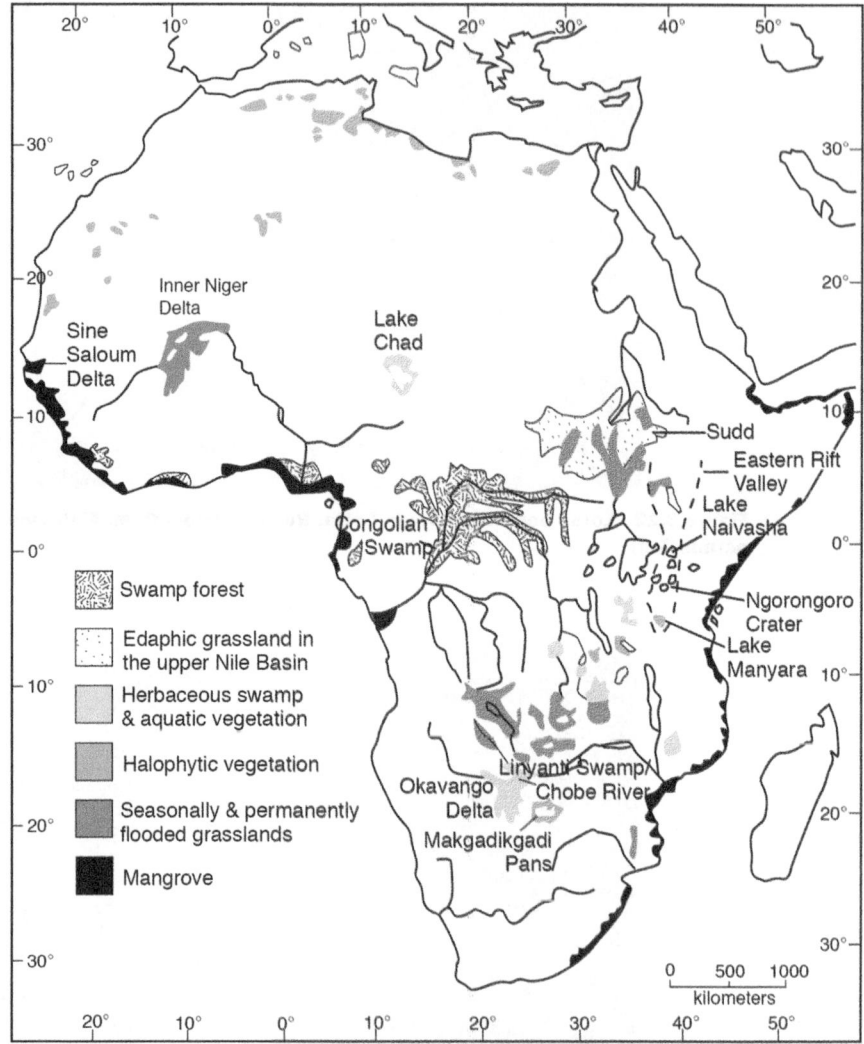

Figure 3.24 Map of major wetland areas of Africa.

the Inner Niger Delta of Mali (320,000 km² when flooded), the Congolian Swamp Forests (190,000 km²), the Sudd of the Upper Nile (more than 30,000 km² when flooded), and the Okavango Delta in Botswana (28,000 km²).

Okavango Delta

One of the great seasonally pulsed inland deltas of the world, the Okavango Delta (28,000 km²) forms at the convergence of the Okavango River and the sands of the Kalahari in Botswana, Africa (Mendelsohn and el Obeid, 2004; Fig. 3.25).

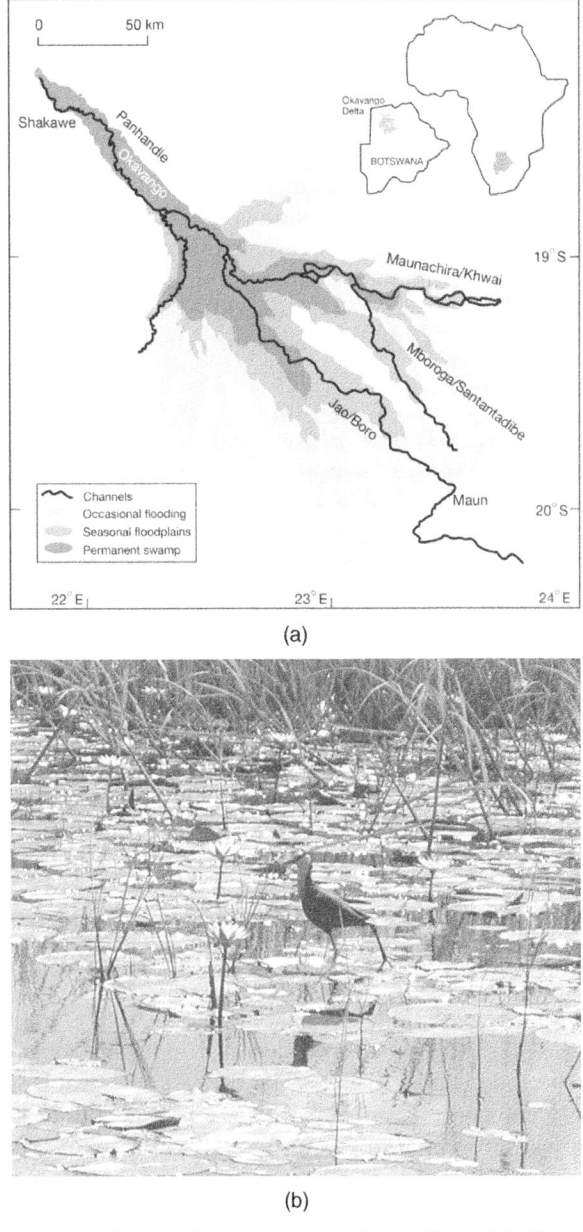

Figure 3.25 The Okavango Delta of Botswana, southern Africa: (a) Map of wetland showing permanent wetlands and seasonal and occasional floodplains; and (b) photo of floating-leaved aquatics, mostly the day water lily *Nymphaea nouchall* var. *caerulea* and the African jacana (*Actophilornis africanus*) in a permanently flooded stream. (Photo by W. J. Mitsch)

The wetland has a dramatic seasonal pulse with the water surface expanding from 2500 to 4000 km^2 in February/March to a peak of 6000 to 12,000 km^2 in August/September (McCarthy et al., 2004; Ramberg et al., 2006a,b; Ringrose et al., 2007; Mitsch et al., 2010). The system is thus divided into three major hydrologic zones: permanent swamp, seasonally flooded floodplains, and occasional floodplains (Fig. 3.25a). There is little to no surface outflow from this inland delta, and infiltration to the groundwater from the seasonally floodplain is very rapid during the 90 to 175 days of flooding (Ramberg et al., 2006a). The wetland has a web of channels, islands, and lagoons supporting crocodiles, elephants, lions, hippos, and water buffalo, and more than 400 bird species (Fig. 3.25b). Several species of tilapia and bream spawn in the Okavango Delta, contributing to the 71 species of fish found in the streams and floodplains of the delta (Ramberg et al., 2006b). Many of northern Botswana's diverse tribes find refuge there. Most of the inhabitants depend on the delta's resources. Like so many other wetlands, however, the Okavango is threatened by the increased burning (fires are natural in the Okavango), clearing associated with crop production and livestock grazing, and possible plans by upstream countries to use some of the Okavango River water. Tourism is an issue here, too, as in many other wetland sites; ecotourism is the largest single employer in Maun, located on the edge of the delta. Maun also benefits economically from the wetland's water lily tubers, bulrush roots, palm hearts, and palm wine, made from the sap of the *Hyphaene* palm. Fencing, roofing, and wall materials are also derived from the wetlands.

Congolian Swamp Forests

The Congolian swamp forests, a 190,000-km^2 region in Congo and the Democratic Republic of Congo (formerly Zaire), is one of the largest yet least studied swamp forests in the world (Campbell, 2005). This freshwater tropical African wetland includes swamp forests, flooded savannas, and floating prairies on its rivers and streams. The Congolian swamp forest is found on the banks of the middle reaches of the Congo River in a large depression in equatorial Africa called the *cuvette centrale congolaise*. The Congo River has the second highest flow of any river in the world and, along with its tributaries, provides the water that supports these forested alluvial swamps. In the wet season, the forests are flooded to a depth of 0.5 to 1.0 m; during the dry season, they often lack standing water. Human population is low in the region, and the people who live in the region are involved in hunting and fishing in the forest and its rivers. The eastern portion of the Congolian swamp forest in the Democratic Republic of Congo is generally thought to be more diverse than the western region in the Congo. Large tracts of the forest remain free of logging because of their relative isolation (Minnemeyer, 2002).

East Africa Tropical Marshes

Several wetlands that form around tropical lakes in Africa are typical of what are called "swamps" in Old World usage of the word but would be "marshes" in New World terminology (see Chapter 2 "Wetland Definitions"). These highly productive wetland margins tend to be dominated by tropical species of cattail (*Typha domingensis*) and papyrus (*Cyperus papyrus*), often with mats of floating plants (*Eichhornia crassipes* and

Figure 3.26 Wildlife is abundant in the Rift Valley lakes and wetlands. This photo, showing wildebeests, monkeys, and yellow-billed storks (*Ibis Ibis*), is along Lake Manyara, Tanzania, one of the southernmost lakes along the Rift Valley. (Photo by W. J. Mitsch)

Salvinia molesta). Many lakes and wetlands are found along the 6,500-km Rift Valley of eastern Africa. Not far from Nairobi on the floor of the Rift Valley, Lake Naivasha is one of the most studied tropical lakes in East Africa. The area provides a home for nearly the entire range of ducks and herons found in eastern Africa. Vegetation changes in Lake Naivasha have been caused by a combination of water-level fluctuations, the introduction of crayfish (*Procambarus clarkii*), and the physical effects of floating rafts of *Eichhornia crassipes* (Harper et al., 1995).

Other vast wetlands are found in the Rift Valley in northern Tanzania, including the wetlands of Ngorongoro Crater and the shorelines of Lake Manyara (Figs. 3.26, 3.27). Two swamps, Mandusi Swamp and Gorigor Swamp, and one lake, Lake Makat, are found in the caldera. The abundant wildlife of Ngorongoro was summarized by the East African/German conservationist Bernhard Grzimek, who stated, "It is impossible to give a fair description of the size and beauty of the Crater, for there is nothing with which one can compare it. It is one of the wonders of the World" (Hanby and Bygott, 1998).

West Africa Mangrove Swamps

Extensive mangrove swamps are found on Africa's tropical and subtropical coastlines. One example on the Atlantic Ocean west coast of Africa, about 150 km south of Dakar, is the Sine Saloum Delta in Senegal (Vidy, 2000), a vast (180,000 ha) almost-untouched expanse of mangrove swamps (Fig. 3.28). These mangrove swamps, as well as similar mangroves at the Senegal River delta around St. Louis to the north and in the coastal reaches of the Gambia River in Gambia to the south, support a wide variety of bird life, mammals, and four species of breeding turtles

Figure 3.27 Wetlands and wildlife of the Rift Valley of northern Tanzania including waterfowl in the wetlands of the Ngorongoro Crater, northern Tanzania, including yellow-billed duck (*Anas undulata*), red-billed duck (*A. erythrorhynchos*), and Egyptian goose (*Alopochen aegyptiaca*). (Photo by W. J. Mitsch)

Figure 3.28 African reef heron (*Egretta gularis*) in mangrove prop roots in Sine Saloum Delta, Senegal. (Photo by W. J. Mitsch)

in what is otherwise an extremely arid climate. Birds included several species of herons and egrets as well as the great white pelican (*Pelecanus onocrotalus*). To the east, where the delta meets the arid uplands, salt pans, or "tannes," develop where little vegetation is supported because of excessive salinity. The mangrove system is distinguished by the lack of permanent river flow related to the Sahelian drought dating to the 1970s. For that reason, the Sine Saloum is termed a *reverse estuary*, meaning that salinity increases going upstream. But desertification is also due to mismanagement of natural resources. The region is only lightly populated, and local people support themselves with fishing, salt production, and peanut farming. The mangrove forests have suffered from overexploitation for the wood they provide for housing and charcoal as well as from conversion to rice fields. UNESCO and other international agencies are encouraging both ecotourism in the region and adaptation of oyster farming techniques to better fit the mangrove system, as well as the creation of village "green belts." Many shell islands built well above the intertidal zone are found throughout the delta and indicate a long human history here.

Middle East

Mesopotamian Marshlands

The crown-jewel wetlands of the Middle East are the Mesopotamian marshlands of southern Iraq and Iran. These wetlands are in an arid region of the world and exist at the confluence of the Tigris and Euphrates rivers. The watersheds of both the Euphrates and the Tigris are predominantly in the countries of Turkey, Syria, and Iraq. The Tigris–Euphrates Basin has had water control projects for over six millennia. The Mesopotamian wetlands are the largest wetland ecosystem in the Middle East, have been the home to the Marsh Arabs for 5,000 years, and support a rich biodiversity (UNEP, 2001). Since 1970, the wetlands have been damaged dramatically. The Mesopotamian wetlands once were 15,000 to 20,000 km^2 in area but were drained in the 1980s and 1990s to less than 10 percent of that extent. There was a 30 percent decline just between 2000 and 2002. The draining of the wetlands was the result of human-induced changes. Upstream dams and drainage systems constructed in the 1980 and 1990s drastically altered the river flows and have eliminated the flood pulses that sustained the wetlands (UNEP, 2001). Turkey alone constructed more than a dozen dams on the upper rivers. But the main cause of the disappearance of the wetlands was water control structures built by Iraq between 1991 and 2002 (Altinbilek, 2004). These marshes, dominated by *Phragmites australis*, are located on the intercontinental flyway of migratory birds and provide wintering and staging areas for waterfowl. Two-thirds of West Asia's wintering waterfowl have been reported to live in the marshes. Globally threatened wildlife, which have been recorded in the marshes, include 11 species of birds, 5 species of mammals, 2 species of amphibians and reptiles, 1 species of fish, and 1 species of insect. The drying of the marshes has had a devastating effect on wildlife (UNEP, 2001). Their restoration is described in Chapter 18 "Wetland Creation and Restoration."

Australia/New Zealand

Eastern Australian Billabongs

Australia's wetlands are distinctive for their seasons of general dryness caused by high evaporation rates and low rainfall. Wetlands do occur on the Australian mainland, but only where the accumulation of water is possible, generally on the eastern and western

Figure 3.29 A billabong of New South Wales, Australia, showing bulrushes, river red gum (*Eucalyptus camaldulensis*) in the background, and invasive *Salvinia molesta* on the water's surface.

portions of the continent. Thus, there are not many permanent wetlands—most are intermittent and seasonal. Furthermore, because of the high evaporation rates, saline wetlands and lakes are not uncommon. A particular feature in eastern Australia is the *billabong* (Shiel, 1994), a semipermanent pool that develops from an overflowing river channel (Fig. 3.29). Although found throughout Australia, billabongs are best concentrated along the Murray and Darling rivers in southeastern Australia. There are about 1,400 wetlands representing 32,000 ha in four watersheds alone in New South Wales. Billabongs support a variety of aquatic plants, are a major habitat for birds and fish, and are often surrounded by one of many species of eucalyptus, especially the river red gum (*Eucalyptus camaldulensis*). The billabongs serve as refuges for aquatic animals during the dry season, when the rivers come close to drying.

Western Australia Wetlands

The Mediterranean-type climate of southwestern Australia favors a wide variety of wetlands, which are especially important to the waterfowl that are separated from the rest of the continent by vast expanses of desert. Swamps are numerous, and many can be found inland or just above the saline wetlands of the tidal rivers and bays of the Swan Coastal Plain, near Perth (Fig. 3.30). Nevertheless, it is estimated that 75 percent of the wetlands in the Swan Coastal Plain in southwestern Australia have been lost (Chambers and McComb, 1994).

Figure 3.30 Freshwater wetland in the Swan Coastal Plain, Western Australia. (Photo by J. Davis, reprinted with permission)

New Zealand Wetlands

For a small country, New Zealand has a wide variety of wetland types (Johnson and Gerbeaux, 2004). However, New Zealand has lost 90 percent of its wetlands, amounting to over 300,000 ha. The western region of South Island, called Westland, is sometimes humorously called "Wetland" because of the enormous amount of rain it receives (2–10 m annually) due to its location between the Tasman Sea to the west and the Southern Alps to its east. It is thus, not surprisingly, the location of a great variety of coastal wetlands. Grand Kahikatea (*Dacrycarpus dacrydiodes*) or "white pine"—forested wetlands (Fig. 3.31), reminiscent of the bald cypress swamps of the southeastern United States, are found throughout Westland and also on North Island. *Pakihi* (peatlands) are found on both North Island and South Island.

One of the largest wetlands in North Island is Whangamarino Wetland (Fig. 3.32), a 7,300-ha peatland and seasonally flooded swamp adjacent to the Waikato River, New Zealand's largest river (Clarkson, 1997; Shearer and Clarkson, 1998). Management issues facing this and other peatlands in the area are reduced inundation by the river, silt deposition from agricultural development, increased fire frequency over presettlement times, and invading willows and other exotics. Flax (*Phormium tenax*) swamps and raupo (*Typha orientalis*) marshes are also common in New Zealand. Willows (*Salix* spp.) are generally considered undesirable woody invaders to many of these wetlands.

Figure 3.31 Kahikatea Swamp in the background with Okarito Lagoon in the foreground in western New Zealand. The Kahikatea tree (*Dacrycarpus dacrydlodes*) is locally called white pine. These white pine forests once dominated both coastlines of New Zealand. (Photo by W. J. Mitsch)

Figure 3.32 Peatland in the lower Waikato River basin, about 60 km south of Aukland, New Zealand. Circular ponds with earthen paths are hunting ponds. (Photo by W. J. Mitsch)

Asia

Western Siberian Lowlands

One of the largest contiguous wetland areas in the world is the region of central Russia bordered by the Kara Sea of the Arctic Ocean to the north, the Ural Mountains to the west, and Kazakhstan to the south. The area is referred to the Western Siberian Lowland and encompasses about 2.7 million km², about 787,000 km² of which is peatland (Solomeshch, 2005). The region also has more than 800,000 lakes. Precipitation is relatively low (<600 mm/yr), but evapotranspiration is even lower (<400 mm/yr), leading to excess moisture that creates the peatlands. Part of this region includes the Bi-Ob region of central Russia, a large floodplain on the Ob River between Kazakhstan to the south and the Ob River's estuary to the north on the Kara Sea. This valley of channels, floodplain lakes, and river distributaries is actually an inland delta caused more by decreased sea levels than by deposited sediments. The region has been described as "the largest single breeding area for waterfowl in Eurasia" (Dugan, 1993). One of the greatest values of these peatlands could be carbon sequestration. It has been estimated that these wetlands alone have an average carbon accumulation of 22.8 Tg/yr (Tg = teragram = 10^{12} g) or about 24 to 35 percent of the global accumulation rate of all northern peatlands (Solomeshch, 2005).

Indian Freshwater Marshes

The world's second most populous nation is India. It is slightly more than one-third the size of the United States but has more than three times as many people. Droughts, soil erosion, overgrazing, and desertification are common. Agriculture employs two-thirds of the labor force, based in and around the alluvial plains and coastal zones on 55 percent of the land. The wetlands are under intense pressure for farm expansion, water control, and urbanization. Flooding cycles on alluvial valleys have been aggravated by these developments, resulting in "natural" disasters to humans and habitat alike. A few conservation wetlands remain under moderate protection, sometimes as remnants of the lands formerly held by the upper-classes. Keoladeo National Park in Bharatpur (Fig. 3.33) is an example, where the former hunting reserve is now a protected area of international significance. About 850 ha of the park are wetlands. The local economy benefits from tourism and also collects or illegally harvests products from the area. The protected wildlife heritage includes migratory species from northern Asia. In all, more than 350 species of birds, 27 mammals, 13 amphibians, 40 fish, and 90 wetland flowering plants are found in the park (Prasad et al., 1996).

Southern Asia River Deltas

More than 80 percent of Asian wetlands are located in seven countries: Indonesia, China, India, Papua New Guinea, Bangladesh, Myanmar, and Vietnam. The diversity of Asia's wetlands is reflected in its intertidal mud flats, swamp forests, natural lakes, open marshes, arctic tundra, and mangrove forests (recognized as one of the most productive ecosystems in the world—yielding over 70 direct and indirect uses

Figure 3.33 Keoladeo National Park, Bharatpur, India, during flooding season. (Photo by B. Gopal, reprinted with permission)

of the forest or its products—but now threatened by logging). The snowfields and glaciers of the Himalayas are the birthplace of many of the world's well-known rivers, including the Ganges, the Indus, the Mekong, and the Yangtze. The Mekong, Southeast Asia's longest river, begins in the Tibetan Plateau, enters its lower basin at the boundary of Myanmar, Laos, and Thailand, and then flows to the ocean through one of the world's great deltas. The basin catchment area is more than 600,000 km^2 and includes Laos, Cambodia, Thailand, and Vietnam. There has been little coordination among these countries concerning the basin's management, especially with regard to the extensive wetlands in the Mekong Delta region. Problems stemming from devegetation and drainage during the war years in southeast Asia have been exacerbated by more recent efforts at agricultural intensification, urbanization, industrialization, and dam and reservoir construction. Even drained soil became acidic (pH < 3) when sulfur-rich soils oxidized, making the soil unsuitable for agriculture. Restoration of a freshwater portion of the Mekong Delta, known as the Dong Thap Muoi (Plain of Reeds), continues with international assistance.

The largest expanse of mangrove swamps in the world is found in the Ganges Delta in Bangladesh and West Bengal in India. This large coastal mangroves area is part of what is referred to as the Sundarbans, which means "beautiful jungle." The Sundarbans were originally about 17,000 km^2 in size but are now only a small fraction of that area, perhaps 4,000 km^2. This estuary is a region of transition between the freshwater of the rivers originating from the Ganges and the saline water of the Bay of Bengal. These wetlands in the Bay of Bengal delta are formed and nourished by the Padma, Brahmaputra, and Meghna rivers in southern Bangladesh. The wetlands and adjacent uplands are the home to a rich diversity of wildlife, including the royal Bengal tiger (*Panthera tigris tigris*), the national animal of Bangladesh and India that was recently declared endangered by the International Union for Conservation of Nature (IUCN). The Sundarbans support a population of the tiger, which swims among mangrove islands hunting prey, estimated to be in the low 100s. The Sundarbans also includes seasonally flooded freshwater marshes and swamps upstream of the mangroves and is both a UNESCO world heritage site and a Ramsar Wetland of International Importance.

Issyk Kul

One of the world's great mountain lakes, the 623,600-ha Issyk Kul (also Ysyk Köl) is a brackish wetland/lake lying in a basin of the Tian Shan mountain chain in eastern Kyrgyzstan. It is an incredibly deep (up to 670 m depth) lake with up to 118 rivers and streams flowing into it but with no obvious outflow. The Issyk Kul State Reserve was acknowledged as a Ramsar site in 1975 and is also a UNESCO Bisphere Reserve. Around the lake are 3 species of amphibians, 11 species of reptiles, 54 species of mammals, and 267 species of birds. From 60,000 to 80,000 migratory water birds (16 species) gather around Lake Issyk Kul for wintering. The slightly salty lake has dropped 2.5 m in depth over the past few decades due partially to water diversions, causing some concern for the protection of habitats.

Wetlands of China

The total area of wetlands in China, estimated 625,000 km^2, ranks as Asia's highest (Lu, 1990; Chen, 1995); 250,000 km^2 are natural wetlands, with the rest artificial wetlands, such as rice paddies and fish ponds. Natural wetlands thus comprise about 2.5 percent of the country. There are several important wetland sites in and around China (Fig. 3.34). Few of the wetlands are preserved in semipristine conditions as is done in the West for habitat conservation; most wetlands in China provide fish, cattle, grain, duck, and other food as well as habitat and recreation benefits, in a symbiotic relationship between humans and nature.

River Deltas   Many of the important wetlands of China are found in the lower and delta regions of the Changjiang (Yangtze) River (Fig. 3.35), the Zhujiang (Pearl) River, and the Liaohe River. Because these regions are among the most populated in the world, very few natural wetlands remain, as most have been converted to rice paddies or fish ponds. But many new wetlands are being created by accretion of sediments, such as those on the downstream (east) and upstream (west) coasts of 1400-km^2 Chongming Island in the Yangtze Delta in Shanghai (Fig. 3.34 and 3.35). Chongming Island is the third-largest island in China and supports a human population of 600,000. In many cases, wetlands and reed (*Phragmites*) fields are connected hydrologically with fish ponds and rice paddies to enhance food and fiber production (Ma et al., 1993).

Yangtze River Wetlands   There are also extensive inland wetlands associated with the Yangtze River, particularly in the Jianghan–Dongting Plain in the middle of the river

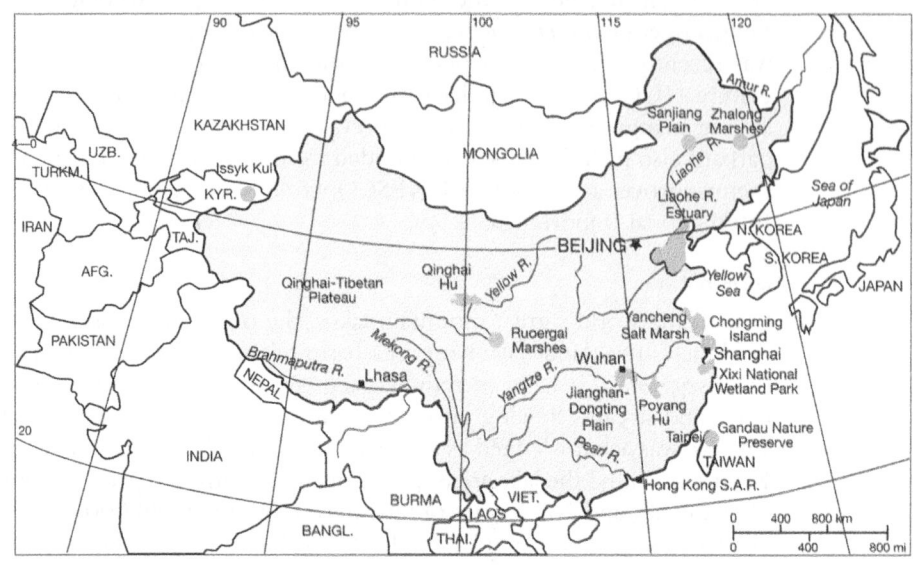

Figure 3.34 Wetlands of China and its neighboring countries discussed in this chapter.

Figure 3.35 Marshes on the eastern extent of Chongming Island in the Yangtze River near Shanghai, China. (Photo by W. J. Mitsch)

valley near Wuhan, Hubei Province, and Poyang Hu, in northern Jiangxi Province. The Jianghan–Dongting Plain is approximately 10,000 km² of former marshes and lakes that has been extensively drained and diked yet consistently suffers crop damage due to excessive water. Integrating these backwater areas with the Yangtze River as they once were is probably impossible, but converting wet areas from rice and other crops to wetland crops such as lotus (*Nelumbo nucifera*) and wild rice stem (*Zizania latifolia*) has been suggested as a viable "ecological" approach (Bruins et al., 1998). Poyang Hu is the largest lake in China but varies considerably with the season. The lake shrinks to less than 1,000 km² in the dry season and grows to 4,000 km² in the late-summer rainy season. The lake is connected to the Yangtze River with a 1-km-long channel that allows natural overflow. The lake's basin is one of China's most important rice-producing regions, but because of regular flooding, Jiangxi Province is among the poorest in China.

Northeastern China Wetlands There are extensive wetlands in northeastern China especially in Jilin Province north of Changchun. The 140,000-ha Momoge National Nature Reserve in northern Jilin Province, fed by the Nen and Tao'er rivers, supports almost 300 species of birds, including 6 species of cranes, among them the protected red-crowned crane (Fig. 3.36).

Qinghai-Tibetan Plateau The sources of several major rivers can also be found in the Qinghai–Tibetan Plateau of China—the Yellow, the Yangtze, the Indus, and the Ganges—along with high-altitude lakes and bogs. Qinghai is thus described as the "water tower of China." Most of the plateau's larger lakes are saline, and, at 458,000 ha and 3,200 m above sea level, Qinghai Lake is the largest (Fig. 3.37).

Figure 3.36 Red-crowned cranes at Momoge National Nature Reserve in Jilin Province in northeastern China. (Photo by W. J. Mitsch)

Figure 3.37 Qinghai Hu, western China: (a) A bird island with cormorants; (b) shallow marshes on one shoreline. This lake is in an arid Tibetan Plateau region of China and is the largest saltwater lake in the country.

As this whole area is experiencing desiccation, the lakes are shrinking, most recently at a rate of 12 cm per year in depth. Nevertheless, these wetlands are habitat for millions of migratory and resident birds comprising over 160 species. Birds common at Qinghai Lake include wild goose, brown-headed gulls, the cormorant, sandpipers, the extremely rare black-necked cranes, and the bar-headed goose (*Anser indicus*).

Wetland Parks China has become interested in establishing urban wetland parks that provide scenery and relaxation for the public in an aquatic setting. Wetland parks are now in many Chinese cities. Xixi National Wetland Park (Fig. 3.38), located in the suburbs of Hangzhou China in eastern China and covering an area of about 350 ha, is

Figure 3.38 Xixi National Wetland Park, Hangzhou, China. (Photo by W. He, reprinted with permission)

but one example. The park includes streams for boating and several marshes, swamps, and ponds. It is the first formal national wetland park in China and provides a semi-natural water-land park opened to the public. Much of the park is restored on former fish ponds and rice paddies.

The 61-ha Hong Kong Wetland Park (Fig. 3.39) provides a green oasis in an otherwise concrete and asphalt dominated megacity. Its mission is to educate the public about wetlands of East Asia. Opened to the public in May 2006, it consists of a 10,000-m^2 visitor center, a Wetland Interactive World, and a 60-ha wetland reserve with plentiful boardwalks and interpretative signage. In 2013, the park had 440,000 visitors, including 61,000 overseas visitors.

Urban Wetland Park, Taiwan

Similar to the emerging urban wetland parks in mainland China, a site worth mentioning where wetlands have been brought before a large urban population is the 57-ha Gandau Nature Park in Taipei, Taiwan (Fig. 3.40). The wetland site, which is very popular with local environmental and bird-watching groups, includes a bird-viewing gallery and several paved and unpaved pathways. It forms along a major bend of the

Figure 3.39 Boardwalk through 61-ha Hong Kong Wetland Park located in otherwise densely populated urban Hong Kong. (Photo by W. J. Mitsch)

Figure 3.40 View of Gandau Wetland Park, Taipei, Taiwan, from its nature center building. (Photo by W. J. Mitsch)

Keelung River in Taipei. The wetland supports mostly created freshwater wetland ponds at the Gandau Nature Park and several hectares of saline mangrove forest in the adjacent Gandau nature reserve along the river.

Recommended Readings

Fraser, L. H., and P. A. Keddy, eds. 2005. The *World's Largest Wetlands: Ecology and Conservation*. Cambridge, UK: Cambridge University Press.

Grunwald, M. 2006. *The Swamp: The Everglades, Florida, and the Politics of Paradise*. New York: Simon & Schuster.

Junk, W., ed. 2013. The World's Wetlands and Their Future under Global Climate Change. Special issue of *Aquatic Sciences* 75 (1): 1–167.

References

Abraham, K. F. and C. J. Keddy. 2005. The Hudson Bay Lowland. In L.A. Fraser and P.A. Keddy, eds., *The World's Largest Wetlands: Ecology and Conservation*, Cambridge University Press, Cambridge, UK, pp. 118–148.

Altinbilek, D. 2004. Development and management of the Euphrates–Tigris basin. *Water Resources Development* 20: 15–33.

Aselmann, I., and P. J. Crutzen. 1989. Global distribution of natural freshwater wetlands and rice paddies, their net primary productivity, seasonality and possible methane emissions. *Journal of Atmospheric Chemistry* 8: 307–358.

Australian Nature Conservation Agency. 1996. *Wetlands Are Important*. Two-page flyer, National Wetlands Program, ANCA, Canberra, Australia.

Bartlett, C. H. 1904. *Tales of Kankakee Land*. Charles Scribner's Sons, New York. 232 pp.

Bazilevich, N. I., L. Ye. Rodin, and N. N. Rozov. 1971. Geophysical aspects of biological productivity. *Soviet Geography* 12: 293–317.

Bruins, R. J. F., S. Cai, S. Chen, and W. J. Mitsch. 1998. Ecological engineering strategies to reduce flooding damage to wetland crops in central China. *Ecological Engineering* 11: 231–259.

Calheiros, D. F., M. D. de Oliveira, and C. R. Padovani. 2012. Hydro-ecological processes and anthropogenic impacts on the ecosystem services of the Pantanal wetland. In A. A. R. Ioris, ed., *Tropical Wetland Management: The South-American Pantanal and the International Experience*. Ashgate Publishing Ltd., Farnham, England, pp. 29–57.

Campbell, D. 2005. The Congo River basin. In L.A. Fraser and P.A. Keddy, eds., *The World's Largest Wetlands: Ecology and Conservation*, Cambridge University Press, Cambridge, UK, pp. 149–165.

Chambers, J. M., and A. J. McComb. 1994. Establishment of wetland ecosystems in lakes created by mining in Western Australia. In W. J. Mitsch, ed., *Global Wetlands: Old World and New*. Elsevier, Amsterdam, The Netherlands, pp. 431–441.

Chen, Y. 1995. *Study of Wetlands in China*. Jilin Sciences Technology Press, Changchun, China. [in Chinese with English summaries.]

Clarkson, B. R. 1997. Vegetation recovery following fire in two Waikato peatlands at Whangamarino and Moanatuatua. *New Zealand Journal of Botany* 35: 167–179.

Comin, F. A., J. A. Romero, V. Astorga, and C. Garcia. 1997. Nitrogen removal and cycling in restored wetlands used as filters of nutrients for agricultural runoff. *Water Science and Technology* 35: 255–261.

Contreras-Espinosa, F., and B. G. Warner. 2004. Ecosystem characteristics and management considerations from coastal wetlands in Mexico. *Hydrobiologia* 511: 233–245.

Costanza, R., W. J. Mitsch, and J. W. Day. 2006. A new vision for New Orleans and the Mississippi delta: Applying ecological economics and ecological engineering. *Frontiers in Ecology and the Environment* 4: 465–472.

Dahl, T. E. 1990. Wetlands losses in the United States, 1780s to 1980s. U.S. Department of Interior, Fish and Wildlife Service, Washington, DC. 21 pp.

Dahl, T. E. 2000. Status and trends of wetlands in the conterminous United States 1986 to 1997. U.S. Department of the Interior, Fish and Wildlife Service, Washington, DC, 82 pp.

Dahl, T. E. 2006. Status and trends of wetlands in the conterminous United States 1998 to 2004. U.S. Department of the Interior, Fish and Wildlife Service, Washington, DC, 112 pp.

Dahl, T. E. 2011. Status and trends of wetlands in the conterminous United States 2004 to 2009. U.S. Department of the Interior, Fish and Wildlife Service, Washington, DC, 108 pp.

Dahl, T. E. 2014. Status and trends of prairie wetlands in the United States 1997 to 2009. U.S. Department of the Interior, Fish and Wildlife Service, Washington, DC, 67 pp.

Dahl, T. E., and C. E. Johnson. 1991. Wetlands status and trends in the conterminous United States mid-1970s to mid-1980s. U.S. Department of Interior, Fish and Wildlife Service, Washington, DC. 28 pp.

da Silva, C. J., and P. Girard. 2004. New challenges in the management of the Brazilian Pantanal and catchment area. *Wetlands Ecology and Management* 12: 553–561.

Davidson, N.C. 2014 How much wetland has the world lost? Long-term and recent trends in global wetland area. *Marine and Freshwater Research* 65: 934–941.

Day, J. W., Jr., J. Barras, E. Clairain, J. Johnston, D. Justic, G. P. Kemp, J.-Y. Ko, R. Lane, W. J. Mitsch, G. Steyer, P. Templet, and A. Yañez-Arancibia. 2005. Implications of global climatic change and energy cost and availability for the restoration of the Mississippi Delta. *Ecological Engineering* 24: 253–265.

Day, J. W., Jr., D. F. Boesch, E. J. Clairain, G. P. Kemp, S. B. Laska, W. J. Mitsch, K. Orth, H. Mashriqui, D. R. Reed, L. Shabman, C. A. Simenstad, B. J. Streever, R. R. Twilley, C. C. Watson, J. T. Wells, and D. F. Whigham. 2007. Restoration of the Mississippi Delta: Lessons From Hurricanes Katrina and Rita. *Science* 315: 1679–1684.

Dennis, J. V. 1988. *The Great Cypress Swamps*. Louisiana State University Press, Baton Rouge. 142 pp.

Douglas, M. S. 1947. *The Everglades: River of Grass*. Ballantine, New York. 308 pp.

Dugan, P. 1993. *Wetlands in Danger*. Michael Beasley, Reed International Books, London. 192 pp.

Ellison, A. M. 2004. Wetlands of Central America. *Wetlands Ecology and Management* 12: 3–55.

Finlayson, M., and N. C. Davidson. 1999. Global Review of Wetland Resources and Priorities for Wetland Inventory. Ramsar Bureau Contract 56, Ramsar Convention Bureau, Gland, Switzerland.

Forsyth, J. L. 1960. *The Black Swamp*. Ohio Department of Natural Resources, Division of Geological Survey, Columbus, OH. 1 p.

Frayer, W. E., T. J. Monahan, D. C. Bowden, and F. A. Graybill. 1983. Status and trends of wetlands and deepwater habitat in the conterminous United States, 1950s to 1970s. Department of Forest and Wood Sciences, Colorado State University, Fort Collins. 32 pp.

Gorham, E. 1991. Northern peatlands: Role in the carbon cycle and probable responses to climatic warming. *Ecological Applications* 1: 182–195.

Grant, M. 1962. *Myths of the Greeks and Romans*. Mentor/New American Library, New York.

Gray, L. C., O. E. Baker, F. J. Marschner, B. O. Weitz, W. R. Chapline, W. Shepard, and R. Zon. 1924. The utilization of our lands for crops, pasture, and forests. In *U.S. Department of Agriculture Yearbook*, 1923. Government Printing Office, Washington. DC, pp. 415–506.

Grunwald, M. 2006. *The Swamp: The Everglades, Florida, and the Politics of Paradise*. Simon & Schuster, New York, 450 pp.

Hanby, J., and D. Bygott. 1998. *Ngorongoro Conservation Area*. Kibuyu Partners, Karatu, Tanzania.

Harper, D. M., C. Adams, and K. Mavuti. 1995. The aquatic plant communities of the Lake Naivasha wetland, Kenya: Pattern, dynamics, and conservation. *Wetlands Ecology and Management* 3: 111–123.

Harris, M. B., W. Tomas, G. Mourao, C. J. DaSilva, E. Guimaraes, F. Sonoda, and E. Fachim. 2005. Safeguarding the Pantanal wetlands: Threats and conservation initiatives. *Conservation Biology* 19: 714–720.

Ioris, A. A. R., ed. 2012. *Tropical Wetland Management: The South-American Pantanal and the International Experience*. Ashgate Publishing Ltd., Farnham, England, 351 pp.

Johnson, D. C. 1942. *The Origin of the Carolina Bays*. Columbia University Press, New York. 341 pp.

Johnson, P., and P. Gerbeaux. 2004. *Wetland Types in New Zealand*. New Zealand Department of Conservation, Wellington, 184 pp.

Junk, W. J. 1993. Wetlands of tropical South America. In D. F. Whigham, D. Dykyjová, and S. Hejny, eds., *Wetlands of the World, I: Inventory, Ecology, and Management*. Kluwer Academic Publishers, Dordrecht, The Netherlands, pp. 679–739.

Junk, W. J., and M. T. F. Piedade. 2004. Status of knowledge, ongoing research, and research needs in Amazonian wetlands. *Wetlands Ecology and Management* 12: 597–609.

Junk, W. J., and M. T. F. Piedade. 2005. The Amazon River basin. In L.A. Fraser and P.A. Keddy, eds., *The World's Largest Wetlands: Ecology and Conservation*, Cambridge University Press, Cambridge, UK, pp. 63–117.

Junk, W. J., and C. Nunes de Cunha. 2005. Pantanal: A large South American wetland at a crossroads. *Ecological Engineering* 24: 391–401.

Kaatz, M. R. 1955. The Black Swamp: A study in historical geography. *Annals of the Association of American Geographers* 35: 1–35.

Kirk, P. W., Jr. 1979. *The Great Dismal Swamp.* University Press of Virginia, Charlottesville. 427 pp.

Lehner, B. and P. Döll. 2004. Development and validation of a global database of lakes, reservoirs, and wetlands. *Journal of Hydrology* 296: 1–22.

Lide, R. F., V. G. Meentemeyer, J. E. Pinder, and L. M. Beatty 1995. Hydrology of a Carolina Bay located on the Upper Coastal Plain of western South Carolina. *Wetlands* 15: 47–57.

Lodge, T. E. 2010. *The Everglades Handbook: Understanding the Ecosystem*, 3rd ed. CRC Press, Boca Raton, FL.

Lu, J., ed. 1990. *Wetlands in China.* East China Normal University Press, Shanghai. 177 pp. [in Chinese].

Lu, J. 1995. Ecological significance and classification of Chinese wetlands. *Vegetatio* 118: 49–56.

Ma, X., X. Liu, and R. Wang. 1993. China's wetlands and agro-ecological engineering. *Ecological Engineering* 2: 291–330.

Maltby, E., and R. E. Turner. 1983. Wetlands of the world. *Geographic Magazine* 55: 12–17.

Matthews, E., and I. Fung. 1987. Methane emissions from natural wetlands: Global distribution, area, and environmental characteristics of sources. *Global Biogeochemical Cycles* 1: 61–86.

Mauchamp, A., P. Chauvelon, and P. Grillas. 2002. Restoration of floodplain wetlands: Opening polders along a coastal river in Mediterranean France, Vistre marshes. *Ecological Engineering* 18: 619–632.

McCarthy, J. M., T. Gumbricht, T. McCarthy, P. Frost, K. Wessels, and F. Seidel. 2004. Flooding patterns in the Okavango wetland in Botswana between 1972 and 2000. *Ambio* 32: 453–457.

Mendelsohn, J. and S. el Obeid. 2004. *Okavango River: The Flow of a Lifeline.* Struik Publishers, Cape Town, South Africa. 176 pp.

Meyer, A. H. 1935. The Kankakee "Marsh" of northern Indiana and Illinois. *Michigan Academy of Science, Arts, and Letters Papers* 21: 359–396.

Michalak, A. M., E. J. Anderson, D. Beletsky, S. Boland et al. 2013. Record-setting algal bloom in Lake Erie caused by agricultural and meteorological trends consistent with expected future conditions. *Proceedings National Academy of Sciences* 110: 6524–6529.

Millennium Ecosystem Assessment, 2005. Ecosystems and Human Well-being: Synthesis. Island Press and World Resources Institute, Washington, DC. 137 pp.

Minnemeyer, S. 2002. *An Analysis of Access into Central Africa's Rainforests.* World Resources Institute, Washington DC.

Mitsch, W. J. 2010. The 2010 Gulf of Mexico oil spill: What would Mother Nature do? *Ecological Engineering* 36: 1607–1610.

Mitsch, W. J., R. H. Mitsch, and R. E. Turner. 1994. Wetlands of the Old and New Worlds—Ecology and Management. In W. J. Mitsch, ed., *Global Wetlands: Old and New*. Elsevier, Amsterdam, The Netherlands, pp. 3–56.

Mitsch, W. J., A. M. Nahlik, P. Wolski, B. Bernal, L. Zhang, and L. Ramberg. 2010. Tropical wetlands: Seasonal hydrologic pulsing, carbon sequestration, and methane emissions. *Wetlands Ecology and Management* 18: 573–586.

Mitsch, W. J, and M. Hernandez. 2013. Landscape and climate change threats to wetlands of North and Central America. *Aquatic Sciences* 75: 133–149.

Nahlik, A. M., and W. J. Mitsch. 2006. Tropical treatment wetlands dominated by free-floating macrophytes for water quality improvement in the Caribbean coastal plain of Costa Rica. *Ecological Engineering* 28: 246–257.

National Wetlands Working Group. 1988. *Wetlands of Canada*. Ecological and Classification Series 24, Environment Canada, Ottawa, Ontario, and Polyscience Publications, Montreal, Quebec. 452 pp.

Niu, Z., H. Zhang, and P. Gong. 2011. More protection for China's wetlands. *Nature* 471: 305.

Novacek, J. M. 1989. The water and wetland resources of the Nebraska Sandhills. In A. G. van der Valk, ed., *Northern Prairie Wetlands*. Iowa State University Press, Ames, pp. 340–384.

Odum, H. T. 1951. The Carolina Bays and a Pleistocene weather map. *American Journal of Science* 250: 262–270.

Pérez-Arteaga, A., K. J. Gaston, and M. Kershaw. 2002. Undesignated sites in Mexico qualifying as wetlands of international importance. *Biological Conservation* 107: 47–57.

Por, F. D. 1995. *The Pantanal of Mato Grosso (Brazil)*. Kluwer Academic Publishers, Dordrecht, The Netherlands. 122 pp.

Prasad, V. P., D. Mason, J. E. Marburger, and C. R. A. Kumar. 1996. *Illustrated Flora of Keoladeo National Park, Bharatpur, Rajasthan*. Bombay Natural History Society, Mumbai, India. 435 pp.

Prigent, C., F. Papa, F. Aires, C. Jiménez, W. B. Rossow, and E. Matthews. 2012. Changes in land surface water dynamics since the 1990s and relation to population pressure. *Geophys. Res. Lett.*, doi:10.1029/2012GL051276

Prouty, W. F. 1952. Carolina Bays and their origin. *Geological Society of America Bulletin* 63: 167–224.

Qiu, J. 2011. China faces up to '"terrible"' state of its ecosystems. *Nature* 471: 19.

Ramberg, L., P. Wolski, and M. Krah. 2006a. Water balance and infiltration in a seasonal floodplain in the Okavango Delta, Botswana. *Wetlands* 26: 677–690.

Ramberg, L., P. Hancock, M. Lindholm, T. Meyer, S. Ringrose, J. Silva., J. Van As, and C. VanderPost. 2006b. Species diversity of the Okavango Delta, Botswana. *Aquatic Sciences* 68: 310–337.

Ramsar Convention Secretariat. 2004. *Ramsar Handbook for the Wise Use of Wetlands*, 2nd ed. Handbook 10, Wetland Inventory: A Ramsar framework for wetland inventory. Ramsar Secretariat, Gland, Switzerland.

Revenga, C., J. Brunner, N. Henninger, K. Kassem, and R. Payne. 2000. *Pilot Analysis of Global Ecosystems: Freshwater Systems*. World Resources Institute, Washington, DC. 65 pp.

Richardson, C. J., R. Evans, and D. Carr. 1981. Pocosins: An ecosystem in transition. In C. J. Richardson, ed., *Pocosin Wetlands*. Hutchinson Ross Publishing, Stroudsburg, PA, pp. 3–19.

Ringrose, S., C. Vanderpost, V. Matheson, P. Wolski, P. Huntsman-Mapila, M. Murray-Hudson, and A. Jellema. 2007. Indicators of desiccation-driven change in the distal Okavango Delta, Botswana. *Journal of Arid Environments* 68: 88–112.

Roe, H. B., and Q. C. Ayres. 1954. *Engineering for Agricultural Drainage*. McGraw-Hill, New York. 501 pp.

Russi D., P. ten Brink, A. Farmer, T. Badura, D. Coates, J. Förster, R. Kumar, and N. Davidson. 2013. The Economics of Ecosystems and Biodiversity for Water and Wetlands. IEEP, Ramsar Secretariat, Gland, London and Brussels.

Sabin, T.J. and V.T. Holliday. 1995. Playas and lunettes on the Southern High Plains: Morphometric and spatial relationships. *Annals of the Association American Geographers* 85: 286–305.

Savage, H. 1983. *The Mysterious Carolina Bays*. University of South Carolina Press, Columbia. 121 pp.

Scavia, D., J. D. Allan, K. K. Arend, S. Bartell et al. 2014. Assessing and addressing the re-eutrophication of Lake Erie: Central basin hypoxia. *Journal of Great Lakes Research* 40: 226–246.

Schmidt, K. F. 2001. A true-blue vision for the Danube. *Science* 294: 1444–1447.

Shaw, S. P., and C. G. Fredine. 1956. *Wetlands of the United States, Their Extent, and Their Value for Waterfowl and Other Wildlife*. Circular 39, U.S. Fish and Wildlife Service, U.S. Department of Interior, Washington, DC. 67 pp.

Shearer, J. C., and B. R. Clarkson. 1998. Whangamarino wetland: Effects of lowered river levels on peat and vegetation. *International Peat Journal* 8: 52–65.

Sheffield, R. M., T. W. Birch, W. H. McWilliams, and J. B. Tansey. 1998. Chamaecyparis thyoides (Atlantic white cedar) in the United States. In A. D. Laderman, ed., *Coastally Restricted Forests*. Oxford University Press, New York, pp. 111–123.

Shiel, R. J. 1994. Death and life of the billabong. In X. Collier, ed., *Restoration of Aquatic Habitats*. Selected Papers from New Zealand Limnological Society 1993 Annual Conference, Department of Conservation, pp. 19–37.

Smith, L.M. 2003. *Playas of the Great Plains*. University of Texas Press, Austin, Texas, 257 pp.

Solomeshch, A. I. 2005. The West Siberian Lowland. In L. A. Fraser and P. A. Keddy, eds., *The World's Largest Wetlands: Ecology and Conservation*, Cambridge University Press, Cambridge, UK, pp. 11–62.

Tiner, R. W. 1984. *Wetlands of the United States: Current Status and Recent Trends*. National Wetlands Inventory, U.S. Fish and Wildlife Service, Washington, DC. 58 pp.

Trama F. A., F. L. Riso-Patrón, A. Kumar, E. González, D. Somma, and M.B. McCoy. 2009. Wetland cover types and plant community changes in response to cattail control activities in the Palo Verde Marsh, Costa Rica. *Ecological Restoration* 27:278–289.

United Nations Environmental Programme (UNEP). 2001. *The Mesopotamian Marshlands: Demise of an Ecosystem* (UNEP/DEWA/TR.01-3 Rev.1), UNEP, Nairobi, Kenya.

Vasander, H., E.-S. Tuittila, E. Lode., L. Lundin, M. Ilomets, T. Sallantaus, R. Heikkila, M.-L. Pitkanen, and J. Laine. 2003. Status and restoration of peatlands in northern Europe. *Wetlands Ecology and Management* 11: 51–63.

Verhoeven, J. T. A. 2014. Wetlands in Europe: Perspectives for restoration of a lost paradise. *Ecological Engineering* 66: 6–9.

Vidy, G. 2000. Estuarine and mangrove systems and the nursery concept: which is which. The case of the Sine Saloum system (Senegal). *Wetlands Ecology and Management* 8: 37–51.

Whooten, H. H., and M. R. Purcell. 1949. *Farm Land Development: Present and Future by Clearing, Drainage, and Irrigation*. Circular 825, U.S. Department of Agriculture, Washington, DC.

Wolff, W. J. 1993. Netherlands—Wetlands. In E. P. H. Best and J. P. Bakker, eds., *Netherlands—Wetlands*. Kluwer Academic Publishers, Dordrecht, The Netherlands, pp. 1–14.

Wright, J. O. 1907. *Swamp and Overflow Lands in the United States*. Circular 76, U.S. Department of Agriculture, Washington, DC.

Zoltai, S. C., S. Taylor, J. K. Jeglum, G. F. Mills, and J. D. Johnson. 1988. Wetlands of boreal Canada. In National Wetlands Working Group, ed., *Wetlands of Canada*. Ecological Land Classification Series 24, Environment Canada, Ottawa, Ontario, and Polyscience Publications, Montreal, Quebec, pp. 97–154.

Part II

The Wetland Environment

Chapter 4

Wetland Hydrology

Hydrologic conditions are extremely important for the maintenance of a wetland's structure and function. They affect many abiotic factors, including soil anaerobiosis, nutrient availability, and, in coastal wetlands, salinity. These, in turn, determine the biota that develops in a wetland. Finally, completing the cycle, biotic components are active in altering the wetland hydrology and other physicochemical features. The hydroperiod, or hydrologic signature of a wetland, is the result of the balance between inflows and outflows of water (called the water budget), the wetland basin geomorphology, and subsurface conditions. The hydroperiod can have dramatic seasonal and year-to-year variations, yet it remains the major determinant of wetland processes. The major components of a wetland's water budget include precipitation, evapotranspiration, surface inflows and outflows including overbank flooding into riparian wetlands, groundwater fluxes, and tides or seiches in coastal wetlands. Simple determinations of the hydroperiod, water budget, and turnover time in wetland studies can contribute to a better understanding of wetland function. Hydrology affects species composition and richness, primary productivity, organic accumulation, and nutrient cycling in wetlands.

The hydrology of a wetland creates the unique physiochemical conditions that make such an ecosystem different from both well-drained terrestrial systems and deepwater aquatic systems. Hydrologic pathways such as precipitation, surface runoff, groundwater, tides, and flooding rivers transport energy and nutrients to and from wetlands. Water depth, flow patterns, and duration and frequency of flooding, which are the result of all of the hydrologic inputs and outputs, influence the biochemistry of the soils and are major factors in the ultimate selection of the biota of wetlands. Biota ranging from microbial communities, to vegetation, to waterfowl are all constrained or enhanced by hydrologic conditions. An important point about wetlands—one that is often missed by ecologists who begin to study these systems—is this: *Hydrology is probably the single most important determinant of the establishment and maintenance of specific types of wetlands and wetland processes.* An understanding of rudimentary hydrology should be in the repertoire of any wetland scientist.

Importance of Hydrology in Wetlands

Wetlands are transitional between terrestrial and open-water aquatic ecosystems. They are transitional in terms of spatial arrangement, for they usually are found between uplands and aquatic systems (see Fig. 2.1a). They are also transitional in the amount of water they store and process and in other ecological processes that result from the water regime. Wetlands form the aquatic boundary of the habitats of many terrestrial plants and animals; they also form the terrestrial edge for many aquatic plants and animals. Hence, small changes in hydrology can result in significant biotic changes.

The starting point for the *hydrology* of a wetland is the climate and basin geomorphology (Fig. 4.1). All things being equal, wetlands are more prevalent in cool or wet climates than in hot or dry climates. Cool climates have less water loss from the land via evapotranspiration, whereas wet climates have excess precipitation. The second important factor is the geomorphology of the landscape and basin. Steep terrain tends to have fewer wetlands than flat or gently sloping landscapes. Isolated basins have different potential for wetlands than do tidal-fed or river-fed environments. When climate, basin geomorphology, and hydrology are considered as one unit, it is referred to as a wetland's *hydrogeomorphology*. Figure 4.1 illustrates that the hydrology of a wetland directly modifies and changes its *physiochemical environment* (chemical and physical properties), particularly oxygen availability and related chemistry, such as nutrient availability, pH, and toxicity (e.g., the production of hydrogen sulfide). Hydrology also transports sediments, nutrients, and even toxic materials into wetlands, thereby further influencing the physiochemical environment. Except in nutrient-poor wetlands such as bogs, water inputs are the major source of nutrients to wetlands. Hydrology also causes water outflows from wetlands that often remove biotic and abiotic material, such as dissolved organic carbon, excessive salinity, toxins, and excess sediments and detritus. Some modifications in the physicochemical environment, such as the buildup of sediments, can modify the hydrology by changing the basin geometry or affecting the hydrologic inflows or outflows (pathway A in Fig. 4.1).

Modifications of the physiochemical environment, in turn, have a direct impact on the biota in the wetland. When hydrologic conditions in wetlands change even slightly, the biota may respond with massive changes in species composition and richness and in ecosystem productivity. Biota such as emergent aquatic plants adapt to the anoxia in the sediments, although the anoxia excludes most vascular plant species. The level of nutrients in the sediments determines productivity and which species will dominate. Animals adapted to shallow water and this vegetation cover will flourish. Microbes able to metabolize in anoxic conditions dominate the reduced sediments, while aerobic microorganisms survive in a thin layer of oxidized sediments and in the water column

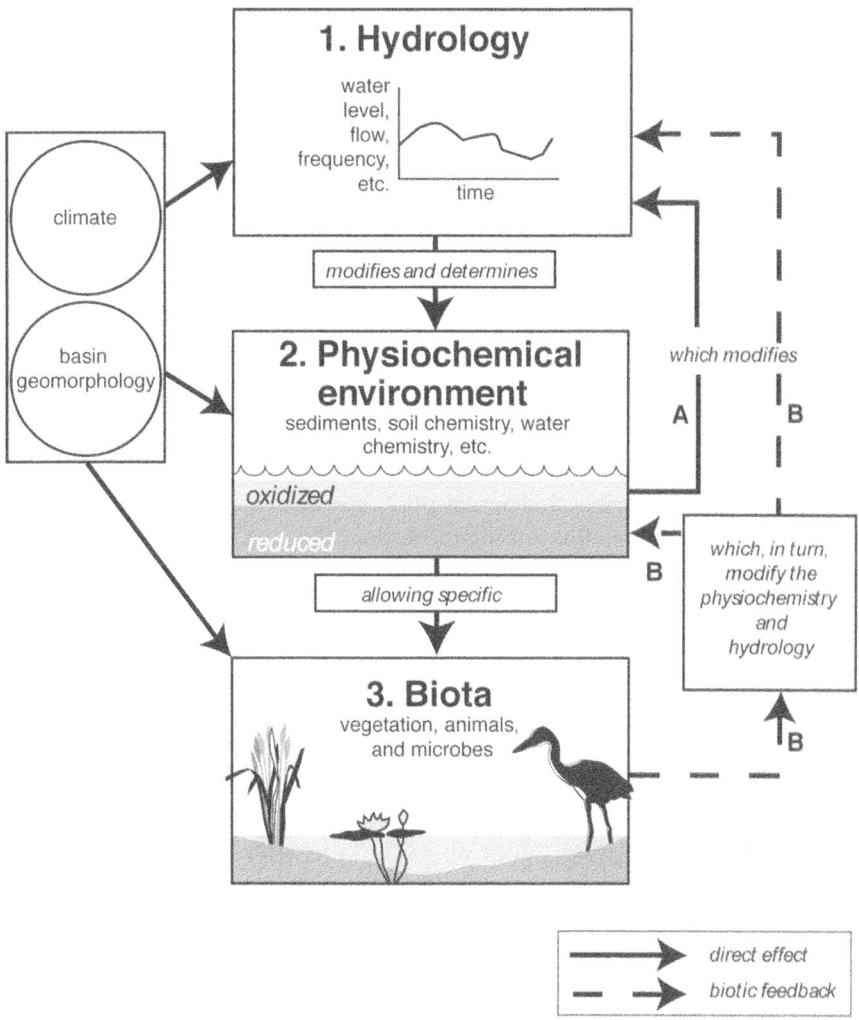

Figure 4.1 Conceptual diagram illustrating the effects of hydrology on wetland function and the biotic feedbacks that affect wetland hydrology. Pathways A and B are feedbacks to the hydrology and physiochemistry of the wetland.

if oxygen is present there. When hydrologic patterns remain similar from year to year, a wetland's biotic structural and functional integrity may persist for many years.

Biotic Control of Wetland Hydrology

Just as many other ecosystems exert feedback (cybernetic) control of their physical environments, wetland biota are not passive to their hydrologic conditions. Pathway B in Figure 4.1 shows that the biotic components of wetlands can control the hydrology and chemistry of their environment through a variety of mechanisms. Microbes, in particular, catalyze virtually all chemical changes in wetland soils and thus control nutrient availability to plants and even the production of phytotoxins, such as sulfides. Plants, animals, and microbes that use these essential biological feedback mechanisms were formally recognized in the ecological literature as *ecosystem engineers*. Plants cause changes in their physical environment through processes such as peat building, sediment trapping, nutrient retention, water shading, and transpiration. Wetland vegetation influences the hydrologic conditions of the physicochemical environment by binding sediments to reduce erosion, by trapping sediments, by interrupting water flows, and by building peat. Accumulated sediments and organic matter, in turn, interrupt water flows and can eventually decrease the duration and frequency by which the wetlands are flooded. Bogs build peat to the point at which they are no longer influenced at the surface by the inflow of mineral waters. Some trees in some southern swamps save water by their deciduous nature, their seasonal shading, and their relatively slow rates of transpiration. In more temperate climates, trees that invade shallow marshes and vernal pools can decrease water levels during the growing season by increasing transpiration, thus allowing even more woody plants to take over. Removal of these trees in what appears to be a dry forest sometimes surprisingly causes standing water and marsh vegetation to reappear.

Several animals are particularly noted for their contributions to hydrologic modifications and subsequent changes in wetlands. The exploits of beavers (*Castor canadensis*) in much of North America in both creating and destroying wetland habitats are well known. They build dams on streams, backing up water across great expanses and creating wetlands where none existed before. In colonial times, beaver populations covered the entire American continent north of Mexico, before fur trappers drastically reduced them. Beavers have been an important causal force in the creation of the Great Dismal Swamp of Virginia and North Carolina. Hey and Philippi (1995) estimated that a population of 40 million beavers could have accounted for 207,000 km^2 of beaver ponds (wetlands) in the upper Mississippi and Missouri River basins before European trappers entered the region and that, with the demise of the beaver, only 1 percent of those beaver ponds exist today.

Muskrats (*Ondatra zibethicus*) burrow through wetlands, changing flow patterns and sometimes water levels directly. They harvest large amounts of emergent vegetation for their food and to build winter lodges, thereby opening up large areas of marshes. Geese, especially Canada geese (*Branta canadensis*) and several varieties of snow geese (*Chen* spp.), cause *eat-outs*, or major wetland vegetation removal by herbivory, in many parts of the world. Newly planted wetlands are particularly susceptible to Canada geese eat-outs in North America. By removing vegetation cover, these herbivores reset the successional status of the wetlands and thus have a major impact on wetland hydrology.

The American alligator (*Alligator mississippiensis*) is known for its role in the Florida Everglades in constructing "gator holes" that serve as oases for fish, turtles, snails, and other aquatic animals during the dry season. In all of these cases, the biota of the ecosystem have contributed to their own survival, to the survival of other species, and to the elimination of others by influencing the ecosystem's hydrology and other physical characteristics.

Wetland Hydroperiod

The *hydroperiod* is the seasonal pattern of the water level of a wetland and is the wetland's hydrologic signature. It characterizes each type of wetland, and the constancy of its pattern from year to year ensures a reasonable stability for that wetland. It defines the rise and fall of a wetland's surface and subsurface water by integrating all of the inflows and outflows. The hydroperiod is also influenced by physical features of the terrain and by proximity to other bodies of water.

Many terms are used to describe qualitatively a wetland's hydroperiod (Table 4.1). These terms such as *seasonally flooded* or *intermittently flooded* are specific in their meaning and should be used with care and with sufficient data in describing a wetland's hydroperiod. For wetlands that are not subtidal or permanently flooded, the amount of time that a wetland is in standing water is called the *flood duration*. The average number of times that a wetland is flooded in a given period is known as the *flood frequency*. Both terms are used to describe periodically flooded wetlands such as coastal salt marshes and riparian wetlands.

Typical hydroperiods for a diverse set of wetlands are shown in Figure 4.2. A coastal salt marsh has a hydroperiod of semidiurnal flooding and dewatering superimposed on a twice-monthly pattern of spring and ebb tides (Fig. 4.2a). Wetlands along coastlines often show some of this same spring-and-ebb pulsing (Fig. 4.2b), whereas others reflect seasonal water-level changes of freshwater inflows and the water levels

Table 4.1 Definitions of wetland hydroperiods

Tidal Wetlands
Subtidal—permanently flooded with tidal water
Irregularly exposed—surface exposed by tides less often than daily
Regularly flooded—alternately flooded and exposed at least once daily
Irregularly flooded—flooded less often than daily

Nontidal Wetlands
Permanently flooded—flooded throughout the year in all years
Intermittently exposed—flooded throughout the year except in years of extreme drought
Semipermanently flooded—flooded during the growing season in most years
Seasonally flooded—flooded for extended periods during the growing season but usually no surface water by end of growing season
Saturated—substrate is saturated for extended periods during the growing season, but standing water is rarely present
Temporarily flooded—flooded for brief periods during the growing season, but water table is otherwise well below surface
Intermittently flooded—surface is usually exposed with surface water present for variable periods without detectable seasonal pattern

Source: After Cowardin et al. (1979)

116 Chapter 4 Wetland Hydrology

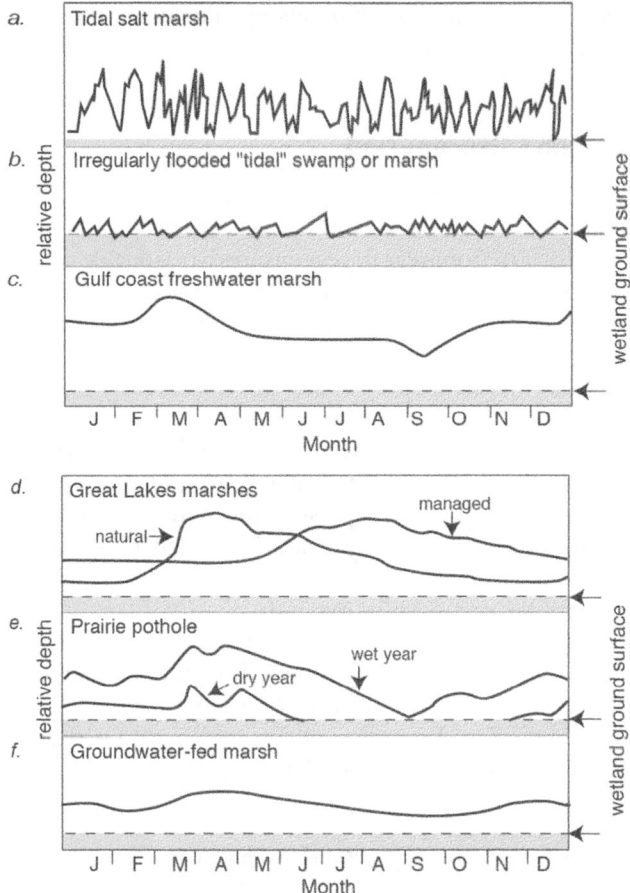

Figure 4.2 Hydroperiods for several different wetlands, presented in approximately the same relative scale: (a) tidal salt marsh, Rhode Island; (b) irregularly flooded "tidal" swamp or marsh; (c) Gulf Coast freshwater marsh, Louisiana; (d) Great Lakes marshes, northern Ohio (natural and managed); (e) prairie pothole marsh with little groundwater flow (dry and wet years); (f) groundwater-fed prairie pothole marsh; (g) vernal pool, California; (h) subtropical cypress dome, Florida; (i) alluvial swamp, North Carolina; (j) bottomland hardwood forest, northern Illinois; (k) mineral soil swamp, Ontario, Canada; (l) rich fen, North Wales; (m) pocosin or Carolina Bay, North Carolina; (n) tropical floodplain forest, Amazon River, Manaus, Brazil. (Data from Nixon and Oviatt, 1973; Mitsch et al., 1979; Gilman, 1982; Junk, 1982; P.H. Zedler, 1987; Mitsch, 1989; van der Valk, 1989; Brinson, 1993; Woo and Winter, 1993)

of the ocean itself (Fig. 4.2c). Hydroperiods of coastal lacustrine wetlands along the Laurentian Great Lakes in the United States and Canada vary considerably, depending on whether pumps and water management are used or whether the marshes are open to the seasonal patterns of river flows and lake levels (Fig. 4.2d). In fact, the hydroperiods of these wetlands, when used as hunting clubs for waterfowl production, actually are managed to be dry when the normal season calls for wet and wet when the seasonal

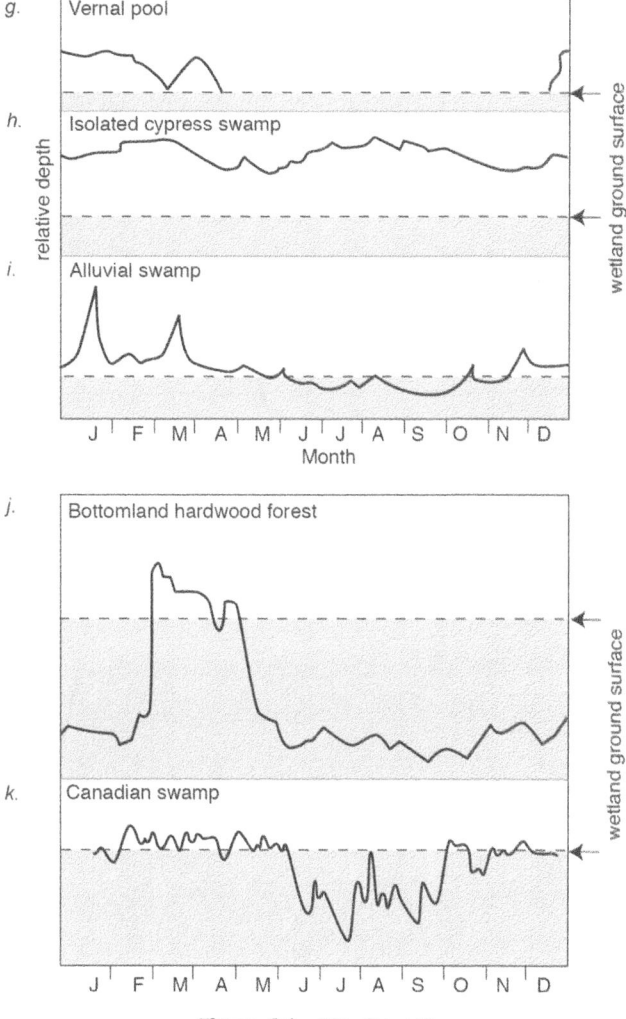

Figure 4.2 *(Continued)*

pattern calls for dry conditions. Water levels for interior wetlands, such as the prairie potholes of North America, vary considerably from year to year (see the next section), with differences depending on climate variability (Fig. 4.2e). Wetlands affected by groundwater tend to have water levels that are less seasonally variable (Fig. 4.2f).

Some of the most seasonally variable wetlands are the vernal pools of central California, where surface water essentially disappears in this Mediterranean-type climate for all but four or five months (Fig. 4.2g). Cypress domes in central Florida have standing water during the wet summer season and dry periods in the late autumn and early spring (Fig. 4.2h). Low-order riverine wetlands, such as the alluvial swamps in the southeastern United States, respond sharply to local rainfall events rather than to

118 Chapter 4 Wetland Hydrology

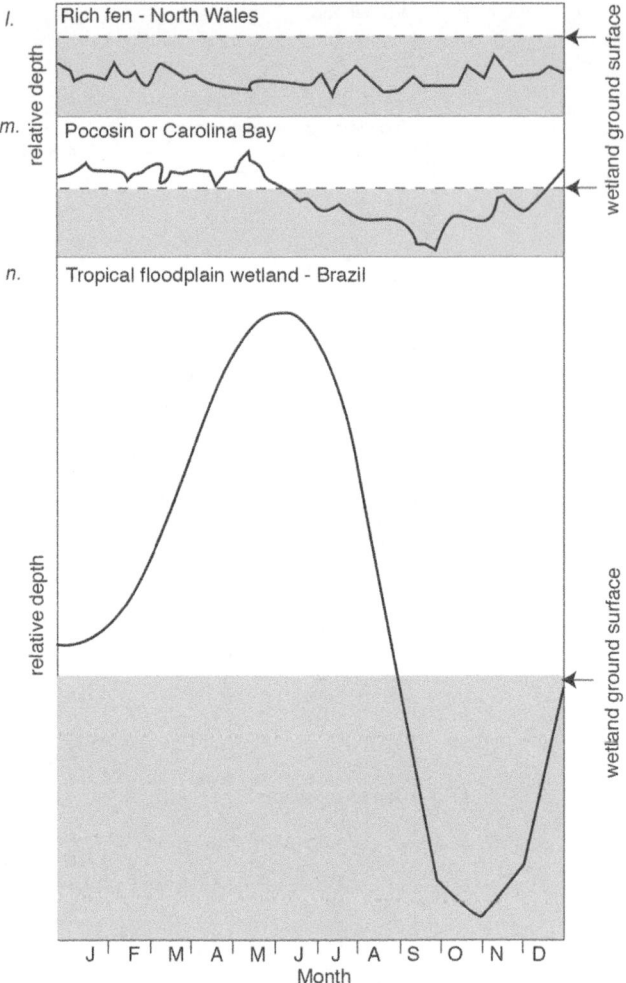

Figure 4.2 *(Continued)*

general seasonal patterns (Fig. 4.2i). The hydroperiods of many bottomland hardwood forests and swamps in colder climates have distinct periods of surface flooding in the winter and early spring due to snow and ice conditions followed by spring floods but otherwise have a water table that can be a meter or more below the surface (Fig. 4.2j and k).

Peatlands in cooler climates can have hydroperiods with little pronounced seasonal fluctuation, as in the fen from North Wales (Fig. 4.2l). If peatlands such as the pocosins of North Carolina are located in regions of warm summers, significant patterns of seasonal water-level change will occur (Fig. 4.2m). The most dramatic hydroperiods result from high-order rivers that are more influenced by seasonal patterns of precipitation throughout a large watershed than by local precipitation, leading to a more predictable and seasonally distinct hydroperiod. For example, the annual fluctuation of water in the tropical floodplain forests along the Amazon River is a predictable seasonal pattern that can include a seasonal fluctuation in water level of 5 to 10 m caused by flooding of upstream rivers (Fig. 4.2n).

Year-to-Year Fluctuations

The hydroperiod is not the same each year but varies according to climate and antecedent conditions. Great variability can be seen from year to year for some wetlands, as illustrated in Figure 4.3 for a prairie pothole regional wetland in Canada and for the Big Cypress Swamp region of south Florida. In the pothole region, a wet-dry cycle of 10 to 20 years is seen; spring is almost always wetter than fall, but depths vary significantly from year to year (Fig. 4.3a). Figure 4.3b illustrates cases of an even seasonal rainfall pattern for the Big Cypress Swamp in Florida between a fairly stable hydroperiod and a year with a significant dry season, which caused the hydroperiod to vary about 1.5 m between high and low water. A three-year study of groundwater levels in a red maple swamp shows dramatically different growing season water levels from year to year (Fig. 4.4). Water is near or at the surface during high precipitation periods (last half of first year and entire second year) while dry low-water conditions are mainly driven by seasonal evapotranspiration in the swamp accelerated by groundwater loss during tree transpiration.

Pulsing Water Levels

Water levels in most wetlands are generally not stable but fluctuate seasonally (riparian wetlands), daily or semidaily (types of tidal wetlands), or unpredictably (wetlands in low-order streams and coastal wetlands with wind-driven tides). Flooding "pulses" that occur seasonally or periodically especially in riverine wetlands nourish the wetlands with additional nutrients and carry away detritus and waste products. Pulse-fed wetlands are often the most productive wetlands and are the most favorable for exporting materials, energy, and biota to adjacent ecosystems. Despite this obvious fact, many wetland managers, especially those who manage wetlands for waterfowl, often attempt to control water levels by isolating formerly open wetlands with levees meant to restrict flooding. A seasonally fluctuating water level, then, is the rule, not the exception, in most wetlands.

Wetland Water Budget

The hydroperiod, or hydrologic state of a given wetland, can be summarized as being a result of these three factors:

1. The balance between the inflows and outflows of water
2. The surface contours of the landscape
3. Subsurface soil, geology, and groundwater conditions

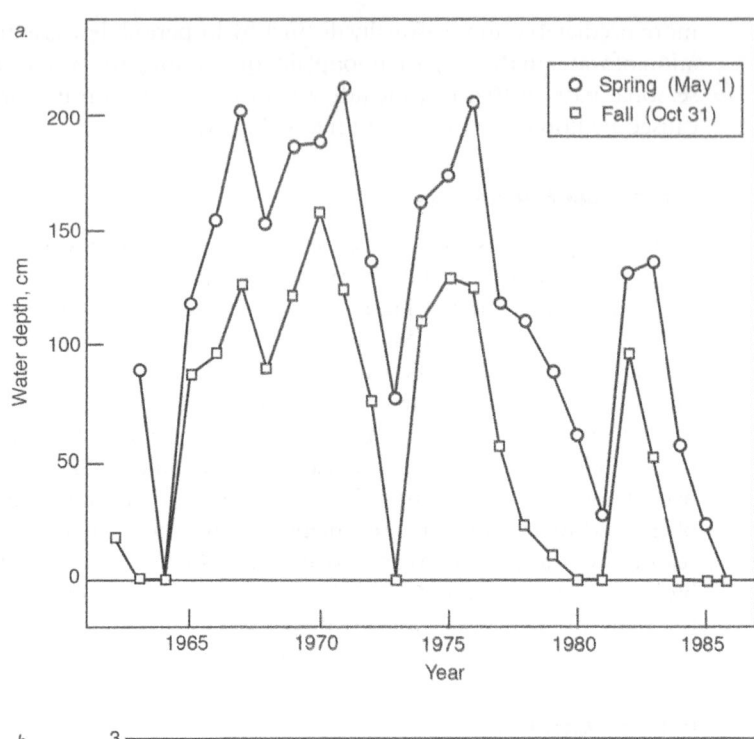
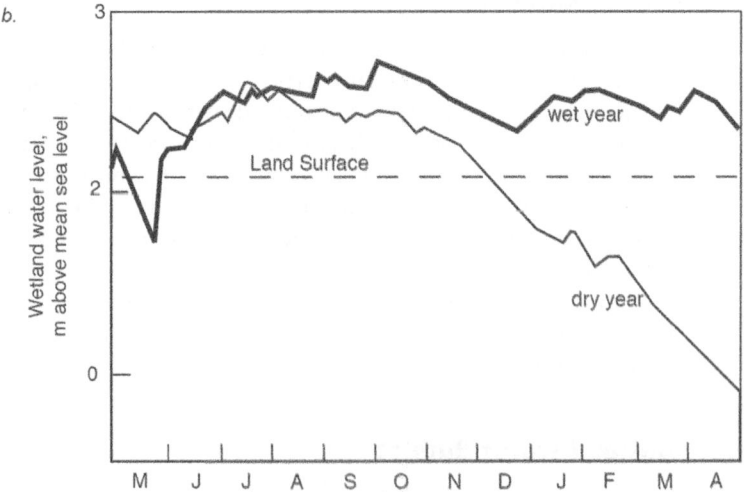

Figure 4.3 Year-to-year fluctuations in wetland water levels in two regions: (a) spring and fall water depths for 25 years in shallow open-water wetlands in the prairie pothole region of southwestern Saskatchewan, Canada; and (b) wet and dry year hydrographs for the Big Cypress Swamp region of the Everglades, southwestern Florida. ((a) After Kantrud et al., 1989 and Millar, 1971; (b) after Freiberger, 1972 and Duever, 1988)

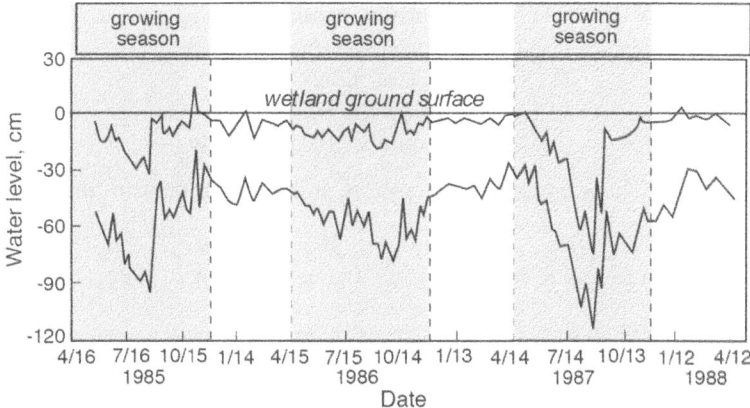

Figure 4.4 Relative water levels in two seasonally saturated red maple swamps in Rhode Island, United States, for 1985 to 1987. Growing season precipitation amounts for 1985, 1986, and 1987 were 104, 76, and 59 cm, respectively. (After Golet et al., 1993)

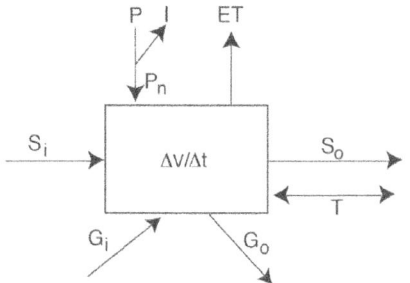

Figure 4.5 Generalized water budget for a wetland with corresponding terms as in Equation 4.1. *P* = precipitation; *ET* = evapotranspiration; *I* = interception; P_n = net precipitation; S_i = surface inflow; S_o = surface outflow; G_i = groundwater inflow; G_o = groundwater outflow; *T* = tide or seiche; Δ*V*/Δ*t* = change in storage per unit time.

The first condition defines the *water budget* of the wetland, whereas the second and the third define the capacity of the wetland to store water. The general balance between water storage and inflows and outflows, illustrated in Figure 4.5, is expressed as

$$\frac{\Delta V}{\Delta t} = P_n + S_i + G_i - ET - S_o - G_o \pm T \qquad (4.1)$$

where

V = volume of water storage in wetlands
$\Delta V/\Delta t$ = change in volume of water storage in wetland per unit time, t
P_n = net precipitation
S_i = surface inflows, including flooding streams

G_i = groundwater inflows
ET = evapotranspiration
S_o = surface outflows
G_o = groundwater outflows
T = tidal inflow (+) or outflow (−)

The average water depth, d, at any one time, can further be described as

$$d = \left(\frac{V}{A}\right) \qquad (4.2)$$

where

A = wetland surface area

Each of the terms in Equation 4.1 can be expressed in terms of depth per unit time (e.g., cm/yr) or in terms of volume per unit time (e.g., m^3/yr).

Examples of Water Budgets

Equation 4.1 and Figure 4.5 serves as useful summaries of the major hydrologic components of any wetland water budget. Examples of hydrologic budgets for several wetlands are illustrated in Figure 4.6. The terms in the equation vary in importance according to the type of wetland observed; furthermore, not all terms in the hydrologic budget apply to all wetlands (Table 4.2). There is a large variability in certain flows, particularly in surface inflows and outflows, depending on the openness of the

Table 4.2 Major components of hydrologic budgets for wetlands

Component	Pattern	Wetlands Affected
Precipitation	Varies with climate, although many regions have distinct wet and dry seasons	All
Surface inflows and outflows	Seasonally, often matched with precipitation pattern or spring thaw; can be channelized as streamflow or nonchannelized as runoff; includes river flooding of alluvial wetlands	Potentially all wetlands except ombrotrophic bogs; riparian wetlands, including bottomland hardwood forests and other alluvial wetlands, are particularly affected by river flooding
Groundwater	Less seasonal than surface inflows and not always present	Potentially all wetlands except ombrotrophic bogs and other perched wetlands
Evapotranspiration	Seasonal with peaks in summer and low rates in winter. Dependent on meteorological, physical, and biological conditions in wetlands	All
Tides	One to two tidal periods per day; flooding frequency varies with elevation	Tidal freshwater and salt marshes; mangrove swamps

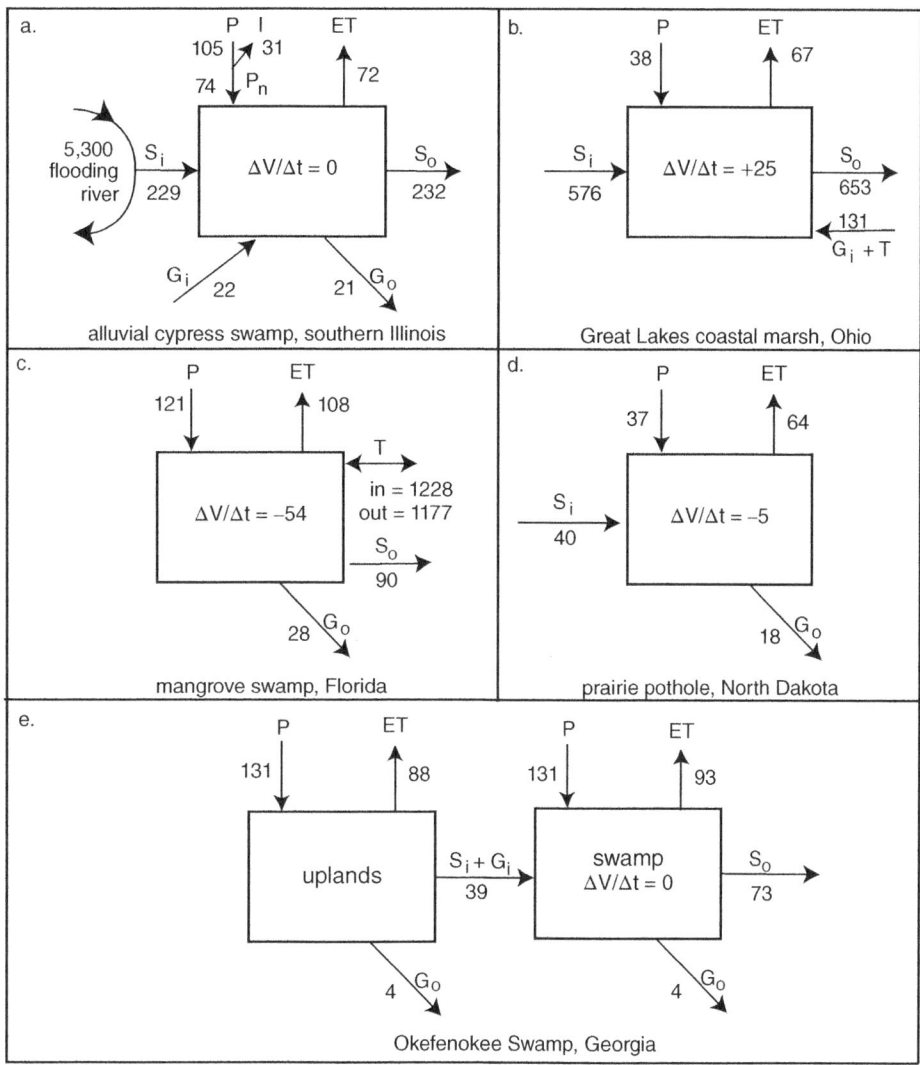

Figure 4.6 Annual water budgets for several wetlands. See Figure 4.5 for symbol definitions. All values are expressed in centimeters per year (cm/yr) except (b), which is March–September only. (Data from Pride et al., 1966; Shjeflo, 1968; Mitsch, 1979; Hemond, 1980; Gilman, 1982; Twilley, 1982; Richardson, 1983; Mitsch and Reeder, 1992; Mitsch et al., 2010)

wetlands. An alluvial cypress swamp in southern Illinois received a gross inflow of floodwater from one flood that was more than 50 times the gross precipitation for the entire year (Fig. 4.6a). Even the net surface inflow from that flood (the water left behind after the flooding river receded) was three times the precipitation input for the entire year. Surface and groundwater inflows to a coastal Lake Erie marsh in northern

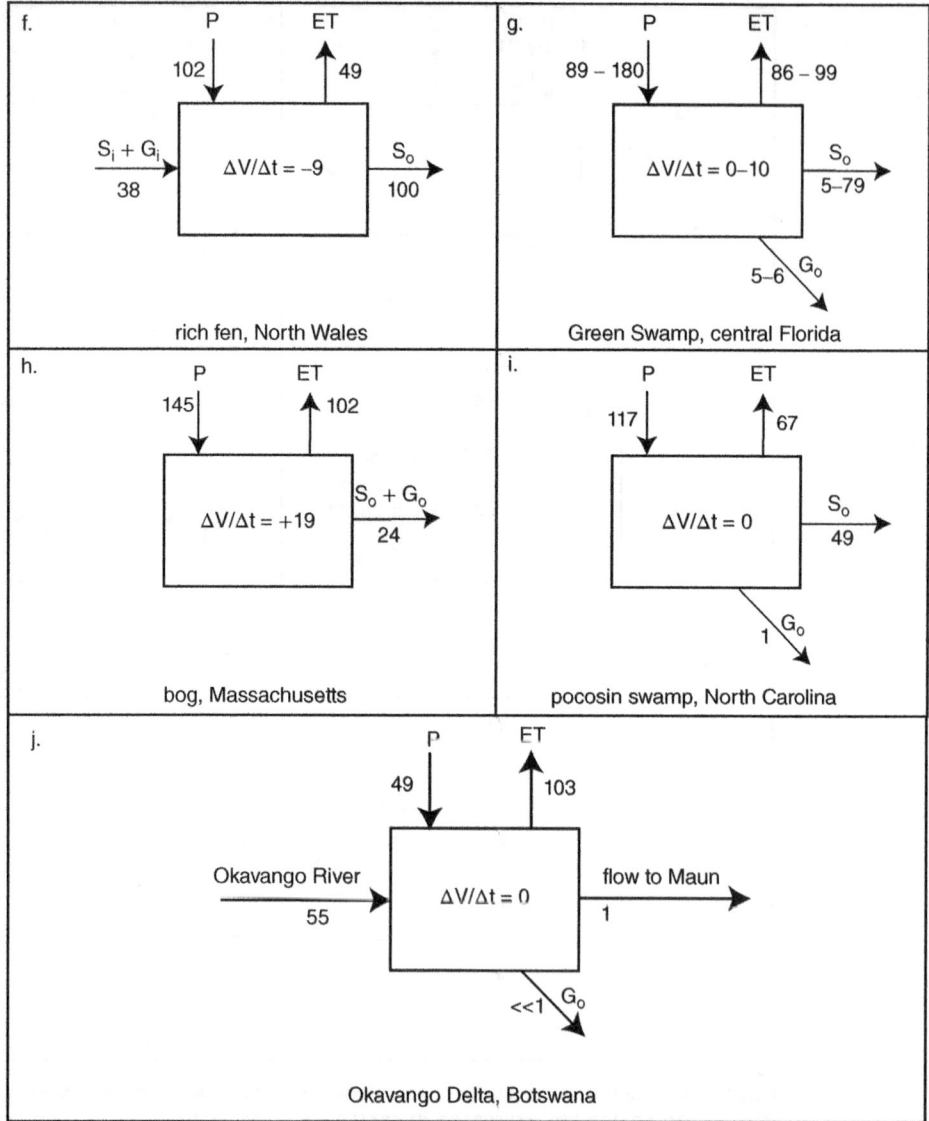

Figure 4.6 *(Continued)*

Ohio were estimated to be almost 20 times the precipitation for a major part of a drought year (Fig. 4.6b), and tides contributed 10 times the precipitation to a black mangrove swamp in Florida (Fig. 4.6c).

In contrast to these inflow-dominated wetlands, surface inflow is approximately equal to the precipitation inflow in the prairie pothole marshes of North Dakota

(Fig. 4.6d), considerably less than the precipitation for the Okefenokee Swamp in Georgia (Fig. 4.6e) and a rich fen in North Wales (Fig. 4.6f), and essentially nonexistent in the upland Green Swamp of central Florida (Fig. 4.6g), a bog in Massachusetts (Fig. 4.6h), and a pocosin wetland of North Carolina (Fig. 4.6i). In most of these examples, the change in storage is small or zero, indicating that the water level at the end of the study period (usually an annual cycle) is close to where it was at the beginning of the study period.

The water budget for the tropical Okavango Delta in southern Africa (Botswana) has been investigated for many years. Figure 4.6j represents the average conditions for the past 36 years. The data show that the Okavango River input, when averaged over the entire delta, is about equivalent to the rainfall over this vast area. Furthermore, the budget shows that essentially all of the inputs are balanced by a loss of evapotranspiration in this semiarid climate, and only about 1 percent of the water now leaves the wetland region to the downstream village of Maun.

Residence Time—How Long Does Water Stay in a Wetland?

A generally useful concept of wetland hydrology is that of the *renewal rate* or *turnover rate* of water, defined as the ratio of throughput to average volume within the system:

$$t^{-1} = \left(\frac{Q_t}{V}\right) \qquad (4.3)$$

where

t^{-1} = renewal rate (time^{-1})
Q_t = total inflow rate (volume/time)
V = average volume of water storage in wetland

Few measurements of renewal rates have been made in wetlands, although the renewal rate is a frequently used parameter in limnological studies. Chemical and biotic properties are often determined by the openness of the system, and the renewal rate is an index of this because it indicates how rapidly the water in the system is replaced. The reciprocal of the renewal rate is the *turnover time* or *residence time* (t, sometimes called *detention time* by engineers for constructed wetlands), which is a measure of the average time that water remains in the wetland. The theoretical residence time, as calculated as the reciprocal of Equation 4.3, is often much longer than the actual residence time of water flowing through a wetland, because of nonuniform mixing. Because there are often parts of wetland where waters are stagnant and not well mixed, the theoretical residence time (t) estimate should be used with caution when estimating the hydrodynamics of wetlands.

Precipitation

Wetlands occur most extensively in regions where *precipitation*, a term that includes rainfall and snowfall, is in excess of losses such as evapotranspiration and surface runoff. The fate of precipitation that falls on a wetland with forested, shrub, or emergent vegetation is shown in Figure 4.7. When some of the precipitation is retained by the vegetation cover, particularly in forested wetlands, the amount that actually passes through

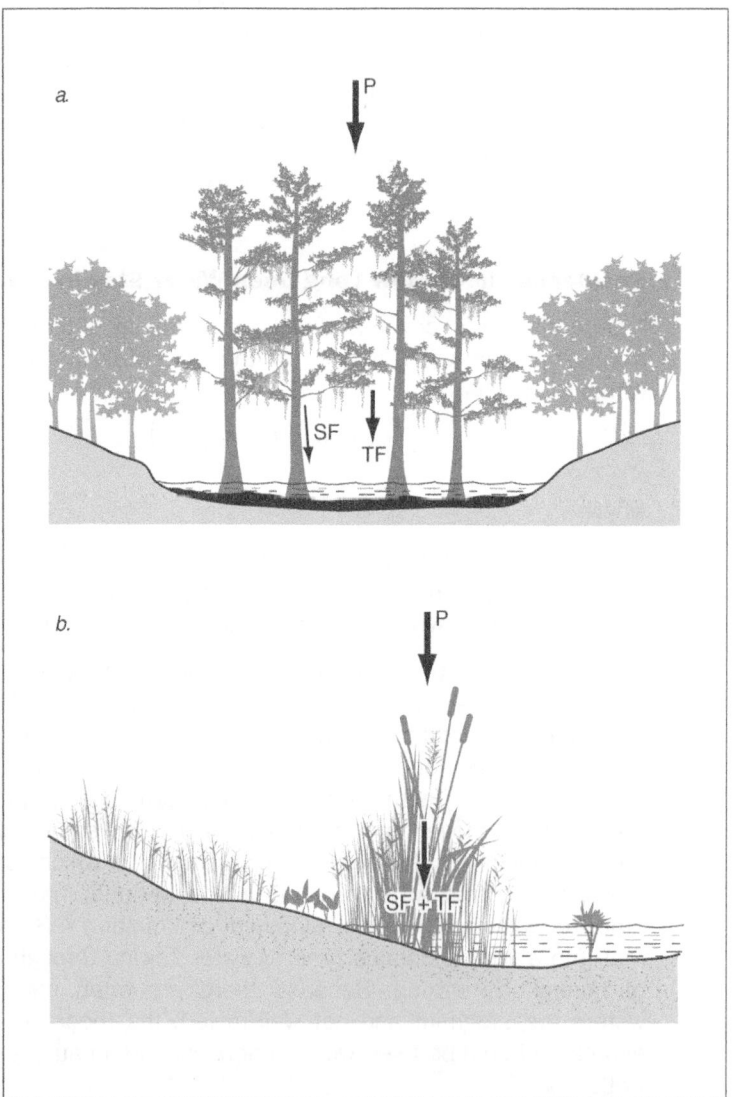

Figure 4.7 Fate of precipitation in (a) a forested wetland and (b) a marsh. *P* = precipitation; *TF* = throughfall; *SF* = stemflow.

the vegetation to the water or substrate below is called *throughfall*. The amount of precipitation that is retained in the overlying vegetation canopy is called *interception*. Interception depends on several factors, such as the total amount of precipitation, the intensity of the precipitation, and the character of the vegetation, including the stage of vegetation development, the type of vegetation (e.g., deciduous or evergreen), and the strata of the vegetation (e.g., tree, shrub, or emergent macrophyte). The percentage of precipitation that is intercepted in forests varies between 8 and 35 percent. The water budget in Figure 4.6a, for example, illustrates that 29 percent of precipitation in a forested wetland was intercepted by a canopy dominated by the deciduous conifer *Taxodium distichum*.

Little is known about the interception of precipitation by emergent herbaceous macrophytes, but it probably is similar to that measured in grasslands or croplands. Essentially, in those systems, interception at maximum growth can be as high as that in a forest (10 to 35 percent of gross precipitation). An interesting hypothesis about interception and the subsequent evaporation of water from leaf surfaces is that, because the same amount of energy is required whether water evaporates from the surface of a leaf or is transpired by the plant, the evaporation of intercepted water is not "lost" because it may reduce the amount of transpiration loss that occurs. This suggests that wetlands with either high or low interception may have similar overall water loss to the atmosphere.

Another term related to precipitation, *stemflow*, refers to water that passes down the stems of the vegetation (Fig. 4.7). This flow is generally a minor component of the water budget of a wetland. For example, Heimburg (1984) found that stemflow was, at maximum, 3 percent of throughfall in cypress dome wetlands in north-central Florida.

These terms are related in a simple water balance as follows:

$$P = I + TF + SF \tag{4.4}$$

where

P = total precipitation
I = interception
TF = throughfall
SF = stemflow

The total amount of precipitation that actually reaches the water's surface or substrate of a wetland is called the net precipitation (P_n) and is defined as

$$P_n = P - I \tag{4.5}$$

Surface Flow

Watersheds and Runoff

The percentage of precipitation that becomes surface flow depends on several variables, with climate being the most important. Humid cool regions such as the Pacific Northwest, western British Columbia, and the northeastern Canadian provinces have 60 to 80 percent of precipitation converted to runoff. In the arid southwestern United States, less than 10 percent of the already low precipitation becomes runoff. This difference is related, in large part, to the higher temperatures in the arid Southwest, which translate into higher evapotranspiration rates, greater soil moisture deficits, and higher soil infiltration rates than in the Pacific Northwest. Even though runoff in arid regions is small relative to that in humid areas, it does contribute *streamflow*, which is an important part of a riparian wetland's water budget. Wetlands can be receiving systems for surface water flows (*inflows*), or surface water streams can originate in wetlands to feed downstream systems (*outflows*). Surface outflows are found in many wetlands that are located in the upstream reaches of a watershed. These wetlands are often important water flow regulators for downstream rivers. Some wetlands have surface outflows that develop only when their water stages exceed a critical level.

Wetlands are subjected to surface inflows of several types. *Overland flow* is non-channelized sheet flow that usually occurs during and immediately following rainfall or a spring thaw, or as tides rise in coastal wetlands. A wetland influenced by a drainage basin may receive channelized streamflow during most or all of the year. Wetlands are often an integrated part of a stream or river, for example, as instream freshwater marshes or riparian bottomland forests. Wetlands that form in wide, shallow expanses of river channels or floodplains adjacent to them are greatly influenced by the seasonal streamflow patterns of the river. Wetlands can also receive surface inflow from seasonal or episodic pulses of flood flow from adjacent streams and rivers that may otherwise not be connected hydrologically with the wetland. Coastal saline and brackish wetlands are also significantly influenced by freshwater runoff and streamflow (in addition to tides) that contribute nutrients and energy to the wetland and often ameliorate the effects of soil salinity and anoxia.

Surface inflow from a drainage basin into a wetland is usually difficult to estimate without a great deal of data. Nevertheless, it is often one of the most important sources of water in a wetland's hydrologic budget. The direct runoff component of streamflow refers to rainfall during a storm that causes an immediate increase in streamflow. An estimate of the amount of precipitation that results in direct runoff, or *quickflow*, from an individual storm can be determined from the following equation:

$$S_i = R_p P A_w \tag{4.6}$$

where

S_i = direct surface runoff into wetland (m³ per storm event)
R_p = hydrologic response coefficient
P = average precipitation in watershed (m)
A_w = area of watershed draining into wetland (m²)

This equation states that the flow is proportional to the volume of precipitation ($P \times A_w$) on the watershed feeding the wetland in question. R_p, which represents the fraction of precipitation in the watershed that becomes direct surface runoff, ranges from 4 to 18 percent for small watersheds in the eastern North America and generally increases with latitude. Slope and type of vegetation appear to have little influence on R_p in a watershed with a mature forest cover. As the following paragraph suggests, land use and soil type can strongly influence runoff.

While Equation 4.6 predicts the volume of direct runoff caused by a storm event, in some cases wetland scientists and managers might be interested in calculating the peak runoff (*flood peak*) into a wetland caused by a specific rainfall event. Although this is generally a difficult calculation for large watersheds, a formula with the unlikely name of the *rational runoff method* is a widely accepted and useful way to predict peak runoff for watersheds less than 80 ha in size. The equation is given by

$$S_{i(pk)} = 0.278 C I A_w \qquad (4.7)$$

where

$S_{i(pk)}$ = peak (pk) runoff into wetland (m³/s)
C = rational runoff coefficient (see Table 4.3)
I = rainfall intensity (mm/h)
A_w = area of watershed draining into wetland (km²)

The coefficient C, which ranges from 0 to 1 (Table 4.3), depends on the upstream land use. Concentrated urban areas have a coefficient ranging from 0.5 to 0.95, and

Table 4.3 Values of the rational runoff coefficient C used to calculate peak runoff

		C
Urban Areas		
Business areas:	high-value districts	0.75–0.95
	neighborhood districts	0.50–0.70
Residential areas:	single-family dwellings	0.30–0.50
	multiple-family dwellings	0.40–0.75
	suburban	0.25–0.40
Industrial areas:	light	0.50–0.80
	heavy	0.60–0.90
Parks and cemeteries		0.10–0.25
Playgrounds		0.20–0.35
Unimproved land		0.10–0.30
Rural Areas		
Sandy and gravelly soils:	cultivated	0.20
	pasture	0.15
	woodland	0.10
Loams and similar soils:	cultivated	0.40
	pasture	0.35
	woodland	0.30
Heavy clay soils; shallow soils over bedrock:	cultivated	0.50
	pasture	0.45
	woodland	0.40

rural areas have lower coefficients that greatly depend on soil type, with sandy soils lowest ($C = 0.1$–0.2) and clay soils highest ($C = 0.4$–0.5).

Channelized Streamflow

Channelized streamflow into and out of wetlands is described simply as the product of the cross-sectional area of the stream (A_x) and the average velocity (v) and can be determined through stream velocity measurements in the field:

$$S_i \text{ or } S_o = A_x v \tag{4.8}$$

where

S_i, S_o = surface channelized flow into or out of wetland (m³/s)
A_x = cross-sectional area of stream (m²)
v = average velocity (m/s)

The velocity can be determined in several ways, ranging from handheld velocity meter readings taken at various locations in the stream cross-section to the floating-orange technique where the velocity of a floating orange or similar fruit (which is 90 percent or more water and therefore floats but just beneath the water surface) is timed as it goes downstream. If a continuous or daily record of streamflow is needed, then a *rating curve* (Fig. 4.8), a plot of instantaneous streamflow (as estimated using Equation 4.8) versus stream elevation or stage, is useful. If this type of rating curve is developed for a stream (the basis of most hydrologic streamflow gauging stations operated by the U.S. Geological Survey), then a simple measurement of the stage in the stream can be used to determine the streamflow. Because hydrographs generally assume a constant water gradient, caution should be taken in using this approach for streams flowing into wetlands to ensure that no "backwater effect" of the wetland's water level will affect the stream stage at the point of measurement.

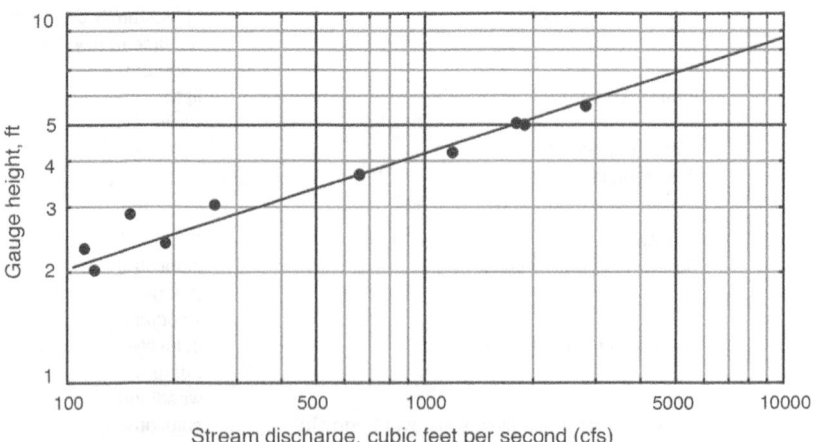

Figure 4.8 Rating curve for streamflow determination as a function of stream stage. 100 cfs = 2.832 m³/s. (After Dunne and Leopold, 1978)

Measuring Streamflow with Weirs

When a weir or other control structure is used at the outflow of a wetland (Fig. 4.9), the outflow of a wetland can be estimated to be a function of the water level in the wetland itself according to the equation:

$$S_o = xL^y \qquad (4.9)$$

where

S_o = surface outflow
L = wetland water level above a control structure crest (level at which flow just begins)
x, y = calibration coefficients

Figure 4.9 Control structures such as the V-notched weir shown here can be used for measuring surface water flow in small streams into or out of wetlands. (Photo by W. J. Mitsch)

> If a control structure such as a rectangular or V-notched weir is used to measure the outflow from a wetland, standard equations of the form of Equation 4.9 can be obtained from water measurement manuals (e.g., U.S. Department of Interior, 2001). Care should be taken to calibrate standard weir equations with actual measurements of streamflow and water level.

When an estimate of surface flow into or out of a riverine wetland is needed and no stream velocity measurements are available, the *Manning equation* often can be used if the slope of the stream and a description of the surface roughness are known:

$$S_i \text{ or } S_o = \frac{A_x R^{2/3} s^{0.5}}{n} \quad (4.10)$$

where

n = roughness coefficient (Manning coefficient; see Table 4.4)
R = hydraulic radius (m) (cross-sectional area divided by the wetted perimeter; this is an estimate of the relative portion of the stream cross section and hence flow volume, in contact with the streambed)
s = channel slope (dimensionless)

The equation states that flow is proportional to stream cross-section, as modified by the roughness of the streambed and the proportion of flow in contact with that bed. Although the potential exists for their use in wetland studies, the roughness coefficients given in Table 4.4 and the Manning equation (Eq. 4.10) have not been used very often. The relationship is particularly useful for estimating streamflow where velocities are too slow to measure directly and to estimate flood peaks from high-water marks on ungauged streams. These circumstances are common in wetland studies.

Floods and Riparian Wetlands

A special case of surface flow occurs in wetlands that are in floodplains adjacent to rivers or streams and are occasionally flooded by those rivers or streams. These ecosystems

Table 4.4 Roughness coefficients (n) for Manning equation used to determine streamflow in natural streams and channels

Stream Conditions	Manning Coefficient, n
Straightened earth canals	0.02
Winding natural streams with some plant growth	0.035
Mountain streams with rocky streambed	0.040–0.050
Winding natural streams with high plant growth	0.042–0.052
Sluggish streams with high plant growth	0.065
Very sluggish streams with high plant growth	0.112

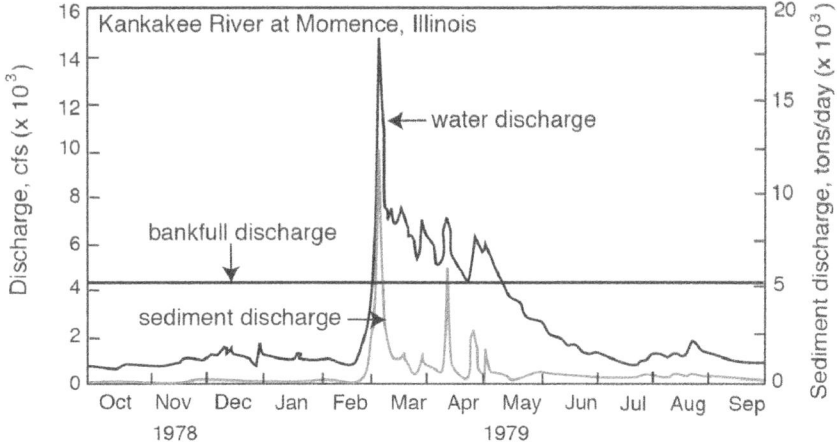

Figure 4.10 River hydrograph from northeastern Illinois, showing discharge and sediment load of the river and discharge at which a riparian wetland is flooded (bankfull discharge). 1,000 cfs = 28.32 m³/s. (After Bhowmik et al., 1980)

are often called *riparian wetlands*. The flooding of these wetlands varies in intensity, duration, and number of floods from year to year, although the probability of flooding is fairly predictable. In the eastern and midwestern United States and in much of Canada, a pattern of winter or spring flooding caused by rains and sudden snowmelt is often observed. When river flow begins to overflow onto the floodplain, the streamflow is referred to as *bankfull discharge*. A hydrograph of a stream that flooded its riparian wetlands above bankfull discharge for several months in the spring is shown in Figure 4.10. There is a remarkable consistency in the hydrographs of rivers in the midwestern United States, in that they tend to overflow their banks (bankfull discharge) at intervals between one and two years or on the average two years out of three (see the next box).

Recurrence Interval

The *recurrence interval* is the average interval between the occurrences of floods at a given or greater stage (depth). The inverse of the recurrence interval is the average probability of flooding in any one year. Figure 4.11 suggests that streams in the midwestern and southern United States will overflow their banks onto the adjacent riparian forest with an average recurrence interval of 1.5 years (or a probability of 1/1.5, or 67 percent, of overbank flooding in any one year). Stated another way, these rivers, on average, overflow their banks in two out of every three years. Figure 4.11 also illustrates that flow that is twice that of bankfull discharge occurs at recurrence intervals of approximately five years; this flow, however, results in only a 40 percent greater river depth over bankfull depth on the floodplain. This predictable relationship suggests that in natural stream systems, the size of a stream channel is related to the hydraulic energy that scours the streambed.

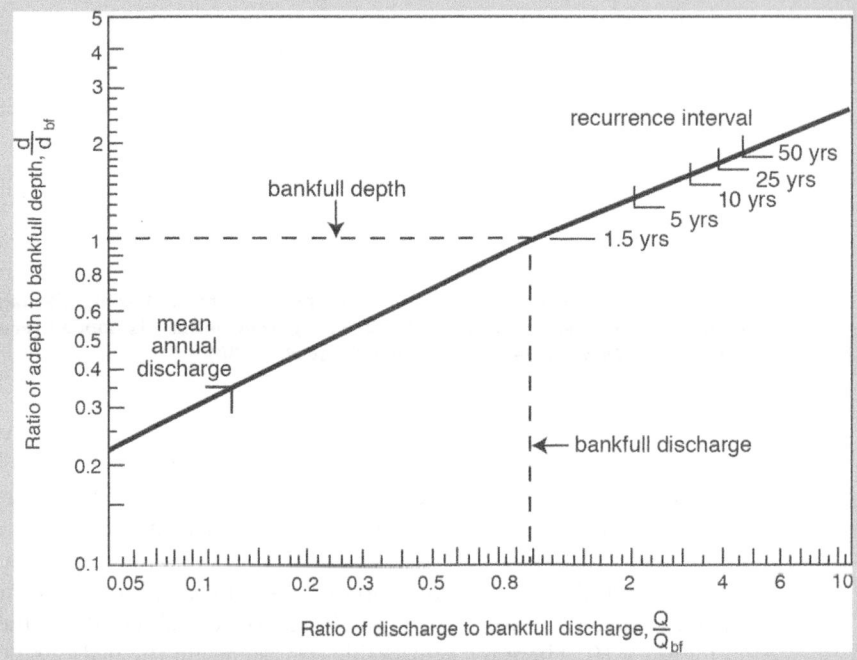

Figure 4.11 Relationships among streamflow (discharge), stream depth, and recurrence interval for streams and rivers in the midwestern and southern United States. Q = stream discharge; Q_{bf} = bankfull discharge; d = stream depth; d_{bf} = bankfull depth (depth of river with floodplain is initially flooded). (After Leopold et al., 1964)

Groundwater

Recharge and Discharge Wetlands

Groundwater can heavily influence some wetlands, whereas in others it may have hardly any effect at all. The influence of wetland recharge and discharge on groundwater resources has often been cited as one of the most important attributes of wetlands, but it does not hold for all wetland types; nor is there sufficient experience with site-specific studies to make many generalizations. Groundwater inflow results when the surface water (or groundwater) level of a wetland is lower hydrologically than the water table of the surrounding land (called a *discharge wetland* by geologists, who generally view their water budget from a groundwater, not from a wetland, perspective). Wetlands can intercept the water table in such a way that they have only inflows and

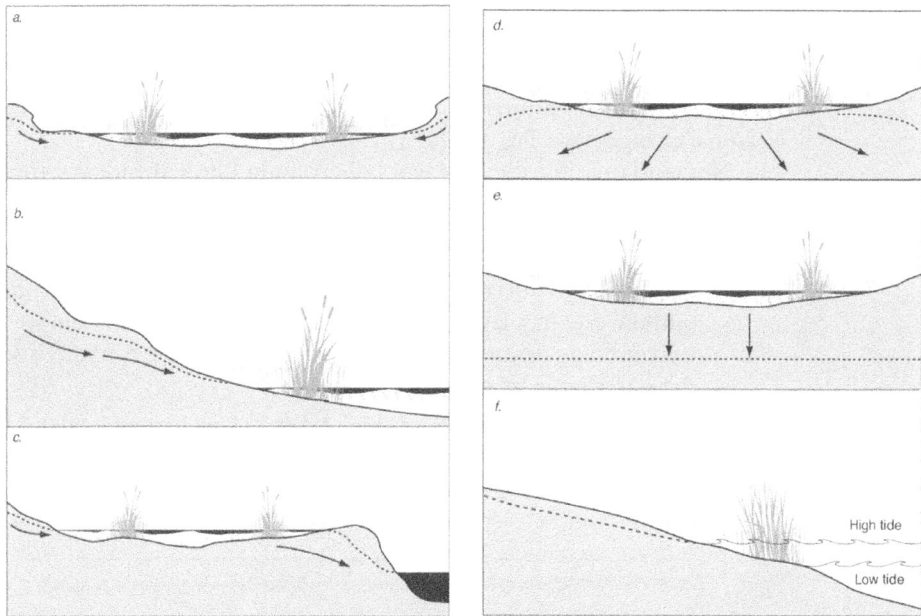

Figure 4.12 Possible discharge–recharge interchanges between wetlands and groundwater systems including (a) marsh as a depression receiving groundwater flow (discharge wetland); (b) groundwater spring or seep wetland or groundwater slope wetland at the base of a steep slope; (c) floodplain wetland fed by groundwater; (d) marsh as a recharge wetland adding water to groundwater; (e) perched wetland or surface water depression wetland; (f) groundwater flow through a tidal wetland. Dashed lines indicate groundwater level.

no outflows, as shown for a prairie marsh in Figure 4.12a. Another type of discharge wetland, called a *spring* or *seep* wetland, is often found at the base of steep slopes where the groundwater surface intersects the land surface (Fig. 4.12b). This type of wetland can be an isolated low point in the landscape; more often, it discharges excess water downstream as surface water or as groundwater, as shown in the riparian wetland in Figure 4.12c.

When the water level in a wetland is higher than the water table of its surroundings, groundwater will flow out of the wetland (called a *recharge wetland*; Fig. 4.12d). When a wetland is well above the groundwater of the area, the wetland is referred to as being *perched* (Fig. 4.12e). This type of wetland, also referred to as a *surface water depression wetland*, loses water only through infiltration into the ground and through evapotranspiration. Tidally influenced wetlands often have significant groundwater inflows that can influence soil salinity and keep the wetland soil wet even during low tide (Fig. 4.12f).

A final type of wetland, one that is fairly common, is little influenced by groundwater inflows. Because wetlands often occur where soils have poor permeability, the major source of water can be restricted to surface water runoff, with losses occurring through evapotranspiration and other outflows. This type of wetland

136 Chapter 4 Wetland Hydrology

often has fluctuating hydroperiods and intermittent flooding (e.g., prairie potholes [Fig. 4.12e] and vernal pools [Fig. 4.12d], with standing water dependent on seasonal precipitation and surface inflows. If, however, such a wetland were to be influenced by groundwater, its water level would be better buffered against dramatic seasonal changes (see Fig. 4.12a, c).

Nomenclature for the four types of groundwater hydrologic settings for freshwater wetlands are illustrated in Figure 4.13 and summarized here:

1. *Surface water depression wetland* (Fig. 4.13a). This type of wetland is dominated by surface runoff and precipitation, with little groundwater outflow due to a layer of low-permeability soils. This is similar to the perched wetland type described in Figure 4.12e, where the wetland is separated from the water table by an unsaturated zone.

2. *Surface water slope wetland* (Fig. 4.13b). This type of wetland is generally found in alluvial soil adjacent to a lake or stream and is fed, to some degree, by precipitation and surface runoff but, more important, by overbank

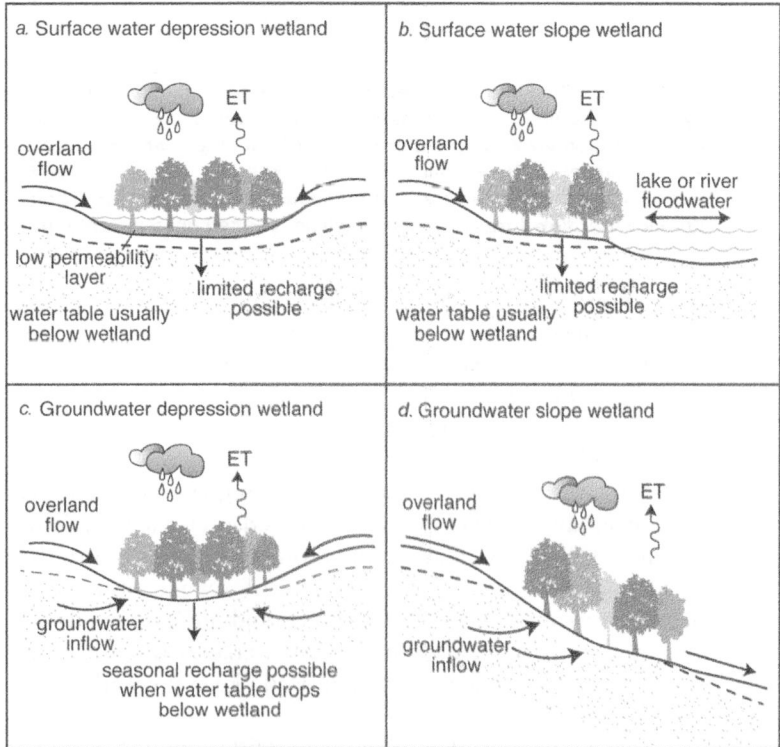

Figure 4.13 Novitski groundwater flow patterns for wetlands: (a) surface water depression, (b) surface water slope, (c) groundwater depression, and (d) groundwater slope. Dashed lines indicate groundwater level. (After Golet et al., 1993)

flooding from the adjacent stream, river, or lake. Hydroperiods of these wetlands match the seasonal patterns of the adjacent bodies of water, with relatively rapid wetting and drying. Some groundwater recharge is possible, but that groundwater soon discharges back to the stream, river, or lake.

3. *Groundwater depression wetland* (Fig. 4.13c). This is the groundwater discharge wetland described previously (Fig. 4.12a), where the wetland is in a depression low enough to intercept the local groundwater table. These kinds of wetlands can occur in coarse-textured glaciofluvial deposits, where the interchange between groundwater and surface water is enhanced by relatively coarse soil material. Water-level fluctuations in these types of wetlands are less dramatic than fluctuations in surface flow wetlands because of the relative stability of the groundwater levels.

4. *Groundwater slope wetland* (Fig. 4.13d). Wetlands often develop on slopes or hillsides where groundwater discharges to the surface as springs and seeps. Groundwater flow into these wetlands can be continuous or seasonal, depending on the local geohydrology and on the evapotranspiration rates of the wetland and adjacent uplands.

Darcy's Law

Darcy's law, an equation familiar to groundwater hydrologists, often describes the flow of groundwater into and out of a wetland. This law states that the flow of groundwater is proportional to (1) the slope of the piezometric surface (the hydraulic gradient) and (2) the hydraulic conductivity, or *permeability*, the capacity of the soil to conduct water flow. In equation form, Darcy's law is given as

$$G = kA_x s \qquad (4.11)$$

where

G = flow rate of groundwater (volume per unit time)
k = hydraulic conductivity or permeability (length per unit time)
A_x = groundwater cross-sectional area perpendicular to the direction of flow
s = hydraulic gradient (slope of water table or piezometric surface)

Despite the importance of groundwater flows in the budgets of many wetlands, there is poor understanding of groundwater hydraulics in wetlands, particularly in those that have organic soils. The hydraulic conductivity of both organic and inorganic wetland soils is discussed in more detail in Chapter 5: "Wetland Soils."

Evapotranspiration

The water that vaporizes from water or soil in a wetland (*evaporation*), together with moisture that passes through vascular plants to the atmosphere (*transpiration*), is called *evapotranspiration*. The meteorological factors that affect evaporation and transpiration are similar as long as there is adequate moisture, a condition that almost always exists in most wetlands. The rate of evapotranspiration is proportional to the difference between the vapor pressure at the water surface (or at the leaf surface) and the vapor pressure in the overlying air. This is described in a version of *Dalton's law*:

$$E = cf(u)(e_w - e_a) \tag{4.12}$$

where

E = rate of evaporation
c = mass transfer coefficient
$f(u)$ = function of wind speed, u
e_w = vapor pressure at surface, or saturation vapor pressure at wet surface
e_a = vapor pressure in surrounding air

Evaporation and transpiration are enhanced by the same meteorological conditions, such as solar radiation or surface temperature, that increase the value of the vapor pressure at the evaporating surface and by factors such as decreased humidity or increased wind speed that decrease the vapor pressure of the surrounding air. This equation assumes an adequate supply of water for capillary movement in the soil or for access by rooted plants. When the water supply is limited (not a frequent occurrence in wetlands), evapotranspiration is limited as well. Transpiration can also be physiologically limited in plants through the closing of leaf stomata despite adequate moisture during periods of stress such as anoxia.

Direct Measurement of Wetland Evapotranspiration

Several direct measurement techniques can be used in wetlands to determine evapotranspiration. The classical reference method is the measurement of evaporation from a water-filled pan, usually by measuring the weight loss, by measuring the volume required to replace lost water over a period of time, or by measuring the drop in water level. This is generally considered a measurement of potential evaporation, since the evaporating surface is saturated. The method is tedious and the results often poorly correlated with actual evaporation from vegetated surfaces, because the transpiration, unsaturated soils, winds, and shading effects of the plant canopy all influence the rate, often in unknown ways. However, pan evaporation provides a reference evaporation rate for comparison with other techniques. Furthermore, because wetland soils tend to be saturated most of the time, the pan method may be more accurate for wetlands than for terrestrial environments.

Wetland evapotranspiration can also be estimated by measuring the change in water level of the water in the wetland itself. This method, illustrated in Figure 4.14, can be calculated as follows:

$$ET = S_y(24h \pm s) \tag{4.13}$$

where

ET = evapotranspiration (mm/day)
S_y = specific yield of aquifer (unitless)
   = 1.0 for standing-water wetlands
   <1.0 for groundwater wetlands
h = hourly rise in water level from midnight to 4:00 A.M. (mm/h)
s = net fall (+) or rise (−) of water table or water surface in one day

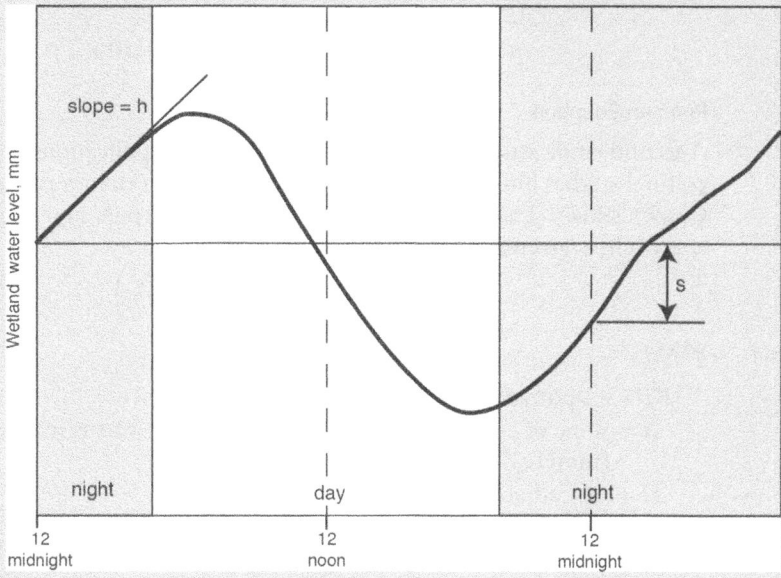

Figure 4.14 Diurnal water fluctuation in some wetlands can be used to estimate evapotranspiration as in Equation 4.13.

The pattern assumes active "pumping" of water by vegetation during the day and a constant rate of recharge equal to the midnight-to-4:00 A.M. rate. This method also assumes that evapotranspiration is negligible around midnight and that the water table around this time approximates the daily mean. The water level is usually at or near the root zone in many wetlands, a necessary condition for this method to measure evapotranspiration accurately.

Empirical Estimates of Wetland Evapotranspiration

Thornthwaite Equation

Evapotranspiration can be determined with any number of empirical equations that use easily measured meteorological variables. One of the most frequently used empirical equations for evapotranspiration from terrestrial ecosystems, which has been applied with some success to wetlands, is the *Thornthwaite equation* for potential evapotranspiration:

$$ET_i = 16(10T_i/I)^a \tag{4.14}$$

where

ET_i = potential evapotranspiration for month i (mm/month)
T_i = mean monthly temperature (°C)
I = local heat index $\sum_{i=1}^{12}(T_i/5)^{1.514}$
$a = (0.675 \times I^3 - 77.1 \times I^2 + 17{,}920 \times I + 492{,}390) \times 10^{-6}$

Penman Equation

A second empirical relationship that has had many applications in hydrologic and agricultural studies but relatively few in wetlands is the *Penman equation* (Penman, 1948; Chow, 1964). This equation, based on both Dalton's law and the energy budget approach, is given as

$$ET = \left(\frac{\Delta H + 0.27 E_a}{\Delta + 0.27}\right) \tag{4.15}$$

where

ET = evapotranspiration (mm/day)
Δ = slope of curve of saturation vapor pressure versus mean air temperature (mmHg/°C)
H = net radiation (cal/cm^2-day)
  = $R_t(1-a) - Rb$
R_t = total shortwave radiation
a = albedo of wetland surface
R_b = effective outgoing longwave radiation = $f(T^4)$
E_a = term describing the contribution of mass transfer to evaporation
  = $0.35(0.5 + 0.00625u)(e_w - e_a)$
u = wind speed 2m above ground (km/day)
e_w = saturation vapor pressure of water surface at mean air temperature (mmHg)
e_a = vapor pressure in surrounding air (mmHg)

The Penman equation was compared with the pan evaporation (multiplied by a factor of 0.8) and other methods at natural enriched fens in Michigan and constructed wetlands in Nevada. The Penman equation, like the Thornthwaite equation, generally underpredicted evapotranspiration from the humid Michigan wetland but agreed within a few percentage points with other measurement techniques for the arid Nevada wetlands.

Because of the many meteorological and biological factors that affect evapotranspiration, none of the many empirical relationships is entirely satisfactory for estimating wetland evapotranspiration. Several comparisons of approaches to measuring evapotranspiration have been attempted (Lott and Hunt, 2001; Rosenberry et al., 2004). One finding has been that empirical estimates of potential evapotranspiration (PET), such as those determined from the Penman equation, generally underestimate true wetland evapotranspiration during the growing season, possibly due to limitation of the equation for describing surface roughness. A comparison of an energy budget method for estimating evapotranspiration at a wetland in North Dakota with 12 empirical evapotranspiration equations found that most of the empirical methods gave reasonable approximations of evapotranspiration (Rosenberry et al., 2004).

The Thornthwaite equation, the simplest method investigated as it only requires air temperature, worked relatively well and may provide the most accurate measurement per instrument cost. It remains one of the more commonly uses empirical equations for estimating wetland evapotranspiration, but it only gives monthly estimates, not daily or hourly rates.

Effects of Vegetation on Wetland Evapotranspiration

A question about evapotranspiration from wetlands that does not elicit a uniform answer in the literature is: "Does the presence of wetland vegetation increase or decrease the loss of water compared to that which would occur from an open body of water?" Data from individual studies are conflicting. Obviously, the presence of vegetation retards evaporation from the water surface, but the question is whether the transpiration of water through the plants equals or exceeds the difference. Eggelsmann (1963) found evaporation from bogs in Germany to be generally less than that from open water except during wet summer months. In studies of evapotranspiration from small bogs in northern Minnesota, Bay (1967) found it to be 88 percent to 121 percent of open-water evaporation. Eisenlohr (1976) reported 10 percent lower evapotranspiration from vegetated prairie potholes than from nonvegetated potholes in North Dakota. Hall et al. (1972) estimated that a stand of vegetation in a small New Hampshire wetland lost 80 percent more water than did the open water in the wetland. In a forested pond cypress dome in north-central Florida, Heimburg (1984) found that swamp evapotranspiration was about 80 percent of pan evaporation during the dry season (spring and fall) and as low as 60 percent of pan evaporation during the wet season (summer). S. L. Brown (1981) found that transpiration losses from pond cypress wetlands were lower than evaporation from an open-water surface even with adequate standing water.

In the arid West, it has been a long-standing practice to conserve water for irrigation and other uses by clearing riparian vegetation from streams. In this environment where groundwater is often well below the surface but within the rooting zone of deep-rooted plants, trees "pump" water to the leaf surface and actively transpire even when little evaporation occurs at the soil surface.

The conflicting measurements and the difficulty of measuring evaporation and evapotranspiration led Linacre (1976) to conclude that neither the presence of wetland vegetation nor the type of vegetation had major influences on evaporation rates, at least during the active growing season. Bernatowicz et al. (1976) also found little difference in evapotranspiration among several species of vegetation. The general unimportance of plant species variation on overall wetland water loss is probably a reasonable conclusion for most wetlands, although it is clear that the type of wetland ecosystem and the season are important considerations. Ingram (1983), for example, found that fens have about 40 percent more evapotranspiration than do treeless bogs and that evaporation from the bogs is less than potential evapotranspiration in the summer and greater than potential evapotranspiration in the winter.

In some cases, the type of vegetation in the wetland does matter. When trees are removed from some forested swamps where the soil is hydric but there is little surface flooding, standing water may return and, with it, herbaceous marsh vegetation. This resets a hydrologic succession; woody plants are able to reinvade the marsh during dry years and reestablish the site back to a forested wetland.

Tides

The periodic and predictable tidal inundation of coastal salt marshes, mangroves, and freshwater tidal marshes is a major hydrologic feature of these wetlands. The tide acts as a stress by causing submergence, saline soils, and soil anaerobiosis; it acts as a subsidy by removing excess salts, reestablishing aerobic conditions, and providing nutrients. Tides also shift and alter the sediment patterns in coastal wetlands, causing a uniform surface to develop.

Typical tidal patterns for several coastal areas of the United States are shown in Figure 4.15a. Seasonal as well as diurnal patterns exist in the tidal rhythms. Annual variations of mean monthly sea level are as great as 25 cm (Fig. 4.15b). Tides also have significant bimonthly patterns, because they are generated by the gravitational pull of the moon and, to a lesser extent, the sun. When the sun and the moon are in line and pull together, which occurs almost every two weeks, *spring tides*, or tides of the greatest amplitude, develop. When the sun and the moon are at right angles, *neap tides*, or tides of least amplitude, occur. Spring tides occur roughly at full and new moons, whereas neap tides occur during the first and third quarters.

Tides vary more locally than regionally. The primary determinant is the coastline configuration. In North America, tidal amplitudes vary from less than 1 m along the Texas Gulf Coast to several meters in the Bay of Fundy in Canada. Tidal amplitude can actually increase as one progresses inland in some funnel-shaped estuaries. Typically, on a rising tide, water flows up tidal creek channels until the channels are bankfull. It overflows first at the upstream end, where tidal creeks break up into small creeks that lack natural levees. The overflowing water spreads back downstream over the marsh surface. On falling tides, the flows are reversed. At low tides, water continues to drain through the natural levee sediments into adjacent creeks because these sediments tend to be relatively coarse; in the marsh interior, where sediments are finer, drainage is poor and water is often impounded in small depressions in the marsh.

Seiches

While inland wetlands are nontidal by definition, periodic water-level fluctuations in wetlands adjacent to large freshwater lakes do occur as a result of short-term

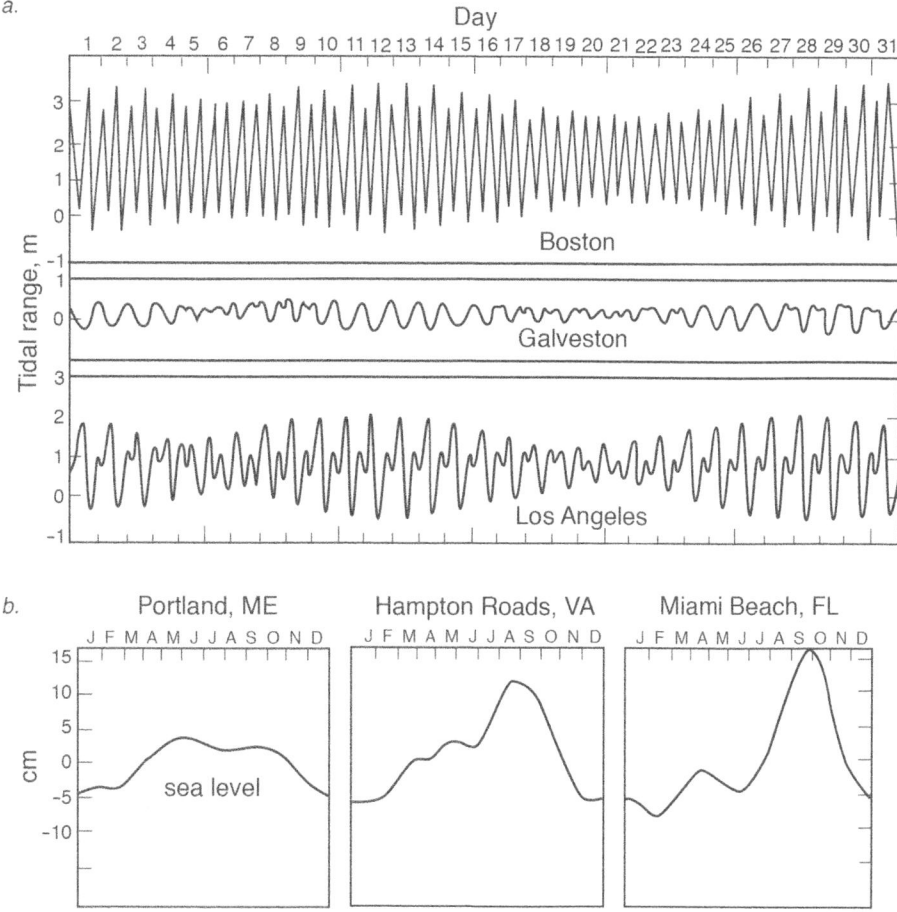

Figure 4.15 Patterns of tides: (a) daily tides for a month and (b) seasonal changes in mean monthly sea level for several locations in North America. (After Emery and Uchupi, 1972)

water-level seiches, or "wind tides." These are a common occurrence in wetlands adjacent to large lakes, such as the Laurentian Great Lakes in the United States and Canada (Fig. 4.16). When wind has a persistent direction, particularly in a long fetch across a lake, water "piles up" on the downwind side of the lake, causing high-water events for wetlands in that location. When the wind shifts or dies down, the high water is released and flows to the opposite shoreline, causing a secondary wind-relaxation seiche there and lower-than-normal water in the original high-water location.

Effects of Hydrology on Wetland Function

The effects of hydrology on wetland structure and function can be described with a complicated series of cause-and-effect relationships. A conceptual model of the general

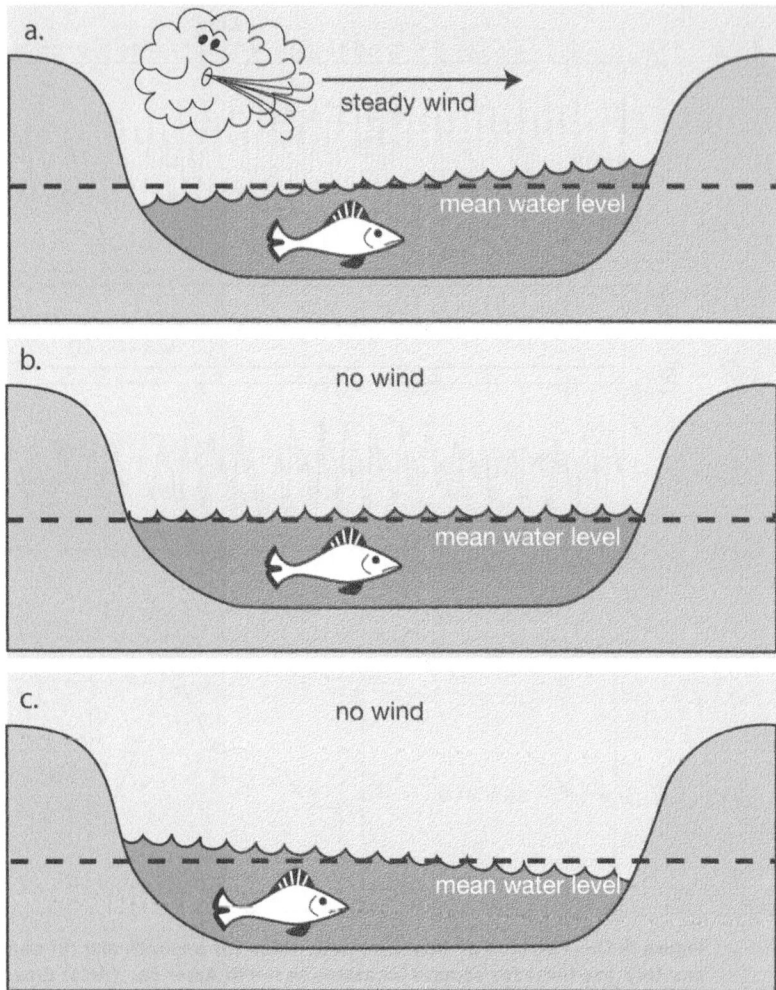

Figure 4.16 Concept of a seiche: a wind-relaxation seiche caused by (a) a steady wind that (b) relaxes or shifts directions from initial wind set and (c) results in an oppositely directed tilt; (d) water levels in Ohio (Toledo and Cleveland) and New York (Buffalo) coastlines of Lake Erie during an April 1979 storm and subsequent wind-relaxation seiche. (After Korgen, 1995)

effects of hydrology in wetland ecosystems was shown in Figure 4.1. The effects are shown to be primarily on the chemical and physical aspects of the wetlands, which, in turn, influence the biotic components of the ecosystem. The biotic components then have a feedback effect on hydrology. Four principles underscoring the importance of hydrology in wetlands can be elucidated from studies that have been conducted to date. These principles are described next.

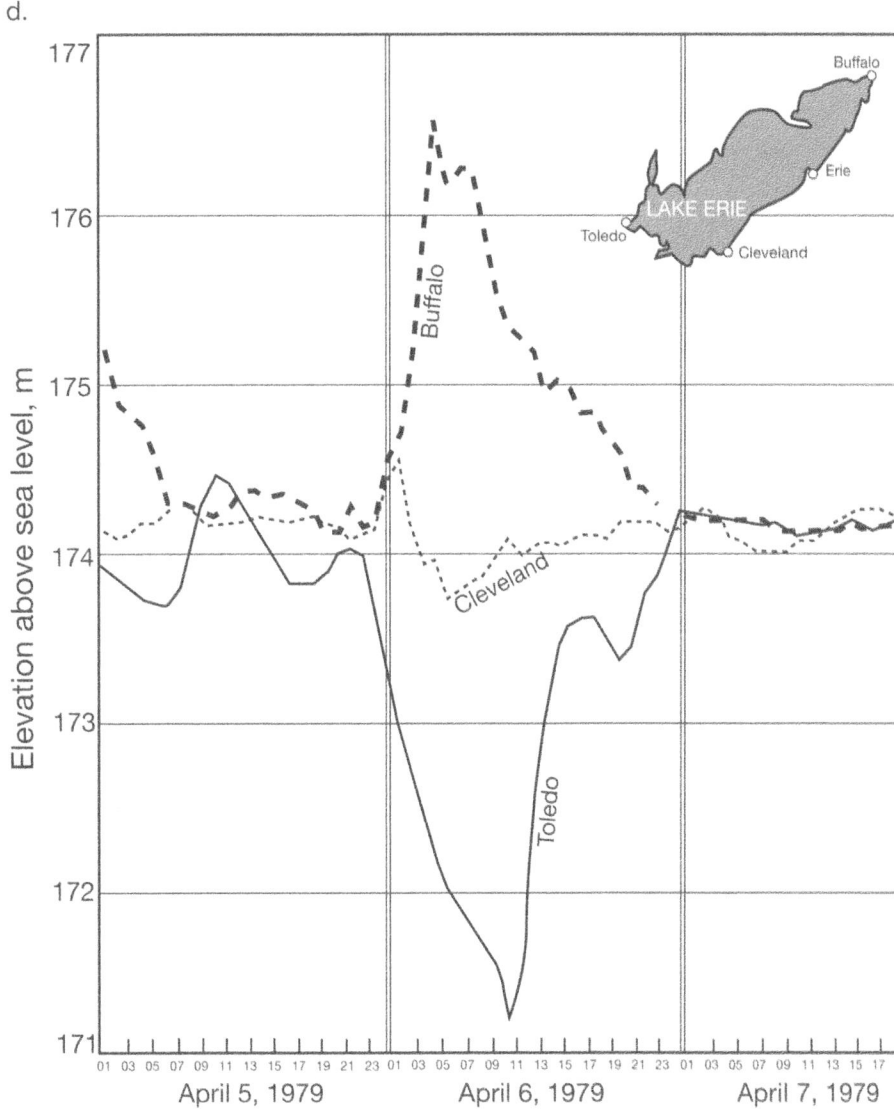

Figure 4.16 *(Continued)*

1. Hydrology leads to a unique vegetation composition but can limit or enhance species richness.

Hydrology is a two-edged sword for species composition and diversity in wetlands. It acts as a limit or a stimulus to species richness, depending on the hydroperiod and physical energies. At a minimum, the hydrology acts to select water-tolerant vegetation in both freshwater and saltwater conditions and to exclude flood-intolerant species. Of the thousands of vascular plants on Earth, relatively few have adapted to waterlogged soils. Although it is difficult to generalize, many wetlands that sustain long flooding durations have lower species richness in vegetation than do less frequently flooded or pulsing areas. Waterlogged soils and the subsequent changes in oxygen content and other chemical conditions significantly limit the number and the types of rooted plants that can survive in this environment.

In general, species richness, at least in the vegetation community, increases as flow-through or pulsing hydrology increases. Flowing water can be thought of as a stimulus to diversity, probably caused by its ability to renew minerals and reduce anaerobic conditions. Hydrology also stimulates diversity when the action of water and transported sediments creates spatial heterogeneity, opening up additional ecological niches. When rivers flood riparian wetlands or when tides rise and fall in coastal marshes, erosion, scouring, and sediment deposition sometimes create niches that allow diverse habitats to develop. However, flowing water can also create a relatively uniform surface that might allow monospecific stands of *Typha* or *Phragmites* to dominate a freshwater marsh or *Spartina* to dominate a coastal marsh. Keddy (1992) likened water-level fluctuations in wetlands to fires in forests. They eliminate one growth form of vegetation (e.g., woody plants) in favor of another (e.g., herbaceous species) and allow regeneration of species from buried seeds.

2. Primary productivity and other ecosystem functions in wetlands are often enhanced by flowing conditions and a pulsing hydroperiod and are often depressed by stagnant conditions.

In general, the "openness" of a wetland to hydrological fluxes is probably one of the most important determinants of potential primary productivity. For example, peatlands that have flow-through conditions (fens) have long been known to be more productive than stagnant raised bogs. Some studies have found that wetlands in stagnant (nonflowing) or continuously deep water have low productivities, whereas wetlands that are in slowly flowing strands or are open to flooding rivers have high productivities.

This relationship between hydrology and ecosystem primary productivity has been investigated most extensively for forested wetlands. Figure 4.17 shows a set of similar typical "Shelford-type" limitation curves that have been suggested in separate studies to explain the importance of hydrology on forested wetland productivity. All of the curves in Figure 4.17 suggest that the highest productivity occurs in systems that are neither very wet nor too dry but that have either average hydrologic conditions or seasonal hydrologic pulsing.

The subsidy-stress model of H. T. Odum (1971) and E. P. Odum (1979), later refined as the *pulse stability* concept by all three Odums (W. E. Odum et al., 1995), includes concepts that potentially apply well to the effects of hydrology on

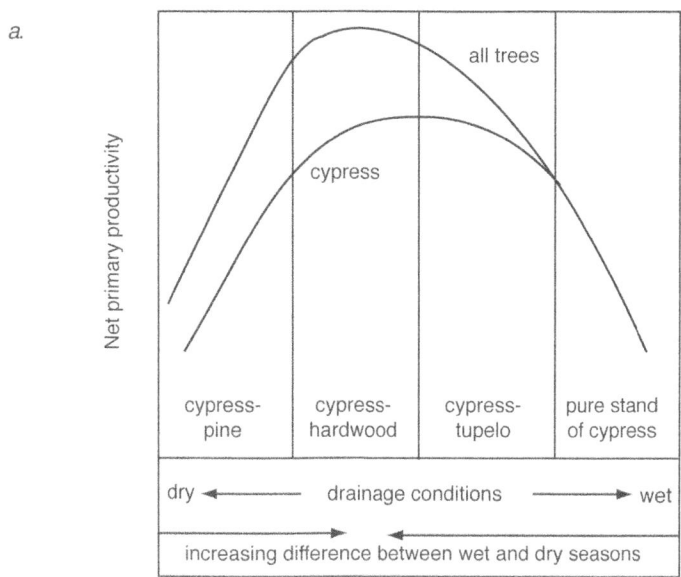

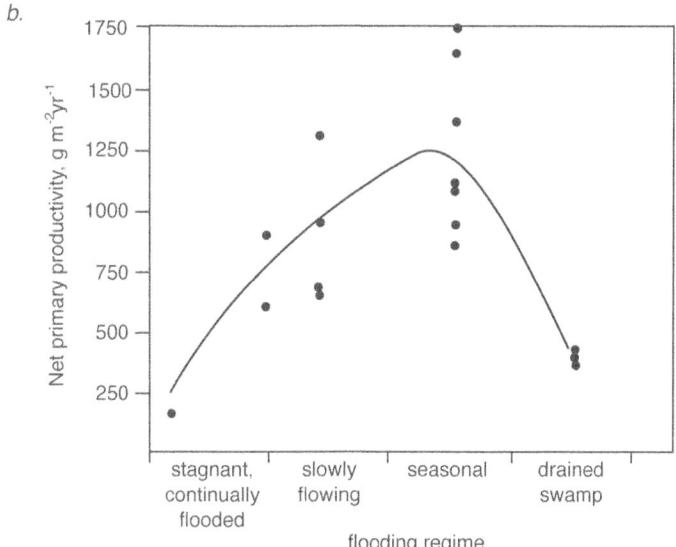

Figure 4.17 Relationships between swamp productivity and hydrologic conditions: (a) for cypress (*Taxodium*) swamps in north-central Florida, (b) between flooding regime and net primary productivity of Louisiana swamps, and (c) between radial growth of red maple (*Acer rubrum*) and annual water level for six Rhode Island red maple swamps over six years. ((a) after Mitsch and Ewel, 1979; (b) after Conner and Day, 1982; (c) after Golet et al., 1993)

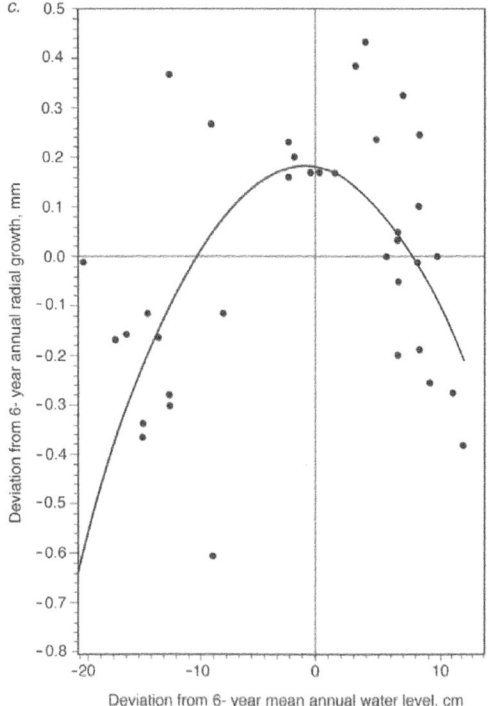

Figure 4.17 *(Continued)*

wetland productivity. Seasonal pulsing of floodwater can be both a subsidy and a stress, whether the wetland is a salt marsh or mangrove swamp subject to twice-per-day flooding or a riparian wetland subject to seasonal river pulses. Pulsing is frequent in nature, and ecosystems such as bottomland forests and salt marshes appear to be well adapted to taking advantage of this subsidy. Despite this clear theoretical basis for understanding the effects of hydrology on productivity, it has been difficult to confirm or deny these theories in practice.

The model shown in Figure 4.18 may explain the difficulty in ascribing a direct relationship between vascular plant productivity and hydrologic conditions. While flood intensity increases available moisture and nutrients, longer flood durations increase stresses caused by an anaerobic root zone and can actually decrease the length of the growing season. In effect, "subsidies and stresses may occur simultaneously and cancel one another" (Megonigal et al., 1997). In this Mitsch-Rust model, flood intensity and duration affect moisture, available nutrients, anaerobiosis, and even length of growing season in a complex and nonlinear "push-pull" arrangement.

The influence of hydrologic conditions on freshwater marsh productivity is less certain. If peak biomass or similar measures are used as indicators of marsh

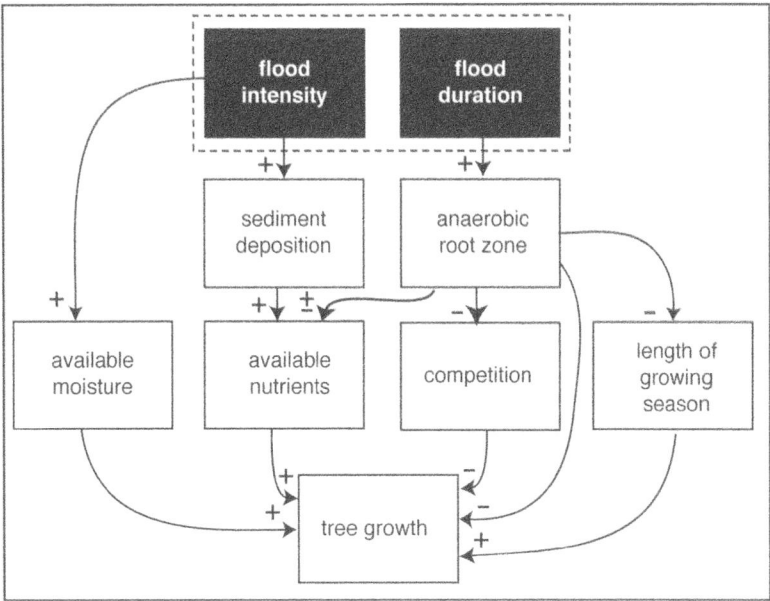

Figure 4.18 Causal model that describes the major causes for increases and decreases in individual tree growth in riparian floodplain forests. Plus (+) sign indicates a positive effect; minus sign (−) indicates a negative effect. (After Mitsch and Rust, 1984)

productivity, some studies have shown the classical stimulation of vegetation along the water's edge, whereas other studies have indicated a higher macrophyte productivity in sheltered, nonflowing marshes than in wetlands that are open to flowing conditions or coastal influences. For example, consistently higher macrophyte biomass was found in wetlands isolated from surface fluxes with artificial dikes than in wetlands that were open to coastal fluxes along Lake Erie. Several explanations are possible: (1) The coastal fluxes may also be serving as a stress as well as a subsidy on the macrophytes; (2) the open marshes may be exporting a significant amount of their productivity; and (3) the diked wetlands have more predictable hydroperiods.

Similar results were found in a hydrologic pulsing experiment in central Ohio, where simulated river floods caused a decrease in macrophyte and water column primary productivity but led to changes in greenhouse gas emissions because of a flushing effect (Mitsch et al., 2005; Altor and Mitsch, 2006, 2008; Hernandez and Mitsch, 2006, 2007; Tuttle et al., 2008; Fig. 4.19a) Conversely, an earlier study in Illinois of the influence of flow-through conditions on water column primary productivity of constructed marshes found that, after two years of experimentation, water column (phytoplankton and submerged aquatics) productivity was higher in high-flow wetlands compared to low-flow wetlands (Fig. 4.19b). While macrophyte productivity may take many years to respond to the difference

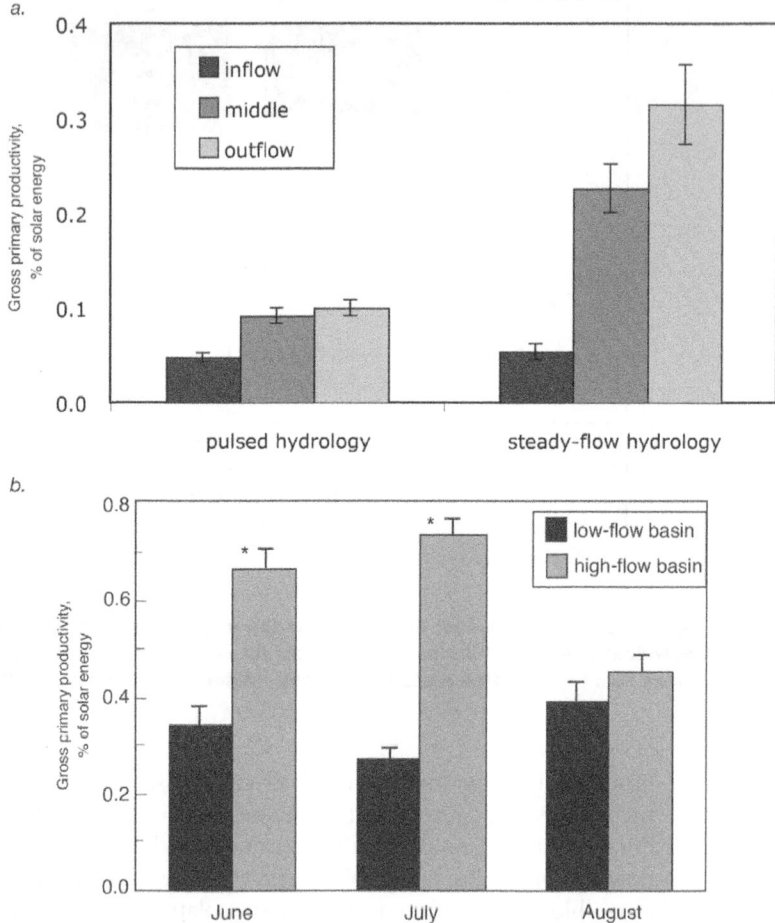

Figure 4.19 Aquatic primary productivity in freshwater marshes as a function of hydrologic conditions: (a) pulsed flooding versus steady-flow hydrology at the Olentangy River Wetland Research Park, central Ohio; (b) high-flow and low-flow conditions at the Des Plaines Wetland Demonstration Project, northeastern Illinois. * indicates statistical differences (0.05_ between low-flow and high-flow conditions. ((a) After Tuttle et al., 2008; (b) after Cronk and Mitsch, 1994)

in hydrology, water column productivity, which is often caused by attached and planktonic algae, responds relatively quickly to changing hydrologic conditions.

Coastal wetlands subject to frequent tidal action are generally more productive than those that are only occasionally inundated. A comparison of several Atlantic Coast salt marshes, for example, showed a direct relationship between tidal range (as a measure of water flux) and end-of-season peak biomass of *Spartina alterniflora* (Fig. 4.20). Apparently, vigorous tides increase the nutrient subsidy and cause a flushing of toxic materials, such as salt. Freshwater tidal

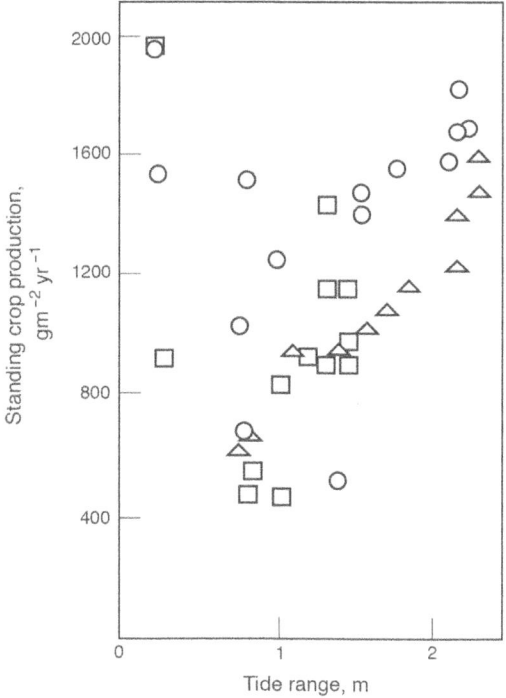

Figure 4.20 Production of *Spartina alterniflora* versus mean tidal range for several Atlantic Coast salt marshes. Different symbols indicate different data sources. (After Steever et al., 1976)

wetlands are even more productive than saline tidal wetlands, because they receive the energy and nutrient subsidy of tidal flushing while avoiding the stress of saline soils.

3. Accumulation of organic material in wetlands is controlled by hydrology through its influence on primary productivity, decomposition, and export of particulate organic matter.

Wetlands can accumulate excess organic matter as a result of either increased primary productivity (as described previously) or decreased decomposition and export. Notwithstanding the discrepancies from short-term litter decomposition studies, peat accumulates to some degree in all wetlands as a result of these processes. The effects of hydrology on decomposition pathways are even less clear than the effects on primary productivity discussed previously. Probably the lack of agreement among the many studies published on the subject results from the complexity of the decomposition process. In general, decomposition of organic detritus requires electron donors (usually oxygen, but alternate chemicals such as sulfate or nitrate may be effective under anoxic conditions), moisture, inorganic nutrients, and microorganisms capable of metabolizing in the specific environment concerned. The observed rate of organic decomposition is also influenced by the ambient temperature and by the activity of macrodetritivores that shred the plant remains and/or repackage it as bacterially inoculated fecal pellets. Hydrology modifies many of these variables; for example, moisture depends on the flooding regime, flowing water carries oxygen and nutrients, while in stagnant water oxygen is rapidly depleted and nutrients are transformed to more or less available forms. Given this complexity, it is not surprising that the results of short-term *in situ* decomposition studies often disagree.

The importance of hydrology for organic carbon export is obvious. A generally higher rate of export is to be expected from wetlands that are open to the flowthrough of water. Riparian wetlands often contribute large amounts of organic detritus to streams, including macrodetritus such as whole trees. There is also considerable evidence that watersheds that drain wetland regions export more organic material but retain more nutrients than do watersheds that do not have wetlands (Fig. 4.21). For example, the slope of the line in Figure 4.21 for wetland-dominated watersheds is much steeper than that for upland watersheds,

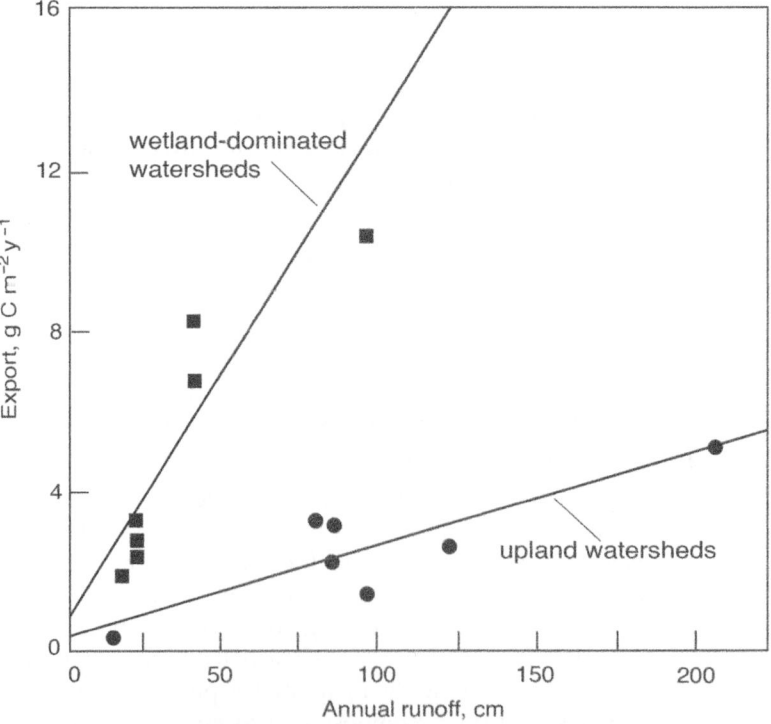

Figure 4.21 Organic carbon export from wetland-dominated watersheds compared with non-wetland watersheds. (From Mulholland and Kuenzler, 1979)

indicating a much greater organic carbon concentration in runoff as well as greater export for a given runoff from the wetland-dominated watersheds. Salt marshes and mangrove swamps are also considered major exporters of their productivity by most, but the generality of this concept is not fully accepted by coastal ecologists. Hydrologically isolated wetlands, such as northern peatlands, have much lower organic export.

4. Nutrient cycling and nutrient availability are both significantly influenced by hydrologic conditions.

Nutrients are carried into wetlands by the hydrologic inputs of precipitation, river flooding, tides, and surface and groundwater inflows. Outflows of nutrients are controlled primarily by the outflow of water. These hydrologic/nutrient flows are also important determinants of wetland productivity and decomposition (see previous sections). Intrasystem nutrient cycling is generally, in turn, tied to pathways such as primary productivity and decomposition. When productivity and decomposition rates are high, as in flowing water or pulsing hydroperiod wetlands, nutrient cycling is rapid. When productivity and decomposition processes are slow, as in isolated ombrotrophic bogs, nutrient cycling is also slow.

The hydroperiod of a wetland has a significant effect on nutrient transformations, on the availability of nutrients to vegetation, and on loss from wetland soils of nutrients that have gaseous forms. Thus, nitrogen availability and loss are affected in wetlands by the reduced conditions that result from waterlogged soil. Typically, a narrow oxidized surface layer develops over the anaerobic zone in wetland soils, causing a combination of reactions in the nitrogen cycle—nitrification and denitrification—that may result in substantial losses of dinitrogen gas to the atmosphere. Furthermore, ammonium nitrogen is usually the form of nitrogen most available to plants in wetland soils, because the anaerobic environment favors the reduced ionic form over the nitrate common in agricultural soils.

Flooding of wetland soil, by altering both the pH and the redox potential of the soil, influences the availability of other nutrients. The pH of both acid and alkaline soils tends to converge on a pH of 7 when they are flooded. The redox potential, a measure of the intensity oxidation or reduction of a chemical or biological system, indicates the state of oxidation (and, hence, availability) of several nutrients. Phosphorus is known to be more soluble under anaerobic conditions, partly because of the hydrolysis and reduction of ferric and aluminum phosphates to more soluble compounds. The availability of major ions, such as potassium and magnesium, and several trace nutrients, such as iron, manganese, and sulfur, is also affected by hydrologic conditions in the wetlands.

Techniques for Wetland Hydrology Studies

It is curious that so little attention has been paid to hydrologic measurements in wetland studies, despite the importance of hydrology in ecosystem function. A great

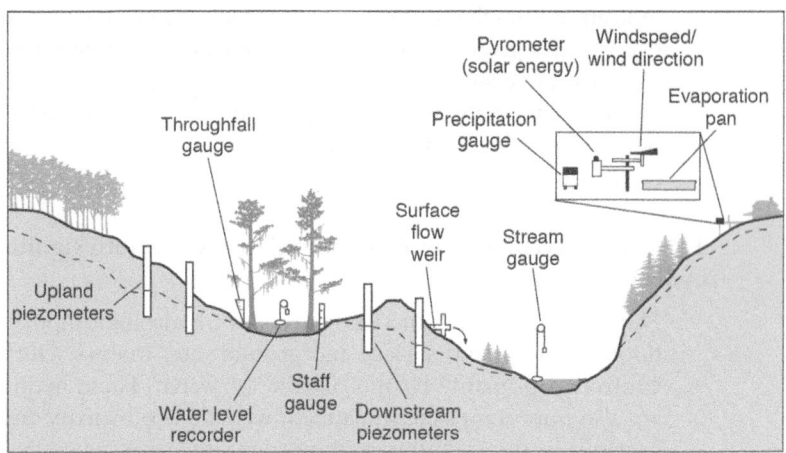

Figure 4.22 Placement of hydrology instruments in the landscape to estimate a water budget for a floodplain wetland.

deal of information can be obtained with only a modest investment in supplies and equipment. A diagram summarizing many of the hydrology measurements typical for developing a wetland's water budget is given in Figure 4.22. Water levels can be recorded continuously with water-level recorders or data loggers or during site visits with a staff gauge. With records of water level, all of the following hydrologic parameters can be determined: hydroperiod, frequency of flooding, duration of flooding, and water depth. Water-level recorders can also be used to determine the change in storage in a water budget, as in Equation 4.1.

Evapotranspiration measurements are more difficult to obtain, but several empirical relationships, such as the Thornthwaite equation, use meteorological variables. Evaporation pans can also be used to estimate total evapotranspiration from wetlands, although pan coefficients are highly variable. Evapotranspiration of continuously flooded nontidal wetlands can also be determined by monitoring the diurnal water-level fluctuation.

Precipitation or throughfall or both can be measured by placing a statistically adequate number of rain gauges in random locations throughout the wetland or by utilizing weather station data. Surface runoff to wetlands can usually be determined as the increase in water level in the wetland during and immediately following a storm after net precipitation has been subtracted. Weirs can be constructed on more permanent streams to monitor surface water inputs and outputs.

Groundwater flows are usually the most difficult and most costly hydrologic flows to measure accurately. In some cases, clusters of shallow monitoring wells, placed around a wetland, will help indicate the direction of groundwater flow and the slope of the water or hydraulic gradient as required in Equation 4.11. The wells are called *piezometers* when they are only partially screened, and thus measure the piezometric head of an isolated part of the groundwater rather than being screened through the entire length of the well and thus measuring the surface water aquifer. Piezometers can be installed by professional well-drilling companies or, for low-budget installations, generally can be installed with augers or as well points. Estimates of permeability or hydraulic conductivity are then required to quantify the flows. Permeability can be estimated through *in situ* pump tests using the wells or through laboratory analysis of intact soil cores. The variability of results among different hydraulic conductivity measuring techniques suggests that caution should be used in taking these numbers.

If a wetland is a perched or a recharge wetland, seepage can be estimated either through a water budget approach (e.g., subtracting evapotranspiration losses from water-level decreases when there are no other inflows or outflows) or by using half-barrel seepage meters. Other methods available to measure groundwater flows in wetlands include the use of stable isotopes, generally $^{18}O/^{16}O$ or $^{2}H/^{1}H$, because of the propensity of the lighter isotope in each case to evaporate more readily, allowing water to be "tagged" according to its source (Hunt et al., 1996). Groundwater flow models have also been used to estimate the flow of groundwater into and out of wetlands with some success (Hunt et al., 1996; Koreny et al., 1999).

The uncertainty in the scientific literature concerning many wetland processes (e.g., the rates of organic matter decomposition discussed earlier) is often closely related to unquantified hydrologic parameters. Thus, careful attention to quantification of pertinent hydrologic parameters in wetland research studies is virtually certain to improve our understanding of the ecological processes that control wetlands.

Recommended Readings

Brooks, K. N., P. F. Ffolliott, and J. A. Magner. 2012. *Hydrology and Management of Watersheds*, 4th ed. Chichester, UK: Wiley-Blackwell.

Winter, T. C., and M. R. Llamas, eds., 1993. *Hydrogeology of Wetlands. Special Issue of Journal of Hydrology* 141:1–269.

References

Altor, A. E., and W. J. Mitsch. 2006. Methane flux from created riparian marshes: Relationship to intermittent versus continuous inundation and emergent macrophytes. *Ecological Engineering* 28: 224–234.

Altor, A. E., and W. J. Mitsch. 2008. Pulsing hydrology, methane emissions, and carbon dioxide fluxes in created marshes: A 2-year ecosystem study. *Wetlands* 28: 423–438.

Bay, R. R. 1967. Groundwater and vegetation in two peat bogs in northern Minnesota. *Ecology* 48: 308–310.

Bernatowicz, S., S. Leszczynski, and S. Tyczynska. 1976. The influence of transpiration by emergent plants on the water balance of lakes. *Aquatic Botany* 2: 275–288.

Bhowmik, N. G., A. P. Bonini, W. C. Bogner, and R. P. Byrne. 1980. Hydraulics of flow and sediment transport to the Kankakee River in Illinois. *Illinois State Water Survey Report of Investigation* 98, Champaign, IL. 170 pp.

Brinson, M. M. 1993. Changes in the functioning of wetlands along environmental gradients. *Wetlands* 13: 65–74.

Brown, S. L. 1981. A comparison of the structure, primary productivity, and transpiration of cypress ecosystems in Florida. *Ecological Monographs* 51: 403–427.

Chow, V. T., ed. 1964. *Handbook of Applied Hydrology*. McGraw-Hill, New York. 1453 pp.

Conner, W. H., and J. W. Day, Jr. 1982. The ecology of forested wetlands in the southeastern United States. In B. Gopal, R. E. Turner, R. G. Wetzel, and D. F. Whigham, eds., *Wetlands: Ecology and Management*. National Institute of Ecology and International Scientific Publications, Jaipur, India, pp. 69–87.

Cowardin, L. M., V. Carter, F. C. Golet, and E. T. LaRoe. 1979. *Classification of Wetlands and Deepwater Habitats of the United States*. FWS/OBS-79/31. U.S. Fish and Wildlife Service, Washington, DC. 103 pp.

Cronk, J. K., and W. J. Mitsch. 1994. Aquatic metabolism in four newly constructed freshwater wetlands with different hydrologic inputs. *Ecological Engineering* 3: 449–468.

Duever, M. J. 1988. Hydrologic processes for models of freshwater wetlands. In W. J. Mitsch, M. Straskraba, and S. E. Jørgensen, eds., *Wetland Modelling*. Elsevier, Amsterdam, The Netherlands, pp. 9–39.

Dunne, T., and L. B. Leopold. 1978. *Water in Environmental Planning*. W. H. Freeman and Company, New York. 818 pp.

Eggelsmann, R. 1963. *Die Potentielle und Aktuelle Evaporation eines Seeklimathochmoores*. Publication 62, International Association of Hydrological Sciences, pp. 88–97.

Eisenlohr, W. S. 1976. Water loss from a natural pond through transpiration by hydrophytes. *Water Resources Research* 2: 443–453.

Emery, K. O., and E. Uchupi. 1972. Western North Atlantic Ocean: Topography, rocks, structure, water, life, and sediments. *Memoirs of the American Association of Petroleum Geologists* 17. 1532 pp.

Freiberger, H. J. 1972. *Streamflow Variation and Distribution in the Big Cypress Watershed During Wet and Dry Periods*. Map Series 45, Bureau of Geology, Florida Department of Natural Resources, Tallahassee.

Gilman, K. 1982. Nature conservation in wetlands: Two small fen basins in western Britain. In D. O. Logofet and N. K. Luckyanov, eds., *Ecosystem Dynamics in Freshwater Wetlands and Shallow Water Bodies, Vol. I*. SCOPE and UNEP Workshop, Center of International Projects, Moscow, pp. 290–310.

Golet, F. C., A. J. K. Calhoun, W. R. DeRagon, D. J. Lowry, and A. J. Gold. 1993. *Ecology of Red Maple Swamps in the Glaciated Northeast: A Community Profile*. Biological Report 12, U.S. Fish and Wildlife Service, Washington, DC. 151 pp.

Hall, F. R., R. J. Rutherford, and G. L. Byers. 1972. *The Influence of a New England Wetland on Water Quantity and Quality*. New Hampshire Water Resource Center Research Report 4. University of New Hampshire, Durham. 51 pp.

Heimburg, K. 1984. Hydrology of north-central Florida cypress domes. In K. C. Ewel and H. T. Odum, eds., *Cypress Swamps*. University Presses of Florida, Gainesville, pp. 72–82.

Hernandez, M. E., and W. J. Mitsch. 2006. Influence of hydrologic pulses, flooding frequency, and vegetation on nitrous oxide emissions from created riparian marshes. *Wetlands* 26: 862–877.

Hernandez, M. E., and W. J. Mitsch. 2007. Denitrification in created riverine wetlands: Influence of hydrology and season. *Ecological Engineering* 30: 78–88.

Hemond, H. F. 1980. Biogeochemistry of Thoreau's Bog, Concord, Mass. *Ecological Monographs* 50: 507–526.

Hey, D. L., and N. S. Philippi. 1995. Flood reduction through wetland restoration: The Upper Mississippi River Basin as a case study. *Restoration Ecology* 3: 4–17.

Hunt, R. J., D. P. Krabbenhoft, and M. P. Anderson. 1996. Groundwater inflow measurements in wetland systems. *Water Resources Research* 32: 495–507.

Ingram, H. A. P. 1983. Hydrology. In A. J. P. Gore, ed. *Ecosystems of the World, Vol. 4A: Mires: Swamp, Bog, Fen, and Moor*. Elsevier, Amsterdam, The Netherlands, pp. 67–158.

Junk, W. J. 1982. Amazonian floodplains: Their ecology, present and potential use. In D. O. Logofet and N. K. Luckyanov, eds., *Ecosystem Dynamics in Freshwater Wetlands and Shallow Water Bodies*, Vol. I. SCOPE and UNEP Workshop, Center of International Projects, Moscow, pp. 98–126.

Kantrud, H. A., J. B. Millar, and A. G. van der Valk. 1989. Vegetation of wetlands of the prairie pothole region. In A. G. van der Valk, ed., *Northern Prairie Wetlands*. Iowa State University Press, Ames, pp. 132–187.

Keddy, P. A. 1992. Water level fluctuations and wetland conservation. In J. Kusler and R. Smandon, eds., *Wetlands of the Great Lakes*. Proceedings of an International Symposium. Association of State Wetland Managers, Berne, NY, pp. 79–91.

Koreny, J. S., W. J. Mitsch, E. S. Bair, and X. Wu. 1999. Regional and local hydrology of a constructed riparian wetland system. *Wetlands* 19: 182–193.

Korgen, B. J. 1995. Seiches. *American Scientist* July-August 1995: 330–341.

Leopold, L. B., M. G. Wolman, and J. E. Miller. 1964. *Fluvial Processes in Geomorphology*. W. H. Freeman, San Francisco. 522 pp.

Linacre, E. 1976. Swamps. In J. L. Monteith, ed., *Vegetation and the Atmosphere*, vol. 2: Case Studies. Academic Press, London, pp. 329–347.

Lott, R. B., and R. J. Hunt. 2001. Estimating evapotranspiration in natural and constructed wetlands. *Wetlands* 21: 614–628.

Megonigal, J. P., W. H. Conner, S. Kroeger, and R. R. Sharitz. 1997. Aboveground production in southeastern floodplain forests: A test of the subsidy–stress hypothesis. *Ecology* 78: 370–384.

Millar, J. B. 1971. Shoreline-area as a factor in rate of water loss from small sloughs. *Journal of Hydrology* 14: 259–284.

Mitsch, W. J. 1979. Interactions between a riparian swamp and a river in southern Illinois. In R. R. Johnson and J. F. McCormick, tech. coords., Strategies for the Protection and Management of Floodplain Wetlands and Other Riparian Ecosystems. Proceedings of the Symposium, Calaway Gardens, GA, December 11–13, 1978. General Technical Report WO-12, U.S. Forest Service, Washington, DC, pp. 63–72.

Mitsch, W. J., and K. C. Ewel. 1979. Comparative biomass and growth of cypress in Florida wetlands. *American Midland Naturalist* 101: 417–426.

Mitsch, W. J., ed. 1989. *Wetlands of Ohio's Coastal Lake Erie: A Hierarchy of Systems*. NTIS, OHSU-BS-007, Ohio Sea Grant Program, Columbus. 186 pp.

Mitsch, W. J., W. Rust, A. Behnke, and L. Lai. 1979. *Environmental Observations of a Riparian Ecosystem during Flood Season*. Research Report 142, Illinois University Water Resources Center, Urbana. 64 pp.

Mitsch, W. J., and W. G. Rust. 1984. Tree growth responses to flooding in a bottomland forest in northeastern Illinois. *Forest Science* 30: 499–510.

Mitsch, W. J., and B. C. Reeder. 1992. Nutrient and hydrologic budgets of a Great Lakes coastal freshwater wetland during a drought year. *Wetlands Ecology and Management* 1(4): 211–223.

Mitsch, W. J., L. Zhang, C. J. Anderson, A. Altor, and M. Hernandez. 2005. Creating riverine wetlands: Ecological succession, nutrient retention, and pulsing effects. *Ecological Engineering* 25: 510–527.

Mitsch, W.J., A.M. Nahlik, P. Wolski, B. Bernal, L. Zhang, and L. Ramberg. 2010. Tropical wetlands: Seasonal hydrologic pulsing, carbon sequestration, and methane emissions. *Wetlands Ecology and Management* 18: 573–586.

Mulholland, P. J., and E. J. Kuenzler. 1979. Organic carbon export from upland and forested wetland watersheds. *Limnology and Oceanography* 24: 960–966.

Nixon, S. W., and C. A. Oviatt. 1973. Ecology of a New England salt marsh. *Ecological Monographs* 43: 463–498.

Odum, E. P. 1979. Ecological importance of the riparian zone. In R. R. Johnson and J. F. McCormick, tech. coords., Strategies for Protection and Management of Floodplain Wetlands and Other Riparian Ecosystems. Proceedings of the Symposium, Callaway Gardens, GA, December 11–13, 1978. General Technical Report WO-12, U.S. Forest Service, Washington, DC, pp. 2–4.

Odum, H. T. 1971. *Environment, Power and Society*. John Wiley & Sons, New York.

Odum, W. E., E. P. Odum, and H. T. Odum. 1995. Nature's pulsing paradigm. *Estuaries* 18: 547–555.

Penman, H. L. 1948. Natural evaporation from open water, bare soil and grass. *Proceedings of the Royal Society of London* 93: 120–145.

Pride, R. W., F. W. Meyer, and R. N. Cherry. 1966. *Hydrology of Green Swamp Area in Central Florida*. Report 42, Florida Division of Geology, Tallahassee. 137 pp.

Richardson, C. J. 1983. Pocosins: Vanishing wastelands or valuable wetlands. *BioScience* 33: 626–633.

Rosenberry, D. O., D. I. Stannard, T. C. Winter, and M. L. Martinez. 2004. Comparison of 13 equations for determining evapotranspiration from a prairie wetland, Cottonwood Lake Area, North Dakota, USA. *Wetlands* 24: 483–497.

Shjeflo, J. B. 1968. Evapotranspiration and the water budget of prairie potholes in North Dakota. Professional Paper 585-B. U.S. Geological Survey, Washington, DC. 49 pp.

Steever, E. Z., R. S. Warren, and W. A. Niering. 1976. Tidal energy subsidy and standing crop production of Spartina alterniflora. *Estuarine, Coastal and Marine Science* 4: 473–478.

Tuttle, C. L., L. Zhang, and W. J. Mitsch. 2008. Aquatic metabolism as an indicator of the ecological effects of hydrologic pulsing in flow-through wetlands. *Ecological Indicators* 8: 795–806.

Twilley, R. R. 1982. Litter dynamics and organic carbon exchange in black mangrove *(Avicennia germinans) basin forests in a Southwest Florida estuary*. Ph.D. dissertation, University of Florida, Gainesville.

U.S. Department of Interior (USDI), Bureau of Reclamation. 2001. *Water Measurement Manual*, 3rd ed., revised reprint. U.S. Government Printing Office, Washington, DC, 327 pp. Available at www.usbr.gov/pmts/hydraulics_lab/pubs/wmm/.

van der Valk, A. G., ed. 1989. *Northern Prairie Wetlands*. Iowa State University Press, Ames. 400 pp.

Woo, M.-K., and T. C. Winter. 1993. The role of permafrost and seasonal frost in the hydrology of nothern wetlands in North America. *Journal of Hydrology* 141: 5–31.

Vernal Pools: A Community Profile. Biological Report 85 (7.11), U.S. Fish and Wildlife Service, Washington, DC.

Chapter **5**

Wetland Soils

Wetland soils, known as hydric soils, are defined as soils that formed under conditions of saturation, flooding or ponding long enough during the growing season to develop anaerobic conditions in the upper part. They occur when oxygen is cut off due to the presence of water, causing a predictable sequence of chemically reduced conditions. Wetland soils can be organic soils or mineral soils. Hydric mineral soils can be identified through hydric soil indicators such as redox concentrations, redox depletions, and reduced soil matrices, often by using a Munsell color chart. Redox potential is a useful measure of the degree to which wetland soils are oxidized or reduced. Various chemical and biological transformations take place as coupled oxidation (e^- donor)–reduction (e^- acceptor) reactions in wetland soils and occur in a predictable sequence within predictable redox ranges.

Types and Definitions

Wetland soils are both the medium in which many of the wetland chemical transformations take place and the primary storage of available chemicals for most wetland plants. They are often described as *hydric soils*, defined by the U.S. Department of Agriculture's Natural Resources Conservation Service (NRCS, 2010) as soils "that formed under conditions of saturation, flooding or ponding long enough during the growing season to develop anaerobic conditions in the upper part." Wetland soils are of two types: (1) *mineral soils* or (2) *organic soils.* All soils have some organic material, but when a soil has less than 20 to 35 percent organic matter (on a dry-weight basis), it is considered a mineral soil.

Organic soils and organic soil materials (*peat, mucky peat,* and *muck*) are defined under either of two conditions of saturation:

1. Soils are saturated with water for long periods or are artificially drained and, excluding live roots, (a) have 18 percent or more organic carbon if the

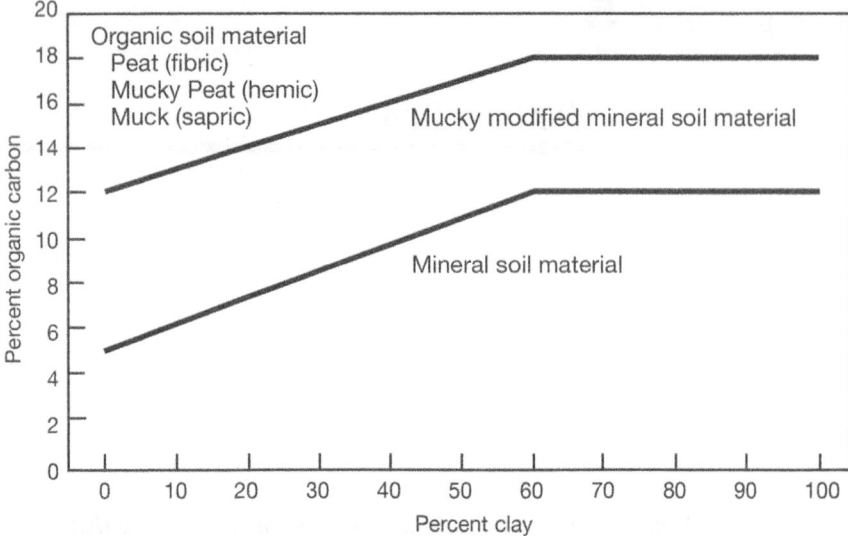

Figure 5.1 Percentage organic carbon required for a soil material to be called organic, mucky modified mineral, or mineral soil material versus clay content. (From NRCS, 2010)

mineral fraction is 60 percent or more clay, (b) have 12 percent or more organic carbon if the mineral fraction has no clay, or (c) have a proportional content of organic carbon between 12 and 18 percent if the clay content of the mineral fraction is between 0 and 60 percent (Fig. 5.1); or

2. Soils are never saturated with water for more than a few days and have 20 percent or more organic carbon.

For an estimate of organic carbon when organic matter content is known,

$$\% C_{org} = \% OM/2 \qquad (5.1)$$

where

$\% C_{org}$ = percentage of organic carbon
$\% OM$ = percentage of organic matter

Any soil material that is not included in the preceding definition is considered mineral soil material. Where mineral soils occur in wetlands, such as in some freshwater marshes or riparian forests, they generally have a soil profile made up of horizons, or layers. The upper layer of wetland mineral soils is often organic peat composed of partially decayed plant materials.

Although the preceding definition of organic soil is applicable to many types of wetlands, particularly to northern peatlands, *peat*, a generic term for relatively undecomposed organic soil material, is not usually that strictly defined. Most peats contain less than 20 percent unburnable inorganic matter (and therefore usually contain more than 80 percent burnable organic material, which is about 40 percent organic carbon). Some soil scientists, however, allow up to 35 percent unburnable inorganic matter (approximately 33 percent organic carbon), and commercial operations sometimes allow 55 percent unburnable material (22 percent organic carbon). *Muck* is defined as sapric organic soil material with plant material so decomposed that identification of plant forms is not possible. Its bulk density is generally greater than $0.2\,\text{g/cm}^3$ and more than peat.

Organic soils are different from mineral soils in four physicochemical features other than the percentage of organic carbon (Table 5.1):

1. *Bulk density and porosity*. Organic soils have lower bulk densities and higher water-holding capacities than do mineral soils. Bulk density, defined as the dry weight of soil material per unit volume, is generally 0.2 to $0.3\,\text{g/cm}^3$ when the organic soil is well decomposed, although peatland soils composed of *Sphagnum* moss can be extremely light, with bulk densities as low as $0.04\,\text{g/cm}^3$. By contrast, mineral soil bulk density generally ranges between 1.0 and $2.0\,\text{g/cm}^3$. Bulk density is low in organic soils because of their high porosity, or percentage of pore spaces. Peat soils generally have at least 80 percent pore spaces and are thus 80 percent water by volume when flooded. Mineral soils generally range from 45 to 55 percent total pore space, regardless of the amount of clay or texture.

2. *Hydraulic conductivity*. Both mineral and organic soils have wide ranges of possible hydraulic conductivities. Organic soils may hold more water than mineral soils, but, given the same hydraulic conditions, they do not necessarily allow water to pass through more rapidly. Hydraulic conductivity can be predicted for some peatland soils from their bulk density or fiber content, both of which can easily be measured (Fig. 5.2). In general, the conductivity of organic peat decreases as the fiber content decreases through the process of decomposition. Water can pass through fibric, or poorly

Table 5.1 Comparison of mineral and organic soils in wetlands

	Mineral Soil	Organic Soil
Organic content (percent)	Less than 20 to 35	Greater than 20 to 35
Organic carbon (percent)	Less than 12 to 20	Greater than 12 to 20
pH	Usually circumneutral	Acid
Bulk density	High	Low
Porosity	Low (45–55%)	High (80%)
Hydraulic conductivity	High (except for clays)	Low to high
Water holding capacity	Low	High
Nutrient availability	Generally high	Often low
Cation exchange capacity	Low, dominated by major cations	High, dominated by hydrogen ion
Typical wetland	Riparian forest, some marshes	Northern peatland

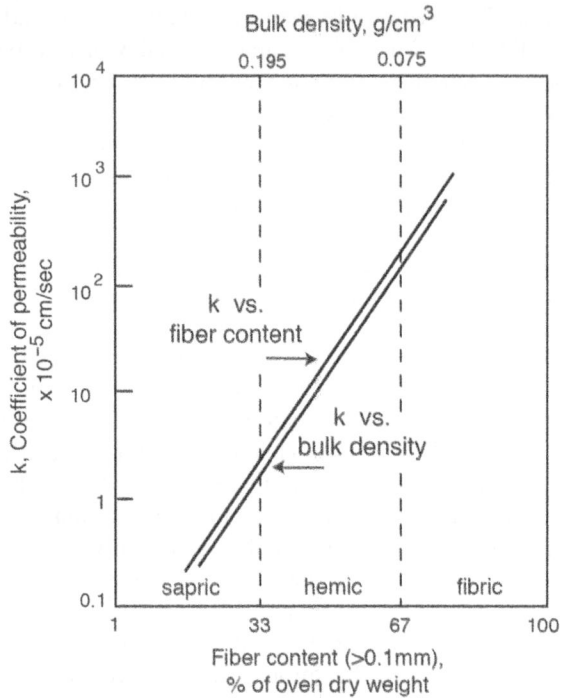

Figure 5.2 Permeability of peatland soil as a function of fiber content and bulk density. (After Verry and Boelter, 1979)

decomposed, peats 1,000 times faster than it can through more decomposed sapric peats. The type of plant material that makes up the peat is also important. Peat composed of the remains of grasses and sedges such as *Phragmites* and *Carex*, for example, is more permeable than the remains of most mosses, including sphagnum. The hydraulic conductivity of peat can vary over several orders of magnitude, showing a range almost as great as the range for mineral soil between clay ($k = 5 \times 10^{-7}$ cm/s) and sand ($k = 5 \times 10^{-2}$ cm/s) (Table 5.2). There has been some disagreement over the appropriate methods for measuring hydraulic conductivity in wetlands and about whether Darcy's law applies to flow through organic peat.

3. *Nutrient availability.* Organic soils generally have more minerals tied up in organic forms unavailable to plants than do mineral soils. This follows from the fact that a greater percentage of the soil material is organic. This does not mean, however, that there are more total nutrients in organic soils; very often, the opposite is true in wetland soils. For example, organic soils can be extremely low in bioavailable phosphorus or iron content—enough to limit plant productivity.

Table 5.2 Typical hydraulic conductivity for wetland soils compared with other soil materials

Wetland or Soil Type	Hydraulic Conductivity, k (cm/s $\times 10^{-5}$)	Reference
Northern Peatlands		
Highly humified blanket bog, UK	0.02–0.006	Ingram (1967)
Fen, Russia		
Slightly decomposed	500	Romanov (1968)
Moderately decomposed	80	
Highly decomposed	1	
Carex fen, Russia		
0–50cm deep	310	Romanov (1968)
100–150cm deep	6	
North American peatlands (general)		
Fibric	>150	Verry and Boelter (1979)
Hemic	1.2–150	
Sapric	<1.2	
Coastal Salt Marsh		
Great Sippewissett Marsh, Massachusetts (vertical conductivity)		Hemond and Fifield (1982)
0–30cm deep	1.8	
High permeability zone	2,600	
Sand–peat transition zone	9.4	
Nonpeat Wetland Soils		
Cypress dome, Florida		
Clay with minor sand	0.02–0.1	Smith (1975)
Sand	30	
Okefenokee Swamp watershed, Georgia	3.4–834	Hyatt and Brook (1984)
Mineral Soils (general)		
Clay	0.05	
Limestone	5.0	
Sand	5000	

4. *Cation exchange capacity.* Organic soils have a greater cation exchange capacity, defined as the sum of exchangeable cations (positive ions) that a soil can hold. Figure 5.3 summarizes the general relationship between organic content and cation exchange capacity of soils. Mineral soils have a cation exchange capacity that is dominated by the major metal cations (Ca^{2+}, Mg^{2+}, K^+, and Na^+). As organic content increases, both the percentage and the amount of exchangeable hydrogen ions increase. For *Sphagnum* moss peat, the high cation capacity may be caused by long-chain polymers of uronic acid (Clymo, 1983).

Organic Wetland Soil

Organic soil is composed primarily of the remains of plants in various stages of decomposition and accumulates in wetlands as a result of the anaerobic conditions

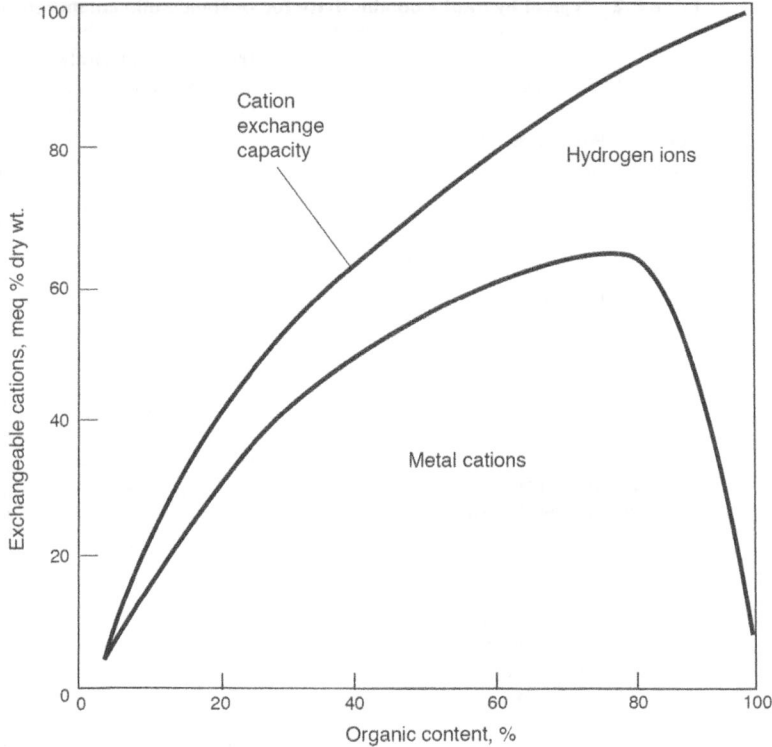

Figure 5.3 Relationship between cation exchange capacity and organic content for wetland soils. For low organic content (mineral soils), the cation exchange capacity is saturated by metal cations; when organic content is high, the exchange capacity is dominated by hydrogen ions. (After Gorham, 1967)

created by standing water or poorly drained conditions. Two of the more important characteristics of organic soil, including soils commonly termed peat and *muck*, are the botanical origin of the organic material and the degree to which it is decomposed. Several of the properties that have been discussed, including bulk density, cation exchange capacity, hydraulic conductivity, and porosity, are often dependent on these characteristics. Therefore, it is often possible to predict the range of the physical properties of an organic soil if the origin and state of decomposition can be observed in the field or laboratory.

Botanical Origin

The botanical origin of the organic material can be (1) mosses, (2) herbaceous material, and (3) wood and leaf litter. For most northern peatlands, the moss is usually *Sphagnum*, although several other moss species can dominate if the peatland is receiving inflows of mineral water. Organic soils can originate from herbaceous grasses such as reed grass (*Phragmites*), wild rice (*Zizania*), and salt marsh cordgrass (*Spartina*), or from sedges such as *Carex* and *Cladium*. Organic soils can also be produced in freshwater marshes by plant fragments from several nongrass and nonsedge plants, including cattails (*Typha*) and water lilies (*Nymphaea*). In forested wetlands, the peat can be a result of woody detritus or leaf material or both. In northern peatlands, the material can originate from birch (*Betula*), pine (*Pinus*), or tamarack (*Larix*), and in southern deepwater swamps, the organic horizon can be composed of material from cypress (*Taxodium*) or water tupelo (*Nyssa*) trees.

Decomposition

The state of decomposition, or *humification*, of wetland soils is the second key characteristic of organic peat. As decomposition proceeds, albeit at a very slow rate in flooded conditions, the original plant structure is changed physically and chemically until the resulting material little resembles the parent material. As peat decomposes, bulk density increases, hydraulic conductivity decreases, and the quantity of larger (>1.5 mm) fiber particles decreases as the material becomes increasingly fragmented. Chemically, the amount of peat "wax," or material soluble in nonpolar solvents, and lignin increase with decomposition, whereas cellulose compounds and plant pigments decrease. When some wetland plants, such as salt marsh grasses, die, the detritus rapidly loses a large percentage of its organic compounds through leaching. These readily soluble organic compounds are thought to be easily metabolized in adjacent aquatic systems.

Classification and Characteristics

Organic soils (*histosols*) are classified into four groups, the first three of which listed here are considered hydric soils:

1. *Saprists (muck)*. Two-thirds or more of the material is decomposed, and less than one-third of plant fibers are identifiable.
2. *Fibrists (peat)*. Less than one-third of material is decomposed, and more than two-thirds of plant fibers are identifiable.
3. *Hemists (mucky peat or peaty muck)*. Conditions fall between saprist and fibrist soil.
4. *Folists*. Organic soils caused by excessive moisture (precipitation > evapotranspiration) that accumulate in tropical and boreal mountains; these soils are not classified as hydric soils because saturated conditions are the exception rather than the rule.

Organic soil is generally dark in color, ranging from the dark black soils characteristic of mucks such as those found in the Everglades in Florida to the dark brown color of partially decomposed peat from northern bogs.

Mineral Wetland Soil

Mineral soils, when flooded for extended periods, develop certain characteristics that allow for their identification. These characteristics are collectively called *redoximorphic features*, defined as features formed by the reduction, translocation, and/or oxidation of iron and manganese oxides (Vepraskas, 1995).

The development of redoximorphic features in mineral soils is mediated by microbiological processes. The rate at which they are formed depends on three conditions, all of which must be present:

1. Sustained anaerobic conditions
2. Sufficient soil temperature (5°C is often considered "biological zero," below which much biological activity ceases or slows considerably; see description of biological zero and its importance to wetland science by Rabenhorst, 2005)
3. Organic matter, which serves as a substrate for microbial activity

Reduced Matrices and Redox Depletions

One characteristic of many hydric mineral soils that are semipermanently or permanently flooded is the development of black, gray, or sometimes greenish or blue-gray color as the result of a process known as *gleying*. This process, also known as *gleization*, is the result of the chemical reduction of iron (see "Iron and Manganese Transformations" in Chapter 6: "Wetland Biogeochemistry"). When soils are not saturated with water, iron (ferric = Fe^{3+}) oxides are the principal chemicals that give the soil its typical red, brown, yellow, or orange color. Manganese (Mn^{3+} or Mn^{4+}) oxides give the soil a black color. When soils are flooded and become reduced, the iron is reduced to a soluble form of iron (ferrous = Fe^{2+}) and the manganese is reduced to its soluble manganous (Mn^{2+}) form. These soluble forms of iron and manganese can be leached out of the soil, leaving the natural (gray or black) color of the parent sand, silt, or clay, called the matrix. A similar term used to describe these reduced soils is *redox depletions*—iron is reduced and then depleted from the soil matrix. In a similar manner, *clay depletions* occur when clay is selectively removed along root channels after iron and manganese oxides have been depleted, only to redeposit as clay coatings on soil particles below the clay depletions (Vepraskas, 1995).

Oxidized Rhizosphere

Another characteristic of some mineral wetland soils is the presence of an *oxidized rhizosphere* (also called *oxidized pore linings*) that results from the capacity of many hydrophytes to transport oxygen through aboveground stems and leaves to below-ground roots (Fig. 5.4). Excess oxygen, beyond the root's metabolic needs, diffuses from the roots to the surrounding soil matrix, forming deposits of oxidized iron along small roots. When a wetland soil is examined, these oxidized rhizosphere deposits can often be seen as thin traces through an otherwise dark matrix.

Redox Concentrations

Mineral soils that are seasonally flooded, particularly by alternate wetting and drying, develop spots of highly oxidized materials called *mottles* or *redox concentrations* (Fig. 5.5). Mottles and redox concentrations are orange/reddish-brown (because of

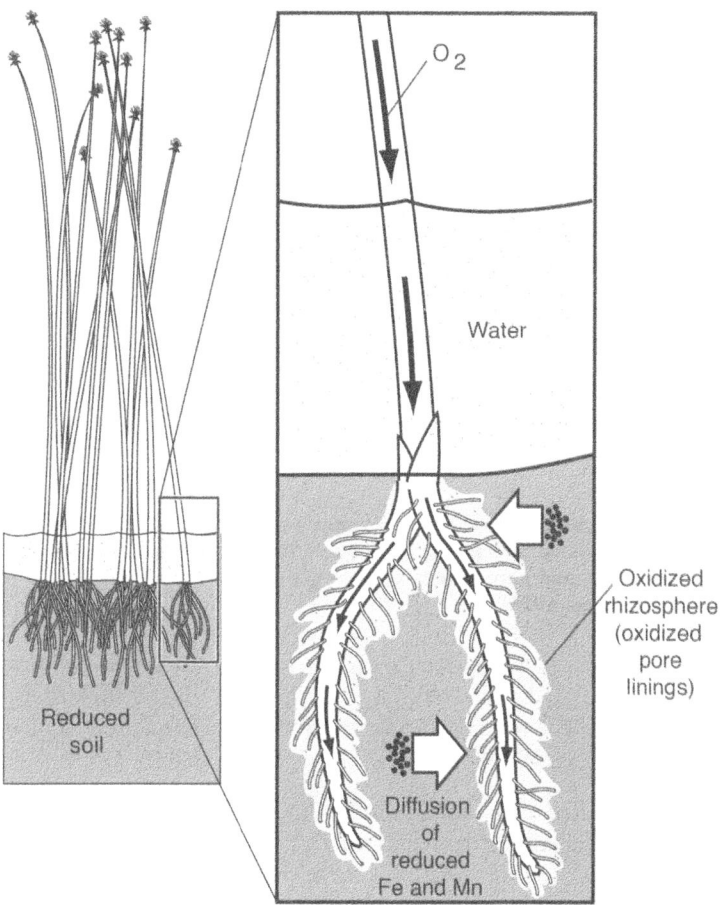

Figure 5.4 Formation of oxidized rhizospheres, or *pore linings*, around the roots of a wetland plant caused by the transport of excess oxygen by wetland plants to their roots. When the plant dies, pore linings of iron and manganese oxides often remain in the soil. (After Vepraskas, 1995)

iron) or dark reddish-brown/black (because of manganese) spots seen throughout an otherwise gray (gleyed) soil matrix and suggest intermittently exposed soils with spots of iron and manganese oxides in an otherwise reduced environment. Mottles are relatively insoluble, enabling them to remain in soil long after it has been drained.

Modern Nomenclature

A revised set of terms defining redoximorphic features has been devised by soil scientists to describe indicators of hydric soils, or more properly, to identify an *aquic condition*—the condition in which soils are saturated with water, are reduced, and display redoximorphic features. The term *aquic condition* was introduced in the early 1990s to better reconcile field techniques that used soil colors (e.g., iron

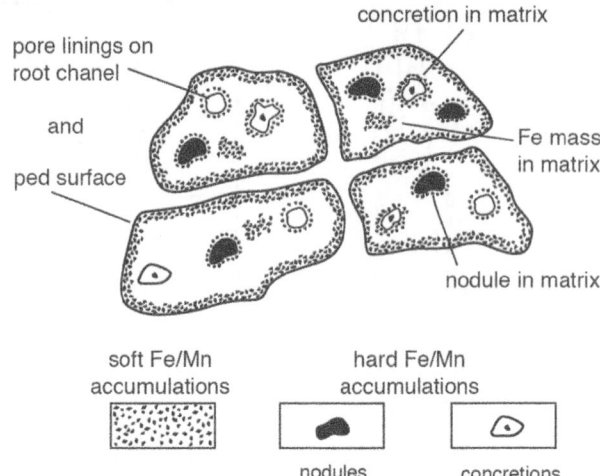

Figure 5.5 Different kinds of *redox concentrations*, or *mottles*, in soil peds (soil macroparticles), including *nodules* and *concretions*, *iron masses* in soil matrix (also called reddish mottles), and *pore linings* on root channel (also called *oxidized rhizospheres*). (After Vepraskas, 1995)

reduction or oxidation) with the former term *aquic moisture regime*—any soil that was saturated with water and chemically reduced such that no dissolved oxygen was present. The redoximorphic features that can be used to identify aquic conditions are (Vepraskas, 1995):

1. *Redox concentrations.* Accumulation of iron and manganese oxides (formerly called mottles) in at least three different structures (Fig. 5.5):
   a. *Nodules and concretions.* Firm to extremely firm irregularly shaped bodies with diffuse boundaries
   b. *Masses.* Formerly called reddish mottles
   c. *Pore linings.* Formerly included oxidized rhizospheres (Figs. 5.4 and 5.5)
2. *Redox depletions.* Low-chroma (≤ 2) bodies with high values (≥ 4) including:
   a. *Iron depletions.* Sometimes called gray mottles or gley mottles; these are low-chroma bodies
   b. *Clay depletions.* Contain less iron, manganese, and clay than adjacent soils
3. *Reduced matrices.* Low-chroma soils (because of presence of Fe^{2+}) *in situ* that change color if exposed to air and iron is oxidized to Fe^{3+}

> **Mineral Hydric Soil Determination**
>
> In practice, the determination of whether a mineral soil is a hydric soil is a complicated process, but it is often done by determining soil color relative to a standard color chart called the Munsell® Soil Color Chart (Fig. 5.6a). Soils

that contain *low chromas* (as indicated by the color chips on the left-hand side of the color chart in Fig. 5.6b) indicate hydric soils. Soils that contain bright reds, browns, yellows, or oranges are nonhydric. In general, a chroma of 2 or less on the Munsell color chart is necessary for a soil to be classified as a hydric soil. These color charts are commonly used in the United States to identify the presence of hydric soils for the delineation of wetlands.

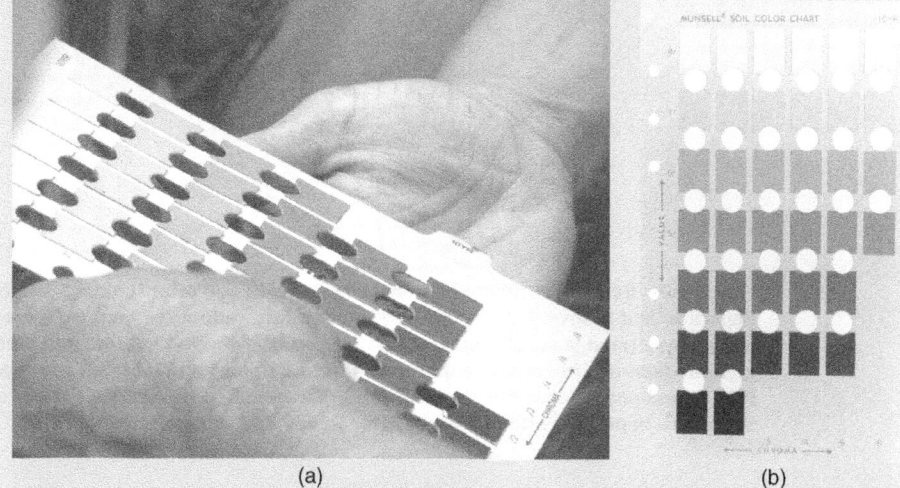

Figure 5.6 (a) Hydric soils can be identified by comparing the soil color with standard soil color charts such as the Munsell Soil Color Chart shown here. (b) A representative Munsell Soil Color Chart (10YR in this case): The hue, given in the upper right-hand corner of the chart, indicates the relation to standard spectral colors, in this case yellow (Y) and red (R). The value notation (vertical scale) indicates the soil lightness (darker with lower value), and the chroma (horizontal scale) indicates the color strength or purity, with grayer soils to the left. Chromas of 2 or less generally indicate hydric soils.

Reduction/Oxidation in Wetland Soil

All soils contain air and water in a mineral/organic matrix. When soils, whether mineral or organic, are inundated with water, anaerobic conditions usually result as water fills the air spaces or soil pores. When water fills the pore spaces, the rate at which oxygen can diffuse through the soil is drastically reduced. Diffusion of oxygen in an aqueous solution has been estimated at 10,000 times slower than oxygen diffusion through a porous medium such as drained soil. This low diffusion rate leads relatively quickly to anaerobic, or reduced, conditions, with the time required for oxygen depletion on the order of several hours to a few days after inundation begins (Fig. 5.7). The rate at which the oxygen is depleted depends on the ambient temperature, the

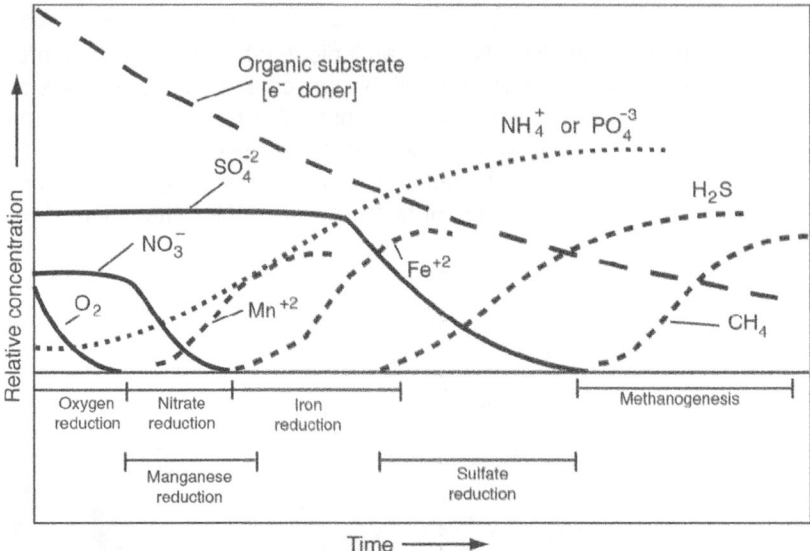

Figure 5.7 Sequence in time of transformations in soil after flooding, beginning with oxygen depletion and followed by nitrate and then sulfate reduction. Increases are seen in reduced manganese (manganous), reduced iron (ferrous), hydrogen sulfide, and methane. Note the gradual decrease in organic substrate (electron donor) and increases in available ammonium (NH_4^+) and phosphate (PO_4^{3-}) ions. The graph can also be interpreted as relative concentrations with depth in wetland soils. (After Reddy and DeLaune, 2008)

availability of organic substrates for microbial respiration, and sometimes the chemical oxygen demand from reductants such as ferrous iron. The resulting lack of oxygen prevents plants from carrying out normal aerobic root respiration and strongly affects the availability of plant nutrients and toxic materials in the soil. As a result, plants that grow in anaerobic soils generally have some specific adaptations to this environment (see Chapter 7).

It is not always true that oxygen is totally depleted from the soil water of wetlands. There is usually a thin layer of oxidized soil, sometimes only a few millimeters thick, at the surface of the soil at the soil–water interface (Fig. 5.8). The thickness of this oxidized layer is directly related to four things:

1. The rate of oxygen transport across the atmosphere–surface water interface
2. The small population of oxygen-consuming organisms present
3. Photosynthetic oxygen production by algae within the water column
4. Surface mixing by convection currents and wind action

Even though the deeper layers of the wetland soils remain reduced, this thin oxidized layer is often very important in the chemical transformations and nutrient cycling that occur in wetlands. Oxidized ions such as Fe^{3+}, Mn^{4+}, NO_3^-, and $SO_4^=$ are found

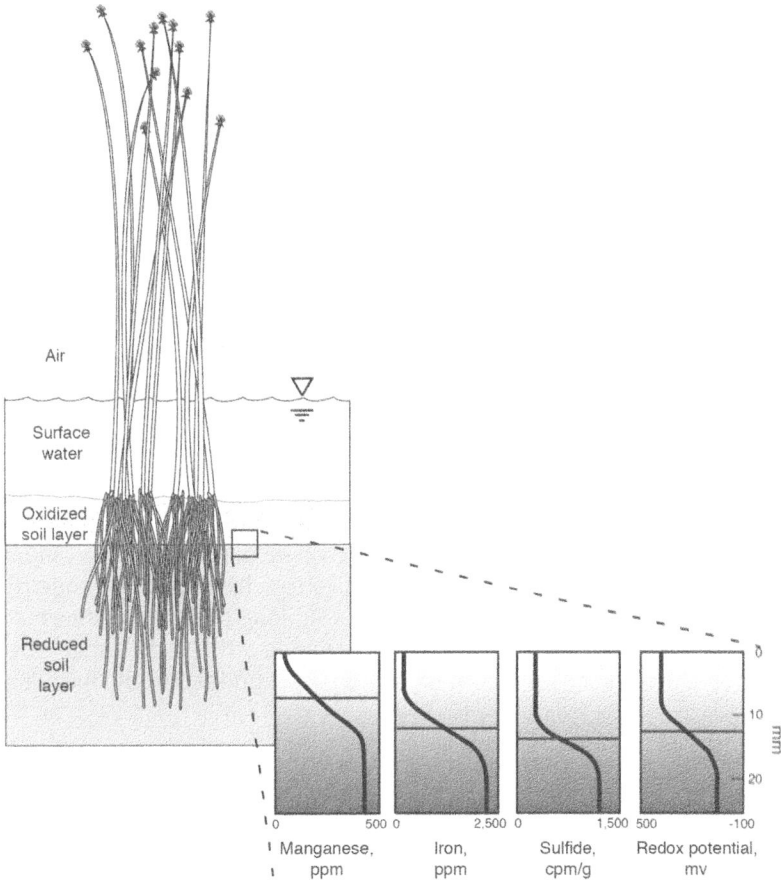

Figure 5.8 Characteristics of many wetland soils showing a shallow oxidized soil layer over a reduced soil layer. Also shown are soil profiles of reduced forms of manganese (sodium acetate–extractable manganese), iron (ferrous iron), and sulfur (sulfide), and redox potential. (After Patrick and Delaune, 1972)

in this microlayer, whereas the lower anaerobic soils are dominated by reduced forms, such as ferrous and manganous salts, ammonia, and sulfides. Because of the presence of oxidized ferric iron (Fe^{3+}) in the oxidized layer, the soil surface often is a brown or brownish-red color, in contrast to the bluish-gray to greenish-gray color of the reduced gleyed sediments, dominated by ferrous iron (Fe^{2+}).

Redox potential, or oxidation–reduction potential, a measure of the electron pressure (or availability) in a solution, is often used to further quantify the degree of electrochemical reduction of wetland soils. *Oxidation* occurs not only during the uptake of oxygen but also when hydrogen is removed (e.g., $H_2S \rightarrow S^{2-} + 2H^+$) or, more generally, when a chemical gives up an electron (e.g., $Fe^{2+} \rightarrow Fe^{3+} + e^-$). *Reduction* is the opposite process of releasing oxygen, gaining hydrogen (hydrogenation), or gaining an electron.

Measuring Redox Potential

Redox potential can be measured in wetland soils and is a quantitative measure of the tendency of the soil to oxidize or reduce substances. When based on a hydrogen scale, redox potential is referred to as E_H and is related to the concentrations of oxidants (*ox*) and reductants (*red*) in a redox reaction by the *Nernst equation*:

$$E_H = E^0 + 2.3[RT/nF]\log[ox]/\{red\} \qquad (5.2)$$

where

E^0 = potential of reference (mV)
R = gas constant = 81.987 cal deg^{-1} mol^{-1}
T = temperature (°K)
n = number of moles of electrons transferred
F = Faraday constant = 23,061 cal/mole-volt

Redox potential can be measured with a platinum electrode (Fig. 5.9a, b), which is easily constructed in the laboratory. Electric potential in units of millivolts (mV) is measured relative to a hydrogen electrode ($H^+ + e \longrightarrow H$) or to a calomel reference electrode. As long as free dissolved oxygen is present in a solution, the redox potential varies little (in the range of +400 to +700 mV). However, it becomes a sensitive measure of the degree of reduction

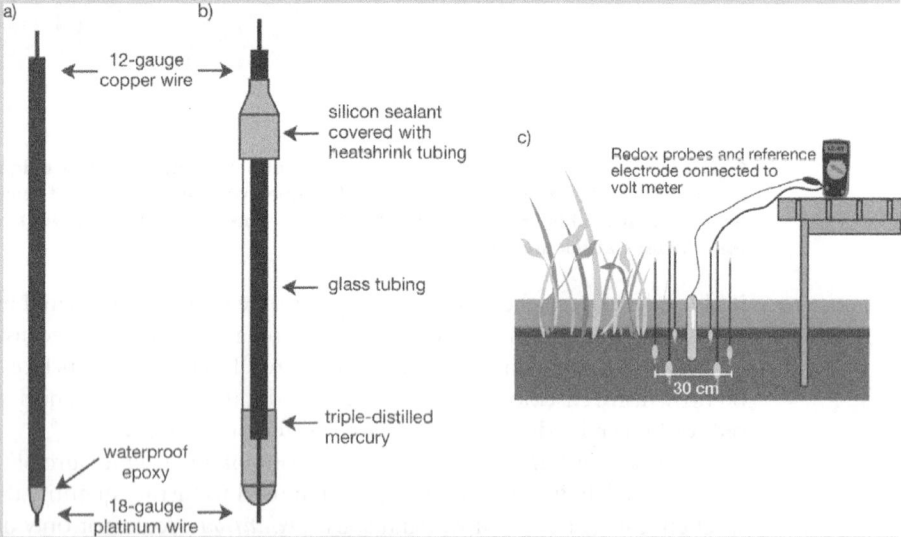

Figure 5.9 Design of (a, b) constructed redox and (c) possible deployment of multiple redox probes and reference electrode in a wetland for redox potential measurements. ((a), (b): After Faulkner et al., 1989)

of wetland soils after oxygen disappears, ranging from +400 mV down to −400mV. Flooding and/or redox conditions in pond and wetland soils can be estimated by constructing platinum electrodes, using microplatinum electrodes or steel rods. Normally a small tip of platinum (Pt) is connected to a copper wire, which is then put in the soil, with the reference electrode also put in the soil at a set distance (Fig. 5.9c). The potential between the Pt tip and the reference electrode can be measured after the system has stabilized, which takes sometimes up to two days.

As organic substrates in a waterlogged soil are oxidized (donate electrons), the redox potential drops as a sequence of reductions (electron gains) takes place. Because organic matter is one of the most reduced of substances, it can be oxidized when any number of terminal electron acceptors is available, including O_2, NO_3^-, Mn^{4+}, Fe^{3+}, or $SO_4^=$. Rates of organic decomposition are most rapid in the presence of oxygen and slower for electron acceptors such as nitrates and sulfates.

The oxidation of organic substrate is described by Equation 5.3, which illustrates the organic substrate as an electron (e^-) donor:

$$[CH_2O]n + nH_2O \rightarrow nCO_2 + 4ne^- + 4nH^+ \quad (5.3)$$

Various chemical and biological transformations take place as coupled oxidation (e^- donor)–reduction (e^- acceptor) reactions. Equations 5.3 and 5.4 make one such coupled reaction.

$$O_2 + 4e^- + 4H^+ \rightarrow 2H_2O \quad (5.4)$$

These transformations occur in a predictable sequence (Fig. 5.7), within predictable redox ranges to provide electron acceptors for this oxidation or decomposition (Table 5.3).

The first and most common transformation is through aerobic oxidation when oxygen itself is the terminal electron acceptor (Eq. 5.4) at a redox potential of between 400 and 600 mV.

Table 5.3 Oxidized and reduced forms of several elements and approximate redox potentials for transformation

Element	Oxidized Form	Reduced Form	Approximate Redox Potential for Transformation (mV)
Nitrogen	NO_3^- (nitrate)	N_2O, N_2, NH_4^+	250
Manganese	Mn^{4+} (manganic)	Mn^{2+} (manganous)	225
Iron	Fe^{3+} (ferric)	Fe^{2+} (ferrous)	+100 to −100
Sulfur	$SO_4^=$ (sulfate)	$S^=$ (sulfide)	−100 to −200
Carbon	CO_2 (carbon dioxide)	CH_4 (methane)	Below −200

One of the first reactions that occur in wetland soils after they become anaerobic (i.e., the dissolved oxygen is depleted) is the reduction of NO_3^- (nitrate) first to NO_2^- (nitrite) and ultimately to N_2O (nitrous oxide) or N_2 (nitrogen gas); nitrate becomes an electron acceptor at a redox potential of approximately 250 mV:

$$2NO_3 + 10e^- + 12H^+ \rightarrow N_2 + 6H_2O \tag{5.5}$$

As the redox potential continues to decrease, manganese is transformed from manganic to manganous compounds at about 225 mV:

$$MnO_2 + 2e^- + 4H^+ \rightarrow Mn^{2+} + 2H_2O \tag{5.6}$$

Iron is transformed from ferric to ferrous form at about +100 to –100, while sulfates are reduced to sulfides at –100 to –200 mV:

$$Fe(OH)_3 + e^- + 3H^+ \rightarrow F^{2+} + 3H_2O \tag{5.7}$$

$$SO_4^= + 8e^- + 9H^+ \rightarrow HS^- + 4H_2O \tag{5.8}$$

Finally, under the most reduced conditions, the organic matter itself (or carbon dioxide) becomes the terminal electron acceptor below –200 mV, producing low-molecular-weight organic compounds and methane gas, as, for example,

$$CO_2 + 8e^- + 8H^+ \rightarrow CH_4 + 2H_2O \tag{5.9}$$

These redox potentials are not precise thresholds, because pH and temperature are also important factors in the rates of transformation. These major chemical transformations and others related to the nitrogen, sulfur, and carbon cycles are discussed in the next chapter.

Recommended Readings

Baskin, Y. 2005. *Under Ground: How Creatures of Mud and Dirt Shape Our World.* Washington, DC: Island Press.

Richardson, J. L., and M. J. Vepraskas. 2001. *Wetland Soils: Genesis, Hydrology, Landscapes, and Classification.* Boca Raton, FL: CRC Press.

Natural Resources Conservation Service (NRCS). 2010. *Field Indicators of Hydric Soils in the United States,* Version 7.0. L. M. Vasilas, G. W. Gurt, C. V. Noble, eds. USDA, NRCS in cooperation with the National Technical Committee for Hydric Soils, 45 pp.

References

Clymo, R. S. 1983. Peat. In A. J. P. Gore, ed., *Ecosystems of the World, Vol. 4A: Mires: Swamp, Bog, Fen, and Moor.* Elsevier, Amsterdam, The Netherlands, pp. 159–224.

Faulkner, S. P., W. H. Patrick, Jr., and R. P. Gambrell. 1989. Field techniques for measuring wetland soil parameters. *Soil Science Society of America Journal* 53: 883–890.

Gorham, E. 1967. *Some Chemical Aspects of Wetland Ecology.* Technical Memorandum 90, Committee on Geotechnical Research, National Research Council of Canada, pp. 2–38.

Hemond, H. F., and J. L. Fifield. 1982. Subsurface flow in salt marsh peat: A model and field study. *Limnology and Oceanography* 27: 126–136.

Hyatt, R. A., and G. A. Brook. 1984. Groundwater flow in the Okefenokee Swamp and hydrologic and nutrient budgets for the period August, 1981 through July, 1982. In A. D. Cohen, D. J. Casagrande, M. J. Andrejko, and G. R. Best, eds., *The Okefenokee Swamp: Its Natural History, Geology, and Geochemistry.* Wetland Surveys, Los Alamos, NM, pp. 229–245.

Ingram, H. A. P. 1967. Problems of hydrology and plant distribution in mires. *Journal of Ecology* 55: 711–724.

Natural Resources Conservation Service (NRCS). 2010. *Field Indicators of Hydric Soils in the United States*, Version 7.0. L. M. Vasilas, G. W. Gurt, C. V. Noble, eds. USDA, NRCS in cooperation with the National Technical Committee for Hydric Soils, 45 pp.

Patrick, W. H., Jr., and R. D. Delaune. 1972. Characterization of the oxidized and reduced zones in flooded soil. *Proceedings of the Soil Science Society of America* 36: 573–576.

Rabenhorst, M. C. 2005. Biological zero: A soil temperature concept. *Wetlands* 25: 616–621.

Reddy, K. R. and R. D. DeLaune. 2008. *Biogeochemistry of Wetlands.* CRC Press, Boca Raton, FL, 774 pp.

Romanov, V. V. 1968. *Hydrophysics of Bogs.* Translated from Russian by N. Kaner; edited by Prof. Heimann. Israel Program for Scientific Translation, Jerusalem. Available from Clearinghouse for Federal Scientific and Technical Information, Springfield, VA. 299 pp.

Smith, R. C. 1975. Hydrogeology of the experimental cypress swamps. In H. T. Odum and K. C. Ewel, eds., *Cypress Wetlands for Water Management, Recycling and Conservation.* Second Annual Report to NSF and Rockefeller Foundation, Center for Wetlands, University of Florida, Gainesville, pp. 114–138.

Vepraskas, M. J. 1995. *Redoximorphic Features for Identifying Aquic Conditions.* Technical Bulletin 301, North Carolina Agricultural Research Service, North Carolina State University, Raleigh. 33 pp.

Verry, E. S., and D. H. Boelter. 1979. Peatland hydrology. In P. E. Greeson, J. R. Clark, and J. E. Clark, eds., *Wetland Functions and Values: The State of Our Understanding.* American Water Resources Association, Minneapolis, MN, pp. 389–402.

Chapter 6

Wetland Biogeochemistry

Wetland biogeochemistry features a combination of many chemical transformations and chemical transport processes. Many transformations of nitrogen, sulfur, iron, manganese, carbon, phosphorus and other chemicals occur in wetlands as a result of the combination of both aerobic and anaerobic conditions in proximity. Wetlands can be sources, sinks, or transformers of nutrients but are valued especially for their ability to be sinks for nutrients. Some transformations cause toxic conditions, as with the production of hydrogen sulfide, whereas others, such as sedimentation, denitrification, and carbon sequestration, improve water quality and improve the planet's carbon balance. Still other processes from wetlands allow emissions of greenhouse gases to the atmosphere. Many transformations in wetlands, especially in the nitrogen, sulfur, and carbon cycles, are mediated by microbial populations that are adapted to the anaerobic environment, while many other processes, such as those in the phosphorus cycle, are chemical and physical. Wetlands are often coupled to adjacent ecosystems such as by exporting vital organic carbon to downstream aquatic ecosystems.

The transport and transformation of chemicals in ecosystems, known as *biogeochemical cycling*, involve a great number of interrelated physical, chemical, and biological processes. The diverse hydrologic conditions in wetlands discussed in Chapter 4 and the soil types described in Chapter 5 both markedly influence biogeochemical processes. These processes result not only in changes in the chemical forms of materials but also in the spatial movement of materials within wetlands, as in water–sediment exchange and plant uptake, and with surrounding ecosystems, as in organic exports. These processes, in turn, determine overall wetland productivity. The interrelationships among hydrology, the physiochemical environment, and wetland biota were already summarized in Figure 4.1.

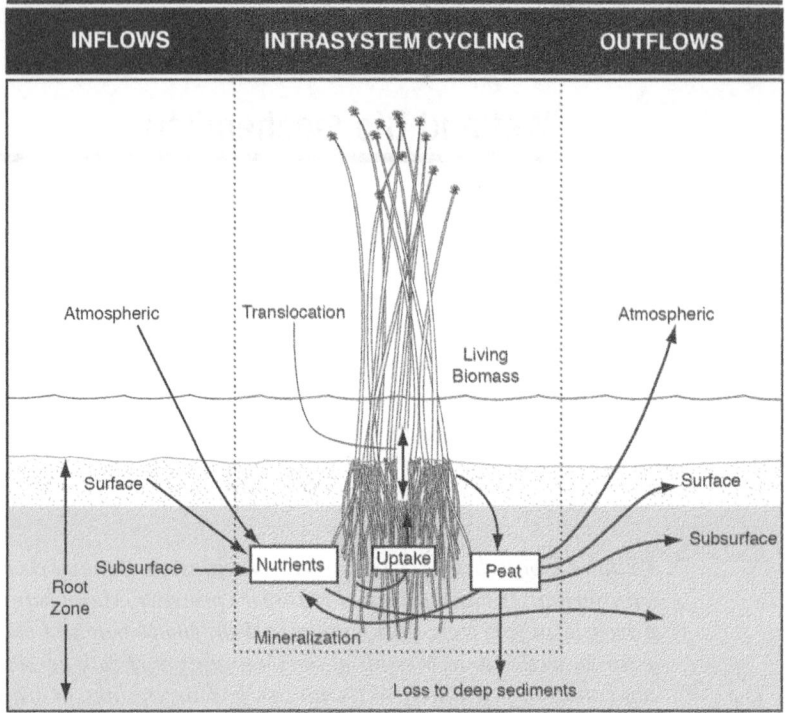

Figure 6.1 Components of a wetland nutrient budget, including inflows, outflows, and intrasystem cycling.

The biogeochemistry of wetlands can be divided into (1) *intrasystem cycling* through various transformation processes and (2) the exchange of chemicals between a wetland and surrounding waters, landscapes, and atmosphere (Figs. 6.1 and 6.2). Although no transformation processes are unique to wetlands, the permanent to intermittent flooding of these ecosystems causes certain processes to be more dominant in wetlands than in either upland or deep aquatic ecosystems. For example, while anaerobic, or oxygen-less, conditions are sometimes found in other ecosystems, they prevail in wetlands. Wetland soils are characterized by waterlogged conditions during part or all of the year, which produce reduced conditions, which, in turn, have a marked influence on several biochemical transformations unique to anaerobic conditions.

This intrasystem cycling, along with hydrologic conditions, influences the degree to which chemicals are transported to or from wetlands. An ecosystem is considered biogeochemically *open* when there is an abundant exchange of materials with its surroundings. When there is little movement of materials across the ecosystem boundary, it is biogeochemically *closed*. Wetlands can fall into either category. For example, wetlands such as bottomland forests and tidal salt marshes have a significant exchange of minerals with their surroundings through river flooding and tidal exchange, respectively. Other wetlands such as ombrotrophic bogs and cypress domes

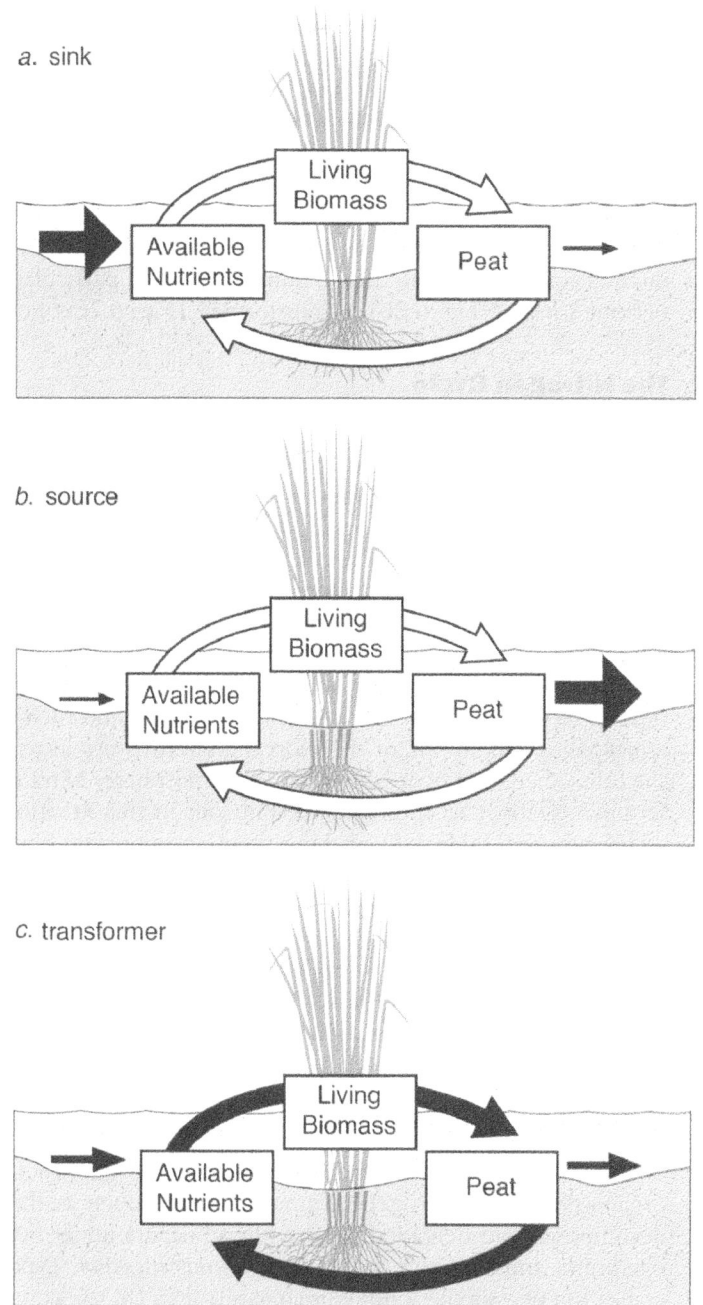

Figure 6.2 A wetland can serve as an (a) inorganic nutrient sink, (b) source of total nutrients, and (c) transformer of inorganic nutrients to organic nutrients.

have little material exchange except for precipitation and gases that pass into or out of the ecosystem. These latter systems depend more on intrasystem cycling than on throughput for their chemical supplies.

Wetlands serve as *sources, sinks,* or *transformers* of chemicals or nutrients, depending on the wetland type, the hydrologic conditions, and the length of time the wetland has been subjected to chemical loadings (Fig. 6.2). When wetlands serve as sinks for certain chemicals (Fig. 6.2a), the long-term sustainability of that situation depends on the hydrologic and geomorphic conditions, the spatial and temporal distribution of chemicals in the wetland, and the ecosystem succession. Wetlands can become saturated in certain chemicals after a number of years, particularly if inflows are high, and become sources (Fig. 6.2b) or transformers (Fig. 6.2c) of chemicals.

The Nitrogen Cycle

The nitrogen cycle (Fig. 6.3) is one of the most important and studied chemical cycles in wetlands. Nitrogen appears in a number of oxidation states in wetlands, several of which are important in a wetland's biogeochemistry. Nitrogen is often the most limiting nutrient in flooded soils, whether the flooded soils are in natural wetlands or on agricultural wetlands, such as rice paddies. Nitrogen is considered one of the major limiting factors in coastal waters, making the nitrogen dynamics in coastal wetlands particularly significant, although this universal belief in nitrogen limitation in coastal wetlands has been challenged (e.g., Sundareshwar et al., 2005). Because of the presence of anoxic conditions in wetlands, microbial denitrification of nitrates to gaseous forms of nitrogen in wetlands and their subsequent release to the atmosphere remain one of the more significant ways in which nitrogen is lost from the lithosphere and hydrosphere to the atmosphere. Nitrates serve as one of the first terminal electron acceptors in wetland soils in this situation after the disappearance of oxygen (see Table 5.3), making them an important chemical in the oxidation of organic matter in wetlands.

Nitrogen transformations in wetlands (Fig. 6.3) involve several microbiological processes, some of which make the nutrient less available for plant uptake. The ammonium ion (NH_4^+), with a nitrogen oxidation state of –3, is the primary form of mineralized nitrogen in most flooded wetland soils, although much nitrogen can be tied up in organic forms in highly organic soils. The presence of an oxidized zone over the anaerobic or reduced zone is critical for several of the pathways.

Nitrogen Mineralization

Nitrogen mineralization refers to a series of biological transformations that converts organically bound nitrogen to ammonium nitrogen as the organic matter is being decomposed and degraded. This pathway occurs under both anaerobic and aerobic conditions and is often referred to as *ammonification*. Typical formulas for the mineralization of a simple soluble organic nitrogen (SON) compound, urea, are given as

$$NH_2CONH_2 + H_2O \rightarrow 2NH_3 + CO_2 \tag{6.1}$$

$$NH_3 + H_2O \rightarrow NH_4 + OH^- \tag{6.2}$$

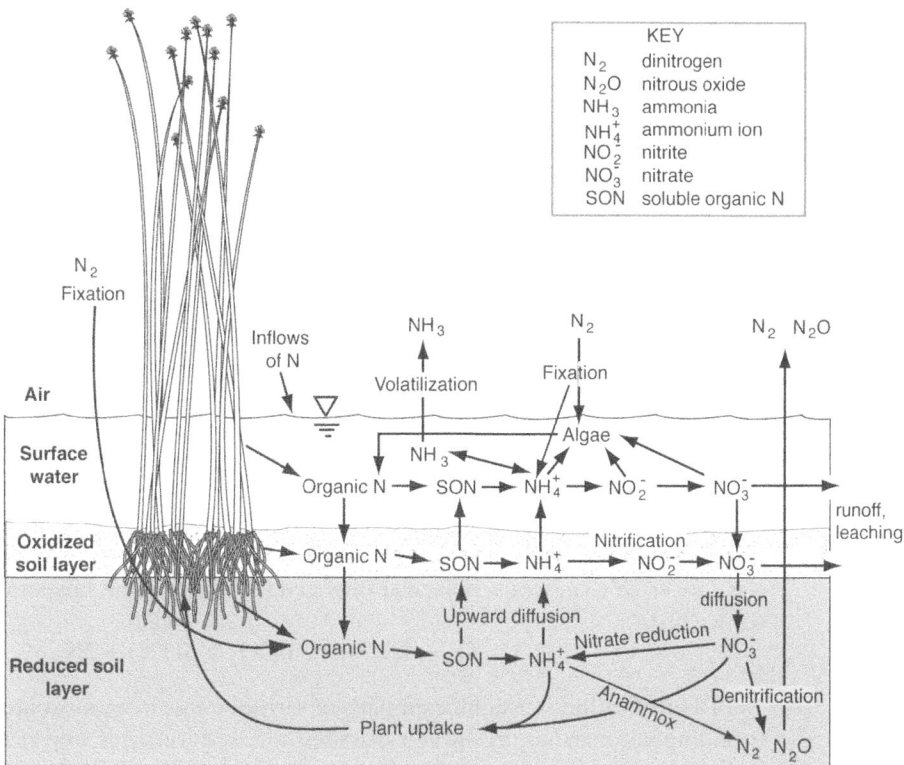

Figure 6.3 The nitrogen cycle in wetlands. Major pathways illustrated are nitrogen fixation, ammonia volatilization, nitrification, denitrification, plant uptake, dissimilatory nitrate reduction to ammonia (DNRA), and anammox (anaerobic ammonium oxidation)

Ammonia Transformations and Nitrification

Once the ammonium ion (NH_4^+) is formed, it can take several possible pathways. It can be absorbed by plants through their root systems or by anaerobic microorganisms and converted back to organic matter. Under high-pH conditions (pH >8), a common occurrence in marsh waters with excessive algal blooms, the ammonium ion can be converted to NH_3, which is then released to the atmosphere through *volatilization*. The ammonium ion can also be immobilized through ion exchange onto negatively charged soil particles. Because of the anaerobic conditions in wetland soils, ammonium would normally be restricted from further oxidation and would build up to excessive levels were it not for the thin oxidized layer at the surface of many wetland soils. The gradient between high concentrations of ammonium in the reduced soils and low concentrations in the oxidized layer causes an upward diffusion of ammonium, albeit very slowly, to the oxidized layer. In this aerobic environment, ammonium nitrogen can be oxidized through the process of *nitrification* in two steps by *Nitrosomonas* sp.:

$$2NH_4^+ + 3O_2 \rightarrow 2NO_2^- + 2H_2O + 4H^+ + energy \qquad (6.3)$$

and by *Nitrobacter* sp.:

$$2NO_2^- + O_2 \rightarrow 2NO_3^- + energy \qquad (6.4)$$

Nitrification can also occur in the oxidized rhizosphere of plants, where adequate oxygen is often available to convert the ammonium nitrogen to nitrate nitrogen.

Nitrate Transformations and Denitrification

Nitrate (NO_3), as a negative ion rather than the positive ammonium ion, is not subject to immobilization by negatively charged soil particles and is thus much more mobile in solution. If it is not assimilated immediately by plants or microbes (*assimilatory nitrate reduction*) or is lost through groundwater flow stemming from its rapid mobility, it has the potential to undergo *dissimilatory nitrogenous oxide reduction*, a term that refers to several pathways of nitrate reduction. It is called dissimilatory because the nitrogen is not assimilated into a biological cell. The most prevalent are reduction to ammonia and *denitrification*.

Denitrification, carried out by facultative bacteria under anaerobic conditions, with nitrate acting as a terminal electron acceptor, results in the loss of nitrogen as it is converted to gaseous molecular nitrogen (N_2) with some small fraction to nitrous oxide (N_2O):

$$C_6H_{12}O_6 + 4NO_3^- \rightarrow 6CO_2 + 6H_2O + 2N_2 \qquad (6.5)$$

Denitrification is a significant path of nitrogen loss from most kinds of wetlands, including salt marshes, freshwater marshes, forested wetlands, and rice paddies. Denitrification is inhibited in acid soils and peat and is therefore thought to be of less consequence in northern peatlands. As illustrated in Figure 6.3, the entire process occurs after (1) ammonium nitrogen diffuses to the aerobic soil layer, (2) nitrification occurs, (3) nitrate nitrogen diffuses back to the anaerobic layer, and (4) denitrification, as described in Equation 6.5, occurs. The diffusion rates of the ammonium ion to the aerobic soil layer and the nitrate ion to the anaerobic layer are governed by the concentration gradients of the ions. There is generally a steep gradient of ammonium between the anaerobic and aerobic layers. Nevertheless, because nitrate diffusion rates in wetland soils are seven times faster than ammonium diffusion rates, ammonium diffusion and subsequent nitrification appear to limit the entire process of nitrogen loss by denitrification.

Given an adequate supply of nitrate-nitrogen, the next most significant factor that affects denitrification appears to be temperature. This pattern is clearly shown in a summary of denitrification measurements in created riverine wetlands in Ohio from 2004 through 2009 (Fig. 6.4). The highest nitrate-nitrogen concentrations in the wetlands are January through June (winter and spring), and the highest water temperatures are in July through September (summer). Denitrification peaks in June, then, after lower numbers in the early summer (due to low concentrations of nitrate nitrogen), it peaks again in September, when temperatures remain high despite continued low levels of nitrate-nitrogen. The strongest regression for denitrification for the multiyear comparison was with water temperature (Fig. 6.5).

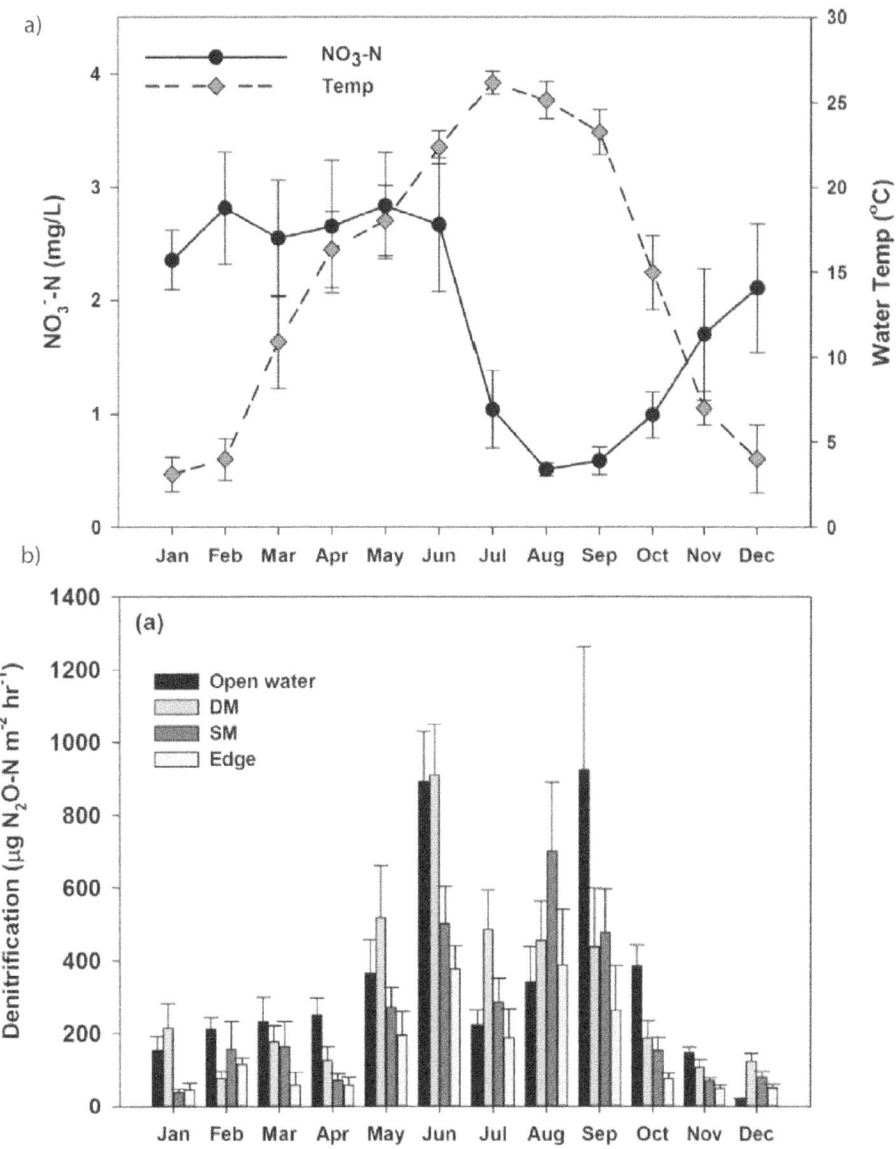

Figure 6.4 Seasonal patterns of denitrification and associated environmental variables for two 1-ha created riverine wetlands in Ohio, summarized for studies in 2004, 2005, 2008, and 2009. (a) Monthly averaged nitrate-nitrogen concentrations and water temperature of inflowing water to the wetlands; (b) average ± standard error of monthly denitrification measurements for open water, deepwater marsh (DM), shallow marsh (SM), and edge of wetland. (From Song et al., 2014; reprinted with permission, Elsevier.)

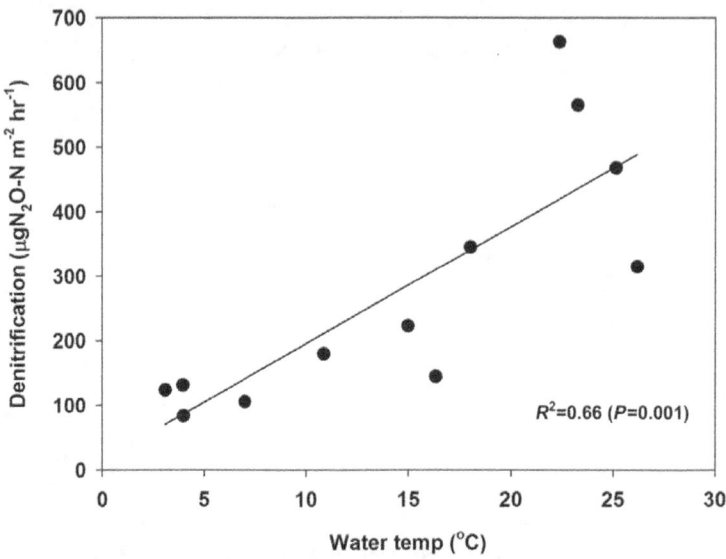

Figure 6.5 Relationship between denitrification and water temperature for the wetland described in Figure 6.4. (From Song et al. 2014); reprinted with permission, Elsevier

There are two gaseous products of denitrification—dinitrogen (N_2) and nitrous oxide (N_2O). The predominant gas that usually results from denitrification in most wetlands is N_2, and that is no environmental issue with an atmosphere already having 80 percent N_2. However, nitrous oxide is one of the so-called greenhouse gases that could cause climate change, so any attempt to design wetlands for nitrate removal should recognize this and understand conditions that minimize nitrous oxide production in favor of dinitrogen production. Hernandez and Mitsch (2006, 2007) found lower nitrous oxide fluxes in the spring in pulse-flooded conditions in the higher marshes compared to steady flow conditions in the same Ohio riverine wetlands the next year (Fig. 6.6). The rate is probably limited by the lack of nitrates in the permanently flooded soils. Nitrous oxide production was highest when soil temperatures were greater than 20°C in the summer months (Fig. 6.6). In addition, wetland plants appeared to increase nitrous oxide emissions when sites were flooded but not when soils were exposed. Overall, the amount of nitrogen emitted as nitrous oxide in these riverine wetlands as a percentage of the total nitrogen released via denitrification was quite small. This study suggests that nitrous oxide emissions and nitrous oxide/dinitrogen gas ratios (N_2O/N_2) in denitrification are higher on the aerobic/anaerobic edges of wetlands than in the more anaerobic middle. It is reasonable to conclude then that if nitrate-nitrogen is denitrified in more aerobic farm fields, ditches, streams and rivers, and even downstream coastal waters rather than in wetlands, higher nitrous oxide emissions would result from those systems than from the wetlands (Hernandez and Mitsch, 2006, 2007). Thus wetlands may not be the

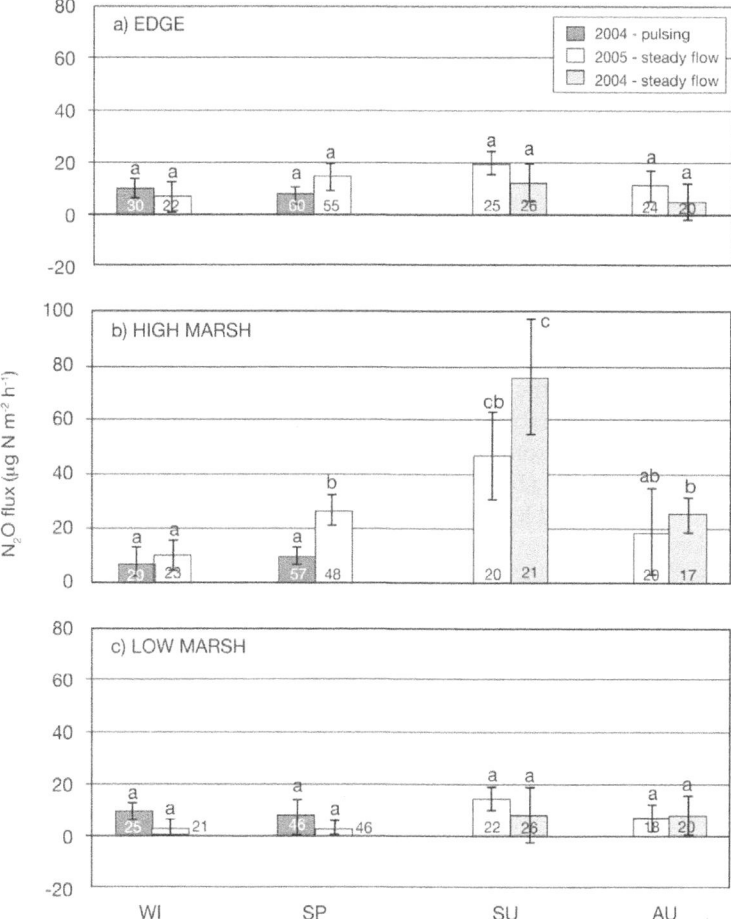

Figure 6.6 Seasonal nitrous oxide fluxes under different hydrologic conditions and along (a) dry edge, (b) high marsh (saturated soils with some standing water), and (c) low marsh (continuous standing water) in the freshwater experimental marshes in central Ohio. Numbers indicate number of flux measurements. (From Hernandez and Mitsch, 2006)

cause of additional nitrous oxide emissions; created and restored wetlands actually may decrease the overall nitrous oxide emissions on a landscape scale.

Nitrogen Fixation

Nitrogen fixation results in the conversion of N_2 gas to organic nitrogen through the activity of certain organisms in the presence of the enzyme nitrogenase. It may be the source of significant nitrogen for some wetlands. Nitrogen fixation, which is carried out by certain aerobic and anaerobic bacteria and blue-green algae, is favored in low oxygen conditions because nitrogenase activity is inhibited by high oxygen.

In wetlands, nitrogen fixation can occur in overlying waters, in the aerobic soil layer, in the anaerobic soil layer, in the oxidized rhizosphere of the plants, and on the leaf and stem surfaces of plants. Bacterial nitrogen fixation can be carried out by nonsymbiotic bacteria, by symbiotic bacteria of the genus *Rhizobium*, or by certain actinomycetes. Bacterial fixation is the most significant pathway for nitrogen fixation in salt marsh soils, while nitrogen-fixing bacteria are virtually absent from the low-pH peat of northern bogs. Cyanobacteria (blue-green algae) are common nitrogen fixers in wetlands, occurring in flooded delta soils in Louisiana, in northern bogs, and in rice cultures.

Dissimilatory Nitrate Reduction to Ammonia

Because conversion of nitrate-nitrogen to dinitrogen and nitrous oxide is considered to be the primary transformation of nitrates in anaerobic soils, an additional process whereby nitrate-nitrogen is transformed in anaerobic conditions is often overlooked (Megonigal et al., 2004). The process—called dissimilatory nitrate reduction to ammonia (DNRA)—occurs as follows, with mobile nitrates as the initial form of nitrogen and less-mobile ammonium as the product.

$$NO_3^- + 4H_2 + 2H^+ \to 3H_2O + NH_4^+ \qquad (6.6)$$

The process yields energy to the many microorganisms capable of carrying out this process. The bacteria can be anaerobic, aerobic, or facultative. In some cases, nitrate reduction can be a more significant pathway than the other dissimilatory nitrate loss—denitrification. Studies have supported the concept that high availability of organic carbon and/or low nitrate concentrations favors DNRA over denitrification (Megonigal et al., 2004).

Anammox

Anammox (for anaerobic ammonium oxidation) involves nitrite-nitrogen (rather than nitrate-nitrogen as originally thought) as the oxidant:

$$NO_2^- + NH_4^+ \to 2H_2O + N_2 \qquad (6.7)$$

Few studies have definitively determined the importance of anammox in the cycling of nitrogen in natural or created wetlands, but it does appear that this process may be more important in wetlands where denitrification is limited by lack of organic carbon (Megonigal et al., 2004). Erler et al. (2008) found anammox contributed up to 24 percent of the dinitrogen production in a surface flow treatment wetland. Ligi et al. (2015) detected bacterial genes that are specific to organisms capable of anammox from soils samples taken from the Ohio created wetlands described above. They suggested that anammox converting ammonium to dinitrogen gas may compensate for the relatively low rates of denitrification reported at these wetlands (Mitsch et al., 2012; Song et al., 2014).

The Nitrogen Cycle, Wetlands, and Hypoxia

Humans have essentially doubled the amount of nitrogen entering the land-based nitrogen cycle through fertilizer manufacturing, increased use of nitrogen-fixing crops, and fossil fuel burning (Galloway et al., 2003; Doering et al., 2011). Significant amounts of this excess nitrogen are transported as nitrate-nitrogen to rivers and streams, leading to eutrophication and episodic and persistent hypoxia (dissolved oxygen <2 mg/L) in coastal waters worldwide. For example, a hypoxic zone that currently averages close to 14,350 km^2 reappears annually in the Gulf of Mexico (Figs. 6.7 and 6.8), caused almost certainly by excessive nitrogen coming from farm fields in from the Mississippi-Ohio-Missouri (MOM) river basin 1,000 km to the north of the gulf. The extent of the hypoxia was much smaller than that area in the late 1980s. The federal government decreed in 2000 and then again in 2008 that the hypoxia should be no larger than 5000 km^2 (Mississippi River/Gulf of Mexico Watershed Nutrient Task Force, 2008).

Many options were investigated for controlling nutrient flow into the gulf by research teams in the late 1990s (e.g., Mitsch et al., 2001). In the end, there were the general approaches that involve either revision of agronomic approaches or wetland creation and riparian restoration that make the most sense (Fig. 6.9). Two million ha of restored and created wetlands and restored riparian buffers were recommended as necessary to provide enough denitrification to substantially reduce the nitrogen entering the Gulf of Mexico (Mitsch et al., 2001, 2005; Mitsch and Day, 2006). The anaerobic process of denitrification in wetlands was a particularly important process recognized in this recommendation. Two million hectares of wetlands is less than 1 percent of the Mississippi River Basin. Interestingly, Hey and Phillipi (1995) found that a similar scale of wetland restoration would be required in the Upper Mississippi River Basin to mitigate the effects of very large and costly floods, such as the one that occurred in the summer of 1993 in the Upper Mississippi River Basin.

Murphy et al. (2013) looked at the trend of nitrate-nitrogen in the Mississippi River for the 30-year period of 1980 to 2010 and found that, although two states (Iowa and Illinois) with the highest nitrate-nitrogen in the 1980s had 11 to 15 percent reductions in nitrate-nitrogen concentrations and loading over the 30 years, other sites on the river had 8 to 55 percent increases in nitrate loading over that same period, essentially overshadowing the modest decreases from Iowa and Illinois. The flux of nitrate-nitrogen to the Gulf of Mexico increased by 14.5 percent over those 30 years and was at an all-time high in 2010; the concentrations of nitrate-nitrogen entering the gulf increased by 19 percent over those years. In the early 2000s, there was

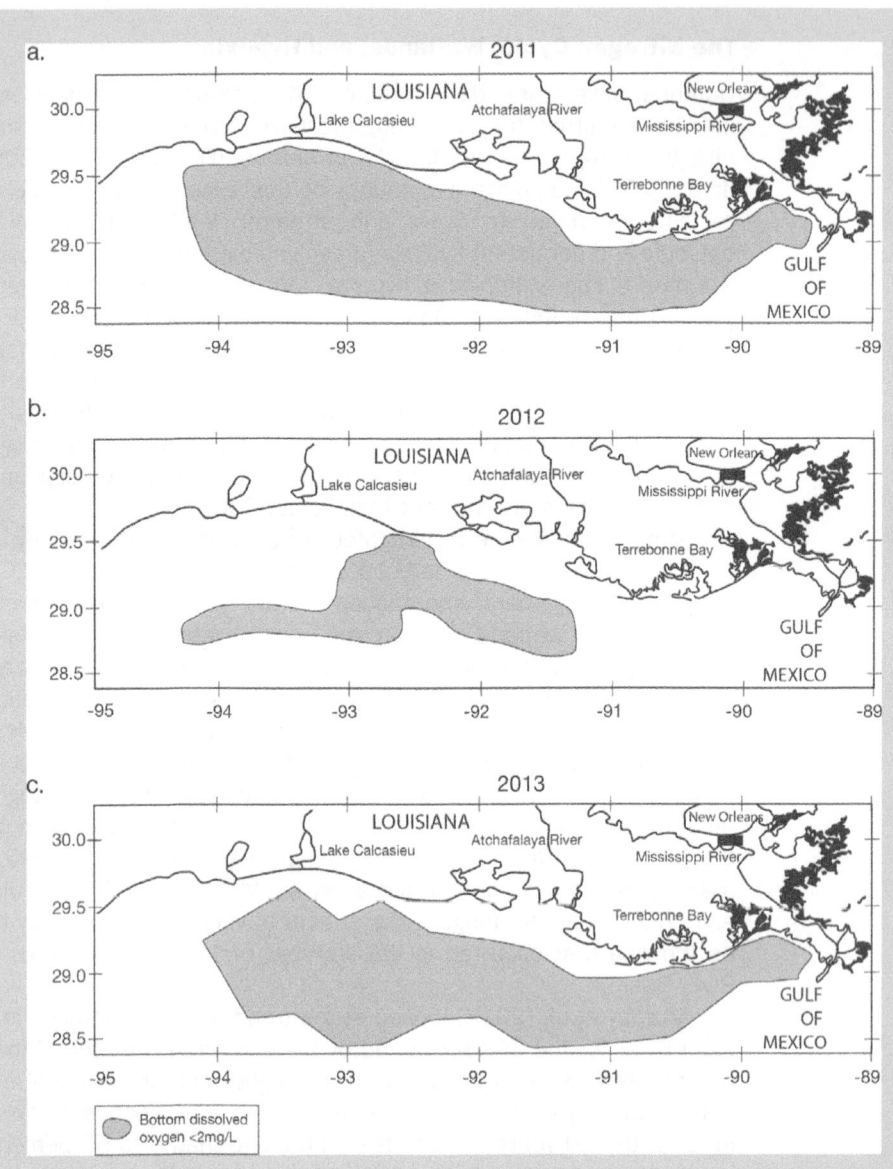

Figure 6.7 Extent of hypoxic conditions in Gulf of Mexico in summers of (a) 2011, (b) 2012, and (c) 2013. Shaded area indicates where gulf waters are less than 2mg/L in dissolved oxygen. The hypoxia covered 17,520, 7,500, and 15,000 km^2 in those three years, respectively. The smaller hypoxia area in 2012 may have been due to an extensive drought in the Midwestern USA that year that led to reduced Mississippi River flows. Source: N. Rabalais, Louisiana Universities Marine Consortium, and NOAA, Center for Sponsored Coastal Ocean Research.

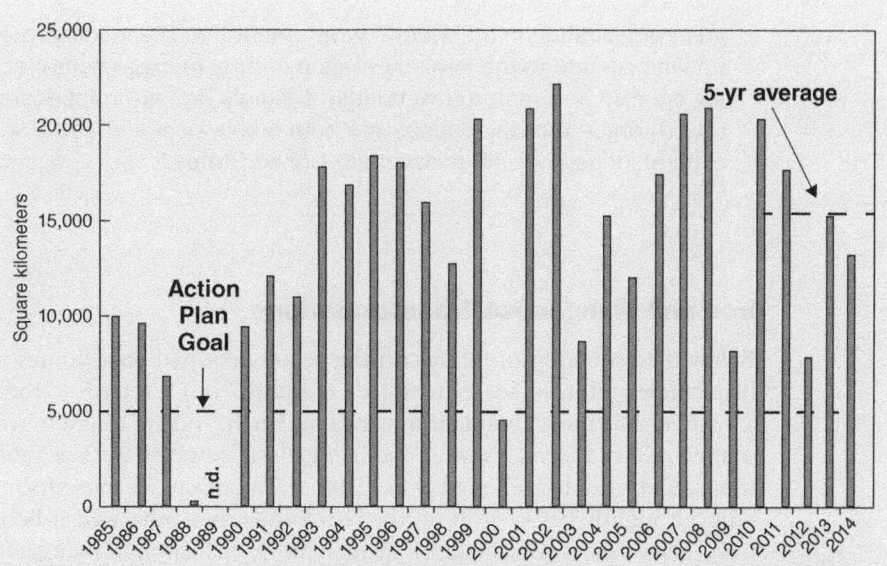

Figure 6.8 The extent of mid-summer Gulf of Mexico hypoxia from 1985 through 2014. The average size over the last five years (2010–2014) is shown to be about 14,350 km². Also shown is the action plan goal of 5,000 km² set by a government task force in 2000 and reaffirmed in its action plan of 2008 (Mississippi River/Gulf of Mexico Watershed Nutrient Task Force, 2008) Source: N. Rabalais, Louisiana Universities Marine Consortium, and NOAA, Center for Sponsored Coastal Ocean Research.

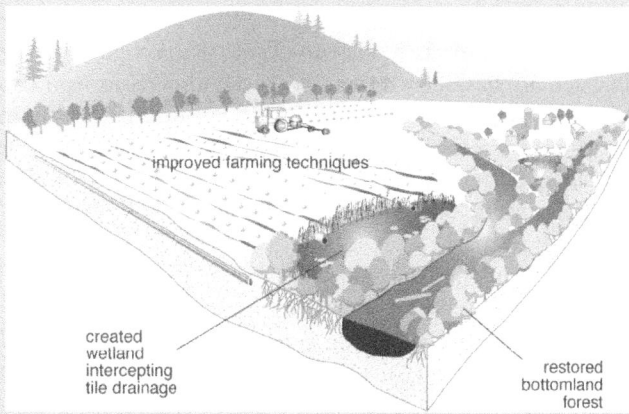

Figure 6.9 Sketch of strategy for wetland restoration and better farming practices in an agricultural setting to improve water quality in the midwestern United States, particularly to control nitrate-nitrogen to protect the downstream Gulf of Mexico. Mitsch et al. (2001, 2005) recommended 2 million ha of created and restored wetlands and riparian forest buffers in the Mississippi River Basin to intercept surface and subsurface drainage coming from agricultural nonpoint sources. (From Mitsch et al., 2001, 2005)

> great anticipation in the federal government that the loading rates were going to diminish due to the implementation of best management practices, including created and restored wetlands. It simply has not happened. David et al. (2013) argue that the causes are both biophysical and social within the agricultural industry of the midwestern United States.

Iron and Manganese Transformations

Below the reduction of nitrate on the redox potential scale comes the reduction of manganese and iron (see Equations 5.6 and 5.7 in Chapter 5). Iron and manganese are among the most abundant minerals on Earth, and are found in wetlands primarily in their reduced forms (ferrous and manganous, respectively; see Table 5.3). Both are more soluble and more readily available to organisms in those forms. Manganese is reduced slightly before iron on the redox scale, but otherwise it behaves similarly to iron. The direct involvement of bacteria in the reduction of manganic oxide (MnO_2) has been questioned by some researchers, although several experiments have shown the generation of energy by the bacterial reduction of oxidized manganese (Laanbroek, 1990).

Iron can be oxidized from reduced ferrous iron to the insoluble ferric form by chemosynthetic bacteria in the presence of oxygen:

$$4Fe^{2+} + O_2(aq) + 4H^+ \rightarrow 4Fe^{3+} + 2H_2O \tag{6.8}$$

Although this reaction can occur nonbiologically at neutral or alkaline pH, microbial activity has been shown to accelerate ferrous iron oxidation by a factor of 10^6 in coal mine drainage water (Singer and Stumm, 1970). A similar type of bacterial process is believed to exist for manganese.

Iron bacteria are thought to be responsible for the oxidation to insoluble ferric compounds of soluble ferrous iron that originated in anaerobic groundwaters in northern peatland areas. These "bog-iron" deposits form the basis of the ore that has been used in the iron and steel industry. Iron in its reduced ferrous form causes a gray-green coloration (gleying) of mineral soils instead of the normal red or brown color in oxidized conditions caused by ferric hydroxide [$Fe(OH)_3$]. This appearance gives a relatively easy field check on the oxidized and reduced layers in a mineral soil profile.

Iron and manganese in their reduced forms can reach toxic concentrations in wetland soils. Ferrous iron, diffusing to the surface of the roots of wetland plants, can be oxidized by oxygen leaking from root cells, immobilizing phosphorus and coating roots with an iron oxide, and causing a barrier to nutrient uptake.

The Sulfur Cycle

Sulfur, as the fourteenth most abundant element in the Earth's surface, occurs in several different states of oxidation in wetlands. Like nitrogen, it is transformed through several pathways that are mediated by microorganisms (Fig. 6.10). Sulfur is rarely present in such low concentrations that it is limiting to plant or animal growth in wetlands. The release of the reduced form of sulfur, sulfide (S^{-2}), when wetland sediments are disturbed causes the odor familiar to those who carry out research in wetlands—the smell of rotten eggs as hydrogen sulfide (H_2S). On the redox scale, sulfur compounds are the next major electron acceptors after nitrates, iron, and manganese, with

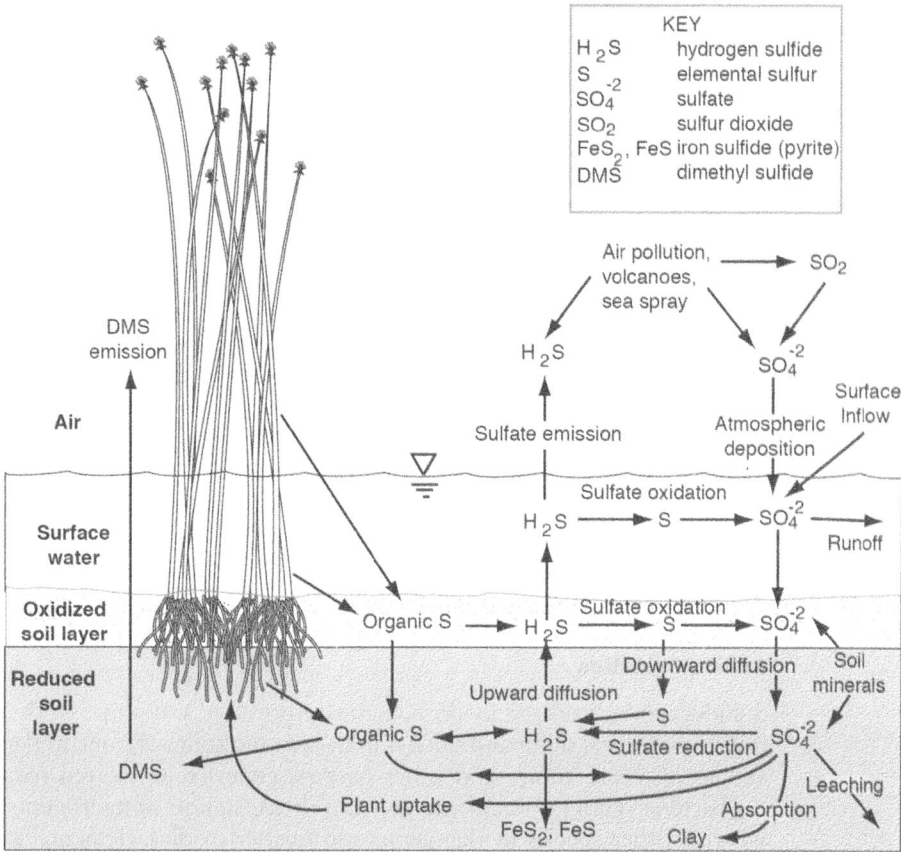

Figure 6.10 The sulfur cycle in wetlands. Major pathways illustrated are sulfur oxidation, sulfate reduction, iron sulfide production, sulfate absorption and leaching, and hydrogen sulfide emissions.

reduction occurring at about −100 to −200mV on the redox scale (see Table 5.3). The most common oxidation states (valences) for sulfur in wetlands are:

Form	Valence
S^{-2} (sulfide)	−2
S (elemental sulfur)	0
S_2O_3 (thiosulfate)	+2
$SO_4^=$ (sulfate)	+6

Sulfate Reduction

Sulfate reduction can take place as *assimilatory sulfate reduction* in which certain sulfur-reducing obligate anaerobes, such as *Desulfovibrio* bacteria, utilize the sulfates as terminal electron acceptors in anaerobic respiration:

$$4H_2 + SO_4^= \rightarrow H_2S + 2H_2O + 2OH^- \tag{6.9}$$

This sulfate reduction can occur over a wide range of pH, with the highest rates prevalent near neutral pH.

There have been a few measurements of the rate at which hydrogen sulfide is produced in and released from wetlands, and those measurements have ranged over several orders of magnitude. It can be safely generalized that saltwater wetlands have higher rates of sulfide emission per unit area than do freshwater wetlands, where sulfate ions are much less abundant (~2700 mg/L in sea water; ~10 mg/L in fresh water). Sulfur can also be released to the atmosphere as organic sulfur compounds, especially as dimethyl sulfide (DMS), $(CH_3)_2S$; this flux is thought by some to be as important as or more important than H_2S emissions from some wetlands. The general consensus, however, is that most DMS comes from oceans as a product of decomposing phytoplankton cells and that the most important loss of sulfur from terrestrial freshwater wetland systems is H_2S.

Sulfide Oxidation

Sulfides can be oxidized by both chemoautotrophic and photosynthetic microorganisms to elemental sulfur and sulfates in the aerobic zones of some wetland soils. Certain species of *Thiobacillus*—and other bacteria collectively referred to as colorless sulfur bacteria (CSB)—obtain energy from the oxidation of hydrogen sulfide to sulfur, whereas other species in this genus can further oxidize elemental sulfur to sulfate. These reactions are summarized in Equations 6.10 and 6.11:

$$2H_2S + O_2 \rightarrow 2S + 2H_2O + energy \tag{6.10}$$

and

$$2S + 3O_2 + 2H_2O \rightarrow 2H_2S_4 + energy \tag{6.11}$$

Under anaerobic conditions, nitrate-nitrogen can be used as the terminal electron acceptor in oxidizing hydrogen sulfides.

Photosynthetic sulfur-oxidizing bacteria, such as the green and purple sulfur bacteria found in salt marshes and mud flats, are capable of producing organic matter in the presence of light according to Equation 6.12:

$$CO_2 + 2H_2S + light \rightarrow CH_2O + 2S + H_2O \qquad (6.12)$$

This reaction, called *anoxygenic photosynthesis*, uses hydrogen sulfide as an electron donor rather than H_2O but is otherwise similar to the more traditional photosynthesis equation. This reaction often takes place under anaerobic conditions where hydrogen sulfide is abundant, but at the surface of sediments where sunlight is also available.

Sulfide Toxicity

Hydrogen sulfide, which is characteristic of anaerobic wetland sediments, can be toxic to rooted higher plants and microbes, especially in saltwater wetlands where the concentration of sulfates is high. The negative effects of sulfides on higher plants include the following:

1. The direct toxicity of free sulfide as it comes in contact with plant roots;
2. The reduced availability of sulfur for plant growth because of its precipitation with trace metals; and
3. The immobilization of zinc and copper by sulfide precipitation.

In wetland soils that contain high concentrations of ferrous iron (Fe^{2+}), sulfides can combine with iron to form insoluble ferrous sulfides (FeS), thus reducing the toxicity of the free hydrogen sulfide. Ferrous sulfide gives the black color characteristic of many anaerobic wetland soils; one of its common mineral forms is pyrite, FeS_2, the form of sulfur commonly found in coal deposits.

The Carbon Cycle

The major processes of carbon transformation under aerobic and anaerobic conditions are shown in Figure 6.11. Photosynthesis (Equation 6.13) and aerobic respiration (Equation 6.14) dominate the aerobic horizons (aerial and aerobic water and soil), with H_2O as the major electron donor in photosynthesis and oxygen as the terminal electron acceptor in respiration:

$$6CO_2 + 12H_2O + light \rightarrow C_6H_{12}O_6 + 6O_2 + 6H_2O \qquad (6.13)$$

$$C_6H_{12}O_6 + 6O_2 \rightarrow 6CO_2 + 6H_2O + 12e^- + energy \qquad (6.14)$$

The degradation of organic matter by aerobic respiration is fairly efficient in terms of energy transfer. However, because of the anoxic nature of wetlands, anaerobic

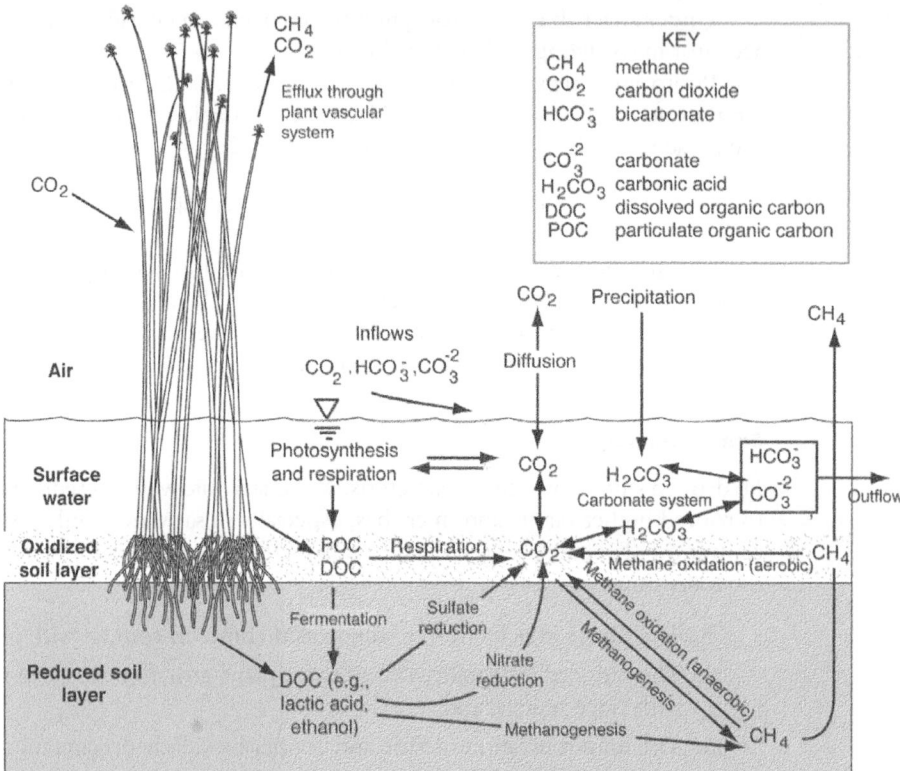

Figure 6.11 The carbon cycle in wetlands. Major pathways include photosynthesis, respiration, fermentation, methanogenesis, and methane oxidation (anaerobic and aerobic). Also indicated are the roles of sulfate and nitrate reduction in the carbon cycle.

processes, less efficient in terms of energy transfer, occur in proximity to aerobic processes. Two of the major anaerobic processes are fermentation and methanogenesis.

Fermentation

The *fermentation* of organic matter, also called *glycolysis* for the substrate involved, occurs when organic matter is the terminal electron acceptor in anaerobic respiration by microorganisms and forms various low-molecular-weight acids and alcohols and CO_2. Examples are lactic acid (Eq. 6.15):

$$C_6H_{12}O_6 \rightarrow 2CH_3CH_2OCOOH \text{ (lactic acid)} \tag{6.15}$$

and ethanol (Eq. 6.16):

$$C_6H_{12}O_6 \rightarrow 2CH_3CH_2OH \text{ (ethanol)} + 2CO_2 \tag{6.16}$$

Fermentation can be carried out in wetland soils by either facultative or obligate anaerobes. Although *in situ* studies of fermentation in wetlands are rare, fermentation plays a central role in providing substrates for other anaerobes, such as methanogens in wetland sediments. Fermentation represents one of the major ways in which high-molecular-weight carbohydrates are broken down to low-molecular-weight organic compounds, usually as dissolved organic carbon, which are, in turn, available to other microbes.

Methanogenesis

Methanogenesis occurs when certain bacteria (*methanogens*) use CO_2 as an electron acceptor for the production of gaseous methane (CH_4), as described in Chapter 5:

$$CO_2 + 8H^+ \rightarrow CH_4 + 2H_2O \qquad (6.17)$$

or, alternatively, use a low-molecular-weight organic compound, such as one from a methyl group:

$$CH_3COOH \text{ (acetic acid)} \rightarrow CH_4 + CO_2 \qquad (6.18)$$

or

$$3CH_3OH \text{ (methanol)} + 6H^+ \rightarrow 3CH_4 + 3H_2O \qquad (6.19)$$

Methane, which can be released to the atmosphere when sediments are disturbed, is often referred to as *swamp gas* or *marsh gas*. Methane production requires extremely reduced conditions, with a redox potential below –200 mV, after other terminal electron acceptors (O_2, NO_3, and $SO_4^=$) have been reduced. Methanogenesis is carried out by *methanogens*—a group of microbes called the *Archaea*. Archaea are prokaryotes that includes several obligate halophiles, and thermophiles in addition to methanogens.

Methane Oxidation

Methane oxidation is carried out by obligate *methanotropic bacteria*, which are from a larger group of eubacteria; they convert methane gas in sequence to methanol (CH_3OH), formaldehyde (HCHO), and finally CO_2:

$$CH_4 \rightarrow CH_3OH \rightarrow HCHO \rightarrow HCOOH \rightarrow CO_2 \qquad (6.20)$$

Nonflooded lands (e.g., forests, agricultural land, grasslands) are normally considered the major biological sinks of methane and are where most *methanotrophs* occur. But wetlands, which have stratified anoxic-oxic horizons, may have a lower anoxic zone dominated by methanogenesis and a surface oxygenated zone with methane oxidation (Fig. 6.11). Thus methane produced in the lower reaches of wetland soils may be "modulated" by methanotrophs that intercept methane from below and convert it to carbon dioxide. Methanotrophs are also able to tolerate extended periods of anoxia, as with temporary flooding, and can resume methane oxidation within a few hours of reexposure to oxygen (Whalen, 2005). Methanogens, however, are extremely sensitive to oxygen; methane production does not continue very long once flooded soils are drained. Roy-Chowdhury et al. (2014) found a high rate of potential methane oxidation (PMO) by methanotrophs in Ohio created wetlands [equivalent to a methane oxidation rate of 104 g-C m^{-2} yr^{-1}] and also concluded that the soil methane concentration had a greater influence than temperature on controlling methanotroph activity in these wetlands.

In addition to methanotrophs, the autotrophic nitrifier communities discussed previously are also able to carry out methane oxidation, because methane and ammonia molecules have a similar size and structure. As a result, the ammonium molecule can also essentially inhibit the methanotrophs from oxidizing CH_4, and CH_4 can substitute for NH_4^+ in nitrifiers and be co-oxidized.

Methane Emissions

Methane emissions, which are the net result of methanogenesis and methane oxidation, have a considerable range from both saltwater and freshwater wetlands as well as from domestic wetlands, such as rice paddies. Comparison of rates of methane production from different studies is difficult, because different methods are used and because the rates depend on both soil temperature (season) and hydroperiod. Methane emissions have clear seasonal patterns in temperate-zone wetlands (Fig. 6.12) and much less seasonality in tropical and subtropical wetlands (Fig. 6.13). Summer rates can be highest in seasonal climates, but estimation of total methane generation requires year-long measurements, particularly in subtropical and tropical regions. The pattern also depends on the degree of flooding and the presence or absence of vegetation. Studies of methane fluxes in temperate zone marshes have shown that methane fluxes are higher in permanently flooded parts of the marshes than in intermittently exposed areas (Altor and Mitsch, 2006; Sha et al. 2011), suggesting that seasonal pulsing rather than permanent flooding minimizes methane emissions. The lower rates of methane generation in the intermittently exposed marshes could be a result of either lower methanogenesis or higher rates of methane oxidation. Methane emissions in tropical and subtropical climes show interesting patterns versus hydrologic conditions (Fig. 6.14). In a series of tropical wetlands in different climates in Costa Rica, Nahlik and Mitsch (2011) found a Shelford curve pattern with highest methane emissions at middle water depths between 30 and 50 cm (Fig. 6.14a). They attributed lower methane emissions in shallow depths to better oxygen diffusion into the entire water column, allowing oxidation of the soil-water interface and lower emissions in deepwater because of stratification patterns typical in tropical bodies of water. Villa and Mitsch (2014) describe a similar pattern for several plant communities in Corkscrew Swamp area of the Greater Florida Everglades. Using a slightly different metric called days after inundation (DAI) for the summer seasonal rains, they found that methane emissions were low just after inundation, increased for about two months of inundation, but began to decrease if flooding lasted longer than two months for a freshwater prairie (Fig. 6.14b). This same pattern was also seen for several other wetland communities at Corkscrew Swamp.

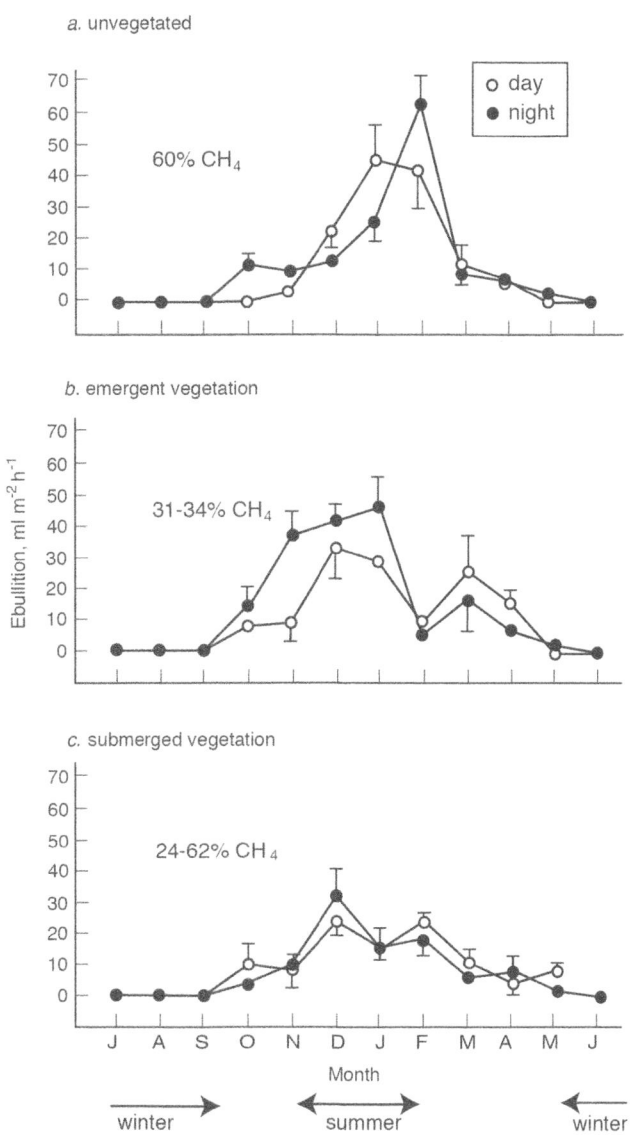

Figure 6.12 Seasonal patterns of gas ebullition (flux of methane-rich bubbles) from three different wetland community types in a floodplain lake (billabong) along the River Murray, New South Wales, Australia: (a) no vegetation; (b) beds of the emergent plant *Eleocharis sphacelata*; and (c) beds of the submerged aquatic plant *Vallisneria gigantea*. Methane concentrations were 60 percent of the emissions from the bare area, 31 to 54 percent of the emergent plant site, and 24 to 62 percent of the submerged aquatic plant site. (After Sorrell and Boon, 1992)

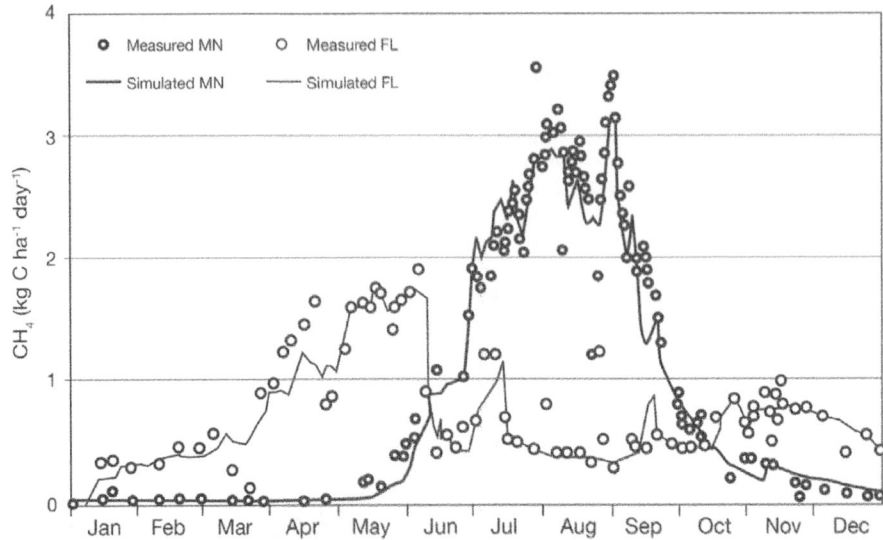

Figure 6.13 Comparison of methane emission rates from Florida (subtropical) and Minnesota (temperate with cold winters), and model results that attempted to simulate both conditions. (After Cui et al., 2005)

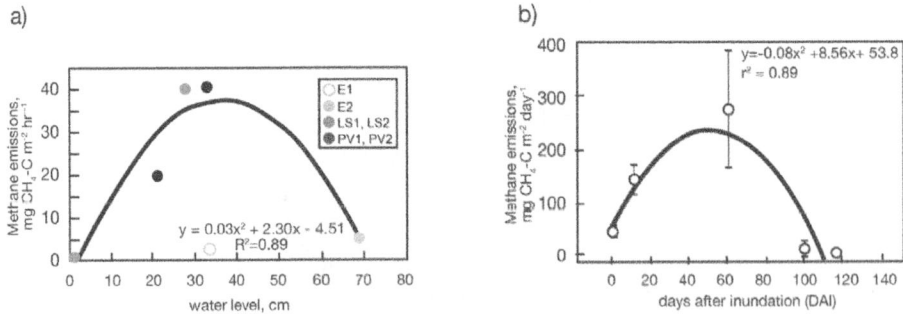

Figure 6.14 Relationships between hydrologic conditions and methane emissions for tropical and subtropical wetlands: (a) mean methane emissions versus water level for six wetland transects in three tropical wetlands in Costa Rica (E = EARTH University campus wetland; LS = La Selva Biological Station; PV = Palo Verde Biological Station); (b) methane emissions (mean ± standard error) versus days after inundation (DAI) for seasonally wet subtropical wet prairie communities in Corkscrew Swamp Sanctuary in southwest Florida. Methane emissions in bald cypress and pond cypress communities showed similar patterns versus DAI. ((a) Nahlik and Mitsch, 2011; (b) Villa and Mitsch, 2014)

Ebullition and Gaseous Transport in Plants

With the exception of CO_2 and O_2, gases emitted from wetlands (1) emanate from the sediment or soil surface through the water column by diffusive flux or diffusion,

(2) bubble to the surface in a process called *ebullitive flux* or *ebullition* and then exit to the atmosphere, or (3) pass through the vascular system of emergent plants (Boon, 1999). Boon and Sorrell (1995) noted that there were substantially more methane fluxes during the day than during the night in chamber studies of Australian wetlands when wetland plants were included in the chambers. They also noted that there was a discrepancy in chambers between the total methane flux and the amount measured by inverted funnels (which capture the ebullitive flux). As a result, the pressures, flows, and gas concentrations were measured within a dominant wetland plant, *Eleocharis sphacelata*, in both "influx" culms, which could generate high pressures, and "efflux" culms, which could not. Methane concentrations were three orders of magnitude greater in the efflux culms than in the influx culms. Carbon dioxide concentrations, as expected, were 50 times higher, whereas dissolved oxygen concentrations decreased 20 percent. These studies and others suggest that between 50 and 90 percent of all methane generated from a vegetated wetland could be passing through the vascular system of emergent plants.

Carbon–Sulfur Interactions

The sulfur cycle is important in some wetlands for the oxidation of organic carbon. This is particularly true in most coastal wetlands where sulfur is abundant. In general, methane is emitted at low concentrations in reduced soils when sulfate concentrations are high. Possible reasons for this phenomenon include (1) competition for substrates that occurs between sulfur and methane bacteria, (2) the inhibitory effects of sulfate or sulfide on methane bacteria, (3) a possible dependence of methane bacteria on products of sulfur-reducing bacteria, and (4) a stable redox potential that does not drop low enough to reduce CO_2 because of an ample supply of sulfate. Other evidence suggests that methane may actually be oxidized to CO_2 by sulfate reducers.

Sulfur-reducing bacteria require an organic substrate, generally of low molecular weight, as a source of energy in converting sulfate to sulfide (Eq. 6.9). The process of fermentation described previously can conveniently supply these necessary low-molecular-weight organic compounds, such as lactate or ethanol (see Eq. 6.15 and 6.16 and and Fig. 6.11). Equations for sulfur reduction, also showing the oxidation of organic matter, are shown in Equations 6.21 and 6.22:

$$2CH_3CHOHCOOH \text{ (lactate)} + SO_4^{-2} \rightarrow 2CH_3COOH + 2CO_2 + H_2S + 2H_2O \qquad (6.21)$$

and

$$CH_3COOH \text{ (acetate)} + SO_4^{-2} \rightarrow 2CO_2 + H_2S + 2H_2O \qquad (6.22)$$

This fermentation–sulfur reduction pathway is particularly important in the oxidation of organic carbon to carbon dioxide in saltwater wetlands, which have an excess of sulfates. Fully 54 percent of the carbon dioxide evolution from the salt marsh in New England was caused by the fermentation–sulfur reduction pathway, with aerobic respiration accounting for another 45 percent. By contrast, most of the carbon flux from freshwater systems is through the methane–methane oxidation pathway. In a freshwater billabong in Australia, Boon, and Mitchell (1995) demonstrated that methanogenesis accounted for 30 to 50 percent of the total benthic carbon flux and that a major portion of the carbon fixed by plants leaves the wetland via methanogenesis.

In general, the release of carbon by methane production is dominant in freshwater wetlands, whereas oxidation of organic carbon by sulfate reduction is dominant in saltwater wetlands.

The Phosphorus Cycle

Phosphorus (Fig. 6.15) is one of the most important limiting chemicals in ecosystems, and wetlands are no exception. It is a major limiting nutrient in northern bogs, freshwater marshes, and southern deepwater swamps. In other wetlands, such as agricultural wetlands and salt marshes, phosphorus is an important mineral, although it is not considered a limiting factor because of its relative abundance and biochemical stability.

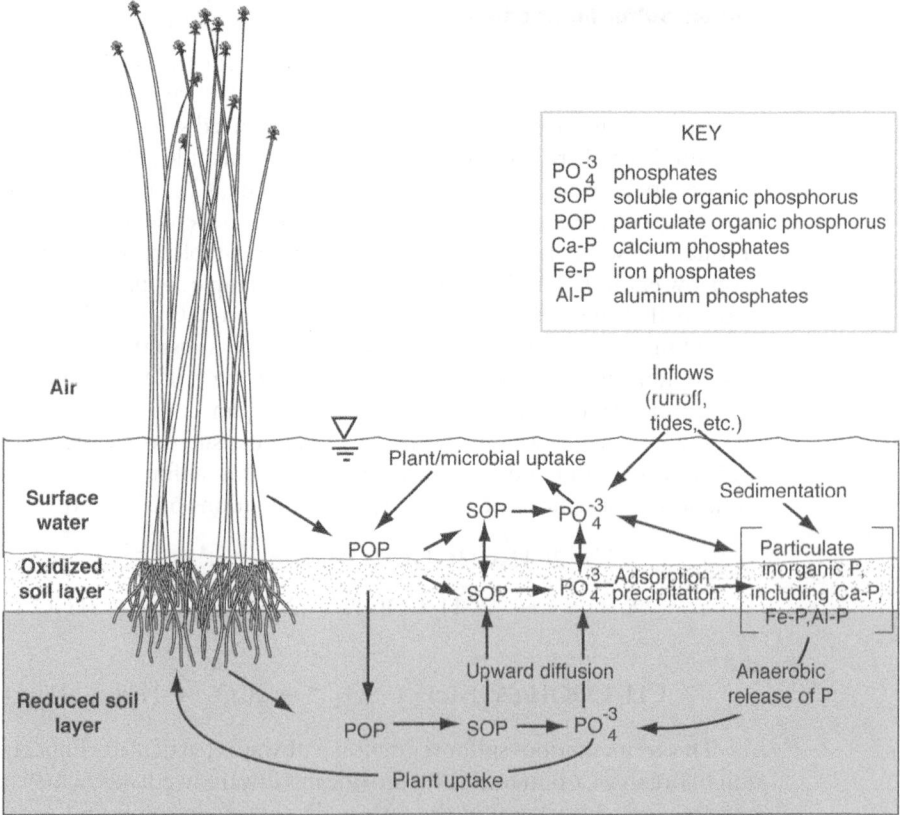

Figure 6.15 The phosphorus cycle in wetlands. Major pathways illustrated are plant/microbial uptake, mineralization, adsorption/precipitation, sedimentation, and anaerobic release.

Phosphorus retention is considered one of the most important attributes of natural and constructed wetlands, particularly those that receive nonpoint source pollution or wastewater.

Phosphorus occurs as soluble and insoluble complexes in both organic and inorganic forms in wetland soils. Inorganic forms include the ions PO_4^{3-}, HPO_4^{2-}, and $H_2PO_4^-$ (collectively referred to as orthophosphates) with the predominant form depending on pH. Phosphorus also has an affinity for calcium, iron, and aluminum, forming complexes with those elements when they are readily available. Phosphorus occurs in a sedimentary cycle rather than in gaseous cycles, such as the nitrogen, sulfur, and carbon cycles described earlier. At any one time, a major proportion of the phosphorus in wetlands is tied up in organic litter and peat and in inorganic sediments, with the former dominating peatlands and the latter dominating mineral soil wetlands.

The analytical measure of biologically available orthophosphates is sometimes called *soluble reactive phosphorus* (SRP), although the equivalence among SRP, exchangeable phosphorus, and orthophosphate is not exact. However, it is often used as indicators of the bioavailability of phosphorus. Dissolved organic phosphorus (DOP) and insoluble forms of organic and inorganic phosphorus are generally not biologically available until they are transformed into soluble inorganic forms.

Although phosphorus is not directly altered by changes in redox potential as are nitrogen, iron, manganese, and sulfur, it is indirectly affected in soils and sediments by its association with several elements, especially iron, that are so altered. Phosphorus is rendered relatively unavailable to plants and microconsumers by:

1. The precipitation of insoluble phosphates with ferric iron, calcium, and aluminum under aerobic conditions;
2. The adsorption of phosphate onto clay particles, organic peat, and ferric and aluminum hydroxides and oxides; and
3. The binding of phosphorus in organic matter as a result of its incorporation into the living biomass of bacteria, algae, and vascular macrophytes.

There are three general conclusions about the tendency of phosphorus to precipitate with selected ions: (1) Phosphorus is fixed as aluminum and iron phosphates in acid soils; (2) phosphorus is bound by calcium and magnesium in alkaline soils; and (3) phosphorus is most bioavailable at slightly acidic to neutral pH (Reddy and DeLaune, 2008). The precipitation of metal phosphates and the adsorption of phosphates onto ferric or aluminum hydroxides and oxides are believed to result from the same chemical forces, namely, those involved in the forming of complex ions and salts.

Co-precipitation of Phosphorus

In many surface water wetlands, high algal productivity can pull CO_2 out of the water, shift the whole carbonate equilibrium, and drive the pH as high as 9 or 10 on a diurnal basis. Under these conditions, co-precipitation of phosphorus as it adsorbs onto calcite and precipitates as calcium phosphate can be significant, just as precipitation of calcium carbonate is also accelerated. In a study of created marshes in central Ohio, calcite and dolomite were found in significant concentrations in the algal mat biomass and wetland sediments but not in the river inflow, indicating that the precipitated calcite was produced within the wetlands in significant amounts. Phosphorus co-precipitating with calcite was up to 47 percent of the total phosphorus contained in the algal mat in these wetlands, suggesting that phosphorus co-precipitation essentially doubled the phosphorus removal capability of the algal mat (Liptak, 2000). Wetlands with high algal productivity thus have two major pathways for phosphorus removal: the assimilation of phosphorus by algal cells and co-precipitation of phosphates caused by high pH created by the algal water column productivity.

The Phosphorus Cycle

The sorption of phosphorus onto clay particles is important in aquatic ecosystems. It is believed to involve both the chemical bonding of the negatively charged phosphates to the positively charged edges of the clay and the substitution of phosphates for silicate in the clay matrix. This clay–phosphorus complex is particularly important for many wetlands, including riparian wetlands and coastal salt marshes, because a considerable portion of the phosphorus brought into these systems by flooding rivers and tides is brought in sorbed to clay particles. Thus, phosphorus cycling in many mineral soil wetlands tends to follow the sediment pathways of sedimentation and resuspension. Because most wetland macrophytes obtain their phosphorus from the soil, sedimentation of phosphorus sorbed onto clay particles is an indirect way in which the phosphorus is made available to the biotic components of the wetland. In essence, the plants transform inorganic phosphorus to organic forms that are then stored in organic peat, mineralized by microbial activity, or exported from the wetland.

The Phosphorus Cycle

When soils are flooded and conditions become anaerobic, several changes in the availability of phosphorus result. A well-documented phenomenon in the hypolimnion of lakes is the increase in soluble phosphorus when the hypolimnion and the sediment–water interface become anoxic. In general, a similar phenomenon often occurs in wetlands on a compressed vertical scale. As ferric (Fe^{3+}) iron is reduced to more soluble ferrous (Fe^{2+}) compounds, phosphorus that is in a specific ferric phosphate (analytically known as reductant-soluble phosphorus) is released into solution. Other reactions that may be important in releasing phosphorus upon flooding are the hydrolysis of ferric and aluminum phosphates and the release of phosphorus sorbed to clays and hydrous oxides by the exchange of anions. Phosphorus can also be released from insoluble salts when the pH is changed either by the production of organic acids or by the production of nitric and sulfuric acids by chemosynthetic bacteria. Phosphorus sorption onto clay particles, however, is highest under acidic to slightly acidic conditions.

Water Chemistry

The inputs of materials to wetlands occur through geologic, biologic, and hydrologic pathways. The geologic input from weathering of parent rock, although poorly understood, may be important in some wetlands. Biologic inputs include photosynthetic uptake of carbon, nitrogen fixation, and biotic transport of materials by mobile animals such as birds. Except for gaseous exchanges such as carbon fixation in photosynthesis and nitrogen fixation, however, elemental inputs to wetlands are generally dominated by hydrologic inputs.

Oceans and Estuaries

Wetlands such as salt marshes and mangrove swamps are continually exchanging tidal waters with adjacent estuaries and other coastal waters. The chemistry of these waters differs considerably from rivers, streams, and lakes. Although estuaries are places where rivers meet the sea, they are not simply places where seawater is diluted with fresh water. Table 6.1 contrasts the chemical makeup of average river water with the average composition of seawater. The chemical characteristics of seawater are fairly constant worldwide compared with the relatively wide range of river water chemistry. Total salinity typically range from 33 to 37 parts per thousand (ppt). Although seawater contains almost every element that can go into solution, 99.6 percent of the salinity is accounted for by 11 ions. In addition to seawater dilution, estuarine waters can also involve chemical reactions when sea and river waters meet, including the dissolution of particulate substances, flocculation, chemical precipitation, biological assimilation and mineralization, and adsorption and absorption of chemicals on and into particles of clay, organic matter, and silt. In most estuaries and coastal wetlands, biologically important chemicals such as nitrogen, phosphorus, silicon, and iron come from rivers, whereas other important chemicals such as sodium, potassium, magnesium, sulfates, and bicarbonates/carbonates come from ocean sources.

Table 6.1 Average chemical concentrations (mg/L) of ocean water and river water

Chemical	Seawater	"Average" River
Na^+	10,773	6.3
Mg^{2+}	1,294	4.1
Ca^{2+}	412	15
K^+	399	2.3
Cl^-	19,340	7.8
SO_4^{2-}	2,712	11.2
HCO_3^-/CO_3^{2-}	142	58.4
B	4.5	0.01
F	1.4	0.1
Fe	<0.01	0.7
SiO_2	$<0.1->10^4$	13.1
N	0–0.5	0.2
P	0–0.07	0.02
Particulate organic carbon	0.01–10	5–10
Dissolved organic carbon	1–5	10–20

Streams, Rivers, and Groundwater

As precipitation reaches the ground in a watershed, it infiltrates into the ground, passes back to the atmosphere through evapotranspiration, or flows on the surface as runoff. When enough runoff comes together, sometimes combined with groundwater flow, in channelized streamflow, its mineral content is different from that of the original precipitation. The "average" concentration of dissolved materials in the world's rivers is compared to seawater in Table 6.1. There is not, however, a typical water quality for surface and subsurface streams and rivers as there is for seawater. Figure 6.16 illustrates the cumulative frequency of the ionic composition of freshwater streams and rivers in the United States. It shows, for example, the average concentrations of the many ions at the 50 percent line. Average NO_3 concentrations are about 1 mg/L, whereas the average for Mg^{2+} about 10 mg/L and the average total dissolved solids is approximately 500 mg/L. The curves demonstrate the wide range over which these chemicals are found in streams and rivers.

The variability in concentrations of chemicals in runoff and streamflow is caused by five factors:

1. *Groundwater influence.* The chemical characteristics of streams and rivers depend on the degree to which the water has previously come in contact with underground formations and on the types of minerals present in those formations. Soil and rock weathering, through dissolution and redox reactions, provides major dissolved ions to waters that enter the ground.

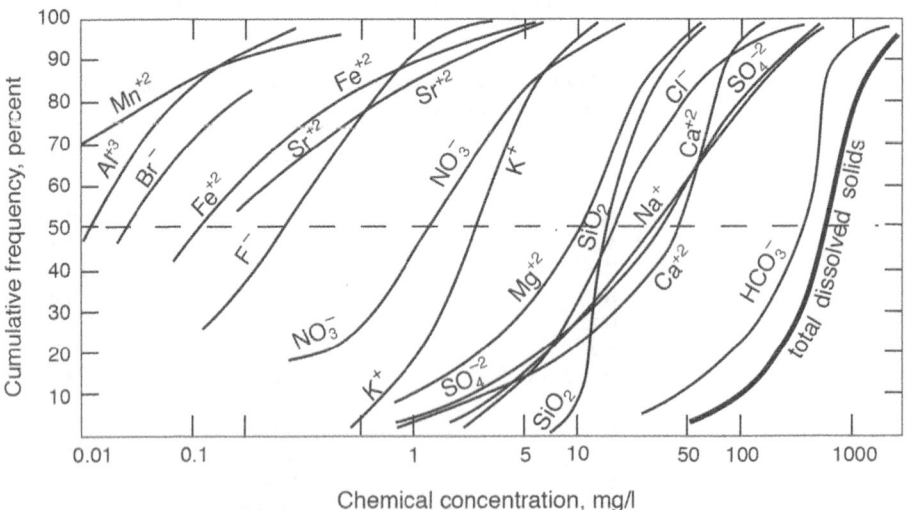

Figure 6.16 Cumulative frequency curves for concentrations of various dissolved minerals in surface waters. Horizontal dashed line indicates median concentrations, 90 percent cumulative frequency indicates the 90th percentile concentration, 50 percent indicates the 50th percentile concentration, and so on. (After Davis and DeWiest, 1966)

The dissolved materials in surface water can range from a few milligrams per liter, found in precipitation, to 500 or even 1,000 mg/L. The ability of water to dissolve mineral rock depends, in part, on its nature as a weak carbonic acid. The rock being mineralized is also an important consideration. Minerals such as limestone and dolomite yield high levels of dissolved ions, whereas granite and sandstone formations are relatively resistant to dissolution.

2. *Climate.* Climate influences surface water quality through the balance of precipitation and evapotranspiration. Arid regions tend to have higher concentrations of salts in surface waters than do humid regions. Climate also has a considerable influence on the type and extent of vegetation on the land, and it therefore indirectly affects the physical, chemical, and biological characteristics of soils and the degree to which soils are eroded and transported to surface waters.

3. *Geographic effects.* The amounts of dissolved and suspended materials that enter streams, rivers, and wetlands also depend on the size of the watershed, the steepness or slope of the landscape, the soil texture, and the variety of topography. Surface waters that have high concentrations of suspended (insoluble) materials caused by erosion are often relatively low in dissolved substances. However, waters that have passed through groundwater systems often have high concentrations of dissolved materials and low levels of suspended materials. The presence of upstream wetlands also influences the quality of water entering downstream wetlands. Johnston et al. (2001) found in a comparison of two riverine wetland areas of different soils and geomorphology that there was nevertheless a seasonal convergence of surface water chemistry caused by the wetlands that overrode the basin differences.

4. *Streamflow/ecosystem effects.* The water quality of surface runoff, streams, and rivers varies seasonally. There is generally an inverse correlation between streamflow and concentrations of dissolved materials and streamflow. During wet periods and storm events, the water is contributed primarily by recent precipitation that becomes streamflow very quickly without coming into contact with soil and subsurface minerals. During low flow, some or much of the streamflow originates as groundwater and has higher concentrations of dissolved materials. The relationship between particulate matter and streamflow is often the opposite. High flow often causes high concentrations of sediments (particulate matter).

5. *Human effects.* Water that has been modified by humans through, for example, sewage effluent, urbanization, and runoff from farms often drastically alters the chemical composition of streamflow and groundwater that reach wetlands. If drainage is from agricultural fields, higher concentrations of sediments and nutrients and some herbicides and pesticides might be expected. Urban and suburban drainage is often lower than that from farmland in those constituents, but it may have high concentrations of trace organics, oxygen-demanding substances, and some toxins.

Nutrient Budgets of Wetlands

A quantitative description of the inputs, outputs, and internal cycling of materials in an ecosystem is called an *ecosystem mass balance*. If the material being measured is one of several elements such as phosphorus, nitrogen, or carbon that are essential for life, then the mass balance is called a *nutrient budget*. In wetlands, mass balances have been developed both to describe ecosystem function and to determine the importance of wetlands as sources, sinks, and transformers of chemicals.

A general mass balance for a wetland, already shown in Figure 6.1, illustrates the major categories of pathways and storages that are important in accounting for materials passing into and out of wetlands. Nutrients or chemicals that are brought into the system are called *inputs* or *inflows*. For wetlands, these inputs are primarily through hydrologic pathways (described in Chapter 4), such as precipitation, surface water and groundwater inflow, and tidal exchange. Biotic pathways of note that apply to the carbon and nitrogen budgets are the fixation of atmospheric carbon through photosynthesis and the capture of atmospheric nitrogen through nitrogen fixation.

Hydrologic *exports*, or *losses* or *outflows*, are by both surface water and groundwater, unless the wetland is an isolated basin that has no outflow, such as a northern ombrotrophic bog. The long-term burial of chemicals in the sediments is also considered a nutrient or chemical outflow, although the depth at which a chemical goes from internal cycling to permanent burial is an uncertain threshold. The depth of available chemicals is usually defined by the root zone of vegetation in the wetland. Biologically mediated exports to the atmosphere are also important in the nitrogen cycle (denitrification) and in the carbon cycle (respiratory loss of CO_2). The significance of other losses of elements to the atmosphere, such as ammonia volatilization and methane and sulfide releases, are potentially important pathways for individual wetlands as well as for the global cycling of minerals.

Intrasystem cycling involves exchanges among various *pools*, or *standing stocks*, of chemicals within a wetland. This cycling includes pathways such as litter production, remineralization, and various chemical transformations discussed earlier. The *translocation* of nutrients from the roots through the stems and leaves of vegetation is another important intrasystem process that results in the physical movement of chemicals within a wetland.

Figure 6.17 illustrates in detail some of the major pathways and storages that investigators should consider when developing nutrient mass balances for wetlands. Few, if any, investigators have developed a complete mass balance for wetlands that includes measurement of all of the pathways shown in the figure, but the diagram remains a useful guide.

A phosphorus budget developed for an alluvial river swamp in southern Illinois showed that 10 times more phosphorus was deposited with sediments during river flooding ($3.6 \text{ g P m}^{-2} \text{ yr}^{-1}$) than was returned from the swamp to the river during the rest of the year (Fig. 6.18). Thus, the swamp was a sink for a significant amount of phosphorus and sediments during that particular year of flooding, although the percentage of retention was low (3–4.5 percent) because a very large volume of phosphorus passed over the swamp ($80.2 \text{ g P m}^{-2} \text{ yr}^{-1}$) during flooding conditions.

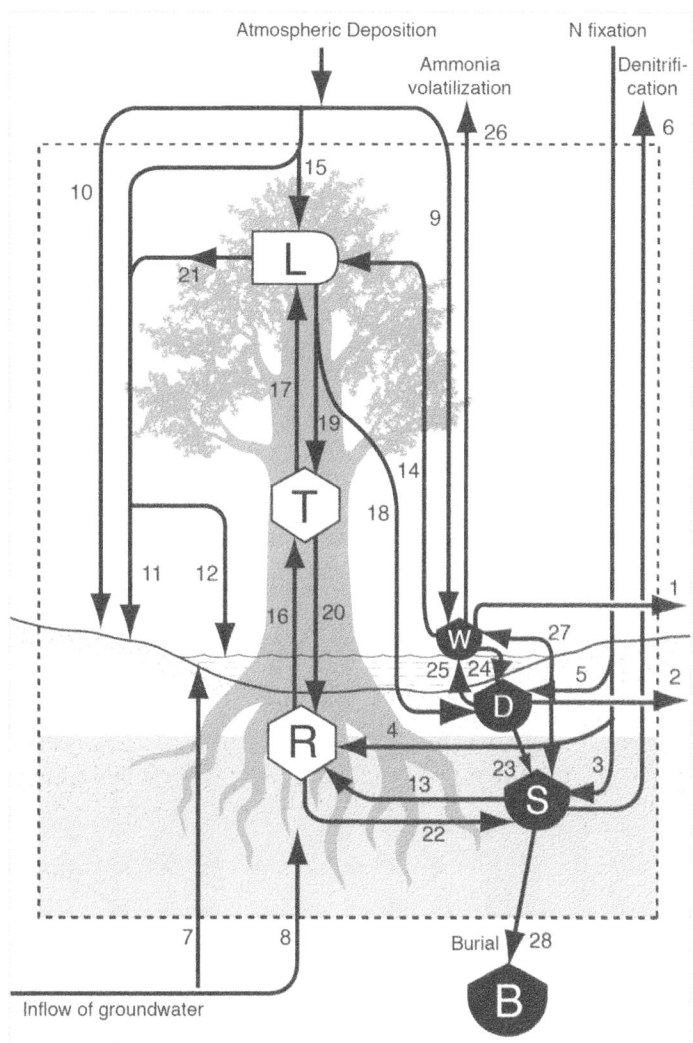

Figure 6.17 Model of major chemical storages and flows in a forested wetland. Storages: L, above-ground shoots or leaves; T, stems, branches, perennial above-ground storage; R, roots and rhizomes; W, surface water; D, litter and detritus; S, near-surface sediments; B, deep sediments essentially removed from internal cycling. Flows: 1 and 2 are exchanges of dissolved and particulate matter with adjacent waters; 3–5 are nitrogen fixation in sediments, rhizosphere microflora, and litter; 6 is denitrification; 7 and 8 are groundwater inputs; 9 and 10 are atmospheric inputs (e.g., precipitation); 11 and 12 are throughfall and stemflow; 13 is uptake by roots; 14 is foliar uptake from surface water; 15 is foliar uptake directly from precipitation; 16 and 17 are translocation from roots through stem to leaves; 18 is litterfall; 19 and 20 are translocation of materials from leaves back to stems and roots; 21 is leaching from leaves; 22 is death/decay of roots; 23 is incorporation of detritus into peat; 24 is adsorption from water to detritus; 25 is release from detritus to water; 26 is volatilization of ammonia; 27 is sediment–water exchange; and 28 is long-term burial of sediments. (After Nixon and Lee, 1986)

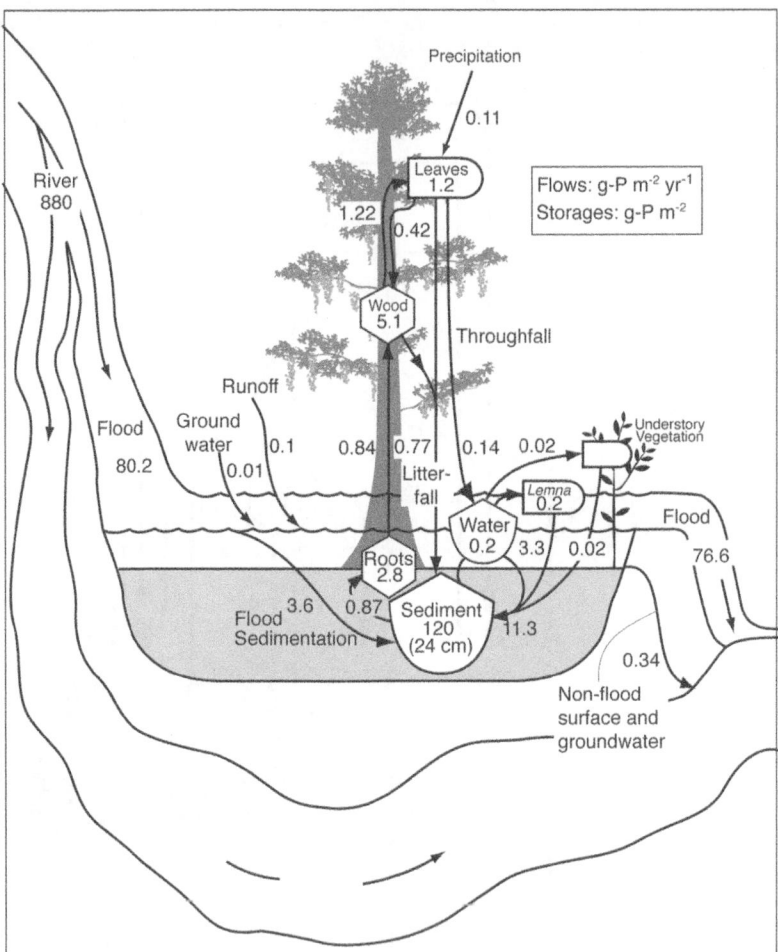

Figure 6.18 Annual phosphorus budget for alluvial cypress swamp in southern Illinois. (After Mitsch et al., 1979)

Detailed nitrogen and carbon budgets for created marshes in Ohio are illustrated in Figure 6.19. Both of these budgets illustrate the importance of accurate hydrologic measurements. Each also shows significant nitrogen and carbon sequestration in the wetland soils.

Generalizations about Nutrient Budgets in Wetlands

Chemical balances that have been developed for various wetlands are extremely variable, but four generalizations have emerged from these studies:

1. *Seasonal patterns of nutrient uptake and release are characteristic of many wetlands.* In temperate climates, retention of certain chemicals, such as nutrients, is greatest during the growing season, primarily because of higher

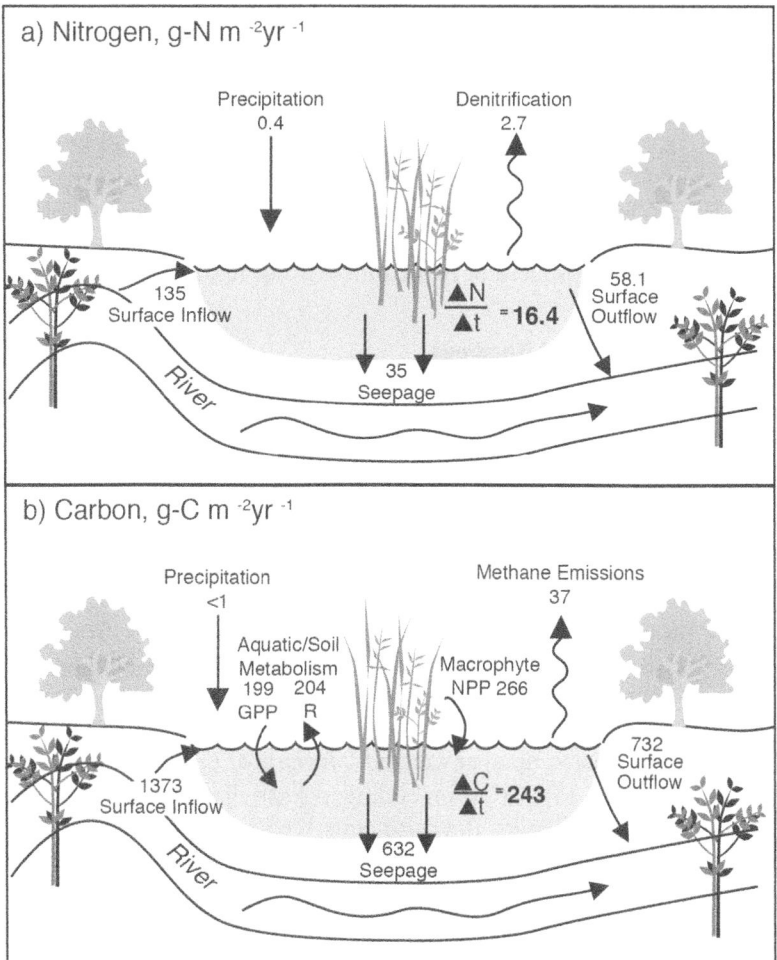

Figure 6.19 Annual wetland nutrient budgets for (a) nitrogen and (b) carbon for two created riparian wetlands in central Ohio. (Data are from Batson et al., 2012 and Waletzko and Mitsch, 2013)

microbial activity in the water column and sediments and secondarily because of greater macrophyte productivity. For example, in cold temperate climates, distinct seasonal patterns of nitrate retention are evident in many cases, with greater retention during the summer months when warmer temperatures accelerate both denitrification microbial activity and algal and macrophyte growth.

2. *Wetlands are frequently coupled to adjacent ecosystems through chemical exchanges that significantly affect both systems.* Ecosystems upstream of wetlands are often significant sources of chemicals to wetlands, whereas downstream aquatic systems often benefit either from the ability of wetlands to retain certain chemicals or from the export of organic materials.

3. *Nutrient cycling in wetlands differs from both deepwater aquatic and terrestrial ecosystem cycling in temporal and spatial dimensions.* More nutrients are tied up in sediments and peat in wetlands than in most terrestrial systems, and deepwater aquatic systems have autotrophic activity more dependent on nutrients in the water column than on nutrients in the sediments.
4. *Anthropogenic changes have led to considerable changes in chemical cycling in many wetlands.* Although wetlands are quite resilient to many chemical inputs, the capacity of wetlands to assimilate anthropogenic wastes from the atmosphere or hydrosphere is not limitless.

Recommended Readings

Reddy, K. R., and R. D. DeLaune. 2008. *Biogeochemistry of Wetlands*. Boca Raton, FL: CRC Press.

Schlesinger, W. H., and E. S. Bernhardt. 2013. *Biogeochemistry*, 3rd ed. Amsterdam, Netherlands: Academic Press/Elsevier.

References

Altor, A. E., and W. J. Mitsch. 2006. Methane flux from created riparian marshes: Relationship to intermittent versus continuous inundation and emergent macrophytes. *Ecological Engineering* 28: 224–234.

Batson, J., Ü. Mander, and W. J. Mitsch. 2012. Denitrification and a nitrogen budget of created riparian wetlands. *Journal of Environmental Quality* 41: 2024–2032.

Boon, P. I. 1999. Carbon cycling in Australian wetlands: The importance of methane. *Verhandlungen Internationale Vereinigung für Limnologie* 27: 1–14.

Boon, P. I., and A. Mitchell. 1995. Methanogenesis in the sediments of an Australian freshwater wetland: Comparison with aerobic decay and factors controlling methanogenesis. *FEMS Microbiology Ecology* 18: 174–190.

Boon, P. I., and B. K. Sorrell. 1995. Methane fluxes from an Australian floodplain wetland: The importance of emergent macrophytes. *Journal of North American Benthological Society* 14: 582–598.

Cui, J., C. Li, and C. Trettin. 2005. Analyzing the ecosystem carbon and hydrologic characteristics of forested wetland using a biogeochemical process model. *Global Change Biology* 11: 278–289.

David, M. B., C. G. Flint, G. F. McIsaac, L. E. Gentry, M. K. Dolan, and G. F. Czapar. 2013. Biophysical and social barriers restrict water quality improvements in the Mississippi River basin. *Environmental Science & Technology* 47: 11928–11929.

Davis, S. N., and R. J. M. DeWiest. 1966. *Hydrogeology*. John Wiley & Sons, New York. 463 pp.

Doering, O. C., J. N. Galloway, T. L. Theis, V. Aneja, E. Boyer, K. G. Cassman, E. B. Cowling, R. R. Dickerson, W. Herz, D. L. Hey, R. Kohn, J. S. Lighy, W. Mitsch, W. Moomaw, A. Mosier, H. Paerl, B. Shaw, and P. Stacey. 2011.

Reactive Nitrogen in the United States: An Analysis of Inputs, Flows, Consequences, and Management Options. U.S. Environmental Protection Agency Science Advisory Board Integrated Nitrogen Committee, EPA-SAB-11-013, Washington, DC, 139 pp.

Erler, D. V., B. D. Eyre, and L. Davison. 2008. The contribution of anammox and denitrification to sediment N_2 production in a surface flow constructed wetland. *Environ. Sci. Technol.* 42(24): 9144–9150.

Galloway, J. N., J. D. Aber, J. W. Erisman, S. P. Seitzinger, R. W. Howarth, E. B. Cowling, and B. J. Cosby. 2003. The nitrogen cascade. *BioScience* 53: 341–356.

Hernandez, M. E., and W. J. Mitsch. 2006. Influence of hydrologic pulses, flooding frequency, and vegetation on nitrous oxide emissions from created riparian marshes. *Wetlands* 26: 862–877.

Hernandez, M. E., and W. J. Mitsch. 2007. Denitrification in created riverine wetlands: Influence of hydrology and season. *Ecological Engineering* 30: 78–88.

Hey, D. L., and N. S. Philippi. 1995. Flood reduction through wetland restoration: The Upper Mississippi River Basin as a case study. *Restoration Ecology* 3: 4–17.

Johnston, C. A., S. D. Bridgham, and J. P. Schubauer-Berigan. 2001. Nutrient dynamics in relation to geomorphology of riverine wetlands. *Soil Science of America Journal* 65: 557–577.

Laanbroek, H. J. 1990. Bacterial cycling of minerals that affect plant growth in waterlogged soils: A review. *Aquatic Botany* 38: 109–125.

Ligi, T., M. Truu, K. Oopkaup, H. Nõlvak, Ü. Mander, W.J. Mitsch, and J. Truu. 2015. The genetic potential of N_2 emission via denitrification and ANAMMOX from the soils and sediments of a created riverine treatment wetland complex. *Ecological Engineering* doi.org/10.1016/j.ecoleng.2014.09.072

Liptak, M. A. 2000. Water column productivity, calcite precipitation, and phosphorus dynamics in freshwater marshes. Ph.D. dissertation, The Ohio State University, Columbus.

Megonigal, J. P., M. E. Hines, and P. T. Visscher. 2004. Anaerobic metabolism: Linkages to trace gases and aerobic processes. In W. H. Schlesinger, ed., *Biogeochemistry*. Elsevier-Pergamon, Oxford, UK, pp. 317–424.

Mississippi River/Gulf of Mexico Watershed Nutrient Task Force. 2008. *Gulf Hypoxia Action Plan 2008 for Reducing, Mitigating, and Controlling Hypoxia in the Northern Gulf of Mexico and Improving Water Quality in the Mississippi River Basin.* Mississippi River/Gulf of Mexico Watershed Nutrient Task Force, Washington, DC.

Mitsch, W. J., C. L. Dorge, and J. R. Wiemhoff. 1979. Ecosystem dynamics: A phosphorus budget of an alluvial cypress swamp in southern Illinois. *Ecology* 60: 1116–1124.

Mitsch, W. J., J. W. Day, Jr., J. W. Gilliam, P. M. Groffman, D. L. Hey, G. W. Randall, and N. Wang. 2001. Reducing nitrogen loading to the Gulf of Mexico from the Mississippi River Basin: Strategies to counter a persistent ecological problem. *BioScience* 51: 373–388.

Mitsch, W. J., J. W. Day, Jr., L. Zhang, and R. Lane. 2005. Nitrate-nitrogen retention by wetlands in the Mississippi River Basin. *Ecological Engineering* 24: 267–278.

Mitsch, W. J., and J. W. Day, Jr. 2006. Restoration of wetlands in the Mississippi-Ohio-Missouri (MOM) River Basin: Experience and needed research. *Ecological Engineering* 26: 55–69.

Mitsch, W. J, L. Zhang, K. C. Stefanik, A. M. Nahlik, C. J. Anderson, B. Bernal, M. Hernandez, and K. Song. 2012. Creating wetlands: Primary succession, water quality changes, and self-design over 15 years. *BioScience* 62: 237–250.

Murphy, J. C., R. M. Hirsch, and L. A. Sprague, 2013. Nitrate in the Mississippi River and its tributaries, 1980–2010—An update: U.S. Geological Survey Scientific Investigations Report 2013–5169, 31 p., http://pubs.usgs.gov/sir/2013/5169/.

Nahlik, A. M. and W. J. Mitsch. 2011. Methane emissions from tropical freshwater wetlands located in different climatic zones of Costa Rica. *Global Change Biology* 17: 1321–1334.

Nixon, S. W., and V. Lee. 1986. Wetlands and Water Quality. Technical Report Y-86-2, U.S. Army Corps of Engineers Waterways Experiment Station, Vicksburg, MS.

Reddy, K. R. and R. D. DeLaune. 2008. *Biogeochemistry of Wetlands*. CRC Press, Boca Raton, FL, 774 pp.

Roy-Chowdhury, T., W. J. Mitsch, and R. P. Dick. 2014. Seasonal methanotrophy across a hydrological gradient in a freshwater wetland. *Ecological Engineering* 72: 116–124.

Sha, C., W. J. Mitsch, Ü. Mander, J. Lu, J. Batson, L. Zhang, and W. He. 2011. Methane emissions from freshwater riverine wetlands. *Ecological Engineering* 37: 16–24.

Singer, P. C., and W. Stumm. 1970. Acidic mine drainage: The rate-determining step. *Science* 167: 1121–1123.

Song, K., M. E. Hernandez, J. A. Batson, and W. J. Mitsch. 2014. Long-term denitrification rates in created riverine wetlands and their relationship with environmental factors. *Ecological Engineering* 72: 40–46.

Sorrell, B. K., and P. I. Boon. 1992. Biogeochemistry of billabong sediments. II. Seasonal variations in methane production. *Freshwater Biology* 27: 435–445.

Sundareshwar, P. V., J. T. Morris, E. K. Koepfler, and B. Fornwalt. 2005. Phosphorus limitation of coastal ecosystem processes. *Science* 299: 563–565.

Villa, J. A. and W. J. Mitsch. 2014. Methane emissions from five wetland plant communities with different hydroperiods in the Big Cypress Swamp region of Florida Everglades. *Ecohydrology & Hydrobiology* 14: 253–266.

Waletzko, E. and W. J. Mitsch. 2013. The carbon balance of two riverine wetlands fifteen years after their creation. *Wetlands* 33: 989–999.

Whalen, S. C. 2005. Biogeochemistry of methane exchange between natural wetlands and the atmosphere. *Environmental Engineering Science* 22: 73–94.

Chapter **7**

Wetland Vegetation and Succession

There are many plants (hydrophytes) adapted to temporary and permanent flooding conditions found in wetlands. To counter anoxia, one important structural adaptation in vascular plants is the development of pore space in the cortical tissues, which allows oxygen to diffuse from the aerial parts of the plant to the roots to supply root respiratory demands. There are many morphological adaptations, such as pneumatophores, fluted trunks, prop roots, and adventitious roots, that assist vascular plants in adapting to having their roots in water. In addition, wetland plants have several physiological and whole plant adaptations.

Wetland ecosystems have traditionally been considered transitional seres between open lakes and terrestrial forests. The accumulation of organic material from plant production was seen to build up the surface until it was no longer flooded and could support flood-tolerant terrestrial forest species (autogenic succession). An alternative theory is that the vegetation found at a wetland site consists of species adapted to the particular environmental conditions of that site (allogenic succession). Current evidence seems to suggest that both allogenic and autogenic forces act to change wetland vegetation. Models used to describe wetland plant development include a functional guild model, an environmental sieve model, and a centrifugal organization concept.

If one looks at ecosystem attributes as indices of succession, wetlands appear to be mature in some respects and young in others. The strategy for ecosystem development in wetlands includes concepts such as pulse stability and self-organization or self-design. At landscape scales, patterns of wetlands, aquatic and upland habitats reflect a complex and dynamic interaction of physical (allogenic) and biotic (autogenic) forces.

Wetland vegetation is generally viewed as consisting of vascular plants adapted to flooding. Of the 325,000 vascular plant species in the world, only a small percentage have enough adaptations to be considered wetland plants. Wetland vegetation, in the strict sense, also includes many unicellular species of algae and cyanobacteria. Because of the significant number of metabolic and structural adaptations that wetland plants do have, there can be a wide diversity of plants in many wetlands. We often see a minimum 100 vascular plant species in most mature wetlands, with perhaps half of those listed as wetland plants. Defining what a wetland plant is exactly is difficult; for example, it has remained one of the more challenging questions when wetlands are defined legally in the United States.

Vascular Plant Adaptations to Waterlogging and Flooding

Wetland environments are characterized by stresses that most organisms are ill equipped to handle. Aquatic organisms are not adapted to deal with the periodic drying that occurs in many wetlands. Terrestrial organisms are stressed by long periods of flooding. Because of the shallow water, temperature extremes on the wetland surface are greater than would ordinarily be expected in aquatic environments. The most severe stress, however, is the absence of oxygen in flooded wetland soils, which prevents organisms from respiring through normal aerobic metabolic pathways. In the absence of oxygen, the supply of nutrients available to plants is also modified, and concentrations of certain elements and organic compounds can reach toxic levels.

Multicellular organization adds another layer of complexity to individuals compared to unicellular organization. This complexity has enabled plants and animals to develop a wider range of adaptations than bacteria to anoxia and to salt. At the same time, some adaptations found in unicellular organisms, such as the ability to use reduced inorganic compounds in the sediment as a source of energy, are not found in multicellular organisms. These adaptations typically develop in specialized tissue and organ systems.

In contrast to flood-sensitive plants, flood-tolerant species (*hydrophytes*) possess a range of adaptations that enable them either to tolerate stresses or to avoid them. Several adaptations by hydrophytes allow them to tolerate anoxia in wetland soils. These adaptations can be grouped into three main categories: structural or morphological adaptations, physiological adaptations, and whole plant strategies (Table 7.1). Details of these adaptations are discussed here and throughout the chapter.

Morphological Adaptations

Aerenchyma

Virtually all hydrophytes have elaborate structural (or morphological) mechanisms to avoid root anoxia. These responses to flooding are mechanisms that increase the oxygen supply to the plant either by growth into aerobic environments or by enabling oxygen to penetrate more freely into the anoxic zone. The primary plant strategy in

Table 7.1 Plant adaptations and responses to flooding and waterlogging

Structural (or Morphological) Adaptations
a. Aerenchyma tissue in roots and stem
b. Adventitious roots
c. Stem hypertrophy (e.g., buttress trunks)
d. Fluted trunks
e. Rapid vertical growth/growth dormancy
f. Shallow root systems/prop roots
g. Lenticles
h. Pneumatophores and cypress knees

Physiological Adaptations
a. Pressurized gas flow
b. Rhizospheric oxygenation
c. Decreased water uptake
d. Altered nutrient absorption
e. Sulfide avoidance
f. Anaerobic respiration

Whole-Plant Strategies
a. Timing of seed production
b. Buoyant seeds and buoyant seedlings (viviparous seedlings)
c. Persistent seed banks
d. Resistant roots, tubers, and seeds

response to flooding is the development of air spaces (*aerenchyma*) in roots and stems, which allow the diffusion of oxygen from the aerial portions of the plant into the roots (Fig. 7.1). Aerenchyma development is not extensive in the absence of flooding and is characteristic of flood-tolerant plant species, not flood-sensitive ones. In plants with well-developed aerenchyma, the root cells no longer depend on the diffusion of oxygen from the surrounding soil, the main source of root oxygen to terrestrial plants. Unlike the plant porosity of normal plants, which is usually a low 2 to 7 percent of volume, up to 60 percent of the volume of the roots of wetland species consists of pore space. Air spaces are formed either by cell separation during maturation of the root cortex or by cell breakdown. They result in a honeycomb structure. Air spaces are not necessarily continuous throughout the stem and roots. The thin lateral cellular partitions within the aerenchyma, however, are not likely to impede internal gas diffusion significantly. The same kind of cell lysis and air space development has been described in submerged stem tissue. Roots of flood-tolerant species, such as rice, form aerenchyma even in aerated apical cells.

Root porosity is the overriding factor governing internal root oxygen concentration. The effectiveness of aerenchyma in supplying oxygen to the roots has been demonstrated in several plant species. For example, the root respiration of flood-tolerant *Senecio aquaticus* was only 50 percent inhibited by root anoxia, whereas that of *S. jacobaea*, a flood-sensitive species, was almost completely inhibited. Greater

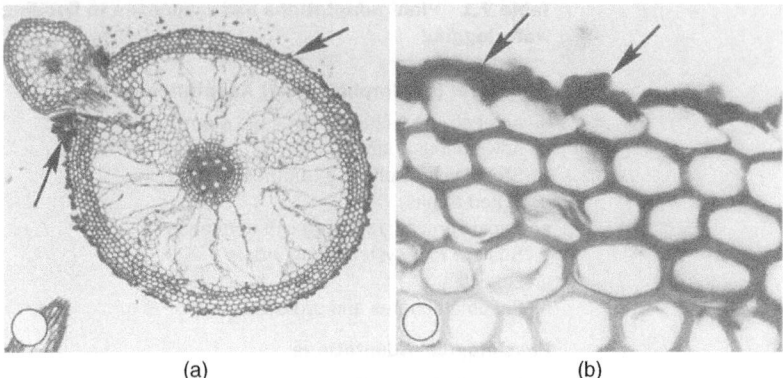

Figure 7.1 Light photomicrographs of *Spartina alterniflora* roots: (a) cross-section of a streamside root; arrows indicate the presence of red ferric deposits on the root epidermis. magnification x192; (b) streamside root cross-section showing the presence of similar materials on the external walls of the epidermal cells. magnification x1,143. Note the extensive pore space (aerenchyma) in the roots. (From Mendelssohn and Postek, 1982)

root porosity in the tolerant species was the primary factor that contributed to the difference. The most extensively studied flood-tolerant plant is rice. Rice plants grown under continuous flooding develop greater root porosity than unflooded plants, and this maintains the oxygen concentration in the root tissues. When deprived of oxygen, rice root mitochondria degraded in the same way as did flood-sensitive pumpkin plants, suggesting that the primary basis of resistance in flooded plants is by the avoidance of root anoxia, not by physiological changes in cell metabolism (Levitt, 1980).

Adventitious Roots

In addition to aerenchyma development, anaerobic conditions result in the formation of certain organs on wetland plants that assist the plant in getting oxygen to its root system. Hormonal changes, especially the concentration of ethylene in hypoxic tissues, initiate some of these structural adaptations. Ethylene has been reported to stimulate the formation of *adventitious roots* in both flood-tolerant trees (e.g., *Salix* and *Alnus*) and flood-tolerant herbaceous species (e.g., *Phragmites*, *Ludwigia*, and *Lythrum salicaria*) and some flood-intolerant plants (e.g., tomato). These roots develop on the stem just above the anaerobic zone when these plants are flooded (Fig. 7.2a). They form as the original roots die and are able to function normally in an aerobic environment above the water line.

Stem Hypertrophy

Stem hypertrophy, a noticeable swelling of the lower stem of vascular plants, is another adaptation of many vascular plants to waterlogged conditions and hence serves as a good indicator or wetland conditions. When this hypertrophy occurs on a tree, it is called a *buttress* (Fig. 7.2b). It is a characteristic of swamp trees, such as bald and pond cypress (*Taxodium* spp.) and water and swamp black gum (*Nyssa* spp.). Hypertrophy is

Figure 7.2 Illustrations of morphological adaptations to flooding and waterlogging by vascular plants: (a) adventitious roots on willow (*Salix*) tree; (b) stem hypertrophy or buttresses on cypress (*Taxodium*) trees in a deepwater swamp; (c) fluted trunk on pin oak tree (*Quercus palustris*) in a freshwater forested wetland; (d) prop roots extending from *Rhizophora* mangrove trees in Costa Rica; and (e) pneumatophores ("knees") of *Taxodium* in a freshwater swamp. (Photo (a) by Ralph Tiner; (b), (c), (d), and (e) by W. J. Mitsch, reprinted with permission)

not caused by the formation of aerenchyma but rather by larger cells and lower density wood, also probably caused by ethylene production. A somewhat similar pattern of trees exhibiting flared or *fluted trunks* (Fig. 7.2c) at the ground surface is common in wetlands with several tree species, such as pin oak (*Quercus palustris*) and American elm (*Ulmus americana*).

Stem Elongation, Root Adaptations, and Lenticels

Another response stimulated by submergence is *rapid stem elongation* in such aquatic and semiaquatic plants as the floating heart (*Nymphoides peltata*), rice (*Oryza sativa*), and bald cypress (*Taxodium distichum*), stimulated by rising water levels. Bald cypress seedlings have rapid vertical growth rates supposedly to get the photosynthetic organs out of harm's way before standing water levels increase. The formation of *shallow root systems* by wetland plants is another clear and common adaptation by vascular plants to avoid anaerobic conditions. Deep taproots, common in upland forests, are almost never found in forested wetlands. Some species are facultative in the regard. Red maple (*Acer rubrum*) develops shallow root systems in wetlands but can have deep taproots in upland forests.

The red mangrove (*Rhizophora* spp.) grows on arched *prop roots* in tropical and subtropical tidal swamps around the world (Fig. 7.2d). These prop roots have numerous small pores, termed *lenticels,* above the tide level, which terminate in long, spongy, air-filled, submerged roots. The oxygen concentration in these roots, embedded in anoxic mud, may remain as high as 15 to 18 percent continuously, but if the lenticels are blocked, this concentration can fall to 2 percent or less in two days. Lenticels are also in the stems of flood-tolerant species, such as *Alnus glutinosa* and *Nyssa sylvatica,* and serve as conduits to the aerenchymatous tissue in the stem.

Pneumatophores

Similarly, the black mangrove (*Avicennia* spp.) tree produces thousands of *pneumatophores* (air roots) about 20 to 30 cm high by 1 cm in diameter, spongy, and studded with lenticels. They protrude out of the mud from the main roots and are exposed during low tides. The oxygen concentration of the submerged main roots has a tidal pulse, rising during low tide and falling during submergence, reflecting the cycle of emergence of the air roots. These pneumatophores are often covered with lenticels that aid in root aeration. The "knees" of bald cypress (*Taxodium distichum*) (Fig. 7.2e) are pneumatophores that improve gas exchange to the root system. Cypress knees generally develop only when the trees are in waterlogged or flooded soils, and their heights were often used as indicators of high water levels in the wetlands.

Physiological Adaptations

Vascular emergent and floating-leaved wetland plants are sessile; only their roots are in an anoxic environment. Typically, if the roots of a flood-sensitive upland plant are inundated, the oxygen supply rapidly decreases. This shuts down the aerobic metabolism of the roots, impairs the energy status of the cells, and reduces nearly all metabolically mediated activities such as cell extension and division and nutrient absorption. Even when cell metabolism shifts to anaerobic glycolysis, adenosine triphosphate (ATP) production is reduced. Toxic metabolic end products of fermentation may accumulate, causing cytoplasmic acidosis and eventually death. Anoxia is soon followed by pathological changes in the mitochondrial structure. The complete destruction of mitochondria and other organelles occurs within 24 hours. Anoxia also changes the chemical environment of the root, increasing the availability of reduced forms of iron, manganese, and sulfur, which may accumulate to toxic levels in the root. Several physiological adaptations of wetland vascular plants attempt to solve the problem of anoxic conditions in the root system.

Pressurized Gas Flow

Dacey (1980, 1981) first described a particularly interesting adaptation that increases the oxygen supply to the roots of the floating-leaved spatterdock (*Nuphar luteum*; currently subdivided into several species including *N. adventa*). Since then, a similar adaptation of pressurized gas flow from the surface to the rhizosphere has been demonstrated for other floating-leaved species. Fourteen emergent plants in southwestern Australia were tested, and eight were found to have significant gas flow ($0.2 \rightarrow 10$ cm^3 min^{-1} culm^{-1}), including *Baumea articulata, Cyperus involucratus, Eleocharis sphacelata, Schoenoplectus validus, Typha domingensis, T. orientalis, Phragmites australis,* and *Juncus ingens* (Table 7.2). These results for such a wide variety of plants suggest that internal pressurization and pressurized gas flow may be common to many hydrophytes.

Table 7.2 Pressurized gas flow in culms or leaves of 13 wetland plants and 1 upland plant in Australia[a]

Water Depth Species	N	ΔP_s(Pa)	Flow Rate (cm^3 min^{-1} culm^{-1})
Potentially Deepwater Plants			
Phragmites australis	12	573 ± 54	5.3 ± 0.4[b]
Typha orientalis	8	1,070 ± 120	4.4 ± 0.3[b]
Typha domingensis	6	780 ± 140	3.4 ± 0.4[b]
Marginal Depth (<1 m) Plants			
Juncus ingens	11	222 ± 24	1.2 ± 0.1[c]
Eleocharis sphacelata	10	1,080 ± 86	0.85 ± 0.02
Schoenoplectus validus	9	1,310 ± 124	0.29 ± 0.05
Baumea articulata	16	494 ± 58	0.23 ± 0.06
Very Shallow Water or Moist-Soil Plants			
Cyperus involucratus	11	903 ± 234	0.33 ± 0.09[c]
Canna sp.	5	27 ± 5	0.06 ± 0.01
Myriophyllum papillosum[d]	6	68 ± 12	0.04 ± 0.01
Cyperus eragrostis	8	111 ± 34	0.02 ± 0.01
Ludwigia pelloides[d]	5	57 ± 1	<0.01
Bolboschoenus medianus	15	2 ± 31	<0.01
Not a True Wetland Plant			
Arundo donax	6	1 ± 10	<0.01

[a]Water depths refer to the potential depths that these plants can grow based on other studies and plant size. ΔP_s refers to the static pressure differential in the plant stem. Plants are listed in order of decreasing gas flow rates. Numbers indicate averages ± standard deviations.
[b]Small specimens or leaves had to be removed to get flow rates within measuring range.
[c]Gas flow measured through detached culms.
[d]Creeping, floating plants that grow in shallow water.
Source: Brix et al. (1992).

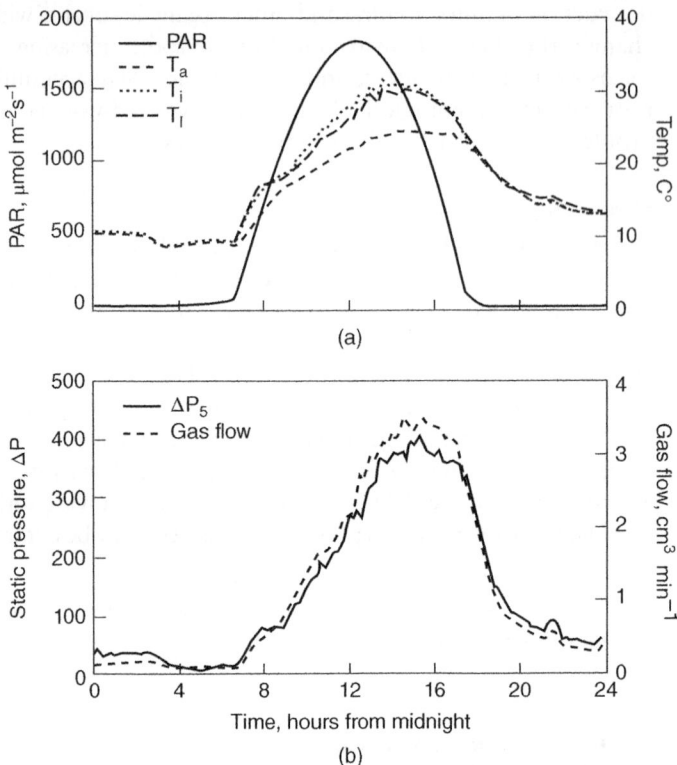

Figure 7.3 [Gas flow.ai] Diel variation of solar energy, temperature, internal pressure, and gas flow for *Typha domingensis*: (a) photosynthetically available solar radiation (PAR), air temperature (T_a), internal leaf temperature (T_i), and leaf surface temperature (T_l); (b) internal static pressure differential (ΔP_s) and gas flow. (From Brix et al., 1992)

T. domingensis had a dramatic diel pattern of convective gas flow related to air and subsequent leaf temperatures (Fig. 7.3). Gas flows of 0.1 to 0.2 cm³ min⁻¹ culm⁻¹ occurred at night but increased to a rate as high as 3 cm³ min⁻¹ culm⁻¹ during the afternoon. The results were interpreted to suggest that humidity-induced pressurization was the dominant driving force for the gas flow in the plants.

The pressure produced by this range of plants matched very nicely the approximate depths at which these plants can potentially occur in wetlands (Table 7.2). *Phragmites*, *Eleocharis*, and the two *Typha* species can grow in water depths up to 2 m; *Schoenoplectus*, *Juncus*, and *Baumea* are found in water less than 1 m deep; the two *Cyperus* species, *Bolboschoenus*, and *Canna*, grow in very shallow water to wet soils. Air moves into the internal gas spaces (the lacunar system) of aerial leaves and is forced down through the aerenchyma of the stem into the roots by a slight pressure (~200–1,300 Pa) generated by a gradient in temperature and water vapor pressure. Older leaves often lose their capacity to support pressure gradients, and so the return flow of gas from the roots is through the older leaves, which are rich in carbon dioxide and methane from root respiration. The gas exchange to the rhizosphere through dead culms of *P. australis* was sufficient to maintain aerobic respiration of the plant roots. The dead culms also provided an escape channel for excess CO_2 and CH_4 in the roots.

Grosse and others (Grosse and Schröder, 1984; Schröder, 1989; Grosse et al., 1992) described a similar process in swamp trees, specifically in common alder (*Alnus glutinosa*), the dominant tree species of European floodplain forests and riverine temperate forests. Seedlings and dormant (leafless) trees of flood-tolerant species show enhanced gas transport from aerial shoots to the roots when the shoots are heated by the sun or incandescent light, compared to plants in the dark. Grosse et al. (1998) called this phenomenon "pressurized gas flow" or "thermo-osmotic" gas flow. This phenomenon occurs when a temperature gradient is established between the exterior ambient air and the interior gas spaces in a plant's cortical tissue. A second requirement is a permeable partition between the exterior and interior with pore diameters "similar to or smaller than the 'mean free path length' of the gas molecules in the system (e.g., 70 nm at room temperature and standard barometric pressure" (Grosse et al., 1998). In alder, meristematic tissue in the lenticels forms such a partition.

When the surface of the stem is mildly heated by sunlight, the mean free path length of gas molecules in the intercellular spaces of the plant increases, preventing the molecules from moving out through the osmotic barrier in the lenticels. The cooler exterior molecules, however, can still diffuse into the plant. This sets up an internal pressure gradient that forces gas down through the plant stem to the roots. This "thermal pump" is not as effective in moving oxygen to the roots in alder as is extensive aerenchyma tissue. For example, alder seedlings grown in flooded soil for two months transported oxygen at eight times the rate of seedlings grown in aerated soil. The difference was because of aerenchyma and lenticel development under flooded conditions. By comparison, the thermo-osmotic effect (in the absence of flooding) led to a fourfold increase in the rate of gas transport. The thermal pump is also not as active in foliated trees as in dormant ones. Therefore, for trees, the adaptation appears to be most effective in enhancing root aeration during seedling establishment in saturated soils before aerenchyma development is accomplished, and in deciduous trees during the dormant season.

Rhizosphere Oxygenation

Secondary effects of adaptations to root aeration influence other parts of the plants or their environment. When anoxia is moderate, the magnitude of oxygen diffusion through many wetland plants into the roots is apparently large enough not only to supply the roots but also to diffuse out, oxidize the adjacent anoxic soil, and produce an oxidized rhizosphere (see Chapter 5: "Wetland Soils"). The brown deposits found around the roots of *Spartina alterniflora* in Figure 7.1 were composed of iron and manganese deposits formed when root oxygen comes in contact with reduced soil ferrous ions. Oxygen diffusion from the roots is an important mechanism that moderates the toxic effects of soluble reduced ions such as manganese in anoxic soil and restores ion uptake and plant growth. These ions tend to be reoxidized and precipitated in the rhizosphere, which effectively detoxifies them. In a similar vein, McKee et al. (1988) determined that soil redox potentials were higher and that pore water sulfide concentrations were three to five times lower in the presence of the aerial prop roots of the red mangrove (*Rhizophora*) or the pneumatophores of the black mangrove (*Avicennia*) than in nearby bare mud soils, in all probability because of the diffusion of oxygen from the mangrove roots into the soil. An interesting possibility is that the root systems of these flood-tolerant plants may modify sediment anoxia enough to allow the survival of nearby nontolerant plants (Ernst, 1990).

The presence of *oxidized rhizospheres* (now called *oxidized pore linings* by soil scientists; see Chapter 5), which form as a result of root oxidation, is an important way in which wetlands can be identified. Long after the plant roots die, residual veins of red and orange, resulting from oxidized iron (Fe^{3+}) deposits, remain in many mineral soils, a telltale sign that hydrophytes had been living in the soil. They are used in wetland delineation practices as one indicator that hydric soils and, thus, wetlands are present.

Lower Water Uptake

Plants intolerant to anaerobic environments typically show decreased water uptake despite the abundance of water, probably as a response to an overall reduction of root metabolism. Decreased water uptake results in symptoms similar to those seen under drought conditions: closing of stomata, decreased carbon dioxide uptake, decreased transpiration, and wilting. The adaptive advantage of these responses is probably the same as for drought-stricken plants—to minimize water loss and accompanying damage to the cytoplasm. An accompanying depression of the photosynthetic machinery is generally seen as an unavoidable corollary.

Sulfide Avoidance

Sulfur as sulfide is toxic to plant tissues. The element is reduced to sulfide in anaerobic soils and accumulates to toxic concentrations, especially in coastal wetlands. Although sulfate uptake is metabolically controlled, sulfide can enter the plant without control and is found in elevated concentrations in many flood-adapted species under highly reduced conditions. In experiments with *Spartina alterniflora*, a salt marsh species, and *Panicum hemitomon*, a freshwater marsh species, Koch et al. (1990) reported that the activity of alcohol dehydrogenase (ADH), the enzyme that catalyzes the terminal step in alcohol fermentation, was significantly inhibited by hydrogen sulfide and that this inhibition may help explain the physiological mechanism of sulfide phytotoxicity often seen in salt marshes. Sulfur tolerance in wetland plants varies widely, probably because of the variety of detoxification mechanisms available. These include the oxidation of sulfide to sulfate through root aeration of the rhizosphere; the accumulation of sulfate in the vacuole; the conversion to gaseous hydrogen sulfide, carbon disulfide, and dimethylsulfide and their subsequent diffusive loss; and a metabolic tolerance to elevated sulfide concentrations.

Anaerobic Respiration

Under conditions of oxygen deprivation, plant tissues respire anaerobically, as described for bacterial cells. In most plants, pyruvate, the end product of glycolysis,

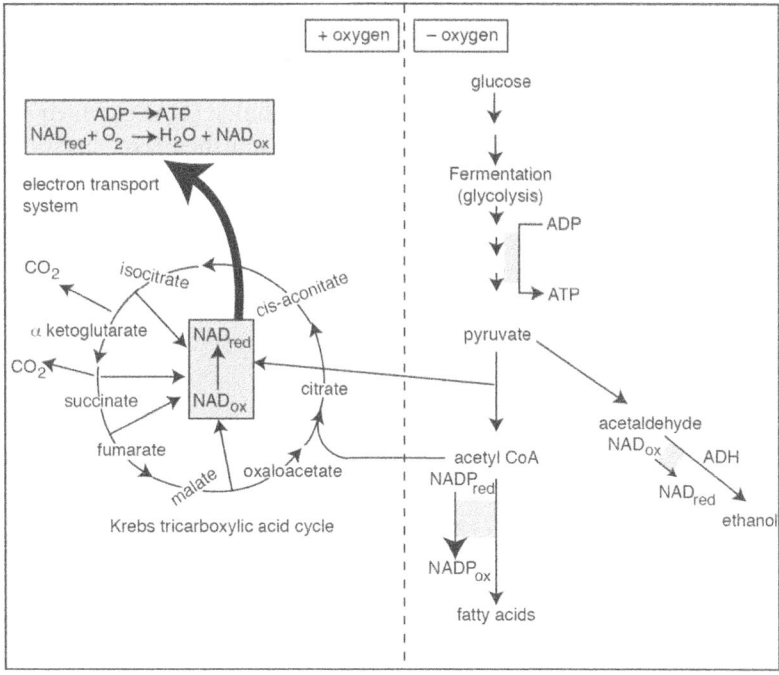

Figure 7.4 Schematic of metabolic respiration pathway in flood-tolerant plants. Left side of diagram is aerobic respiration; right side is anaerobic respiration (fermentation/glycolsis), which yields pyruvate, acetaldehyde, ethanol, and fatty acids such as malic acid.

ADH: alcohol dehydrogenase.
NAD: nicotinamide adenine dinucleotide.
NADP: NAD phosphate.
Subscripts refer to oxidized (ox) and reduced (red) forms.

is decarboxylated to acetaldehyde, which is reduced to ethanol (see right side of Fig. 7.4). Both of these compounds are potentially toxic to root tissues. Flood-tolerant plants often have adaptations to minimize this toxicity. For example, under anaerobic conditions, *S. alterniflora* roots show much increased activity of ADH, the inducible enzyme that catalyzes the reduction of acetaldehyde to ethanol. The increase in the enzyme indicates a switch to anaerobic respiration, and it explains why acetaldehyde does not accumulate in the root tissue. Ethanol does not accumulate either, although its production is apparently stimulated. It diffuses from rice roots during anaerobiosis, thus preventing a toxic buildup, and the same probably occurs in other wetland plants. Another metabolic strategy reduces the production of alcohol by shifting the metabolism to accumulate nontoxic organic fatty acids instead (Fig. 7.4). At one time, it was suggested that malic acid (malate) accumulation may be a characteristic feature of wetland species. The accumulation of malate cannot easily be interpreted, however, in part because malate is an intermediate in several metabolic pathways.

The metabolic problem encountered by plants deprived of oxygen is the loss of the electron acceptor that enables normal energy metabolism through ATP formation and use. The metabolic bottleneck in this process is often the electron-accepting coenzyme nicotinamide adenine dinucleotide (NAD), which is reduced in the oxidative steps of carbohydrate metabolism and then reoxidized in the mitochondria by molecular oxygen to yield the biological energy currency ATP (see shaded boxes in left side of Fig. 7.4). In the absence of oxygen, reduced NAD (NAD_{red}) accumulates and "jams" the metabolic system, blocking ATP generation. In the process of fermentation, acetaldehyde replaces oxygen, reoxidizing reduced NAD. Malate acts in the same way through the tricarboxylic acid cycle. Thus, glycolysis can occur as long as NAD is reoxidized to the oxidized form, NAD_{ox}.

Whole Plant Strategies

Many plant species have evolved avoidance or escape strategies by life-history adaptations. The five most common of these strategies are listed next:

1. The timing of seed production in the non–flood season by either delayed or accelerated flowering
2. The production of buoyant seeds that float until they lodge on high, unflooded ground
3. The germination of seeds while the fruit is still attached to the tree (*vivipary*), as in the red mangrove
4. The production of a large, persistent seed bank
5. The production of tubers, roots, and seeds that can survive long periods of submergence

In many riparian wetlands, flooding occurs primarily during the winter and early spring, when trees are dormant and much less susceptible to anoxia than they are during the active growing season. The *viviparous seedlings* that germinate live in the canopy of red mangrove (*Rhizophora*) trees fall in the water from the canopy after germination and are transported, sometimes great distances. (See Chapter 9: "Mangrove Swamps"). The seedling rights itself to a vertical position and develops roots if the water is shallow until it lodges in shallow sediments, allowing the seedling to then grow to a tree.

The freshwater aquatic monocot *Sagittaria latifolia* has a similar way of distributing its seeds. (The plant is sometimes called duck potato because its seeds resemble potatoes.) The seed floats through a wetland until it lodges in a shallow area or amid other emergent macrophytes, after which it germinates.

Mutualism and Commensalism

The close interactions among members of an ecological community reflect the high degree of adaptability of members of the community, not only to their physical environment but also to their biological environment. This chapter has documented adaptations to the physical wetland environment, but positive interactions among organisms are also predicted to play a significant role in ecosystem dynamics, especially in marginal or stressed environments such as wetlands. Two such possibilities of reactions between wetland populations are *mutualism*, when there are positive and obligatory benefits to both populations, and *commensalism*, where one population benefits and the other does not have either a positive or negative effect (i.e., it is neutral).

Documentations of these effects in wetland environments are relatively few. Many appear to involve nutrients that are limiting in these environments. For example, Grosse et al. (1990) reported a commensalism or mutualism between alder (*Alnus*) trees and fungi. Increased levels of nitrogen fixation by the symbiotic fungus *Frankia alni* presumably occurred because of thermal pumping of oxygen through and out of the root system of common alder into the rhizosphere. Ellison et al. (1996) reported a mutualistic interaction between root-fouling sponges (*Todania ignis* and *Haliclona implexiformis*) and the red mangroves (*Rhizophora mangle*) on which they grow. Fine, adventitious mangrove rootlets ramify throughout the sponges. They absorb dissolved ammonium from the sponges, which stimulates additional root growth. The sponges also protect the roots from isopod attack. Mangrove roots, in turn, provide the only hard substrate for sponges in this habitat, and they stimulate sponge growth by leaking carbon.

These examples of the positive interactions among wetland species point to an extremely interesting line of neglected research that may lead to important new insights into the complexity of mutualistic adaptations in wetland ecosystems and their importance not only to the organisms involved but also to the energetic dynamics of the entire community.

Wetland Succession

Allogenic versus Autogenic Succession

The beginning and subsequent development of a plant community is characterized by the initial conditions at the site and by subsequent events, including the availability of viable seeds or other propagules, appropriate environmental conditions for germination and subsequent growth, and replacement by plants of the same or different species as site conditions change in response to both abiotic and biotic factors. The concept of succession (i.e., the replacement of plant species in an orderly sequence of development), in particular, has exerted a strong influence on plant ecology for more than a century. Ecological theories of plant succession were advanced by H. C. Cowles in his classic work on plant succession based on the sequential exposure of sand dunes on the southern and eastern shores of Lake Michigan (Cowles, 1899). In that study, dunes left bare from a retreating Lake Michigan were shown through a series of successional ecosystems over thousands of years to go in an orderly primary succession to a climax beech-maple forest (see Case Study 1).

Autogenic succession was further enunciated by Clements (1916) and applied to wetlands by the English ecologist W. H. Pearsall in 1920 and by an American, L. R. Wilson, in 1935. E. P. Odum (1969) adapted and extended the ideas of those early ecologists to include ecosystem properties such as productivity, respiration, and diversity. This classical use of the term *succession* involves three fundamental concepts: (1) vegetation occurs in recognizable and characteristic *communities*; (2) community change through time is brought about by the biota (i.e., changes are *autogenic*); and (3) changes are linear and *directed* toward a mature, stable *climax* ecosystem. Although this concept of autogenic succession was a dominating paradigm of great importance in terrestrial ecology, the concept has been challenged and altered for almost a century. Gleason (1917) enunciated an *individualistic* hypothesis to explain the distribution of plant species. His ideas have developed into the *continuum* concept, which holds that the distribution of a species is governed by its response to its environment (*allogenic succession*). Because each species responds differently to its environment, no two occupy exactly the same zone. The observed invasion/replacement sequence is also influenced by the chance occurrence of propagules at a site. The result is a continuum of overlapping sets of species, each responding to subtly different environmental cues. In this view, no communities exist in the sense used by Clements, and although ecosystems change, there is little evidence that this is directed or that it leads to a particular climax.

A key issue in discussions of ecosystem development is whether biota determine their own future by modifying their own environment, or if the development of an ecosystem is simply a response to the external environment. In the classical view of succession, wetlands are considered transient stages in the *hydrarch development* of a terrestrial forested climax community from a shallow lake (Fig. 7.5). In this view, lakes and open water gradually fill in as organic material from dying plants accumulates and minerals are carried in from upslope. At first, change is slow because the source of organic material is single-celled plankton. When the lake becomes shallow enough to support rooted aquatic plants, however, the pace of organic deposition increases. Eventually, the water becomes shallow enough to support emergent marsh vegetation, which continues to build a peat mat. Shrubs and small trees appear. They continue to transform the site to a terrestrial one, not only by adding organic matter to the soil but also by drying it through enhanced evapotranspiration. Eventually, a climax terrestrial forest occupies the site (Fig. 7.5a). The important point in this description of hydrarch succession is that most of the change is brought about by the plant community as opposed to externally caused environmental changes. A second important feature, shown in Figure 7.5b, is that the process can reverse if the environmental conditions, particularly the hydrology, change.

How realistic is this concept of succession? It is certainly well documented that forests do occur on the sites of former lakes, but the evidence that the successional sequence leading to these forests was autogenic is not clear. Because peat building is crucial to filling in a lake and its conversion to dry land, key questions involve the conditions for peat accumulation and the limits of that accumulation. Peat underlies many wetlands, often in beds 10 m or more deep. In coastal marshes, peat has accumulated at

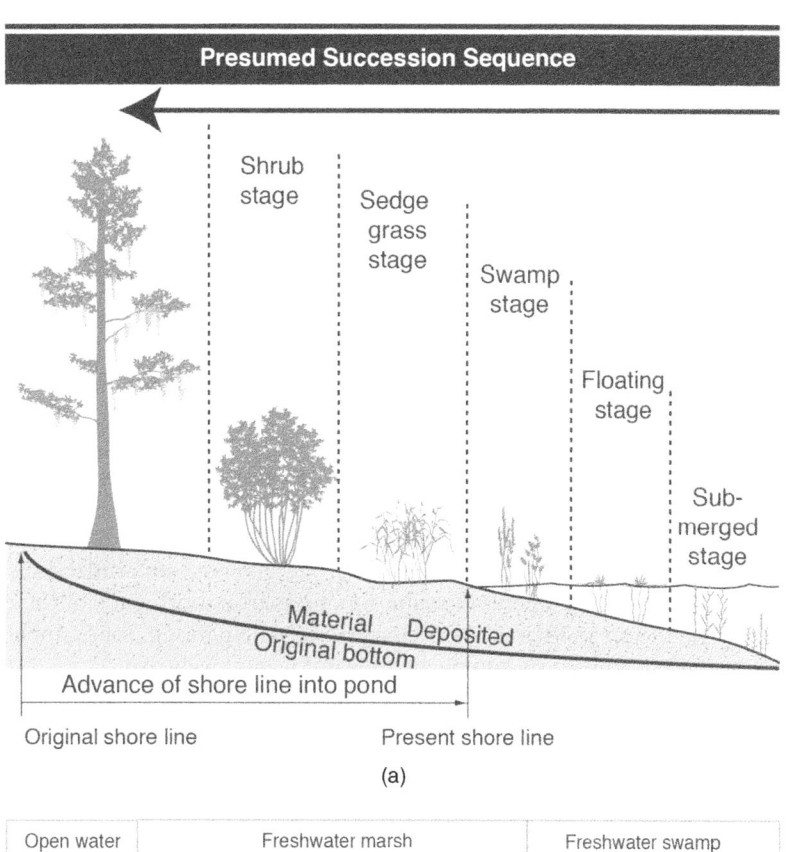

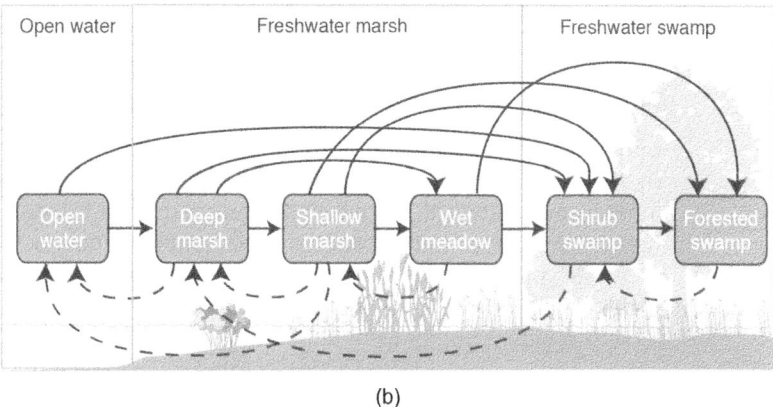

Figure 7.5 Classical hydrarch succession of freshwater wetlands: (a) succession from a pond to a terrestrial forest at the edge of a pond, and (b) general succession to mineral soil forested wetlands in glaciated regions of North America. ((b) after Golet et al., 1993)

rates varying from less than 1 to up to 15 mm/yr. Most of this accumulation seems to be associated with rising sea levels (or submerging land). By contrast, northern inland bogs accumulate peat at rates of 0.2 to 2 mm/yr. (See Chapter 12: "Peatlands.")

In general, accumulation occurs only in anoxic sediment. When organic peats are drained, they rapidly oxidize and subside, as farmers who cultivate drained marshes have discovered. As the wetland surface accretes and approaches the water surface or at least the upper limit of the saturated zone, peat accretion in excess of subsidence must cease. It is difficult to see how this process can turn a wetland into a dry habitat that can support terrestrial vegetation unless there is a change in hydrologic conditions that lowers the water table. For example, Cushing (1963) used paleoecological techniques to show that most of the peatlands in the Lake Agassiz plain (Minnesota and south-central Canada) formed during the mid-Holocene (beginning about 4,000 years ago) during a moist climatic period when surface water levels rose about 4 m.

Wetlands are at the center of the dispute about the importance of autogenic versus allogenic processes because of their transitional nature. In addition to being *seres*, wetlands are often described as being *ecotones*—that is, transitional spatial gradients between adjacent aquatic and terrestrial environments. Thus, wetlands can be considered transitional in both space and time. As ecotones, wetlands usually interact strongly to varying (allogenic) forcing functions from both ends of the ecotone. These forces may push a wetland toward its terrestrial neighbor if, for example, regional water levels fall, or toward its aquatic neighbor if water levels rise.

Alternately, plant production of organic matter may raise the level of the wetland, resulting in a drier environment in which different species succeed. Because these environmental changes can be subtle, it is often difficult to determine whether the observed ecosystem response is autogenic or allogenic. Without careful measurements, the causes of the response are often obscure.

CASE STUDY 1: Revisiting the Lake Michigan Dunes

In the early twentieth century, H. C. Cowles (1899, 1901, 1911) and Victor Shelford (1907, 1911, 1913) studied ponds of different ages in the Indiana dunes region along the southern shore of Lake Michigan. The ponds were thought to represent an autogenic successional sequence. Along this age gradient, the young ponds were deep and dominated by aquatic vegetation. Older ponds were shallower and supported emergent vascular plants along their borders. The oldest ponds were shallowest and contained the most "terrestrial" vegetation. This sequence was interpreted as evidence of classical autogenic succession.

Wilcox and Simonin (1987) and Jackson et al. (1988) revisited the Indiana dunes ponds. Using modern quantitative methods of ordination, they found the same progression of plant species from young to old ponds, supporting the sequence observed by earlier workers. In addition to the current vegetation, they also examined pollen and macrofossils in the sediments of a 3,000-year-old pond to determine whether the sediments support the presumed successional sequence found in the modern-day chronosequences. Pollen and macrofossil data older than 150 B.P. (before the present) consisted of a diverse assemblage of submersed, floating-leaved, and emergent macrophyte groups. The data indicated a major and rapid vegetation change after 150 B.P., which the authors attributed to post-European settlement, such as railroad construction and forest clearing.

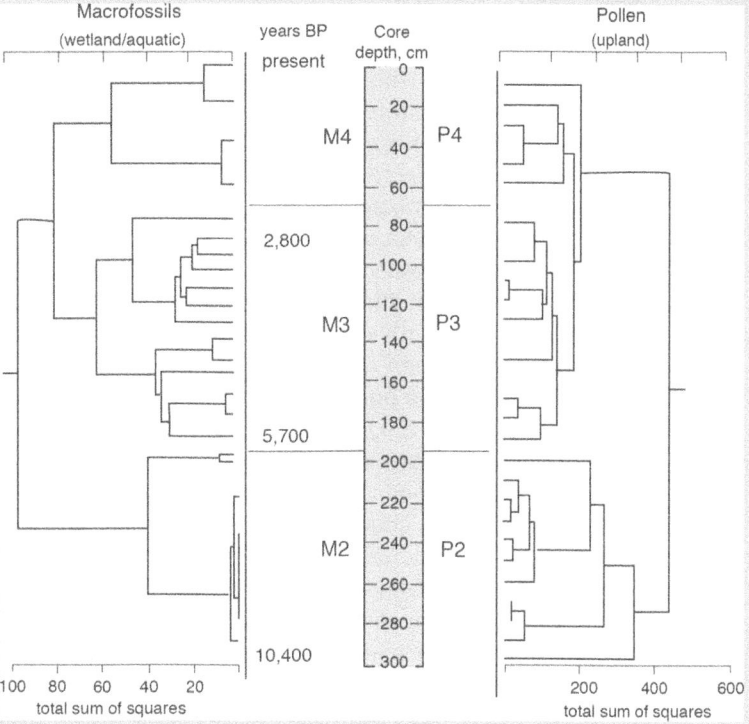

Figure 7.6 **Stratigraphically constrained cluster diagrams for macrofossil and pollen data from an Indiana dunes pond in northern Indiana adjacent to Lake Michigan. Macrofossil zonation is based on the presence or absence of aquatic and wetland taxa; pollen zonation is based on percentages of selected upland pollen types. Close individual taxa indicate close occurrence in fossil record. Thus the macrofossil record indicates three different groups of organisms. (After Singer et al., 1996)**

To further evaluate the historical changes in the Indiana dunes ponds, Singer et al. (1996) examined the sediment pollen and macrofossil record of aquatic and emergent plants in one of the old Indiana dunes pond sites and compared it with the regional terrestrial pollen record (of airborne pollen found in the same cores). The latter tracks long-term climate changes in the region. If aquatic and emergent paleotaxonomic remains showed changes in species dominance that mirrored the terrestrial pollen record, then the changes in the ponds could be attributed to regional climate change rather than to autogenic processes. From their 10,000-year record, Singer et al. (1996) determined that historic changes in pond vegetation did correspond to regional climate change (Fig. 7.6). Between 10,000 and 5,700 B.P., the sampled area was a shallow lake; the regional climate was mesic (a pine/oak/elm terrestrial assemblage). A rapid increase in oak and hickory pollen around 5,700 B.P. signaled a regional climate shift to a drier environment. At the same level in the sediment record, the pond macrofossil record showed a rapid shift to a peat-forming marsh environment. After about 3,000 B.P., modest increases in beech and birch pollen suggested a trend toward a cooler, moister climate. The concomitant pond vegetation remained dominated by emergents, but transitions among several taxa suggest that water-level fluctuations and occasional fires were characteristic of the period.

These studies, taken together, provide a fuller, more complex picture of plant development than the autogenic succession process proposed in Cowles's and Shelford's earlier studies. The picture that emerges is one of an interaction between allogenic and autogenic processes, with allogenic forces driving the development of the biotic system, but modified by autogenic processes. Over the 10,000-year span of the fossil record, changes in the plant assemblage correlated well with regional climate change. However, during the same period, the lake was slowly filling with organic sediments, first 100 cm of gyttja, characteristic of open freshwater systems; then 200 cm of fibrous peat, characteristic of vascular aquatic plants. During the period of a slow climate shift to a less xeric environment after approximately 3,000 B.P., the pond environment remained a marsh, although the species assemblage changed. This fits well with the idea that the organic sediments moderated the climatic influence on the local water levels. Finally, during the modern period after about 150 B.P., human activities, which probably altered water levels locally, resulted in rapid vegetation changes.

The Community Concept and the Continuum Idea

The Indiana dunes ponds example just described in Case Study 1 is only a small part of the extensive literature concerning questions about plant and ecosystem development in wetlands. The idea of the community is particularly strong in wetland literature. Historic names for different kinds of wetlands—marshes, swamps, carrs, fens, bogs, reedswamps—often used with the name of a dominant plant (*Sphagnum* bog, leatherleaf bog, cypress swamp)—signify our recognition of distinctive associations of plants that are readily recognized and at least loosely comprise a community. One reason these associations are so clearly identified is that zonation patterns in wetlands often tend to be sharp, having abrupt boundaries that call attention to vegetation change and, by implication, the uniqueness of each zone. The plant community is central to the historic idea of succession because the mature climax resulting from succession was presumed to be a predictable group of plant species, with each group dependent on the regional climate.

The identification of a community is also, to some extent, a conceptual issue that is confused by the scale of perception. Field techniques are adequate to describe the vegetation in an area and its variability. However, its homogeneity—one index of community—may depend on size. For example, Louisiana coastal marshes have been classified into four zones, or communities, based on the dominant vegetation. If the size of the sampling area is large enough, any sample within one of these zones will always identify the same species. If smaller grids are used, however, differences appear within a zone. The intermediate marsh zone is dominated on a broad scale by *S. patens*, but aerial imagery shows patterns of vegetation within the zone, and intensive sampling and cluster analysis of the vegetation reveal five subassociations that are characteristic of intermediate marshes. Is the intermediate marsh a community? Are the subassociations communities? Or is the community concept a pragmatic device to reduce the bewildering array of plants and possible habitats to a manageable number of groups within which there are reasonable similarities of ecological structure and function?

Supporters of the continuum concept would argue that the scale dependence of plant associations illustrates that individual species are simply responding to subtle environmental cues, implying little, if anything, about communities, and that plant zonation simply indicates an environmental gradient to which individual species are responding. The reason zonation is so sharp in many wetlands, they argue, is that environmental gradients are "ecologically" steep, and groups of species have fairly similar tolerances that tend to group them on these gradients.

One major difference between classical community ecologists and proponents of the continuum idea is the greater emphasis put on allogenic processes by the latter. In some wetlands, abiotic environmental factors often seem to overwhelm biotic forces. In coastal areas, plants can do little to change the tidal pulse of water and salt. Tidal energy may be modified by vegetation as stems create friction that slows currents or as dead organic matter accumulates and changes the surface elevation. These effects are limited, however, by the overriding tides. These wetlands are often in dynamic equilibrium with the abiotic forces, an equilibrium that is sometimes called *pulse stability* (see discussion later in this chapter).

In the low-energy environment of northern peatlands, in contrast to tidal marshes, hydrologic flows can be dramatically changed by biotic forces, resulting in distinctive patterned landscapes. Thus, changes in wetlands may be autogenic but are not necessarily directed toward a terrestrial climax. In fact, wetlands in dynamically stable environmental regimes seem to be extremely stable, contravening the central idea of succession. Pollen profiles were used to determine the successional sequence in British northern peatlands. Sequences were variable and there were reversals and skipped

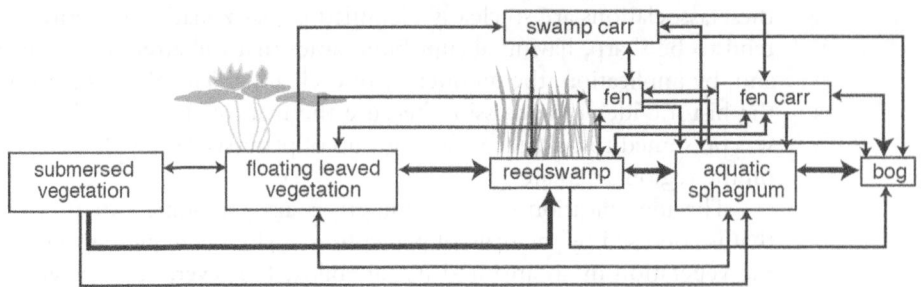

Figure 7.7 Successional sequences reconstructed from stratigraphic and palynological studies of postglacial British peatlands. Thicker lines indicate the more common transitions. (After Walker, 1970)

stages that may have been influenced by the dominant species first reaching a site (Fig. 7.7). A bog, not some type of terrestrial forest that hydrarch succession would have predicted, was the most common endpoint in most of the sequences described.

Linear Directed Change

If plant species development on a site is determined by allogenic processes and is, therefore, simply a response to environmental forcing processes, then the successional concept of linear directed change makes little sense. Although the scientific literature is replete with schematic diagrams showing the expected successional sequence from wetland to terrestrial forest, most of these are based on observed zonation patterns (or chronosequences), assuming that these spatial patterns presage the temporal pathway of change.

However, paleological analyses of soil profiles (such as those discussed earlier for the Indiana dunes ponds) provide the best evidence to evaluate the concept. These records, mostly from northern peat bogs, suggest two generalizations: (1) In some sites, the current vegetation has existed for several thousands of years; and (2) climatic change and glaciation had major impacts on plant species composition and distribution; generally, bogs expanded during warm, wet periods and contracted during cool, drier periods. Pollen sequences, however, are generally consistent across Europe and North America, indicating a response to similar global climate shifts. West (1964), as quoted in McIntosh (1985), wrote tellingly: "We may conclude that our present plant communities have no long history in the Quaternary, but are merely temporary aggregations under given conditions of climate, other environmental factors, and historical factors."

Seed Banks

Seed banks, referred to as buried reserves of viable seeds (Keddy, 2010), are an important component of wetland succession. Many studies have documented the role of chance in the development of plant communities, especially in the early stages. The chance development can be the result of the availability of a seed bank and a changing environmental condition (e.g., the flooding of a site after years of dry conditions).

In this respect, studies of seed banks and their role in the introduction and invasion of plant species have been important. If the development of plants on a site can be explained only in terms of the response of individual species to local conditions, then the previous history of the site is important because it determines what propagules are present for future invasion. This—the sediment seed bank—has been found to be extremely variable—both in space and in time. Pederson and Smith (1988) made these five generalizations about freshwater marsh seed banks:

1. Marshes with drawdowns produce the greatest number of seeds.
2. Seed banks are dominated by the seeds of annual plants and flood-intolerant species. Areas that contain emergent plants have greater seed densities than mud flats. Perennials generally produce fewer seeds that have shorter viability than annuals. They are more likely to reproduce by asexual means such as rhizomes.
3. Seed distribution decreases exponentially with the depth of the sediment.
4. Water is a major factor in seed banks. Seeds are concentrated along drift lines. The kinds of seeds produced depend on the flooding regime—by submergents when deep flooded, by emergents when periodically flooded, and by flood-intolerant annuals during drawdowns.
5. Saline zones produce few seeds. A salt marsh is an example of a perennial-dominated system in which most reproduction is asexual.

The germination of seedlings from a seed bank is similarly influenced by many factors that vary in space and time. Environmental factors, such as flooding, temperature, soil chemistry, soil organic content, pathogens, nutrients, and allelopathy, have been shown to influence recruitment. Water, in particular, is a critical variable, because most wetland plant seeds require moist but not flooded conditions for germination and early seedling growth. As a result of this restrictive moisture requirement, it is common to find even-aged stands of trees at low elevations in riparian wetlands, reflecting seed germination during relatively uncommon years when water levels were unusually low during the spring and summer.

Postrecruitment processes play a major role in the distribution of adult plants at a site, leading to plant assemblages that cannot be predicted from the seed bank alone. Thus, in coastal areas where the dominant plant, *Spartina alterniflora*, occurs in large monotypic stands, it is often the pioneer species and remains dominant throughout the life of the marsh. In contrast, in tidal and nontidal freshwater marshes, the seed bank is much larger and richer, and the first species to invade a site may later be replaced by other species.

Models of Wetland Community Succession

Plant Species Functional Groups

Historically, although the community concept has been of immense value in ecology, it has been criticized for being imprecise and not subject to accurate predictive models for ecological communities. Some ecologists have addressed this problem in different ways. One approach is to describe vegetation communities in terms of *guilds* or the more recent term *functional groups* that can be defined by measurable traits (Keddy, 2010). A functional group is defined as a group of functionally similar species in a vegetation community. This approach has two advantages: (1) It collapses the large number of vegetation species in a wetland to a manageable subset; and (2) species are defined in terms of measurable functional properties. Boutin and Keddy (1993) illustrated a functional classification of 43 wetland plant species in eastern North America according to 27 functional traits (Table 7.3). Figure 7.8 summarizes the results, which groups the species according to their traits into three groups: (1) ruderal annuals, (2) interstitial perennials, and (3) matrix perennials. These are further split into seven guilds, ranging from *obligate annuals* that flowered in the first year and then die at the end of the growing season to tall *clonal dominant* species with deep roots such as *Typha* x *glauca*) that reproduce vegetatively with extensive lateral spread. Most of the functional groups appear to fall along a continuum of life histories adjusted to different light regimes, which is consistent with the results of other studies.

Environmental Sieve Model

Van der Valk's (1981) environmental sieve model of wetland succession (Fig. 7.9) is also a Gleasonian model and is similar to Keddy's model in several ways. The presence and the abundance of each species depend on its life history and its adaptation to

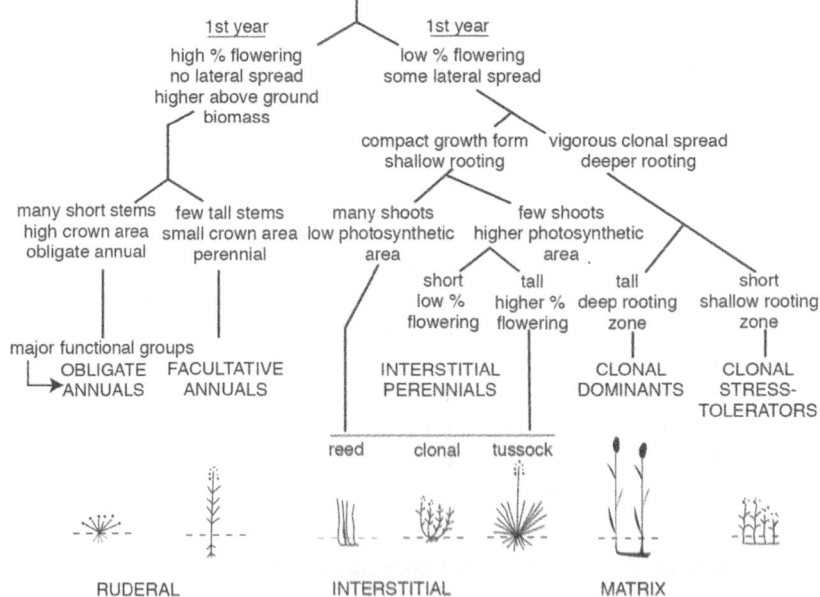

Figure 7.8 Functional classification of 43 species of plants from various wetland habitats in eastern North America, based on 27 plant traits displayed in Table 7.3. (After Boutin and Keddy, 1993).

Table 7.3 Traits measured on wetland plant species for functional guild classification

A. Traits Measured on 1-Year-Old Plants in the Garden

1	Life span: 1 = annuals 2 = facultative annuals (100% flowering) 3 = partly facultative annuals (>50<100 % flowering) 4 = perennials (<50% flowering)
2	Percentage flowering first year
3	Final height or highest height (cm)
4	Rate of shoot extension (cm/day): $\dfrac{\log_n \text{height at day 94} - \log_n \text{height at day 36}}{\text{Day 94} - \text{Day 36}}$
5	Total biomass at harvest (g)
6	Aboveground biomass (g)
7	Belowground biomass (g)
8	Ratio belowground/aboveground biomass
9	Photosynthetic area (cm^2); includes leaves and green stems
10	Photosynthetic area/total biomass (cm^2/g)
11	Photosynthetic area/total volume occupied by a plant (cm^2/ml) measured by displacement of water in graduated cylinder
12	Total biomass/total volume (g/ml)
13	Total number of tillers or shoots
14	Crown cover (cm^2): $((D_1 + D_2)/4)^2$ where D_1 = first measure of crown diameter D_2 = second measurement at right angle to first
15	Stem diameter at ground level (cm)
16	Depth to belowground system (cm)
17	Diameter of belowground system, i.e., rhizome or main roots (cm)
18, 19	Shortest (18) and longest (19) distances between two shoots or tillers (measure of degree of clumping of aerial stems) (cm)

B. Traits Measured on Plants in Natural Wetlands (Adult Traits)

20	Total height (cm)
21	Total number of tillers or shoots
22	Stem diameter at ground level (cm)
23, 24	Shortest (23) and longest (24) distances between two shoots or tillers (cm)
25	Diameter of belowground system, i.e., rhizome or main roots (cm)
26	Depth to belowground system (cm)

C. Trait Measured under Greenhouse Conditions

27	Relative growth rate (RGR) (day^{-1}) between days 10 and 30

Source: Boutin and Keddy (1993).

the environment of a site. In van der Valk's model, all plant species are classified into life-history types, based on potential life span, propagule longevity, and propagule establishment requirements. Each life-history type has a unique set of characteristics and, thus, potential behavior in response to controlling environmental factors such as water-level changes. These environmental factors comprise the "environmental sieve" in van der Valk's model. As the environment changes, so does the sieve and, hence, the species present.

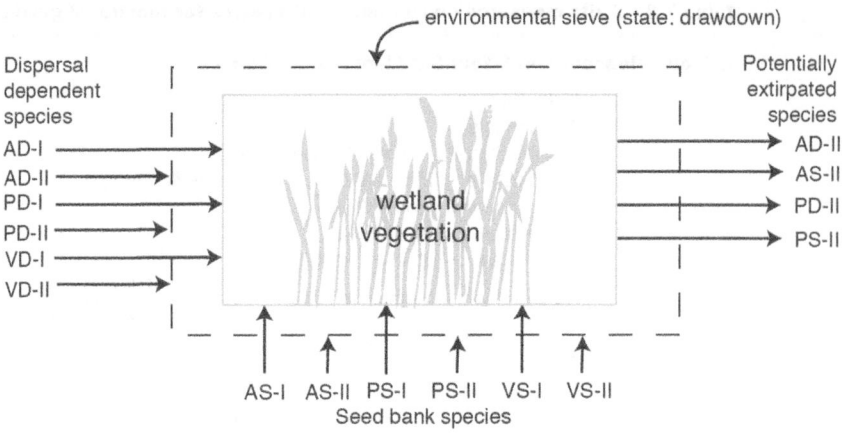

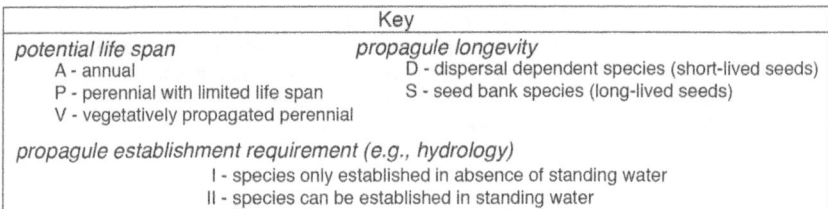

Figure 7.9 General sieve model of Gleasonian wetland (freshwater marsh) succession proposed by van der Valk (1981).

Centrifugal Organization Concept

Several other models of community change have been developed, although few have been applied to wetlands. Grime (1979) proposed that changes in species composition and richness of herbaceous plants was related to the gradients of disturbance and stress factors, which reduced biomass and determined which functional plant strategies would work best. Tilman (1982) suggested that competition among plants controlled community plant distribution, with each species limited by a different ratio of resources and spatial heterogeneity of the resources.

Wisheu and Keddy (1992) combined aspects of both Grime's and Tilman's models to propose a model of centrifugal organization of plant communities (Fig. 7.10a). Centrifugal organization describes the distribution of species and vegetation types along standing-crop gradients caused by combinations of environmental constraints. Wisheu and Keddy (1992) summarize the concept as follows:

> Gradients radiate outwards from a single core habitat to many different peripheral habitats. The assumed mechanism is a competitive hierarchy where weaker competitors are restricted to the peripheral end of the gradient as a result of a trade-off between competitive ability and tolerance limits. The benign ends of the gradients comprise a core habitat, which is dominated by the same species. At the peripheral end of each axis, species with specific adaptations to particular sources of adversity occur.

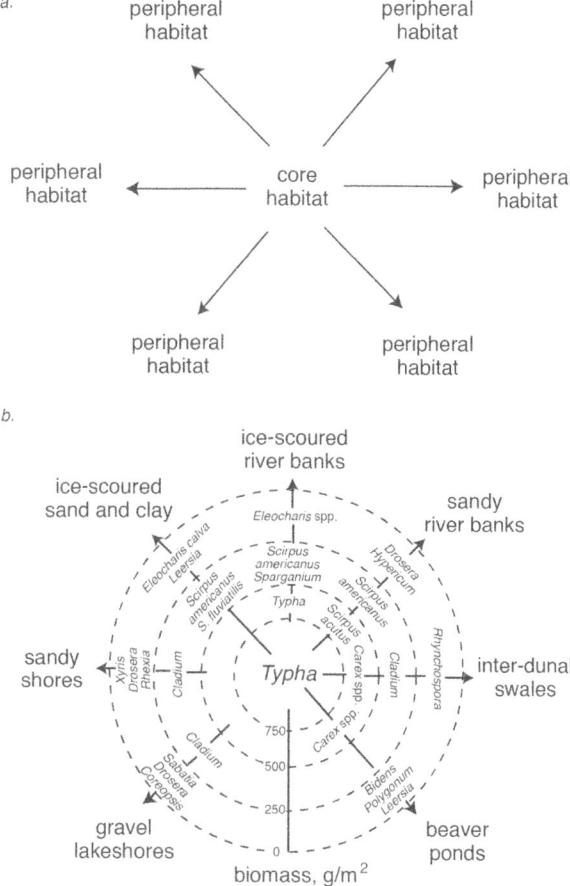

Figure 7.10 Centrifugal organization models illustrating (a) transitions from core habitat to peripheral habitats along resource or stress gradients (general model) and (b) freshwater wetland pattern for eastern North America, where large, leafy species such as cattail (*Typha* spp.) occupy the core habitat, while several different species and communities occupy peripheral habitats stressed by infertile sand, ice scouring, and beaver activity. (From Wisheu and Keddy, 1992).

The core habitat in wetlands has low disturbance and high fertility and is dominated by species that form dense canopies, such as *Typha* in eastern North America (Fig. 7.10b). Peripheral habitats represent different kinds and combinations of stresses (infertility, disturbance) and support distinctive plant associations. The model allows one to predict how changes in gradients and, hence, peripheral habitats will change community composition. In the case of the *Typha*-core centrifugal model shown in Figure 7.10b, ice scouring, infertile sandy soils, flooding by beavers, and open shorelines are among the stresses that shift communities to less productive, albeit possibly more diverse, assemblages. In this model, rare species are restricted to the peripheral habitats that may contain most of the biological diversity of a landscape, suggesting that the model could be useful for the protection of rare and endangered plant species and the conservation of biodiversity (Keddy, 2010).

So far in this chapter, we have discussed vegetation changes in wetlands. We summarize this discussion with a statement by Bill Niering (1989):

> Traditional successional concepts have limited usefulness when applied to wetland dynamics. Wetlands typically remain wet over time exhibiting a wetland aspect rather than succeeding to upland vegetation. Changes that occur may not necessarily be directional or orderly and are often not predictable on the long term. Fluctuating hydrologic conditions are the major factor controlling the vegetation pattern. The role of allogenic factors, including chance and coincidence, must be given new emphasis. Cyclic changes should be expected as water levels fluctuate. Catastrophic events such as floods and droughts also play a significant role in both modifying yet perpetuating these systems.

Ecosystem Development

E. P. Odum (1969) described the maturation of ecosystems as a whole (as distinct from plants, communities and species) in an article entitled "The Strategy of Ecosystem Development." The concepts, in general, have withstood the test of time, and are republished with some update in Odum and Barrett (2005), published three years after Odum's death. In ecosystem development, species composition in immature to mature (climax) stages are less important than are ecosystem functions, such as those described in Table 7.4. Immature ecosystems, Odum had observed, are characterized,

Table 7.4 Selected attributes for ecosystem development

Ecosystem Type	Community Energetics				Community Structure			
	P:R Ratio*	P:B Ratio*	Net Community Production	Food Chains	Total Biomass and Nonliving Organic Matter	Species Diversity	Organism Size	
Developing	<1 or >1	High	High	Linear, grazing	Low	Increases Initially	Small	
Mature (Climax)	1	Low	Low	Weblike, detrital	High	High or Declines	Large	

	Natural Selection		Biogeochemical Cycles		Regulation	
Ecosystem Type	Growth Form	Life Cycle	Mineral Cycles	Internal Cycling	Resilience	Resistance
Developing	r-selection	Short, simple	Open	Not important	High	Low
Mature (Climax)	K-selection	Long, complex	Closed	Important	Low	High

*P = gross primary productivity; R = respiration; B = biomass
Sources: E. P. Odum (1969, 1971) and E. P. Odum and Barrett (2005)

in general, by high production to biomass ($P{:}B$) ratios; an excess of production over community respiration ($P{:}R$ ratio >1); simple, linear, grazing food chains; low species diversity; small organisms; simple life cycles; and open mineral cycles. In contrast, mature ecosystems such as old-growth forests, tend to use all of their production to maintain themselves and therefore have $P{:}R$ ratios about equal to 1 and little, if any, net community production. Production may be lower than in immature systems, but the quality is better; that is, plant production tends to be high in fruits, flowers, tubers, and other materials that are rich in protein. Because of the large structural biomass of trees in forested ecosystems, the $P{:}B$ ratio is small. Food chains are elaborate and detrital based, species diversity is high, space is well organized into many different niches, organisms are larger than in immature systems, and life cycles tend to be long and complex. Nutrient cycles are closed; nutrients are efficiently stored and recycled within the ecosystem.

It is instructive to see how wetland ecosystems fit into this scheme of ecosystem development. Do their ecosystem-level characteristics fit the classical view that all wetlands are immature transitional seres? Or do they resemble the mature features of a terrestrial forest? Five conclusions can be made:

1. *Wetland ecosystems have properties of both immature and mature ecosystems.* For example, nearly all of the nonforested wetlands have $P{:}B$ ratios intermediate between developing and mature systems and $P{:}R$ ratios greater than 1. Primary production tends to be very high compared with most terrestrial ecosystems. These attributes are characteristic of immature ecosystems. However, all of the ecosystems are detrital based, with complex food webs characteristic of mature systems.

2. *The Odum model used live biomass as an index of structure or "information" within an ecosystem.* This relationship is reflected in the high $P{:}B$ ratios (immature) of nonforested wetlands and the low $P{:}B$ ratios (mature) of forested wetlands. In a real sense, however, peat should be considered a structural element of wetlands because it is a primary autogenic factor modifying the flooding characteristic of a wetland site. If peat is included in biomass, herbaceous wetlands would have the high biomass and low $P{:}B$ ratios characteristic of more mature ecosystems. For example, a salt or fresh marsh has a live peak biomass of less than 2 kg/m^2. However, the organic content of a meter depth of peat (peats are often many meters deep) beneath the surface is on the order of 45 kg/m^2. This is comparable to the above-ground biomass of the most dense wetland or terrestrial forest. As a structural attribute of a marsh, peat is an indication of a maturity far greater than the live biomass alone would signify.

3. *Mineral cycles vary widely in wetlands.* They range from extremely open riparian systems in which surface water (and nutrients) may be replaced thousands of times each year to bogs in which nutrients are derived from precipitation alone and are almost quantitatively retained. An open nutrient cycle is a juvenile characteristic of wetlands, directly related to the large flux

of water through these ecosystems. Even in a system as open as a salt marsh that is flooded daily, however, about 80 percent of the nitrogen used by vegetation during a year is recycled from mineralized organic material.

4. *Spatial heterogeneity is generally well organized in wetlands along allogenic gradients.* The sharp, predictable zonation patterns and abundance of land–water interfaces are examples of this spatial organization. In forested wetlands, vertical heterogeneity is also well organized. This organization is an index of mature ecosystems. In most terrestrial ecosystems, however, the organization results from autogenic factors in ecosystem maturation. In wetlands, most of the organization seems to result from allogenic processes, specifically hydrologic and salinity gradients created by slight elevation changes across a wetland. Thus, the "maturity" of a wetland's spatial organization consists of a high level of adaptation to prevailing microhabitat differences.

5. *Life cycles of wetland consumers are usually relatively short but are often exceedingly complex.* The short cycle is characteristic of immature systems, although the complexity is a mature attribute. Once again, the complexity of the life cycles of many wetland animals seems to be as much an adaptation to the physical pattern of the environment as to the biotic forces. Many animals use wetlands only seasonally or only during certain life stages. For example, small marsh fish and shellfish make daily excursions into wetlands during high tides, retiring to adjacent ponds during ebb tides. Many fish and shellfish species migrate from the ocean to coastal wetlands to spawn or for use as a nursery. Waterfowl use northern wetlands to nest and southern wetlands to overwinter, migrating thousands of miles between the two areas each year.

Strategy of Wetland Ecosystem Development

In the previous sections, we showed that wetlands possess attributes of both immature and mature systems and that both allogenic and autogenic processes are important. Allogenic processes are important as forcing functions, which include factors such as hydrology and propagule introduction, change. Autogenic processes are important as the biota begin to control some of the physics and chemistry, as illustrated in Figure 4.1. In this section, we suggest that in all wetland ecosystems there is a common theme: Development insulates the ecosystem from its environment.

At the level of individual species, this occurs through genetic (structural and physiological) adaptations to anoxic conditions. At the ecosystem level, it occurs primarily through peat production, which tends to stabilize the flooding regime and shifts the main source of nutrients to recycled material within the ecosystem. In forests, shading is important in regeneration following disturbance.

Turnover Rates and Nutrient Influxes

The intensity of water flow over and through a wetland can be described by the water renewal rate (t^{-1}), the ratio of throughflow to the volume stored on the site (see

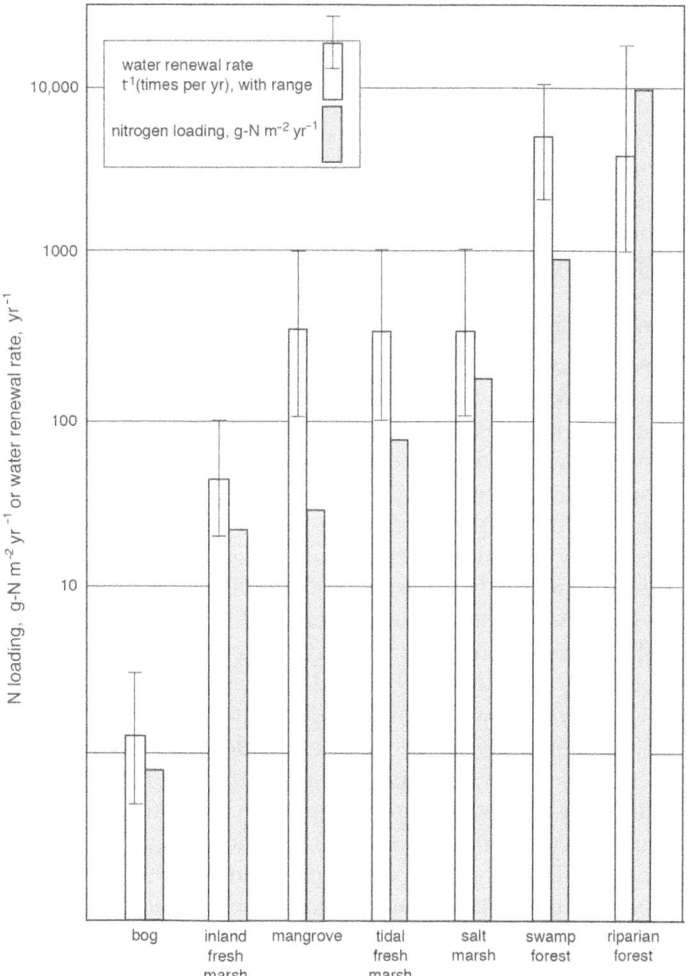

Figure 7.11 Renewal rates of water and nitrogen loading of major wetland types. This figure illustrates that wetlands have five orders of magnitude differences of hydrologic inflow and nutrient inflow.

Chapter 4: "Wetland Hydrology"). In wetlands, t^{-1} varies by five orders of magnitude (Fig. 7.11), ranging from about 1 per year in northern bogs to almost 10,000 per year in swamp forests. The nutrient input to a wetland follows closely the water renewal rate, because nutrients are carried to a site by water. The amount of nitrogen delivered to a wetland site, for example, also varies by five orders of magnitude, ranging from less than 1 g m^{-2} yr^{-1} in a northern bog to perhaps 10,000 g m^{-2} yr^{-1} in a riparian forested wetland (Fig. 7.11). Of course, not all of this nitrogen is available to plants in the ecosystem because, in many cases, it is flowing through much faster than it can be immobilized, but these figures indicate the potential nutrient supply to the ecosystem.

Yet despite the extreme variability in these outside (allogenic) forces of hydrology and nutrient inflows varying over 10,000 times, as illustrated in Figure 7.11, wetland ecosystems are remarkably similar in many respects. Total stored biomass, including peat to 1 m depth, ranges from 40 to 60 kg m^{-2}—less than twofold. Soil nitrogen similarly varies only about threefold, from about 500 to 1,500 g m^{-2}. Net primary production (NPP), a key index of ecosystem function, ranges for wetlands from 400 (peat bogs) to 2,000 (forested wetlands) g m^{-2} yr^{-1}, a factor of five.

Wetland Insularity

So allogenic forces vary by 10,000 times (five orders of magnitude) but functions of wetlands vary by two to five times (not even one order of magnitude). Although some studies of individual species (e.g., *Spartina alterniflora*) or ecosystems (e.g., cypress swamps) have concluded that productivity is directly proportional to the water renewal rate, when different wetland ecosystems that constitute greatly different water regimes are compared, the relationship breaks down or at least is logarithmic. The apparent contradiction may be explained primarily by the role of stored nutrients within the ecosystem. As the large store of organic nutrients in the sediment mineralizes, it provides a steady source of inorganic fertilizer for plant growth. As a result, much of the nutrient demand is satisfied by recycling, even in systems as open as salt marshes and riparian wetlands. External nutrient inputs provide a subsidy to this basic supply. Therefore, growth is often apparently limited by the mineralization rate, which, in turn, is strongly temperature and hydroperiod dependent. Temperatures during the growing season are uniform enough to provide a similar nitrogen supply to plants in different wetland systems, except probably in northern bogs. There, low temperatures and short growing seasons limit mineralization and restrict nutrient input. The combination of the two factors limits productivity. Thus, as wetland ecosystems develop, they become increasingly insulated from the variability of the environment by storing nutrients. Often the same process that stores nutrients (i.e., peat accumulation) also reduces the variability of flooding, further stabilizing the system. The surface of marshes, in general, is built up by the deposition of peats and waterborne inorganic sediments. As the elevation increases, flooding becomes less frequent, and sediment input decreases. In the absence of overriding factors, coastal wetland marshes in time reach a stable elevation somewhere around local mean high water. The surfaces of riparian wetlands similarly rise until they become flooded only infrequently. Northern bogs grow by peat deposition above the water table, stabilizing at an elevation that maintains saturated peat by capillarity. Prairie potholes may be exceptions to these generalizations. They appear to be periodically "reset" by a combination of herbivore activity and long-term precipitation cycles and to achieve stability only in some cyclic sense.

Pulse Stability

In contrast to the lack of evidence for community succession to a stable set of species, which was summarized earlier, the concept of a progression toward a mature ecosystem has greater merit. The attributes of a mature stable ecosystem place it in dynamic equilibrium with its environment, and although individual species may come and go, a mature ecosystem is stable in the sense that it has built-in mechanisms (species diversity, nutrient storage, and recycling) that resist short-term environmental fluctuations. In fact, the three Odums (W. E. Odum et al., 1995) suggested that natural processes pulse regularly, and the mature ecosystem responds in a pulsing steady state. In wetlands, examples of this phenomenon include salt marshes, tidal freshwater marshes, riverine forests, and seasonally flooded freshwater marshes, all of which are functionally similar despite marked differences in species composition, diversity, and community structure. Odum et al. (1995) suggest that natural pulses such as tides pump energy into ecosystems and enhance productivity. Biotic events are geared to and take advantage of these pulses; for example, the influx of small fish into flooded marshes to feed during high tide or the capturing of young fish in backwater oxbows and billabongs during flooding, with the captured fish serving as food for wading birds during periods of low water. This concept is referred to as *pulse stability*.

Self-Organization and Self-Design

Most wetland ecosystems are continually open to atmospheric, hydrologic, and biotic inputs of propagules of plants, animals, and microbes. *Self-organization*, as discussed by Howard T. Odum (1989), manifests itself in both microcosms and newly created ecosystems, "showing that after the first period of competitive colonization, the species prevailing are those that reinforce other species through nutrient cycles, aids to reproduction, control of spatial diversity, population regulation, and other means." Self-organization is further defined as "the process whereby complex systems consisting of many parts tend to organize to achieve some sort of stable, pulsing state in the absence of external interference" (E. P. Odum and Barrett, 2005).

Self-design, defined as "the application of self-organization in the design of ecosystems" (Mitsch and Jørgensen, 2004), relies on the self-organizing ability of ecosystems; natural processes (e.g., wind, rivers, tides, biotic inputs) contribute to species introduction; selection of those species that will dominate from this gene inflow is then nature's manifestation of ecosystem design (Mitsch and Wilson, 1996; Mitsch and Jørgensen, 2004; Mitsch et al., 2012). In self-design, the presence and survival of species resulting from the continuous introduction of them and their propagules is the essence of the successional and functional development of an ecosystem. This can be thought of as analogous to the continuous production of mutations necessary for evolution to proceed. In the context of ecosystem restoration and creation, self-design means that, if an ecosystem is open to allow "seeding," through human or natural means, of enough species' propagules, then the system will optimize its design by selecting for the assemblage of plants, microbes, and animals that is best adapted to the existing conditions. It is an important process to be investigated, particularly in view of the interest in restoring and creating wetlands.

In contrast to the self-design approach, the wetland restoration approach that is still used today involves the introduction of organisms (often plants), the survival of which becomes the measure of success of the restoration. This has sometimes been referred to as the "designer wetland" approach (Mitsch, 1998; van der Valk, 1998). This latter approach, while understandable because of the natural human tendency to control events, may be less sustainable than an approach that relies more on nature being involved in the design.

Case Study 2: Wetland Primary Succession—A Wetland Experiment in Self-Design

In a 20-year, whole-ecosystem experiment in two created freshwater marsh basins at the Wilma H. Schiermeier Olentangy River Wetland Research Park at The Ohio State University in central Ohio, Mitsch et al. (1998, 2005a, 2005b, 2012, 2014) describe how 2,500 individual wetland plants representing 13 species were introduced to one 1-ha flowthrough wetland basin while an adjacent identical wetland basin remained an unplanted control, essentially testing the self-design capabilities of nature with and without human intervention. Both basins have had identical inflows of river water and hydroperiods from 1994 through 2012. These experimental wetlands have allowed simultaneous long-term study of three different questions related to wetland development: (1) How important is wetland plant introduction on long-term ecosystem function? (2) How long does it take for hydric soils and other wetland features to develop at a site where no hydric soils previously existed? and (3) What are the long-term patterns of biogeochemical changes of flowthrough wetlands as they develop from open ponds of water to vegetated, hydric-soil marshes?

For the first six years of the wetland experiment, 17 different biotic and abiotic functional indicators of wetland function were measured, and similarities of the wetland basins were estimated from these indicators. Indices were in six different categories, including macrophytes, algal communities, water quality changes, nutrient changes, benthic invertebrate diversity, and bird use. After only three years, there appeared to be a convergence of wetland function of the planted and unplanted basins, with a 71 percent similarity between the two basins after only one year of divergence. By year 3, over 50 species of macrophytes, 130 genera of algae, over 30 taxa of aquatic invertebrates, and dozens of bird species found their way naturally to both wetlands to supplement the 13 introduced plant species (Mitsch et al., 1998). This convergence in year 3 followed the second year (1995), which showed only a 12 percent similarity in the wetlands, probably because the planted wetland had macrophytes, but the unplanted wetland did not.

The pattern of vegetation succession in the two experimental wetlands for 17 years is shown in Figure 7.12a. Not surprisingly, the clonal dominant *Typha* spp. began to dominate the naturally colonizing wetland (unplanted basin) quickly because there was little competition there. But *Typha* had minimal cover in the planted wetland until about the seventh year (2000) of the experiment. *Typha* was held at bay in the planted wetland by competition

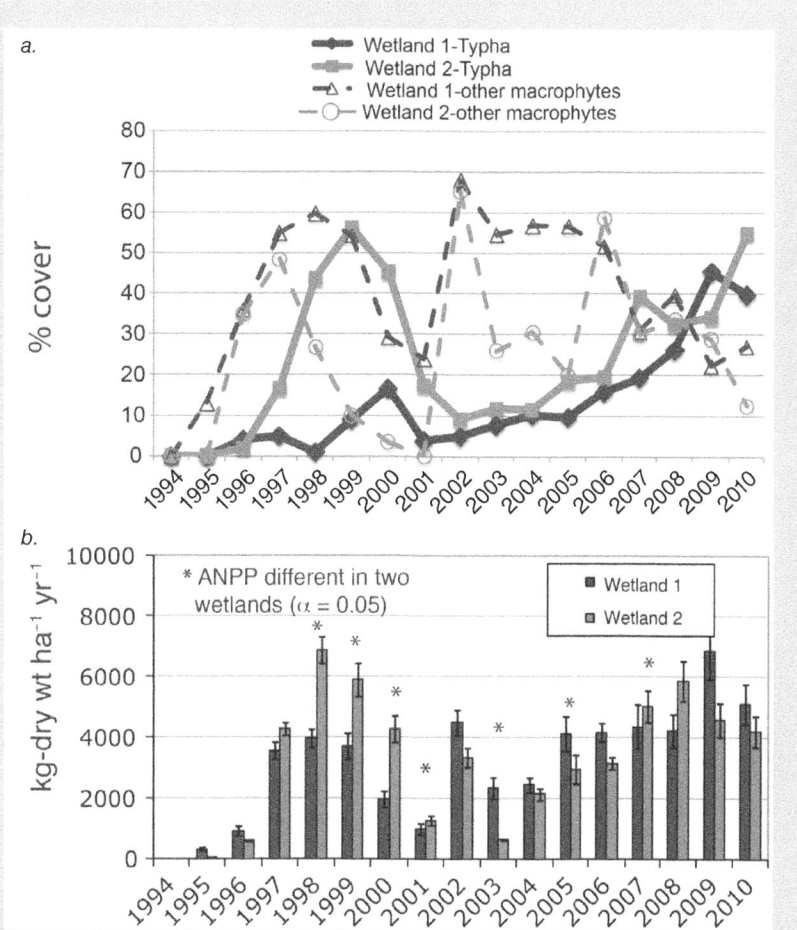

Figure 7.12 Patterns of emergent macrophyte vegetation structure and function in two experimental wetlands in Ohio for 17 years (1994–2010) after wetlands were created and one of the wetlands (Wetland 1) was planted in May 1994. (a) Percentage cover of *Typha* spp. and other emergent macrophytes in planted Wetland 1 and naturally colonizing Wetland 2. (b) Above-ground net primary productivity (ANPP; average ± std error) of planted Wetland 1 and naturally colonizing Wetland 2. (Updated from Mitsch et al., 2012)

by several planted species, most notably *Sparganium eurycarpum* (bur reed) and *Schoenoplectus tabernaemontani* (soft-stemmed bulrush). A muskrat "eat-out" in 2000–2001 eliminated most of the *Typha* and other macrophytes in both wetlands, but a seed bank of one of the planted plants—*S. tabernaemontani*—caused an explosion in cover in 2002, especially in the wetland where it was planted and a year after the eat-out. After that, a

10-year pattern of a slowly developing domination of both wetlands by *Typha* occurred until both wetlands had an average cover of about 40 percent *Typha* in 2009–2010 (years 16 and 17 since the wetlands were created). The pattern of above-ground net primary productivity (ANPP) (Fig. 7.12b) showed four years early in the experiment (1998–2001) when the so-called unplanted wetland had higher productivity than did the planted wetland, mostly because of *Typha* dominance. Then productivity decreased in both wetlands

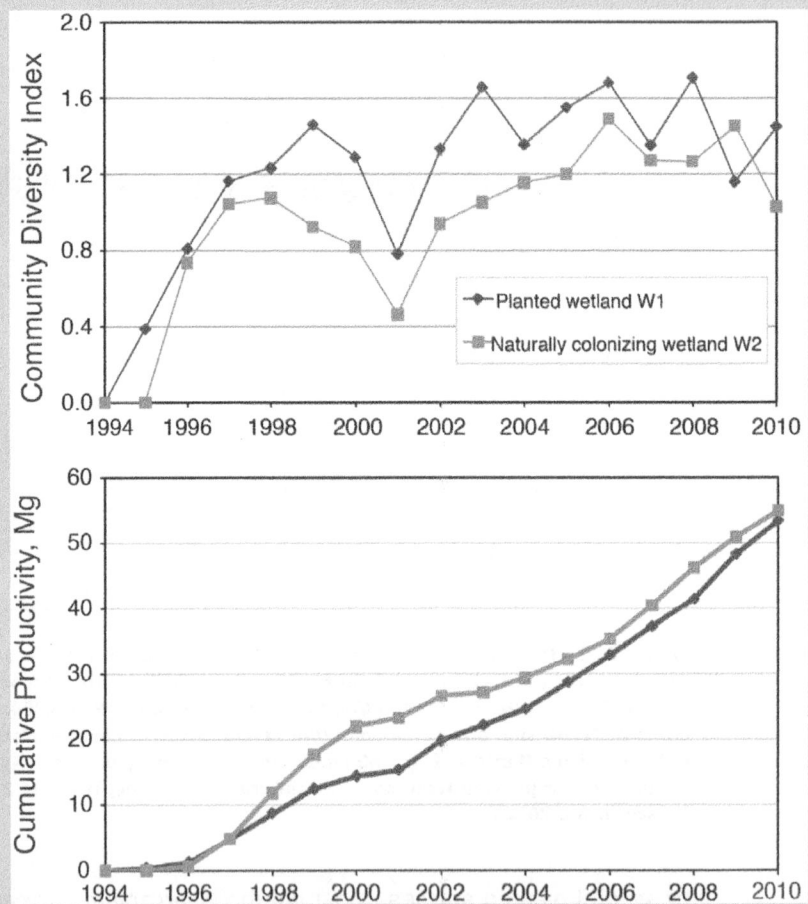

Figure 7.13 (a) Emergent macrophyte community diversity and (2) cumulative organic productivity in two created experimental wetlands in Ohio for 17 years (1994–2010). Wetland 1 (W1) was planted in May 1994; Wetland 2 (W2) remained an unplanted control. Hydrologic conditions were identical over that period in the two wetlands. (Updated from Mitsch et al., 2012)

because of the muskrat eat-out but recovered quicker in the planted wetland, which had higher productivity in 2003 and 2005.

In general, the planted wetland maintained a higher spatial macrophyte diversity throughout much the study until the end (Fig. 7.13a) while the naturally colonizing wetland has been more susceptible to disturbances such as muskrat herbivory and hydrologic pulses than has the more diverse planted wetland. More details of the vegetation richness over the 17 years are presented in Chapter 18: "Wetland Creation and Restoration."

The naturally colonizing wetland had benthic invertebrate diversity and amphibian populations similar to the planted wetland for several of the early years and had greater accumulated productivity after 17 years (Fig. 7.13b) because of the several years when *Typha* dominated the unplanted wetland.

The continual introduction of species, whether introduced through flooding and other abiotic and biotic pathways, appeared to have a much longer-lasting effect in development of these ecosystems than the few species of plants that were introduced to one of the wetlands in the beginning. This 17-year study of wetland development showed that the planting had some long-term effect on functions such as nutrient retention (Mitsch et al., 2012, 2014; see also Chapter 19: "Wetlands and Water Quality") but none on denitrification (Hernandez and Mitsch, 2007; Song et al., 2014). But the planting did appear to have had a significant effect on decreasing carbon accumulation in the soil (Anderson and Mitsch, 2006; Mitsch et al., 2012; Mitsch et al., 2013) and methane emissions (Altor and Mitsch, 2006; Nahlik and Mitsch, 2010; Sha et al. 2011). After 17 years, the two created wetlands may have converged on plant cover structure, but the residual effect of the years when they were different in productivity may still be influencing some ecosystem functions.

Ecosystem Engineers

Another way to describe the importance of autogenic successional processes involves the introduced term *ecosystem engineer*—a term that is used to describe organisms that have dramatic and important effects on an ecosystem (Jones et al., 1997; Alper, 1998). [This concept, also discussed in Chapter 4, should not be confused with the field of ecological engineering, described by Mitsch and Jørgensen (2004) and Mitsch (2012)]. In wetlands, examples of ecosystem engineers could be muskrats and beavers, both of which can have dramatic effects on vegetation cover and ecosystem hydrology in freshwater marshes. In these cases, the biota does show a dramatic feedback to the many features of the wetland (e.g., water levels or vegetation productivity). One could argue that these ecosystem engineers "set back" succession to an earlier stage; an alternative argument is that they are part of the ecosystem development and that their behavior and its effects should be both expected and appreciated as normal ecosystem behavior.

250 Chapter 7 Wetland Vegetation and Succession

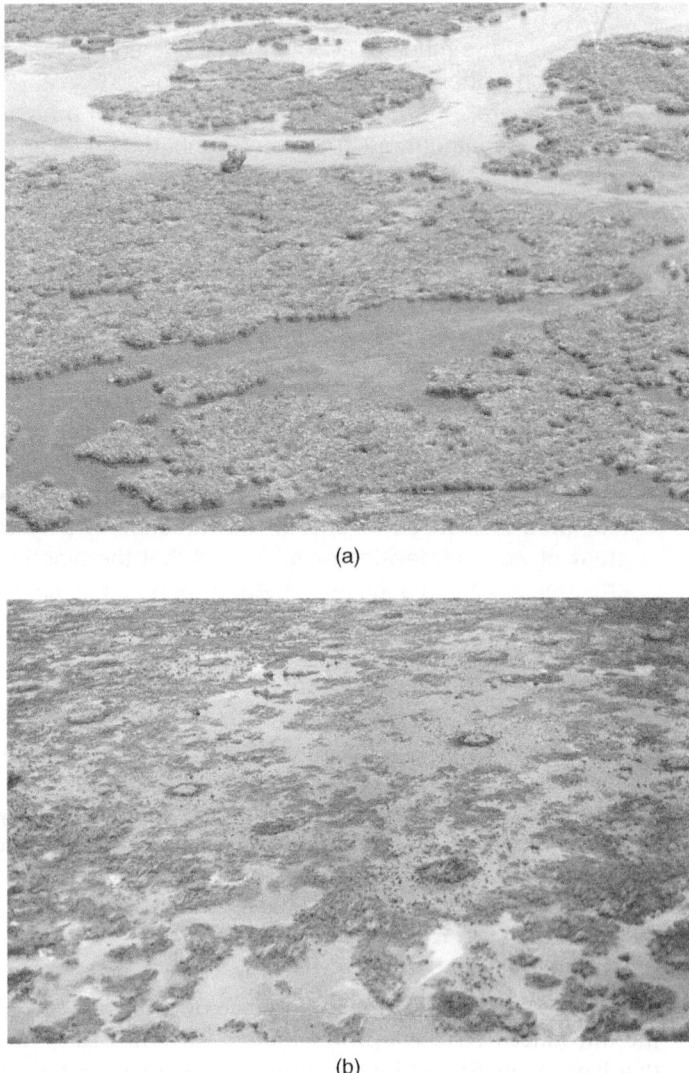

(a)

(b)

Figure 7.14 Landscape patterns in Louisiana wetlands: (a) a physically controlled pattern of vegetation in tidal creeks in a Louisiana salt marsh; (b) a biologically controlled pattern of vegetation caused by a muskrat eat-out in a brackish marsh. Note the high density of muskrat lodges in (b)

Landscape Patterns

Many large wetland landscapes develop predictable and often complex patterns of aquatic, wetland, and terrestrial habitats or ecosystems. In high-energy environments, these patterns appear to reflect abiotic forces, but they are largely controlled by biotic processes in low-energy environments. At the high-energy end of the spectrum, the microtopography and sediment characteristics of mature floodplains—complex mosaics of river channels, natural levees, back swamps, abandoned first and second terrace flats, and upland ridges—reflect the flooding pattern of the adjacent river. The vegetation responds to the physical topography and sediments with typical zonation patterns. Salt marshes similarly develop a characteristic pattern of tidal creeks, creekside levees, and interior flats that determine the zonation pattern and vigor of the vegetation (Fig. 7.14a).

At the low-energy end of the spectrum, the characteristic pattern of strings and flarks stretching for miles across northern peatlands appears to be controlled primarily by biotic processes. Similarly, in many freshwater marshes, herbivores can be major actors in the development of landscape patterns (Fig. 7.14b). In actuality, both physical (climatic, topographic, hydrologic) and biotic (production rates, root binding, herbivory, peat accumulation) processes combine in varying proportions and interact to produce observed wetland landscape patterns.

Recommended Readings

Keddy, P. A. 2010. *Wetland Ecology: Principles and Conservation*, 2nd ed. Cambridge, UK: Cambridge University Press.

van der Valk, A. G. 2012. *The Biology of Freshwater Wetlands*, 2nd ed. Oxford, UK: Oxford University Press.

References

Alper, J. 1998. Ecosystem "engineers" shape habitats for other species. *Science* 280: 1195–1196.

Altor, A. E., and W. J. Mitsch. 2006. Methane flux from created riparian marshes: Relationship to intermittent versus continuous inundation and emergent macrophytes. *Ecological Engineering* 28: 224–234.

Anderson, C. J., and W. J. Mitsch. 2006. Sediment, carbon, and nutrient accumulation at two 10-year-old created riverine marshes. *Wetlands* 26: 779–792.

Bernal, B. and W. J. Mitsch. 2013. Carbon sequestration in two created riverine wetlands in the Midwestern United States. *Journal of Environmental Quality* 42: 1236–1244.

Boutin, C., and P. A. Keddy. 1993. A functional classification of wetland plants. *Journal of Vegetation Science* 4: 591–600.

Brix, H., B. K. Sorrell, and P. T. Orr. 1992. Internal pressurization and convective gas flow in some emergent freshwater macrophytes. *Limnology and Oceanography* 37: 1420–1433.

Clements, F. E. 1916. *Plant Succession*. Publication 242. Carnegie Institution of Washington. 512 pp.

Cowles, H. C. 1899. The ecological relations of the vegetation on the sand dunes of Lake Michigan. *Botanical Gazette 27*: 95–117, 167–202, 281–308, 361–369.

Cowles, H. C. 1901. The physiographic ecology of Chicago and vicinity. *Botanical Gazette* 31: 73–108, 145–182.

Cowles, H. C. 1911. The causes of vegetative cycles. *Botanical Gazette* 51: 161–183.

Cushing, E. J. 1963. *Late-Wisconsin pollen stratigraphy in East-Central Minnesota.* Ph.D. dissertation, University of Minnesota, Minneapolis.

Dacey, J. W. H. 1980. Internal winds in water lilies: An adaption for life in anaerobic sediments. *Science* 210: 1017–1019.

Dacey, J. W. H. 1981. Pressurized ventilation in the yellow waterlily. *Ecology* 62: 1137–1147.

Ellison, A. M., E. J. Farnsworth, and R. R. Twilley. 1996. Facultative mutualism between red mangroves and root-fouling sponges in Belizean mangal. *Ecology* 77: 2431–2444.

Ernst, W. H. O. 1990. Ecophysiology of plants in waterlogged and flooded environments. *Aquatic Botany* 38: 73–90.

Gleason, H. A. 1917. The structure and development of the plant association. *Torrey Botanical Club Bulletin* 44: 463–481.

Golet, F. C., A. J. K. Calhoun, W. R. DeRagon, D. J. Lowry, and A. J. Gold. 1993. *Ecology of Red Maple Swamps in the Glaciated Northeast: A Community Profile.* Biological Report 12, U.S. Fish and Wildlife Service, Washington, DC. 151 pp.

Grime, H. H. 1979. *Plant Strategies and Vegetation Processes.* John Wiley & Sons, New York.

Grosse, W., and P. Schröder. 1984. Oxygen supply of roots by gas transport in alder trees. *Zeitschrift für Naturforsch.* 39C: 1186–1188.

Grosse, W., S. Sika, and S. Lattermann. 1990. Oxygen supply of roots by thermo-osmotic gas transport in Alnus-glutinosa and other wetland trees. In D. Werner and P. Müller, eds. *Fast Growing Trees and Nitrogen Fixing Trees.* Gustav Fischer Verlag, New York, pp. 246–249.

Grosse, W., J. Frye, and S. Lattermann. 1992. Root aeration in wetland trees by pressurized gas transport. *Tree Physiology* 10: 285–295.

Grosse, W., H. B. Büchel, and S. Lattermann. 1998. Root aeration in wetland trees and its ecophysiological significance. In A. D. Laderman, ed., *Coastally Restricted Forests.* Oxford University Press, New York, pp. 293–305.

Hernandez, M. E., and W. J. Mitsch. 2007. Denitrification in created riverine wetlands: Influence of hydrology and season. *Ecological Engineering* 30: 78–88.

Jackson, S. T., R. P. Rutyma, and D. A. Wilcox. 1988. A paleoecological test of a classical hydrosere in the Lake Michigan dunes. *Ecology* 69: 928–936.

Jones, C. G., J. H. Lawton, and M. Shachak. 1997. Positive and negative effects of organisms as physical ecosystem engineers. *Ecology* 78: 1946–1957.

Keddy, P. A. 2010. *Wetland Ecology: Principles and Conservation*, 2nd ed. Cambridge University Press, Cambridge, UK. 497 pp.

Koch, M. S., I. A. Mendelssohn, and K. L. McKee. 1990. Mechanism for the hydrogen sulfide-induced growth limitation in wetland macrophytes. *Limnology and Oceanography* 35: 399–408.

Levitt, J. 1980. *Responses of Plants to Environmental Stresses*, Vol. II: Water, Radiation, Salt, and Other Stresses. Academic Press, New York. 607 pp.

McIntosh, R. P. 1985. *The Background of Ecology, Concept and Theory*. Cambridge University Press, Cambridge, UK. 383 pp.

McKee, K. L., I. A. Mendelssohn, and M. W. Hester. 1988. Reexamination of pore water sulfide concentrations and redox potentials near the aerial roots of Rhizophora mangle and Avicennia germinans. *American Journal of Botany* 75: 1352–1359.

Mendelssohn, I. A., and M. L. Postek. 1982. Elemental analysis of deposits on the roots of *Spartina alterniflora* Loisel. *American Journal of Botany* 69: 904–912.

Mitsch, W. J. 1998. Self-design and wetland creation: Early results of a freshwater marsh experiment. In A. J. McComb and J. A. Davis, eds. *Wetlands for the Future*. Contributions from INTECOL's Fifth International Wetlands Conference. Gleneagles Publishing, Adelaide, Australia, pp. 635–655.

Mitsch, W. J. 2012. What is ecological engineering? *Ecological Engineering* 45: 5–12.

Mitsch, W. J., and R. F. Wilson. 1996. Improving the success of wetland creation and restoration with know-how, time, and self-design. *Ecological Applications* 6: 77–83.

Mitsch, W. J., X. Wu, R. W. Nairn, P. E. Weihe, N. Wang, R. Deal, and C. E. Boucher. 1998. Creating and restoring wetlands: A whole-ecosystem experiment in self-design. *BioScience* 48: 1019–1030.

Mitsch, W. J., and S. E. Jørgensen. 2004. *Ecological Engineering and Ecosystem Restoration*. John Wiley & Sons, Hoboken, NJ. 411 pp.

Mitsch, W. J., N. Wang, L. Zhang, R. Deal, X. Wu, and A. Zuwerink. 2005a. Using ecological indicators in a whole-ecosystem wetland experiment. In S. E. Jørgensen, F-L. Xu, and R. Costanza, eds. *Handbook of Ecological Indicators for Assessment of Ecosystem Health*. CRC Press, Boca Raton, FL, pp. 211–235.

Mitsch, W. J., L. Zhang, C. J. Anderson, A. Altor, and M. Hernandez. 2005b. Creating riverine wetlands: Ecological succession, nutrient retention, and pulsing effects. *Ecological Engineering* 25: 510–527.

Mitsch, W. J, L. Zhang, K. C. Stefanik, A. M. Nahlik, C. J. Anderson, B. Bernal, M. Hernandez, and K. Song. 2012. Creating wetlands: Primary succession, water quality changes, and self-design over 15 years. *BioScience* 62: 237–250.

Mitsch, W. J., B. Bernal, A. M. Nahlik, U. Mander, L. Zhang, C. J. Anderson, S. E. Jørgensen, and H. Brix. 2013. Wetlands, carbon, and climate change. *Landscape Ecology* 28: 583–597.

Mitsch, W. J., L. Zhang, E. Waletzko, and B. Bernal. 2014. Validation of the ecosystem services of created wetlands: Two decades of plant succession, nutrient retention, and carbon sequestration in experimental riverine marshes. *Ecological Engineering* 72: 11–24.

Nahlik, A. M., and W. J. Mitsch. 2010. Methane emissions from created riverine wetlands. *Wetlands* 30: 783-793 with Erratum in *Wetlands* (2011) 31: 449–450.

Niering, W. A. 1989. Wetland vegetation development. In S. K. Majumdar, R. P. Brooks, F. J. Brenner, and J. R. W. Tiner, eds. *Wetlands Ecology and Conservation: Emphasis in Pennsylvania*. Pennsylvania Academy of Science, Easton, pp. 103–113.

Odum, E. P. 1969. The strategy of ecosystem development. *Science* 164: 262–270.

Odum, E. P. 1971. *Fundamentals of Ecology*, 3rd ed. W. B. Saunders, Philadelphia. 544 pp.

Odum, E. P., and G. W. Barrett. 2005. *Fundamentals of Ecology*, 5th ed. Thomson Brooks/Cole, Belmont, CA, 598 pp.

Odum, H. T. 1989. Ecological engineering and self-organization. In W. J. Mitsch and S. E. Jørgensen, eds. *Ecological Engineering*. John Wiley & Sons, New York, pp. 79–101.

Odum, W. E., E. P. Odum, and H. T. Odum. 1995. Nature's pulsing paradigm. *Estuaries* 18: 547–555.

Pearsall, W. H. 1920. The aquatic vegetation of the English lakes. *Journal of Ecology* 8: 163–201.

Pederson, R. L., and L. M. Smith. 1988. Implications of wetland seed bank research: A review of Great Britain and prairie marsh studies. In D. A. Wilcox, ed., *Interdisciplinary Approaches to Freshwater Wetlands Research*. Michigan State University Press, East Lansing, pp. 81–95.

Schröder, P. 1989. Characterization of a thermo-osmotic gas transport mechanism in *Alnus glutinosa* (L.) Gaertn. *Trees* 3: 38–44.

Sha, C., W. J. Mitsch, Ü. Mander, J. Lu, J. Batson, L. Zhang and W. He. 2011. Methane emissions from freshwater riverine wetlands. *Ecological Engineering* 37: 16–24.

Shelford, V. E. 1907. Preliminary note on the distribution of the tiger beetle (Cicindela) and its relation to plant succession. *Biological Bulletin* 14.

Shelford, V. E. 1911. Ecological succession. II. Pond fishes. *Biological Bulletin* 21: 127–151.

Shelford, V. E. 1913. *Animal Communities in Temperate America as Illustrated in the Chicago Region*. University of Chicago Press, Chicago.

Singer, D. K., S. T. Jackson, B. J. Madsen, and D. A. Wilcox. 1996. Differentiating climatic and successional influences on long-term development of a marsh. *Ecology* 77: 1765–1778.

Song, K., M. E. Hernandez, J. A. Batson, and W. J. Mitsch. 2014. Long-term denitrification rates in created riverine wetlands and their relationship with environmental factors. *Ecological Engineering* 72: 40–46.

Tilman, D. 1982. *Resource Competition and Community Structure*. Princeton University Press, Princeton, NJ. 296 pp.

van der Valk, A. G. 1981. Succession in wetlands: A Gleasonian approach. *Ecology* 62: 688–696.

van der Valk, A. G. 1998. Succession theory and wetland restoration. In A. J. McComb and J. A. Davis, eds., *Wetlands for the Future. Contributions from INTECOL's Fifth International Wetland Conference*. Gleneagles Publishing, Adelaide, Australia, pp. 657–667.

Walker, D. 1970. Direction and rate in some British post-glacial hydroseres. In D. Walker and R. G. West, eds. *Studies in the Vegetational History of the British Isles*. Cambridge University Press, Cambridge, UK, pp. 117–139.

West, R. G. 1964. Inter-relations of ecology and quaternary paleobotany. *Journal of Ecology* (Supplement) 52: 47–57.

Wilcox, D. A., and H. A. Simonin. 1987. A chronosequence of aquatic macrophyte communities in dune ponds. *Aquatic Botany* 28: 227–242.

Wilson, L. R. 1935. Lake development and plant succession in Vilas County, Wisconsin. 1. The medium hard water lakes. *Ecological Monographs* 5: 207–247.

Wisheu, I. C., and P. A. Keddy. 1992. Competition and centrifugal organization of plant communities: Theory and tests. *Journal of Vegetation Science* 3: 147–156.

Part III

Wetland Ecosystems

Tidal marshes: (a) tidal salt marsh in Louisiana; (b) tidal freshwater marsh in Maryland (tidal freshwater photo courtesy of A. Baldwin)

Chapter 8

Tidal Marshes

The salt marsh, distributed worldwide along coastlines in middle and high latitudes, flourishes wherever the accumulation of sediments is equal to or greater than the rate of land subsidence and where there is adequate protection from destructive waves and storms. The important physical and chemical variables that determine the structure and function of the salt marsh include tidal flooding frequency and duration, soil salinity, soil permeability, and nutrient limitation, particularly by nitrogen. The vegetation of the salt marsh, primarily salt-tolerant grasses and rushes, develops in identifiable zones in response to these and possibly other factors. Mud and epiphytic algae are also often an important component of the autotrophic community. Heterotrophic communities are dominated by detrital food chains, with the grazing food chain being much less significant except during marsh die-off episodes.

Freshwater tidal marshes combine many features of both salt marshes and inland marshes. They act in many ways like salt marshes, but the biota reflect the increased diversity made possible by the reduction of the salt stress. Plant diversity is high, and more birds use these marshes than any other marsh type. Because they are inland from the saline parts of the estuary, they are often close to urban centers. This makes them more prone to human impact than coastal salt marshes. Along coastal rivers, tidal freshwater swamps tend to occupy a narrow range at the furthest extent of the tidal range. They occur where tidal waters are normally fresh and shallow enough for tree establishment.

Several types of wetlands in coastal areas are influenced by alternating floods and ebbs of tides. Coastal wetlands include tidal salt marshes, tidal freshwater wetlands (marshes

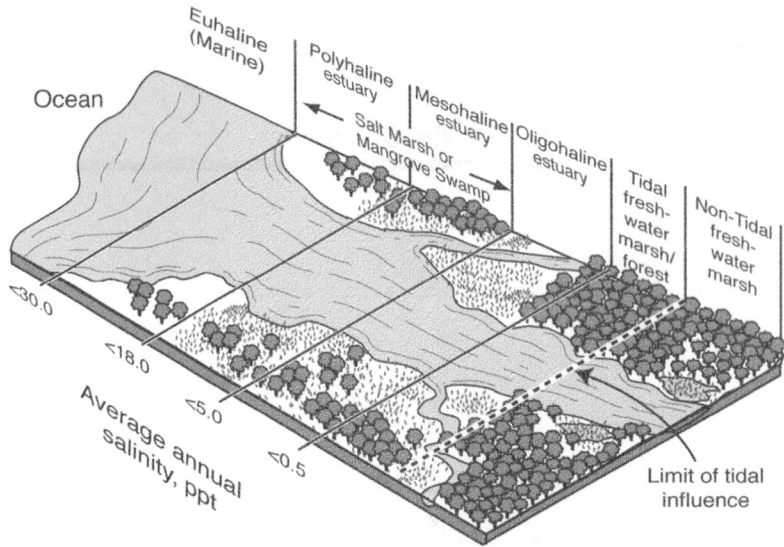

Figure 8.1 Coastal wetlands lie on gradients of increasing salinity from inland to the ocean in an estuary. Where salinity is sufficient, salt marshes (in temperate zone) and mangroves (in tropics) are found. Tidal freshwater marshes and tidal freshwater forests still experience tides but are above the salt boundary. Farther inland are marshes and forested swamps that experience neither salt nor tides.

and forests), and mangrove swamps. Salt marshes and tidal freshwater marshes are discussed in this chapter. Mangrove swamps are discussed in the next chapter.

Near coastlines, the salinity of the water approaches that of the ocean (35 ppt), whereas farther inland, the tidal effect can remain significant even when the salinity is that of fresh water (Fig. 8.1). Tidal freshwater wetlands are found upstream of salt water (0.5 ppt = 500 ppm and lower salinity), while salt marshes and mangroves are found downstream (salt marshes in temperate and boreal zones; mangroves in tropics) in polyhaline and mesohaline estuarine waters greater than 5 ppt (=5,000 ppm) in salinity. These coastal wetlands are found in abundance in the river deltas and estuaries of the world—where large rivers debouch onto low-energy coasts (Fig. 8.2). These river deltas and estuaries span the world's latitudes and climatic zones. In the tropics, tidally influenced wetlands of these deltas are mangroves. Above 25° latitude, mangroves give way to salt marshes. In North America, large deltas are restricted to the coasts of the South Atlantic and the Gulf of Mexico. The Mississippi River deltaic marshes are the major example of this type of development and support the most extensive coastal marshes in the United States.

We estimate that there are about 270,000 km² of coastal wetlands in the world, representing about 3 to 4 percent of all the wetlands in the world. About 150,000 km² of those coastal wetlands are mangroves (see Chapter 9: "Mangrove Swamps") with tidal marshes (freshwater and salt) probably covering slightly less than that amount of area worlwide (Mitsch et al., 2009). For the United States, the total area of

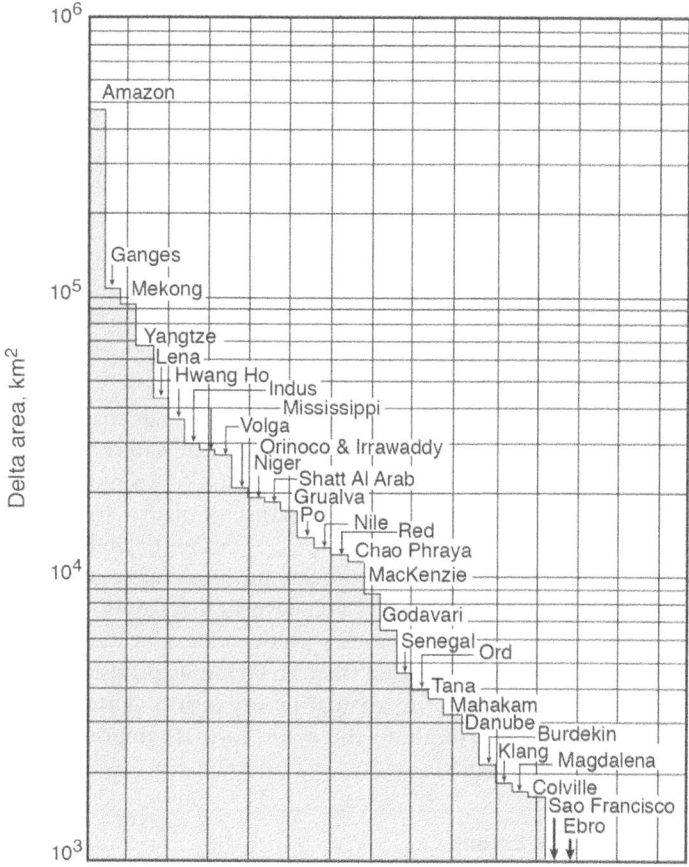

Figure 8.2 The area of deltaic plains of selected major river systems of the world. (After Coleman and Roberts, 1989)

wetlands considered coastal or estuarine wetlands, including Alaska, is approximately 32,000 km², with about 19,000 km² as salt marsh, 8,000 km² as tidal freshwater marshes, and 5,000 km² as mangrove swamps (Table 8.1). Almost 40 percent of the salt marshes in continental United States are found in the Mississippi River Delta in Louisiana (Ibánez et al., 2013).

Tidal Salt Marshes

Salt marshes are found throughout the world along protected coastlines in the middle and high latitudes (Fig. 8.3a). Salt marshes can be narrow fringes on steep shorelines or expanses that are several kilometers wide. They are found near river mouths, in bays, on protected coastal plains, and around protected lagoons. Different plant associations dominate different coastlines, but the ecological structure and function

Table 8.1 Estimated area of coastal wetlands in the United States (x 1,000 ha)

	Salt Marsh[a]	Freshwater Tidal Marsh[b]	Mangrove[b]	Total
Atlantic Coast	669	400		1,069
Gulf of Mexico	1,011	362	506	1,879
Pacific Coast	49	57		106
Alaska[c]	146			146
Total	1,875	819	506	3,200

[a]Watzin and Gosselink (1992)
[b]Field et al. (1991)
[c]Hall et al. (1994)

of salt marshes is similar around the world. Salt marshes, dominated by rooted vegetation that is alternately inundated and dewatered by the rise and fall of the tide, appear from afar to be vast fields of grass of a single species. In reality, salt marshes have a complex zonation and structure of plants, animals, and microbes, all tuned to the stresses of salinity fluctuations, alternate drying and submergence, and extreme daily and seasonal temperature variations. A maze of tidal creeks with plankton, fish, nutrients, and fluctuating water levels crisscrosses the marsh, forming conduits for energy and material exchange with the adjacent estuary. Studies of a number of different salt marshes have found them to be highly productive and to support the spawning and feeding habits of many marine organisms. Thus, salt marshes and tropical mangrove swamps throughout the world form an important interface between terrestrial and marine habitats.

Geographic Extent

Salt marshes are found near river mouths, in bays, on protected coastal plains, and around protected lagoons. Different plant associations dominate different coastlines, but the ecological structure and function of salt marshes is similar around the world. Based the classification system developed by Valentine Chapman (1960, 1976), the world's salt marshes can be divided into the following eight major geographical groups:

1. *Arctic.* This group includes marshes of northern Canada, Alaska, Greenland, Iceland, northern Scandinavia, and Russia. Probably the largest extent of marshes in North America, as much as 300,000 km², occurs along the southern shore of the Hudson Bay. These marshes, influenced by ice, extreme low temperatures, a positive water balance, and numerous inflowing streams, can be generally characterized as brackish rather than saline. Various species of the sedge *Carex* and the grass *Puccinellia phryganodes* often dominate. Parts of the southwestern coast of Alaska are dominated by species of *Salicornia* and *Suaeda*.

2. *Northern Europe.* This group includes marshes along the west coast of Europe from the Iberian Peninsula to Scandinavia, including Great Britain and the Baltic

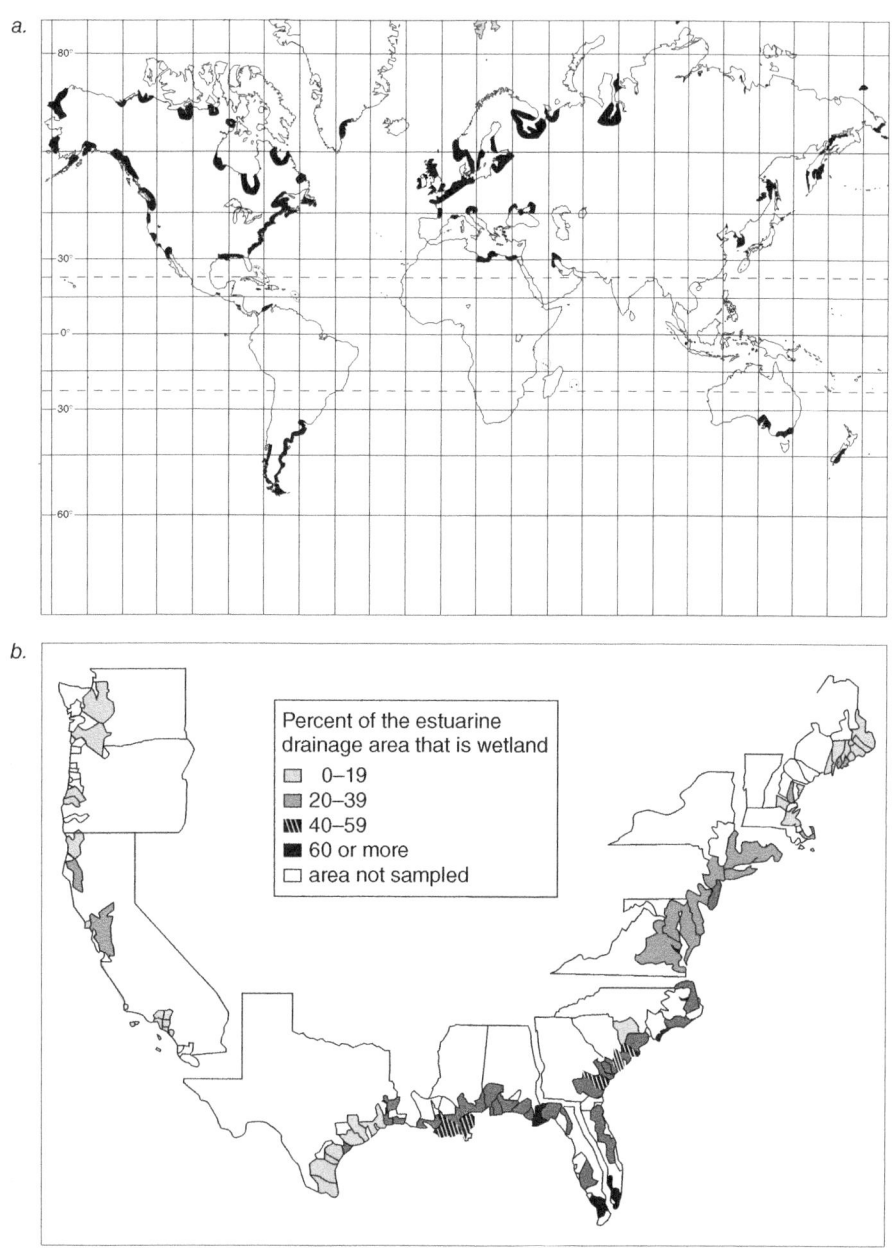

Figure 8.3 Distribution of (a) salt marshes of the world and (b) wetlands in coastal drainage areas of the United States, including freshwater tidal wetlands and mangroves as well as tidal salt marshes. (After Chapman, 1977 and Field et al., 1991)

Sea coast. Most of the western European coastal environment is characterized by a moderate climate with sufficient precipitation but high salinities toward the southern extremes. Dominant species include *Puccinellia maritima, Juncus gerardi, Salicornia* spp., *Spartina anglica,* and *S. townsendii.* The west coast of Great Britain and parts of the Scandinavian and Baltic Sea coasts, where substrates are dominated by sand and salinities are low, are populated by *Festuca rubra, Agrostis stolonifera, Carex paleacea, Juncus bufonius, Desmoschoenus bottanica,* and *Scripus* spp. The muddy coast of the English Channel is dominated by *Spartina townsendii.* Salt marshes in northern Europe are often characterized by a lack of vegetation in the intertidal zone, in contrast to North American marshes.

3. *Mediterranean.* This group includes the arid, rocky-to-sandy, high-salinity coasts of the Mediterranean Sea. The salt marshes are dominated by low shrubby vegetation, *Arthrocnemum, Limonium, Juncus* spp., and the halophyte *Salicornia* spp.

4. *Eastern North America.* These marshes, mostly dominated by *Spartina* and *Juncus* species, are found along the eastern coasts of the United States and Canada and the Gulf Coast of the United States. Salt marshes are most prevalent along the eastern coast of the United States from Maine to Florida and on into Louisiana and Texas along the Gulf of Mexico (Fig. 8.3b). The Eastern North American group is further divided into three subgroups:

   a. *Bay of Fundy.* River and tidal erosion is high in the soft rocks of this region, producing an abundance of reddish silt. The tidal range, as exemplified at the Bay of Fundy, is large, leading to a few marshes in protected areas and considerable depth of deposited sediments. *Puccinellia americana* dominates the lower marsh, and *Juncus balticus* is found on the highest levels.

   b. *New England.* Marshes are built mainly on marine sediments and marsh peat, and there is little transport of sediment from the hard-rock uplands. These marshes range from Maine to New Jersey and are dominated by *Spartina alterniflora* in the low marsh, with *S. patens* mixed with *Distichlis spicata* in the high marsh.

   c. *Coastal Plain.* These marshes extend southward from New Jersey along the southeastern coast of the United States to Texas along the Gulf of Mexico. Major rivers supply an abundance of silt from the recently elevated Coastal Plain. The tidal range is relatively small. The marshes are laced with tidal creeks. Mangrove swamps replace salt marshes along the southern tip of Florida. Because of the extensive delta marshes built by the Mississippi River, the Gulf Coast contains about 60 percent of the coastal salt and fresh marshes of the United States (Fig. 8.3b). Dominant species are *Spartina alterniflora, S. patens, Juncus roemerianus,* and *Distichlis spicata.*

5. *Western North America.* Compared with the Arctic and the eastern coast of North America, salt marshes are far less developed along the western coasts of the United States and Canada because of the geomorphology of the coastline. On this rugged coast with its Mediterranean-type climate is found a narrow belt of *Spartina*

foliosa, often bordered by broad belts of *Salicornia* and *Suaeda*. *Spartina alterniflora* is a nonnative invasive plant in coastal marshes north of California.

6. *Australasia*. Salt marshes are frequently found in river deltas along the temperate coastlines of eastern Asia, Australia, and New Zealand on the Pacific Ocean, Indian Ocean, and Tasman Sea.

   a. *Eastern Asia*. The coasts of China, Japan, Russia, and Korea are generally rugged and uplifted, with moderate precipitation but limited marsh development. These marshes are dominated by *Triglochin maritima*, *Limonium japonicum*, *Salicornia*, and *Zoysia macrostachya*. Major areas of salt marsh rehabilitation have occurred on China's eastern coastline, owing to the introduction of *Spartina anglica* and *S. alterniflora*, although now both species, especially *S. alterniflora*, are considered invasive and are being eliminated where they compete with the native *Phragmites australis*.

   b. *Australia*. This group also includes New Zealand and Tasmania. It is characterized by high rainfall and geographic isolation. Cosmopolitan species in Australian salt marshes include *Sporobolus virginicus*, *Sarcocornia quinqueflora*, and *Suaeda australis*. However, invasion of salt marshes of Australia and New Zealand by several species is common. In New Zealand and other temperate-zone salt marshes of the region, *Spartina anglica* is a major invasive species. Even with less rainfall and a clearly defined seasonal pattern of wet and dry on the western coast of Australia, salt marshes can be found, particularly around Shark Bay and the Peel–Harvey estuaries. In contrast to the general case around much of the world, a majority of the salt marshes of Australia are found in tropical regions (Adam, 1998).

7. *South America*. South American coasts too far south and too cold for mangroves are rugged and geographically isolated. They are dominated by unique species of *Spartina, Limonium, Distichlis, Juncus, Heterostachys*, and *Allenrolfea*.

8. *Tropics*. Although mangroves generally dominate tropical coastlines, salt marshes are found in the tropics on high-salinity flats that mangroves cannot tolerate. *Spartina* spp. and the halophytic genera *Salicornia* and *Limonium* often dominate.

Hydrogeomorphology

The physical features of tides, sediments, freshwater inputs, and shoreline structure determine the development and extent of salt marsh wetlands within their geographical range. Coastal salt marshes are predominantly intertidal; that is, they are found in areas at least occasionally inundated by high tide but not flooded during low tide. A gentle, rather than steep, shoreline slope allows for tidal flooding and the stability of the vegetation. Adequate protection from wave and storm energy is also a physical requirement for the development of salt marshes. Sediments that build salt marshes originate from upland runoff, marine reworking of the coastal shelf sediments, or organic production within the marsh itself.

Table 8.2 Hydrologic demarcation between low marsh and high marsh in salt marshes

Marsh	Submergences per Day in Daylight	Per Year	Maximum Period of Continuous Exposure (days)
High marsh	<1	<360	≥10
Low marsh	>1.2	>360	≤9

Source: Chapman (1960)

Hydrology

Tidal energy represents a subsidy to the salt marsh that influences a wide range of physiographic, chemical, and biological processes, including sediment deposition and scouring, mineral and organic influx and efflux, flushing of toxins, and the control of sediment redox potential. These physical factors in turn influence the species that occur on the marsh and their productivity. The lower and upper limits of the marsh are generally set by the tide range. The lower limit is set by the depth and the duration of flooding and by the mechanical effects of waves, sediment availability, and erosional forces. The upland side of the salt marsh generally extends to the limit of flooding on extreme tides, normally between mean high water and extreme high water of spring tides. Based on marsh elevation and flooding characteristics, the marsh is often divided into two zones, the upper marsh (*high marsh*) and the intertidal lower marsh (*low marsh*) (Table 8.2). The high marsh is flooded irregularly and can experience at least 10 days of continuous exposure to the atmosphere, whereas the low marsh is flooded almost daily, and there are never more than 9 continuous days of exposure. In the Gulf Coast marshes of the United States, the terms *streamside marshes* and *inland marshes* generally replace low and high marsh, respectively, because in these flat, expansive marshes the streamside levees are actually the highest marsh elevations.

Marsh Development

Although a number of different patterns of development can be identified, salt marshes can be classified broadly into two classes: (1) those that were formed from reworked marine sediments on marine-dominated coasts; and (2) those that were formed in deltaic areas where the main source of mineral sediment is riverine.

Marine-dominated marshes are typical of most of the world's coastlines. On marine-dominated coasts, salt marsh development requires sufficient shelter to ensure sedimentation and to prevent excessive erosion from wave action. Marshes can develop at the mouths of estuaries where sediments are deposited by the river, behind spits and bars, and in bays that offer protection from waves and long-shore currents. A spit is neck of land that acts to trap sediment on its lee side and protects the marsh from the full forces of the open sea. The most extensive examples of this type of coastal salt marsh in the United States have developed behind outer barrier reefs along the Georgia–Carolina coast. Several large bays, such as Chesapeake Bay,

Hudson Bay, the Bay of Fundy, and San Francisco Bay, are also protected adequately from storms and waves so that they can support extensive salt marshes. These salt marshes in bays have features of both marine and deltaic origins. They occur on the shores of estuaries where shallow water and low gradients lead to river sediment deposition in areas protected from destructive wave action. Tidal action must be strong enough to maintain salinities above about 5 ppt; otherwise, the salt marsh will be replaced by reeds, rushes, and other freshwater aquatic plants.

Major rivers carrying large sediment loads can build marshes in shallow estuaries or out onto the shallow continental shelf where the ocean is fairly quiet. The size of a delta increases with the size of the inflowing river's drainage basin and its discharge, but is modified by such factors as the slope of the ocean shelf into which the river drains and the tidal range. In coasts with shallow slopes and low wave energies, deltas can build out onto the shelf. These deltas tend to have long shorelines relative to their straight-line width. The interaction of river discharge and tidal energy determines the salinity of the delta wetlands, with fresh river water reducing salinities and tidal action extending the zone of marine-riverine interactions. One of the most dynamic and expansive river-fed salt marshes is found in Mississippi River Delta in Louisiana. Typically, the first marshes developing on newly deposited sediments are dominated by freshwater species. However, the river course shifts through geologic time as the delta lobe extends and the river loses efficiency. The abandoned marshes, no longer supplied with fresh river water, become increasingly marine influenced. In the Mississippi River Delta, these marshes undergo a 5,000-yr cycle of growth as fresh marshes, transition to salt marshes, and finally degradation back to open water under the influence of subsidence and marine transgression. During the last stage, the seaward edges of the marshes are reworked into barrier islands and spits in the same way as marine-fed coastal marshes on the Atlantic Coast.

Tidal Creeks

A notable physiographic feature of salt marshes, especially low marshes, is the development of *tidal creeks* in the marsh itself (Fig. 8.4). These creeks develop, as do rivers, "with minor irregularities sooner or later causing the water to be deflected into definite channels" (Chapman, 1960). The creeks serve as important conduits for material and energy transfer between the marsh and its adjacent body of water. A tidal creek has salinity similar to that of the adjacent estuary or bay, and its water depth varies with tide fluctuations. Its microenvironments include different vegetation zones along its banks that have aquatic food chains important to the adjacent estuaries. Because the flow in tidal channels is bidirectional, the channels tend to remain fairly stable; that is, they do not meander as much as streams that are subject to a unidirectional flow. As marshes mature and sediment deposition increases elevation, however, tidal creeks tend to fill in and their density decreases (Fig. 8.4).

Pannes

A distinctive feature of many salt marshes is the occurrence of pannes (pans). The term *panne* is used to describe bare, exposed, or water-filled depressions in the marsh,

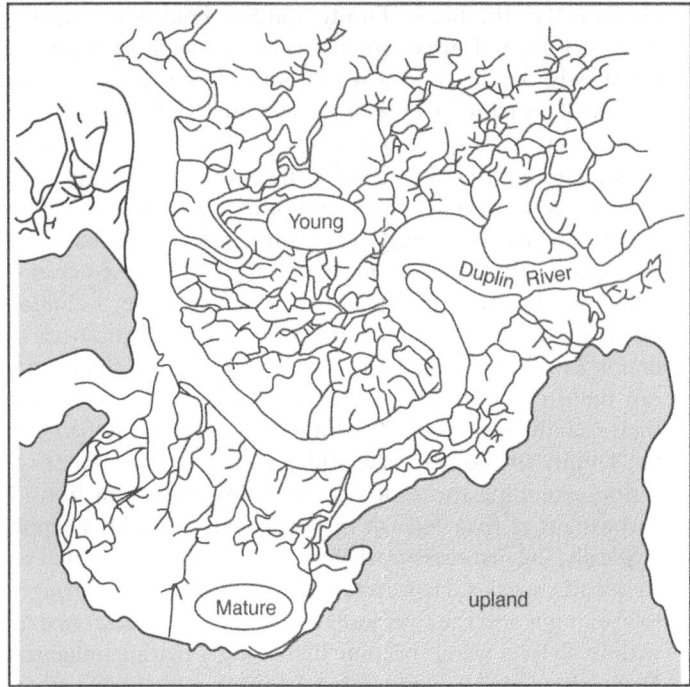

Figure 8.4 Drainage patterns of tidal creeks in young and mature *Spartina alterniflora* salt marshes in the Duplin River drainage, Doboy Sound, Georgia. (After Wiegert and Freeman, 1990, and Wadsworth, 1979)

which may have different sources. In the higher reaches of the marsh, inundated by only the highest tides, *sand barrens* appear where evaporation concentrates salts in the substrate, killing the rooted vegetation. These exposed barrens are often covered by thin films of blue-green algae. *Mud barrens* are naturally occurring depressions in the marsh that are intertidal and retain water even during low tide. Pannes are often devoid of vascular vegetation or support submerged or floating vegetation because of the continuous standing water and the elevated salinities when evaporation is high and are continually forming and filling due to shifting sediments and organic production. The vegetation that develops in a mud panne, for example, wigeon grass (*Ruppia* sp.), is tolerant of salt at high concentrations in the soil water. Relatively permanent ponds are formed on some high marshes and are flooded infrequently by tides. Because of their shallow depth and their support of submerged vegetation, they are used heavily by migratory waterfowl. Pannes are a common feature due to human intervention, occurring where free tidal movement has been blocked by roads or levees, where spoil deposits have elevated a site, or where soil excavation, for example, for highway construction, has occurred in a marsh.

Soil and Salinity

The sediment source and tidal current patterns determine the sediment characteristic of the marsh. Salt marsh sediments can come from river silt, organic productivity in the marsh itself, or reworked marine deposits. As a tidal creek rises out of its banks, water flowing over the marsh slows and drops its coarser-grained sediment load near the stream edge, creating a slightly elevated streamside levee. Finer sediments drop out farther inland, giving rise to the well-known "streamside" effect, characterized by the greater productivity of grasses along tidal channels than inland, a result of the slightly larger nutrient input, higher elevation, and better drainage.

Salinity

Salt marshes that experience a large tide range (e.g., the Wash, England) tend to approximate the ambient marine water salinity even though rainfall may be significant. In coastal marshes adjacent to large rivers, in contrast (e.g., the north coast of the Gulf of Mexico), fresh water dilutes marine sources, and the marshes are brackish or even fresh. Extreme salinities can be found in subtropical areas, such as the Texas Gulf Coast, where rivers and rainfall supply little fresh water and tides have a narrow range so that flushing is reduced. As a result, marine water is concentrated by evapotranspiration, often to double seawater strength or even higher.

Lateral salinity gradients develop as a function of flooding frequency and subsequently influence vegetation productivity (Fig. 8.5). Near the adjacent tidal creek, frequent tidal inundation keeps sediment salinity at or below sea strength. As the marsh

Tidal creek	*Spartina alterniflora*		Salt flats	*Salicornia Batis*	*Juncus Borrichia*	Vegetation zones
	Tall	Short				
100	80–100	40–80	5–10	4–8	2–5	Frequency of flooding
20.0	23.3	33.2	127.0	41.0	24.5	Interstitial salinity, ppt

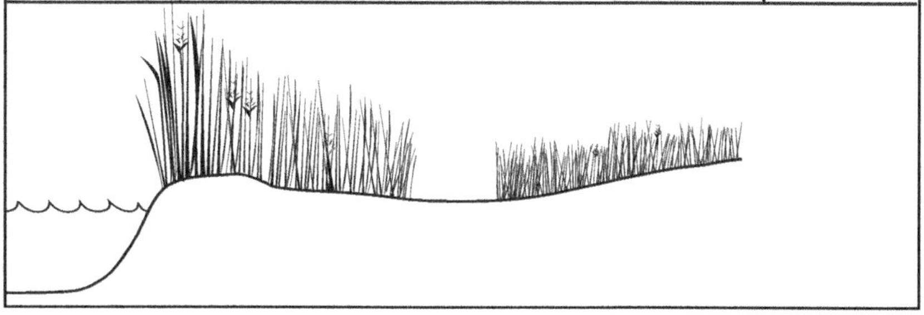

Figure 8.5 The relation of a salt flat's interstitial soil salinity and its vegetation. (After Antlfinger and Dunn, 1979 and Wiegert and Freeman, 1990)

elevation increases, the inundation frequency decreases and the finer sediments drain poorly. At the salt flat zone shown in Figure 8.5, infrequent spring tides bring in salt water that is concentrated by evaporation. Flushing is not frequent enough to remove these salts, so they accumulate to lethal levels. Above this elevation, tidal flooding is so infrequent that salt input is restricted, and flushing by rainwater is sufficient to prevent salt accumulation. In this way, the salt gradient set up by the interaction of marsh elevation, tides, and rain often controls the general zonation pattern of vegetation and its productivity. Within the salt marsh zone itself, however, all plants are salt tolerant, and it is misleading to account for plant zonation and productivity on the basis of salinity alone. Salinity, after all, is the net result of many hydrodynamic factors, including slope and elevation, tides, rainfall, freshwater inputs, and groundwater. Thus, when *Spartina* flourishes in the intertidal zone, it is also responding to tides that reduce the local salinity, remove toxic materials, supply nutrients, and modify soil anoxia. All of these factors collectively contribute to different productivities and different growth forms in the intertidal and high marshes.

Vegetation

The salt marsh ecosystem has diverse biological components, which include vegetation and animal and microbe communities in the marsh and plankton, invertebrates, and fish in the tidal creeks, pannes, and estuaries. The discussion here will be limited to the biological structure of the marsh itself. Plants and animals in these systems have adapted to the stresses of salinity, periodic inundation, and extremes in temperature.

The vegetation of salt marshes can be divided into zones that are related to the high and low marshes described previously but that also reflect regional differences. Figure 8.6 shows a typical New England vegetation zonation pattern from streamside to upland. The intertidal zone or low marsh next to the estuary, bay, or tidal creek is dominated by the tall form of *S. alterniflora* Loisel (smooth cordgrass). In the high marsh, *S. alterniflora* gives way to extensive stands of *S. patens* (saltmeadow cordgrass) mixed with *Distichlis spicata* (spikegrass) and occasional patches of the shrub *Iva frutescens* (marsh elder) and various forbs. Beyond the *S. patens* zone and at normal high tide, *Juncus gerardi* (blackgrass) forms pure stands. At the upper edge of a marsh inundated only by spring tides, two groups of species are common, depending on the local rainfall and temperature. Where rainfall exceeds evapotranspiration, salt-tolerant species give way to less tolerant species, such as *Panicum virgatum* (switchgrass), *Phragmites australis* (common reed), *Limonium carolinianum* (sea lavender), *Aster* spp. (asters), and *Triglochin maritima* (arrow grass). On the southeastern New England coast where evapotranspiration may exceed rainfall during the summer, salts can accumulate in these upper marshes, and salt-tolerant halophytes such as *Salicornia* spp. (saltwort) and *Batis maritima* flourish. Bare areas with salt efflorescence are common. Other features of New England salt marshes include well-flushed mosquito ditches lined with tall *S. alterniflora* and salt pannes containing short-form *S. alterniflora*.

Crain et al. (2004) used greenhouse and field transplants along a New England coast to compare biotic and abiotic factors influencing salt marsh plants. Salt marsh

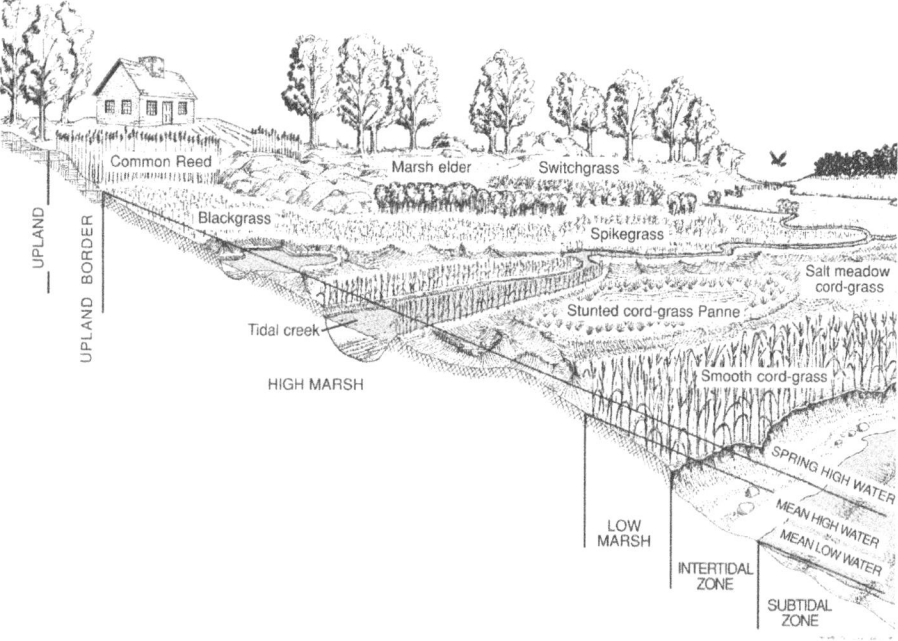

Figure 8.6 Idealized zonation of communities on a typical North Atlantic salt marsh. The location of the different plant associations is strongly influenced by small differences in elevation above the mean high water level. (After Dreyer and Niering, 1995)

plants transplanted into freshwater marsh conditions without competitors grew better than in the salt marsh. However, when salt marsh plants were transplanted into a freshwater marsh with neighboring plants, they were outcompeted by the freshwater marsh plants. The authors surmised that plants in environmental extremes, such as salt marshes, are determined by their tolerance to physiological stress while inland wetland plant occurrence is dictated by the plant's competitive ability.

Characteristic patterns of vegetation found in other salt marshes are shown in Figure 8.7. South of the Chesapeake Bay along the Atlantic Coast, salt marshes typical of the Coastal Plains appear (Fig. 8.7a). These marshes are similar in zonation to those in New England except that (1) tall *S. alterniflora* often forms only in very narrow bands along creeks, (2) the short form of *S. alterniflora* occurs more commonly in the wide middle zone, and (3) *Juncus roemerianus* (black rush) replaces *J. gerardi* in the high marsh. At maturity, low and high marsh areas are approximately equal. The low marsh is almost entirely *S. alterniflora*, tall on the creek bank and shorter behind the natural levee as elevation gradually increases in an inland direction. It may contain small vegetated or unvegetated ponds and mud barrens. The high marsh is much more diverse, containing short *S. alterniflora* intermixed with associations of *Distichlis spicata*, *Juncus roemerianus*, and *Salicornia* spp.

a. Southeast Atlantic coast salt marsh

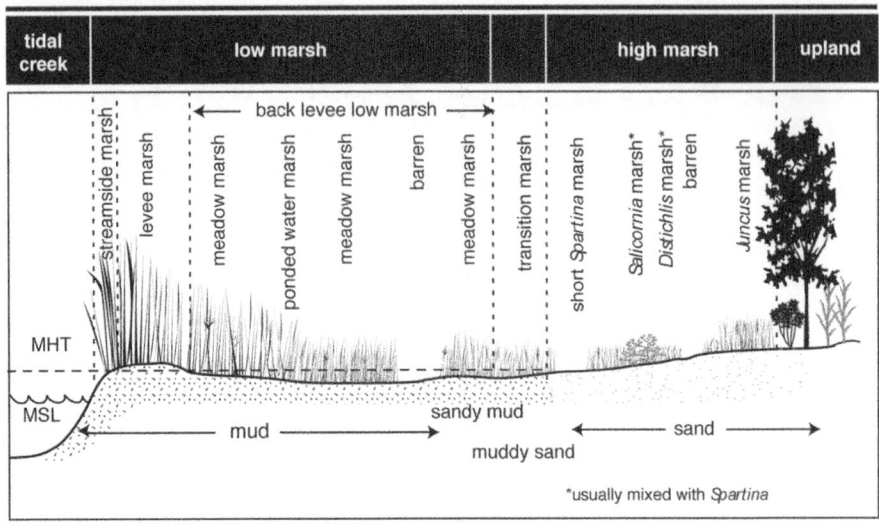

b. Eastern Gulf of Mexico coast salt marsh

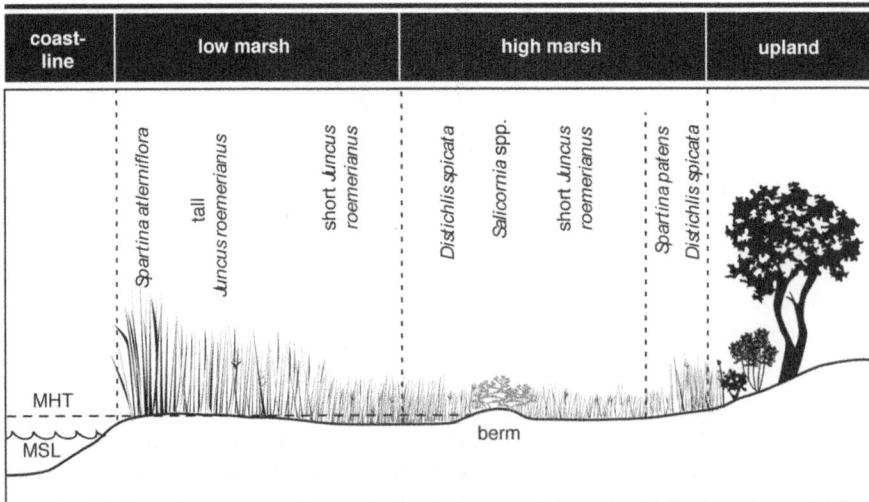

Figure 8.7   Zonation of vegetation in typical salt marshes: (a) southeastern U.S. Atlantic Coast; (b) eastern and northern Gulf of Mexico; (c) northern France. MHT = mean high tide; MSL = mean sea level. ((a) After Wiegert and Freeman, 1990; (b) after Montague and Wiegert, 1990; (c) after LeFeuvre and Dame, 1994)

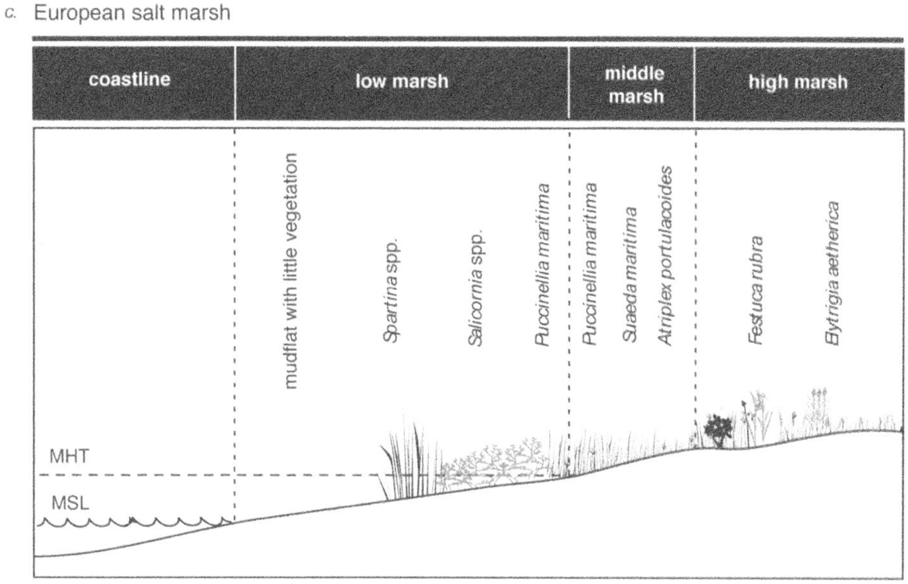

c. European salt marsh

Figure 8.7 (Continued)

Along the Mississippi and northwest Florida coasts, *J. roemerianus* is found in extensive monocultures (Fig. 8.7b). There is often a fringe of *S. alterniflora* along the seaward margin, followed, in an inland direction, by large areas of tall and short *J. roemerianus*. Mixtures of *S. patens* and *D. spicata* line the marsh on the landward edge, and *Salicornia* spp. can be found in small areas such as berms where salt accumulates. Along the northern Gulf Coast, *S. patens* is the dominant species, occurring in a broad zone inland of the more salt-tolerant *S. alterniflora*. More than 200,000 ha of coastal marsh in Louisiana are dominated by *S. patens*.

In Europe, a totally different salt marsh is found, at least compared to the eastern United States marshes (Fig. 8.7c). One of the most notable features is that the intertidal zone between high tide and mean high tide is sparsely covered if it is vegetated at all in Europe, whereas it is dominated by *S. alterniflora* in the United States (Lefeuvre and Dame, 1994). So much of what would be called the low marsh in Europe is, in fact, a mud flat or sparsely vegetated. In Europe, the salt marshes that are studied are mostly between mean high tide and spring tide. The cordgrass found in Europe is generally *S. anglica* or *S. townsendiii*, and it is found in a relatively narrow band. In the last few decades, the clonal grass *Elymus athericus* has spread into the middle and low marshes of many European salt marshes. This invasive plant, because of its large size, traps macrodetritus on the marsh, limiting its export to the adjacent estuary (Bouchard and Lefeuvre, 2000; Lefeuvre et al., 2003; Valéry et al., 2004) and has a significant impact on salt marsh biodiversity (Pétillon et al., 2005).

Salt marshes closer to the polar regions are less well understood. Funk et al. (2004) found that elevation, conductivity, and soil ion composition all contributed to plant cover and species composition in an Alaskan salt marsh. As elevation increased, salinity decreased, resulting in increased plant species richness. At the lowest marsh elevations, only *Puccinella phryganodes* occurred. Midelevation sites were dominated by *Carex subspathaceae*, and high-elevations sites had the highest cover with as many as 16 species, including *Dupontia fischeri* and *Eriophorum angustifolium*. Zhu et al. (2008) reported that biomass in coastal tundra marshes in eastern Antarctica were dominated by a combination of algae, moss, cyanobacteria, and bacteria.

Consumers

Salt marshes, whose features are characteristic of both terrestrial (aerobic) and aquatic (anoxic) environments, provide a harsh environment for consumers. Salt is an additional stress with which they must contend. In addition, the variability of the environment through time is extreme. The dominant plant food source for marsh consumers is generally a marsh grass, which is usually limited in its nutritional value. Considering all these limitations, the number of consumers in the salt marsh is surprisingly diverse.

Many faunal species, particularly vertebrate taxa, utilize tidal marshes (both salt and freshwater) as a component of a larger set of coastal ecosystems. There are, however, many reported species that are considered endemic to tidal marshes. In a global review of terrestrial vertebrates and their occurrence in tidal marshes, Greenberg et al. (2006) found 25 species (or subspecies) that were endemic to tidal marshes. Interestingly, nearly all of the species were restricted to North America. Sampling bias may explain some of this; however, very few records of tidal endemics were reported in well-studied regions, such as Europe and Australia. Another possible factor is the large area of tidal marshes present in North America, and the higher occurrence of endemism is a reflection of a species-area relationship. While these factors or others may contribute to the high North American endemism, currently there is no comprehensive theory for this phenomenon.

It is convenient to classify consumers according to the type of marsh habitat they occupy, although the animals, especially in the higher trophic levels, move from one habitat to another. The marsh can be divided into three major habitats: an *aerial habitat*—the above-ground portion of the macrophytes, which is seldom flooded; a *benthic habitat*—the marsh surface and lower portions of the living plants; and an *aquatic habitat*—the marsh pools and creeks (Fig. 8.8).

Aerial Habitat

The aerial habitat is similar to a terrestrial environment and is dominated by insects and spiders that live in and on the plant leaves. This is the grazing portion of the salt marsh food web. The most common leaf-chewing organisms in salt marshes in the eastern United States are the arthropod *Orchelimum*, the weevil *Lissorhoptrus*, and the squareback crab *Sesarma*. In addition, there are abundant sap-sucking insects (*Prokelisia marginata*, *Delphacodes detecta*) that ingest material translocated through the plant's vascular tissue or empty the contents of mesophyll cells. Numerous carnivorous insects are also found in this habitat. Pfeiffer and Wiegert (1981) listed 81 species of spiders and insects in North Carolina, South Carolina, and Georgia *Spartina* marshes.

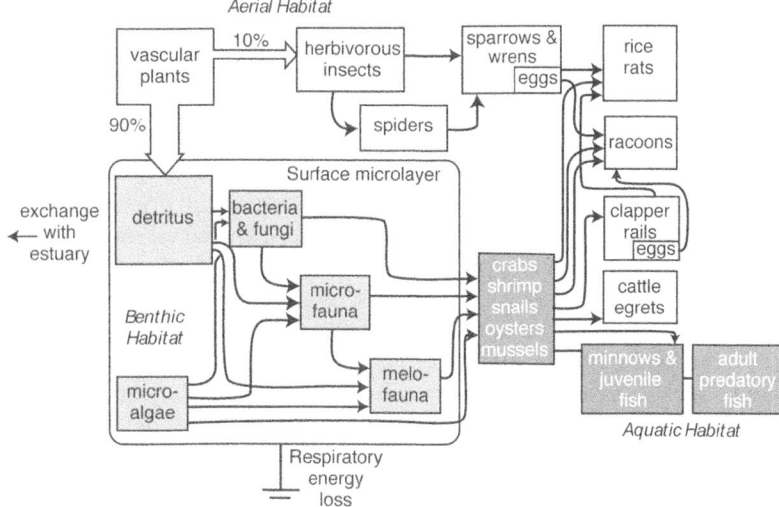

Figure 8.8 Salt marsh food web, showing the major producer and consumer groups of the aerial habitat, benthic habitat, and aquatic habitat. (After Montague and Wiegert, 1990)

Salt marshes support large populations of wading birds, including egrets, herons, willets, and even wood stork and roseate spoonbills. Coastal marshes also support vast populations of migratory waterfowl, including the mallard (*Anas platyrhynchos*), American wigeon (*Anas americana*), gadwall (*Anas strepena*), redheads (*Aythya americana*), and teals (*Anas discors* and *A. crecca*). Black duck (*Anas rubripes*) is a permanent resident in many marshes, as are a number of songbirds.

A number of birds, including the marsh wren (*Cistothorus palustris*) and the seaside sparrow (*Ammodramus maritimus*), laughing gulls (*Larus atricilla*), and Forster's (*Sterna forsteri*) and common terns (*S. hirundo*), feed and nest in the marsh grasses. Wrens feed primarily on insects, and the sparrows apparently feed on the marsh surface, eating worms, shrimp, small crabs, grasshoppers, flies, and spiders. The clapper rail (*Rallus longirostris*) is another permanent marsh resident, feeding primarily on cutworm moths and small crabs. Many nonpermanent insectivorous birds forage in the salt marsh periodically, entering from adjacent fresher marshes, beaches, and upland habitats or migrating through. These include the sharptailed sparrow (*Ammodramus caudacutus*), swallows (*Tachycineta bicolor*, *Hirundo rustica*, and *Stelgidopteryx serripennis*), red-winged blackbirds (*Agelaius phoeniceus*), and various gulls.

Migratory waterfowl use coastal marshes extensively, mostly as wintering grounds, but also as stopover areas during fall and spring migrations. In some areas, geese or duck flocks numbering in the hundreds of thousands denude coastal marshes. Repeated and intense herbivory, especially when followed by high water levels, salinity extremes, or extended drought, may result in the formation of mud flats or shallow open-water ponds.

Benthic Habitat

Probably less than 10 percent of the above-ground primary production of the salt marsh is grazed by aerial consumers. Most plant biomass dies and decays on the marsh surface, and its energy is processed through the detrital pathway. The primary consumers are microbial fungi and bacteria. These organisms, in turn, are preyed on by meiofauna in the decaying grass, the surface microfilm of the marsh, and the decaying bases of plant shoots. Most of these microscopic organisms are protozoa, nematodes, harpacticoid copepods, annelids, rotifers, and larval stages of larger invertebrates. The larger invertebrates on the marsh surface are of two groups, foragers (deposit feeders) and filter feeders. In a general sense, they are considered aquatic because most have some kind of organ to filter oxygen out of water. Foragers include polychaetes, gastropod mollusks such as *Littorina irrorata* and *Melampus bidentatus*, and crustaceans such as *Uca* spp., the blue crab (*Callinectes sapidus*), and amphipods. These organisms browse on the sediment surface, ingesting algae, detritus, and meiofauna. The filter feeders, such as the ribbed mussel (*Geukensia demissus*) and the oyster (*Crassostrea virginica*), filter particles out of the water column.

Aquatic Habitat

Animals classified as aquatic overlap with those in the benthic habitat. For convenience, we include in this group animals in higher trophic levels (mostly vertebrates) and migratory organisms that are not permanent residents of the marsh. Few fish species are permanent residents of the marsh. Most feed along the marsh edges and in small, shallow marsh ponds and move up into the marsh on high tides. Werme (1981) found 30 percent of silverside (*Menidia extensa*) and mummichog (*Fundulus heteroclitus*) in a North Atlantic estuary up in the marsh at high tide. Fish common in small salt marsh ponds in Louisiana include sheepshead minnow (*Cyprinodon variegatus variegatus*), diamond killifish (*Adinia xenica*), tidewater silverside (*Menidia beryllina*), Gulf killifish (*Fundulus grandis*), and sailfin molly (*Poecilia latipinna*). Shrimp (*Penaeus* spp.) and blue crabs (*Callinectes sapidus*) are also common. Most other species use the marsh intermittently for shelter and for food but range widely. Many fish and shellfish spawn offshore or upstream and, as juveniles, migrate into the salt marsh, which offers an abundant food supply and shelter. As subadults, they migrate back into the estuary or offshore. This group of migratory organisms includes more than 90 percent of the commercially important fish and shellfish of the southeastern Atlantic and Gulf coasts.

Mammals

Two mammals in North American salt marshes deserve attention because of their impact on the marshes. The muskrat (*Ondatra zibethicus*) is native to North America; the coypu or nutria (*Myocastor coypus*) is an exotic species introduced from South America. Both prefer fresh marshes but are also found in salt marshes. In Louisiana, the muskrat appears to have been displaced by the nutria from its preferred freshwater habitat into saline marshes. Both mammals are voracious herbivores that consume plant leaves and shoots during the growing season and dig up tubers during the winter. They destroy far more vegetation than they ingest and are responsible for eat-outs that degrade large areas of marsh. These areas recover extremely slowly, especially in the subsiding environment of the northern Gulf Coast. In European marshes, it is common to have domestic animals (e.g., cattle, sheep, or goats) grazing in coastal salt marshes (Bouchard et al., 2003). This grazing has a profound effect on the plant communities and zonation that develops in these marshes.

Ecosystem Function

Major points that have been demonstrated in several studies about the functioning of salt marsh ecosystems include the following five:

1. Primary productivity of macrophytes is high in much of the salt marsh—almost as high as in subsidized agriculture. This high productivity is a result of subsidies in the form of tides, nutrient import, and abundance of water that offset the stresses of salinity, widely fluctuating temperatures, and alternate flooding and drying.
2. Although the biomass of edaphic algae is small, algal production sometimes can be as high as or higher than that of the community's macrophytes, especially in hypersaline marshes.
3. Direct grazing of vascular plant tissue is a minor energy flow in the salt marsh, but grazing on edaphic and epiphytic algae is a significant source of high-quality food energy for meio- and macro-invertebrates.
4. Fungi and bacteria are primary consumers that break down and transform indigestible plant cellulose (detritus) into protein-rich microbial biomass for consumers. This detrital pathway is a major flow of energy utilization in the salt marsh.
5. Salt marshes have been shown at times to be both sources and sinks of nutrients, particularly nitrogen.

Primary Productivity

Tidal marshes are among the most productive ecosystems in the world, annually producing up to 80 metric tons per hectare of plant material ($8,000 \text{ g-wet weight m}^{-2} \text{ yr}^{-1}$) in the southern Coastal Plain of North America. The three major autotrophic units of the salt marsh are marsh grasses, mud algae, and phytoplankton of the tidal creeks. Extensive studies of the net primary production have been conducted in salt marshes, especially along the Atlantic and Gulf coasts of the United States. A comparison of some of the measured values of net above-ground and below-ground production is given in Table 8.3. Above-ground production varies widely, from as little as $410 \text{ g m}^{-2} \text{ yr}^{-1}$ in a Normandy salt marsh to a high of $4,200 \text{ g m}^{-2} \text{ yr}^{-1}$ in a Louisiana *Spartina patens* marsh. Below-ground production is difficult to measure and can be much higher than above-ground production. Productivity of salt marshes is often higher along creek channels and in low or intertidal marshes than in high marshes because of the increased exposure to tidal and freshwater flow. These conditions also

Table 8.3 Net primary productivity estimates of salt marshes and dominant plant species

Species	Aboveground Net Primary Production (g m^{-2} yr^{-1})	Belowground Net Primary Production (g m^{-2} yr^{-1})	Source
Louisiana			
Distichlis spicata	1,162–1,291		White et al. (1978)
Juncus roemerianus	1,806–1,959		
Spartina alterniflora	1,473–2,895		
Spartina patens	1,342–1,428		
Distichlis spicata	1,967		Hopkinson et al. (1980)
Juncus roemerianus	3,295		
Spartina alterniflora	1,381		
Spartina cynosuroides	1,134		
Spartina patens	4,159		
Alabama			
Juncus roemerianus	3,078	7,578	Stout (1978)
Spartina alterniflora	2,029	6,218	
Mississippi			
Juncus roemerianus	1,300		de la Cruz (1974)
Distichlis spicata	1,072		
Spartina alterniflora	1,089		
Spartina patens	1,242		
Normandy (France)			
Spartina anglica/	1,080 (nongrazed)		Lefeuvre et al. (2000)
Salicornia/Suaeda	410 (grazed)		
maritima low marsh	1,990 (nongrazed)		
high marsh	550 (grazed)		
Mediterranean Sea			
Rhone River Delta (France)			Ibánez et al. (1999)
Sarcocornia fruticosa	1123–1262		
Ebro Delta (Spain)			Curco et al. (2002)
Arthocnemum macrostachyum	190	50	
Sarcocornia fruticosa	580	950	
A. macrostachyum, S. fruticosa	840	340	

produce the taller forms of *Spartina*, as discussed earlier. Below-ground production is sizable—often greater than aerial production (Table 8.3). Under unfavorable soil conditions, plants seem to put more of their energy into root production. Hence, roots:shoot ratios seem to be generally higher inland than at streamside locations.

The productivity of edaphic algae was summarized by Sullivan and Currin (2000). Annual benthic algal production, as measured in a number of studies, ranges from 28 g C m^{-2} yr^{-1} in a Gulf of Mexico coast *Juncus roemerianus* marsh to 341 g C m^{-2} yr^{-1} in a southern California *Jaumea carnosa* marsh (Table 8.4). Benthic algal production increases in a southerly direction along the Atlantic coast, but is lowest on the

Table 8.4 Comparison of annual benthic microalgal production (g C m^{-2} yr^{-1}) and ratio of annual benthic microalgal to vascular plant net aerial production (BMP/VPP) in different salt marshes of the United States

State	Algal productivity, g C m^{-2} yr^{-1}	BMP/VPP, %	Reference
Massachusetts	105	25	Van Raalte (1976)
Delaware	61–99	33	Gallagher and Daiber (1974)
South Carolina	98–234	12–58	Pinckney and Zingmark (1993)
Georgia	200	25	Pomeroy (1959)
Georgia	150	25	Pomeroy et al. (1981)
Mississippi	28–151	10–61	Sullivan and Moncreiff (1988)
Texas	71	8–13	Hall and Fisher (1985)
California	185–341	76–140	Zedler (1980)

Source: Sullivan and Currin (2000)

Gulf coast. Much of the algal production on the east and west coasts of the United States occurs when the overstory plants are dormant. On the Atlantic and Gulf coasts, the productivity of algae is 10 to 60 percent of vascular plant productivity. Zedler (1980), however, found that algal net primary productivity in southern California was 76 to 140 percent of vascular plant productivity. She hypothesized that the arid and hypersaline conditions of southern California favor algal growth over vascular plant growth. Algae are important components of the salt marsh food web, so much so that Kreeger and Newell (2000) stated:

> We question the paradigm that salt marshes have "detritus-based food webs" (Odum, 1980), considering that the bulk of secondary production by metazoans could actually be linked to primary production by the microphytobenthos rather than through either direct (herbivory) or indirect (detrivory) linkages to primary production by vascular plants.

Variations in productivity on the local scale result from complex interactions of soil anoxia, soluble sulfide, and salinity (Mendelssohn and Morris, 2000). Although water appears plentiful, the concentration of dissolved salt makes the salt marsh environment similar in many respects to a desert. The "normal" water gradient is from plant to substrate. To overcome the osmotic influence of salt, plants must expend energy to increase their internal osmotic concentration in order to take up water. As a result, numerous studies confirm that plant growth is progressively inhibited by increasing salt concentrations in the soil. This is true even for the salt-tolerant species of the salt marsh, and the salinity effects may be subtle. For example, Morris et al. (1990) showed that the year-to-year variation in marsh production at a single site on the East Coast of the United States was correlated with the mean summer water level, which they equated with soil salinity. (Soil salinity was inversely correlated with the frequency of marsh flooding in this study.)

Another factor limiting production is the degree of anaerobiosis of the substrate. Vascular plants, even those that have developed adaptations to anaerobic conditions, grow best in aerobic soils. Many effects of anaerobiosis have been documented: reduced energy availability as the aerobic respiratory pathway is blocked, reduced nutrient uptake, the accumulation of toxic sulfides in the substrate, and changes in the availability of nutrients. Salt inhibition and oxygen depletion frequently occur together. *Spartina* grows shorter in the inland marsh because its drainage is poor; hence oxygen deficits are severe. Salt may concentrate in this environment. The primary result of poor drainage in inland salt marshes, however, is apparently a dramatically lower soil redox potential, which in turn leads to elevated sulfide concentrations. Although *S. alterniflora* is able to mitigate the toxic effects of sulfide to some extent through its ability to transport oxygen through the root system to the rhizosphere and by the enzymatic oxidation of sulfides, its growth is inhibited when the interstitial soluble sulfide concentration exceeds 1 mM sulfide (Bradley and Dunn, 1989; Koch et al., 1990).

Primary productivity also contributes to sediment accretion in salt marshes, and there is increasing interest in the ability of salt marshes to withstand relative rises in sea level. Because of their location, salt marshes are constantly adjusting to maintain an equilibrium near mean sea level. Morris et al. (2002) demonstrated the importance of primary productivity for increased sediment accretion. By experimentally increasing productivity in a South Carolina salt marsh, sediment accretion in the marsh was also enhanced. Salt marshes tend to be most productive at elevations just below mean high tide (Fig. 8.9). However, in terms of long-term response to rapidly rising sea levels, these lower marshes may be incapable of accreting sediment quickly enough to

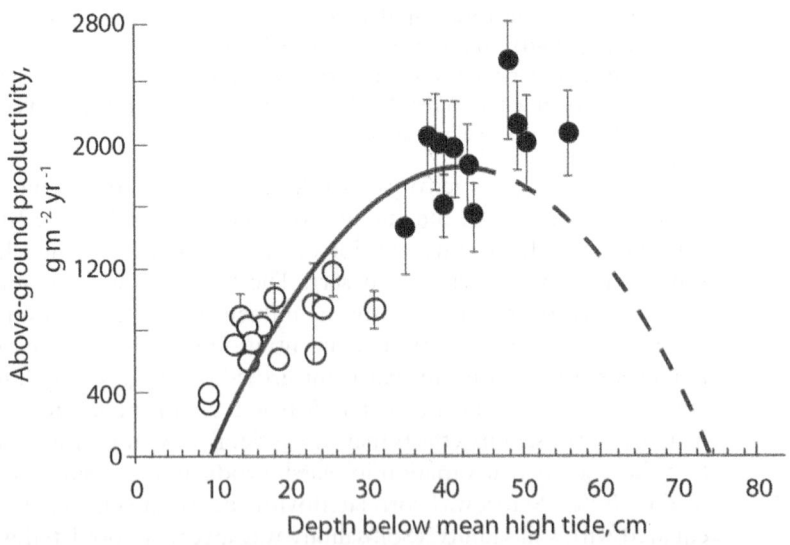

Figure 8.9 Above-ground net primary productivity of salt marshes as a function of elevation below mean high tide during peak growing season of June and July. Data are for high (open circles) and low (solid circles) *Spartina alterniflora* salt marshes. (After Morris et al., 2002)

keep pace with rising water levels. Marshes at slightly higher elevations with abundant supplies of sediment are the most likely ones to acclimate to rapidly rising sea levels and maintain their position.

Decomposition and Consumption

Since John Teal's seminal publication on energy flow in the salt marsh system (Teal, 1962), salt marshes have been considered detrital systems. Almost three-quarters of the primary production in the salt marsh ecosystem is broken down by bacteria and fungi. In his study of energy flow within the salt marsh environment, Teal (1962) estimated that 47 percent of the total net primary productivity was lost through respiration by microbes. It was largely assumed that the rich secondary productivity of estuaries was fueled by a detritus food web. With the development of new techniques, such as multiple stable isotope fractionation, these early assumptions have been questioned, and a refined and quite different picture of decomposition and secondary production has emerged. With a few exceptions such as salt marshes in Mediterranean-type climates, primary production is dominated by emergent spermatophytes, usually grasses. When they senesce, the soluble organic contents are rapidly flushed from their tissues. This labile soluble organic matter from both living and decomposing salt marsh vegetation (which may be as much as 25 percent of the initial dry weight of the dying grass) is an important energy source for microorganisms in the marsh and the adjacent estuary (Wilson et al., 1986; Newell and Porter, 2000). The remaining 75 percent of dead vegetation biomass is largely composed of refractory structural lignocellulose that is indigestible by all but a few metazoans. Ideas about the fate of the vegetation biomass have changed. Two key conclusions about the process of decomposition are described next.

1. The initial secondary producers, or decomposers, on epibenthic marsh grass stems are ascomycetous fungi. These fungi may reach a biomass equal to 3 (summer) to 28 (winter) percent of live *Spartina alterniflora* standing crop. Most of this biomass occurs in standing dead grass or on the marsh surface. In South Atlantic coastal marshes, fungal productivity is 10 times greater in winter than in summer. In contrast, most of the bacterial biomass is found in the sediment surface microlayer. Productivity of bacteria is twice fungal productivity in summer but only one-tenth as great in winter. The conversion efficiency of grass biomass to fungal biomass can be as high as 50 percent (Newell and Porter, 2000).

2. There appear to be at least three decomposer groups. (a) Fungi are the major decomposers of the epibenthic standing dead grass; (b) aerobic bacteria in the surface microlayer decompose the decayed grass leaf shoots that are shredded by gastropods and amphipods and fall to the marsh surface; and (c) anaerobic bacteria, a third group of decomposers in deeper anoxic sediments, are able to use electron acceptors other than oxygen to metabolize. Prime among these are sulfate reducers, which may oxidize a major proportion of the underground senescent root and rhizome biomass.

During the decomposition process, the nitrogen content of the grass/fungal/bacterial brew increases. This is due, in part, to the low C:N ratio of bacterial decomposers compared to raw grass tissue. It was assumed for many years that nitrogen enrichment made the decaying plant material a nutritionally better food supply for consumers; in more recent studies, however, it was determined that much of the nitrogen is bound in refractory compounds in the decaying grass (Teal, 1986). The nutritious bacterial population is kept at low concentrations by metazoan grazing.

These discoveries about the decomposition process in marsh macrophytes have led to a reevaluation of the source of energy for the abundant consumer population found in tidal marshes and their associated tidal creeks. Although much of the change consists of elaboration and clarification of the detrital process, a major shift has been toward a much greater role for algae, both phytoplankton and especially edaphic algae, as major flows of energy in the salt marsh food web. Vascular plants are still the major source of organic carbon, but few metazoans can assimilate this cellulose-rich material. Hence, direct grazers are limited to several species of herbaceous insects, which collectively consume less than 10 percent of plant production. The ribbed mussel *Geukensia demissa* is an exception to this generalization. It has been shown to assimilate aseptic detrital cellulose with an efficiency of up to 15 percent. Kreeger and Newell (2000) suggest that the mussel must either possess endogenous cellulases or contain a vigorous gut flora capable of cellulose breakdown.

Decay begins with fungal decomposition of the aerial parts of the senescent vascular plants. Epiphytic algae growing on the lower parts of the grass culms are also a part of this detrital brew. The complex is ingested and shredded by gastropod snails, such as *Littorina irrorata* and perhaps amphipods. In a microcosm experiment, the snails had the capacity to ingest 7 percent of their weight of naturally decayed leaves per day and assimilate it with an efficiency of about 50 percent. The epiphytic algae are also ingested and assimilated by amphipods and other organisms grazing on the dead leaf surfaces.

The finely shredded grass/fungal/algal material that falls to the marsh surface is infected by aerobic bacteria that continue the process of decomposition. Also part of this mixture is the algal community, largely diatoms, growing on the marsh surface. This complex is consumed by benthic meiofaunal and macrofaunal deposit feeders. Primary among the meiofauna are nematodes; also feeding on the surface are harpacticoid copepods, amphipods, polychaetes, turbellarians, ostracods, foraminifera, and gastroliths. The larger consumers in this group include fiddler crabs, snails, polychaetes, oligochaetes, and some bivalves.

Finally, some of the finely decomposed organic material on the surface microlayer is periodically suspended by winds and currents, where it mixes with the phytoplankton growing in the water. For example, as much as 25 percent of the suspended algae have been found to be edaphic species (MacIntyre and Cullen, 1995). This sestonic mix of bacterial/organic fragments, free-living bacteria, and algae is consumed by suspension feeders, especially benthic suspension feeders such as bivalve mollusks and oligochaete annelids. Also active are zooplankton, although they probably do not process as much material as the benthic bivalves. The meio- and macrofauna feeding on algae, fungi, and bacteria are in turn consumed by animals in the higher trophic levels (Fig. 8.8).

Organic Export

A central paradigm of salt marsh ecology has long been the *outwelling hypothesis*, which was first enunciated by E. P. Odum in 1968. The hypothesis was based, in part, on a salt marsh energy flow analysis presented by John Teal at the first salt marsh conference, held in 1958 at the University of Georgia Marine Laboratory on Sapelo Island, Georgia (published as Teal, 1962). Odum (1968) described salt marshes as "primary production pumps" that feed large areas of adjacent waters, and he compared the flow of organic material and nutrients from salt marshes to the *upwelling* of deep ocean water, which supplies nutrients to some coastal waters. Teal (1962) hypothesized that salt marshes exported organic material and energy primarily as detritus from the marsh surface. In the intervening years, there have been many attempts to measure this export. Teal's (1962) energy flow analysis estimated that about 45 percent of net primary production was exported from the salt marsh. Nixon's (1980) summary agreed in that most studies showed an export of dissolved and particulate material, in an amount that could account for about 10 to 50 percent of phytoplankton production in coastal and estuarine waters.

Childers et al. (2000) pointed out that the original hypothesis was ambiguous in that it equated salt marsh export to coastal ocean import. In reality, the flows from a salt marsh are into nearby tidal creeks, and fluxes to the coastal ocean depend on the geomorphology of the estuary and the distance from the marsh to the coast. Hence, salt marshes interact with nearby tidal creeks and the inner estuary, which, in turn, exchange flows with the greater estuarine basin, which, finally, interacts with the coastal ocean (Fig. 8.10). Failure to take these spatial factors into account has

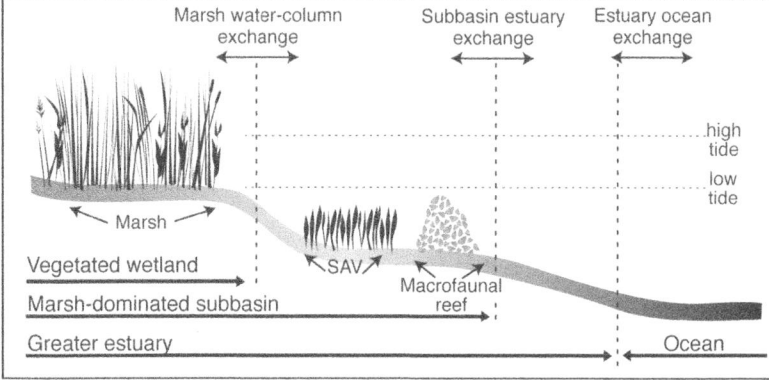

Figure 8.10 **Hierarchy of estuarine-coastal landscape that includes estuarine subbasins nested within the greater estuary, and vegetated wetland ecosystems nested within both. SAV = submerged aquatic vegetation. (After Childers et al., 2000)**

made it difficult to compare studies and is one reason for the lack of agreement in study results.

The evidence for outwelling rests on more than organic flux data, such as reported in the studies summarized by Nixon (1980) and Childers (1994). Hopkinson (1985) reported that water column respiration offshore of the Georgia barrier islands exceeded *in situ* production; that is, the zone was heterotrophic, implying that organic matter was being imported from the inshore estuaries and marshes. Turner et al. (1979) reported that offshore, within 10 km of the coastal estuaries, primary productivity measurements were often 10 times greater than that farther offshore. They attributed this high productivity to outwelling of nutrients from the estuaries.

Other evidence of outwelling comes from fishery studies. Turner (1977) found a close correlation worldwide between commercial yield of shrimp (which are harvested both in the estuary and offshore) and the area of estuarine intertidal vegetation. Teal and Howes (2000) analyzed fish catch statistics dating back to 1880 from the Long Island Sound, New York, and determined that fish catch was closely related to marsh edge length. Since edge length is an index of accessibility to the marsh, the result implicated salt marsh production in commercial fishery catch.

Several general factors affect the outwelling hypothesis. First, material and energy usually flow from concentrated hot spots to lower concentration areas. Salt marshes are hot spots of production, so it is logical to expect an outwelling of production and food energy (E. P. Odum, 2000). Second, outwelling can be expected to be modified by the geomorphology of the estuary and the location of a salt marsh in the estuary. Thus, open estuaries with salt marshes close to the coast are expected to export more material than estuaries with small coastal passes and distant marshes. Finally, salt marshes and coastal estuaries are pulsing systems, with daily tidal variation, seasonal variations in rainfall and river flow, and periodic severe storms. Extreme events often lead to import or export that overwhelms the normal daily fluxes.

Salt Marsh Die-off

For the first several years at the turn of the century (2000–2005), *Spartina* salt marshes in the southeastern and Gulf coasts of the United States were experiencing major die-off, totaling more that 100,000 ha and affecting 1,500 km of coastline. One theory presented to explain this die-off was described by Silliman et al. (2005) as having the following sequence: a protracted and intense drought that occurred for three to four years (bottom-up effect) followed by snails (*Littoraria irrorata*) concentrating on the die-off borders to prolong the effect (top-down effect). In addition, declines in blue crab populations, a major predator of the snails, of 40 to 85 percent provided synergy for the snail grazing. Essentially Silliman et al. (2005) suggested that "drought-induced soil stress can amplify top-down control by grazers and initiate marsh plant die-off.... These disturbances then stimulate the formation of consumer fronts, leading to waves of salt marsh destruction resulting from runaway consumption." Such epidemic ecosystem die-offs that combine bottom-up and top-down stresses on coastal ecosystems in a synergistic way are another example of an undesirable positive feedback that could occur on coastal ecosystems with any significant climate change.

Tidal Freshwater Wetlands

Tidal freshwater wetlands are interesting because they receive the same "tidal subsidy" as mangroves and salt marshes but without the salt stress. One would expect, therefore, that these ecosystems might be very productive and also more diverse than their saltwater counterparts. As tides attenuate upstream, the wetlands assume more of the characteristics of inland freshwater wetlands (see Chapter 10: Freshwater Marshes). The distinction between freshwater tidal and inland wetlands is not clear-cut because on the coast they form a continuum (Fig. 8.1). Inland from the tidal salt marshes but still close enough to the coast to experience tidal effects, tidal freshwater marshes are dominated by a variety of grasses and by annual and perennial broad-leaved aquatic plants. In the United States, they are found primarily along the Middle and South Atlantic coasts and along the coasts of Louisiana and Texas. Tidal freshwater swamps tend to be most abundant along the farthest tidal extent of coastal rivers, particularly those rivers with low gradients and high discharge. Most of the extensive tidal freshwater forests in the United States occur along the southeastern coastline (Maryland to Texas). Estimates of tidal freshwater wetlands in the United States range from 400,000 ha along the Atlantic Coast to 819,000 ha for the conterminous United States (Table 8.1). The extent of tidal freshwater swamps in the United States is less certain, but 200,000 ha has been is estimated for the southeast U.S. coastline (Field et al., 1991). The uncertainty in the estimates is related to where the line is drawn between tidal and nontidal areas. Tidal freshwater marshes can be described as intermediate on the continuum from coastal salt marshes to freshwater marshes. Because they are tidally influenced but lack the salinity stress of salt marshes, often tidal freshwater marshes have been reported to be very productive ecosystems, although a considerable range in their productivity has been measured. Elevation differences across a freshwater tidal marsh correspond to different plant associations. These associations are not discrete enough to call communities, and the species involved change with latitude. Nevertheless, they are characteristic enough to allow some generalizations.

Vegetation

Marsh Vegetation

On the Atlantic Coast of the United States (Fig. 8.11a, b), submerged vascular plants, such as *Nuphar advena* (spatterdock), *Elodea* spp. (waterweed), *Potamogeton* spp. (pondweed), and *Myriophyllum* spp. (water milfoil), grow in the streams and permanent ponds. The creek banks are scoured clean of vegetation each fall by the strong tidal currents, and they are dominated during the summer by annuals, such as *Polygonum punctatum* (water smartweed), *Amaranthus cannabinus* (water hemp), and *Bidens laevis* (bur marigold). The natural stream levee is often dominated by *Ambrosia trifida* (giant ragweed). Behind this levee, the low marsh is populated with

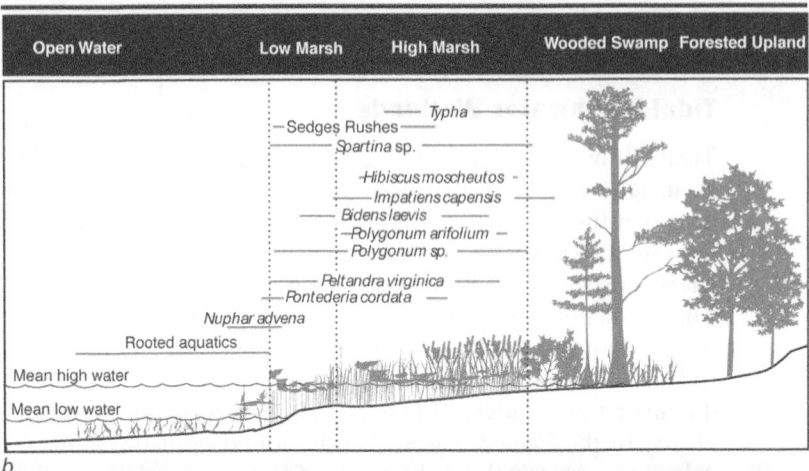
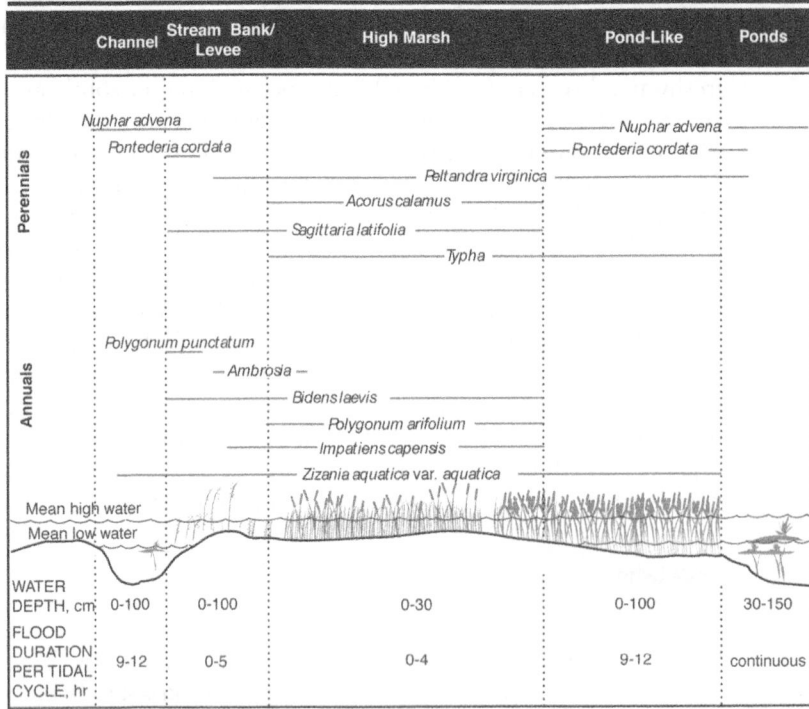

Figure 8.11 Cross sections across typical freshwater tidal marshes, showing elevation changes and typical vegetation: (a) and (b) Atlantic Coast marshes; (c) new marsh in the Atchafalaya Delta, Louisiana. ((a) After W. E. Odum et al., 1984; (b) after Simpson et al., 1983; (c) after Gosselink et al., 1998)

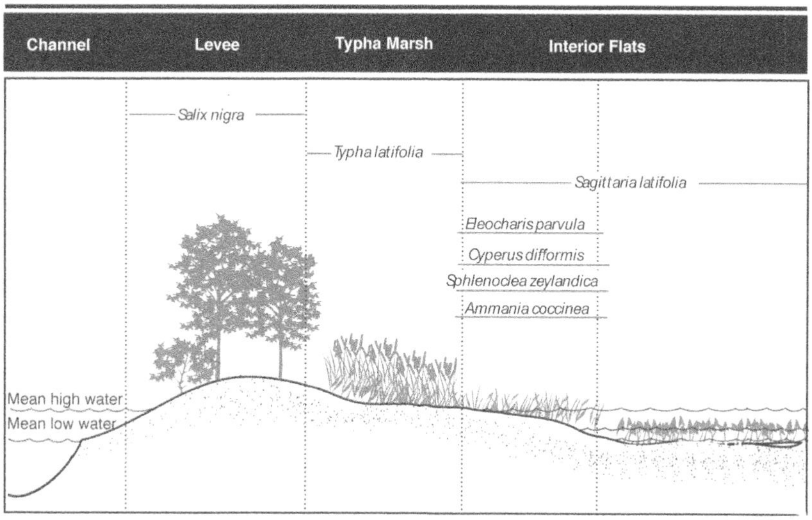

Figure 8.11 *(Continued)*

broad-leaved monocotyledons, such as *Peltandra virginica* (arrow arum), *Pontederia cordata* (pickerelweed), and *Sagittaria* spp. (arrowhead).

Typically, the high marsh has a diverse population of annuals and perennials. W. E. Odum et al. (1984) called this the "mixed aquatic community type" in the Mid-Atlantic region. Leck and Graveline (1979) described a "mixed annual" association in New Jersey while Caldwell and Crow (1992) described the vegetation of a tidal freshwater marsh in Massachusetts. Generally, the areas were dominated early in the season by perennials, such as arrow arum. A diverse group of annuals—*Bidens laevis, Polygonum arifolium* (tear-thumb) and other smartweeds, *Pilea pumila* (clearweed), *Hibiscus coccineus* (rose mallow), *Acnida cannabina*, and others—assumed dominance later in the season. In addition to these associations, there are often almost pure stands of *Zizania aquatica* (wild rice), *Typha* spp. (cattail), *Zizaniopsis miliacea* (giant cutgrass), and *Spartina cynosuroides* (big cordgrass). In the northern Gulf of Mexico, arrowheads (*Sagittaria* spp.) replace arrow arum (*Peltandra* spp.) and pickerelweed (*Pontedaria cordata*) at lower elevations. Visser et al. (1998) described three vegetation associations in this area:

1. Bulltongue (*Sagittaria lancifolia*) occurs with co-dominants maidencane (*Panicum hemitomon*) and spikerush (*Eleocharis* spp.). Commonly the ferns *Thelypteris palustris* and *Osmunda regalis*, wax myrtle (*Myrica cerifera*), and pennywort (*Hydrocotyi* spp.) are also present. Fifty-two different species occur in this association.

2. A maidencane-dominated association is widespread across the delta and includes 55 species.

3. Cutgrass (*Zizaniopsis miliacea*) occurring with co-dominant maidencane is relatively uncommon. It includes 20 other species.

Interestingly, rising sea level and/or surface subsidence on both the Gulf Coast and the Atlantic Coast has resulted in vegetation shifts. Although the previously dominant species are still present, in a Chesapeake Bay tidal freshwater wetland, for example, the oligohaline species *Spartina cynosuroides*, which was not among the dominant species in 1974, is now second in peak biomass and fourth in importance value (Perry and Hershner, 1999). Similarly, Visser et al. (1999) reported that the maidencane association has decreased from 51 percent coverage of the tidal wetlands of Terrebonne Bay (in the Mississippi River delta) in 1968 to only 14 percent in 1992. It has been replaced by *Eleocharis baldwinii*–dominated marshes, which were uncommon in 1968 (3 percent coverage) but in 1992 covered 42 percent of the area.

Floating Marshes

Floating marshes in the tidal reaches of the northern Gulf of Mexico are similar to the nontidal riverine and lacustrine marshes found extensively around the globe. Large expanses of floating marshes (*Phragmites communis* marshes) have been found in the Danube Delta for at least a century (Pallis, 1915), along the lower reaches of the Sud in Africa (papyrus swamps; Beadle, 1974), in South America (floating meadows in lakes of the *varzea*; Junk, 1970), and in Tasmania (floating islands in the Lagoon of Islands; Tyler, 1976). Floating marshes have also been reported in Germany, the Netherlands (Verhoeven, 1986), England (Wheeler, 1980), and North Dakota and Arkansas in the United States (Eisenlohr, 1972; Huffman and Lonard, 1983). In Louisiana, floating marshes are usually floristically diverse, but different stands are dominated by *Panicum hemitomon* with ferns and vines such as *Vigna luteola* and *Ipomoea sagittata*; *Sagittaria lancifolia* with *Eleocharis* spp., *Panicum dichotomiflorum*, *Bacopa monnieri*, and *Spartina patens*; and *Eleocharis baldwinii* and *Eleocharis parvula* with *Ludwigia leptocarpa*, *Phyla nodiflora*, and *Bidens laevis* (Sasser et al., 1996). The marsh substrate is composed of a thick organic mat, entwined with living roots, that rises and falls (all year or seasonally) with the ambient water level (Swarzenski et al., 1991). This type of marsh is interesting in a successional sense because it appears to be an endpoint in development; it is freed from normal hydrologic fluctuation and mineral sediment deposition. Hence, in the absence of salinity intrusions, it appears to support a remarkably stable community (Sasser et al., 1995).

New Marshes

The active deltas of the Mississippi and Atchafalaya rivers are the sites of the largest newly progading coastal deltas in the continental United States. Fresh tidal marshes (Fig. 8.11c) that have formed in the last 25 years on emergent islands are dominated on the natural levees by black willow (*Salix nigra*), and the extensive back-island mud flats are dominated by common three-square sedge (*Scirpus deltarum*) or by arrowhead (*Sagittaria latifolia*), with areas of cattail (*Typha latifolia*) and a seasonally variable annual/perennial mix in between.

Swamp Vegetation

The richness of canopy trees tends to be low in tidal freshwater forests, such as those found in southeastern United States. At the lowest elevations, canopy species tend to be dominated by those that can withstand long periods of inundation, such as cypress (*Taxodium distichum*), water tupelo (*Nyssa aquatica*), and swamp tupelo (*Nyssa biflora*). (See Chapter 11: "Freshwater Swamps and Riparian Ecosystems," for more descriptions of these forested wetlands.) At slightly higher elevations, other species can become dominant, including ash (*Fraxinus* spp.), red maple (*Acer rubrum*), sweetgum (*Liquidambar styraciflua*), American hornbeam (*Carpinus caroliniana*), and sweetbay (*Magnolia virginiana*). Tidal freshwater swamps are often characterized by a distinctive hummock and hollow topography in which most of the trees are limited to the hummocks and hollows are sparsely vegetated to unvegetated by trees. Because these forests can have fairly open canopies, subcanopy and understory vegetation can be extensive and species rich. In tidal swamps along the Pamunkey River in Virginia, Rheinhardt (1992), found spicebush (*Lindera benzoin*), winterberry (*Ilex verticillata*), *C. caroliniana*, and *Ilex opaca* to be among the most dominant subcanopy species. Understory dominants included halberd-leaved tearthumb (*Polygonum arifolium*), lizard's tail (*Saururus cernuus*), and sedges (*Carex* spp.). Along the forested tidal reaches of the Suwannee River on the Gulf Coast of Florida, dominant subcanopy vegetation was pumpkin ash (*Fraxinus profunda*), Carolina ash (*Fraxinus caroliniana*), and wax myrtle (*Morella cerifera*). Common understory vegetation included variable panic grass (*Dicanthelium commutatum*), string lily (*Crinum americanum*), *S. cernuus*, and *Carex* spp.

Seed Banks

The species composition of a tidal freshwater wetland does not appear to depend on the availability of seed in particular locations. Seeds of most species are found in almost all habitats, although the most abundant seed reserves are generally from species found in that vegetation zone (Whigham and Simpson, 1975; Leck and Simpson, 1987, 1995; Baldwin et al., 1996). They differ, however, in their ability to germinate under the local field conditions and in seedling survival. Flooding is one of the main controlling physical factors. Many of the common plant species seem to germinate well even when submerged—for example, *Peltandra virginica* and *Typha latifolia*—whereas others—such as *Impatiens capensis, Cuscuta gronovii*, and *Polygonum arifolium*—show reduced germination. Baldwin et al. (2001) manipulated the hydrology in a series of experiments examining freshwater tidal marshes along the Patuxent River in Maryland and found that increasing flood depths by 3 to 10 cm can significantly reduce species richness and plant growth. In particular, shallow flooding early in the growing season reduced the germination of annuals.

Competitive factors also play a role in vegetation assemblages. Arrow arum and cattail, for example, produce chemicals that inhibit the germination of seed; and shading by existing plants is apparently responsible for the inability of arrow arum plants to become established anywhere except along the marsh fringes. Some species (*Impatiens capensis, Bidens laevis*, and *Polygonum arifolium*) are restricted to the high marsh because the seedlings are not tolerant of extended flooding. Seed bank strategies differ in different zones of the marsh. The seeds of most of the annuals in the high marsh germinate each spring so that there is little carryover in the soil. In contrast, perennials tend to maintain seed reserves. The seeds of most species, however, appear to remain in the soil for a restricted period. In one study, 31 to 56 percent of the seeds were present only in surface samples, and 29 to 52 percent germinated only in sediment samples taken in early spring (Leck and Simpson, 1987). The complex interaction of all these factors has not been elucidated to the extent that it is possible to predict what species will be established where on the marsh.

In addition to vascular plants, phytoplankton and epibenthic algae abound in freshwater tidal marshes, but relatively little is known about them. In one study of Potomac River marshes, diatoms (bacillariophytes) were the most common phytoplankton, with green algae (chlorophytes) comprising about one-third of the population and blue-green algae (cyanobacteria) present in moderate numbers. The same three taxa accounted for most of the epibenthic algae. Indeed, many of the algae in the water column are probably entrained by tidal currents off the bottom. In a study of New Jersey tidal freshwater marsh soil algae, Whigham et al. (1980) identified 84 species exclusive of diatoms. Growth was better on soil that was relatively mineral and coarse, compared with growth on fine organic soils. Shading by emergent plants reduced algal populations in the summer months. In nontidal freshwater marshes, algae epiphytic on emergent plants and litter made important contributions to invertebrate consumers (Campeau et al., 1994). Algal biomass is probably two to three orders of magnitude less than peak biomass of the vascular plants, but the turnover rate is much more rapid.

Consumers

Coastal freshwater wetlands are used heavily by wildlife. The consumer food chain is predominantly detrital, and benthic invertebrates are an important link in the food web. Bacteria and protozoa decompose litter, gaining nourishment from the organic material. It appears unlikely that these microorganisms concentrate in large enough numbers to provide adequate food for macroinvertebrates. Meiobenthic organisms, primarily nematodes, comprise most of the living biomass of anaerobic sediments. They probably crop the bacteria as they grow, packaging them in bite-sized portions for slightly larger macrobenthic deposit feeders. In coastal freshwater marshes, the microbenthos is composed primarily of amoebae (thecamoebinids, a group of amoebae with theca casings). This is in sharp contrast to more saline marshes in which foraminifera predominate. The slightly larger macrobenthos is composed of amphipods, especially *Gammarus fasciatus*, oligochaete worms, freshwater snails, and insect larvae. Copepods and cladocerans are abundant in the tidal creeks. The Asiatic clam (*Corbicula fluminaea*), a species introduced into the United States in the twentieth century, has spread throughout the coastal marshes of the southern states. Caridean shrimp, particularly *Palaemonetes pugio*, are common, as are freshwater shrimp, *Macrobrachium* spp. The density and diversity of these benthic organisms are reported to be low compared with those in nontidal freshwater wetlands, perhaps because of the lack of diverse bottom types in the tidal reaches of the estuary.

Where coastal forests transition from tidal to nontidal, shifts in invertebrate species have been shown to correspond with hydrological and species shifts. At the tidal transition of the Suwannee River, Wharton et al. (1982) reported that faunal associations changed from a brackish water snail-fiddler crab community (*Neretina-Uca*) to a freshwater snail-crayfish (*Vivipara-Cabarus*) community. Salinity and not vegetation was the primary factor dictating infauna and epifauna taxa along the lower reaches of the Cape Fear River in North Carolina (Hackney et al., 2007). Common faunal groups in tidal swamps include oligochaetes (especially Tubificidae and Lumbriculidae), fiddler crabs (*Uca* spp.) and grass shrimp (*Palaemontes pugio*).

Nekton

Coastal freshwater wetlands are important habitats for many nektonic species that use the area for spawning, year-round food and shelter, and a nursery zone and juvenile habitat. Fish of coastal freshwater marshes can be classified into five groups (Fig. 8.12). Most of them are freshwater species that spawn and complete their lives within freshwater areas. The three main families of these fish are cyprinds (minnows, shiners, carp), centrarchids (sunfish, crappies, bass), and ictalurids (catfish). Juveniles of all species are most abundant in the shallows, often using submerged marsh vegetation for protection from predators. Predator species—the bluegill (*Lepomis macrochirus*), largemouth bass

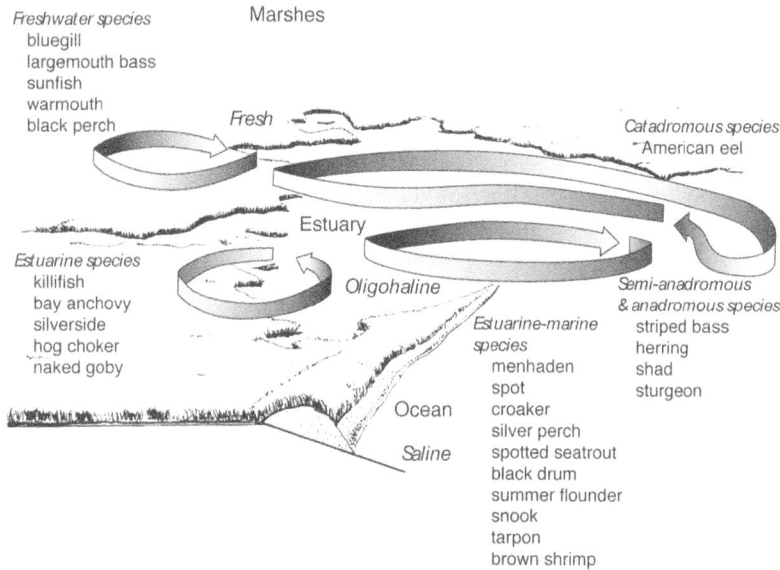

Figure 8.12 Fish and shellfish that use tidal freshwater marshes and other coastal systems can be classified into five groups: freshwater, estuarine, anadromous, catadromous, and estuarine–marine.

(*Micropterus salmoides*), sunfish (*Lepomis* spp.), warmouth (*Lepomis gulosus*), and black crappie (*Pomoxis nigromaculatus*)—are all important for sport fishing. Gar (*Lepisosteus* spp.), pickerel (*Esox* spp.), and bowfin (*Amia calva*) are other common predators often found in both coastal marshes and tidal freshwater creeks.

Some oligohaline or estuarine fish and shellfish that complete their entire life cycle in the estuary extend their range to include the freshwater marshes. Killifish (*Fundulus* spp.), particularly the banded killifish (*F.* diaphanus) and the mummichog (*F. heteroclitus*), are abundant in schools in shallow freshwater marshes, where they feed opportunistically on any available food. The bay anchovy (*Anchoa mitchilli*) and tidewater silverside (*Menidia beryllina*) are also often abundant in freshwater areas. The latter breed in this habitat more than in saltwater areas. Juvenile hog chokers (*Trinectes anadensi*) and naked gobies (*Gobiosoma bosci*) use tidal freshwater areas as nursery grounds (W. E. Odum et al., 1984).

Anadromous species of fish, which live as adults in the ocean, or *semianadromous species*, whose adults remain in the lower estuaries, pass through coastal freshwater marshes on their spawning runs to freshwater streams. For many of these species, the tidal freshwater areas are major nursery grounds for juveniles. Along the Atlantic Coast, herrings (*Alosa* spp.) and shads (*Dorosoma* spp.) fit into this category. The young of all of these species, except the hickory shad (*A. mediocris*), are found in peak abundance in tidal fresh waters, where they feed on small invertebrates and, in turn, are an important forage fish for striped bass (*Morone saxatilis*), white perch (*Morone anadensi*), catfish (*Ictalurus* spp.), and others (W. E. Odum et al., 1984). As they mature late in the year, they migrate downstream to saline waters and offshore. Two species of sturgeon (*Acipenser brevirostrum* and *A. oxyrhynchus*) were formerly important commercially in East Coast estuaries but were seriously overfished and currently are rare. Both species spawn in nontidal and tidal fresh waters, and juveniles may spend several years there before migrating to the ocean.

The striped bass is perhaps the most familiar semianadromous fish of the Mid-Atlantic Coast because of its importance in both commercial and sport fisheries. Approximately 90 percent of the striped bass on the East Coast are spawned in tributaries of the Chesapeake Bay system. They spawn in spring in tidal fresh and oligohaline waters; juveniles remain in this habitat along marsh edges, moving gradually downstream to the lower estuary and nearshore zone as they mature. Because the critical period for survival of the young is the larval stage, conditions in the tidal fresh marsh area where these larvae congregate are important determinants of the strength of the year class.

The only *catadromous species* in Atlantic Coast estuaries is the American eel (*Anguilla rostrata*). It spends most of its life in fresh or brackish water, returning to the ocean to spawn in the region of the Sargasso Sea. Eels are common in tidal and nontidal coastal freshwater areas, in marsh creeks, and even in marshes.

The juveniles of a few species of fish that are marine spawners move into freshwater marshes, but most remain in the oligohaline reaches of the estuary. Species whose range extends into tidal freshwater marshes are menhaden (*Brevoortia tyrannus*), spot

(*Leiostomus xanthurus*), croaker (*Micropogonias undulatus*), silver perch (*Bairdiella chrysoura*), spotted seatrout (*Cynoscion nebulosus*), black drum (*Pogonias cromis*), summer flounder (*Paralichthys dentatus*), snook (*Centropomus undecimalis*), and tarpon (*Megalops atlanticus*). Along the northern Gulf Coast, juvenile brown and white shrimp (*Penaeus* spp.) and male blue crabs (*Callinectes sapidus*) may also move into freshwater areas. These juveniles emigrate to deeper, more saline waters as temperatures drop in the fall.

Birds

Of all wetland habitats, coastal freshwater marshes may support the largest and most diverse populations of birds. W. E. Odum et al. (1984), working from a number of studies, compiled a list of 280 species of birds that have been reported from tidal freshwater marshes. They stated that although it is probably true that this environment supports the greatest bird diversity of all marshes, the lack of comparative quantitative data makes it difficult to test this hypothesis. Bird species include: waterfowl (44 species); wading birds (15 species); rails and shorebirds (35 species); birds of prey (23 species); gulls, terns, kingfishers, and crows (20 species); arboreal birds (90 species); and ground and shrub birds (53 species). A major reason for the intense use of these marshes is the structural diversity of the vegetation provided by broad-leaved plants, tall grasses, shrubs, and interspersed ponds.

Dabbling ducks (family Anatidae) and Canada geese actively select tidal freshwater areas on their migratory flights from the North. They use the Atlantic Coast marshes in the late fall and early spring, flying farther south during the cold winter months. Most of these species winter in fresh coastal marshes of the northern Gulf of Mexico, but some fly to South America. Their distribution in apparently similar marshes is variable; some marshes support dense populations, others few birds. For example, Fuller et al. (1988) found extensive use of new Atchafalaya River delta marshes by many species of ducks. Although the vegetation was dominated by arrowhead throughout the newly created islands, duck populations were twice as dense in the western islands of the delta compared to the east and central islands. On the central islands, ducks preferentially selected stands of three-square sedge over arrowhead; on the western islands where there was not three-square sedge, they frequented stands with mixed grass species over arrowhead. The reason for this selectivity is unclear. In the Atchafalaya Delta, it may be because the western islands remain fresh year-round, whereas the other islands sometimes experience saltwater encroachment (Holm, 1998). The birds feed in freshwater marshes on the abundant seeds of annual grasses and sedges, the rhizomes of perennial marsh plants, and also in adjacent agricultural fields. They are opportunistic feeders, on the whole, ingesting from the available plant species. An analysis by Abernethy (1986) suggests that many species that frequent the fresh marsh early in the winter move seaward to salt marshes before beginning their northward migration in the spring. The reason for this behavior pattern is not known, but Abernethy speculated that the preferred foods of the freshwater marshes are depleted by early spring and the birds move into salt marshes that have not been previously grazed.

The wood duck (*Aix sponsa*) is the only duck species that nests regularly in coastal freshwater tidal marshes, although an occasional black duck (*Anas rubripes*) or mallard (*A. platyrhynchos*) nest is found in Atlantic Coast marshes.

Wading birds are common residents of coastal freshwater marshes. They are present year-round in Gulf Coast marshes but only during the summer along the Atlantic Coast. An exception is the great blue heron (*Ardea herodias*), which is seen throughout the winter in the northern Atlantic states. Nesting colonies are common throughout the southern marshes, and some species, such as green-backed herons (*Butorides striatus*) and bitterns (*Ixobrychus exilis* and *Botaurus lentiginosus*), nest along the Mid-Atlantic Coast. They feed on fish and benthic invertebrates, often flying long distances each day from their nesting areas to fish.

Rails (*Rallus* spp.) and shorebirds, including the killdeer (*Charadrius vociferus*), sandpipers (Scolopacidae), and the American woodcock (*Scolopax minor*), are common in coastal freshwater marshes. They feed on benthic macroinvertebrates and diverse seeds. Gulls (*Larus* spp.), terns (*Sterna* spp.), belted kingfishers (*Ceryle alcyon*), and crows (*Corvus* spp.) are also common. Some are migratory; some are not. A number of birds of prey are seen hovering over freshwater marshes, including the northern harrier (*Circus cyaneus*), the American kestrel (*Falco sparverius*), falcons (*Falco* spp.), eagles, ospreys (*Pandion haliaetus*), owls (Tytonidae), vultures (Cathartidae), and the loggerhead shrike (*Lanius ludovicianus*). Swallow-tailed kites have been found to use the lower tidal reach of forested wetlands for nesting (Sykes et al., 1999). Arboreal birds use the coastal freshwater marshes intensively during short periods of time on their annual migrations. Flocks of tens of thousands of swallows (Hirundinidae) have been reported over the upper Chesapeake freshwater marshes. Flycatchers (Tyrannidae) are also numerous. They often perch on trees bordering the marsh, darting out into the marsh from time to time to capture insects. Although coastal marshes may be used for only short periods of time by a migrating species, they may be important temporary habitats. For example, the northern Gulf coastal marshes are the first landfall for birds on their spring migration from South America. Often they reach this coast in an exhausted state, and the availability of forested barrier islands for refuge and marshes for feeding is critical to their survival.

Sparrows, finches (Fringillidae), juncos (*Junco* spp.), blackbirds (Icteridae), wrens (Troglodytidae), and other ground and shrub birds are abundant residents of coastal freshwater marshes. W. E. Odum et al. (1984) indicated that 10 species breed in Mid-Atlantic Coast marshes, including the ring-necked pheasant (*Phasianus colchicus*), red-winged blackbird (*Agelaius phoeniceus*), American goldfinch (*Carduelis tristis*), rufous-sided towhee (*Pipilo erythrophthalmus*), and a number of sparrows. The most abundant are the red-winged blackbirds, dickcissels (*Spiza americana*), and bobolinks (*Dolichonyx oryzivorus*), which can move into and strip a wild rice marsh in a few days.

Amphibians and Reptiles

Although W. E. Odum et al. (1984) compiled a list of 102 species of amphibians and reptiles that frequent coastal freshwater marshes along the Atlantic Coast, many are poorly understood ecologically, especially with respect to their dependence on this type of habitat. None is specifically adapted for life in tidal freshwater marshes. Instead, they are able to tolerate the special conditions of this environment. River turtles, the most conspicuous members of this group, are abundant throughout the southeastern United States. Three species of water snakes (*Nerodia*) are common. *Agkistrodon piscivorus* (the cottonmouth) is found south of the James River in Virginia. In the South, especially along the Gulf Coast, the American alligator's preferred habitat is the tidal freshwater marsh. These large reptiles used to be listed as threatened or endangered, but they have come back so strongly in most areas that currently they are harvested legally (under strict control) in Louisiana and Florida. They nest along the banks of coastal freshwater marshes, and the animal, identified by its high forehead and long snout, is a common sight gliding along the surface of marsh streams.

Mammals

The mammals most closely associated with coastal freshwater marshes are all able to get their total food requirements from the marsh, have fur coats that are more or less impervious to water, and are able to nest (or hibernate, in northern areas) in the marsh. These include the river otter (*Lutra canadensis*), muskrat (*Ondatra zibethicus*), nutria (*Myocastor coypus*), mink (*Mustela vison*), raccoon (*Procyon lotor*), marsh rabbit (*Silvilagus palustris*), and marsh rice rat (*Oryzomys palustris*). In addition, the opossum (*Didelphis virginiana*) and white-tailed deer (*Odocoileus virginianus*) are locally abundant. The nutria was introduced from South America some years ago and has spread steadily in the Gulf Coast states and into Maryland, North Carolina, and Virginia. It is not likely to spread farther north because of its intolerance to cold, but the South Atlantic marshes would seem to provide an ideal habitat. The nutria is more vigorous than the muskrat and has displaced it from the freshwater marshes in many parts of the northern Gulf. As a result, muskrat density is highest in oligohaline marshes. The muskrat, for some reason, is not found in coastal Georgia and South Carolina or in Florida, although it is abundant farther north along the Atlantic Coast. Muskrat, nutria, and beaver (*Castor canadensis*) can influence the development of a marsh. The first two species destroy large amounts of vegetation with their feeding habits (they prefer juicy rhizomes and uproot many plants when digging for them), their nest building, and their underground passages. Beavers have been observed in tidal freshwater marshes in Maryland and Virginia. Their influence on forested habitats is well known, but their impact on tidal freshwater marshes needs to be studied more closely.

Ecosystem Function

Primary Productivity

Many production estimates have been made for freshwater coastal marshes. Productivity is generally high, usually falling in the range of 1,000 to 3,000 g m^{-2} yr^{-1} (Table 8.5). The large variability reported from different studies stems, in part, from

Table 8.5 Peak standing crop and annual net primary production (NPP) estimates for tidal freshwater marsh associations in approximate order from highest to lowest productivity[a]

Vegetation Type[b]	Peak Standing Crop (g m^{-2})	Annual NPP (g m^{-2} yr^{-1})
Extremely High Productivity		
Spartina cynosuroides (big cordgrass)	2,311	—
Lythrum salicaria (spiked loosestrife)	1,616	2,100
Zizaniopsis miliacea (giant cutgrass)	1,039	2,048
Panicum hemitomon (maidencane)	1,160	2,000
Phragmites communis (common reed)	1,850	1,872
Moderate Productivity		
Zizania aquatica (wild rice)	1,218	1,578
Amaranthus cannabinus (water hemp)	960	1,547
Typha sp. (cattail)	1,215	1,420
Bidens spp. (bur marigold)	1,017	1,340
Polygonum sp./Leersia oryzoides (smartweed/rice cutgrass)	1,207	—
Ambrosia tirifida (giant ragweed)	1,205	1,205
Acorus calamus (sweet flag)	857	1,071
Sagittaria latifolia (duck potato)	432	1,071
Low Productivity		
Peltandra virginica/Pontederia cordata (arrow arum/pickerelweed)	671	888
Hibiscus coccineus (rose mallow)	1,141	869
Nuphar adventa (spatterdock)	627	780
Rosa palustris (swamp rose)	699	—
Scirpus deltarum	—	523
Eleocharis baldwinii	130	—

[a]Values are means of 1 to 8 studies.
[b]Designation indicates the dominant species in the association.
Sources: W. E. Odum et al. (1984); Sasser and Gosselink (1984); Visser (1989); White (1993); and Sasser et al. (1995)

a lack of standardization of measurement techniques, but real differences can be attributed to three factors:

1. *Type of plant and its growth habit.* Fresh coastal marshes, in contrast to saline marshes, are floristically diverse, and productivity is determined, at least to some degree, by genetic factors that regulate the species' growth habits. Tall perennial grasses, for example, appear to be more productive than broad-leaved herbaceous species such as arrow arum and pickerelweed.
2. *Tidal energy.* The stimulating effect of tides on production has been shown for salt marshes and appears to be true for tidal freshwater marshes as well.
3. *Other factors.* Soil nutrients, grazing, parasites, and toxins are other factors that can limit production in tidal freshwater marshes.

The elevation gradient across a fresh coastal marsh and the resulting differences in vegetation and flooding patterns account for three broad zones of primary production. The low marsh bordering tidal creeks, dominated by broad-leaved perennials, is characterized by apparently low production rates. Biomass peaks early in the growing season. Turnover rates, however, are high, suggesting that annual production may be much higher than can be determined from peak biomass. Much of the production is stored in below-ground biomass (root : shoot $\gg$ 1) in mature marshes; this biomass is mostly rhizomes rather than fibrous roots. Decomposition is rapid, the litter is swept from the marsh almost as fast as it forms, the soil is bare in winter, and erosion rates are high. The parts of the high marsh dominated by perennial grasses and other erect, tall species are characterized by the highest production rates of freshwater species, and root : shoot ratios are approximately 1. Because tidal energy is not as strong and the plant material is not so easily decomposed, litter accumulates on the soil surface, and little erosion occurs. The high-marsh mixed-annual association typically reaches a large peak biomass late in the growing season. Most of the production is above-ground (root : shoot <1), and litter accumulation is common.

Few estimates of primary productivity have been conducted in tidal freshwater swamps; however, they are expected to benefit from the same tidal nutrient subsidy. Relative elevation can have an effect of tree composition, geomorphology, flooding duration, and tidal exchange—all of which may influence productivity. There is likely a broad range of productivity and contribution for tidal freshwater wetlands. Tree growth in forests near their downstream threshold can be stunted by higher salinities and inundation frequency. Upstream, however, forests may be highly productive, benefiting from the nutrient subsidy provided by the tides and diminished saltwater intrusion. Ozalp et al. (2007) found that above-ground net primary productivity (ANPP) at a tidal forest along the lower Pee Dee River in South Carolina ranged between 477 and $1,117 \text{ g m}^{-2} \text{ yr}^{-1}$. Along the tidal reaches of the Pamunkey River, Fowler (1987) found that forest ANPP at $1,230 \text{ g m}^{-2} \text{ yr}^{-1}$, with 40 percent of this production from herbaceous plants. Along Lake Maurepas in southeastern Louisiana, where flooding durations and depths have increased due to anthropogenic alterations, Effler et al. (2007) reported that annual tree production of tidal swamps was low, ranging between 220 and $700 \text{ g m}^{-2} \text{ yr}^{-1}$.

Energy Flow

There are three major sources of organic carbon to tidal freshwater marshes. The largest source is probably the vascular marsh vegetation, but organic material brought from upstream (terrestrial carbon) may be significant, especially on large rivers and where domestic sewage waters are present. Phytoplankton productivity is a largely unknown quantity. Most of the organic energy flows through the detrital pool and is distributed to benthic fauna and deposit-feeding omnivorous nekton. These groups feed fish, mammals, and birds at higher trophic levels. The magnitude of the herbivore food chain, in comparison to the detritus one, is poorly understood. Insects are more abundant in fresh marshes than in salt marshes but most do not appear to be herbivorous. Marsh mammals apparently can "eat out" significant areas of vegetation (Evers et al., 1998), but direct herbivory is probably small in comparison to the flow of organic energy from destroyed vegetation into the detrital pool. Nevertheless, these rodents may exert strong control on species composition and on primary production (Evers et al., 1998). Herbivores also act in synergy with other stresses, for example, saltwater intrusion, flooding, and fire (Taylor et al., 1994; Grace and Ford, 1996).

The phytoplankton–zooplankton–juvenile fish food chain in fresh marshes is of interest because of its importance to humans. Zooplankton are an important dietary component for a variety of larval, postlarval, and juvenile fish of commercial importance that are associated with tidal freshwater marshes (Van Engel and Joseph, 1968).

Birds are major seasonal or year-round consumers in all types of tidal freshwater marshes. In addition, they move materials out of the system, processing it into guano, which in some areas may be a significant source of nutrients. They also modify plant composition and production by eat-outs (Smith and Odum, 1981).

Organic Import and Export

In mature tidal freshwater marshes, most organic production is decomposed to litter and peat within the marsh system, and nutrients are extracted and recycled. Floating tidal freshwater marshes may have even more closed cycles. Because they float, no surface flows export or import organic material. This limits fluxes to subsurface dissolved materials. The largest loss of organic energy in these mature marshes is probably to deep peats in the case of anchored marshes or to an organic sludge layer under the water column in floating marshes (Sasser et al., 1991). The magnitude of this loss was measured as 145 to 150 g C m^{-2} yr^{-1} in a Gulf Coast freshwater marsh (Hatton, 1981).

Other losses of organic carbon from marshes occur through flushing from the marsh surface, conversion to methane that escapes as a gas, and export as biomass in the bodies of consumers that feed on the marsh. In highly reduced freshwater sediments, where, in contrast to salt marshes, little sulfur is available as an electron acceptor, it is expected that methanogenesis from carbon dioxide and fermentation should be a dominant pathway of respiratory energy flow. However, there is evidence that macrophyte biomass may regulate methanogenesis. Neubauer et al. (2005) examined the anaerobic metabolism in soils from tidal (freshwater and salt) marshes along the Patuxent River, Maryland. In the freshwater tidal marsh, they found that anaerobic metabolism was dominated by iron (III) reduction early in the growing season (Fig. 8.13). Later, when plant biomass declined, methanogenesis became the dominant pathway for anaerobic metabolism. The authors attributed the enhanced ferric iron (Fe^{+3}) reduction to radial oxygen loss from plants during their peak growing period that allow the replenishment of Fe^{+3} oxides back into the rhizosphere. Rates of anaerobic metabolism were lower in the salt marsh and were dominated almost equally by both Fe reduction and sulfate reduction when plants were most productive and almost exclusively by sulfate reduction late in the growing season. The relationship between plant biomass and iron reduction was less clear in the salt marsh (Fig. 8.13) and may have been confounded by shifts in seasonal flooding.

Carbon dynamics in tidal freshwater forests can be variable and likely change along the estuarine gradient. Tidal export of detritus matter is likely significant; however, there is little information and few estimates available for these wetlands. Soil organic matter tends to be high in these wetlands and is often linked to hydrology. A review of

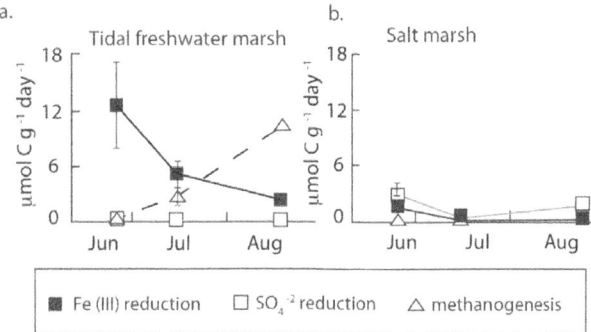

Figure 8.13 Seasonal changes in three anaerobic metabolism processes (iron reduction, sulfate reduction, and methanogeneis) in Maryland tidal freshwater and salt marshes. In the tidal freshwater marsh, iron reduction dominated early in the growing season (coinciding with peak plant biomass) and methanogenesis later. In the salt marsh, rates were lower and were dominated by iron and sulfate reduction early and by only sulfate reduction later in the growing season. (After Neubauer et al., 2005)

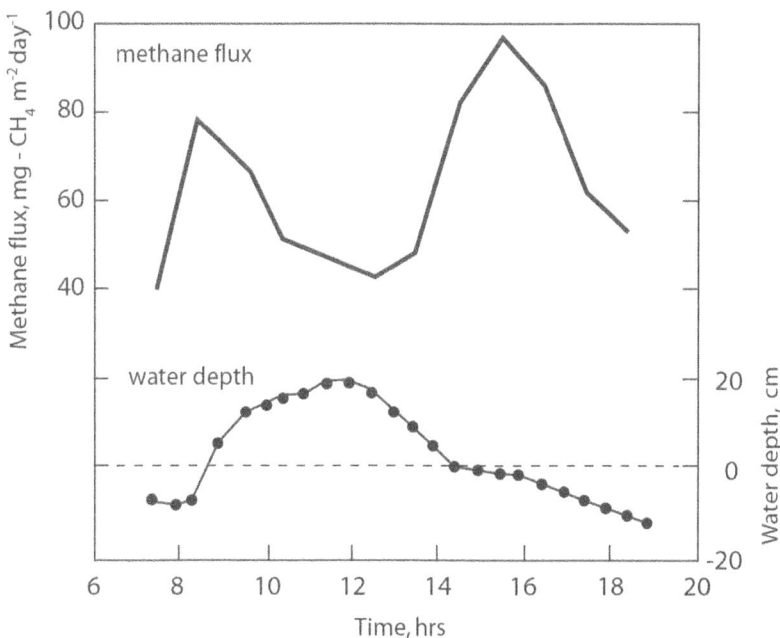

Figure 8.14 Methane fluxes from static chambers and concurrent tidal water level changes in a tidal freshwater swamp in White Oak River Estuary, North Carolina. The highest methane emission occurred while water levels coincided with the soil surface. When water levels were below the soil surface, methane oxidation increased in an aerobic surface layer, resulting in decreased emission. When water levels exceeded the soil surface, methane emission was probably reduced by a water diffusion barrier. (After Kelley et al., 1995)

reported soil conditions in the southeastern United States found that the percentage of soil organic matter at the surface ranged from 9 to 77 percent with the highest concentrations reported for blackwater rivers (Anderson and Lockaby, 2007). For comparison, surface soils in a tidal freshwater shrub wetland in the Netherlands had 35 percent organic matter (Verhoeven et al., 2001). At the lower river reaches where waters can be more brackish, carbon mineralization and anaerobic metabolism may alternate between sulfate reduction and methanogenesis, although iron reduction may be important as well. During low tides, water levels in these swamps can drop below the sediment surface, and the increased aerobic conditions can increase methane oxidation and decrease emissions from these wetlands; during high tides, inflowing water can both cool soil temperatures and provide a medium for methane oxidation, reducing methane generation then as well (Fig. 8.14). Consequently, methane flux rates in tidal freshwater marshes and swamps tend to be lower than in comparable nontidal freshwater marshes and swamps (Anderson and Lockaby, 2007).

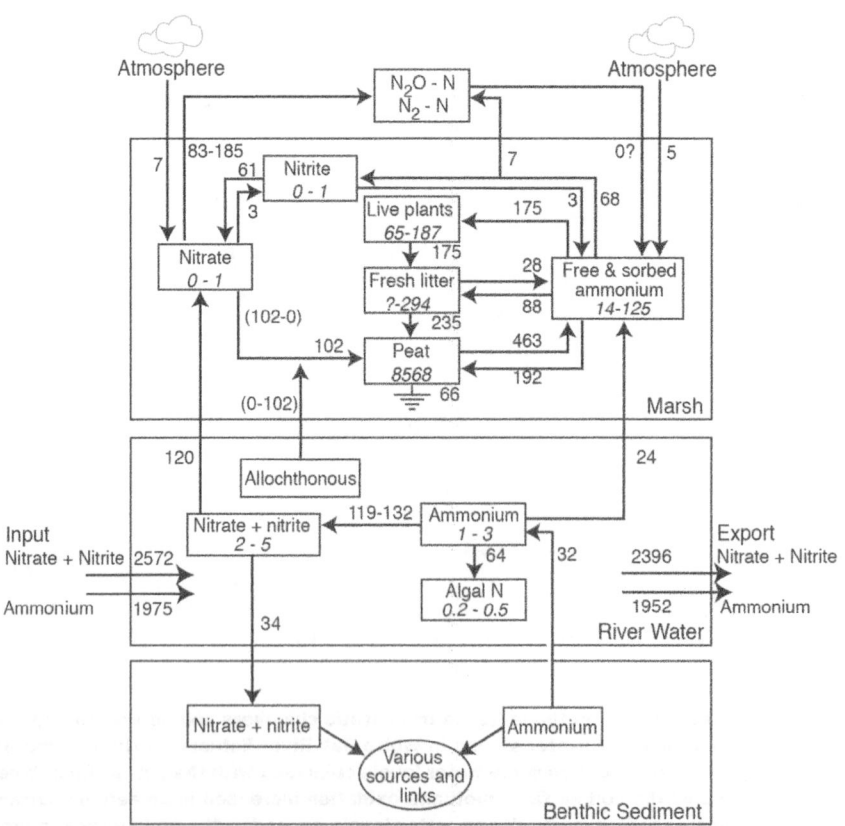

Figure 8.15 Nitrogen budget for a 23-ha tidal freshwater marsh in coastal Massachusetts. Pool sizes are in kmoles N, fluxes in kmoles N yr^{-1}. (After Bowden et al., 1991)

Nutrient Budgets

In general, nutrient cycling and nutrient budgets in coastal freshwater wetlands appear to be similar to salt marshes and mangrove swamps: They are fairly open systems that have the capacity to act as long-term sinks, sources, or transformers of nutrients. Even though these marshes generally are vigorously flooded by tides, they recycle a major portion of the nitrogen requirements of the vegetation. Figure 8.15 illustrates nitrogen cycling in a tidal freshwater marsh near Boston, Massachusetts. In this budget, most nutrient inputs are inorganic from the North River, and the marsh and river nutrient cycles are mostly independent. Within the marsh itself, the major cycle is from peat to ammonium-nitrogen to live plants. Some of the ammonium-nitrogen is nitrified to nitrate-nitrogen and denitrified. Overall nitrate loss always exceeded denitrification measurements by the acetylene block method, suggesting that other sinks of nitrate, such as assimilatory nitrate reduction, may be important (Bowden et al., 1991). Peat mineralization is sufficient to satisfy the nitrogen demands of the vegetation, and nearly all of the nitrogen flowing across the marsh from the adjacent river is reexported. Mineralized litter and peat are conserved in the marsh by plant uptake and by microbial and litter immobilization. Despite this closed mineral cycle, the small uptake of nitrogen from the river may be important.

Recommended Readings

Day, J. W., B. C. Crump, W. M. Kemp, and A. Yáñez-Arancibia, eds. 2013. *Estuarine Ecology*. Hoboken, NJ: John Wiley & Sons.

Perillo, G., E. Wolanski, D. Cahoon, and M. Brinson. 2009. *Coastal Wetlands: An Integraded Ecosystem Approach*. Amsterdam: Elsevier.

Tiner, R. W. 2013. *Tidal Wetlands Primer: An Introduction to Their Ecology, Natural History, Status, and Conservation*. Amherst: University of Massachusetts Press.

References

Abernethy, R. K. 1986. *Environmental Conditions and Waterfowl Use of a Backfilled Pipeline Canal*. M.S. thesis, Louisiana State University, Baton Rouge. 125 pp.

Adam, P. 1998. Australian saltmarshes: A review. In A. J. McComb and J. A. Davis, eds., *Wetlands for the Future*. Gleneagles Publishing, Adelaide, Australia, pp. 287–295.

Anderson, C. J. and B. G. Lockaby. 2007. Soils and biogeochemistry of tidal freshwater forested wetlands. In W. H. Conner, T. W. Doyle, and K. W. Krauss, eds., *Ecology of Tidal Freshwater Forested Wetlands of the Southeastern United States*. Springer, Inc., Dordrecht, Netherlands, pp. 65–88.

Antlfinger, A. E., and E. L. Dunn. 1979. Seasonal patterns of CO_2 and water vapor exchange of three salt-marsh succulents. *Oecologia* (Berl.) 43: 249–260.

Baldwin, A. H., K. L. McKee, and I. A. Mendelssohn. 1996. The influence of vegetation, salinity, and inundation on seed banks of oligohaline coastal marshes. *American Journal of Botany* 83: 470–479.

Baldwin, A. H., M. S. Egnotovish, and E. Clarke. 2001. Hydrologic change and vegetation of tidal freshwater marshes: Field, greenhouse, and seed-bank experiments. *Wetlands* 21: 519–531.

Beadle, L. C. 1974. *The Inland Waters of Tropical Africa*. Longman, London.

Bouchard, V. and J. C. Lefeuvre. 2000. Primary production and macro-detritus dynamics in a European salt marsh: carbon and nitrogen budgets. *Aquatic Botany* 67: 23–42.

Bouchard, V., M. Tessier, F. Digaire, J. P. Vivier, L. Valery, J. C. Gloaguen, and J. C. Lefeuvre. 2003. Sheep grazing as management tool in Western European saltmarshes. *C.R. Biologies* 326: S148–S157.

Bowden, W. B., C. J. Vorosmarty, J. T. Morris, B. J. Peterson, J. E. Hobbie, P. A. Steudler, and B. Moore. 1991. Transport and processing of nitrogen in a tidal freshwater wetland. *Water Resources Research* 27: 389–408.

Bradley, P. M., and E. L. Dunn. 1989. Effects of sulfide on the growth of three salt marsh halophytes of the southeastern United States. *American Journal of Botany* 76: 1707–1713.

Caldwell, R., and Crow. 1992. A floristic and vegetation analysis of a tidal freshwater marsh on the Merrimac River, West Newberry, MA. *Rhodora* 94: 63–97.

Campeau, S., H. R. Murkin, and R. D. Titman. 1994. Relative importance of algae and emergent plant litter to freshwater marsh invertebrates. *Canadian Journal of Fisheries and Aquatic Sciences* 51: 681–692.

Chapman, V. J. 1960. *Salt Marshes and Salt Deserts of the World*. Interscience, New York. 392 pp.

Chapman, V. J. 1976. *Coastal Vegetation*, 2nd ed., Pergamon Press, Oxford, UK. 292 pp.

Chapman, V. J. 1977. *Wet Coastal Ecosystems*. Elsevier, Amsterdam, 428 pp.

Childers, D. L. 1994. Fifteen years of marsh flumes: a review of marsh-water column interactions in southeastern USA estuaries. In W. J. Mitsch, ed., *Global Wetlands: Old World and New*. Elsevier, Amsterdam, pp. 277–293.

Childers, D. L., J. W. Day, Jr., and H. N. McKellar, Jr. 2000. Twenty more years of marsh and estuarine flux studies: revisiting Nixon (1980). In M. P. Weinstein and D. A. Kreeger, eds., *Concepts and Controversies in Tidal Marsh Ecology*. Kluwer Academic Publishers, Netherlands, pp. 389–421.

Coleman, J. M., and H. H. Roberts. 1989. Deltaic coastal wetlands. *Geologie en Mijnbouw* 68: 1–24.

Crain, C. M., B. R. Silliman, S. L. Bertness, and M. D. Bertness. 2004. Physical and biotic drivers of plant distribution across estuarine salinity gradients. *Ecology* 85: 2539–2549.

Curco, A., C. Ibánez, and J. W. Day, Jr. 2002. Net primary production and decomposition of salt marshes of the Ebro Delta (Catalonia, Spain). *Estuaries* 25: 309–324.

de la Cruz, A. A. 1974. Primary productivity of coastal marshes in Mississippi. *Gulf Research Reports* 4: 351–356.

Dreyer, G. D., and W. A. Niering, eds. 1995. *Tidal Marshes of Long Island Sound: Ecology, History and Restoration.* The Connecticut College Arboretum, Bulletin No. 34, New London, CT.

Effler, R. S., G. P. Shaffer, S. S. Hoeppner, and R. A. Goyer. 2007. Ecology of the Maurepas Swamp: Effects of salinity, nutrients, and insect defoliation. Soils and biogeochemistry of tidal freshwater forested wetlands. In W. H. Conner, T. W. Doyle, and K. W. Krauss, eds., *Ecology of Tidal Freshwater Forested Wetlands of the Southeastern United States.* Springer, Dordrecht, Netherlands, pp. 349–384.

Eisenlohr, W. S. 1972. Hydrologic Investigations of Prairie Potholes in North Dakota, 1958-1969, Professional Paper 585, U.S. Geological Survey, Washington, DC.

Evers, D. E., C. E. Sasser, J. G. Gosselink, D. A. Fuller, and J. M. Visser. 1998. The impact of vertebrate herbivores on wetland vegetation in Atchafalaya Bay, Louisiana. *Estuaries* 21: 1–13.

Field, D. W., A. Reyer, P. Genovese, and B. Shearer. 1991. Coastal wetlands of the United States- An accounting of a valuable national resource. Strategic Assessment Branch, Ocean Assessments Division, Office of Oceanography and Marine Assessments, National Ocean Service, National Oceanic and Atmosphere Administration, Rockville, MD.

Fowler, K. 1987. Primary production and temporal variation in the macrophyte community of a tidal freshwater swamp. M.A. thesis, Virginia Institute of Marine Science, College of William and Mary, Williamsburg, Virginia.

Fuller, D. A., G. W. Peterson, R. K. Abernethy, and M. A. LeBlanc. 1988. The distribution and habitat use of waterfowl in Atchafalaya Bay, Louisiana. In C. E. Sasser and D. A. Fuller, eds., *Vegetation and Waterfowl Use of Islands in Atchafalaya Bay.* Coastal Ecology Institute, Louisiana State University, Baton Rouge, LA, pp. 73–103.

Funk, D. W., L. E. Noel, and A. H. Freedman. 2004. Environmental gradients, plant distribution, and species richness in artic salt marsh near Prudhoe Bay, Alaska. *Wetlands Ecology and Management* 12: 215–233.

Gallagher, J. L. and F. C. Daiber. 1974. Primary production of edaphic algae communities in a Delaware salt marsh. *Limnology & Oceanography* 19: 390–395.

Gosselink, J. G., Coleman, J. M., and R. E. Stewart, Jr. 1998. *Coastal Louisiana.* In M. J. Mac, P. A. Opler, C. E. Puckett Haecker, and P. D. Doran. eds. *Status and Trends of the Nation's Biological Resources,* vol. 1. U.S. Department of the Interior, U.S. Geological Survey, Reston, VA, pp. 385–436.

Grace, J. B., and M. A. Ford. 1996. The potential impact of herbivores on the susceptibility of the marsh plant Sagittaria lancifolia to saltwater intrusion in coastal wetlands. *Estuaries* 19: 13–20.

Greenberg, R., J. E. Maldonado, S. Droege, and M. V. McDonald. 2006. Tidal marshes: A global perspective on the evolution and conservation of their terrestrial vertebrates. *BioScience* 56: 675–685.

Hackney, C. T., G. B. Avery, L. A. Leonard, M. Posey, and T. Alphin. 2007. Biological, chemical, and physical characteristics of tidal freshwater swamp forests of the lower Cape Fear River/Estuary, North Carolina. In W. H. Conner, T. W. Doyle, and K. W. Krauss, eds., *Ecology of Tidal Freshwater Forested Wetlands of the Southeastern United States*. Springer, Inc., Dordrecht, Netherlands. pp. 183–221.

Hall, S. L., and F. M. Fisher. 1985. Annual productivity and extracellular release of dissolved organic compounds by the epibenthic algal community of a brackish marsh. *Journal of Phycology* 21: 277–281.

Hall, J. V., W. E. Frayer, and B. O. Wilen. 1994. *Status of Alaska Wetlands*. U.S. Fish & Wildlife Service, Alaska Region, Anchorage, AK, 32 pp.

Hatton, R. S. 1981. *Aspects of Marsh Accretion and Geochemistry: Barataria Basin, La*. Master's thesis, Louisiana State University, Baton Rouge, LA.

Holm, G. O., Jr. 1998. Comparisons of the Atchafalaya and West Lake Delta salinity incursions and the loss of Sagittaria latifolia Willd. M.S. thesis. Louisiana State University, Baton Rouge, Louisiana.

Hopkinson, C. S. 1985. Shallow water benthic and pelagic metabolism: Evidence for heterotrophy in the nearshore. *Marine Biology* 87: 19–32.

Hopkinson, C. S., J. G. Gosselink, and F. T. Parronndo. 1980. Production of coastal Louisiana marsh plants calculated from phenometric techniques. *Ecology* 61: 1091–1098.

Huffman, R. T., and R. E. Lonard. 1983. Successional patterns on floating vegetation mats in a southwestern Arkansas bald cypress swamp. *Castanea* 48:73 78.

Ibánez, C., J. W. Day, Jr., and D. Pont. 1999. Primary production and decomposition in wetlands of the Rhone Delta, France: Interactive impacts of human modifications and relative sea level rise. *Journal Coastal Research* 15: 717–731.

Ibánez, C., J. T. Morris, I. A. Mendelssohn, and J. W. Day, Jr. 2013. Coastal marshes. In J. W. Day, Jr., B. C. Crump, W. M. Kemp, and A. Yanez-Arancibia, eds., *Estuarine Ecology*, 2nd ed. Wiley-Blackwell, Hoboken, NJ, pp. 129–163.

Junk, W. J. 1970. Investigations on the ecology and production biology of the "floating meadows" (Paspalo echinochloetum) on the middle Amazon: 1. The floating vegetation and its ecology. *Amazonia* 2: 449–495.

Kelley C. A., C. S. Martens, and W. Ussler III. 1995. Methane dynamics across a tidally influenced riverbank margin. *Limnology & Oceanography* 40: 1112–1129.

Koch, M. S., I. A. Mendelssohn, and K. L. McKee. 1990. Mechanism for the hydrogen sulfide-induced growth limitation in wetland macrophytes. *Limnology and Oceanography* 35: 399–408.

Kreeger, D. A., and R. I. E. Newell. 2000. Trophic complexity between producers and invertebrate consumers in salt marshes. In M. P. Weinstein and D. A. Kreeger, eds., *Concepts and Controversies in Tidal Marsh Ecology*. Kluwer Academic Publishers, Netherlands, pp. 187–220.

Leck, M. A., and K. J. Graveline. 1979. The seed bank of a freshwater tidal marsh. *American Journal of Botany* 66: 1006–1015.

Leck, M. A., and R. L. Simpson. 1987. Seed bank of a freshwater tidal wetland: turnover and relationship to vegetation change. *American Journal of Botany* 74: 360–370.

Leck, M. A., and R. L. Simpson. 1995. Ten-year seed bank and vegetation dynamics of a tidal freshwater marsh. *American Journal of Botany* 82: 1547–1557.

Lefeuvre, J. C., and R. F. Dame. 1994. Comparative studies of salt marsh processes in the New and Old Worlds: An introduction. In W. J. Mitsch, ed., *Global Wetlands: Old World and New*. Elsevier, Amsterdam, pp. 169–179.

Lefeuvre J. C., V. Bouchard, E. Feunteun, S. Grare, P. Lafaille, and A. Radureau. 2000. European salt marshes diversity and functioning: The case study of the Mont Saint-Michel Bay, France. *Wetlands Ecology and Management* 8: 147–161.

Lefeuvre, J. C., P. Laffaille, E. Feunteun, V. Bouchard, and A. Radureau. 2003. Biodiversity in salt marshes: From patrimonial value to ecosystem functioning. The case study of Mont-Saint-Michel Bay. *C.R. Biologies* 326: S125–S131.

MacIntyre, H. L., and J. J. Cullen. 1995. Fine-scale vertical resolution of chlorophyll and photosynthetic parameters in shallow-water benthos. *Marine Ecology Progress Series* 122: 227–237.

Mendelssohn, I. A., and J. T. Morris. 2000. Eco-physiological controls on the productivity of Spartina alterniflora Loisel. In M. P. Weinstein and D. A. Kreeger, eds. *Concepts and Controversies in Tidal Marsh Ecology*. Kluwer Academic Publishers, The Netherlands, p. 59–80.

Mitsch, W. J., J. G. Gosselink, C. J. Anderson, and L. Zhang. 2009. *Wetland Ecosystems*. John Wiley & Sons, Inc., Hoboken, NJ, 296 pp.

Montague, C. L., and R. G. Wiegert. 1990. Salt marshes. In R. L. Myers and J. J. Ewel, eds. *Ecosystems of Florida*. University of Central Florida Press, Orlando, FL, pp. 481–516.

Morris, J. T., B. Kjerfve, and J. M., Dean. 1990. Dependence of estuarine productivity on anomalies in mean sea level. *Limnology & Oceanography* 35: 926–930.

Morris, J. T., P. V. Sundareshwar, C. T. Nietch, B. Kjerfve, and D. R. Cahoon. 2002. Response of coastal wetlands to rising sea level. *Ecology* 83: 2869–2877.

Newell, S. Y., and D. Porter. 2000. Microbial secondary production from saltmarsh grass shoots, and its known and potential fates. In M. P. Weinstein and D. A. Kreeger, eds., *Concepts and Controversies in Tidal Marsh Ecology*. Kluwer Academic Publishers, Dordrecht, Netherlands, pp. 159–185.

Neubauer, S. C., K. Givler, S. Valentine, and J. P. Megonigal. 2005. Seasonal patterns and plant-mediated controls of subsurface wetland biogeochemistry. *Ecology* 86: 3334–3344.

Nixon, S. W. 1980. Between coastal marshes and coastal waters—a review of twenty years of speculation and research on the role of salt marshes in estuarine productivity and water chemistry. In PP. Hamilton and K. B. MacDonald, eds., *Estuarine and Wetland Processes*. Plenum, New York, pp. 437–525.

Odum, E. P. 1968. A research challenge: Evaluating the productivity of coastal and estuarine water. *Proceedings Second Sea Grant Conference*, University of Rhode Island, pp. 63–64.

Odum, E. P. 1980. The status of three ecosystem-level hypotheses regarding salt marsh estuaries: tidal subsidy, outwelling, and detritus-based food chains. In V. S. Kennedy, ed., *Estuarine Perspectives*. Academic Press, New York, pp. 485–495.

Odum, E. P. 2000. Tidal marshes as outwelling/pulsing systems. In M. P. Weinstein and D. A. Kreeger, eds., *International Symposium: Concepts and Controversies in Tidal Marsh Ecology*. Kluwer Academic Publishers, Dordrecht, Netherlands, pp. 3–7.

Odum, W. E, T. J. Smith in, J. K. Hoover, and C. C. McIvor. 1984. *The Ecology of Freshwater Marshes of the United States East Coast: A Community Profile*. U.S. Fish and Wildlife Service, FWS/OBS-87/17, Washington, DC, 177 pp.

Ozalp, M., W. H. Conner, and B. G. Lockaby. 2007. Above-ground productivity and litter decomposition in a tidal freshwater forested wetland on Bull Island, SC, USA. *Forest Ecology and Management* 245: 31–43.

Pallis, M. 1915. The structural history of Plav: The floating fen of the delta of the Danube. *J. Linn. Soc. Bot.* 43: 233–290.

Perry, J. E., and C. H. Hershner. 1999. Temporal changes in the vegetation pattern in a tidal freshwater marsh. *Wetlands* 19: 90–99.

Pétillon, J., F. Ysnel, A. Canard, and J. C. Lefeuvre. 2005. Impact of an invasive plant (Ellymus athericus) on the conservation value of tidal salt marshes in western France and implications for management: Responses of spider populations. *Biological Conservation* 126: 103–117.

Pfeiffer, W. J., and R. G. Wiegert. 1981. Grazers on *Spartina* and their predators. In L. R. Pomeroy and R. G. Wiegert, eds. *The Ecology of a Salt Marsh*. Springer-Verlag, New York, pp. 87–112.

Pinckney, J., and R. G. Zingmark. 1993. Modeling the annual production of intertidal benthic microalgae in estuarine ecosystems. *Journal of Phycology* 29: 396–407.

Pomeroy, L. R. 1959. Algae productivity in salt marshes of Georgia. *Limnology & Oceanography* 4: 386–397.

Pomeroy, L. R., W. M. Darley, E. L. Dunn, J. L. Gallagher, E. B. Haines, and D. M. Witney. 1981. Primary production. In L. R. Pomeroy and R. G. Wiegert, eds., *Ecology of a Salt Marsh*. Springer-Verlag, New York, pp. 39–67.

Rheinhardt, R. 1992. A multivariate analysis of vegetation patterns in tidal freshwater swamps of lower Chesapeake Bay, USA. *Bulletin of the Torrey Botanical Club* 119: 192–207.

Sasser, C. E. and J. G. Gosselink. 1984. Vegetation and primary production in a floating wetland marsh in Louisiana. *Aquatic Botany* 20: 245–255.

Sasser, C. E., J. G. Gosselink, and G. P. Shaffer. 1991. Distribution of nitrogen and phosphorus in a Louisiana freshwater floating marsh. *Aquatic Botany* 41: 317–331.

Sasser, C. E., J. M. Visser, D. E. Evers, and J. G. Gosselink. 1995. The role of environmental variables on interannual variation in species composition and biomass in a subtropical minerotrophic floating marsh. *Canadian Journal of Botany* 73: 413–424.

Sasser, C. E., J. G. Gosselink, E. M. Swenson, C. M. Swarzenski, and N. C. Leibowitz. 1996. Vegetation, substrate and hydrology in floating marshes in the Mississippi river delta plain wetlands, USA. *Vegetatio* 122: 129–142.

Silliman, B. R., H. van de Koppel, M. D. Bertness, L. E. Stanton, and I. A. Mendelssohn. 2005. Drought, snails, and large-scale die-off of southern U.S. salt marshes. *Science* 310: 1803–1806.

Simpson, R. L., R. E. Good, M. A. Leck and D. F. Whigham. 1983. The ecology of freshwater tidal wetlands. *BioScience* 33: 255–259.

Smith, T. J., and W. E. Odum. 1981. The effects of grazing by snow geese on coastal salt marshes. *Ecology* 62: 98–106.

Stout, J. P. 1978. *An Analysis of Annual Growth and Productivity of Juncus roemerianus Scheele and Spartina alterniflora Loisel in Coastal Alabama*. Ph.D. dissertation, University of Alabama, Tuscaloosa.

Sullivan, M. J., and C. A. Moncreiff. 1988. Primary production of edaphic algal communities in a Mississippi salt marsh. *Journal of Phycology* 24: 49–58.

Sullivan, M. J., and C. A. Currin. 2000. Community structure and functional dynamics of benthic microalgae in salt marshes. In M. PP. Weinstein and D. A. Kreeger, eds., *Concepts and Controversies in Tidal Marsh Ecology*. Kluwer Academic Publishers, The Netherlands.

Swarzenski, C., E. M. Swenson, C. E. Sasser, and J. G. Gosselink. 1991. Marsh mat flotation in the Louisiana Delta Plain. *Journal of Ecology* 79: 999–1011.

Sykes, P. W. Jr, C. B. Kepler, K. L. Litzenberger, H. R. Sansing, E. T. R. Lewis, and J. S. Hatfield. 1999. Density and habitat of breeding swallow-tailed kites in the lower Suwannee ecosystem, Florida. *Journal of Field Ornithology* 70: 321–336.

Taylor, K. L., J. B. Grace, G. R. Guntenspergen, and A. L. Foote. 1994. The interactive effects of herbivory and fire on an oligohaline marsh, Little Lake, Louisiana, USA. *Wetlands* 14: 82–87.

Teal, J. M. 1962. Energy flow in the salt marsh ecosystem of Georgia. *Ecology* 43: 614–624.

Teal, J. M. 1986. *The Ecology of Regularly Flooded Salt Marshes of New England: A Community Profile*. U.S. Fish and Wildlife Service, Washington, DC, Biol. Repp. 85(7.4), 61pp.

Teal, J. M., and B. L. Howes. 2000. Salt marsh values: Retrospection from the end of the century. In M. P. Weinstein and D. A. Kreeger, eds., *Concepts and Controversies in Tidal Marsh Ecology*. Kluwer Academic Publishers, The Netherlands, pp. 9–19.

Turner, R. E. 1977. Intertidal vegetation and commercial yields of penaeid shrimp. *American Fisheries Society Transactions* 106: 411–416.

Turner, R. E., W. Woo, and H. R. Jitts. 1979. Estuarine influences on a continental shelf plankton community. *Science* 206: 218–220.

Tyler, P. A. 1976. Lagoon of Islands, Tasmani—Deathknell for a unique ecosystem? *Biological Conservation* 9: 1–11.

Van Engel, W. A., and E. B. Joseph. 1968. *Characterization of Coastal and Estuarine Fish Nursery Grounds as Natural Communities*. Final Report, Bureau of Commercial Fisheries, Virginia Institute of Marine Science, Glocester Point, VA, 43 pp.

Valéry, L., V. Bouchard, and J. C. Lefeuvre. 2004. Impact of the invasive native species Elymus athericus on carbon pools in a salt marsh. *Wetlands* 24: 268–276.

Van Raalte, C. D. 1976. Production of epibenthic salt marsh algae: light and nutrient limitation. *Limnology and Oceanography* 21: 862–872.

Verhoeven, J. T. A. 1986. Nutrient dynamics in minerotrophic peat mires, *Aquatic Botany* 25: 117–137.

Verhoeven, J. T. A., D. F. Whigham, R. van Logtestijn, and J. O'Neill. 2001. A comparative study of nitrogen and phosphorus cycling in tidal and non-tidal river wetlands. *Wetlands* 21: 210–222.

Visser, J. M. 1989. *The Impact of Vertebrate Herbivores on the Primary Production of Sagittaria Marshes in the Wax Lake Delta, Atchafalaya Bay, Louisiana*. Ph.D. dissertation, Louisiana State University, Baton Rouge, 88 pp.

Visser, J. M., C. E. Sasser, R. G. Chabreck, and R. G. Linscombe. 1998. Marsh vegetation types of the Mississippi River deltaic plain. *Estuaries* 21: 818–828.

Visser, J. M., C. E. Sasser, R. H. Chabreck, and R. G. Linscombe. 1999. Long-term vegetation change in Louisiana tidal marshes. *Wetlands* 19: 168–175.

Wadsworth, J. R., Jr. 1979. Duplin River Tidal System. Sapelo Island, Georgia. Map reprinted in Jan. 1982 by the University of Georgia Marine Institute, Sapelo Island, GA.

Watzin, M. C., and J. G. Gosselink. 1992. *Coastal Wetlands of the United States*. Louisiana Sea Grant College Program, Baton Rouge, LA, and U.S. Fish and Wildlife Service, Lafayette, LA, 15 pp.

Werme, C. E. 1981. *Resource Partitioning in the Salt Marsh Fish Community*. Ph.D. dissertation, Boston University, Boston, 126 pp.

Wharton, C. H., W. M. Kitchens, E. C. Pendleton, and T. W. Sipe. 1982. The ecology of bottomland hardwood swamps of the Southeast: A community profile. U.S. Fish and Wildlife Service, Biological Services Program FWS/OBS-81/37, 133 pp.

Wheeler, B. D. 1980. Plant communities of rich fen systems in England and Wales. *Journal of Ecology* 68: 365–395.

Whigham, D. F., and R. L. Simpson. 1975. *Ecological Studies of the Hamilton Marshes*. Progress report for the period June 1974-January 1975, Rider College, Biology Department, Lawrenceville, NJ.

Whigham, D. F., R L. Simpson, and K. Lee. 1980. *The Effect of Sewage Effluent on the Structure and Function of a Freshwater Tidal Wetland Water Resources*. Research Institute Report, Rutgers University, New Brunswick, NJ, 160 pp.

White, D. A. 1993. Vascular plant community development on mudflats in the Mississippi River delta, Louisiana, USA. *Aquatic Bot.* 45: 171–194.

White, D. A., T. E. Weiss, J. M. Trapani, and L. B. Thien. 1978. Productivity and decomposition of the dominant salt marsh plants in Louisiana. *Ecology* 59: 751–759.

Wiegert, R. G., and B. J. Freeman. 1990. *Tidal Salt Marshes of the Southeast Atlantic Coast: A Community Profile*. U. S. Department of Interior, Fish and Wildlife Service, Biological Report 85 (7.29), Washington, DC.

Wilson, J. O., R. Buchsbaum, I. Valiela, and T. Swain. 1986. Decomposition in salt marsh ecosystems: Phenolic dynamics during decay of litter of Spartina alterniflora. *Marine Ecology Progress Series* 29: 177–187.

Zedler, J. B. 1980. Algae mat productivity: Comparisons in a salt marsh. *Estuaries* 3: 122–131.

Zhu, R., Y. Liu, J. Ma, H. Xu, and L. Sun. 2008. Nitrous oxide flux to the atmosphere from two coastal tundra wetlands in eastern Antarctica. *Atmospheric Environment* 42: 2437–2447.

Mangrove swamp

Chapter 9

Mangrove Swamps

Mangrove swamps replace salt marshes as the dominant coastal ecosystems in subtropical and tropical regions. An estimated 140,000 to 170,000 km^2 of mangrove wetlands are found throughout the world. Mangrove wetlands are limited in the United States (where there are approximately 5,000 km^2 of mangroves) mostly to Florida coastlines and to emerging mangrove swamps on the Louisiana coastline. Mangrove wetlands have been classified according to their hydrodynamics and topography as fringe mangroves, riverine mangroves, basin mangroves, and dwarf or scrub mangroves. The dominant mangrove plant species are known for several adaptations to the saline wetland environment, including prop roots, pneumatophores, salt exclusion, salt excretion, and the production of viviparous seedlings. Their productivity and organic export are closely related to their hydrogeomorphic setting.

The coastal salt marsh of temperate middle and high latitudes gives way to its analog, the mangrove swamp, in tropical and subtropical regions of the world. The mangrove swamp is an association of halophytic trees, shrubs, and other plants growing in brackish to saline tidal waters of tropical and subtropical coastlines. This coastal, forested wetland (called a *mangal* by some researchers) is infamous for its impenetrable maze of woody vegetation, its unconsolidated peat that seems to have no bottom, and its many adaptations to the double stresses of flooding and salinity. The word *mangrove* comes from the Portuguese word *mangue* for "tree" and the English word *grove* for "a stand of trees" and refers to both the dominant trees and the entire plant community.

Many myths have surrounded the mangrove swamp. It was described at one time or another in history as a haven for wild animals, a producer of fatal "mangrove root gas," and a wasteland of little or no value. Researchers, however, have established the importance of mangrove swamps in exporting organic matter to adjacent coastal food chains, in providing physical stability to certain shorelines to prevent erosion, in protecting inland areas from severe damage during hurricanes and tidal waves, and in serving as sinks for nutrients and carbon. The extensive literature on the mangrove swamp on a worldwide basis grows exponentially. This interest probably stems from the worldwide scope of these ecosystems, the many unique features that they possess, and their role in climate change—both as being on the front line of sea-level rise in the tropics and their high productivity that leads to significant carbon sequestration (popularly called "blue carbon"). Much of the early literature on mangroves concerned floristic and structural topics. Beginning in the early 1970s, the focus was on hydrogeomorphology and the functional aspects of mangrove swamps. Since that time, a significant literature on ecophysiology, primary productivity, stressors, food chains, and detritus dynamics of mangrove ecosystems has been produced, along with new work on nutrient cycling, mangrove restoration, valuation of mangrove resources, blue carbon sequestration, and responses of mangroves to sea level changes.

Geographical Extent

Mangrove swamps are found along tropical and subtropical coastlines throughout the world, usually between 25° N and 25° S latitude (Fig. 9.1a). Their limit in the Northern Hemisphere generally ranges from 24° to 32° N latitude, depending on the local climate and the southern limits of freezing weather. There are an estimated 138,000 to 170,000 km^2 of mangrove swamps in the world (Giri et al., 2011; Twilley and Day, 2013; Krauss et al., 2014), with more than half of those swamps found in the latitudinal belts between 0° and 10° (Fig. 9.1b). Mangroves are divided into two groups—the Old World mangrove swamps and the New World and West African mangrove swamps. Over 50 species of mangroves exist, and their distribution is thought to be related to continental drift in the long term and possibly to transport by early humans in the short term. The distribution of these species, however, is uneven. The swamps are particularly dominant in the Indo-West Pacific region (part of the Old World group), where they contain the greatest diversity of species. There are 36 species of mangroves in that region, whereas there are only about 10 mangrove species in the Americas (Fig. 9.1c). It has been argued, therefore, that the Indo-Malaysian region was the original center of distribution for the mangrove species (Chapman, 1976). Certainly some of the most intact mangrove forests in the world are found in Malaysia and in Micronesia, in the small islands east of the Philippines in the western Pacific. Studies have illustrated how important these mangrove swamps are to local economies in these regions (Ewel et al., 1998; Cole et al., 1999). Several mangrove species, not native to the Hawaiian Archipelago despite its appropriate climate and coastal geomorphology, invaded the islands in the early twentieth century and are now permanent fixtures on coastlines there (Allen, 1998).

There is also a great deal of segregation between the mangrove vegetation found in the Old World region and that found in the New World of the Americas and West Africa. Two of the primary genera of mangrove trees, *Rhizophora* (red mangrove) and *Avicennia* (black mangrove), contain separate species in the Old and New Worlds, suggesting "that speciation is taking place independently in each region" (Chapman, 1976).

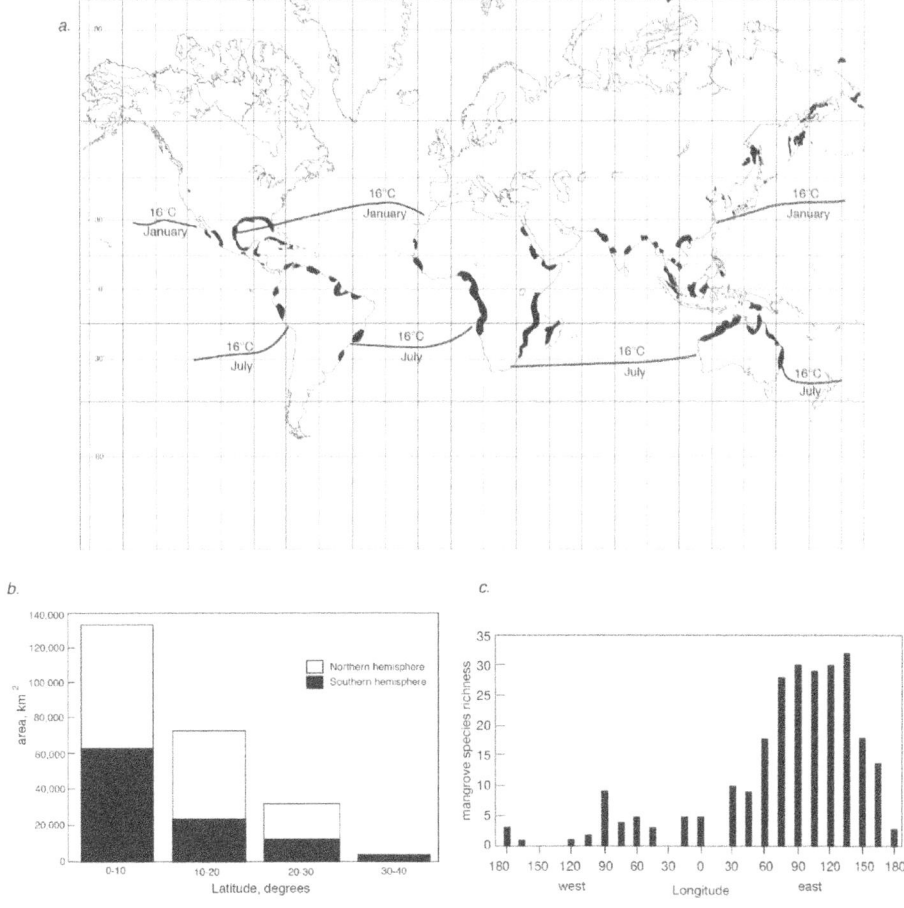

Figure 9.1 Distribution of mangrove wetlands (a) in the world and (b) by latitude, and (c) mangrove species richness by longitude. (After Chapman, 1977; Twilley et al., 1992; Ellison et al. 1999)

Most of the mangrove swamps in the United States, estimated to cover over 5,000 km², are found in Florida (see Table 8.1). The best development of mangroves in Florida is along the southwest coast, where the Everglades and the Big Cypress Swamp drain to the sea. Mangroves extend up to 30 km inland along water courses on this coast. The area includes Florida's Ten Thousand Islands, one of the largest mangrove swamps in the world at 600 km². Because of development pressure, a significant fraction of the original mangroves on these islands has been lost or altered. Patterson (1986) reported that there was a loss of 24 percent of mangroves on one of the most developed islands in this region, Marco Island, from 1952 to 1984. Mangroves are now protected in Florida, and it is illegal to remove them from the shoreline.

Mangrove swamps are also common farther north along Florida's coasts, north of Cape Canaveral on the Atlantic coast and to Cedar Key on the Gulf of Mexico, where mixtures of mangrove and salt marsh vegetation appear. One species of mangrove (*Avicennia germinans*) is found in Louisiana and in the Laguna Madre of Texas, and it has been spreading extensively for the last 40 years, a possible sign of climate shift. Extensive mangrove swamps are also found throughout the Caribbean Islands, including Puerto Rico. Lugo (1988) estimated that there were originally 120 km^2 of mangroves in Puerto Rico, although only half of those remained by 1975.

Geographic Limitation and Recent Expansion

The frequency and severity of frosts are the main factors that limit the extension of mangroves beyond tropical and subtropical climes. For example, in the United States, mangrove wetlands are found primarily along the Atlantic and Gulf coasts of Florida up to 27° to 29° N latitude, north of which they are replaced by salt marshes. The red mangrove can survive temperatures as low as –2° to –4° C for 24 hours, whereas the black mangrove can withstand several days at this temperature, allowing black mangroves to extend farther north on Florida's east coast than red mangroves (as far north as 30° N). Similarly, Schaeffer-Novelli et al. (1990) described mangroves as extending to 28° to 30° S latitude along the Brazilian coast. Three to four nights of a light frost are sufficient to kill even the hardiest mangrove species. Lugo and Patterson-Zucca (1977) showed that mangroves survived approximately five nonconsecutive days of frost in January 1977 in Sea Horse Key Florida on the Gulf of Mexico shoreline (latitude 29° N), but estimated that it would take 200 days for the forest to recover from frost damage. They also hypothesized that soil salinity stress could modify frost stress on mangroves, suggesting that the latitudinal limit of mangroves reflects a number of stresses rather than one factor.

Cavanaugh et al. (2014) investigated recent changes of this latitudinal limit of mangroves in Florida by reviewing 28 years of satellite imagery on the northern extreme of mangrove extent on the Atlantic Ocean shoreline of Florida. They found a poleward expansion of mangroves along this coastline from 1984 to 2011 (Fig. 9.2) and a strong correlation between that expansion and a reduction in the frequency of extreme cold events (days colder than –4°C). They concluded that this poleward expansion of mangroves is not related to mean annual temperatures but rather inversely related to the frequency of cold extreme temperatures, mostly brought to the Florida coastline by polar fronts in the winter.

Hydrogeomorphology

There are several different types of mangrove wetlands, each having a unique set of topographic and hydrodynamic conditions. A classification scheme of five geomorphological settings where mangrove forests occur, as developed by Thom (1982), includes systems dominated by waves, tides, and rivers or, most often, by combinations of these three energy sources. Like the coastal salt marsh, the mangrove swamp can develop only where there is adequate protection from high-energy wave action. A number of physiographic settings favor the protection of mangrove swamps, including (1) protected shallow bays, (2) protected estuaries, (3) lagoons, (4) the leeward sides of peninsulas and islands, (5) protected seaways, (6) behind spits, and (7) behind

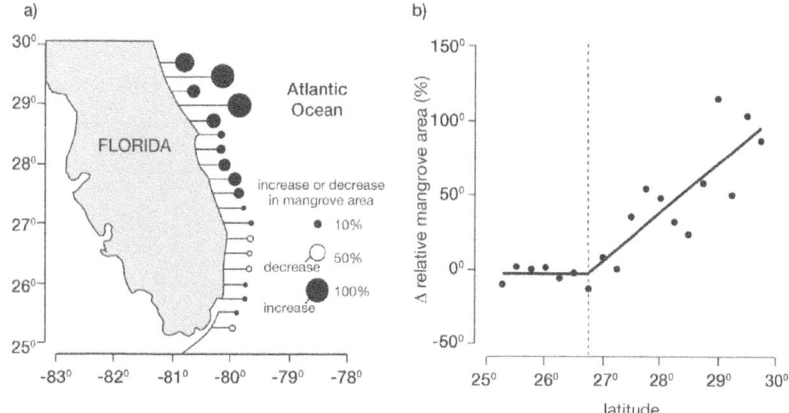

Figure 9.2 (a) The Florida, U.S., peninsula showing the long-term increase (solid) or long-term decrease (gray) in mangrove cover from 1980s to 2007–2011 for each 0.25° latitude on the Atlantic Ocean coastline; (b) relationship between mangrove change (increase or decrease) and latitude. Vertical line on (b) indicates breakpoint of latitude of 26.75° where increase in mangrove cover begins. (After Cavanaugh et al., 2014)

offshore shell or shingle islands. Unvegetated coastal and barrier dunes usually develop where this protection does not exist, and mangroves are also often found behind these dunes.

In addition to the required physical protection from wave action, the range and duration of the flooding of tides exert a significant influence over the extent and functioning of the mangrove swamp. Tides constitute an important subsidy for the mangrove swamp, importing nutrients, aerating the soil water, and stabilizing soil salinity. Salt water is important to the mangroves in eliminating competition from freshwater species. Tides provide a subsidy for the movement and distribution of the seeds of several mangrove species. They also circulate organic sediments in some fringe mangroves for the benefit of filter-feeding organisms, such as oysters, sponges, and barnacles, and for deposit feeders, such as snails and crabs. Like salt marshes, mangrove swamps are intertidal, although a large tidal range is not necessary. Most mangrove wetlands are found in tidal ranges of 0.5 to 3 m or more. Mangrove tree species can also tolerate a wide range of inundation frequencies. *Rhizophora* spp., the red mangrove, is often found growing in continually flooded coastal waters below normal low tide.

At the other extreme, mangroves can be found several kilometers inland along riverbanks where there is less tidal action. These mangroves depend on river discharge and are nourished by river flooding in addition to infrequent tidal inundation and the stability of groundwater and surface water levels near the coast.

Hydrodynamic Classification

The development of mangrove swamps is the result of topography, substrate, and freshwater hydrology as well as tidal action. A classification of mangrove wetland ecosystems according to their physical hydrologic conditions was developed in the

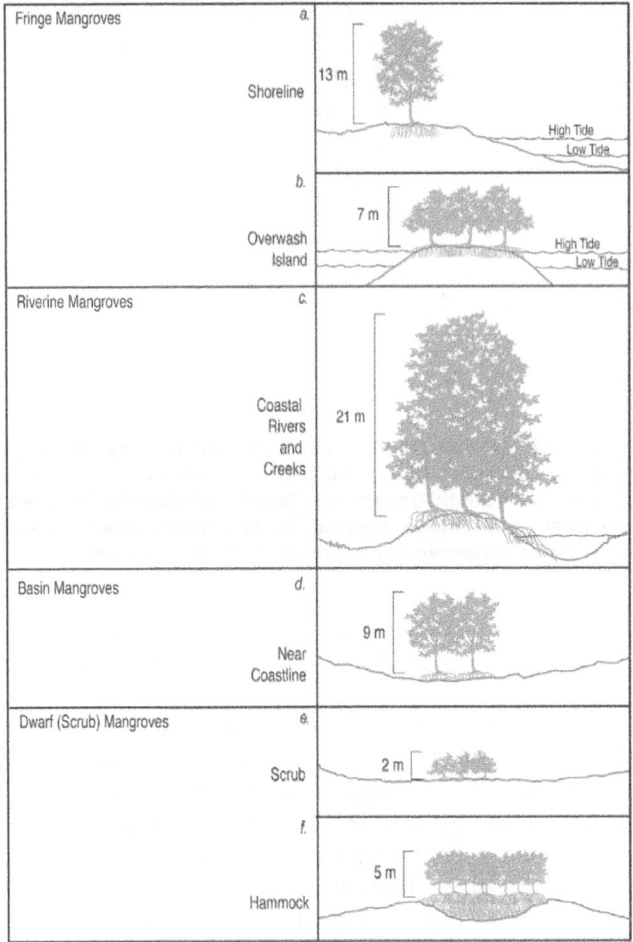

Figure 9.3 Classification of mangrove wetlands according to four hydrogeomorphic classes (and six types overall): (a) and (b) fringe mangroves; (c) riverine mangroves; (d) basin mangroves; (e) and (f) scrub (dwarf) mangroves. (After Wharton et al., 1976; Lugo, 1980; Cintrón et al. 1985)

1970s by Ariel Lugo, Sam Snedaker, and others at the University of Florida. The four major classes of mangrove wetlands, based on their hydrogeomorphology, are shown in Figure 9.3 and are discussed below.

1. *Fringe mangroves.* Fringe mangrove wetlands are found along protected shorelines, on narrow berms along the coastline or in wide expanses along gently sloping beaches, and along some canals, rivers, and lagoons (Fig. 9.3a). If a berm is present, the mangroves may be isolated from freshwater runoff and then have to depend completely on rainfall, the sea, and groundwater for their nutrient supply. A special case of fringe mangroves are small islands and narrow extensions of larger and masses (spits) that are "overwashed" on a daily basis during high tide. These are sometimes called

overwash mangrove islands (Fig. 9.3b). The forests are dominated by the red mangrove (*Rhizophora*) and a prop root system that obstructs the tidal flow and dissipates wave energy during periods of heavy seas. Tidal velocities are high enough to wash away most of the loose debris and leaf litter into the adjacent bay. The islands often develop as concentric rings of tall mangroves around smaller mangroves and a permanent, usually hypersaline, pool of water. These wetlands are abundant in the Ten Thousand Islands region of Florida and along the southern coast of Puerto Rico. They are particularly sensitive to the effects of ocean pollution.

2. *Riverine mangroves*. Tall, productive riverine mangrove forests are found along the edges of coastal rivers and creeks, often several miles inland from the coast (Fig. 9.3c). These wetlands may be dry for a considerable time, although the water table is generally just below the surface. In Florida, freshwater input is greatest during the wet summer season, causing the highest water levels and the lowest salinity in the soils during that time. Riverine mangrove wetlands export a significant amount of organic matter because of their high productivity. These wetlands are affected by freshwater runoff from adjacent uplands and from water, sediments, and nutrients delivered by the adjacent river. Hence they can be significantly affected by upstream activity or stream alteration. The combination of adequate fresh water and high inputs of nutrients from both upland and estuarine sources causes these systems to be generally very productive, supporting large (16–26 m) mangrove trees. Salinity varies but is usually lower than that of the other mangrove types described here. The flushing of fresh water during wet seasons causes salts to be leached from the sediments.

3. *Basin mangroves*. Basin mangrove wetlands occur in inland depressions, or basins, often behind fringe mangrove wetlands, and in drainage depressions where water is stagnant or slowly flowing (Fig. 9.3d). These basins are often isolated from all but the highest tides and yet remain flooded for long periods once tide water does flood them. Because of the stagnant conditions and less frequent flushing by tides, soils have high salinities and low redox potentials. These wetlands are often dominated by black mangroves (*Avicennia* spp.) and white mangroves (*Laguncularia* spp.), and the ground surface is often covered by pneumatophores from these trees.

These hydrogeomorphic classes of mangrove are broad categorizations, and within these mangrove types, there are likely to be subtypes that can be defined by specific hydrologic conditions. Knight et al. (2008) characterized three basin forest subtypes for *A. marina* forests using hydrology and forest structure data from the Coombabah Lake region in Southeast Queensland, Australia. These subtypes included a "deep basin" (characterized by ~50 cm standing water, ~3 tides yr^{-1}, and mature tree development), a "medium depth basin" (characterized by 15–30 cm standing water, 20–40 tides yr^{-1}, and intermediate tree development), and a "shallow basin" (characterized by 5–15 cm standing water, ~80 tides yr^{-1}, and recent tree establishment). Mangrove subtypes could be categorized in the other hydrogeomorphic classes as well.

4. *Dwarf mangroves*. There are several examples of isolated, low-productivity scrub mangrove wetlands that are usually limited in productivity because of the lack of nutrients or freshwater inflows. Dwarf mangrove wetlands are dominated by scattered, small (often less than 2 m tall) mangrove trees growing in an environment that is

probably nutrient poor (Fig. 9.3e). The nutrient-poor environment can be a sandy soil or limestone marl. Hypersaline conditions and cold at the northern extremes of the mangrove's range can also produce "scrub," or stressed mangrove trees, in riverine, fringe, or basin wetlands. True dwarf mangrove wetlands, however, are found in the coastal fringe of the Everglades and the Florida Keys and along the northeastern coast of Puerto Rico. Some of these wetlands in the Everglades are inundated by seawater only during spring tides or storm surges and are often flooded by freshwater runoff in the rainy season. These types of wetlands are actually an intermix of small red mangrove trees with marsh vegetation such as sawgrass (*Cladium jamaicense*) and rush (*Juncus roemarianus*). Hammock mangrove wetlands also occur as isolated, slightly raised tree islands in the coastal fringe of the Florida Everglades and have characteristics of both basin and scrub mangroves. They are slightly raised as a result of the buildup of peat in what was once a slight depression in the landscape (Fig. 9.3f). The peat has accumulated from many years of mangrove productivity, actually raising the surface from 5 to 10 cm above the surrounding landscape.

Soils and Salinity

Soil salinity in mangrove ecosystems varies from season to season and with mangrove type (Table 9.1). In riverine mangrove systems, the soil salinity is less than that of normal seawater because of the influx of fresh water. In basin mangroves, however, salinity can be well above that of seawater because of evaporative losses (>50 ppt). Noted Florida naturalist John Henry Davis (1940) summarized four major points about salinity in mangrove wetlands from his studies in Florida years ago that still hold today:

1. There is a wide annual variation in salinity in mangrove wetlands.
2. Saltwater is not necessary for the survival of any mangrove species but only gives mangroves a competitive advantage over salt-intolerant species.

Table 9.1 Soil salinity ranges for major mangrove types

Hydrodynamic Type	Soil salinity, ppt
Fringe mangroves	
Avicennia	59
Rhizophora	39
Riverine mangroves	10–20[a]
Basin mangroves	
Avicennia	>50
Laguncularia	low salinity
mixed forest	30–40

[a]Higher in dry season when less freshwater streamflow is available.
Source: Cintron et al., 1985

3. Salinity is usually higher and fluctuates less in interstitial soil water than in the surface water of mangroves.
4. Saline conditions in the soil extend farther inland than normal high tide because of the slight relief, which prevents rapid leaching.

Seasonal oscillations in salinity in mangrove wetlands are a function of the height and duration of tides, the seasonality and intensity of rainfall, and the seasonality and amount of fresh water that enters the mangrove wetlands through rivers, creeks, and runoff. In Florida, summer wet-season convective storms and associated freshwater flow in streams and rivers as well as an occasional hurricane in the late summer or early fall lead to the dilution of saltwater and the lowest salinity concentrations. Salinity is generally highest during the dry season, which occurs in the winter and early spring.

Soil Acidity

Mangrove soils are often acidic, although, in the presence of carbonate, as is often the case in south Florida, the soil pore water can be close to neutrality. The soils are often highly reduced, with redox potentials ranging from -100 to -400 mV. The highly reduced conditions and the subsequent accumulation of reduced sulfides in mangrove soils cause extremely acidic soils in many mangrove areas. Dent (1986, 1992) reported a measured accumulation of 10 kg S m^{-3} of sediment per 100 years in mangroves. When these soils are drained and aerated for conversion to agricultural land, the reduced sulfides, generally stored as pyrites, oxidize to sulfuric acid, producing what are known as cat clays. These highly acidic soils make traditional agriculture difficult and are one of the reasons that, when mangrove swamps are converted to fishponds, the ponds have a short lifetime before they are abandoned. Dent (1992) argued that the "dereclamation" of some previously "reclaimed" marginal coastal soils back to mangroves and salt marshes may be the best strategy for these acidic soils.

Vegetation

As is evident in coastal salt marshes, the stresses of waterlogged soils and salinity lead to a relatively simple flora in most mangrove wetlands, particularly when compared to their upland neighboring ecosystem, the tropical rain forest. There are more than 50 species of mangroves throughout the world (Stewart and Popp, 1987; Twilley and Day, 2013), representing 12 genera in 8 families. Fewer than 10 species of mangroves are found in the New World, and only 3 species are dominant in the south Florida mangrove swamps—the red mangrove (*Rhizophora mangle* L.), the black mangrove (*Avicennia germinans* L., also named *A. nitida* Jacq.), and the white mangrove (*Laguncularia racemosa* L. Gaertn.). Buttonwood (*Conocarpus erecta* L.), although strictly not a mangrove, is occasionally found growing in association with mangroves or in the transition zone between the mangrove wetlands and the drier uplands. Each of the hydrologic types of mangrove wetlands described previously is dominated by different associations of mangrove plants. Fringe mangrove wetlands are dominated

Table 9.2 Structural characteristics of canopy vegetation for major mangrove types[a]

Hydrodynamic Type	Number of Tree Species	Number of Trees (#/(ha))		Basal Area (m²/ha)		Stand Height (m)	Aboveground Biomass (kg/m²)
		>2.5 cm dbh	>10 cm dbh	>2.5 cm dbh	>10 cm dbh		
Fringe mangroves	1.7 ± 0.1 (33)	4005 ± 642 (33)	852 ± 115 (31)	22.2 ± 1.5 (33)	14.6 ± 1.9 (31)	13.3 ± 2.6 (32)	0.8–15.9 (8)
Riverine mangroves	1.9 ± 0.1 (36)	1979 ± 209 (28)	661 ± 71 (32)	30.4 ± 3.5 (5)	32.6 ± 4.7 (32)	21.2 ± 4.8 (26)	1.6–28.7 (8)
Basin mangroves	2.3 ± 0.1 (31)	3599 ± 400 (31)	573 ± 102 (21)	18.5 ± 1.6 (31)	10.6 ± 2.2 (21)	9.0 ± 0.7 (31)	—

[a] Data are based on mangrove sites in Florida, Mexico, Puerto Rico, Brazil, Costa Rica, Panama, and Ecuador. Values are the average ± standard error (number of observations) except for above-ground biomass, which is the range (number of observations).
Source: Cintrón et al. (1985).

by red mangroves (*Rhizophora*) that contain abundant and dense prop roots, particularly along the edges that face the open sea. Riverine mangrove wetlands are also numerically dominated by red mangroves, although they are straight trunked and have relatively few, short prop roots. Black (*Apicennia* spp.) and white (*Laguncularia* spp.) mangroves also frequently grow in these wetlands. Basin wetlands support all three species of mangroves, although black mangroves are the most common in basin swamps and hammock wetlands are mostly composed of red mangroves. Scrub mangrove wetlands are typically dominated by widely spaced, short (less than 2 m tall) red or black mangroves.

A comparison of the structural characteristics of the major hydrodynamic types of mangrove wetlands is provided in Table 9.2. These data were compiled from more than 100 mangrove research sites throughout the New World. Fringe mangroves generally have a greater density of large trees (>10 cm diameter at breast height (dbh)) compared to riverine and basin mangroves. Riverine wetlands, however, have the largest trees and, hence, a much greater basal area and tree height than do fringe or basin mangroves. The biomass of riverine mangroves is generally the highest, although data are difficult to compare because of the different methods and sample sizes used in various observations. Cintrón et al. (1985) reported a range of aboveground biomass for the Florida mangroves of 9 to 17 kg/m² for riverine mangroves and 0.8 to 15 kg/m² for fringe mangroves. Single measures of 0.8 kg/m² for a dwarf mangrove wetland and 9.8 kg/m² for a hammock mangrove (both in Florida) were also reported.

Zonation

In trying to understand the vegetation of mangrove wetlands, most early researchers were concerned with describing plant zonation and successional patterns. Some attempts were made to equate the plant zonation found in mangrove wetlands with successional seres, but Lugo (1980) warned that "zonation does not necessarily recapitulate succession because a zone may be a climax maintained by a steady or recurrent environmental condition." J. H. Davis (1940) is generally credited with the best early description of plant zonation in Florida mangrove swamps, especially in fringe and basin mangrove wetlands (Fig. 9.4). He hypothesized that the

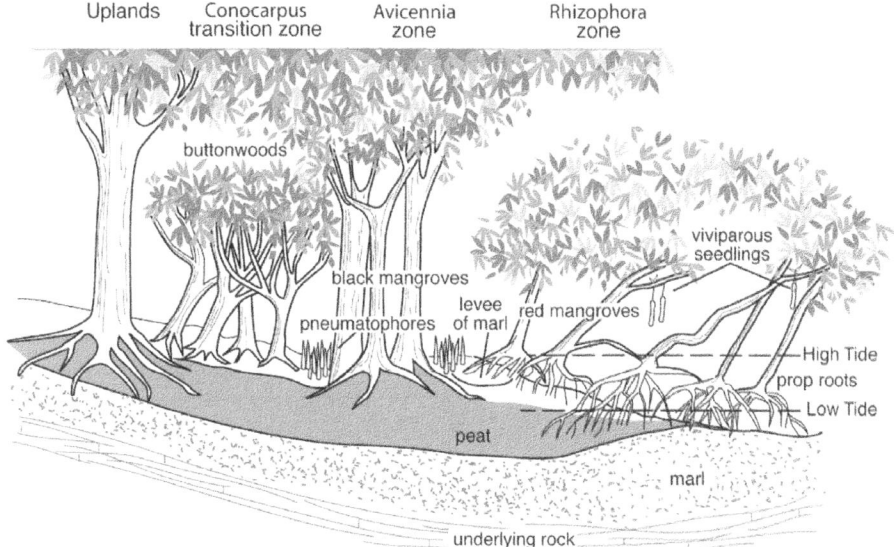

Figure 9.4 Classic zonation pattern of Florida mangrove swamp with illustrations of mangrove adaptations, such as prop roots, viviparous seedlings, and pneumatophores.

entire ecosystem was accumulating sediments and was therefore migrating seaward. Typically, *Rhizophora mangle* is found in the lowest zone, with seedlings and small trees sprouting even below the mean low tide in marl soils. Above the low-tide level but well within the intertidal zone, full-grown *Rhizophora* with well-developed prop roots predominate. There tree height is approximately 10 m. Behind these red mangrove zones and the natural levee that often forms in fringe mangrove wetlands, basin mangrove wetlands, dominated by black mangroves (*Avicennia*) with numerous pneumatophores, are found. Flooding occurs only during high tides. Buttonwood (*Conocarpus erecta*) often forms a transition between the mangrove zones and upland ecosystems. Flooding occurs there only during spring tides or during storm surges, and soils are often brackish to saline.

Thibodeau and Nickerson (1986) suggested that red mangroves have a much lower ability to tolerate high sulfides typical of extremely reduced conditions than do black mangroves, and so red mangroves occur in regions that are frequently flushed by tides, whereas black mangroves are found in isolated basin settings where strongly reduced substrate containing high sulfides are found and pneumatophores can be of the greatest use (see the following section).

Mangroves as Land Builders?

The zonation of plants in mangrove wetlands led some researchers (e.g., J. H. Davis, 1940) to speculate that each zone is a step in an autogenic successional process that leads to freshwater wetlands and, eventually, to tropical upland forests or pine forests. Other researchers, led by Egler (1952), considered each zone to be controlled by its physical environment to the point that it is in a steady state or at least a state of arrested succession (allogenic succession). For example, with a rising sea level, the mangrove zones migrate inland; during periods of decreasing sea level, the mangrove zones move seaward. Egler thought that the impact of fire and hurricanes made conventional succession impossible in the mangroves of Florida. Another theory, advanced by Chapman (1976), is that mangrove succession may be a combination of both autogenic and allogenic strategies, or a "succession of successions." If that is the case, successional stages could be repeated a number of times before the next successional level is attained.

Lugo (1980) reviewed mangrove succession in light of E. P. Odum's criteria (1969; see Chapter 7: Wetland Vegetation and Succession) and found that, except for mangroves on accreting coastlines, traditional successional criteria do not apply. He concluded that mangroves are true steady-state systems in the sense that they are the optimal and self-maintaining ecosystems in low-energy tropical saline environments. In such a situation, high rates of mortality, dispersal, germination, and growth are the necessary tools of survival. Unfortunately, these attributes could lead many to identify mangroves as successional systems.

It is no longer accepted dogma that mangrove wetlands are "land builders" that are gradually encroaching on the sea, as was suggested by J. H. Davis (1940). Yet Lee et al. (2014) state that "few of Davis's critics offered contradictory data as extensively detailed or impressive as those presented in his [Davis's] classic works." In many cases, mangrove vegetation plays a passive role in the accumulation of sediments, and the vegetation usually follows, not leads, the land building that is caused by current and tidal energies. It is only after the substrate has been established that the vegetation contributes to land building by slowing erosion and by increasing the rate of sediment accretion.

The mangrove's successional dynamics appear to involve a combination of (1) peat accumulation balanced by tidal export, fire, and hurricanes over years and decades; and (2) advancement or retreat of zones according to the fall or rise of sea level over centuries. Some researchers (Alongi, 2008; Lee et al., 2014) refer to mangroves as "land stabilizers" rather than "land builders." When peat accumulation is added to the stabilization of marine and riverine sediments, some mangroves can have both horizontal movement as well as vertical movement upward (Fig. 9.5). As Figure 9.5 illustrates, with climate changes in sea level, temperature, atmospheric carbon dioxide and rainfall all at the same time, the stability and/or expansion of mangroves is a complicated matter.

Mangrove Adaptations

Mangrove vegetation, particularly the dominant trees, has several adaptations that allow it to survive in an environment of high salinity, occasional harsh weather, and anoxic soil conditions. (Chapter 7 gives an overview of wetland plant adaptations.) These physiological and morphological adaptations have been of interest to researchers and are among some of the most distinguishing features that the laypeople notice when first viewing these wetlands. Some of the morphological adaptations are shown

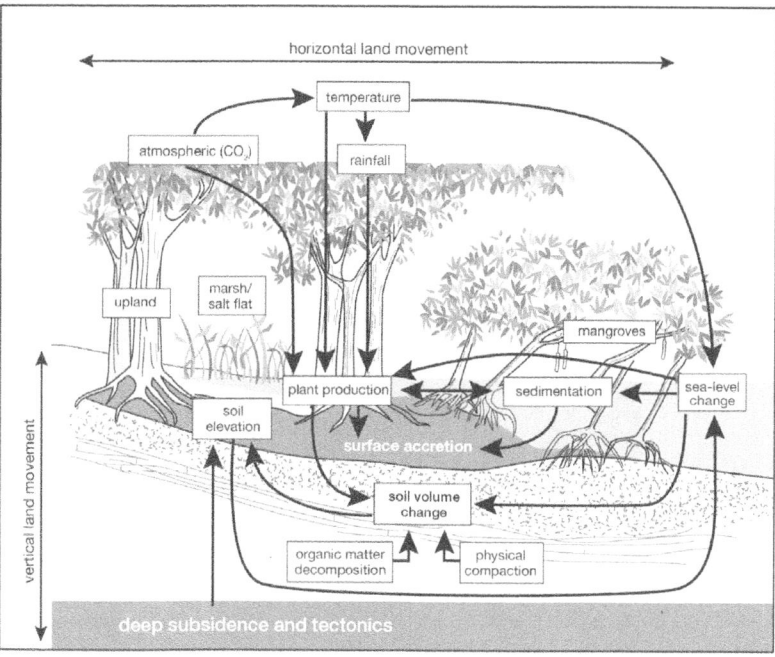

Figure 9.5 Model showing how climate change in sea level, temperature, atmospheric carbon dioxide (CO_2), and rainfall, along with sedimentation, soil erosion, and tectonic uplift, all influence the horizontal and vertical movement of coastal mangroves and possible land building. (After Lee et al., 2014)

in Figure 9.6. Overall, physiological and morphological adaptations of mangroves include (1) salinity control; (2) prop and drop roots, pneumatophores, and lenticles; and (3) viviparous seedlings.

Salinity Control

Mangroves are facultative halophytes; that is, they do not require saltwater for growth but are able to tolerate high salinity and thus outcompete vascular plants that do not have this salt tolerance. The ability of mangroves to live in saline soils depends on their ability to control the concentration of salt in their tissues. In this respect, mangroves are similar to other halophytes. Mangroves have the ability both to prevent salt from entering the plant at the roots (*salt exclusion*) and to excrete salt from the leaves (*salt secretion*). Salt exclusion at the roots is thought to be a result of reverse osmosis, which causes the roots to absorb only freshwater from saltwater. The root cell membranes of mangroves species of *Rhizophora*, *Avicennia*, and *Laguncularia*, among others, may act as ultrafilters that exclude salt ions. Water is drawn into the root through the filtering membrane by the negative pressure in the xylem developed through transpiration at the leaves; this action counteracts the osmotic pressure caused by the solutions in the external root medium. There are also a number of mangrove species (e.g., *Avicennia* and *Laguncularia*) that have salt-secreting glands on the

Figure 9.6 Adaptations of mangroves, including (a) prop roots of red mangroves (*Rhizophora*) in south Florida, (b) drop roots of mangroves in western Costa Rica, (c) pneumatophores of black mangroves (*Avicennia*) in southwest Florida, (d) viviparous germinated seedlings hanging in red mangrove canopy, and (e) a red mangrove seedling floating vertically in the water column, perhaps kilometers from where it fell into the water. (Photos by W. J. Mitsch)

leaves to rid the plant of excess salt. The solutions that are secreted often contain high concentrations of sodium chloride (NaCl) and salt crystals may form on the leaves. Another possible way in which mangroves discharge salt is through leaf fall, although the importance of this method is questioned.. This leaf fall may be significant because mangroves produce essentially two crops of leaves per year.

Prop Roots and Pneumatophores

Some of the most notable features of mangrove wetlands are the *prop roots* and *drop roots* of the red mangrove (Fig. 9.6a,b) and the numerous, small *pneumatophores* of the black mangrove (*Avicennia*) (reaching 10–20 cm above the sediments and sometimes considerably more; Fig. 9.6c). At a distance, drop roots resemble prop roots; they extend from branches and other upper parts of the stem directly down to the ground, hovering above or rooting only a few centimeters into the sediments. They adsorb freshwater from rainfall into the plant, an especially important function in seasonally dry climates. Oxygen enters the plant through small pores, called *lenticles*, which are found on both pneumatophores and prop and drop roots. When lenticels are exposed to the atmosphere during low tide, oxygen is absorbed from the air, and some of it is transported to and diffuses out of the roots through a system of aerenchyma tissue. This maintains an aerobic microlayer around the root system. When prop roots or pneumatophores of mangroves are continuously flooded by stabilizing the water levels, those mangroves that have submerged pneumatophores or prop roots soon die.

In an interesting experiment to determine the importance of oxygen transport from the aerial organs to the sediments, Thibodeau and Nickerson (1986) "capped" with plastic tubing the pneumatophores of *Avicennia germinans* in a fringe mangrove forest in the Bahamas. They observed that the aerobic zone surrounding the roots was reduced in the area by capping, indicating that the pneumatophores help the plant produce an oxidized rhizosphere. They also found that the greater the number of pneumatophores present in a given area, the more oxidized the soil. They described the relationship as

$$E_H = -307 + 1.1 pd \tag{9.1}$$

where

E_H = redox potential (mV)
pd = pneumatophore density (number per 0.25 m²)

Viviparous Seedlings

Red mangroves (and related genera in other parts of the world) have seeds that germinate while they are still in the parent tree; a long, cigar-shaped hypocotyl (*viviparous seedling*) develops while hanging from the tree (Fig. 9.6d). This is apparently an adaptation for seedling success where shallow anaerobic water and sediments would otherwise inhibit germination. The seedling (or propagule) eventually falls and often will root if it lands on sediments or will float and drift in currents and tides if it falls into the sea. After a time, if the floating seedling becomes stranded and the water is shallow enough, it will attach to the sediments and root. Often the seedling becomes heavier with time, rights itself to a vertical position in the water column (Fig. 9.6e), and develops roots if the water is shallow. It is not well understood whether contact with the sediments stimulates root growth or if the soil contains some chemical compound that promotes root development. The value of the floating seedlings for mangrove dispersal and for invasion of newly exposed substrate is obvious. Rabinowitz (1978) reported that the obligate dispersal time (the time required during propagule dispersal for germination to be completed) was 40 days for the red mangrove and 14 days for the black mangrove propagules. She also estimated that red and black mangrove propagules could survive floating in the water for 110 and 35 days, respectively.

Consumers

W. E. Odum et al. (1982) reported the following data from the literature describing faunal use of mangroves in Florida in terms of the number of species: 220 fish; 181 birds, including 18 wading birds, 29 water birds, 20 birds of prey, and 71 arboreal birds; 24 reptiles and amphibians; and 18 mammals. In general, a wide diversity of animals is found in mangrove wetlands; their distribution sometimes parallels the plant zonation described previously. Many of the animals that are found in mangrove wetlands are filter feeders or detritivores, and the wetlands are just as important as a shelter for most of the resident animals as they are a source of food. Some of the important filter feeders found in Florida mangroves include barnacles (*Balanus eburneus*), coon oysters (*Ostrea frons*), and the eastern oyster (*Crassostrea virginica*). These organisms often attach themselves to the stems and prop roots of the mangroves within the intertidal zone, filtering organic matter from the water during high tide.

Crabs are among the most important animal species in mangrove wetlands around the world, and they appear to play an important role in maintaining biodiversity in mangrove ecosystems. Taking into account the role that they have in seedling survival, carbon cycling, sediment microtopography, and soil chemistry, Smith et al. (1991) suggested that crabs are the keystone species of the mangrove ecosystem. They burrow in the sediments, prey on mangrove seedlings, facilitate litter decomposition, and are the key transfer organism for converting detrital energy to wading birds and fish in the mangrove forest itself and to offshore estuarine systems. Mangroves around the world are dominated by 6 of the 30 families of Brachyura that collectively make up about 127 species. *Uca* and *Sesarma* are the most abundant crab genera in mangrove wetlands. *Uca* spp. (fiddler crabs) are particularly abundant in mangrove wetlands in Florida, living on the prop roots and high ground during high water and burrowing in the sediments during low tide. *Sesarma* (sesarmids) is the most abundant genus of crabs in the world, with dozens of species in the Indo-Malaysian and East African mangroves and many fewer in tropical America.

One of the most significant ways in which crabs may influence the distribution of mangroves is by selective predation of mangrove propagules. Yet this effect is not common to all mangrove wetlands. In comparing the effect of this predation on mangroves around the world, Smith et al. (1989) found that crabs consumed up to 75 percent of the mangrove propagules in Australian mangrove swamps but very little of the litterfall or propagules in Panamanian and Florida mangrove swamps.

The role of crabs in leaf litter removal (burial and consumption) has been illustrated in a number of studies. Robertson and Daniel (1989) estimated that in some mangrove forests, leaf processing by sesarmid crabs alone was over 75 times the rate as microbial degradation and that crabs removed more that 70 percent of the litter of the high-intertidal *Ceriops* and *Bruguiera* mangroves in tropical Australia. Smith et al. (1991) developed an experiment where crabs were removed from experimental plots in *Rhizophora* forests. They found that the removal of crabs caused significantly higher concentrations of sulfide and ammonium in the mangrove soils, due primarily to the absence of burrowing, which oxygenates the soil.

Many other invertebrates, including snails, sponges, flatworms, annelid worms, anemone, mussels, sea urchins, and tunicates, are found growing on roots and stems in and above the intertidal zone. Wading birds frequently found in Florida mangroves include the wood stork (*Mycteria americana*), white ibis (*Eudocimus albus*), roseate spoonbill (*Ajaia ajaja*), cormorant (*Phalacrocorx* spp.), brown pelican (*Pelicanus occidentalis*), egrets, and herons. Vertebrates that inhabit mangrove swamps include alligators, crocodiles, turtles, bears, wildcats, pumps, and rats.

Ecosystem Function

Certain functions of mangroves, such as primary productivity, organic export, and outwelling, have been studied extensively. A picture of the dynamics of mangrove wetlands has emerged from several key studies. These studies have demonstrated the importance of the physical conditions of tides, salinity, and nutrients to these wetlands and have shown where natural and human-induced stresses have caused the most effect.

Primary Productivity

Based on a global assessment of mangrove productivity, Bouillon et al. (2008) conservatively estimated worldwide mangrove productivity at 218 ± 72 teragram (Tg = 10^{12} g)-C yr^{-1} and added that over half of the carbon is unaccounted for, based on various estimates of mangrove carbon sinks (organic export, burial, and mineralization). A wide range of productivity has been measured in mangrove wetlands due to the wide variety of hydrodynamic and chemical conditions encountered. Table 9.3 presents a balance of carbon flow in several fringe and basin mangrove swamps in Florida and Puerto Rico. Net primary productivity ranges from 570 to 2,700 g-C m^{-2} yr^{-1} (equivalent to 1,200 to 6,000 g-dry wt m^{-2} yr^{-1}). Gross and net primary productivity are highest in riverine mangrove wetlands, lower in fringe mangrove wetlands, and lowest in basin mangrove wetlands. The highest productivity in riverine mangrove wetlands is due to the greater influence of nutrient loading and freshwater turnover at the riverine site.

The important factors that control mangrove function in general and primary productivity in particular are: (1) tides and storm surges; (2) freshwater discharge; (3) parent substrate; and (4) water–soil chemistry, including salinity, nutrients, and turbidity. These factors are not mutually exclusive, as tides influence water chemistry and hence productivity by transporting oxygen to the root system, by removing the buildup of toxic materials and salt from the soil water, by controlling the rate

Table 9.3 Mass balance of carbon flow (g-C m^{-2} yr^{-1}) in mangrove forests in Florida and Puerto Rico

	Rookery Bay, Florida[a]		Puerto Rico Fringe[b]	Fahkahatchee Bay, Florida[c]		
	Fringe	Basin		Basin	Fringe	Fringe
Gross primary productivity						
Canopy	2,055	3,292	3,004	3,760	4,307	5,074
Algae	402	26	276			
Total	2,457	3,318	3,280			
Respiration (plants)						
Leaves, stems	671	2,022	1,967	1,172	1,416	3,084
Roots, above-ground	22	197	741	146	182	215
Roots, below-ground	?	?	?	?	?	?
Total plant respiration	693	2,219	2,708	1,318	1,598	3,299
Net primary production	1,764	1,099	572	2,442	2,709	1,775
Growth		186	153			
Litterfall		318	237			
Respiration (heterotrophs)		197				
Respiration (total)		2,416	2,843			
Export		64	500			
Net ecosystem production		838	−63			
Burial		?	?			
Growth		186	153			

[a] Lugo et al. (1975), Twilley (1982, 1985), and Twilley et al. (1986).
[b] Golley et al. (1962).
[c] Carter et al. (1973).
Source: Twilley (1988)

of sediment accumulation or erosion, and by indirectly regenerating nutrients lost from the root zone. The principal chemical conditions that affect primary productivity are soil water salinity and the concentration of major nutrients. High soil salinities, which, in turn, are a function of the local hydrology and geomorphology, appear to be the most important variable that influences the productivity of mangroves in a given region. For example, one study of mangroves in Puerto Rico (Cintrón et al., 1978) found that tree height of the mangroves, as a measure of productivity, was inversely related to soil salinity, according to the following relationship:

$$h = 16.6 - 0.20 C_s \tag{9.2}$$

where

h = tree height (m)
C_s = soil salinity concentration (ppt)

In a similar analysis, J. W. Day et al. (1996) were able to demonstrate a relationship between total litterfall in an *Avicennia*-dominated basin mangrove forest in Mexico and soil salinity as

$$L = 3.915 - 0.039 C_s \tag{9.3}$$

where

$$L = \text{litterfall (g m}^{-2}\text{ yr}^{-1})$$

Lovelock (2008) compared mangrove productivity from 11 forests around the world and found very good correlations between soil respiration and measures of leaf production and biomass. However, the highest below-ground carbon allocation per unit litterfall was found in scrub mangrove forests. It seems that mangroves allocate more growth below-ground during environmentally stressful conditions.

Nutrient availability has also been shown to have a major influence on mangrove productivity. Feller et al. (2007) experimentally relieved nutrient deficiencies in mangroves forests that were phosphorus limited (in Twin Cays, Belize) and nitrogen limited (Indian River Lagoon, Florida). When nutrient deficiency was relieved, black mangroves (*A. germinans*) at both forests responded with enhanced stem growth with the greatest response coming from the N-limited forest. Nutrient enrichment was found to influence more ecological processes in the P-limited Twin Cays forest. The authors concluded that eutrophication is more likely to shift nutrient limitations there than the N-limited Indian River Lagoon.

Hurricane Effects

Hurricanes (and typhoons) and mangrove swamps could be described as having a turbulent tropical love-hate relationship. Either by genetic design or chance, the time required for the attainment of a steady state in mangrove forest in Florida is approximately the same as the average period between tropical hurricanes (approximately 20 to 24 years for Caribbean systems). This match suggests that mangroves may have adapted or evolved to go through one life cycle, on average, between major tropical storms. On average, a mangrove forest reaches maturity just as the next hurricane or typhoon hits.

When Hurricane Andrew passed over south Florida in the late summer of 1992, an opportunity existed for detailed studies of the immediate impact of hurricanes on mangroves. Major damage to mangroves in the vicinity of the Everglades occurred due to trunk snapping and uprooting rather than due to any storm surge (T. J. Smith et al., 1994). Mortality was greatest for red mangroves in the 15- to 30-cm dbh age class and over a wider band of 10 to 35 cm dbh for black and white mangrove. T. J. Smith et al. (1994) made two other interesting observations:

1. Gaps that developed in the mangrove forests due to lightning prior to the hurricane were, after the hurricane, small green patches amid the gray matrix of dead mangroves. Apparently, the small-sized mangrove trees that were spared in these patches would now serve as propagule regeneration sites for the area of catastrophic disturbance, which was of much larger scale. Small-scale disturbance nested in large-scale disturbance provides a positive feedback for more rapid mangrove recovery.

2. The loss of mangrove trees in a hurricane removes a major source of aeration of mangrove soils, the trees themselves. As a result of the loss of this aeration, soils in hurricane-impacted mangrove wetlands might become even more reduced, producing even more toxic hydrogen sulfide that could preclude mangrove regeneration for a number of years. Because of this negative feedback, recovery of the mangrove forest to a system similar to that prior to the disturbance is not assured.

Other research after Hurricane Andrew has looked at the vegetative response. Monitoring forest plots inside and outside the eye-wall path of the hurricane between 1995 and 2005, Ward et al. (2006) found that turnover rates (mortality and recruitment) between the two sections differed (dynamic turnover rates were greater inside the eye-wall). However, both sections also exhibited a steady rate of forest turnover dynamics, suggesting that ecological conditions and not structural conditions primarily control productivity. In the case of a large climatic disturbance such as hurricanes, the deposition of coarse woody debris may constitute a substantial nutrient flux.

Response to Rising Sea Levels

Sea-level rise has been documented around the world, and many coastal ecosystems are expected to be adversely affected. The extent that sea-level rise may displace mangrove forests is unclear, however, because these ecosystems are naturally adapted to be resilient to disturbance (Krauss et al., 2014). Alongi (2008) pointed out that mangroves have several characteristics which make them resilient to disturbances that are acute (e.g., hurricanes, tsunamis) or chronic (changing sea level). These characteristics include: below-ground reservoirs of nutrients, rapid nutrient flux and decomposition, complex and efficient biotic controls, and often rapid reconstruction following disturbance due to the self-design and simple architecture of these forests.

It appears that some mangroves forests may be more capable of acclimating to a rising sea level than others. Many of the world's mangrove forests are vulnerable to climate change, including those in the Caribbean and the Pacific Islands. Alongi (2008) predicted that the most susceptible mangrove forests will be slower-growing forests at the thresholds of their habitat range, such as mangrove forests in arid regions where mangroves grow more slowly due to higher salinities, lower humidity, and extreme light conditions. Other susceptible mangroves will include forests in carbonate environments where growth is slow and the input of terrestrial sediment is limited.

Based on current projections of climate change and sea-level rise, Alongi (2008) predicted a decline of 10 to 15 percent of the global mangrove area by 2100. Although this decline is considerable, the author pointed out that the threat may be moot if current mangrove deforestation trends continue worldwide.

Organic Material Storage and Export

Donato et al. (2011) made an interesting comparison that tropical mangrove are among the most carbon-dense ecosystems in the world. They found that tropical

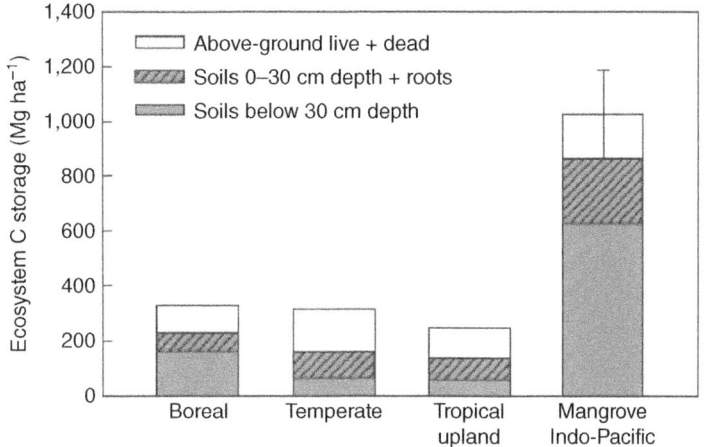

Figure 9.7 Carbon storage in Indo-Pacific tropical mangrove forests compared to boreal, temperate, and tropical forests. (After Donato et al., 2011)

mangroves had carbon storage of over 1,000 Mg ha^{-1} (10,000 g-C m^{-2}) when soil carbon below the 30 cm root zone is included, far more than the carbon storage in tropical, temperate, or boreal forests (Fig. 9.7). They extrapolated, using a world coverage of 140,000 km^2 of mangroves, that mangrove wetlands store on the order of 4 to 20 petagrams ($\times\,10^{15}$ g) of carbon (Pg-C) globally.

Mangrove swamps are important exporters of organic material to the adjacent estuary through the same *outwelling* discussed for coastal marshes in Chapter 8: "Tidal Marshes." In one of the first studies on outwelling from mangroves, Heald (1971) estimated that about 50 percent of the aboveground productivity of a mangrove swamp in southwestern Florida was exported to the adjacent estuary as particulate organic matter (POM). From 33 to 60 percent of the total POM in the estuary came from *Rhizophora* (red mangrove) material. The production of organic matter was greater in the summer (the wet season in Florida) than in any other season, although detrital levels in the swamp waters were greatest from November through February, which is the first four months of a typical seven month dry season. Thirty percent of the yearly detrital export occurred during November. Heald also found that as the debris decomposed, its protein content increased. The apparent cause of this enrichment, also noted in salt marsh studies, is the increase of bacterial and fungal populations.

Since those early studies, an abundance of studies have been undertaken on the outwelling from mangroves to adjacent estuaries. Almost all of the studies agree that there is export of particulate organic carbon from mangroves to the adjacent estuary. A summary of several studies on carbon export from mangrove wetlands suggested an average of about 200 g-C m^{-2} yr^{-1}, about double that exported from salt marshes (Twilley, 1998). In a comparison of leaf litter production and organic export for riverine, fringe, and basin mangrove systems (Fig. 9.8), riverine mangrove systems exported a majority of their organic litter (94 percent, or 470 g-C m^{-2} yr^{-1}), whereas basin mangroves exported much less (21 percent, or 64 g-C m^{-2} yr^{-1}), leaving the

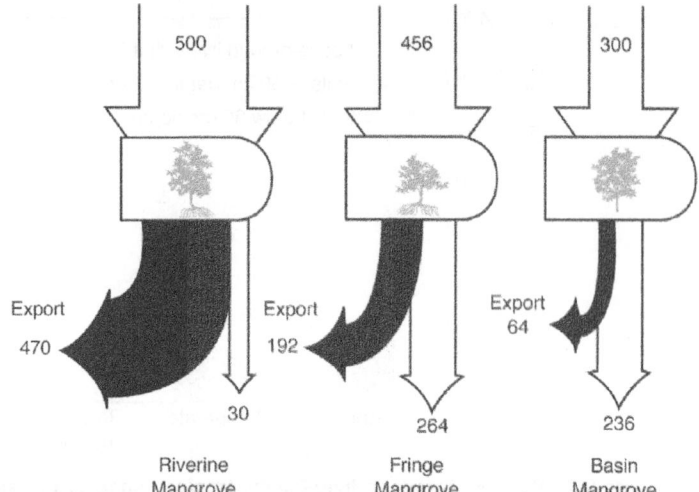

Figure 9.8 Organic carbon fluxes through mangrove swamps: inflows (litterfall), export to adjacent aquatic systems, and other losses (decomposition and peat production). Width of each pathway is proportional to flow (in g-C m^{-2} yr^{-1}). (After Twilley et al., 1986)

leaf litter to decompose or accumulate as peat. The proportion of litterfall production that is exported and the total amount of litter that is exported increase as the tidal influence increases.

Carbon Sequestration

A great deal of attention in the last decade has been paid to the sequestration or burial of carbon by coastal wetlands, even to the point where that carbon is now referred to as "blue carbon" (Mcleod et al., 2011). Carbon accumulation in mangrove soils is a significant part of this coastal wetland carbon accumulation. The most recent meta estimates for carbon sequestration by mangrove swamps range from 160 ± 40 g-C m^{-2} yr^{-1} (Breithaupt et al., 2012) to 226 ± 39 g-C m^{-2} yr^{-1} (Mcleod et al., 2011). These rates are 30 to 50 times the rate of carbon burial estimated for terrestrial uplands forests (Mcleod et al., 2011). These mangrove estimates, when multiplied by area of mangroves in the world described earlier, suggest that mangroves could be responsible for 26 to 34 Tg-C/yr, a small fraction of the total carbon sequestered by the world's wetlands as described in Chapter 17: "Wetlands and Climate Change." Breithaupt et al. (2014) estimated a slightly lower rate of organic carbon sequestration by mangroves of 123 ± 19 (std dev.) g-C m^{-2} yr^{-1} in the southern Florida Everglades. Because this is an inorganic carbon-rich region, it would not be surprising if the total carbon sequestration into the soil is significantly higher than this value.

Effects on the Estuary

The role of mangrove wetlands as both a habitat and a source of food for estuarine fisheries is one of the most often-cited functions of these ecosystems. The fact that

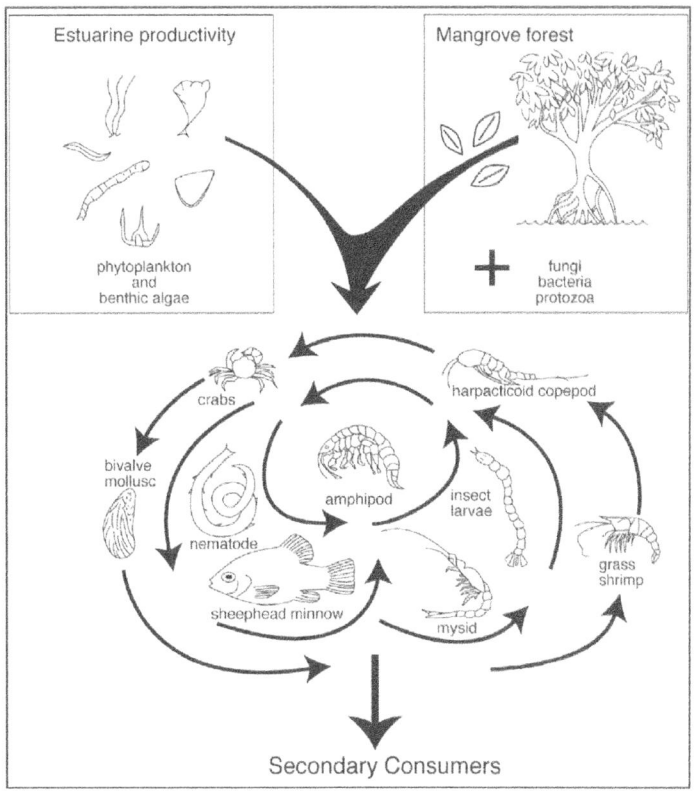

Figure 9.9 Detritus-based food web in south Florida estuary showing major contribution of mangrove detritus to fisheries and the estuarine food chain. (After W. E. Odum and Heald, 1972)

organic carbon is exported from mangrove wetlands does not guarantee that it enters estuarine food chains. Yet several independent studies have verified that mangrove wetlands are important nursery areas and sources of food for sport and commercial fisheries (Fig. 9.9 and 9.10).

Early studies by W. E. Odum (1970) and W. E. Odum and Heald (1972) established that detrital export is important to sport and commercial fisheries (Fig. 9.9). Through the examination of the stomach contents of more than 80 estuarine animals, Odum found that mangrove detritus, particularly from *Rhizophora*, is the primary food source in the estuary. Important consumers include the spiny lobster (*Panulirus argus*), pink shrimp (*Penaeus duorarum*), mullet (*Mugil cephalus*), tarpon (*Megalops atlanticus*), snook (*Centropomus undecimalis*), and mangrove snapper (*Lutjanus apodus*). The primary consumers also used the mangrove estuarine waters during their early life stages as protection from predators and as a source of food.

b.

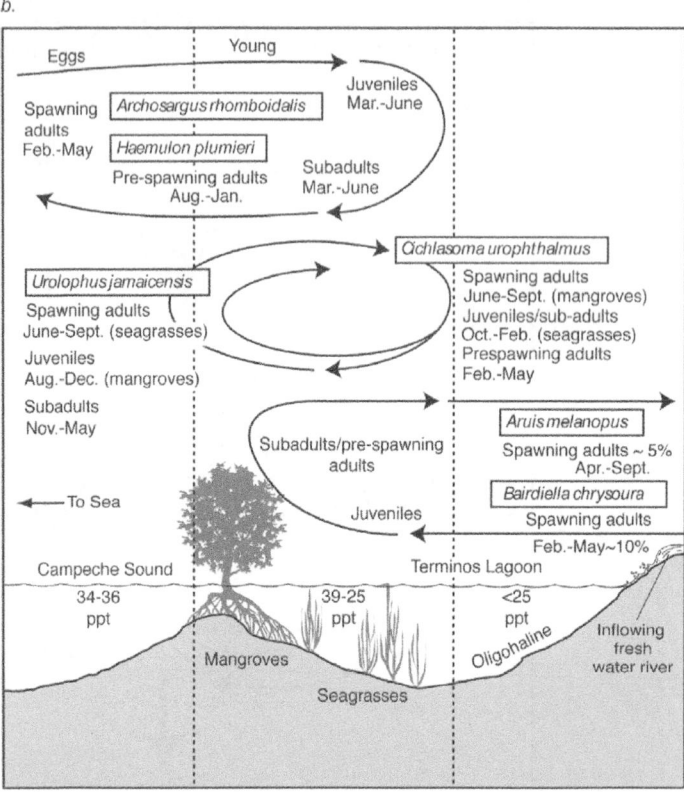

Figure 9.10 Life histories and habitat utilization of six fish species including marine–estuarine spawners, estuarine spawners, and freshwater spawners in a mangrove lagoon on the Gulf of Mexico, Mexico. (After Yáñez-Arancibia et al., 1988)

In a series of studies in Terminos Lagoon in Mexico (Yáñez-Arancibia et al., 1988, 1993), the use of mangrove forests by estuarine fish was clearly illustrated (Fig. 9.10). There are clear connections among seasonal pulsing of mangrove detrital production, adjacent planktonic and seagrass productivity, and fish movement and secondary productivity. It is reasonable to extrapolate from this and similar studies that the removal of mangrove wetlands would cause a significant decline in sport and commercial fisheries in adjacent open waters.

Organic carbon export from mangroves includes not only POM, which is most often measured, but also dissolved organic matter, which has not received as much attention as POM in export studies. Wafar et al. (1997) measured all of these fluxes as well as those of particulate and dissolved nitrogen and phosphorus in a *Rhizophora* swamp on the west coast of India. They made the following three conclusions:

1. Mangrove production is important only for the carbon budget of the adjacent estuary, not for the nitrogen or phosphorus budget.
2. The energy flux coming from the mangroves is more important for sustaining microbial food chains than for sustaining particulate food chains.

3. The influence of the mangrove forest on the adjacent estuary relative to phytoplankton production in the estuary is a function the size of the adjacent estuary relative to the size of the mangrove forests.

Recommended Readings

Lee, S. Y., J. H. Primavera, F. Dahdouh-Guebas, K. McKee, J. O. Bosire, S. Cannicci, K. Diele, F. Fromard, N. Koedam, C. Marchand, I. Mendelsohn, N. Mukherjee, and S. Record. 2014. Ecological role and services of tropical mangrove ecosystems: A reassessment. *Global Ecology and Biogeography* 23: 726–743.

Twilley, R. R., and J. W. Day. 2013. *Mangrove Wetlands* (pp. 165–202). In J. W. Day, B. C. Crump, W. M. Kemp, and A. Yáñez-Arancibia, eds., *Estuarine Ecology*. Hoboken, NJ: Wiley-Blackwell.

References

Allen, J. A. 1998. Mangroves as alien species: The case of Hawaii. *Global Ecology and Biogeography Letters* 7: 61–71.

Alongi, D. M. 2008. Mangrove forests: resilience, protection from tsunamis, and responses to global climate change. *Estuarine, Coastal and Shelf Science* 76: 1–13.

Bouillon, S., A. V. Borges, E. Castañeda-Moya, K. Diele, T. Dittmar, N. C. Duke, E. Kristensen, S. Y. Lee, C. Marchand, J. J. Middelburg, V. H. Rivera-Monroy, T. J. Smith III, and R. R. Twilley. 2008. Mangrove production and carbon sinks: A revision of global budget estimates. *Global Biogeochemistry Cycles* 22:GB2013.

Breithaupt, J. L., J. M. Smoak, T. J. Smith, C. J. Sanders, and A. Hoare 2012, Organic carbon burial rates in mangrove sediments: Strengthening the global budget. *Global Biogeochem. Cycles* 26: GB3011.

Breithaupt, J. L., J. M. Smoak, T. J. Smith, C. J. Sanders. 2014. Temporal variability of carbon and nutrient burial, sediment accretion, and mass accumulation over the past century in a carbonate platform mangrove forest of the Florida Everglades. *J. Geophysical Research: Biogeosciences* 119: 2032–2048.

Carter, M. R., L. A. Bums, T. R. Cavinder, K. R. Dugger, P. L. Fore, D. B. Hicks, H. L. Revells, and T. W. Schmidt. 1973. *Ecosystem Analysis of the Big Cypress Swamp and Estuaries*, U.S. EPA 904/9-74[sim]]02, Region IV, Atlanta.

Cavanaugh, K. C., J. R. Kellner, A. J. Forde, D. S. Gruner, J. D. Parker, W. Rodriguez, and I. C. Feller. 2014. Poleward expansion of mangroves is a threshold response to decreased frequency of extreme cold events. *Proc. Nat. Acad. Sci.* 111: 723–727.

Chapman, V. J. 1976. *Mangrove Vegetation*, J. Cramer, Vaduz, Germany. 447 pp.

Chapman, V. J. 1977. *Wet Coastal Ecosystems*. Elsevier, Amsterdam, 428 pp.

Cintrón, G., A. E. Lugo, D. J. Pool, and G. Morris. 1978. Mangroves of arid environments in Puerto Rico and adjacent islands. *Biotropica* 10: 110–121.

Cintrón, G., A. E. Lugo, and R. Martinez. 1985. Structural and functional properties of mangrove forests. In W. G. D'Arcy and M. D. Corma, eds., *The Botany and Natural History of Panama, IV Series: Monographs in Systematic Botany*, vol. 10. Missouri Botanical Garden, St. Louis, pp. 53–66.

Cole, T. G., K. C. Ewel, and N. N. Devoe. 1999. Structure of mangrove trees and forests in Micronesia. *Forest Ecology and Management* 117: 95–109.

Davis, J. H. 1940. The ecology and geologic role of mangroves in Florida. Publication 517. *Carnegie Institution of Washington*, pp. 303–412.

Day, J. W., Jr., C. Coronado-Molina, F. R. Vera-Herrera, R. Twilley, V. H. Rivera-Monroy, H. Alvarez-Guillen, R. Day, and W. Conner. 1996. A 7-year record of above-ground net primary production in an southeastern Mexican mangrove forest. *Aquatic Botany* 55: 39–60.

Dent, D. L. 1986. *Acid Sulphate Soils: A Baseline for Research and Development.* ILRI Publication 39, Wageninggen, Netherlands.

Dent, D. L. 1992. Reclamation of acid sulphate soils. In R. Lal and B. A. Stewart, eds. *Soil Restoration, Advances in Soil Science*, vol. 17. Springer-Verlag, New York, pp. 79–122.

Donato, D. C., J. B. Kauffman, D. Murdiyarso, S. Kurnianto, M. Stidham, and M. Kanninen. 2011. Mangroves among the most carbon-rich forests in the tropics. *Nature Geoscience Letters* 4: 293–297.

Egler, F. E. 1952. Southeast saline Everglades vegetation, Florida, and its management. *Vegetatio Acta Geobotica* 3: 213–265.

Ewel, K. C., R. R. Twilley, and J. E. Ong. 1998. Different kinds of mangrove forests provide different goods and services. *Global Ecology and Biogeography Letters* 7: 83–94.

Ellison, A. M., E. J. Farnsworth, and R. E. Merkt. 1999. Origins of mangrove ecosystems and the mangrove biodiversity anomaly. *Global Ecology and Biogeography* 8: 95-115.

Feller, I. C., C. E. Lovelock, and K. L. McKee. 2007. Nutrient addition differentially affects ecological processes of *Avicennia germinans* in nitrogen versus phosphorus limited mangrove ecosystems. *Ecosystems* 10: 347–359.

Giri, C., E. Ochieng, L.L. Tieszen, Z. Zhu, m A. Singh, T. Loveland, J. Masek, and N. Duke. 2011 Status and distribution of mangrove forests of the world using earth observation satellite data. Global Ecology and Biogeography 23: 154–159.

Golley, F. B., H. T. Odum, and R. F. Wilson. 1962. The structure and metabolism of a Puerto Rican red mangrove forest in May. *Ecology* 43: 9–19.

Heald E. J. 1971. *The Production of Organic Detritus in a South Florida Estuary.* University of Miami Sea Grant Technical Bulletin No. 6, Coral Gables, FL, 110 pp.

Knight, J. M., P. E. R. Dale, R. J. K. Dunn, G. J. Broadbent, and C. J. Lemckert. 2008. Patterns of tidal flooding within a mangrove forest: Coombabah Lake, Southeast Queensland. *Estuarine, Coastal and Shelf Science* 76:580–593.

Krauss, K. W., K. L. McKee, C. E. Lovelock, D. R. Cahoon, N. Saintilan, R. Reef, and L. Chen. 2014. How mangrove forests adjust to rising sea level. *New Phytologist* 202: 19-34.

Lee, S. Y., J. H. Primavera, F. Dahdouh-Guebas, K. McKee, J. O. Bosire, S. Cannicci, K. Diele, F. Fromard, N. Koedam, C. Marchand, I. Mendelssohn, N. Mukherjee, and S. Record. 2014. Ecological role and services of tropical mangrove ecosystems: A reassessment. *Global Ecology and Biogeography* 23: 726–743.

Lovelock, C. E. 2008. Soil respiration and belowground carbon allocation in mangrove forests. *Ecosystems* 11: 342–354.

Lugo, A. E., G. Evink, M. M. Brinson, A. Broce, and J. C. Snedaker. 1975. Diurnal rates of photosynthesis, respiration, and transpiration in mangrove forests in South Florida. In F. B. Golley and E. Medina, eds. *Tropical Ecological Systems—Trends in Terrestrial and Aquatic Research*. Springer-Verlag, New York, pp. 335–350.

Lugo, A. E. 1980. Mangrove ecosystems: successional or steady state? *Biotropica* (supplement) 12: 65–72.

Lugo, A. E. 1988. The mangroves of Puerto Rico are in trouble. *Acta Científica* 2: 124.

Lugo, A. E., and C. Patterson-Zucca. 1977. The impact of low temperature stress on mangrove structure and growth. *Tropical Ecology* 18: 149–161.

Mcleod, E., G. L. Chmura, S. Bouillon, R. Salm, M. Björk, C. M. Duarte, C. E. Lovelock, W. H. Schlesinger, and B. R Silliman. 2011. A blueprint for blue carbon: toward an improved understanding of the role of vegetated coastal habitats in sequestering CO_2. *Frontiers of Ecology and the Environment* 9: 552–560.

Odum, E. P. 1969. The strategy of ecosystem development. *Science* 164: 262–270.

Odum, W. E. 1970. *Pathways of Energy Flow in a South Florida Estuary*. Ph.D. dissertation, University of Miami, Coral Gables, Florida, 62 pp.

Odum, W. E., and E. J. Heald. 1972. Trophic analyses of an estuarine mangrove community. *Bulletin of Marine Science* 22: 671–738.

Odum, W. E., C. C. McIvor, and T. J. Smith. 1982. *The Ecology of the Mangroves of South Florida: A Community Profile*. U.S. Fish & Wildlife Service, Office of Biological Services, Technical Report FWS/OBS 81-24, Washington, DC.

Patterson, S. G. 1986. *Mangrove Community Boundary Interpretation and Detection of Areal Changes in Marco Island, Florida: Application of Digital Image Processing and Remote Sensing Techniques*. Biological Services Report 86 (10), U.S. Fish and Wildlife Service, Washington, DC.

Rabinowitz, D. 1978. Dispersal properties of mangrove propagules. *Biotropica* 10: 47–57.

Robertson, A. I., and P. A. Daniel. 1989. The influence of crabs on litter processing in high intertidal mangrove forests in tropical Australia. *Oceologia* 78: 191–198.

Schaeffer-Novelli, Y., G. Cintron-Molero, R. R. Adaime, and T. M. de Camargo. 1990. Variability of mangrove ecosystems along the Brazilian coast. *Estuaries* 13: 204–218.

Smith, T. J., H-T. Chan, C. C. McIvor, and M. B. Robblee. 1989. Comparisons of seed predation in tropical, tidal forests on three continents. *Ecology* 70: 146–151.

Smith, T. J., K. G. Boto, S. D. Frusher, and R. L. Giddins. 1991. Keystone species and mangrove forest dynamics: The influence of burrowing by crabs on soil nutrient status and forest productivity. *Estuarine, Coastal, and Shelf Science* 33: 419–432.

Smith, T. J., M. B. Robblee, H. R. Wanless, and T. W. Doyle. 1994. Mangroves, hurricanes, and lightning strikes. *BioScience* 44: 256–262.

Stewart, G.R., and M. Popp. 1987. The ecophysiology of mangroves. In R. M. M. Crawford, ed., *Plant Life in Aquatic and Amphibious Habitats*, Spec Publ. British Ecological Society, vol. 5, pp. 333–345.

Thibodeau, F. R., and N. H. Nickerson. 1986. Differential oxidation of mangrove substrate by *Avicennia germinas* and *Rhizophora mangle*. *American Journal of Botany* 73: 512–516.

Thom, B. G. 1982. Mangrove ecology: A geomorphological perspective. In Clough, B.F., ed., *Mangrove Ecosystems in Australia*. Australian National University Press, Canberra, pp. 3–17.

Twilley, R. R. 1982. Litter dynamics and organic carbon exchange in black mangrove (*Avicennia germinans*) basin forests in a Southwest Florida estuary. Ph.D. dissertation, University of Florida, Gainesville.

Twilley, R. R. 1985. The exchange of organic carbon in basin mangrove forests in a southwest Florida estuary. *Estuarine, Coastal and Shelf Science* 20: 543–557.

Twilley, R. R. 1998. Mangrove wetlands. In M. G. Messina and W. H. Conner, eds., *Southern Forest Wetlands: Ecology and Management*. Lewis Publishers, Boca Raton, FL, pp. 445–473.

Twilley, R. R., A. E. Lugo, and C. Patterson-Zucca. 1986. Litter production and turnover in basin mangrove forests in southwest Florida. *Ecology* 67: 670–683.

Twilley, R. R., R. H. Chen, and T. Hargis. 1992. Carbon sinks in mangroves and their implications to carbon budget of tropical coastal ecosystems. *Water, Air, and Soil Pollution* 64: 265–288.

Twilley, R. R., and J. W. Day, Jr. 2013. Mangrove wetlands. In. J. W. Day, Jr., B. C. Crump, W. M. Kemp, and A. Yanez-Arancibia, eds., *Estuarine Ecology*, 2nd ed. Wiley-Blackwell, Hoboken, NJ, pp. 165–202.

Wafer, S., A. G. Untawale, and M. Wafar. 1997. Litter fall and energy flux in a mangrove ecosystem. *Estuarine, Coastal and Shelf Science* 44: 111–124.

Ward, G. A., T. J. Smith III, K. R. T. Whalen, and T. W. Doyle. 2006. Regional process in mangrove ecosystems: spatial scaling relationships, biomass, and turnmover rates following catastrophic disturbance. *Hydrobiologia* 569: 517–527.

Wharton, C. H., H. T. Odum, K. Ewel, M. Duever, A. Lugo, R. Boyt, J. Bartholomew, E. DeBellevue, S. Brown, M. Brown, and L. Duever. 1976. *Forested Wetlands of Florida—Their Management and Use*. Center for Wetlands, University of Florida, Gainesville, 421 pp.

Yánez-Arancibia, A., A. L. Lara-Dominguez, J. L. Rojan-Galaviz, P. Sánchez-Gil, J. W. Day, and C. J. Madden. 1988. Seasonal biomass and diversity of estuarine fishes coupled with tropical habitat heterogeneity (southern Gulf of Mexico). *Journal of Fish Biology* 33 (Suppl. A): 191–200.

Yánez-Arancibia, A., A. L. Lara-Dominguez, and J. W. Day. 1993. Interactions between mangrove and seagrass habitats mediated by estuarine nekton assemblages: Coupling of primary and secondary production. *Hydrobiologia* 264: 1–12.

Freshwater marsh in Wisconsin

Chapter 10

Freshwater Marshes

Approximately 90 to 95 percent of the world's wetlands are inland, or nontidal. Inland freshwater marshes are perhaps the most diverse of the wetland types discussed in this book and include the pothole marshes of the north-central United States and south-central Canada, the Florida Everglades, many wetland expanses in the Pantanal in Brazil, and the floodplains of the Okavango Delta in Botswana. Vegetation in freshwater marshes is characterized by tall graminoids such as Typha *and* Phragmites, *the grasses* Panicum *and* Cladium, *the sedges* Scirpus, Schoenoplectus, Cyperus, *and* Carex, *broad-leaved monocots such as* Sagittaria *spp., and floating aquatic plants such as* Nymphaea *and* Nelumbo. *Some inland marshes, such as the prairie pothole marshes, follow a cycle that includes drought, reflooding, and herbivory. In contrast to bogs, mineralsoil–based inland marshes have high-pH substrates, high available soil calcium, medium or high loading rates for nutrients, often high productivity, and high soil microbial activity that leads to rapid decomposition, recycling, and nitrogen fixation. Most of the primary productivity is routed through detrital pathways, but herbivory can be seasonally important, particularly by muskrats and geese. Inland marshes are valuable as wildlife islands in the middle of agricultural landscapes and have been tested extensively as sites for assimilating nutrients.*

Most of the wetlands of the world are not located along the coastlines but are found in interior regions. (These wetlands are called "nontidal" in coastal regions to distinguish them from coastal wetlands.) We estimate that there are about 5.5 million km^2 of inland wetlands in the world (Table 10.1); in other words, they make up about 95 percent of all the world's wetlands. About 415,000 km^2 (about 95 percent) of the total wetlands in the conterminous United States are inland. This estimate includes

Table 10.1 Estimated area of inland wetlands in the world and North America (x 1,000 ha)

	Peatlands	Freshwater Marshes	Freshwater Swamps	Total
World	350,000[a]	95,000[b]	109,000[c]	554,000
North America				
Conterminous U.S.[d]	3,700	9,600	28,200	41,500
Alaska[f]	51,800	17,000	[e]	68,800
Canada[g]	110,000	15,900	[e]	125,900

[a] Bridgham et al. (2001).
[b] Average of several independent estimates.
[c] Matthews and Fung (1987).
[d] Dahl (2006); freshwater marshes includes freshwater emergent and freshwater nonvegetated wetlands.
[e] All palustrine forested wetlands in Alaska and Canada are assumed to be peatlands.
[f] Hall et al. (1994).
[g] Zoltai (1988).

about 26,000 km² of nonvegetated freshwater ponds. Including Alaska, there are 1.1 million km² of inland wetlands in the United States.

It is difficult to put these inland wetlands into simple categories. Our simplified scheme divides them into three groups: freshwater marshes (this chapter), freshwater swamps (Chapter 11: "Freshwater Swamps and Riparian Ecosystems"), and boreal peatlands (Chapter 12: "Peatlands"). These divisions roughly parallel the divisions that persist in both the scientific literature and the specializations of wetland scientists.

Terminology for wetlands, especially inland freshwater wetlands, can be confusing and contradictory. In Europe, for example, the term *reed swamp* is often used to describe one type of freshwater marsh dominated by *Phragmites* spp., whereas in the United States, the word *swamp* usually refers to a forested wetland. We consider *Phragmites* reed swamps to be marshes or marshlike, and they are covered in this chapter. Although the use of classifying terms connotes clear boundaries between different wetland types, in reality they form a continuum. The extremes of freshwater marshes are clearly different, but at the boundaries between two wetland types (e.g., marsh and bog), the distinction is not always clear. *Marshes* (and reed swamps) have mineral soils rather than peat soils. American terminology has developed without much regard to whether the system is peat forming. The fact is that most freshwater marshes and swamps, regardless of where they are located and regardless of their geological origins, accumulate some peat.

Freshwater marshes includes a diverse group of wetlands characterized by (1) emergent soft-stemmed aquatic plants, such as cattails, arrowheads, pickerelweed, reeds, and several other species of grasses and sedges; (2) a shallow-water regime; and (3) generally shallow to nonexistent peat deposits. There are few accurate measures of how many freshwater marshes there are in the world for several reasons. First, they are often ephemeral or convert to other types of wetlands, such as unvegetated flats or forested wetlands, over a relatively short period of time. Second, they can be confused with and miscounted with peatlands, particularly fens. Third, freshwater marshes have such a wide number of possible dominant vegetation covers and water depths (from saturated soils to 1 m of water depth) that their classification and inventory is very difficult. We estimate that there are about 950,000 km² million ha of freshwater marshes in the world (Table 10.1), less than 20 percent of the total amount of wetlands in the world. The Okavango Delta in Botswana, the Danube and Volga Deltas in Eastern Europe, the Mesopotamian Marshlands in Iraq, the Everglades in Florida, the Prairie Pothole region of the United States and Canada, and the Pantanal in South America are all examples of regions with extensive expanses of freshwater marshes. Freshwater marshes are estimated to cover about 96,000 km² in the coterminous United States (Table 10.1).

Hydrology

As with any other wetland, the flooding regime, or hydroperiod, of freshwater marshes determines their ecological character. The critical factors that determine the character of these wetlands are the presence of excess water and sources of water other than direct precipitation. The hydroperiods of several freshwater marsh systems were illustrated in Chapter 4: "Wetland Hydrology". Along seacoasts, water levels tend to be stable over the long term because of the influence of the ocean. Water levels in inland marshes, in contrast, are much more controlled by the balance between precipitation and evapotranspiration, especially for marshes in small watersheds that are affected by restricted throughflow. Water levels of marshes, such as those found along the Laurentian Great Lakes, are generally stable but are influenced by the year-to-year variability of lake levels and by whether the wetland is diked or open to the lake. Many marshes, such as wet meadows, sedge meadows, vernal pools, and even prairie potholes, dry down seasonally, but the plant species found there reflect the hydric conditions that exist during most of the year. The seasonality of these marshes is due to the fact that they are fed primarily by runoff and precipitation. Some marshes intercept groundwater supplies. Their water levels, therefore, reflect the local water table, and the hydroperiod is less erratic and seasonal. These types of marshes, such as those found in the prairie pothole region of North America, can be either recharge or discharge wetlands. Other marshes collect surface water and entrained nutrients from watersheds that are large enough to maintain hydric conditions most of the time. For example, overflowing lakes and rivers supply water and nutrients to adjacent riparian or littoral marshes. Because river and stream discharge, lake levels, and precipitation are often notoriously variable due to weather shifts from year to year, the water regime of most inland marshes also varies in a way that is predictable only in a statistical sense.

Even in the same region, water levels can respond differently to shifts in the balance between precipitation and evapotranspiration yearly (Fig. 10.1). Terms such as *ephemeral, temporary, seasonally semipermanent,* and *permanent* can be used to describe freshwater marshes. In addition, marshes can move through several of these classes over the span of a few years. Thus, a marsh that would ordinarily be considered permanent might be in a "drawdown" phase that gives the appearance of an ephemeral marsh.

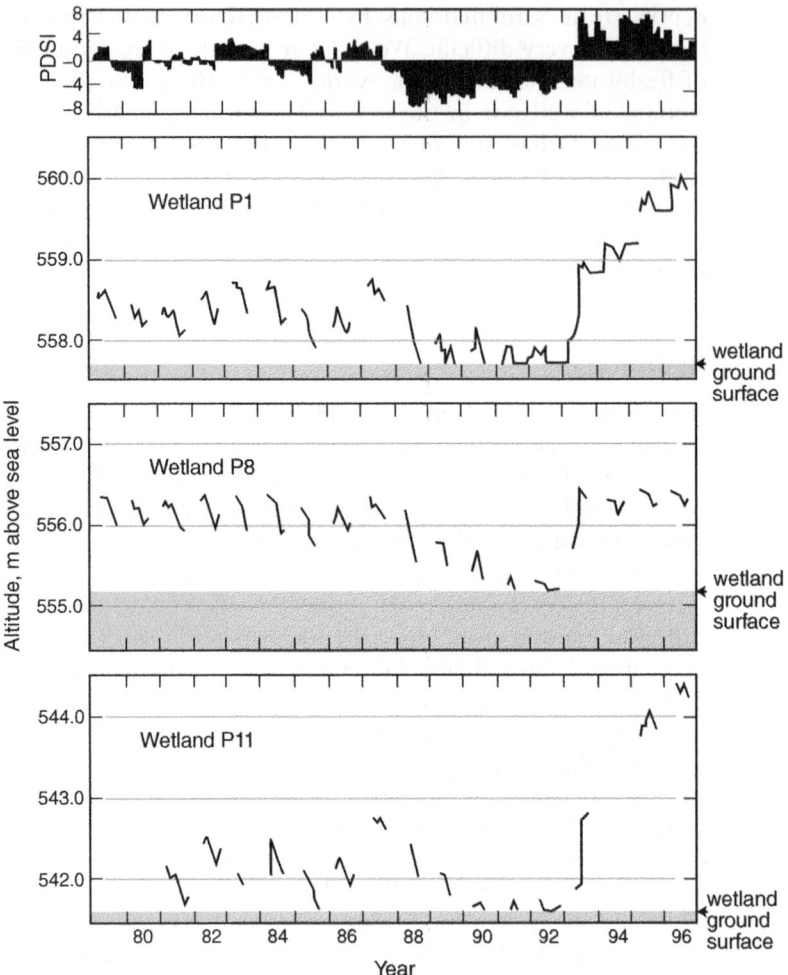

Figure 10.1 Water-level patterns in three wetlands in the prairie pothole region of North America, illustrating the uneven effect that climate has on supposedly similar wetlands in the same region, probably due to nonuniform groundwater effects on the wetlands. PDSI is the Palmer Drought Severity Index and is a relative measure of climatic "wetness." Its value decreases during drought conditions. (After LaBaugh et al., 1996)

Biogeochemistry

The water and soil chemistry of freshwater marshes is dominated by a combination of mineral rather peat soils, overlain with autochthonous inputs of organic matter from the productivity of the vegetation. Given these conditions, there is still a wide range of chemical possibilities for the water and soil in freshwater marshes (Fig. 10.2). Conductivity as a measure of general salinity can range from less than

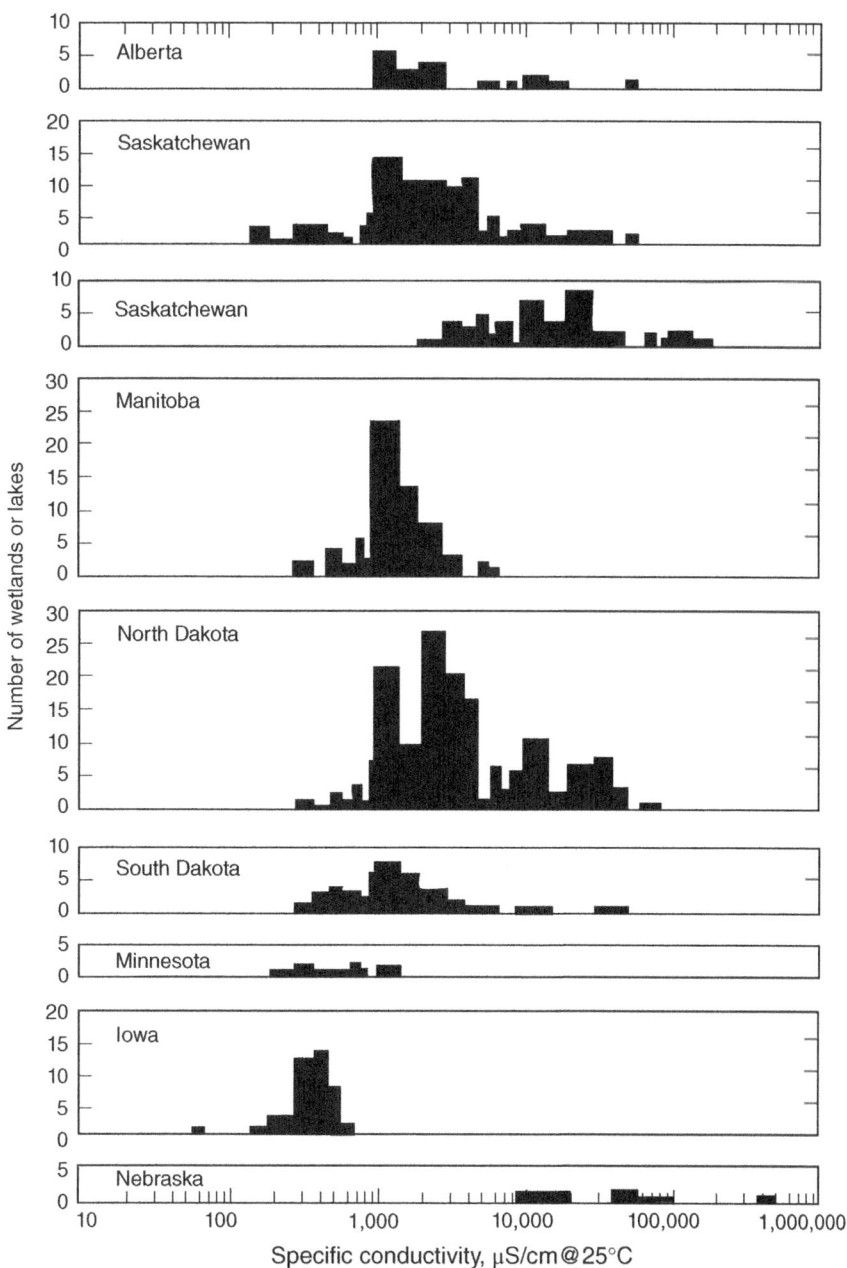

Figure 10.2 Frequency distribution of salinity, as measured by specific conductance, for prairie lakes and wetlands in the United States and Canada as measured by several investigators. Concentrations shown for Nebraska are typical of inland salt marshes. (After LaBaugh, 1989)

100 microSiemens per cm ($\mu S\ cm^{-1}$) in soft-water freshwater marshes dominated by rainfall to over 300,000 $\mu S\ cm^{-1}$ in "inland salt marshes" dominated by saline seeps in semiarid climates. Differences are related to the magnitude of dissolved salts, nutrients, and other chemical inputs and to the relative importance of groundwater and surface water inflow. Inland marshes are generally minerotrophic in contrast to bogs. That is, the inflowing water has higher amounts of dissolved materials, including nutrients, resulting from the presence of dissolved cations in streams, rivers, and groundwater compared to bogs that are fed simply by rainfall. The organic substrate of freshwater marshes, while shallow compared to bogs and fens, is saturated with bases, and as a result the pH is close to neutral. Because nutrients are usually plentiful, productivity is higher in freshwater marshes than it is in bogs, bacteria are active in nitrogen fixation and litter decomposition, and turnover rates are high. The accumulation of organic matter that does occur results from high production rates, not from the inhibition of decomposition by low pH (as occurs in bogs).

Wetland hydrology has a tremendous influence on biogeochemistry and, coupled with a diverse geomorphology, can provide considerable spatial heterogeneity within the wetland. On the St. Louis River system in Minnesota and Wisconsin, Johnston et al. (2001) found higher variability in available nutrients within wetlands than between two wetlands with very different soil conditions (silty versus clayey). The significant intra-wetland variability was attributed to the variety of geomorphic features (levees, backwater zones) that provided very different hydrologic conditions within both wetlands.

Nutrient concentrations reported for sediments in inland freshwater marshes vary widely, depending on the substrate, parent material, open or closed nature of the basin, connection with groundwater, and even nutrient uptake by plants. Ion concentrations of freshwater marshes are high, and water is generally in a pH range of 6 to 9. Organic matter can vary from a very high content (75 percent), as can be found in freshwater marshes in coastal Louisiana, to a low content (10–30 percent) in marshes fed by inorganic sediments from agricultural watersheds or open to organic export. Concentrations of total (as distinguished from available) nutrients are reflections of the kinds of sediments in the marsh. Mineral sediments are often associated with high phosphorus content, for example, whereas total nitrogen is closely correlated to organic content. Dissolved inorganic nitrogen and phosphorus—the elements that most often limit plant growth—often vary seasonally from very low concentrations in the summer, when plants take them up as rapidly as they become available, to high concentrations in the winter, when plants are dormant but mineralization continues in the soil.

In many parts of the world where arid climates persist, inland marshes can be saline rather than freshwater. These marshes, then, have characteristics of both coastal salt marshes (because of the salinity) and inland marshes (because they are not tidal). Good examples of these kinds of marshes are the fringe marshes around the Great Salt Lake in Utah, the Salton Sea in California, and the Caspian Sea in Eastern Europe. There are also inland "salt marshes," some as large as 15 to 20 ha, still

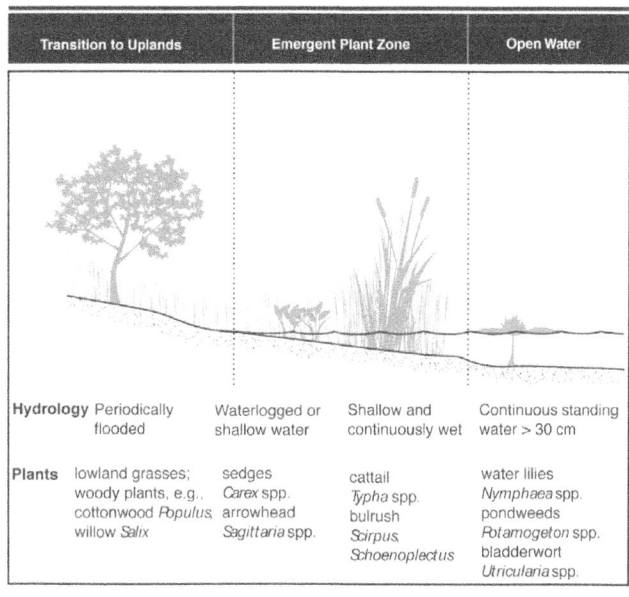

(a)

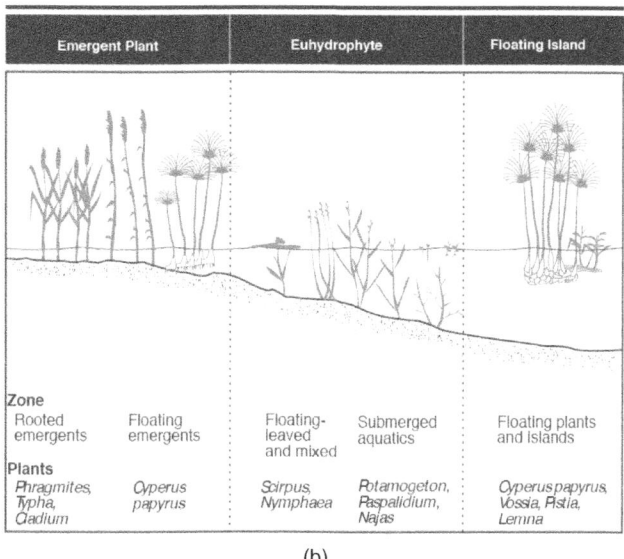

(b)

Figure 10.3 Cross sections of vegetation through freshwater marshes, indicating plant zones according to water depth and typical plants found in each zone for (a) temperate-zone midwestern North America and (b) sub-Saharan Africa. Note floating marsh islands in African wetlands. ((b) After Denny, 1993)

found northwest of Lincoln, Nebraska, along Salt Creek and its tributaries. These marshes are the result of saline seeps from deep groundwater and the consistently high evapotranspiration/precipitation ratio of the region.

Vegetation

The vegetation of fresh inland marshes has been detailed in many studies. The dominant species vary from place to place, but the number of genera common to all locations in the temperate zone is quite remarkable. Common species include the graminoids *Phragmites australis* (= *P. communis*; reed grass), *Typha* spp. (cattail), *Sparganium eurycarpum* (bur reed), *Zizania aquatica* (= *Z. palustris*; wild rice), *Panicum hemitomon*, *Cladium jamaicense*; sedges *Carex* spp., *Schoenoplectus tabernaemontani* (= *Scripus validus*; bulrush), *Scirpus fluviatilis* (river bulrush), and *Eleocharis* spp. (spike rush). In addition, broad-leaved monocotyledons such as *Pontederia cordata* (pickerelweed) and *Sagittaria* spp. (arrowhead) are frequently found in freshwater marshes. Herbaceous dicotyledons are represented by a number of species, typical examples of which are *Ambrosia* spp. (ragweed) and *Polygonum* spp. (smartweed). Frequently represented also are such ferns as *Osmunda regalis* (royal fern) and *Thelypteris palustris* (marsh fern), and the horsetail, *Equisetum* spp. One of the most productive species in the world is the tropical sedge *Cyperus papyrus*, which flourishes in marshes and on floating mats in southern and eastern Africa.

Marsh Vegetation Zonation

These typical plant species do not occur randomly mixed together in marshes. Each has its preferred habitat. Different species often occur in rough zones on slight gradients, especially flooding gradients. Figure 10.3a illustrates the typical distribution of species along an elevation gradient in a midwestern North American freshwater marsh. Sedges (e.g., *Carex* spp., *Scirpus* spp.), rushes (*Juncus* spp.), and arrowheads (*Sagittaria* spp.) typically occupy the shallowly flooded edge of a pothole. Two species of cattail (*Typha latifolia* and *T. angustifolia*) are common. The narrow-leaved species (*T. angustifolia*) is more flood tolerant than the broad-leaved cattail (*T. latifolia*) and may grow in water up to 1 m deep. The deepest zone of emergent plants is typically vegetated with hardstem bulrush (*Scirpus acutus*) and softstem bulrush (*Schoenoplectus tabernaemontani*). Beyond these emergents, floating-leaved and submersed vegetation will grow, the latter to depths dictated by light penetration. Typical floating-leaved aquatic hydrophytes include rhizomatous plants, such as water lilies (*Nymphaea tuberosa* or *N. odorata*), water lotus (*Nelumbo lutea*), and spatterdock (*Nuphar advena*), and stoloniferous plants, such as water shield (*Brasenia schreberi*) and smartweed (*Polygonum* spp.). Submersed hydrophytes include coontail (*Ceratophyllum demersum*), water millfoil (*Myriophyllum* spp.), pondweed (*Potamogeton* spp.), wild celery (*Vallisneria americana*), naiad (*Najas* spp.), bladderwort (*Utricularia* spp.), and waterweed (*Elodea canadensis*).

A unique structural feature of prairie pothole marshes is the 5- to 20-year cycle of dry marsh, regenerating marsh, degenerating marsh, and lake that is related to periodic droughts. During drought years, standing water disappears. Buried seeds in the exposed mud flats germinate to grow a cover of annuals (*Bidens, Polygonum, Cyperus, Rumex*) and perennials (*Typha, Scirpus, Sparganium, Sagittaria*). When rainfall returns to normal, the mud flats are inundated. Annuals disappear, leaving only the perennial emergent species. Submersed species (*Potamogeton, Najas, Ceratophyllum, Myriophyllum, Chara*) also reappear. For the next year or more, during the regenerating stage, the emergent population increases in vigor and density. After a few years, however, these populations begin to decline. The reasons are poorly understood, but often muskrat populations explode in response to the vigorous vegetation growth. Their nest and trail building can decimate a marsh. Whatever the reason, in the final stage of the cycle, there is little emergent marsh; most of the area reverts to an open shallow lake or pond, setting the stage for the next drought cycle. Wildlife use of these wetlands follows the same cycle. The most intense use occurs when there is good interspersion of small ponds with submersed vegetation and emergent marshes with stands diverse in height, density, and potential food.

Figure 10.3b shows the plant zonation of a freshwater marsh/littoral zone in sub-Saharan Africa. The shallow flooded emergent zone is dominated by *Typha, Phragmites,* and *Cyperus papyrus. Typha* taxonomy in Africa has been somewhat confused, but it now appears that there are two distinct species—*T. domingensis* Pers. *sensu lato* in tropical and warm-temperate climates and *T. capensis* Rohrb. in more temperate climates of northern and southern Africa. *Cyperus papyrus* grows in the emergent zone but also develops floating islands when it breaks free from the shoreline. The euhydrophyte (true water plant) zone shown in Figure 10.3b refers to the zone with rooted, floating-leaved, and submerged macrophytes (Denny, 1985). In Africa, examples of other plants in this zone are *Chara, Fontinalis, Nymphaea, Ceratophyllum, Vallisneria, Potamogeton,* and *Paspalidium.* Thus, despite these differences and vastly different climates, freshwater marshes around the world share some common species and many common genera and are functionally much the same.

Inland Salt Marshes

Where evapotranspiration exceeds precipitation and/or saline groundwater seeps occur, inland salt marshes are often found. For example, the Nebraska salt marshes located in eastern Nebraska near Lincoln (Fig. 10.4) support many plant genera familiar to coastal salt marsh ecologists. The most saline parts of these marshes are dominated by salt-tolerant macrophytes such as saltwort (*Salicornia rubra*), sea blight (*Suaeda depressa*), and inland saltgrass (*Distichlis spicata*), whereas the open ponds and their fringes are dominated by plants such as sago pondweed (*Potamogeton pectinatus*), wigeon grass (*Ruppia maritima*), prairie bulrush (*Scirpus maritimus* var. *paludosus*), and even cattails (*Typha angustifolia* and *T. latifolia*). In California, the major species in brackish marshes include pickleweed (*Salicornia virginica*) and alkali bulrush (*Scirpus robustus*).

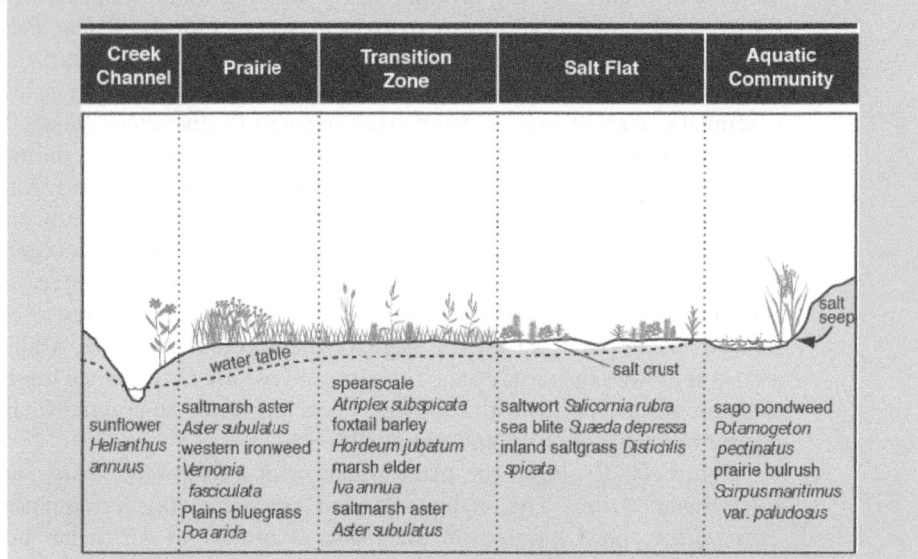

Figure 10.4 Cross section through an inland salt marsh in Nebraska, indicating plant zones and typical plants found in each zone. (After Farrar and Gersib, 1991)

Vegetation Seed Banks and Diversity

Seed banks and fluctuating water levels interact in complicated ways to produce vegetation communities in freshwater marshes (see Chapter 7: "Wetland Vegetation and Succession"). As a general rule, seed germination is maximized under shallow water or damp soil conditions, after which many perennials can reproduce vegetatively into deeper water. For example, fluctuating water levels along the Laurentian Great Lakes allowed greater diversity of plant types and species in the coastal marshes, and these marshes sometimes have a density of buried seeds an order of magnitude greater than that of inland prairie marshes (Keddy, 2010).

An Experiment in Hydroperiods, Seed Banks, and Marsh Vegetation Diversity

The importance of the timing of flooding and drying was illustrated in an experiment in Ohio involving uniform seed banks subjected to several hydroperiods. Although plant density and above-ground biomass were not affected by the different hydroperiods, species composition, diversity, and richness were affected (Fig. 10.5). Highest richness and diversity occurred in continuously moist soils. Flooding followed by a drawdown to moist soils, as is a typical hydroperiod in the midwestern United States, encouraged obligate wetland species, whereas moist soil conditions followed by flooding encouraged the growth of fewer wetland species and more annual species.

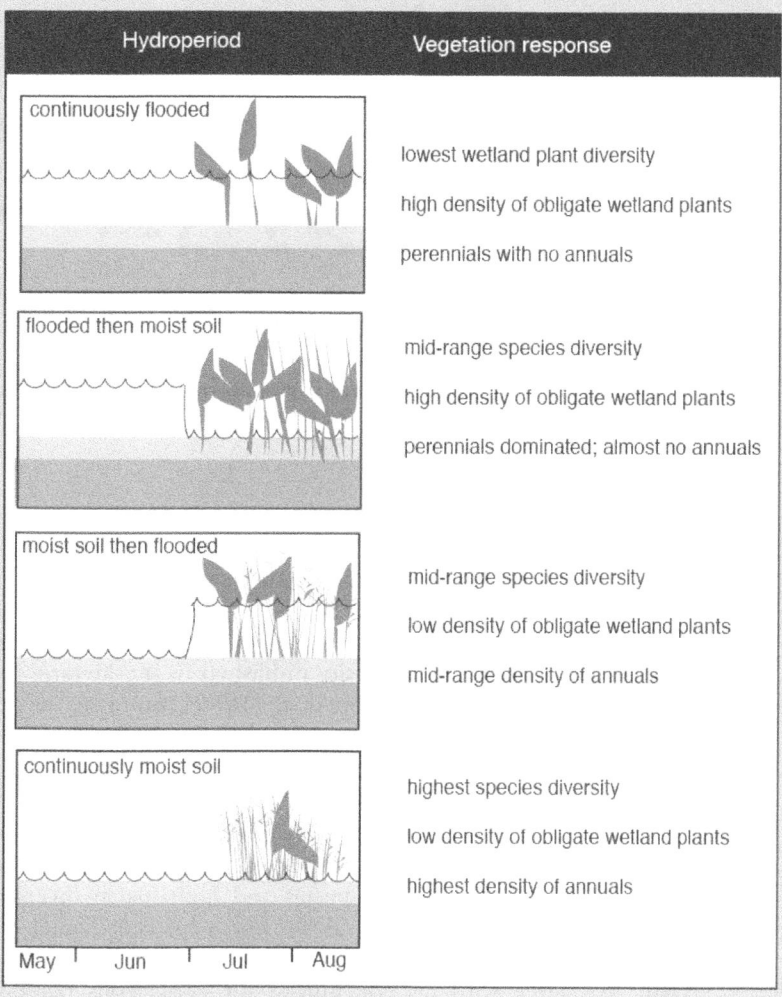

Figure 10.5 **Experimental results of four different growing season hydroperiods on a common seed bank of annual and perennial freshwater marsh plants in central Ohio. Continuously deep water favored low density of obligate wetland perennials; flooded then moist, typical of natural hydroperiod for the midwestern United States, favored perennial-dominated and diverse wetland plants; moist then flooded, typical of some managed marshes around the Laurentian Great Lakes, favored fewer obligate wetland plants and higher density of annuals; continuously flooded soil had highest diversity but fewest obligate wetland plants and highest density of annuals. (From Johnson, 1998)**

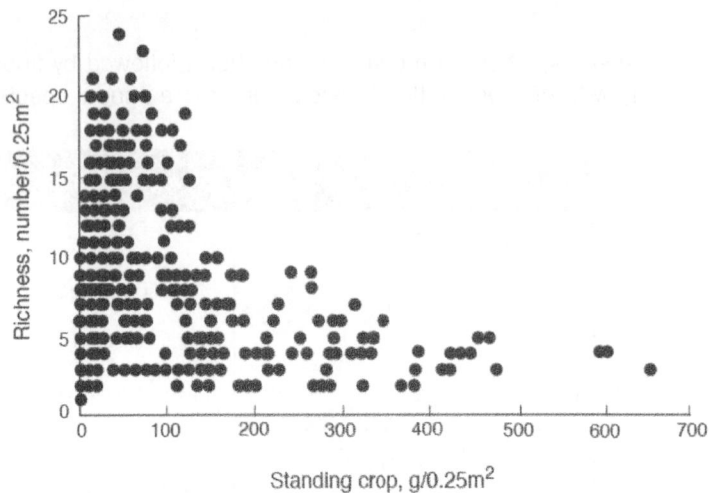

Figure 10.6 Species richness versus vegetation biomass in 0.25-m² quadrants from three wetland areas in Ontario, Quebec, and Nova Scotia. (After Moore et al., 1989)

The particular species found in freshwater wetlands are also determined by many other environmental factors. Nutrient availability determines, to a large degree, whether a wetland site will support mosses or angiosperms (i.e., whether it is a bog or a marsh) and what the species diversity will be. For freshwater marshes, it is not obvious that highly fertile wetlands are highly diverse. In fact, most studies of freshwater marsh plant diversity published in the literature suggest the opposite conclusion. As an example, Moore et al. (1989) contrasted several fertile and infertile sites (as measured by the plant standing crop) in eastern Ontario and found the greatest species richness in marshes that had peak biomass between 60 and 400 g/m² and much less plant richness at higher plant standing crop (>600 g/m²) (Fig. 10.6). They also found rare species only at the infertile sites, suggesting that the conservation of infertile wetlands should be part of overall wetland management strategies. Keddy (2010) pointed out that rare species and high marsh diversity are now routinely found in "peripheral" habitats, as shown in his centrifugal model for herbaceous plants as described in Chapter 7 and shown for a *Typha* core habitat in Figure 7.10b. We agree with his conclusions for herbaceous plant wetlands (freshwater marshes) that:

- Peripheral habitats contain more biological diversity and higher number of rare species.
- The core habitat is dominated by a few species (e.g., *Typha*, *Phragmites*, *Scirpus*, *Papyrus*).
- Any factor that increases fertility or decreases disturbance will force peripheral habitats into the usually undiverse core habitat.

Because many inland marshes are potholes that collect water that leaves only by evaporation, salts may become concentrated during periods of low precipitation, adversely affecting the growth of salt-intolerant species. In a review of 90 emergent wetlands along the U.S. Great Lakes, Johnston et al. (2007) revealed that plant forms were often indicators of wetland soil types. They found that submerged aquatic vegetation tended to indicate silty soils, free-floating plants indicated clay soils, and graminoids indicated sandy soils.

Invasive Species

Nonnative plant species are often a part of the vegetation of freshwater marshes, particularly in areas that have been disturbed. It has been hypothesized that tropical regions are more susceptible to invasion than temperate regions because invading plants grow much more rapidly and are more noticeable in the tropics than in temperate latitudes. Plants such as *Eichhornia crassipes* (water hyacinth), *Salvinia molesta* (salvinia), and *Alternanthera philoxeroidses* (alligator weed) have invaded tropical and subtropical regions of the world. *E. crassipes* can double the area that it covers in two weeks and has choked many waterways that have received high nutrient loads for almost a century. Although there are many theories about alien aquatic plants, there is some validity to the concept that disturbed ecosystems are most susceptible to biological invasions. Werner and Zedler (2002) found that sediment accumulation within Wisconsin sedge meadows reduced tussock microtopography, promoted invasion by *Typha* spp. or *Phalaris arundinacea* (reed canary grass), and reduced wetland species richness.

In the freshwater marshes of the St. Lawrence and Hudson River valleys and in the Great Lakes region of North America, *Lythrum salicaria* (purple loosestrife), a tall purple-flowered emergent hydrophyte, spread at an alarming rate in the twentieth century, causing much concern to those who manage these marshes for wildlife (Stuckey, 1980; Balogh and Bookhout, 1989). The plant is aggressive in displacing native grasses, sedges, rushes, and even *Typha* spp. Many freshwater marsh managers have implemented programs designed to control purple loosestrife by chemical and mechanical means. Other aquatic aliens, such as the submersed *Hydrilla verticillata*, a plant native to Africa, Asia, and Australia, and *Myriophyllum spicatum*, have invaded open, shallow-water marshes in the United States (Steward, 1990; Galatowitsch et al., 1999) but rarely compete well with emergent vegetation.

Phragmites australis (Cav.) Trin. Ex. Steud., common reed, is considered an invasive species in eastern North America, particularly in along the Atlantic coastline and around the Laurentian Great Lakes, even though the plant has been in North America for more than 3,000 years, because of its aggressive expansion through brackish (salinity <5 ppt) and freshwater marshes, especially in the past 50 years. Its spread is attributed to increased disturbances, spread of more aggressive varieties from other parts of the world, including Europe, and changes in hydrology and salinity patterns in coastal estuarine systems (Philipp and Field, 2005). Saltonstall (2002) confirmed the existence of a total of 27 haplotypes of which 11 (types A–H, S, Z, AA) are native to

North America. Within North America, Types AA, F, Z and S are known historically from the Northeast; types E, G, and H from the Midwest; and types A to D from the South and Intermountain West. Two haplotypes, I and M, show worldwide distribution, with M the most common type in North America, Europe, and Asia. A new subspecies, *Phragmites australis* subsp. *americanus*, has been identified and has been shown to be distinctly different from the introduced and Gulf Coast lineages of *P. australis* (Saltonstall et al., 2004). This creates a difficult problem in wetland management and restoration, because wetland managers must be able to distinguish between the invasive "bad" *Phragmites* and the native "good" ones. Major resources are used to control this plant from spreading in eastern North America with techniques such as burning and herbicide application. There is irony in the fact that the plant is taking over North American wetlands while reed dieback of the same species in Europe is the main concern for this species there.

Consumers

Perhaps one reason that small marshes of the prairie region and the western high plains harbor such a rich diversity of organisms and wildlife is that they are often natural islands in a sea of farmland. Cultivated land does not provide a diversity of either food or shelter, and many animals must retreat to the marshes, which have become their only natural habitats. In cases where flow from watersheds is seasonal, freshwater marshes can serve as biological and hydrologic "oases" during low-flow and drought conditions.

Invertebrates

Invertebrates, similar to amphibians, are the links between plants and their detritus, on one hand, and animals such as fish, ducks and other birds, and even several mammals, on the other. Insects make up much of the invertebrate taxa in freshwater marshes, and their composition is often dictated by wetland hydrology and vegetation. Temporary pools tend to be diverse with beetle and midge communities. Insect communities are often productive because of the alternating wet-dry conditions, and many communities are regulated by biotic interactions. As marshes become more perennial, vegetation for habitat structure and as a decaying substrate for becomes important. In open-water sections, certain benthos and nektonic insects may also be important.

The most conspicuous invertebrates are the true flies (Diptera), which often make one's life miserable in the marsh. These include midges, mosquitoes, and crane flies. However, in the larval stage, many of the insects are benthic. Midge larvae, which are called bloodworms because of their rich red color, "are found submerged in bottom soils and organic debris, serving as food for fish, frogs, and diving birds. When pupae surface and emerge as adults, they are exploited as well by surface-feeding birds and fish" (Weller, 1994). Odonata, represented by dragonflies and damselflies, are a notable feature of freshwater marshes; their very presence generally indicates good water quality. Crustaceans such as crayfish and mollusks such as snails can be common in some freshwater marshes. The former are food for large fish and mammals alike, whereas the latter are often found grazing on mats of filamentous algae.

Temporal cycles and spatial patterns of invertebrate species and concentrations reflect the natural seasonal cycle of insect growth and emergence superimposed on the vegetation cycles. McLaughlin and Harris (1990) investigated insect emergence from diked and undiked marshes along Lake Michigan and found more insects, more insect biomass, and a greater number of taxa in diked marshes and the greatest numbers and biomass in the sparsely vegetated zones of the wetlands rather than in open water or dense vegetation. Kulesza and Holomuzki (2006) examined growth and survival of the detritivorous amphipod *Hyalella azteca* from a Lake Erie marsh. They compared its use of *Typha angustifolia* and *Phragmites australis* as substrate and found that both plants supported adequate fungi growth, and the amphipods performed equally well.

Amphibians

Amphibians are an important group of organisms in freshwater marshes, often serving as the link between insect populations and wading birds, mink, raccoons, and some fish in complex food webs. Larval tadpoles, which can be quite abundant in some freshwater marshes, eat small plants and animals and are, in turn, eaten by large fish and wading birds. The adult frogs feast on emerging insects. Even terrestrial toads use freshwater marshes as mating and breeding grounds in the spring. There has been concern about declining amphibian populations; one of the causes that has been suggested has been the loss of wetland habitat. Richter and Azous (1995) investigated the relationships between amphibian richness and variables such as wetland size, vegetation type, presence of competitors and predators, hydrologic characteristics, hydroperiod fluctuations, and land use. The variables that explained the highest correlation with amphibian richness were water-level fluctuations and percentage of the watershed that was urbanized. These data do not explain the exact cause of the loss of amphibians, but urban pollution and stream and hydrological modifications appear to be likely causes.

Porej (2004) compared several created and restored wetlands in central Ohio and found that the presence of a shallow-sloped littoral zone, the absence of fish (often caused by flooding limited to seasonal patterns in the wetland), and a high ratio of edge to area of wetlands (optimized where there are many small basins rather than one large basin of the same area) are among the key physical and biological features that support a diversity of amphibians. American toads (*Bufo americanus*), northern leopard frogs (*Rana pipiens*), western chorus frogs (*Pseudacris triseriata*), gray tree frogs (*Hyla versicolor*) and smallmouth salamanders (*Ambystoma texanum*) were positively correlated with the presence of shallow littoral zones in these freshwater marshes and ponds. Porej (2004) also found higher salamander richness in forested wetlands compared to freshwater marshes (natural or created) while frogs and toads had similar richness in forested wetlands and marshes that was lower than the richness in newly created marshes (Table 10.2). He found a strong association between the presence of forest cover within 200 m of freshwater wetlands and amphibian diversity in agricultural landscapes, particularly for spotted salamanders (*Ambystoma maculatum*),

Table 10.2 Occurrence (percentage of wetlands occupied) of pond-breeding amphibians in 54 natural (emergent and forested) and 42 created wetlands located in the Till Plains and Glaciated Plateau ecoregions of central Ohio

Species	Natural Emergent Wetlands	Natural Forested Wetlands	Created Wetlands
Bufo americanus/Bufo fowleri (American/Fowler's toad)	15	20	50
Rana clamitans melanota (Green frog)	60	59	74
R. pipiens (Northern leopard frog)	74	46	76
R. catesbeiana (American bullfrog)	33	26	55
R. sylvatica (Wood frog)	0	56	0
Pseudacris crucifer (Spring peeper)	87	67	52
P. triseriata (Western chorus frog)	27	31	23
Hyla versicolor (Gray treefrog)	20	26	48
Acris crepitans blanchardii (Blanchard's cricket frog)	0	0	12
Frogs and toads (ave ± st error), species per wetland	**3.2 ± 0.3**	**3.0 ± 0.3**	**3.9 ± 0.3**
Ambystoma tigrinum (Tiger salamander)	43	47	5
A. maculatum (Spotted salamander)	7	43	5
A. texanum (Smallmouth salamander)	15	64	14
A. jeffersonianum complex (Jefferson's salamander complex)	8	57	0
A. opacum (Marbled salamander)	0	7	0
Notophthalamus viridescens v. (Red-spotted newt)	0	22	2
Salamanders (ave ± st error), species per wetland	**1.0 ± 0.3**	**2.4 ± 0.2**	**0.3 ± 0.1**
Total amphibians, species per wetland	**4.2 ± 0.5**	**5.4 ± 0.3**	**4.2 ± 0.3**

Source: Porej (2004)

Jefferson's salamander complex (*A. jeffersonianum* complex), smallmouth salamanders (*A. texanum*), and wood frogs (*Rana sylvatica*).

Fish

One of the most difficult issues about which to generalize is whether freshwater marshes support much fish life or indeed if they should. As a general rule, the deeper the water in the marsh and the more open the system is to large rivers or lakes, the more variety and abundance of fish that can be supported. The positive aspect of freshwater marshes as habitats and nurseries for fish was investigated by Derksen (1989) for a large Manitoba marsh complex and by Stephenson (1990) for Great Lakes marshes. Derksen (1989) found extensive use of the marshes by northern pike (*Esox lucius*) with emigration from the marsh occurring primarily in the autumn. Stephenson (1990) found a total of 36 species of fish in marshes connected to Lake Ontario, including spawning adults of 23 species and the young-of-the-year of 31 species, indicating the importance of these marshes for fish reproduction in the lake. Eighty-nine percent of the species encountered were using the marshes for reproduction.

Common carp (*Cyprinus carpio*) are able to withstand the dramatic seasonal and diel fluctuations of water temperature and dissolved oxygen typical of shallow marshes and are thus abundant in many inland wetlands. They affect marsh vegetation by direct grazing, uprooting vegetation while searching for food, and causing severe turbidity in the water column. For these reasons, carp are not considered desirable by many freshwater marsh managers.

Mammals

A number of mammals inhabit inland marshes. The most noticed is probably the muskrat (*Ondatra zibethicus*). This herbivore reproduces rapidly and can attain population densities that decimate the marsh, causing major changes in its character. Like plants, each mammalian species has preferred habitats. For example, muskrats are found in the most aquatic areas, the water vole (e.g., *Microtus richardsoni* in North America) in overlapping but higher elevations, and other voles in the relatively terrestrial parts of the reed marsh. Most of the mammals are herbivorous. The effects that beavers (*Castor canadensis*) have on hydrology of wetlands and ponds are well known, including their effects on ecosystem functions, such as methane emissions (see Chapters 4 and 17: "Wetland Hydrology" and "Wetlands and Climate Change").

Birds

Waterfowl are plentiful in almost all wetlands, probably because of the food richness and the diversity of habitats for nesting and resting. Migratory waterfowl nest in northern freshwater marshes, winter in southern marshes, and rest in other marshes during their migrations. In a typical freshwater marsh, different species distribute themselves along an elevation gradient according to how well they are adapted to water (Fig. 10.7). In northern marshes, the loon (*Gavia immer*) usually uses the deeper water of marsh ponds, which may hold fish populations. Grebes (*Podilymbus* sp. and *Podiceps* sp.) prefer marshy areas, especially during the nesting season. Some ducks (dabblers), such as mallards (*Anas platyrhynchos*), nest in upland sites, feeding along the marsh–water interface and in shallow marsh ponds. Others (diving ducks), such as the ruddy duck (*Oxyura jamaicensis*), nest over water and fish by diving. For example, the black duck (*Anas rubripes*), one of the most popular ducks for naturalists and hunters alike, uses the emergent marsh as its preferred habitat. The northern shoveler (*A. clypeata*), the "whale of the waterfowl," uses its large bill and laternal lamellae to filter plankton. Geese (*Branta canadensis* and *Chen* sp.) and swans (*Cygnus* sp.), the "cattle of the waterfowl," along with canvasback ducks (*Aythya valisineria*) and the wigeon (*Anas americana*), are major marsh herbivores. Wading birds, such as the great blue heron (*Ardea herodias*) and the great egret (*Casmerodius albus*), usually nest colonially in wetlands and fish along the shallow ponds and streams. The least bittern (*Ixobrychus exilis*) builds nests a meter or less above the water in *Typha* or *Scirpus/Schoenoplectus* stands. Rails live in the whole range of wetlands; many of them are solitary birds that are seldom seen. Marsh wrens (*Cistohorus plaustris*), Virginia rails (*Rallus limicola*), soras (*Porzana carolina*), and swamp sparrows (*Melospiza georgiana*) live amid the dense vegetation of freshwater marshes, often heard but not

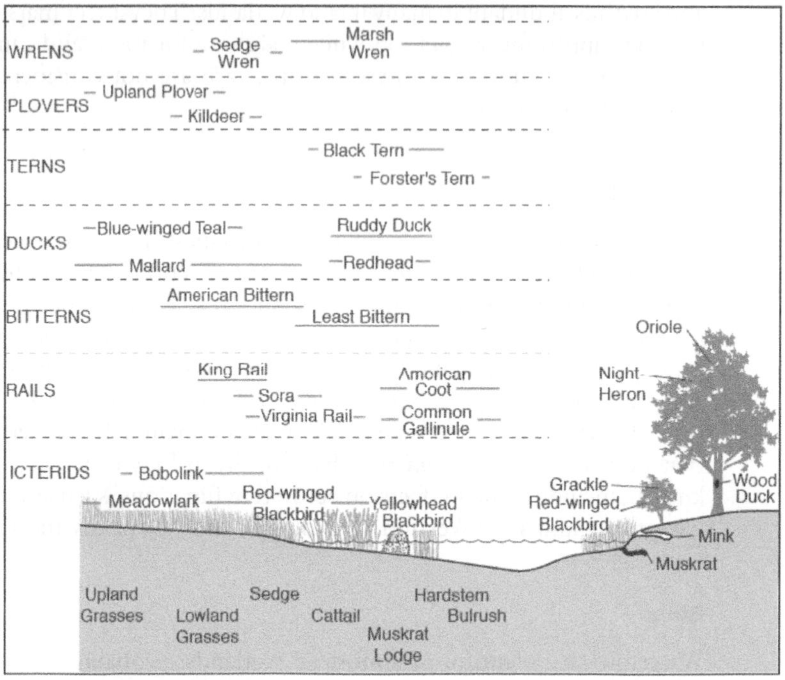

Figure 10.7 Typical distribution of birds across a freshwater marsh from open water edge across shallow water to upland grasses. Placement of muskrat and mink is also illustrated. (After Weller and Spatcher, 1965)

seen. Songbirds are also abundant in and around marshes. They often nest or perch in adjacent uplands and fly into the marsh to feed. Swallows (*Riparia riparia* and *Stelgidopteryx serripennis*) and swifts (e.g., chimney swifts *Chaetura pelagica*) are common around freshwater marshes, flying above the marsh, with their mouths ever open to capture emerging insects, often in swarms of dozens or even hundreds of birds.

One of the most conspicuous wetland birds in United States freshwater marshes is the blackbird, represented by the yellow-headed blackbird (*Xanthocephalus xanthocephalus*) in parts of the midwestern United States and the red-winged blackbird (*Aglaius phoniceus*) in the eastern United States. The red-winged blackbird is a very social species and is quite territorial, especially during the nesting season.

Ecosystem Function

Primary Productivity

The above-ground net primary productivity of inland marshes has been reported in a number of studies (Table 10.3). Estimates are generally quite high, ranging upward

Table 10.3 Selected primary production estimates for inland freshwater marshes

Dominant Species	Location	Net Primary Productivity (g m^{-2} yr^{-1})	Reference
Reeds and Grasses			
Glyceria maxima	Lake, Czech Republic	900–4,300[a]	Kvet and Husak (1978)
Phragmites communis	Lake, Czech Republic	1,000–6,000[a]	Kvet and Husak (1978)
P. communis	Denmark	1,400[a]	Anderson (1976)
Panicum hemitomon	Floating coastal marsh, Louisiana	1,700[b]	Sasser et al. (1982)
Schoenoplectus lacutstris	Lake, Czech Republic	1,600–5,500[a]	Kvet and Husak (1978)
Sparganium eurycarpum	Prairie pothole, Iowa	1,066[b]	van der Valk and Davis (1978)
Typha glauca	Prairie pothole, Iowa	2,297[b]	van der Valk and Davis (1978)
T. latifolia	Oregon	2,040–2,210[a]	McNaughton (1966)
Typha spp.	Lakeside, Wisconsin	3,450[a]	Klopatek (1974)
Typha spp.	central Ohio created marshes (12 yrs; starting 3 years after planting)	627±75[b] planted wetland; 772±91[b] unplanted wetland	Mitsch et al. (2012)
Sedges and Rushes			
Carex atheroides	Prairie pothole, Iowa	2,858[b]	van der Valk and Davis (1978)
Carex lacustris	Sedge meadow, New York	1,078–1,741[a]	Bernard and Solsky (1977)
Juncus effusus	South Carolina	1,860[a]	Boyd (1971)
Scirpus fluviatilis	Prairie pothole, Iowa	943[a]	van der Valk and Davis (1978)
Broad-Leaved Monocots			
Acorus calamus	Lake, Czech Republic	500–1,100[a]	Kevt and Husak (1978)

[a]Above- and below-ground vegetation.
[b]Above-ground vegetation.

from about 1,000 g m^{-2} yr^{-1}. Some of the best estimates, which take into account underground production as well as that above-ground, come from studies of fishponds in the Czech Republic. (These are small artificial lakes and bordering marshes used for fish culture.) These estimates, some indicating values of over 6,000 g m^{-2} yr^{-1}, are high compared with most of the North American estimates and even higher than the productivity of intensively cultivated farm crops.

The emergent monocotyledons *Phragmites* and *Typha*, two of the dominant plants in freshwater marshes, have high photosynthetic efficiency. For *Typha*, efficiency is highest early in the growing season, gradually decreasing as the season

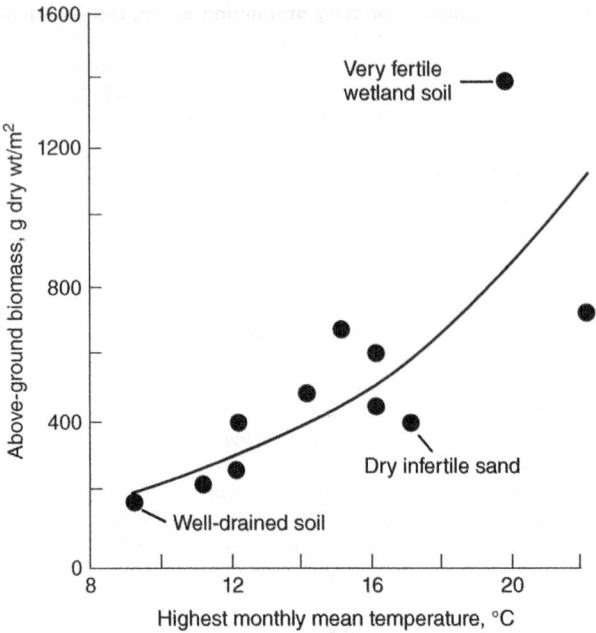

Figure 10.8 Relationship between highest mean monthly temperature and above-ground standing crop of various sedges in freshwater wetlands and uplands. Data points are for wetlands except where noted otherwise. (After Gorham, 1974)

progresses. *Phragmites*, in contrast, has a fairly constant efficiency rate throughout most of the growing season. The efficiencies of conversion by these plants in optimum environments of 4 to 7 percent of photosynthetically active radiation are comparable to those calculated for intensively cultivated crops such as sugar beets, sugarcane, and corn.

Productivity variation is undoubtedly related to a number of factors, including summer air temperatures (Fig. 10.8). Innate genetic differences among species account for part of the variability. For example, in one study that used the same techniques of measurement (Kvet and Husak, 1978), *Typha angustifolia* production was determined to be double that of *T. latifolia*.

The dynamics of underground growth are much less studied than those of aboveground growth. Annuals generally use small amounts of photosynthate to support root growth, whereas species with perennial roots and rhizomes often have root : shoot ratios well in excess of 1. This relationship also appears to hold true for inland freshwater marshes. Perennial species in freshwater marshes generally have more below-ground than above-ground biomass (Fig. 10.9). Even when biomass root : shoot ratios are usually greater than 1, ratios of root production to shoot biomass are generally less than 1. Since above-ground production is often approximated by above-ground biomass, this latter ratio is an index of the allocation of resources by the plant, and it indicates that less than one-half of the photosynthate is translocated to the roots. The coexistence of large root biomass and relatively small

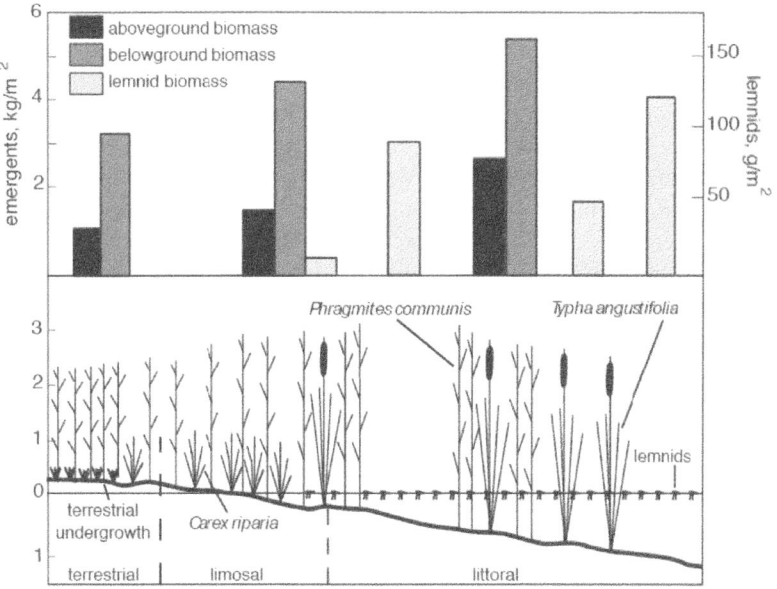

Figure 10.9 Distribution of above-ground and below-ground biomass of emergent vegetation and lemnids across a reed bed (*Phragmites*) transect, showing relation to elevation and flooding. (After Kvet and Husak, 1978)

root production suggests that the root system is generally longer lived (i.e., it renews itself more slowly) than the shoot.

Decomposition and Herbivory

With some notable exceptions, such as muskrat and geese grazing, herbivory is considered fairly minor in inland marshes where most of the organic production decomposes before entering the detrital food chain. The decomposition process is much the same for all wetlands. Variations stem from the quality and resistance of the decomposing plant material, the temperature, the availability of inorganic nutrients to microbial decomposers, and the flooding regime of the marsh.

Consumers play a significant role in detrital cycles. Most litter decomposition studies in freshwater marshes were done with senesced plant material during the winter, and generally low ($k = 0.002$–0.007 day^{-1}) rates were measured. However, in a comparison of the decay of fresh biomass and senesced wetland plant leaves, Nelson et al. (1990a,b) found that samples of freshly harvested wetland plant material (*Typha glauca*) decomposed more than twice as fast ($k = 0.024$ day^{-1}) as did naturally senesced material ($k = 0.011$ day^{-1}). This comparison illustrates the more rapid decomposition that results when animals such as muskrats harvest live plant material.

Muskrats also may play a positive role in the energy flow of a marsh as they harvest aquatic plants and standing detritus for their muskrat mounds. Wainscott et al. (1990)

found in culturing experiments that litter from muskrat mounds supports substantially higher densities of microbes than litter from the marsh floor does. They suggested that these muskrat mounds may act like "compost piles," as they accelerate the decomposition and microbial growth that have become familiar to organic gardeners.

Muskrat eat-outs and the resulting open water contribute to structural and biogeochemical heterogeneity within marshes. In an Iowa prairie pothole marsh, Rose and Crumpton (2006) found that as vegetated zones transitioned into open water, there was a predictable decrease in dissolved oxygen and increase in methane concentrations. They suggested that these conditions were regulated by the presence of emergent vegetation and its influence on aerobic and anaerobic metabolism.

Food Webs

Even though food chains begin in the detrital material of freshwater marshes, they develop into detailed webs that are still poorly understood. Benthic communities that feed on detritus form the basis of food for fish and waterfowl in the marshes. DeRoia and Bookhout (1989) found that chironomids made up 89 percent of the diet of blue-winged teal (*Anas discors*) and 99 percent of the diet of green-winged teal (*A. crecca*) in a Great Lakes marsh. The direct grazing of freshwater marsh vegetation has occasionally been reported in the literature. Crayfish are often important consumers of macrophytes, particularly of submersed aquatic plants, in freshwater marshes. The red swamp crayfish (*Procambarus clarkii*) was shown to have effectively grazed on *Potamogeton pectinatus* in a freshwater marsh in California, where the plant decreased from 70 to 0 percent cover of the marsh while the crayfish population almost doubled (Feminella and Resh, 1989). The direct consumption of marsh plants by geese, muskrats, and other herbivores is common in some parts of the world. Eat-outs causing large expanses of open water are the result of the inability of plants to survive after being clipped below the water surface by animals (Middleton, 1999).

In one study that disputed the low-herbivory assumption of inland marshes, deSzalay and Resh (1997) found that herbivores represented about 27 percent of the benthic community in brackish inland marshes. These herbivores fed primarily on filamentous algae and diatoms. Yet herbivory on marsh macrophytes generally remains low except for large-animal grazing from time to time.

Nutrient Budgets

Vegetation traps nutrients in biomass, but the storage of these nutrients is seasonally partitioned in above-ground and below-ground stocks. For example, nutrient stocks in the roots and rhizomes of macrophytes are mobilized into the shoots early in the growing season and increase to as much as $4\,g\,P\,m^{-2}$ during the summer. In the fall, some nutrients in the shoots are translocated into the below-ground organs before the shoots die, but most nutrients are lost by leaching and in the litter. Nitrogen and carbon budgets for two created flowthrough marshes in Ohio dominated by *Typha* spp. were presented in Chapter 6: "Wetland Biogeochemistry" (Fig. 6.19). In these

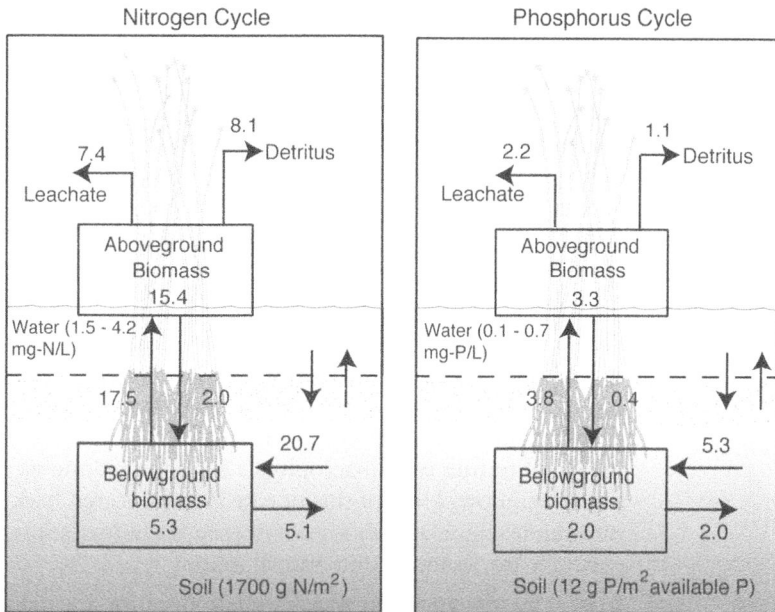

Figure 10.10 Fluxes of nitrogen and phosphorus through a river bulrush (*Scirpus fluviatilis*) stand in Wisconsin. Flows are in g m^{-2} y^{-1}, and storages are in g/m^2 of nitrogen and phosphorus, respectively. (From Klopatek, 1978)

flowthrough wetlands, hydrologic fluxes dominated the wetlands (inflow, outflow, and subsurface seepage), but the sequestration of nitrogen and carbon into the soil was one of the largest fluxes. The nitrogen and phosphorus budgets for a freshwater *Scirpus* marsh in Wisconsin (Fig. 10.10) show a peak biomass storage of 20.7 g N/m^2 and 5.3 g P/m^2. This plant storage is small compared to the nutrients that are stored within the root zone of the peat and mineral soils (shown in Fig. 10.10 to be 1,700 g N/m^2 and 12 g P/m^2 for total nitrogen and available phosphorus, respectively).

Studies such as these lead to five generalizations about nutrient cycling in freshwater marshes:

1. The size of the plant stock of nutrients in freshwater marshes varies widely in contrast to the much more abundant storage of nutrients in marsh soils. More nitrogen and phosphorus are retained in above-ground plant parts in mineral substrate wetlands (freshwater marshes) than in peatlands due to the higher productivity and higher concentrations of nutrients. The above-ground stock of nitrogen ranges from as low as 3 to as high as 30 g-N m^{-2} in freshwater marshes.

2. The biologically inactivated stock of nutrients in plants is only a temporary storage that is released to flooding waters and sediments when the plant shoots die in autumn. Where this occurs, the marsh may retain nutrients during the summer and release them in the winter.

3. Nutrients retained in biomass are often a small portion of nutrients that flow into the marshes, and that percentage decreases with increased nutrient input. Thus, as more nutrients become available to a freshwater marsh, the marsh becomes more "leaky." Nutrients are lost from the system, and nutrient turnover in the vegetation increases. Even if the uptake rate of nutrients is high in wetlands, some of those nutrients are returned via detrital decomposition to the nutrient pool in the sediments and overlying waters. If wetlands are used for nutrient removal (see Chapter 19 "Wetlands and Water Quality"), then it is common for only 10 to 20 percent of the nutrient inflow to transfer temporarily into plant biomass.

4. Marsh vegetation often acts as a nutrient pump, taking up nutrients from the soil, translocating them to the shoots, and releasing them on the marsh surface during senescence. The effect of this pumping mechanism may be to mobilize nutrients that have been sequestered in the soil. In some cases, the uptake of nutrients by macrophytes from the sediments is considerably higher than the inflow. Most of this uptake is translocated back to the roots or lost through leaching and shoot senescence, so biomass storage of nutrients is generally low compared to annual inflow.

5. In general, precipitation and dryfall account for less than 10 percent of plant nutrient demands in freshwater marshes. Similarly, groundwater flows are usually small sources of phosphorus, but, in agricultural settings with artificial drainage, nitrate–nitrogen inflow can be high. Surface inflow is usually a major source of phosphorus because of its ability to sorb onto sediments, particularly clay. Considering all of these variables, it is not surprising that each marsh seems to have its own unique nutrient budget. In low-nutrient wetlands like the Florida Everglades, the marsh system is accustomed to relying primarily on nutrient inflow from precipitation and dry fallout from fires.

Nutrient Limitations

Koerselman and Meuleman (1996), in a study of several wetlands in Europe, found that the nitrogen:phosphorus (N:P) ratios in wetland plant tissues were correlated with the N:P supply ratio and that any N:P ratio less than 14:1 suggests nitrogen limitation. This is twice the often-used Redfield ratio (N:P = 7.2 by weight) that is used in planktonic systems to indicate relative nutrient limitation. As part of an extensive literature review on temperate North American wetlands, Bedford et al. (1999) found that only marshes were consistently N-limited as indicated by leaf-tissue and soil N:P ratios <14 (although swamp tended to have soils with N:P ratios <14 as well). Based on their leaf tissue N:P ratio, other wetland types (swamps, bogs, fens) tended to be co-limited by N and P or just P-limited based on the thresholds derived by Koerselman and Meuleman (1996).

McJannet et al. (1995) investigated the nitrogen and phosphorus content of 41 freshwater marsh plants after they were grown in excess fertilizer for one growing season. There was a wide range of nitrogen (0.25–2.1 percent N) and phosphorus

(0.13–1.1 percent P) that was not related to where the plants came from. However, plants that were from ruderal life histories (i.e., annuals or functional annuals) did have significantly lower nitrogen and phosphorus tissue concentrations than did perennials.

For a Manitoba *Scirpus acutus* marsh, Neill (1990) found that neither nitrogen nor phosphorus increased net productivity when applied alone but that above-ground biomass nearly doubled when nitrogen and phosphorus were applied together. Similar studies of a nearby marsh showed nitrogen limitation, indicating that differences in limiting factors are possible even in the same region (Neill, 1990). Under conditions in which water levels are more stable, such as Louisiana's Gulf Coast, the addition of nitrogen fertilizer at a rate of $10\,g\ NH_4^+\text{--}N/m^2$ caused approximately a 100 percent increase in the growth of *Sagittaria lancifolia* (Delaune and Lindau, 1990).

Experiments by Svengsouk and Mitsch (2001) support the multiple-nutrient limitation of some freshwater marsh plants. Their study investigated the relative limitations of nitrogen and phosphorus in mesocosms planted with both bulrush (*Schoenoplectus tabernaemontani*) and cattail (*Typha* sp.) together. Results suggested that, when both nitrogen and phosphorus are available, *Typha* competed well with *Schoenoplectus*. When only one of the nutrients was in abundance, *Schoenoplectus* did much better than *Typha*.

In contrast, enrichment studies by Craft et al. (1995) on the low-nutrient sawgrass (*Cladium jamaicense*) and other macrophyte communities illustrated that the most important limiting factor in the Florida Everglades is phosphorus. Nitrogen additions had no effect on biomass production, nutrient uptake rates, or nitrogen enrichment of peat. Phosphorus enrichment from agricultural sources has attributed to substantial ecological change in the Everglades, most notably the transition of large areas from sawgrass to cattail (*Typha* spp.). Restoring the Everglades requires reducing P-enriched agricultural stormwater entering the region. (See Chapter 19: "Wetlands and Water Quality," Case Study 2).

Greenhouse Gas Emissions

The anaerobic conditions in marshes and other wetlands give them the potential to emit considerable amounts of nitrous oxide (N_2O) and methane (CH_4) both of which are considered important greenhouse gases. Temperature and diffusion rates through water influence the net emission of these gases, and therefore shallow wetlands often have greater emissions than open water bodies. In a boreal lake in Finland, Huttunen et al. (2003) estimated that the littoral zone, which consisted of 26 percent of the total lake surface area, was responsible for most of the N_2O emissions from the lake. Brix et al. (2001) examined whether a *Phragmites* marsh in Europe could be considered a net source or sink for greenhouse gases given that marshes assimilate carbon dioxide (CO_2) and store carbon while emitting CH_4. They found that when these marshes are evaluated over a shorter time period (<60 years), these wetlands could be considered a net source of greenhouse gases based on their emission of CH_4 and CO_2 relative to carbon fixation. However, CH_4 does not persist in the atmosphere indefinitely. If these marshes are evaluated over a longer time period (>100 years), the balance shifts and the marshes are a net a sink for greenhouse gases. Mitsch et al. (2013) found similar results in a comparison of 21 wetlands from around the world, many of them freshwater marshes. Several recent methane emissions studies from created and natural wetlands in Ohio and Costa Rica by Altor and Mitsch (2006, 2008), Nahlik and Mitsch (2010, 2011), Sha et al., (2011), Mitsch et al. (2013), and Waletzko and Mitsch (2014) suggest that methane emissions from created wetlands are lower in their first two decades than are comparable natural reference wetlands.

Recommended Readings

Coburn, E. A. 2004. *Vernal Pools*. Blacksburg, VA: McDonald & Woodward.
van der Valk, A. G. 2012. *The Biology of Freshwater Wetlands*, 2nd ed. Oxford, UK: Oxford University Press

References

Altor, A. E., and W. J. Mitsch. 2006. Methane flux from created riparian marshes: Relationship to intermittent versus continuous inundation and emergent macrophytes. *Ecological Engineering* 28: 224–234.

Altor, A. E., and W. J. Mitsch. 2008. Pulsing hydrology, methane emissions, and carbon dioxide fluxes in created marshes: A 2-year ecosystem study. *Wetlands* 28: 423–438.

Anderson, F. O. 1976. Primary productivity in a shallow water lake with special reference to a reed swamp. *Oikos* 27: 243–250.

Balogh, G. R., and T. A. Bookhout. 1989. Purple loosestrife (Lythrum salicaria) in Ohio's Lake Erie marshes. *Ohio Journal of Science* 89: 62–64.

Bedford, B. L., M. R. Walbridge, and A. Aldous. 1999. Patterns of nutrient availability and plant diversity of temperate North American wetlands. *Ecology* 8: 1251–1269.

Bernard, J. M., and B. A. Solsky. 1977. Nutrient cycling in a *Carex lacustris* wetland. *Canadian Journal of Botany* 55: 630–638.

Boyd, C. E. 1971. The dynamics of dry matter and chemical substances in a *Juncus effusus* population. *American Midland Naturalist* 86: 28–45.

Bridgham, S. D., K. Updegraff, and J. Pastor. 2001. A comparison of nutrient availability indices along an ombrotrophic-minerotrophic gradient in Minnesota wetlands. *Soil Science Society of America Journal* 65: 259–269.

Brix, H., B. K. Sorrell, and B. Lorenzen. 2001. Are Phragmites-dominated wetlands a net source or net sink of greenhouse gases? *Aquatic Botany* 69: 313–324.

Craft, C. B., J. Vymazal, and C. J. Richardson. 1995. Response of Everglades plant communities to nitrogen and phosphorus additions. *Wetlands* 15: 258–271.

Dahl, T. E. 2006. *Status and trends of wetlands in the conterminous United States 1998 to 2004*. U.S. Department of the Interior, Fish and Wildlife Service, Washington, DC, 112 pp.

Delaune, R. D., and C. W. Lindau. 1990. Fate of added 15N labeled nitrogen in a *Sagittaria lancifolia* L. Gulf Coast marsh. *Journal of Freshwater Ecology* 5: 429–431.

Denny, P. 1985. Wetland plants and associated plant life-forms. In P. Denny, ed., *The Ecology and Management of Wetland Vegetation*. Junk, Dordrecht, the Netherlands, pp. 1–18.

Denny, P. 1993. Wetlands of Africa: Introduction. In D. F. Whigham, D. Dykyjová, and S. Hejny, eds. *Wetlands of the World, I: Inventory, Ecology, and Management*. Kluwer Academic Publishers, Dordrecht, the Netherlands, pp. 1–31.

Derksen, A. J. 1989. Autumn movements of underyearling northern pike, *Esox lucius*, from a large Manitoba marsh. *Canadian Field Naturalist* 103:429–431.

DeRoia, D. M., and T. A. Bookhout. 1989. Spring feeding ecology of teal on the Lake Erie marshes (abstract). *Ohio Journal of Science* 89(2): 3.

deSzalay, F. A., and V. H. Resh. 1997. Responses of wetland invertebrates and plants important in waterfowl diets to burning and mowing of emergent vegetation. *Wetlands* 17: 149–156.

Farrar, J., and R. Gersib. 1991. Nebraska salt marshes: Last of the least. Nebraska Game and Park Commission, Lincoln. 23 pp.

Feminella, J. W., and V. H. Resh. 1989. Submersed macrophytes and grazing crayfish: An experimental study of herbivory in a California freshwater marsh. *Holarctic Ecology* 12: 1–8.

Galatowitsch, S. M., N. O. Anderson, and P. D. Ascher. 1999. Invasiveness in wetland plants in temperate North America. *Wetlands* 19: 733–755.

Gorham, E. 1974. The relationship between standing crop in sedge meadows and summer temperature. *Journal of Ecology* 62: 487–491.

Hall, J. V., W. E. Frayer, and B. O. Wilen. 1994. Status of Alaska wetlands. U.S. Fish & Wildlife Service, Alaska Region, Anchorage. 32 pp.

Huttunen, J. T., H. Nykänen, J. Turunen, and P. J. Martikainen. 2003. Methane emissions from natural peatlands in the northern boreal zone in Finland, Fennoscandia. *Atmospheric Environment* 37: 147–151.

Johnson, S. A. 1998. Effects of hydrology and plant introduction techniques on first-year macrophyte growth in an newly created wetland. Master's thesis, Ohio State University, Columbus.

Johnston, C. A., S. D. Bridgham, and J. P. Schubauer-Berigan. 2001. Nutrient dynamics in relation to geomorphology of riverine wetlands. *Soil Science of America Journal* 65: 557–577.

Johnston, C. A., B. Bedford, M. Bourdaghs, T. Brown, C. Frieswyk, M. Tulbere, L. Vaccaro, and J. B. Zedler. 2007. Plant species indicators of physical environment in Great Lakes coastal wetlands. *Journal of Great Lakes Research* 33: 106–124.

Keddy, P. A. 2010. *Wetland Ecology: Principles and Conservation*, 2nd ed. Cambridge University Press, Cambridge, UK. 497 pp.

Klopatek, J. M. 1974. Production of emergent macrophytes and their role in mineral cycling within a freshwater marsh. Master's thesis, University of Wisconsin, Milwaukee.

Klopatek, J. M. 1978. Nutrient dynamics of freshwater riverine marshes and the role of emergent macrophytes. In R. E. Good, D. F. Whigham, and R. L. Simpson, eds.,

Freshwater Wetlands: Ecological Processes and Management Potential. Academic Press, New York, pp. 195–216.

Koerselman, W., and A. F. M. Meuleman. 1996. The vegetation N:P ratio: A new tool to detect the nature of nutrient limitation. *Journal of Applied Ecology* 33: 1441–1450.

Kulesza, A. E., and J. R. Holomuzki. 2006. Amphipod performance responses to decaying leaf litter of *Phragmites australis* and *Typha angustifolia* from a Lake Erie coastal marsh. *Wetlands* 26: 1079–1088.

Kvet, J., and S. Husak. 1978. Primary data on biomass and production estimates in typical stands of fishpond littoral plant communities. In D. Dykyjova and J. Kvet, eds., *Pond Littoral Ecosystems*. Springer-Verlag, Berlin, pp. 211–216.

LaBaugh, J.W. 1989. Chemical characteristics of water in northern prairie wetlands. In A. van der Valk, ed., *Northern Prairie Wetlands*. Iowa State University Press, Ames, IA.

LaBaugh, J. W., T. C. Winter, G. A. Swanson, D. O. Rosenberry, R. D. Nelson, and N. H. Euliss. 1996. Changes in atmosphereic circulation patterns affect midcontinent wetlands sensitive to climate. *Limnology & Oceanography* 41: 864–870.

Matthews, E., and I. Fung. 1987. Methane emissions from natural wetlands: Global distribution, area, and environmental characteristics of sources. *Global Biogeochemical Cycles* 1: 61–86.

McJannet, C. L., P. A. Keddy, and F. R. Pick. 1995. Nitrogen and phosphorus tissue concentrations in 41 wetland plants: A comparison across habitats and functional groups. *Functional Ecology* 9: 231–238.

McLaughlin, D. B., and H. J. Harris. 1990. Aquatic insect emergence in two Great Lakes marshes. *Wetlands Ecology and Management* 1: 111–121.

McNaughton, S. J. 1966. Ecotype function in the Typha community-type. *Ecological Monographs* 36: 297–325.

Middleton, B. A. 1999. *Wetland Restoration: Flood Pulsing and Disturbance Dynamics*. John Wiley & Sons, New York, 388 pp.

Mitsch, W. J., L. Zhang, K. C. Stefanik, A. M. Nahlik, C. J. Anderson, B. Bernal, M. Hernandez, and K. Song. 2012. Creating wetlands: Primary succession, water quality changes, and self-design over 15 years. *BioScience* 62: 237–250.

Mitsch, W. J., B. Bernal, A. M. Nahlik, U. Mander, L. Zhang, C. J. Anderson, S. E. Jørgensen, and H. Brix. 2013. Wetlands, carbon, and climate change. *Landscape Ecology* 28: 583–597.

Moore, D. R. J., P. A. Keddy, C. L. Gaudet, and I. C. Wisheu. 1989. Conservation of wetlands: do infertile wetlands deserve a higher priority? *Biological Conservation* 47: 203–217.

Nahlik, A. M., and W. J. Mitsch. 2010. Methane emissions from created riverine wetlands. *Wetlands* 30: 783–793. *Erratum in Wetlands* (2011) 31: 449–450.

Nahlik, A. M., and W. J. Mitsch. 2011. Methane emissions from tropical freshwater wetlands located in different climatic zones of Costa Rica. *Global Change Biology* 17: 1321–1334.

Neill, C. 1990. Nutrient limitation of hardstem bulrush (*Scirpus acutus* Muhl.) in a Manitoba interlake region marsh. *Wetlands* 10: 69–75.

Nelson, J. W., J. A. Kadlec, and H. R. Murkin. 1990a. Seasonal comparisons of weight loss of two types of *Typha glauca* Godr. leaf litter. *Aquatic Biology* 37: 299–314.

Nelson, J. W., J. A. Kadlec, and H. R. Murkin. 1990b. Response by macroinvertebrates to cattail litter quality and timing of litter submergence in a northern prairie marsh. *Wetlands* 10: 47–60.

Philipp, K. R., and R. T. Field. 2005. *Phragmites australis* expansion in Delaware Bay salt marshes. *Ecological Engineering* 25: 275–291.

Porej, D. 2004. Faunal aspects of wetland creation and restoration. Ph.D. dissertation, Evolution, Ecology, and Organismal Biology Department, Ohio State University, Columbus.

Richter, K. O., and A. L. Azous. 1995. Amphibian occurrence and wetland characteristics in the Puget Sound Basin. *Wetlands* 15: 305–312.

Rose, C., and W. G. Crumpton. 2006. Spatial patterns in dissolved oxygen and methane concentrations in prairie pothole wetlands in Iowa, USA. *Wetlands* 26: 1020–1025.

Saltonstall, K. 2002. Cryptic invasion by a non-native genotype of the common reed, *Phragmites australis*, into North America. *Proceedings of the National Academy of Sciences* 99: 2445–2449.

Saltonstall, K., P. M. Peterson, and R. J. Soreng. 2004. Recognition of *Phragmites australis* subsp. americanus (Poaceae: Arundinoideae) in North America: Evidence from morphological and genetic analyses. *Brit. Org/SIDA* 21: 683–692.

Sasser, C. E., G. W. Peterson, D. A. Fuller, R. K. Abernethy, and J. G. Gosselink. 1982. *Environmental Monitoring Program, Louisiana Offshore Oil Port Pipeline, 1981*. Annual Report, Coastal Ecology Laboratory, Center for Wetland Resources, Louisiana State University, Baton Rouge, 299 pp.

Sha, C., W. J. Mitsch, Ü. Mander, J. Lu, J. Batson, L. Zhang and W. He. 2011. Methane emissions from freshwater riverine wetlands. *Ecological Engineering* 37: 16–24.

Stephenson, T. D. 1990. Fish reproductive utilization of coastal marshes of Lake Ontario near Toronto. *Journal of Great Lakes Research* 16: 71–81.

Steward, K. K. 1990. Aquatic weed problems and management in the eastern United States. In A. H. Pietersen, and K. J. Murphy, eds. *Aquatic Weeds: The Ecology and Management of Nuisance Aquatic Vegetation*. Oxford University Press, New York, pp. 391–405.

Stuckey, R. L. 1980. Distributional history of *Lythrum salicaria* (purple loosestrife) in North America. *Bartonia* 47: 3–20.

Svengsouk, L. M., and W. J. Mitsch. 2001. Dynamics of mixtures of *Typha latifolia* and *Schoenoplectus tabernaemontani* in nutrient-enrichment wetland experiments. *American Midland Naturalist* 145: 309–324.

van der Valk A. G., and C. B. Davis. 1978. Primary production of prairie glacial marshes, In: *Freshwater Wetlands: Ecological Processes and Management Potential*,

R. E. Good, D. F. Whigham, and R L. Simpson, eds., Academic Press, New York, pp. 21–37.

Wainscott, V. J., C. Bardey, and P. P. Kangas. 1990. Effect of muskrat mounds on microbial density on plant litter. *American Midland Naturalist* 123: 399–401.

Waletzko, E., and W. J. Mitsch. 2014. Methane emissions from wetlands: An in situ side-by-side comparison of two static accumulation chamber designs. *Ecological Engineering* 72: 95–102.

Weller, M. W. 1994. *Freshwater Marshes*, 3rd ed. University of Minnesota Press, Minneapolis. 192 pp.

Weller, M. W., and C. S. Spatcher. 1965. Role of habitat in the distribution and abundance of marsh birds. Iowa State University Agriculture and Home Economics Experiment Station Special Report 43, Ames, IA, 31 pp.

Werner, K. J., and J. B. Zedler. 2002. How sedge meadow soils, microtopography, and vegetation respond to sedimentation. *Wetlands* 22: 451–466.

Zoltai, S. C. 1988. Wetland environments and classification. In National Wetlands Working Group, ed., *Wetlands of Canada*. Ecological Land Classification Series 24, Environment Canada, Ottawa, Ontario, and Polyscience Publications, Montreal, Quebec, pp. 1–26.

Freshwater swamp in Florida

Chapter 11

Freshwater Swamps and Riparian Ecosystems

Freshwater forested wetlands in North America range from deepwater swamps dominated by bald cypress–tupelo (Taxodium distichum–Nyssa aquatica), pond cypress–black gum (Taxodium distichum var. imbricarium–Nyssa sylvatica var. biflora), and Atlantic white cedar (Chamaecyparis thyoides) swamps found along the eastern seaboard of the United States, to less wet red maple (Acer rubrum) swamps found throughout New England and the Mid-Atlantic states. Riparian ecosystems have soils and soil moisture influenced by the adjacent stream or river and are unique because of their linear form along rivers and streams and because they process large fluxes of energy and materials from upstream systems. Riparian ecosystems include bottomland hardwood forests found along rivers in mesic climates everywhere. Trees in forested wetlands have developed several unique adaptations to the wetland environment, including knees, wide buttresses, adventitious roots, fluted trunks, and gas transport to the rhizosphere. Forested swamp primary productivity is closely tied to hydrologic conditions with lower productivity whenever conditions are either too wet or too dry. The function of riparian ecosystems is much better explained by a generalized theory called the flood pulse concept than by a previous theory of streams referred to as the river continuum concept.

In the nomenclature used in this book, swamps are forested wetlands. We discussed saltwater swamps in Chapter 9: "Mangrove Swamps." There are an estimated 1.1 million km^2 of freshwater swamps in the world, representing about 20 percent of the inland wetlands of the world (Table 10.1).

Very few trees flourish in standing water. Exceptions are found in the southeastern United States, where cypress (*Taxodium* sp.) and tupelo/gum (*Nyssa* sp.) swamps are found in deepwater forested wetlands and are characterized by bald cypress–water tupelo communities with permanent or near-permanent standing water. These so-called deepwater swamps were defined by Penfound (1952) as having "fresh water, woody communities with water throughout most or all of the growing season" and include isolated cypress domes and alluvial cypress swamps along rivers. Along the middle-eastern seaboard of the United States and along the Florida Panhandle, the cypress swamp partially gives way to another forested wetland, the Atlantic white cedar (*Chamaecyparis thyoides*) swamp. Farther northeast through New England and well into the Midwest, other types of freshwater forested wetlands occur, although they are not as wet as the cypress–tupelo swamps, nor are the tree species coniferous as are cypress. These broad-leaved deciduous forested wetlands include forest found along river floodplains (riparian forests or bottomland hardwood forests) and a multitude of forested wetlands that are found in isolated upland depressions.

Extensive tracts of riparian wetlands, which occur along rivers and streams, are occasionally flooded by those bodies of water but are otherwise dry for varying portions of the growing season. Riparian forests and freshwater swamps combined constitute the most extensive class of wetlands in the United States, covering an estimated 280,000 km^2. In the southeastern and midwestern United States, riparian ecosystems are often referred to as *bottomland hardwood forests*. They contain diverse vegetation that varies along gradients of flooding frequency. Riparian wetlands also occur in arid and semiarid regions of the United States, where they are often a conspicuous feature of the landscape in contrast to the surrounding arid grasslands and desert. Riparian ecosystems are generally considered to be more productive than the adjacent uplands because of the periodic inflow of nutrients, especially when flooding is seasonal rather than continuous.

Geographic Extent

Cypress–Tupelo Swamps

Bald cypress (*Taxodium distichum* [L.] Rich.) swamps are found as far north as southern Illinois and western Kentucky in the Mississippi River embayment and southern New Jersey along the Atlantic Coastal Plain in the United States (Fig. 11.1a). Pond cypress (*Taxodium distichum* var. *imbricarium* [Nutt.] Croom), described variously as either a different species or a subspecies of bald cypress (Denny and Arnold, 2007), has a more limited range than bald cypress and is found primarily in Florida and southern Georgia; it is not present along the Mississippi River floodplain except in southeastern Louisiana. There is a third cypress, Montezuma cypress (*T. distichum* var. *mexicanum* Gordon), that is found in Mexico and southern Texas. Another species indicative of the deepwater swamp is the water tupelo (*Nyssa aquatica* L.), which has a range similar to that of bald cypress along the Atlantic Coastal Plain and the Mississippi River, although it is generally absent from Florida except for the western peninsula. Water tupelo occurs in pure stands or is mixed with bald cypress in floodplain swamps.

Geographic Extent **375**

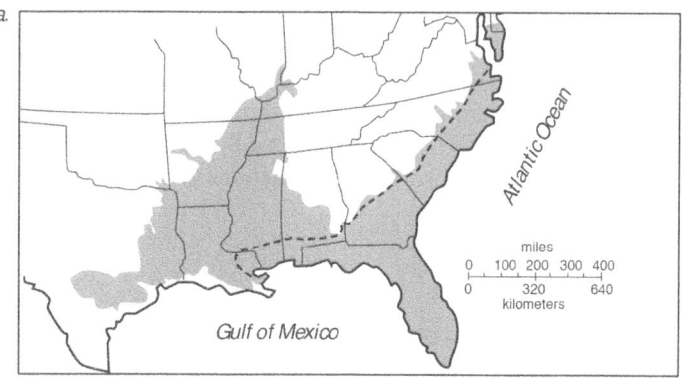

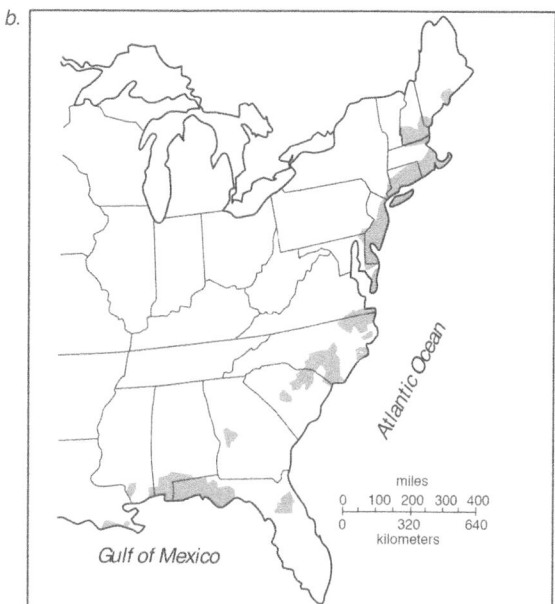

Figure 11.1 Distribution of dominant forested wetland trees in the southeastern United States: (a) bald cypress (*Taxodium distichum*) and pond cypress (*Taxodium distichum* var. *imbricarium*) (with dotted line indicating northern extent of pond cypress) and (b) white cedar (*Chamaecyparis thyoides*). (After Little, 1971; Laderman, 1989)

White Cedar Swamp

White cedar swamps, dominated by Atlantic white cedar (*Chamaecyparis thyoides* [L.] BSP), were once abundant along the Atlantic and Gulf of Mexico coastlines of the United States as far north as southeastern Maine (Fig. 11.1b). These wetlands are not nearly as plentiful as are cypress–tupelo swamps. White cedar occurs in about 2,150 km² of forestland, but the species accounts for a majority of the trees in only about 442 km² (Sheffield et al., 1998). Only 53 km² of Atlantic white cedar swamps remain in the glaciated northeastern United States, with red maple (*Acer rubrum* L.) swamps now more prevalent there. The three states that have the most area of timberland with Atlantic white cedar are North Carolina, Florida, and New Jersey. The regions with the highest concentrations of Atlantic white cedar are the Pinelands of southeastern New Jersey; the Dismal Swamp of Virginia and North Carolina; and the floodplains of the Escambia, Apalachicola, and Blackwater rivers in Florida (Sheffield et al., 1998).

Red Maple Swamps

One of the most common of the broad-leaved deciduous forested wetlands in the northeastern United States is the red maple (*Acer rubrum*) swamp. Toward the west into Pennsylvania and Ohio, red maple swamps are replaced by swamps dominated by trees such as ash (*Fraxinus* spp.), American elm (*Ulmus americana*), swamp white oak (*Quercus bicolor*), and a number of other species, but in the northeastern United States, the red maple swamp is the most common swamp. Using an approximation that all broad-leaved deciduous forested wetlands in several of the coastal states of the northeastern United States are red maple swamps (this approximation would not apply west or south of New York), Golet et al. (1993) estimated that there were $3,530 \text{ km}^2$ of red maple swamps in these six states. Red maple forests also occur in the Upper Peninsula of Michigan and northeastern Wisconsin. The range of the species *Acer rubrum* extends westward to the Mississippi River and northward through much of Ontario and parts of Manitoba and Newfoundland, but the tree can grow in both wetlands and dry, sandy or rocky uplands. Thus, the presence of red maple does not always indicate wetlands, as would the presence of cypress, tupelo, or white cedar.

Riparian Ecosystems

In general terms, riparian ecosystems are found wherever streams or rivers at least occasionally cause flooding beyond their channel confines or where new sites for vegetation establishment and growth are created by channel meandering (e.g., point bars). In arid regions, riparian vegetation may be found along or in ephemeral streams as well as on the floodplains of perennial streams. In most nonarid regions, floodplains and hence riparian zones tend to appear first along a stream "where the flow in the channel changes from ephemeral to perennial—that is, where groundwater enters the channel in sufficient quantity to sustain flow through nonstorm periods" (Leopold et al., 1964).

Riparian ecosystems can be broad alluvial valleys several tens of kilometers wide or narrow strips of streambank vegetation in the arid regions. The "abundance of water and rich alluvial soils" (Brinson et al., 1981) are the factors that make riparian ecosystems different from upland ecosystems. Three major features separate riparian ecosystems from other ecosystem types:

1. Riparian ecosystems generally have a linear form as a consequence of their proximity to rivers and streams.

2. Energy and material from the surrounding landscape converge and pass through riparian ecosystems in much greater amounts than those of any other wetland ecosystem; that is, riparian systems are open systems.

3. Riparian ecosystems are functionally connected to upstream and downstream ecosystems and are laterally connected to upslope (upland) and downslope (aquatic) ecosystems.

Mesic Riparian Ecosystems

Mesic riparian ecosystems, commonly called *bottomland hardwood forests* or bottomland hardwoods in the United States, are one of the dominant types of riparian ecosystems. Historically the term *bottomland hardwood forest* has been used to describe the vast forests that occur on river floodplains of the eastern and central United States, especially in the Southeast. Bottomland hardwood forests are particularly notable wetlands because of the large areas that they cover in the southeastern United States and because of the rapid rate at which they are being converted to other uses, such as agriculture and human settlements. This ecosystem is particularly prevalent in the lower Mississippi River alluvial valley as far north as southern Illinois and western Kentucky and along many streams that drain into the Atlantic Ocean on the south Atlantic Coastal Plain. The Nature Conservancy (1992) estimated that before European settlement, the Mississippi River alluvial plain supported about 21 million ha of riparian forests; about 4.9 million ha remained as of 1991. The Atlantic Coastal Plain from Maryland to Florida is another area of dense riparian forests lining the many rivers that flow into the ocean.

Arid Riparian Ecosystems

Along high-order rivers, the contrast in elevation and vegetation between mesic riparian ecosystems and upland forests is often subtle and the gradients are gradual, whereas for arid riparian ecosyems, the gradients are usually sharp and the visual distinctions are usually clear. In the western United States and many other arid parts of the world, these narrow riparian zones have been extensively modified by human activity. Conversion to housing or agriculture is widespread. Damage from grazing animals is almost ubiquitous. In an area where vegetation is generally limited by the lack of water, riparian vegetation and the availability of water inevitably draw and concentrate cattle. Grazing along these primarily low-order streams results in increased erosion and channel downcutting while higher-order streams have been modified for water use.

Geomorphology and Hydrology

Cypress Swamps

Southern cypress–tupelo swamps occur under a variety of geologic and hydrologic conditions, ranging from the extremely nutrient-poor dwarf cypress communities of southern Florida to the rich floodplain swamps along many tributaries of the lower

Mississippi River basin. A useful classification of deepwater swamps in terms of their geological and hydrological conditions includes five types (Fig. 11.2):

1. *Cypress domes.* Cypress domes (sometimes called cypress ponds or cypress heads) are poorly drained to permanently wet depressions dominated by pond cypress. They are generally small in size, usually 1 to 10 ha, and are numerous in the upland pine flatwoods of Florida and southern Georgia. Cypress domes are found in both sandy and clay soils and usually have several centimeters of organic matter that has accumulated in the wetland depression. These wetlands are called *domes* because of their appearance when viewed from the side: The larger trees are in the middle, and smaller trees are toward the edges (Fig. 11.2a). Ewel and Wickenheiser (1988) confirmed that trees grow slowest at the edges and fastest near the center of the domes but found no significant differences in tree growth among small, medium, and large cypress domes. This dome phenomenon, it has been suggested, is caused by deeper peat deposits in the middle of the dome, fire that is more frequent around the edges of the dome, or a gradual increase in the water level that causes the dome to "grow" from the center outward (Vernon, 1947; Kurz and Wagner, 1953; Watts et al., 2012). A definite reason for this profile has not been determined, nor do all domes display the characteristic shape. An example of a water budget for a cypress dome in north-central Florida is shown in Figure 11.3a.

2. *Dwarf cypress swamps.* Dwarf cypress swamps are major areas in southwestern Florida, primarily in the Big Cypress Swamp and the Everglades, where pond cypress is the dominant tree, but it grows stunted and scattered in a herbaceous understory marsh (Fig. 11.2b). The trees generally do not grow more than 6 or 7 m high and are more typically 3 m in height. The poor growing conditions are caused primarily by the lack of suitable substrate overlying the bedrock limestone that is found in outcrops throughout the region. The hydroperiod includes a relatively short period of flooding as compared with other deepwater swamps, and fire often occurs. The cypress, however, are rarely killed by fire because of the lack of fuel buildup and litter accumulation.

3. *Lake-edge swamps.* Bald cypress swamps are also found as margins around many lakes and isolated sloughs in southeastern United States, ranging from Florida to southern Illinois (Fig. 11.2c). Tupelo and water-tolerant hardwoods such as ash (*Fraxinus* spp.) often grow in association with the bald cypress. A seasonally fluctuating water level is characteristic of these systems and is necessary for seedling survival. The trees in these systems receive nutrients from the lake as well as from upland runoff. The lake-edge swamp can be a filter that receives overland flow from the uplands and allows sediments to settle out and chemicals to adsorb onto the sediments before the water discharges into the open lake. The importance of this filtering function, however, has not been adequately investigated.

4. *Slow-flowing cypress strands.* Cypress strands (Fig. 11.2d) are found primarily in southwest Florida, where the topography is slight, and rivers are replaced by slow-flowing strands with little erosive power. The substrate is primarily sand, and

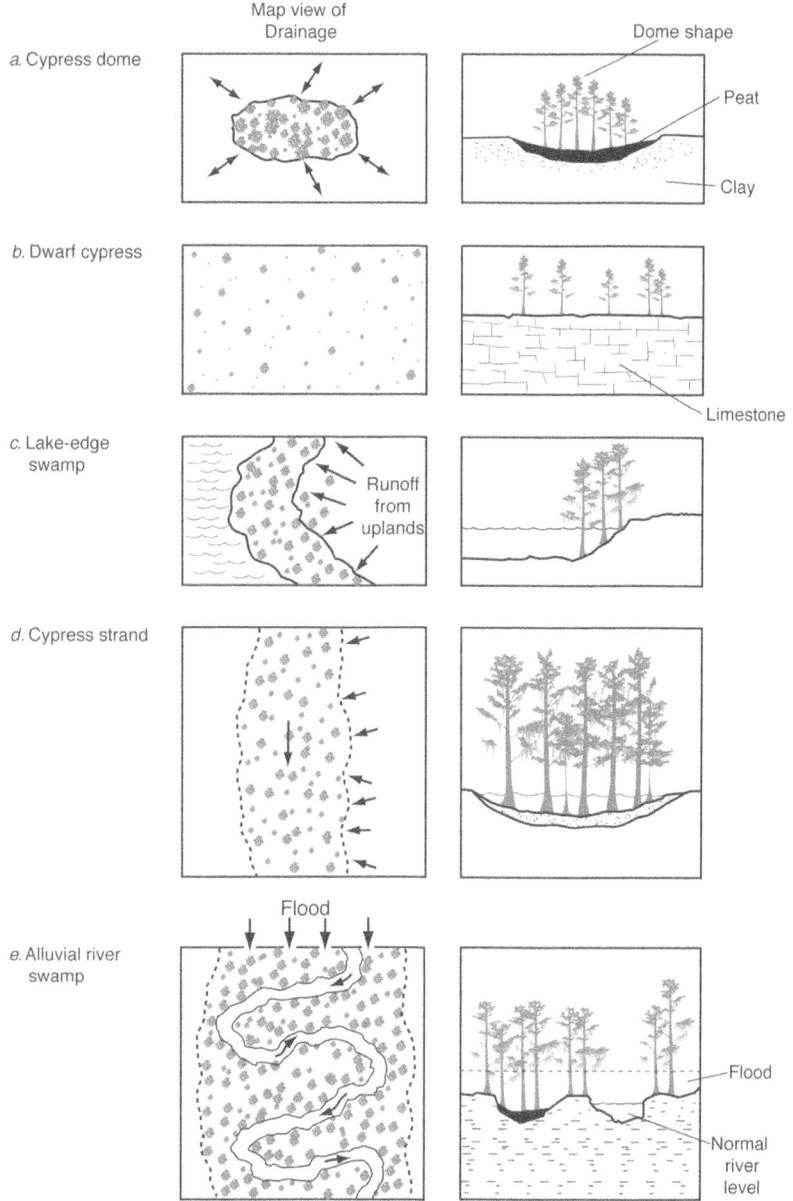

Figure 11.2 General profile and flow pattern of major types of deepwater swamps, showing (a) cypress dome, (b) dwarf cypress, (c) lake-edge swamp, (d) cypress strand, and (e) alluvial river swamp. (After H. T. Odum, 1982)

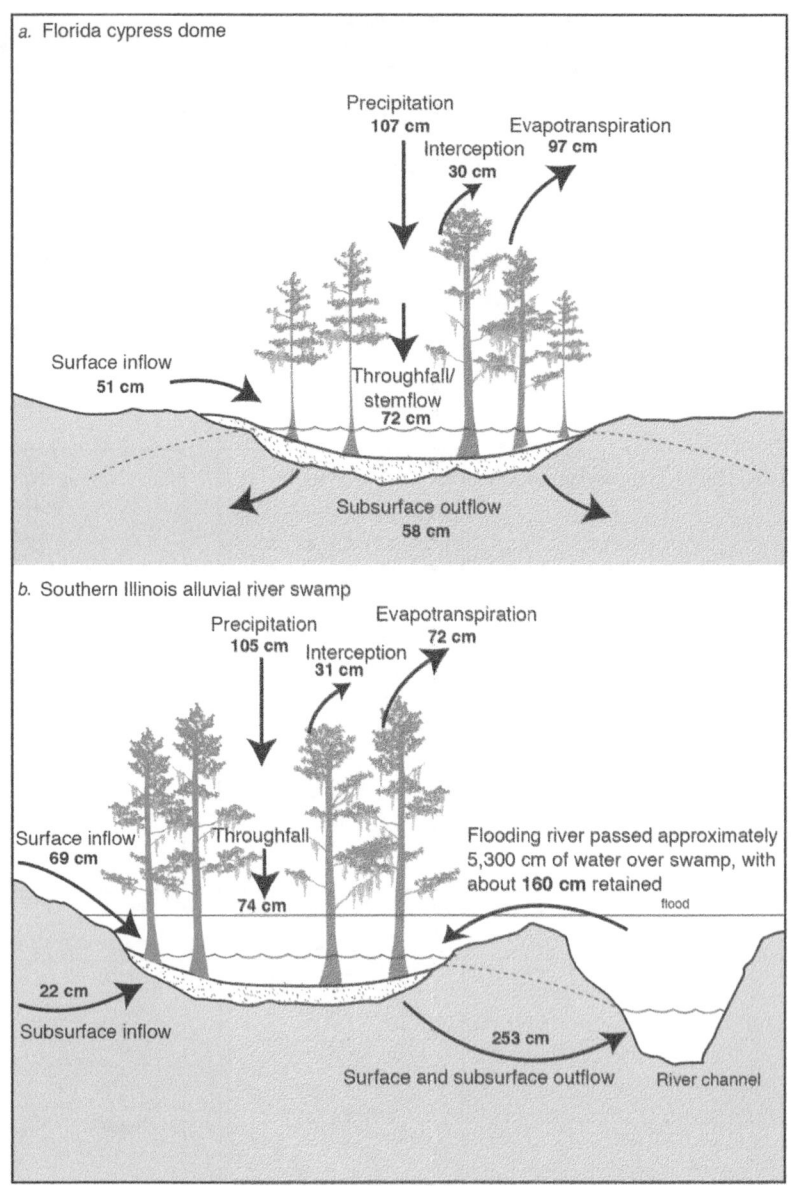

Figure 11.3 Annual water budgets for (a) Florida cypress dome and (b) southern Illinois cypress–tupelo alluvial cypress swamp. ((a) After Heimburg, 1984; (b) After Mitsch et al., 1979)

there is some mixture of limestone and remnants of shell beds. Peat deposits are shallow on higher ground and deeper in the depressions. The hydroperiod has a seasonal wet-and-dry cycle. The deeper peat deposits usually retain moisture even in extremely dry conditions. Much is known about cypress strands from many studies done in Fakahatchee Strand and Corkscrew Swamp (e.g., Carter et al., 1973; Duever et al., 1984; Villa and Mitsch, 2014, 2015).

5. *Alluvial river swamps.* The broad alluvial floodplains of rivers and creeks in humid climates support a vast array of forested wetlands. In the southeastern and lower Mississippi River basin, some of these are permanently flooded deepwater swamps as part of a seasonally flooded forest (Fig. 11.2e). Alluvial river swamps, dominated by bald cypress or water tupelo or both in the southeastern United States, are confined to permanently flooded depressions on floodplains such as abandoned river channels (*oxbows* or *billabongs* in Australia) or elongated swamps that usually parallel the river (*sloughs*). Alluvial river swamps are continuously or almost continuously flooded. The hydrologic inflows are dominated by runoff from the surrounding uplands and by overflow from the flooding rivers. A water budget for an alluvial cypress–tupelo swamp in southern Illinois is shown in Figure 11.3b, and a phosphorus budget for the same swamp, showing the importance of the river input, is shown in Figure 6.18.

White Cedar Swamps

White cedar swamps occupy a narrow hydrologic niche generally between deepwater cypress–tupelo swamps and moist-soil red maple swamps. The hydrologic regime of cedar swamps can be classified as seasonally flooded, with flooding for an extended period during the growing season. Golet and Lowry (1987) found that a group of swamps in Rhode Island had a wide variability in annual water-level fluctuations, ranging from 17 to 75 cm in amplitude and averaging 42 cm over a seven-year period. The percentage of wetland flooded during the growing season ranged from 18 to 76 percent.

Red Maple Swamps

Red maple swamps and mineral-soil forested wetlands occur, in general, in several different hydrogeomorphic regimes, the most common being isolated basins in glacial till or glaciofluvial deposits left behind by glaciations. The hydroperiod for two red maple swamps in Rhode Island is shown in Figure 4.4, and the different hydrologic settings for these types of wetlands are illustrated in Figure 4.13. These wetlands are heavily influenced by regional and local groundwater patterns.

Riparian Ecosystems

Riparian ecosystems are influenced by river flood pulses, usually in the wet winter/spring season, and dry conditions during much of the growing season. They may or may not be jurisdictional wetlands as determined in the United States (see Chapter 15:

"Wetland Laws and Protection") because of the lack of sufficient root-zone flooding in the growing season. Riparian vegetation along a stream or river is determined by the cross-sectional morphology, including braiding of the stream, width of the floodplain, soil type, and elevation and moisture gradients. These are all determined in part by larger scale (continental, basin, stream system) processes that are modified by local biotic and physical processes. The riparian soil moisture regime, in large part, explains the plant associations. The relationship, however, is seldom that simple. Soil moisture and depth to groundwater are not the only factors governing plant establishment. Low floodplain elevations are often swept clean of plants by floods, so that seedlings do not survive, and the vegetation is limited to annuals and perennials that survive until the next flood. Trees mature only at elevations above moderate floods where they can become well enough established to withstand severe floods. There are distinct differences in riparian ecosystems and floodplains in mesic and arid climates.

Mesic Riparian Ecosystems

Most of the extensive riparian ecosystems of south and eastern United States, which include the alluvial cypress swamps described above, are characterized as zones of deposition. These river systems are dominated by spring floods and late-summer flow minima. A typical broad floodplain in mesic climates such as eastern North America contains eight major features (Fig. 11.4):

1. The river channel meanders through the area, transporting, eroding, and depositing alluvial sediments.
2. Natural levees adjacent to the channel are composed of coarse materials that are deposited when floods flow over the channel banks. Natural levees, sloping sharply toward the river and more gently away from the floodplain, are often the highest elevation on the floodplain.
3. *Point bars* are areas of sedimentation on the convex sides of river curves. As sediments are deposited on the point bar, the meander curve of the river tends to increase in radius and migrate downstream. Eventually, the point bar begins to support vegetation that stabilizes it as part of the floodplain.
4. *Meander scrolls* are depressions and ridges on the convex side of bends in the river. They are formed from point bars as the stream migrates laterally across the floodplain. This type of terrain is often referred to as ridge and swale topography.
5. *Oxbows*, oxbow lakes, or *billabongs* (in Australia) are bodies of permanently standing water that result from the cutoff of meanders. Deepwater swamps or freshwater marshes often develop in oxbows.
6. *Sloughs* are areas of dead water that form in meander scrolls and along valley walls. Deepwater swamps can also form in the permanently flooded sloughs.
7. *Backswamps* are deposits of fine sediments that occur between the natural levee and the valley wall or terrace.

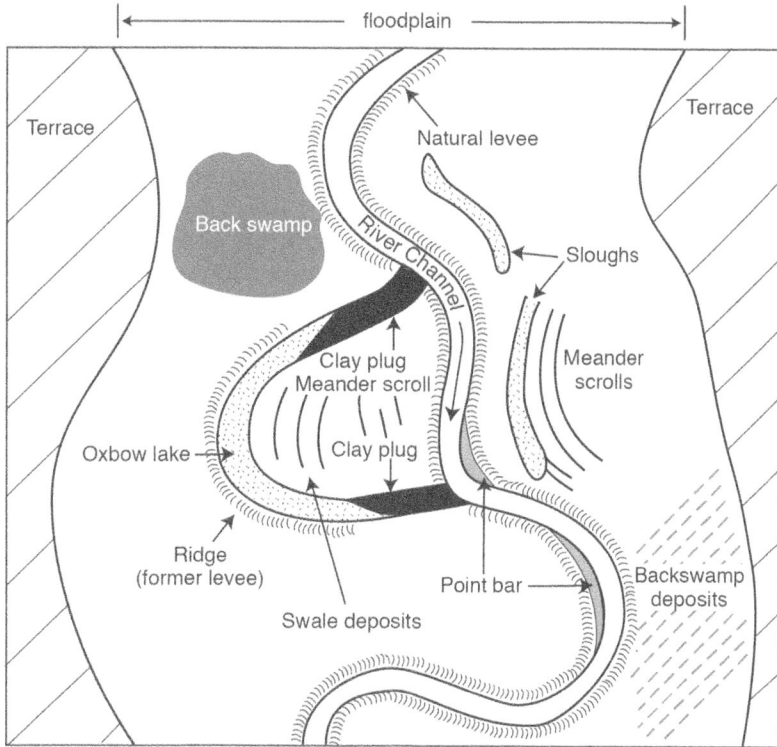

Figure 11.4 Major river geomorphic features of mesic riparian ecosystems (floodplains) including natural levees, meander scrolls, oxbow lakes, back swamp, and river meanders with point bars. (From Mitsch and Jørgensen, 2004)

8. *Terraces* are "abandoned floodplains" that may have been formed by the river's alluvial deposits but are not hydrologically connected to the current river.

Two major aggradation processes are thought to be responsible for the formation of most floodplains: deposition on the inside curves of rivers (point bars) and deposition from overbank flooding. "As a river moves laterally, sediment is deposited within or below the level of the bankfull stage on the point bar, while at overflow stages the sediment is deposited on both the point bar and over the adjacent flood plain" (Leopold et al., 1964). The resulting floodplain is made up of alluvial sediments (or alluvium) that can range from 10 to 80 m thick. Degradation (downcutting) of floodplains occurs when the supply of sediments is decreases, a condition that could be caused naturally with a shift in climate or with the construction of an upstream dam. These processes are difficult to observe over short periods; both aggradation and degradation can be inferred only from the study of floodplain stratigraphy or long-term mapping.

Arid Riparian Ecosystems

Structurally, the temporal and spatial stability of these riparian ecosystems and rivers in arid regions is fundamentally different from those of rivers in mesic climates. In contrast to the broad, flat, expansive southeastern U.S. riparian forests, for example, riparian ecosystems in arid regions, such as western United States, tend to be narrow, linear features of the landscape, often lining streams with steep gradients and narrow floodplains. Because of the dramatic differences in peak floods compared to mean flows, the temporally unpredictable nature of flooding, and the coarseness of most sedimentary material in this region, arid-region channels are time-dependent systems that seldom reach any kind of equilibrium. As with mesic rivers, the two primary factors governing channel morphology are the sediment supply and flow variability.

Biogeochemistry

Lockaby and Walbridge (1998) described the biogeochemistry of forested wetlands as "the most complex and difficult to study with any forest ecosystem type." Forested wetlands have soil and water chemistry that varies from the rich sediments of alluvial cypress swamps to the extremely low mineral and acidic waters of surface water depression red maple swamps and cypress domes. Wide ranges of pH, dissolved substances, and nutrients are found in the soils and waters of these swamps. Three facts should be noted from this wide range of soil and water chemistry:

1. Swamps are generally acidic to circumneutral, depending on the accumulation of peat and the degree to which precipitation dominates the hydrology.
2. Nutrient conditions vary from nutrient- and mineral-poor conditions in rainwater-fed swamps to nutrient- and mineral-rich conditions in alluvial river swamps and groundwater discharge swamps.
3. An alluvial river swamp often has water quality very different from that of the adjacent river. Swamps in alluvial settings are generally fed by both groundwater discharge and flooding rives and can have water chemistry quite different from either source.

Many freshwater swamps, particularly alluvial river swamps, are "open" to river flooding and other inputs of neutral and generally well-mineralized waters. The pH of many alluvial swamps in the southeastern United States is 6 to 7, and there are high concentrations of dissolved ions. Cypress domes and perched-basin swamps, in contrast, are fed primarily by rainwater and have acidic waters, usually in the pH range of 3.5 to 5.0, caused by humic acids produced within the swamp. Colloidal humic substances contribute to both the low pH and the tea-colored or "blackwater" appearance of the standing water in many forested wetlands. Isolated swamps, such as cypress domes, have much in common with the oligotrophic or ombrotrophic peatlands described in Chapter 12: "Peatlands." Swamps open to major surface water and groundwater inputs, however, are generally rich in alkalinity, dissolved ions, and nutrients. For example, conductivity of surface water ranges from only 60 µS/cm in

Table 11.1 Soil chemistry of Atlantic white cedar (*Chamaecyparis thyoides*) and red maple (*Acer rubrum*) swamps in Maryland compared to nonforested peatlands

Soil Parameters (top 50 cm)	White Cedar Swamp	Red Maple Swamp	Nonforested Peatlands
pH	5.34	4.23	4.54
Organic matter (percent)	59 ± 5	67 ± 3	68 ± 2
Nitrogen (percent)	1.6 ± 0.1	1.5 ± 0.1	1.7 ± 0.1
Phosphorus (percent)	0.07 ± 0.01	0.24 ± 0.03	0.10 ± 0.01
NO_3–N (µg/g)	0.8 ± 0.1	0.3 ± 0.1	0.5 ± 0.1
NH_4–N (µg/g)	67 ± 4	72 ± 19	76 ± 10
Ca^{2+} (µg/g)	1,810	339	710
Mg^{2+} (µg/g)	1,420	493	477
K^+ (µg/g)	1,054	1,622	857
Na^+ (µg/g)	841	134	383
Fe (mg/g)	6.3	5.9	5.4
Al (mg/g)	8.0	5.4	7.6

Source: Whigham and Richardson (1988)

cypress domes in Florida to 200 to 400 µS/cm in alluvial cypress swamps in Kentucky and Illinois.

In a comparison of an Atlantic white cedar swamp with adjacent forested wetlands and nonforested peatlands, Whigham and Richardson (1988) found cedar swamp soils to be significantly higher in pH, calcium, and magnesium than the other sites (Table 11.1), suggesting a groundwater or brackish-water source might be important for Atlantic white cedar to compete with other swamp trees. Phosphorus was lowest in the white cedar swamp, suggesting this was the most significant limiting nutrient. The high pH measured in this study suggests that Atlantic white cedar may do best in sites with high pH, although these swamps have been reported to occur under low-pH (3.2–4.4) conditions in the Great Dismal Swamp (F. Day, 1984).

In riparian forest soils, phosphorus availability has been shown to increase during floods although the exact reason for this is often unclear. Wright et al. (2001) examined the availability of P after experimentally flooding plots in a Georgia floodplain forest and found that flooding did release P; however they found that there was no change in Fe/Al phosphates. The reduction of Fe^{+3} phosphates and hydrolysis of Al phosphates has often been credited with increased P availability after soils have been flooded. While this may occur when upland soils are flooded, the authors attributed the increased available P during floods to biological processes such as the release of P from microbial biomass and the suppression of biological P demand during anaerobic conditions.

Vegetation

Cypress Swamps

Southern deepwater swamps, particularly cypress wetlands, have plant communities that either depend on or adapt to the almost continuously wet environment. There are

Table 11.2 Distinction between bald cypress and pond cypress swamps

Characteristic	Bald Cypress Swamp	Pond Cypress Swamp
Dominant cypress	*Taxodium distichum*	*Taxodium distichum* var. *imbricarium*
Dominant tupelo or gum (when present)	*Nyssa aquatica* (water tupelo)	*Nyssa sylvatica* var. *biflora* (black gum)
Tree physiology	Large, old trees, high growth rate, usually abundance of knees and spreading buttresses	Smaller, younger trees, low growth rate, some knees and buttresses but not as pronounced
Location	Alluvial floodplains of Coastal Plain, particularly along Atlantic seaboard, Gulf seaboard, and Mississippi embayment	"Uplands" of Coastal Plain, particularly in Florida and southern Georgia
Chemical status	Neutral of slightly acid, high in dissolved ions, usually high in suspended sediments and rich in nutrients	Low pH, poorly buffered, low in dissolved ions, poor in nutrients
Annual flooding from river	Yes	No
Types of deepwater swamps	Alluvial river swamp, cypress strand, lake-edge swamp	Cypress dome, dwarf cypress swamp

several distinctions between bald cypress and pond cypress swamps. The dominant canopy vegetation found in alluvial river swamps of the southeastern United States includes bald cypress (*Taxodium distichum*) and water tupelo (*Nyssa aquatica*). The trees are often found growing in association in the same swamp, although pure stands of either bald cypress or water tupelo are also frequent in the southeastern United States. Many of the pure tupelo stands may have been the result of the selective logging of bald cypress. The pond cypress–black gum (*Taxodium distichum* var. *imbricarium* [Nutt.]–*Nyssa sylvatica* var. *biflora* [Walt.] Sarg.) swamp is more commonly found on the uplands of the southeastern Coastal Plain, usually in areas of poor sandy soils without alluvial flooding (Table 11.2). These same conditions are usually found in cypress domes.

One of the main features that distinguishes bald cypress trees from pond cypress trees is the leaf structure (Fig. 11.5). Bald cypress has needles that spread from the twig in a flat plane, whereas pond cypress needles are appressed to the twig. Both species are intolerant of salt and are found only in freshwater areas. Pond cypress is limited to sites that are poor in nutrients and are relatively isolated from the effects of river flooding or large inflows of nutrients.

When deepwater swamps are drained or when their dry period is extended dramatically, they can be invaded by pine (e.g., *Pinus elliottii*) or hardwood species. In north-central Florida, a cypress–pine association indicates a drained cypress dome (Mitsch and Ewel, 1979). Hardwoods that characteristically are found in cypress domes include swamp red bay (*Persea palustris*) and sweet bay (*Magnolia virginiana*). In lake-edge and alluvial river swamps, several species of ash (*Fraxinus* sp.) and maple (*Acer* sp.) often grow as subdominants with the cypress or tupelo or both. In the

Figure 11.5 Distinction of leaves between top: bald cypress (*Taxodium distichum*) and bottom: pond cypress (*Taxodium distichum* var. *imbricarium*; formerly known as *T. distichum* var. *nutans*).

Deep South, Spanish moss (*Tillandsia usneoides*) is found in abundance as an epiphyte on the stems and branches of the canopy trees.

The abundance of understory vegetation in cypress–tupelo swamps depends on the amount of light that penetrates the tree canopy. Many mature swamps appear as quiet, dark cathedrals of tree trunks devoid of any understory vegetation. Even when enough light is available for understory vegetation, it is difficult to generalize about its composition. There can be a dominance of woody shrubs, of herbaceous vegetation, or of both. Fetterbush (*Lyonia lucida*), wax myrtle (*Myrica cerifera*), and Virginia willow (*Itea virginica*) are common as shrubs and small trees in nutrient-poor cypress domes. Understory species in higher-nutrient river swamps include buttonbush (*Cephalanthus occidentalis*) and Virginia willow. Some continually flooded cypress swamps that have high concentrations of dissolved nutrients in the water develop dense mats of duckweed (e.g., *Lemna* spp., or *Spirodela* spp., or *Azolla* spp.) on the water surface during most of the year. Floating logs and old tree stumps often provide substrate for understory vegetation to attach and to flourish.

White Cedar Swamps

Cedar swamps occur within a wide climatic range along the East Coast of the United States and in an intermediate hydrology between deepwater cypress swamps in the South and forested swamps such as red maple swamps in the North. Often these swamps are monospecific, even-aged stands with tightly spaced *Chamaecyparis thyoides* trees, and no subcanopy, few shrubs, and minimal herbaceous plants. However, the tree is often found in mixed stands, with co-dominants such as *Betula populifolia* (gray birch), *Picea mariana* (black spruce), *Pinus strobus* (Eastern white pine), and *Tsuga canadensis* (Eastern hemlock) (Laderman, 1989). In the South co-dominant trees include *Gordonia lasianthus* (loblolly bay), *Persea borbonia* (red bay), *P. palustris* (swamp red bay), and *Taxodium distichum* (bald cypress).

The shrub layer in cedar swamps with relatively open canopies includes many ericaceous shrubs, such as *Aronia arbutifolia* (red chokeberry), *Clethra alnifolia* (sweet pepperbush), *Ilex glabra* (gallberry), *Leucothoe racemosa* (fetterbush), and *Vaccinium corymbosum* (highbush blueberry) (Laderman, 1989).

Red Maple Swamps

The canopy of red maple swamps is obviously dominated by *Acer rubrum* L. Canopy cover generally exceeds 80 percent, although trees in these northern swamps tend to be shorter with less biomass than those in southern swamps. Although up to 50 tree species have been found in a red maple swamp, the red maple can account for up to 90 percent of the stem density and basal area (Golet et al., 1993). In general, a specific site will have about four species of trees in the canopy/subcanopy, depending on which region of the glaciated Northeast these red maple swamps occur.

Shrubs include *Ilex vertucillata* (winterberry), *Vaccinium corymbosum* (highbush blueberry), *Lindera benzoin* (spicebush), *Viburnum* spp. (arrowwood), *Alnus rugosa* (speckled alder), *Cephalanthus occidentalis* (buttonbush), *Corylus cornuta* (hazelnut), and *Rhododendron viscosum* (swamp azalea), with dominance depending on the region in which the swamps are found. Shrub cover is generally greater than 50 percent, although some red maple swamps have shrub cover as low as 6 percent. One of the most interesting features of many red maple swamps is the predominance of a great variety of ferns in the herbaceous layer, including *Osmunda cinnamomea* (cinnamon fern), *Onoclea sensibilis* (sensitive fern), *Osmunda regalis* (royal fern), *Thelypteris thelypteroides* (marsh fern), *Matteuccia struthiopteris* (ostrich fern), *Osmunda claytoniana* (interrupted fern), and various *Dryopteris* spp. (wood ferns). Other common herbaceous plants include *Symplocarpus foetidus* (skunk cabbage), *Caltha palustris* (marsh marigold), several species of *Glyceria* (manna grass), and several of more than 32 species of *Carex*.

Riparian Ecosystems

Southeastern U.S. Bottomland Forests

The vegetation of high-order southeastern riparian ecosystems is dominated by diverse trees that are adapted to the wide variety of environmental conditions on the floodplain. The most important local environmental condition is the hydroperiod, which determines the "moisture gradient," or—as Wharton et al. (1982) prefer—the "anaerobic gradient," which varies in time and space across the floodplain. The plant species found along this gradient respond to elevation relative to the river's flooding regime (Fig. 11.6). The lowest parts of the bottomland, nearly always flooded, form cypress–tupelo gum swamps above. At slightly higher bottomland elevations than the deep swamps, the soils are semipermanently inundated or saturated and support an association of black willow (*Salix nigra*), silver maple (*Acer saccharinum*), and sometimes cottonwood (*Populus deltoides*) in the pioneer stage. A more common

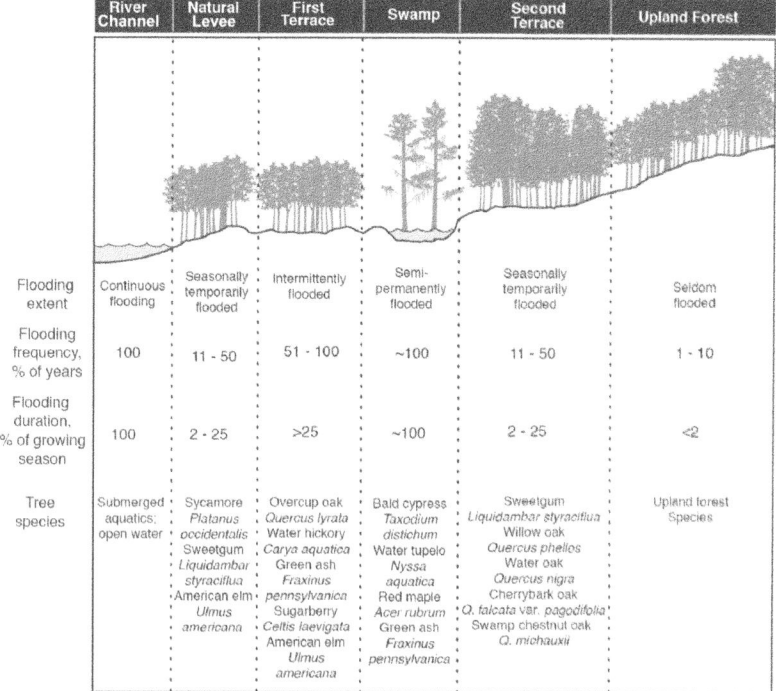

Figure 11.6 **General relationship between vegetation associations and floodplain topography, flood frequency, and flood duration of a southeastern United States bottomland hardwood forest. (From Mitsch and Gossslink, 2000)**

association in this zone includes overcup oak (*Quercus lyrata*) and water hickory (*Carya aquatica*), which often occur in relatively small depressions on floodplains. Also found in this zone are green ash (*Fraxinus pennsylvanica*), red maple (*Acer rubrum*), and river birch (*Betula nigra*). Higher still on the bottomland floodplain in areas flooded or saturated one to two months during the growing season are found an even wider array of hardwood trees, including laurel oak (*Quercus laurifolia*), green ash (*Fraxinus pennsylvanica*), American elm (*Ulmus americana*), and sweetgum (*Liquidambar styraciflua*) as well as sugarberry (*Celtis laevigata*), red maple (*Quercus rubra*), willow oak (*Quercus phellos*), and sycamore (*Platanus occidentalis*). Pioneer successional communities in this zone can consist of monotypic stands of river birch or cottonwood.

Temporarily or infrequently flooded terraces at the highest elevations of the floodplain (second terrace in Fig. 11.6) are flooded for less than a week to about a month during each growing season and are often dominated by several oaks, tolerant of occasionally wet soils such as swamp chestnut oak (*Quercus michauxii*), cherrybark oak (*Quercus falcata* var. *pagodifolia*), and water oak (*Quercus nigra*) and hickories (*Carya* spp.).

Plant zonation is not linear topographically, nor is it vegetationally discrete. Figure 11.6 is a cross section of the microtopography of an alluvial floodplain in the southeastern United States. In reality, the complex microrelief does not show a smooth change from one zone to the next. The natural levee next to the stream (Fig. 11.6), in fact, is often one of the most diverse parts of the floodplain because of fluctuations in its elevation.

Arid and Semiarid Riparian Forests

The vegetation in riparian forests of the semiarid grasslands and arid western United States differ from those found in the humid eastern and southern United States. The natural upland ecosystems of this region are grasslands, deserts, or other nonforested ecosystems, and so the riparian zone is a conspicuous feature of the landscape. Western U.S. riparian ecosystem tree species are *phreatophytes*, that is, they are plants that obtain their water from *phreatic* sources (i.e., groundwater or the capillary fringe of the groundwater table). Many species use surface water supplies when seedlings (hence the general germination requirement of bare, moist soil) but put down long, deep roots that later supply water requirements from groundwater. Cottonwoods (*Populus* spp.) are considered obligate phreatophytes, while both *Prosopis pubescens* (mesquite) and alien *Tamarix ramisissima* (salt cedar) are facultative. Salt cedar is an introduced species that is rapidly replacing cottonwood in many areas.

Swamps of Glaciated Regions

Forested swamps occur throughout the glaciated midwestern United States; in fact, most of the wetlands remaining in states such as Ohio, Indiana, and Illinois are forested wetlands that occur in isolated basins or floodplains amid agricultural fields (Table 11.3). They were the fields that were too wet to plant and gradually were

Table 11.3 Typical vegetation in a hardwood swamp forest in central Ohio[a]

Trees	Wetland indicator Status[b]
Trees	
Quercus palustris (pin oak)	FACW
Quercus bicolor (swamp white oak)	FACW
Acer saccharinum (silver maple)	FACW
Acer rubrum (red maple)	FAC
Ulmus americana (American elm)	FACW
Fraxinus pennsylvanica (green ash)	FACW
Shrubs/Understory	
Lindera benzoin (spicebush)	FACW
Cephalanthus occidentalis (buttonbush)	OBL
Rosa multiflora[c] (multiflora rose)	FACU
Carpinus caroliniana (hornbeam, ironwood)	FAC
Herbs	
Polygonum spp. (smartweed)	FAC/OBL
Symplocarpus foetidus (skunk cabbage)	OBL
Lemna spp. (duckweed)	OBL
Alisma plantago-aquatica (water plantain)	OBL
Aster spp. (asters)	FAC/FACW
Carex spp. (sedges)	FACW/OBL
Ranunculus septentrionalis (swamp buttercup)	OBL
Saxifraga pennsylvanica (swamp saxifrage)	OBL
Onoclea sensibilis (sensitive fern)	FACW
Bidens comosa (leafy-bracted beggar-ticks)	FACW
Bidens frondosa (devil's beggar-ticks)	FACW
Scirpus atrovirens (green bulrush)	OBL
Scirpus cyperinus (wool grass)	FACW

[a]Wetland species at Gahanna Woods Nature Preserve, Franklin County, Ohio.
[b]Use of wetland indicator status for the northeastern United States. In order of wet to dry: OBL = obligate wetland plant; FACW = facultative wet plant; FAC = facultative plant; FACU = facultative upland plant.
[c]Nonnative species.

invaded by tree species. They often are a remnant of a gradual process of ponds of glacial origin slowly infilling and becoming forested (a true hydrarch succession). However, they may also occur in wet basins on mineral hydric soils rather than peat deposits. As with red maple swamps, the trees generally have replaced herbaceous marshes that once occupied those sites, because of natural succession or because of artificial drainage. The succession of these systems is poorly understood.

Tree Adaptations

Vascular plants, particularly trees, have a difficult time surviving under continuously flooded conditions. Only a handful of species of trees in North America can stay viable in continuous flooding, and, even then, their growth is generally slowed; trees that

> **Fires in Swamps?**
>
> Fire is generally infrequent in swamps because of standing water or saturated soil conditions, but it can be a significant ecological factor during droughts or in swamps that have been artificially drained. In general, fire is more frequent in the forested swamps of Florida than anywhere else, because of the more frequent lightning storms and because of a predictable dry season. For example, from 1970 to 1977, there were four fires in the Big Cypress National Preserve in southern Florida, each affecting an average of 500 ha. In April–May 2009, a fire burned a much larger area (12,000 ha) of the northwestern part of the preserve (Watts et al., 2012). Fire is rare in most alluvial river swamps but can be more frequent in cypress domes or dwarf cypress swamps—as frequent as several times per century.
>
> Fire had a "cleansing" effect on the trees in a cypress dome in north-central Florida in the sense that the fire selectively killed almost all of the upland pine and hardwoods that had invaded the cypress dome but left the cypress unharmed (Ewel and Mitsch, 1978). This suggests a possible advantage of fire to some shallow cypress ecosystems in eliminating competition that is less water tolerant. Casey and Ewel (2006) identified fire severity as a key factor influencing tree succession in Florida pond cypress swamps. In their generalized succession model, the exclusion of fire (due to geomorphic conditions) tends to promote mixed bay–cypress communities, while periodic moderate fires tend to promote monotypic cypress or cypress–tupelo forests. Severe fires can lead to shrub or marsh conditions.
>
> Fire can also be an influential factor on white cedar swamps. If water is low, fire can be quite destructive, killing cedar trees and burning the peat deeply. If water levels are high, light fire can have a cleansing effect, eliminating shrubs and brush and favoring cedar seedling germination (Laderman, 1989). In *C. thyoides* swamps of the Atlantic Coast, the highly flammable cedar foliage burned frequently (five fires per each 100- to 200-year interval) during pre-European settlement time; when fires became more rare after European settlement, stands of cedar became the familiar dense monospecific systems that are common today (Motzkin et al., 1993).
>
> Watts (2013) and Watts and Kobziar (2013) described fires in wetlands such as cypress swamps as often being *smoldering combustion* or ground fires, as opposed to flaming combustion typical of fires in upland forests. These fires can continue for many days or even months, are much more difficult to control than are flaming fires, and produce an additional human hazard of abundant smoke, day and night.

are found in freshwater swamps are stressed with the wet conditions but have found ways to adapt. The most conspicuous adaptations specific to the major tree species in forested swamps are discussed here.

Knees and Pneumatophores

Cypress (bald and pond), water tupelo, and black gum are among a number of wetland plants that produce pneumatophores. In deepwater swamps, these organs extend from the root system to well above the average water level (Fig. 11.7a). On cypress, these "knees" are conical and typically less than 1 m in height, although some cypress knees are as tall as 3 to 4 m. Knees are generally much more prominent on cypress than on tupelo. Pneumatophores on black gum in cypress domes are actually arching or "kinked" roots that approximate the appearance of cypress knees. The functions of the knees have been speculated about for more than a century. It was thought that knees might be adaptations for anchoring trees because of the appearance of a secondary root system beneath knees that is similar to and smaller than the trees' main root system. Observations of swamp and upland damage in South Carolina following Hurricane Hugo in 1990 showed that cypress trees often remained standing while hardwoods and pines did not, supporting the tree-anchoring theory for cypress root, knee, and buttress systems (K. Ewel, personal communication).

Figure 11.7 Among the several features of vegetation in cypress swamps are (a) cypress knees, (b) cypress tree butresses, (c) large size and long life of cypress trees.

Other discussions of cypress knee function have centered on their possible use as sites of gas exchange for the root systems. Penfound (1952) argued that cypress knees are often absent where they are most needed—in deep water—and that the wood of the cypress knee is not aerenchymous; that is, there are no intercellular gas spaces capable of transporting oxygen to the root system. However, gas exchange does occur at the knees. S. L. Brown (1981) estimated that gas evolution from knees accounted for 0.04 to 0.12 g C m^{-2} yr^{-1} of the respiration in a cypress dome, and 0.23 g C m^{-2} yr^{-1} in an alluvial river swamp. This accounted for 0.3 to 0.9 percent of the total tree respiration but 5 to 15 percent of the estimated woody tissue (stems and knees) respiration. The fact that carbon dioxide (CO_2) is exchanged at the knee, however, does not prove that oxygen transport is taking place there or that the CO_2 was the result of oxidation of anaerobically produced organic compounds in the root system.

Buttresses

Taxodium and *Nyssa* species and, to a lesser degree, *Chamaecyparis thyoides* often produce swollen bases or buttresses (stem hypertrophy) when they grow in flooded conditions (Fig. 11.7b). The basal swelling can extend from less than 1 m above the soil to several meters, depending on the hydroperiod of the wetland. Swelling generally occurs along the part of the tree that is flooded at least seasonally, although the duration and frequency of the flooding necessary to cause the swelling are unknown. One theory described the height of the buttress as a response to aeration: The greatest swelling occurs where there is a continual wetting and soaking of the tree trunk but where the trunk is also above the normal water level (Kurz and Demaree, 1934). The value of the buttress swelling to ecosystem survivability is unknown; it may simply be a relict response that is of little use to the plant.

Seed Germination and Dispersal

The seeds of swamp trees require oxygen for germination. For example, cypress seeds and seedlings require moist but not flooded soil for germination and survival. Occasional drawdowns, if only at relatively infrequent intervals, are therefore necessary for the survival of trees in these swamps unless floating mats develop. Otherwise, continuous flooding will ultimately lead to an open-water pond.

The dispersal and survival of the seeds of many swamp trees depend on hydrologic conditions. Schneider and Sharitz (1986) found a relatively low number of viable seeds in a seed bank study of a cypress–tupelo swamp in South Carolina. An average of 127 seeds/m^2 were found for woody species (88 percent cypress or tupelo) in the swamp compared to a seed density of 233 seeds/m^2 from an adjacent bottomland hardwood forest. The authors speculated that the continual flooding in the cypress–tupelo swamp leads to reduced seed viability. Huenneke and Sharitz (1986) elaborated further on the importance of *hydrochory* (seed dispersal by water) in these swamps. Hydrologic conditions, particularly scouring by flooding waters, are important factors in determining the composition, dispersal, and survival of seeds in riverine settings. Seeds are transported relatively long distances; the highest seed densities accumulate near obstructions such as logs, tree stumps, cypress knees, and tree stems, and the lowest seed densities occur in open-water areas.

Longevity

Some swamp trees may live for centuries and achieve great sizes (Fig. 11.7c). One individual bald cypress tree in Corkscrew Swamp in southwestern Florida was determined to be about 700 years old. Laderman (1998) reported that the maximum age of *Taxodium* is 1,000 years. By contrast, *Chamaecyparis thyoides* lives to a maximum of 300 years (Clewell and Ward, 1987). Mature bald cypress trees are typically 30 to 40 m in height and 1 to 1.5 m in diameter. Anderson and White (1970) reported a very large cypress tree in a cypress–tupelo swamp in southern Illinois that measured 2.1 m in diameter. C. A. Brown (1984) summarized several reports that documented bald cypress as large as 3.6 to 5.1 m in diameter.

Shallow or Adventitious Roots

Some species, such as *Acer rubrum*, develop very shallow root systems in response to flooding, in all likelihood because the surface soil is closest to the atmospheric source of oxygen. In aerated soils, the same species will develop deep roots. Other swamp species, such as willows (*Salix* sp.), green ash (*Fraxinus pennsylvanica*), and cottonwoods (*Populus deltoides*), develop adventitious roots above-ground from the stem in response to flooding.

Gaseous Diffusion

Woody trees have a particular problem getting oxygen to their rhizosphere when they are flooded, and few species do it well enough to survive continual flooding. The swamp trees, including *Taxodium*, *Nyssa*, *Alnus*, and *Fraxinus*, among others, have the ability to supply oxygen to their root systems in amounts adequate for rhizospheric demands. Solar radiation, which heats up the tree stems by a couple of degrees, causes a light-induced gas flow that can be considerably greater in selected swamp seedlings than in the same trees in the dark: This thermally induced flow of air through vascular plants is called *thermo-osmosis* by some (Grosse et al., 1998) and is enhanced by the development of aerenchymous stem and root tissue (see Chapter 7: "Wetland Vegetation and Succession").

Consumers

Invertebrates

Invertebrate communities, particularly benthic macroinvertebrates, have been analyzed in several cypress–tupelo swamps. A wide diversity and high number of invertebrates have been found in permanently flooded swamps. Species include crayfish, clams, oligochaete worms, snails, freshwater shrimp, midges, amphipods, and various immature insects. Batzer and Wissinger (1996) reported that insects, particularly midges, can dominate forested wetlands and that midges are most likely to reach high densities. Many of these invertebrates are highly dependent, either directly or indirectly, on the abundant detritus found in these systems.

Oligochaetes and midges (Chironomidae), both of which can tolerate low-dissolved-oxygen conditions, and amphipods such as *Hyalella azteca*, which occur in abundance amid aquatic plants such as duckweed, usually dominate alluvial river swamp invertebrate communities. In nutrient-poor cypress domes, the benthic fauna are dominated by Chironomidae, although crayfish, isopods, and other Diptera are also found there. Stresses stemming from low dissolved oxygen and periodic drawdowns account for the low diversity and number in these domes.

The production of wood in deepwater swamps results in an abundance of substrate for invertebrates to colonize, although few studies have documented the importance of this substrate in swamps for invertebrates. Thorp et al. (1985) found that suspended *Nyssa* logs had three times as many invertebrates and twice as many taxa when they were placed in a swamp-influent stream than in the swamp itself or by its outflow stream. The swamp inflow had the highest number of mayflies (Ephemeroptera), stoneflies (Plecoptera), midges (Chironomids), and caddie flies (Trichoptera), whereas Oligochaetes were greatest in the swamp itself, supposedly because of anoxic, stagnant conditions. Understory plants within swamps have also been shown to be important to various invertebrate groups.

Fish

Fish are both temporary and permanent residents of alluvial river swamps. Several studies have noted the value of sloughs and backswamps for fish and shellfish spawning and feeding during the flooding season. Forested swamps often serves as a reservoir for fish when flooding ceases, although the backwaters are less than optimum for aquatic life because of fluctuating water levels and occasional low-dissolved-oxygen levels. Some fish such as bowfin (*Amia calva*), gar (*Lepisosteus* sp.), and certain top minnows (e.g., *Fundulus* spp. and *Gambusia affinis*) are better adapted to periodic anoxia through their ability to utilize atmospheric oxygen. Several species of forage minnows often dominate alluvial river swamps, where larger fish are temporary residents of the wetlands. Fish are sparse to nonexistent in the shallow cypress domes, white cedar swamps, and red maple swamps because of the lack of continuous standing water.

Reptiles and Amphibians

Reptiles and amphibians are prevalent in swamps because of their ability to adapt to fluctuating water levels. Nine or 10 species of frogs are common in many southeastern cypress–gum swamps. Two of the most interesting reptiles in southeastern deepwater swamps are the American alligator (*Alligator mississippiensis*) and the cottonmouth moccasin (*Agkistrodon piscivorus*). The alligator ranges from North Carolina through Louisiana, where alluvial cypress swamps and cypress strands often serve as suitable habitats. The cottonmouth, or water moccasin, a poisonous water snake that has a white inner mouth, is found throughout much of the range of cypress wetlands and is the topic of many a "snake story" of those who have been in these swamps. Other water snakes, particularly several species of *Nerodia*, however, are often more important in terms of number and biomass and often are mistakenly identified as cottonmouth. The snakes feed primarily on frogs, small fish, salamanders, and crayfish.

Red maple swamps are important areas in the forested northeastern United States for the breeding and feeding of reptiles and amphibians. DeGraaf and Rudis (1986) found that 45 species of reptiles and amphibians required forest cover sometime during the year in New England and that of the 11 types of forests studied, red maple swamps were actually the preferred habitat of 12 of those 45 species. In a later study, DeGraaf and Rudis (1990) found that red maple swamps with streams supported twice as many individuals of reptiles and amphibians as did red maple swamps without streams, with wood frog (*Rana sylvatica*), redback salamander (*Plethodon cinereus*), and American toad (*Bufo americanus*) accounting for 90 percent of the abundance.

Ecosystem Function

Four generalizations about the ecosystem function of freshwater swamps will be discussed in this section:

1. Swamp productivity is closely tied to its hydrologic regime.
2. Nutrient inflows, often coupled with hydrologic conditions, are major sources of influence on swamp productivity.
3. Swamps can be nutrient sinks whether the nutrients are a natural source or are artificially applied.
4. Decomposition of woody and nonwoody material in swamps is affected by the water regime and the subsequent degree of anaerobiosis.

Primary Productivity

The importance of flood pulsing (the flood stability concept of W. E. Odum et al., 1995) to the productivity of swamps is illustrated in Figure 11.8a, where the basal-area growth of bald cypress in an alluvial river swamp in southern Illinois was strongly correlated with the annual discharge of the adjacent river. This graph suggests that higher tree productivity in this wetland occurred in years when the swamp was flooded more frequently than average or for longer durations by the nutrient-rich river. Similar correlations were also obtained when other independent variables that indicate degree of flooding were used.

The importance of nutrient inflows as well as hydrologic conditions to productivity in cypress swamps in general is illustrated in Fig. 11.8b. Hydrologic inflows and nutrient inflows are coupled in most swamps, so both charts in Figure 11.8 reflect the same phenomenon. There is a wide range of productivity reported for forested swamps, with almost all of the studies carried out in the southeastern United States (Table 11.4). Primary productivity depends on hydrologic and nutrient conditions

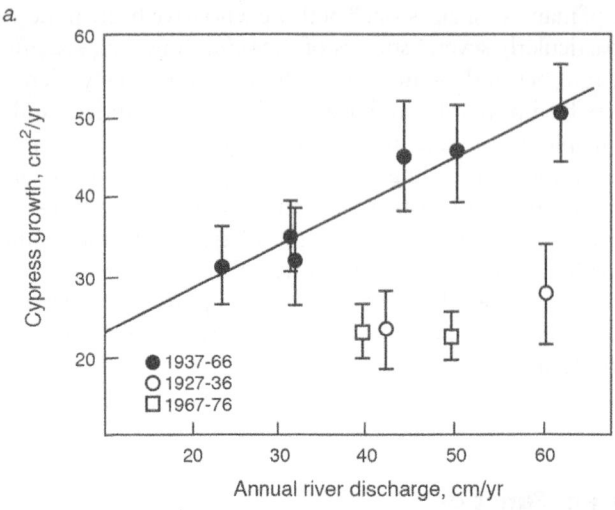
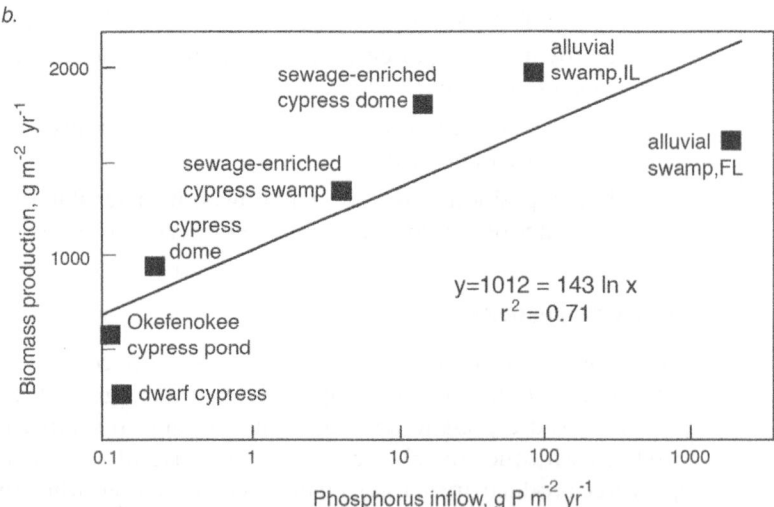

Figure 11.8 Relationships between hydrologic conditions and tree productivity in cypress swamps: (a) increase in basal area of bald cypress trees in southern Illinois alluvial swamp as a function of river discharge for five-year periods, and (b) biomass production as a function of phosphorus inflow for several cypress swamps. Data points in (a) indicate mean; bar indicates 1 standard error. ((a) After Mitsch et al., 1979; (b) after S. L. Brown, 1981)

Table 11.4 Biomass and net primary productivity of deepwater swamps in the southeastern United States

Location/Forest Type	Tree Standing Biomass (kg/m²)	Litterfall (g m⁻² yr⁻¹)	Stem Growth (g m⁻² yr⁻¹)	Above-ground NPP[a] (g m⁻² yr⁻¹)	Reference
Louisiana					
Bottomland hardwood	16.5[b]	574	800	1,374	Conner and Day (1976)
Cypress–tupelo	37.5[b]	620	500	1,120	Conner and Day (1976)
Impounded managed swamp	32.8[b, c]	550	1,230	1,780	Conner et al. (1981)
Impounded stagnant swamp	15.9[b, c]	330	560	890	Conner et al. (1981)
Tupelo stand	36.2[b]	379	—	—	Conner and Day (1982)
Cypress stand	27.8[b]	562	—	—	Conner and Day (1982)
Cypress forests (n = 7)		425 ± 164	269 ± 117	765 ± 245	Megonigal et al. (1997)
Bottomland hardwood forests (n = 7)		670 ± 109	440 ± 135	1,110 ± 177	Megonigal et al. (1997)
North Carolina					
Tupelo swamp	—	609–677	—	—	Brinson (1977)
Floodplain swamp	26.7[d]	523	585	1,108	Mulholland (1979)
South Carolina					
Cypress forests (n = 2)		385 ± 98	242 ± 64	625 ± 35	Megonigal et al. (1997)
Bottomland hardwood forests (n = 10)		720 ± 98	488 ± 122	1,208 ± 198	Megonigal et al. (1997)
Virginia					
Cedar swamp	22.0[b]	758	441	1,097[j]	Dabel and Day (1977), Gomez and Day (1982), Megonigal and Day (1988)
Maple gum swamp	19.6[b]	659	450	1,050[j]	Megonigal and Day (1988)
Cypress swamp	34.5[b]	678	557	1,176[j]	Megonigal and Day (1988)
Mixed hardwood swamp	19.5[b]	652	249	831[j]	Megonigal and Day (1988)
Georgia					
Nutrient-poor cypress swamp	30.7[e]	328	353	681	Schlesinger (1978)
Illinois					
Floodplain forest	29.0	—	—	1,250	F. L. Johnson and Bell (1976)
Floodplain forest		491	177	668	S. L. Brown and Peterson (1983)
Cypress–tupelo swamp	45[d]	348	330	678	Mitsch (1979), Dorge et al. (1984)

(continued)

Table 11.4 (Continued)

Location/Forest Type	Tree Standing Biomass (kg/m²)	Litterfall (g m⁻² yr⁻¹)	Stem Growth (g m⁻² yr⁻¹)	Above-ground NPP[a] (g m⁻² yr⁻¹)	Reference
Ohio					
Bottomland hardwood forest (before and after hydrologic restoration)				531–641 (pre-restoration) 807 ± 86 (post-restoration)	Cochran (2001); Anderson and Mitsch (2008)
Kentucky					
Cypress–ash slough	31.2	136	498	634	Mitsch et al. (1991)
Cypress swamp	10.2	253	271	524	Mitsch et al. (1991)
Stagnant cypress swamp	9.4	63	142	205	Mitsch et al. (1991)
Bottomland forest	30.3	420	914	1,334	Mitsch et al. (1991)
Bottomland forest	18.4	468	812	1,280	Mitsch et al. (1991)
Florida					
Cypress–tupelo (6 sites)	19 ± 4.7[f]	—	289 ± 58[f]	760[g]	Mitsch and Ewel (1979)
Cypress–hardwood (4 sites)	15.4 ± 2.9[f]	—	336 ± 76[f]	950[g]	Mitsch and Ewel (1979)
Pure cypress stand (4 sites)	9.5 ± 2.6[f]	—	154 ± 55[f]	—	Mitsch and Ewel (1979)
Cypress–pine (7 sites)	10.1 ± 2.1[f]	—	117 ± 27[f]	—	Mitsch and Ewel (1979)
Floodplain swamp	32.5	521	1,086	1,607	S. L. Brown (1978)
Natural dome[h]	21.2	518	451	969	S. L. Brown (1978)
Sewage dome[i]	13.3	546	716	1,262	S. L. Brown (1978)
Scrub cypress	7.4	250	—	—	S. L. Brown and Lugo (1982)
Drained strand	8.9	120	267	387	Carter et al. (1973)
Undrained strand	17.1	485	373	858	Carter et al. (1973)
Larger cypress strand	60.8	700	196	896	Duever et al. (1984)
Small tree cypress strand	24.0	724	818	1,542	Duever et al. (1984)
Sewage enriched cypress strand	28.6	650	640	1,290	Nessel (1978)

[a] NPP = net primary productivity = litterfall + stem growth.
[b] Trees defined as >2.54 cm diameter at breast height (DBH).
[c] Cypress, tupelo, ash only.
[d] Trees defined as >10 cm DBH.
[e] Trees defined as >4 cm DBH.
[f] Average ± standard error for cypress only.
[g] Estimated.
[h] Average of five natural domes.
[i] Average of three domes; domes were receiving high-nutrient wastewater.
[j] Above-ground NPP is less than sum of litterfall plus stem growth because some stem growth is measured as litterfall.

and pulsing hydrology supports more productive systems than does permanent flooding or lack of flooding. Several other studies have reported the importance of flooding to forests by linking annual tree growth with flood occurrence (Conner and Day 1976, Robertson et al. 2001, Stromberg, 2001, Anderson and Mitsch, 2008).

Based on the subsidy-stress model (E. P. Odum et al., 1979), floodplain tree growth should be maximized where flooding is frequent or long enough to subsidize nutrients and enhancing growing conditions but not so much that floods become a physiological stress to trees. Attempts to demonstrate this model by comparing forest communities along a wetness gradient have often been inconclusive. Megonigal et al. (1997) investigated productivity in floodplain swamps throughout the southeastern United States and concluded that while permanently flooded floodplain swamps did have lower productivity, there was no evidence that sites that were seasonally pulsed were any more productive than sites that were clearly upland (Fig. 11.9). They suggested that the Mitsch and Rust (1984) model (see Fig. 4.18 in Chapter 4) may be a more appropriate description of the productivity of forested wetlands.

In almost all of these studies, only above-ground productivity was estimated. Powell and Day (1991) made direct measurements of below-ground productivity and found that it was highest in a mixed hardwood swamp ($989 \text{ g m}^{-2} \text{ yr}^{-1}$) and much lower in a more frequently flooded cedar swamp ($366 \text{ g m}^{-2} \text{ yr}^{-1}$), a cypress swamp ($308 \text{ g m}^{-2} \text{ yr}^{-1}$), and a maple–gum swamp ($59 \text{ g m}^{-2} \text{ yr}^{-1}$). These results suggest that the allocation of carbon to the root system decreases with increased flooding.

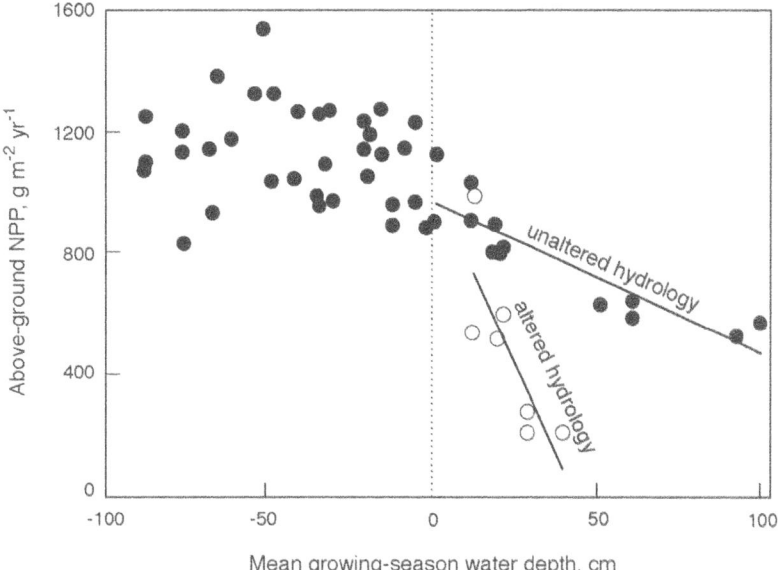

Figure 11.9 **The relationship between net primary productivity of floodplain forests and mean growing season water depth in bottomland hardwood forests of the southeastern United States. (After Megonigal et al., 1997)**

In forested wetlands that have been unaltered, annual mortality rates of trees are often low. Conner et al. (2002) monitored annual changes in forested wetland structure between 1987 and 1999 in South Carolina and Louisiana. Tree mortality in unaltered areas was low (~2 percent); however, higher annual mortality (up to 16 percent) was observed at Louisiana sites where severe water-level rise has occurred. The authors also found that severe windstorms increased short-term mortality, but these events can also lead to elevated long-term mortality rates as damaged trees eventually succumb.

Energy Flow

The energy flow of deepwater swamps is dominated by primary productivity of the canopy trees. Energy consumption is accomplished primarily by detrital decomposition. Significant differences exist, however, between the energy flow patterns in low-nutrient swamps, such as dwarf cypress swamps and cypress domes, and high-nutrient swamps, such as alluvial cypress swamps (Table 11.5). All of the cypress wetlands are autotrophic—productivity exceeds respiration. Gross primary productivity, net primary productivity, and net ecosystem productivity are highest in the alluvial river swamp that receives high-nutrient inflows. Buildup and/or export of organic matter are characteristic of all of these deepwater swamps but are most characteristic of alluvial swamps. There are few allochthonous inputs of energy to the low-nutrient wetlands, and energy flow at the primary producer level is relatively low. The alluvial cypress–tupelo swamp depends more on allochthonous inputs of nutrients and energy, particularly from runoff and river flooding. In alluvial deepwater swamps, productivity of aquatic plants is often high, whereas aquatic productivity in cypress domes is usually low.

Nutrient Budgets

The functioning of forested wetlands as nutrient sinks was first suggested by Kitchens et al. (1975) in a preliminary winter–spring survey of an alluvial river swamp complex in South Carolina. They found a significant reduction in phosphorus as the waters

Table 11.5 Estimated energy flow (kcal m^{-2} day^{-1}) in selected Florida cypress swamps[a]

Parameter	Dwarf Cypress Swamp	Cypress Dome	Alluvial River Swamp
Gross primary productivity[b]	27	115	233
Plant respiration[c]	18	98	205
Net primary productivity	9	17	28
Soil or water respiration	7	13	18
Net ecosystem productivity	2	4	10

[a] Assume 1 g C = 10 kcal.
[b] Assumes gross primary productivity (GPP) = net daytime photosynthesis + nighttime leaf respiration.
[c] Plant respiration = 2 × (nighttime leaf respiration) + stem respiration + knee respiration.
Source: S. L. Brown (1981)

passed over the swamp and assumed this to be the result of biological uptake by aquatic plant communities. In a similar study in Louisiana, J. W. Day et al. (1977) found that nitrogen was reduced by 48 percent and phosphorus decreased by 45 percent as water passed through a lake–swamp complex of Barataria Bay to the lower estuary. They attributed this decrease in nutrients to sediment interactions, including nitrate storage/denitrification and phosphorus adsorption to the clay sediments. Beginning in 1973, H.T. Odum et al. (1977) and colleagues and students investigated recycling of treated sewage applied to cypress domes and other swamps in northcentral Florida. Much of that work was later summarized in Ewel and Odum (1984). Since then, countless studies have illustrated the potential of forested wetlands for nutrient removal, including several studies in Louisiana (Mitsch and Day, 2004; Day et al., 2004; Rivera-Monroy et al., 2013).

Nutrient budgets of deepwater swamps vary from "open" alluvial river swamps that receive and export large quantities of materials to "closed" cypress domes that are mostly isolated from their surroundings (Table 11.6). Mitsch et al. (1979) developed a nutrient budget for an alluvial river swamp in southern Illinois and found that 10 times more phosphorus was deposited with sediments during river flooding ($3.6 \text{ g-P m}^{-2} \text{ yr}^{-1}$) than was returned from the swamp to the river during the rest of the year (see Fig. 6.18 in Chapter 6, "Wetland Biogeochemistry"). The swamp was a sink for a significant amount of phosphorus and sediments during that particular year of flooding, although the percentage of retention was low (3–4.5 percent) because a very large volume of water passed over the swamp during flooding conditions. Noe and Hupp (2005) evaluated net nutrient accumulation in floodplain forests along rivers contributing to the Chesapeake Bay. Mean accumulation rates for C ranged from 61 to $212 \text{ g-C m}^{-2} \text{ yr}^{-1}$, N ranged from 3.5 to $13.4 \text{ g-N m}^{-2} \text{ yr}^{-1}$, and P ranged from 0.2 to $4.1 \text{ g-P m}^{-2} \text{ yr}^{-1}$. Watershed land use was a significant factor in their study. The greatest accumulation of sediment and nutrients occurred along the Chickahominy River, downstream from the urban metropolitan area of Richmond, Virginia.

Table 11.6 Phosphorus inputs to forested swamps ($\text{g-P m}^{-2} \text{ yr}^{-1}$)

Swamp	Rainfall	Surface Inflow	Sediments from River Flooding	Reference
Florida				
Dwarf cypress	0.11	—	0	S. L. Brown (1981)
Cypress dome	0.09	0.12	0	
Alluvial river swamp	—	—	3.1	S. L. Brown (1981)
Southern Illinois				
Alluvial river swamp	0.11	0.1	3.6	Mitsch et al. (1979)
North Carolina				
Alluvial tupelo swamp	0.02–0.04	0.01–1.2	0.2	Yarbro (1983)
Virginia				
Floodplain forests	–	–	0.2–4.1	Noe and Hupp (2005)

Riparian Ecosystems and River Exchanges

Ecologists have reviewed river systems in terms of their ecological function and have developed two different ways of describing flowing water systems. The river continuum system clearly is related to the general differences in ecology along streams and rivers, going *longitudinally* along the river itself. The concepts were developed mostly in low-order streams in the United States. Little attention is paid to lateral connections or two floodplains. The flood pulse concept, however, based on research done in the Amazon River and its tributaries, features the importance of seasonal patterns of stream flow and the importance of *lateral* exchange between the river and its riparian ecosystems.

River Continuum Concept

The *river continuum concept* (RCC) is a theory developed in the early 1980s to describe the longitudinal patterns of biota found in streams and rivers (Vannote et al., 1980; Minshall et al., 1983, 1985). According to the RCC, most organic matter is introduced to streams from terrestrial sources in headwater areas (Fig. 11.10). The production/respiration (P/R) ratio is < 1 (i.e., the stream is heterotrophic), and invertebrate shredders and collectors dominate the fauna. Biodiversity is limited by low temperatures, low light, and low nutrients. In river midreaches, more light

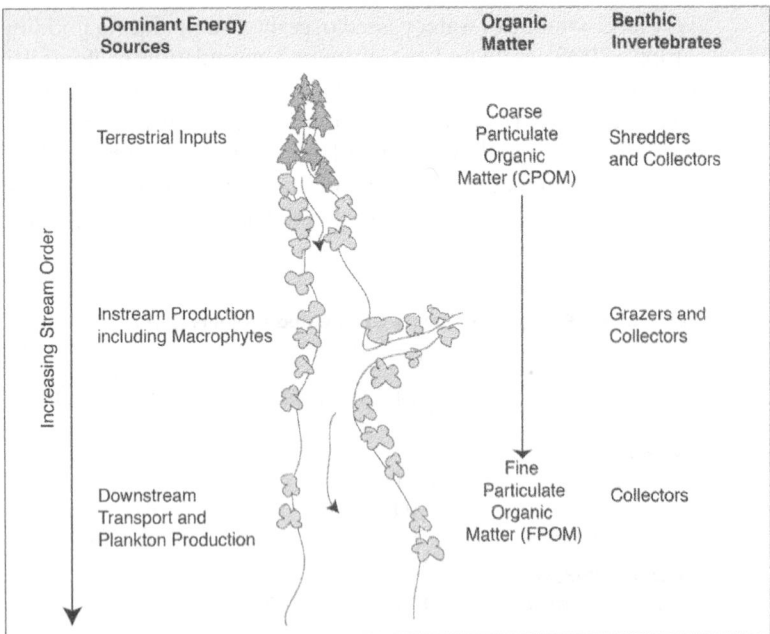

Figure 11.10 The river continuum concept showing transition from small first-order stream to very large eleventh-order river. Charts on the left indicate relative importance of terrestrial, in-stream, or upstream energy sources to the aquatic food chain. Charts on the right indicate the relative importance of different feeding groups of invertebrates. (From Mitsch and Jørgensen, 2004, after Johnson et al., 1995)

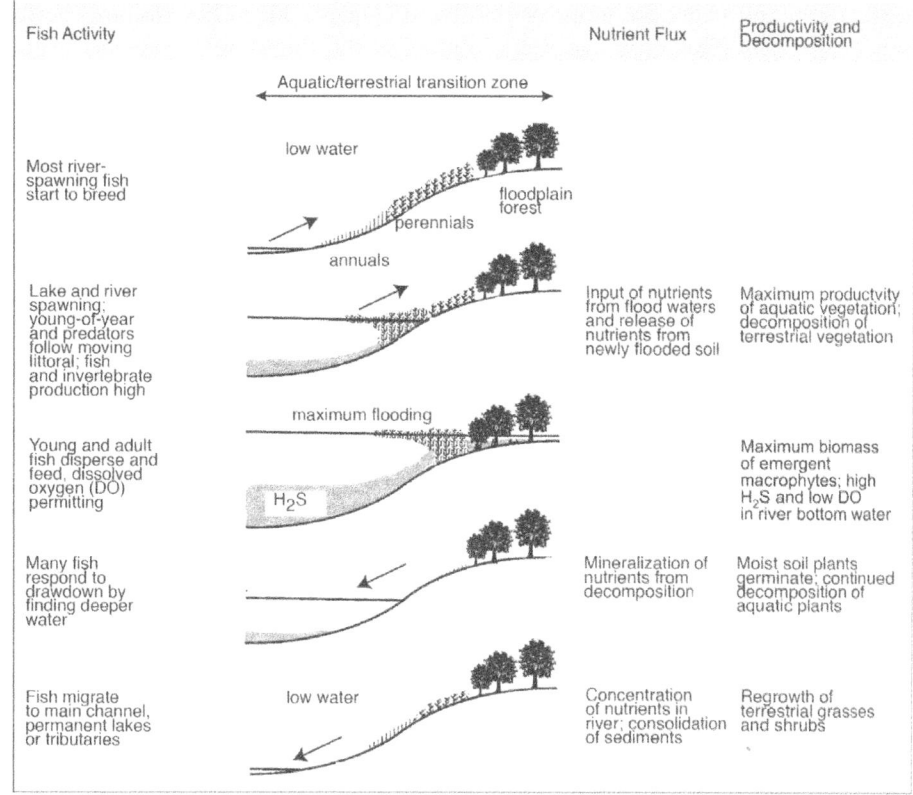

Figure 11.11 The flood pulse concept for a river and its floodplain, illustrating five periods over the wet and dry seasons of a river. (From Mitsch and Jørgensen, after Bayley, 1995, and Junk et al., 1989)

is available, phytoplankton prospers, and biodiversity is highest. The P/R ratio is >1. Organic matter input from upstream is fine; filter feeders dominate the flora. In braided reaches or where the floodplain is broad, however, the bank habitat is a major source of snags and logs that lead to debris dams that slow water flow and increase stream habitat diversity. The increased input of riparian coarse debris increases food diversity and increases heterotrophy. The productivity/respiration (P/R) ratio is <1. Finally, in the highest-order streams, riparian litter inputs are minor and turbidity reduces primary productivity. Hence the system is heterotrophic again (P/R <1), and diversity is often low. The importance of backwaters, oxbows, and floodplains to river ecosystem function are virtually ignored in the RCC.

Flood Pulse Concept

The RCC considers the importance of the riparian zone only in an indirect way by noting that small low-order streams are influenced by shading and abundant contributions of allochthonous organic matter. Junk et al. (1989) developed a *flood pulse concept* (FPC) for floodplain–large river systems based on their experience in both temperate and tropical regions of the world (Fig. 11.11). They dispute the RCC as a generalizable theory because: (1) most of the theory was developed from experience on low-order temperate streams, and (2) the concept is mostly restricted to habitats that are permanent and lotic. In the FPC, the pulsing of the river discharge is the major force controlling biota in river floodplains, and lateral exchange between the floodplain and river channel and nutrient cycling within the floodplain "have more direct impact on biota than nutrient spiraling discussed in the RCC" (Junk et al., 1989). The FPC thus considers the river-floodplain exchange to be of enormous importance in determining the productivity of both the river and the adjacent riparian zone. Alternating dry and wet cycles optimize productivity of the littoral zone and the adjacent forest, decomposition of all that is produced, and fish spawning and feeding.

Recommended Readings

Messina, M. G., and W. H. Conner, eds. 1998. *Southern Forested Wetlands*. Boca Raton, FL: Lewis Publishers.

References

Anderson, C. J., and W. J. Mitsch. 2008. The influence of flood connectivity on bottomland forest productivity in central Ohio, USA. *Ohio Journal of Science* 108 (2): 2–8.

Anderson, R. C., and J. White. 1970. A cypress swamp outlier in southern Illinois. *Illinois State Acad. Sci. Trans.* 63: 6–13.

Batzer, D. P., and S. A. Wissinger. 1996. Ecology of insect communities in nontidal wetlands. *Annual Review of Entomology* 41: 75–100.

Bayley, P. B. 1995. Understanding large river-floodplain ecosystems. *BioScience* 45:153–158.

Brinson, M. M. 1977. Decomposition and nutrient exchange of litter in an alluvial swamp forest. *Ecology* 58: 601–609.

Brinson, M. M., B. L. Swift R. C. Plantico, and J. S. Barclay. 1981. Riparian Ecosystems: Their Ecology and Status, U.S. Fish and Wildlife Service, *Biol. Serv. Prog.*, FWS/OBS-81/17, Washington, DC, 151pp.

Brown, C. A. 1984. Morphology and biology of cypress trees. In K. C. Ewel and H. T. Odum, eds., *Cypress Swamps*. University Presses of Florida, Gainesville.

Brown, S. L. 1978. A Comparison of Cypress Ecosystems in the Landscape of Florida. Ph.D. dissertation, University of Florida, 569 pp.

Brown, S. L. 1981. A comparison of the structure, primary productivity, and transpiration of cypress ecosystems in Florida. *Ecological Monographs* 51: 403–427.

Brown, S. L., and A. E. Lugo. 1982. A comparison of structural and functional characteristics of saltwater and freshwater forested wetlands. In B. Gopal, R. E. Turner, R. G. Wetzel, and D. F. Whigham, eds. *Wetlands: Ecology and Management*. National Institute of Ecology and International Scientific Publications, Jaipur, India, pp. 109–130.

Brown, S. L., and D. L. Peterson. 1983. Structural characteristics and biomass production of two Illinois bottomland forests. *American Midland Naturalist* 110: 107–117.

Carter, M. R., L. A. Bums, T. R. Cavinder, K. R. Dugger, P. L. Fore, D. B. Hicks, H. L. Revells, and T. W. Schmidt. 1973. *Ecosystem Analysis of the Big Cypress Swamp and Estuaries.* U.S. Environmental Protection Agency 904/9–74, Region IV, Atlanta.

Casey, W. P., and K. C. Ewel. 2006. Patterns of succession in forested depressional wetlands in North Florida, USA. *Wetlands* 26: 147–160.

Clewell, A. F., and D. B. Ward. 1987. White cedar in Florida and along the northern Gulf Coast. In A.D. Laderman, ed., *Atlantic White Cedar Wetlands.* Westview Press, Boulder, CO, pp. 69–82.

Cochran, M. 2001. Effect of hydrology on bottomland hardwood forest productivity in central Ohio (USA). Master's thesis, Natural Resources, The Ohio State University, Columbus.

Conner, W. H., and J. W. Day, Jr. 1976. Productivity and composition of a bald cypress-water tupelo site and a bottomland hardwood site in a Louisiana swamp. *American Journal of Botany* 63: 1354–1364.

Conner, W. H., J. G. Gosselink, and R. T. Parrondo. 1981. "Comparison of the vegetation of three Louisiana swamp sites with different flooding regimes." American Journal of Botany 68: 32–331.

Conner, W. H., and J. W. Day, Jr. 1982. The ecology of forested wetlands in the southeastern United States. In B. Gopal, R. E. Turner, R. G. Wetzel, and D. F. Whigham, eds., *Wetlands: Ecology and Management.* National Institute of Ecology and International Scientific Publications, Jaipur, India, pp. 69–87.

Conner, W. H., I. Mihalia, and J. Wolfe. 2002. Tree community structure and changes from 1987 to 1999 in three Louisiana and three South Carolina forested wetlands. *Wetlands* 22: 58–70.

Dabel, C. V., and F. P. Day, Jr. 1977. Structural composition of four plant communities in the Great Dismal Swamp, Virginia. *Torrey Botanical Club Bulletin* 104: 352–360.

Day, F. P. 1984. Biomass and litter accumulation in the Great Dismal Swamp. In K. C. Ewel and H. T. Odum, eds., *Cypress Swamps*, University of Florida Press, Gainesville, pp. 386–392.

Day, J. W., Jr., T. J. Butler, and W. G. Conner. 1977. Productivity and nutrient export studies in a cypress swamp and lake system in Louisiana. In M. Wiley, ed., *Estuarine Processes*, vol. II, Academic Press, New York, pp. 255–269.

Day, J. W., Jr., J. Ko, J., Rybczyk, D. Sabins, R. Bean, G. Berthelot, C. Brantley, L. Cardoch, W. Conner, J. N. Day, A. J. Englande, S. Feagley, E. Hyfield, R. Lane, J. Lindsey, J. Mistich, E. Reyes, and R. Twilley. 2004. The use of wetlands in the Mississippi delta for wastewater assimilation: A review. *Ocean and Coastal Management* 47: 671–691.

DeGraaf, R. M. and D. D. Rudis. 1986. New England Wildlife: Habitat Natural History, and Disturbance. General Technical Report NE-108, Northeastern Forest Experiment Station, Forest Service, U.S. Department of Agriculture, Washington, DC, 491 pp.

DeGraaf, R. M., and D. D. Rudis. 1990. Herpetofaunal species composition and relative abundance among three New England forest types. *Forest Ecology and Management* 32: 155–165.

Denny, G. C., and M. A. Arnold. 2007. Taxonomy and nomenclature of baldcypress, pondcypress, and Montezuma cypress: One, two, or three species? *HortTechnology* (Jan-Mar 2007) 17(1): 125–127.

Dorge, C. L., W. J. Mitsch, and J. R. Wiemhoff. 1984. Cypress wetlands in southern Illinois. In K. C. Ewel and H. T. Odum, eds., *Cypress Swamps*, University Presses of Florida, Gainesville, pp. 393–404.

Duever, M. J., J. E. Carlson, and L. A. Riopelle. 1984. Corkscrew Swamp: A virgin cypress strand. In K. C. Ewel and H. T. Odum, eds., *Cypress Swamps*. University Presses of Florida, Gainesville, pp. 334–348.

Ewel, K. C., and W. J. Mitsch. 1978. The effects of fire on species composition in cypress dome ecosystems. *Florida Scientist* 41: 25–31.

Ewel, K. C., and H. T. Odum, eds. 1984. *Cypress Swamps*. University Presses of Florida, Gainesville, FL.

Ewel, K C., and L. P. Wickenheiser. 1988. Effects of swamp size on growth rates of cypress (*Taxodium distichum*) trees. *American Midland Naturalist* 120: 362–370.

Golet, F. C., and D. J. Lowry. 1987. Water regimes and tree growth in Rhode Island Atlantic white cedar swamps. In A. D. Laderman, ed., *Atlantic White Cedar Wetlands*. Westview Press, Boulder, CO, pp. 91–110.

Golet, F. C., A. J. K. Calhoun, W. R. DeRagon, D. J. Lowry, and A. J. Gold. 1993. *Ecology of Red Maple Swamps in the Glaciated Northeast: A Community Profile*. Biological Report 12, U.S. Fish and Wildlife Service, Washington, DC. 151 pp.

Gomez, M. M., and F. P. Day, Jr. 1982. Litter nutrient content and production in the Great Dismal Swamp. *American Journal of Botany* 69: 1314–1321.

Grosse, W., H. B. Büchel, and S. Lattermann. 1998. Root aeration in wetland trees and its ecophysiological significance. In A. D. Laderman, ed., *Coastally Restricted Forests*. Oxford University Press, New York, pp. 293–305.

Heimburg, K. 1984. Hydrology of north-central Florida cypress domes. In K. C. Ewel and H. T. Odum, eds., *Cypress Swamps*. University Presses of Florida, Gainesville, pp. 72–82.

Huenneke, L. F., and R. R. Sharitz. 1986. Microsite abundance and distribution of woody seedlings in a South Carolina cypress-tupelo swamp. *American Midland Naturalist* 115: 328–335.

Johnson, F. L., and D. T. Bell. 1976. Plant biomass and net primary production along a flood-frequency gradient in a streamside forest. *Castanea* 41: 156–165.

Johnson, B. L., W. B. Richardson, and T. J. Naimo. 1995. Past, present, and future concepts in large river ecology. *BioScience* 45: 134–141.

Junk, W. J., P. B. Bayley, and R. E. Sparks. 1989. The flood pulse concept in river-floodplain systems. In D. P. Dodge, ed., *Proceedings of the International Large River Symposium. Special issue of the Journal of Canadian Fisheries and Aquatic Sciences* 106: 11–127.

Kitchens, W. M., Jr., J. M. Dean, L. H. Stevenson, and J. M. Cooper. 1975. The Santee Swamp as a nutrient sink. In F. G. Howell, J. B. Gentry, and M. H. Smith, eds., *Mineral Cycling in Southeastern Ecosystems*, ERDA Symposium Series 740513. U.S. Government Printing Office, Washington, DC, pp. 349–366.

Kurz, H., and D. Demaree. 1934. Cypress buttresses in relation to water and air. *Ecology* 15: 36–41.

Kurz, H., and K. A. Wagner. 1953. Factors in cypress dome development. *Ecology* 34: 17–164.

Laderman, A. D. 1989. The ecology of the Atlantic white cedar wetlands: A community profile. *Biology Reports* 8S (7.21). U.S. Fish and Wildlife Service, Washington, DC. 114 pp.

Laderman, A. D., ed. 1998. *Coastally Restricted Forests*. Oxford University Press, Oxford, UK, 334 pp.

Leopold, L. B., M. G. Wolman, and J. E. Miller. 1964. *Fluvial Processes in Geomorphology*. W. H. Freeman, San Francisco. 522 pp.

Little, E. L., Jr. 1971. *Atlas of United States Trees*, vol. 1, Conifers and important hardwoods, Misc. Pub. No. 1146, U.S. Department of Agriculture—Forest Service, USGPO, Washington, DC.

Lockaby, B. G., and M. R. Walbridge. 1998. Biogeochemistry. In M. G. Messina and W. H. Conner, eds., *Southern Forested Wetlands: Ecology and Management*. Lewis Publishers, Boca Raton, FL, pp. 149–172.

Megonigal, J. P., and F. P. Day, Jr. 1988. Organic matter dynamics in four seasonally flooded forest communities of the Dismal Swamp. *American Journal of Botany* 75: 1334–1343.

Megonigal, J. P., W. H. Conner, S. Kroeger, and R. R. Sharitz. 1997. Aboveground production in southeastern floodplain forests: A test of the subsidy–stress hypothesis. *Ecology* 78: 370–384.

Minshall, G. W., R. C. Peterson, K. W. Cummins, T. L. Bott, J. R. Sedall, C. E. Cushing, and R. L. Vannote. 1983. Inter-biome comparison of stream ecosystem dynamics. *Ecological Monographs* 53–1–25.

Minshall, G. W., K. W. Cummins, R. C. Peterson, C. E. Cushing, D. A. Bruins, J. R. Sedall, and R. L. Vannote. 1985. Development in stream ecosystem theory. *Canadian Journal of Fisheries and Aquatic Sciences* 37:130–137.

Mitsch, W. J. 1979. Interactions between a riparian swamp and a river in southern Illinois. In R. R. Johnson and J. F. McComlick, tech. coords., *Strategies for the Protection and Management of Floodplain Wetlands and Other Riparian Ecosystems*, Proceedings of a Symposium, Calaway Gardens, U.S. Forest Service General Technical Report WO-12, Washington, DC, pp. 63–72.

Mitsch, W. J., and K. C. Ewel. 1979. Comparative biomass and growth of cypress in Florida wetlands. *American Midland Naturalist* 101: 417–426.

Mitsch, W. J., C. L. Dorge, and J. R. Wiemhoff. 1979. Ecosystem dynamics: A phosphorus budget of an alluvial cypress swamp in southern Illinois. *Ecology* 60: 1116–1124.

Mitsch, W. J., and W. G. Rust. 1984. Tree growth responses to flooding in a bottomland forest in northeastern Illinois. *Forest Science* 30: 499–510.

Mitsch, W. J., J. R. Taylor, and K. B. Benson. 1991. Estimating primary productivity of forested wetland communities in different hydrologic landscapes. *Landscape Ecology* 5: 75–92.

Mitsch, W. J., and J. G. Gosselink. 2000. *Wetlands*, 3rd ed. John Wiley & Sons, New York, 920 pp.

Mitsch, W. J., and J. W. Day, Jr. 2004. Thinking big with whole ecosystem studies and ecosystem restoration—A legacy of H.T. Odum. *Ecological Modelling* 178: 133–155.

Mitsch, W. J., and S. E. Jørgensen. 2004. *Ecological Engineering and Ecosystem Restoration*. John Wiley and Sons, Hoboken, NJ. 411pp.

Motzkin, G., W. A. Patterson, and N.E.R. Drake. 1993. Fire history and vegetation dynamics of a *Chamaecyparis thyoides* wetland on Cape Cod, Massachusetts. *Journal of Ecology* 81: 391–402.

Mulholland, P. J. 1979. Organic Carbon in a Swamp-Stream Ecosystem and Export by Streams in Eastern North Carolina. Ph.D. dissertation, University of North Carolina, Chapel Hill.

The Nature Conservancy. 1992. The forested wetlands of the Mississippi River: An ecosystem in crisis. The Nature Conservancy, Baton Rouge, LA, 25 pp.

Nessel, J. K. 1978. Distribution and Dynamics of Organic Matter and Phosphorus in a Sewage Enriched Cypress Strand. Masters thesis, University of Florida, Gainesville, FL, 159 pp.

Noe, G. P., and C. R. Hupp. 2005. Carbon, nitrogen, and phosphorus accumulation in floodplains of Atlantic Coastal Plain rivers, USA. *Ecological Applications* 15: 1178–1190.

Odum, E. P., J. T. Finn, and E. H. Franz. 1979. Perturbation theory and the subsidy-stress gradient. *BioScience* 29: 349–352.

Odum, H. T. 1982. Role of wetland ecosystems in the landscape of Florida in Ecosystem Dynamics in Freshwater Wetlands and Shallow Water Bodies, vol. 11. D. O. Logofet and N. K Luckyanov, eds., SCOPE and UNEP Workshop, Center of International Projects, Moscow, USSR, pp. 33–72.

Odum, H. T., K. C. Ewel, W. J. Mitsch, and J. W. Ordway. 1977. Recycling treated sewage through cypress wetlands in Florida. In F. M. D'Itri, ed., *Wastewater Renovation and Reuse*. Marcel Dekker, Inc., New York, pp. 35–67.

Odum, W. E., E. P. Odum, and Odum, H. T. 1995. Nature's pulsing paradigm. *Estuaries* 18: 547–555.

Penfound, W. T. 1952. Southern swamps and marshes. *Botanical Review* 18: 413–446.

Powell, S. W., and F. P. Day. 1991. Root production in four communities in the Great Dismal Swamp. *American Journal of Botany* 78: 288–297.

Rivera-Monroy, V. H., B, Branoff, E, Meselhe, A. McCorquodale, M. Dortch, G. D. Steyer, J. Visser, and H. Wang. 2013. Landscape-level estimation of nitrogen removal in coastal Louisiana wetlands: Potential sinks under different restoration scenarios. *Journal of Coastal Research* 67(sp1): 75–87.

Robertson, A. I., P. Y. Bacon, and G. Heagney. 2001. The response of floodplain primary production to flood frequency and timing. *Journal of Applied Ecology* 38: 126–136.

Schlesinger, W. H. 1978. Community structure, dynamics, and nutrient ecology in the Okefenokee Cypress Swamp-Forest. *Ecological Monographs* 48: 43–65.

Schneider, R. L., and R. R. Sharitz. 1986. Seed bank dynamics in a southeastern riverine swamp. *American Journal of Botany* 73: 1022–1030.

Sheffield, R. M., T. W. Birch, W. H. McWilliams, and J. B. Tansey. 1998. Chamaecyparis thyoides (Atlantic white cedar) in the United States. In A. D. Laderman, ed., *Coastally Restricted Forests*. Oxford University Press, New York, pp. 111–123.

Stromberg, J. C. 2001. Influence of stream flow regime and temperature on growth rate of the riparian tree, *Platanus wrightii*, in Arizona. *Freshwater Biology* 46: 227–239.

Thorp, J. H., E. M. McEwan, M. F. Flynn, and F. R. Hauer. 1985. Invertebrate colonization of submerged wood in a cypress-tupelo swamp and blackwater stream. *American Midland Naturalist* 113: 56–68.

Vannote, R. L., G. W. Minshall, K W. Cummins, J. R. Sedell, and C. E. Cushing. 1980. The river continuum concept. *Canadian Journal of Fisheries and Aquatic Sciences* 37:130–137.

Vernon, R. O. 1947. Cypress domes. *Science* 105: 97–99.

Villa, J. A. and W. J. Mitsch. 2014. Methane emissions from five wetland plant communities with different hydroperiods in the Big Cypress Swamp region of Florida Everglades. *Ecohydrology & Hydrobiology* 14: 253–266.

Villa, J. A., and W. J. Mitsch 2015. Carbon sequestration in different wetland plant communities in Southwest Florida. *International Journal for Biodiversity Science, Ecosystems Services and Management*. doi.org/10.1080/21513732.2014.973909

Watts, A. C. 2013. Organic soil combustion in cypress swamps: Moisture effects and landscape implications for carbon release. *Forest Ecology and Management* 294C: 178–187.

Watts, A. C., L. N. Kobziar, and J. R. Snyder. 2012. Fire reinforces structure of pondcypress (*Taxodium distichum* var. *imbricarium*) domes in a wetland landscape. *Wetlands* 32: 430–448.

Watts, A. C., and L. N. Kobziar. 2013. Smoldering combustion and ground fires: Ecological effects and multi-scale significance. *Fire Ecology* 9: 124–132.

Wharton, C. H., W. M. Kitchens, E. C. Pendleton, and T. W. Sipe. 1982. *The Ecology of Bottomland Hardwood Swamps of the Southeast: A Community Profile*. U.S. Fish and Wildlife Service, Biological Services Program, FWS/OBS-81/37. 133 pp.

Whigham, D. F., and C. J. Richardson. 1988. Soil and plant chemistry of an Atlantic white cedar wetland on the Inner Coastal Plain of Maryland. *Canadian Journal of Botany* 66: 568–576.

Wright, R. B., B. G. Lockaby, and M. R. Walbridge. 2001. Phosphorus availability in an artificially flooded southeastern floodplain forest soil. *Soil Science Society of America Journal* 65: 1293–1302.

Yarbro, L. A. 1983. The influence of hydrologic variations on phosphorus cycling and retention in a swamp stream ecosystem. In T. D. Fontaine and S. M. Bartell, eds., *Dynamics of Lotic Ecosystems*. Ann Arbor Science, Ann Arbor, MI, pp. 223–245.

Northern peatland in Estonia

Chapter **12**

Peatlands

Peatlands include bogs and fens distributed primarily in the cool boreal zones of the world where excess moisture is abundant. Bogs and fens can be formed in several ways, originating either from aquatic systems, as in flowthrough succession or quaking bogs, or from terrestrial systems, as with blanket bogs. Although many types of peatlands are identifiable, classification according to chemical conditions usually defines three types: (1) minerotrophic (true fens), (2) ombrotrophic (raised bogs), and (3) transition (poor fens). Features of many peatlands include acidity caused by cation exchange with mosses, oxidation of sulfur compounds, and organic acids, low nutrients and primary productivity, slow decomposition, adaptive nutrient-cycling pathways, and peat accumulation. Several energy and nutrient budgets have been developed for peatlands, with the 1942 energy budget by Lindeman one of the first in ecological sciences. Peatlands collectively are the largest terrestrial storage of carbon on the planet and are seen as potential sources of carbon to the atmosphere if they are disturbed hydrologically or if climate shifts.

As defined here, peatlands include the deep peat deposits of the boreal regions of the world. Bogs and fens, the two major types of peatlands, occur as thick peat deposits in old lake basins or as blankets the landscape. Many of these lake basins were formed by the last glaciation, and the peatlands are considered to be a late stage of a filling-in process. *Bogs* are acid peat deposits with no significant inflow or outflow of surface water or groundwater and support *acidophilic* (acid-loving) vegetation, particularly mosses. *Fens*, in contrast, are open peatland systems that generally receive some drainage from surrounding mineral soils and are often covered by grasses, sedges, or reeds. They are in many respects transitional between marshes and bogs. Fens are important as a successional stage in the development of bogs and will be considered in that context here.

Bogs and fens have been studied and described on a worldwide basis more extensively than any other type of freshwater wetland; European and North American ecology literatures are particularly rich in peatland studies. Peatlands have been studied because of their vast area in temperate climates, their unique biota and successional patterns, their economic importance of peat as a fuel and soil conditioner, and, recently, their importance in the global atmospheric carbon balance. Bogs have intrigued and mystified many cultures for centuries because of such discoveries as the Iron Age "bog people" of Scandinavia, who were preserved intact for up to 2,000 years in the nondecomposing peat (see, e.g., Glob, 1969; Coles and Coles, 1989).

Because bogs and other peatlands are ubiquitous in northern Europe and North America, many definitions and words, some unfortunate, that now describe wetlands in general originated from bog terminology; there is also considerable confusion in the use of terms such as *bog, fen, swamp, moor, muskeg, heath, mire, marsh, highmoor, lowmoor,* and *peatland* to describe these ecosystems. The words *peatlands* in general and *bogs* and *fens* in particular will be used in this chapter to include deep peat deposits, mostly of the cold, northern, forested regions of North America and Eurasia. Peat deposits also occur in warm temperate, subtropical, or tropical regions, and we refer briefly to a major example of these, specifically the *pocosins* of the southeastern Coastal Plain of United States.

Geographic Extent

Bogs and fens are distributed in cold temperate climates of high humidity, mostly in the Northern Hemisphere (Fig. 12.1), where precipitation exceeds evapotranspiration, leading to moisture accumulation. There are also some peatlands in the Southern Hemisphere in southern South America and in New Zealand. But the most extensive areas of bogs and fens occur in Scandinavia, eastern Europe, western Siberia, Alaska, and Canada. Major areas where a very large percentage of the landscape is peatland include the Hudson Bay lowlands in Canada, the Fennoscandian Shield in northern Europe, and the western Siberian lowland around the Ob and Irtysh rivers.

There are about 3.5 million km^2 of peatlands in the world (Gorham, 1991). This total includes 1.6 million km^2 in the former Soviet Union (Botch et al., 1995), 900,000 km^2 of which are in the Western Siberian lowlands (Kremenetski et al., 2003). Fennoscania has another 220,000 km^2. In North America, Canada has approximately 1.10 million km^2 of peatlands. Combining Canada's total with an estimated 0.55 million km^2 of peatlands in the United States (including Alaska) (see Table 10.1), the 1.65 million km^2 of northern peatlands in North America represents almost one-half of the world's peatlands.

Some peatlands, not illustrated in Figure 12.1, are found in the Southern Hemisphere in southern South America and in New Zealand, but the size of these peatlands collectively is small compared to those in the Northern Hemisphere. The New Zealand Land Resource Inventory (Cromarty and Scott, 1996) lists 3,113 km^2 of wetlands in the entire country, many of which are peatlands. Included in that estimate are 439 km^2

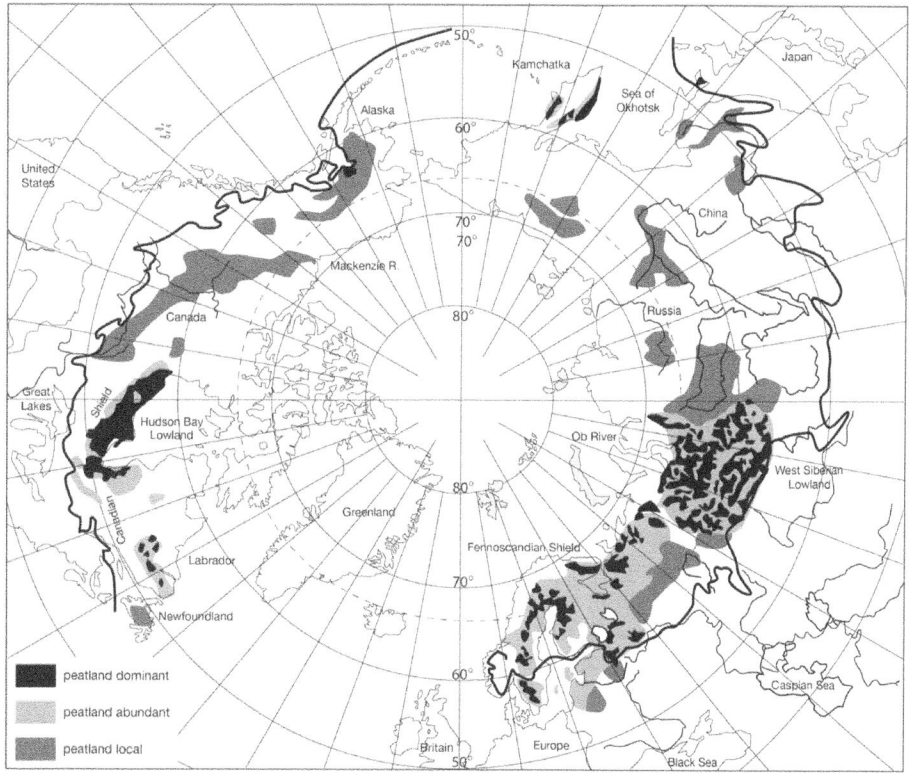

Figure 12.1 Area of abundant peatlands in the boreal zone (taiga) of the Northern Hemisphere. Peatlands are associated with boreal regions and their subalpine equivalents in mountainous regions. South of the tree line (solid line), woodland tundra or subalpine areas extend to the northern broken line. (After Wieder et al., 2006)

of *pakihi* (shallow-peat heathland) and another 356 km² of forest–pakiha associations, some of which support *Sphagnum* moss (Buxton et al., 1996). Raised bogs in New Zealand are not characterized by *Sphagnum* moss or ericaceous species common in the Northern Hemisphere but by rushlike plants co-dominated by restiad (coming from the family Restionaceae) bog species.

In the United States, peatlands dip into the conterminous United States from northern Minnesota to northern Maine. In the northeast United States, bogs and fens are common in Maine, New York, and Vermont. Peatlands are also fairly common in the unglaciated Appalachian Mountains in West Virginia, such as in the *Dolly Sods Wilderness*, which has plant species otherwise found only in sea-level eastern Canada. Bogs and fens are found as far south as Illinois, Indiana, and Ohio in basins scoured out by the Pleistocene glaciers in the north-central United States. The Middle Atlantic Coastal Plain supports an expansive area of poorly drained peatlands, called *pocosins*, that are similar to more northern peatlands in that they are nutrient poor and dominated by evergreen woody plants, such as bilberry, whortleberry, cranberry, heather, and Labrador tea belonging to the Ericaceae or heath family. Pocosins once covered 12,000 km^2, 70 percent of which were in North Carolina. Thirty-three percent of the pocosins in North Carolina have been destroyed (Richardson, 2003).

Hydrology and Peatland Development

Two primary processes necessary for peatland development are a positive water balance and peat accumulation. First, a positive water balance, meaning that precipitation is greater than evapotranspiration, is essential for peatland development and survival. Water budgets for a fen, bog, and pocosin (see Chapter 4: "Wetland Hydrology") show that evapotranspiration is generally only 50 to 70 percent of precipitation. The seasonal distribution of precipitation and excess water is important because peatlands require a humid environment year-round. In seasonally wet climates with cold winters, such as in the midwestern United States south of Minnesota, Wisconsin, and Michigan, peatlands are not common where hot, dry summers persist. The southern limit to bog species and, hence, to bogs is thought to be determined by the intensity of solar radiation in the summer months when precipitation and humidity are otherwise adequate to support bogs.

Some peatlands also depend on local river systems to maintain their moisture regime. Banaszuk and Kamocki (2008) reported on impaired fluvial peatlands of the Narew River valley in northeast Poland that have been impacted by a lower water table, expansion of *Phragmites*, and soil subsidence. Reduced river discharges, linked to a milder and drier climate over the last few decades, may be the cause.

A second requirement for peatland development is a surplus of peat production over decomposition, or accumulation greater than decomposition ($A > D$). Although primary production is generally low in northern peatlands compared to other ecosystems, decomposition is even more depressed, so peat accumulates. This is a necessary condition for the development of ombrotrophic bogs (see description later in this section). The continued development of the ecosystem is directly related to the amount of surplus water and peat. For example, in a cool, moist maritime climate, peatlands can develop over almost any substrate, even on hill slopes. In contrast, in warm climates where both evapotranspiration and decomposition are elevated, ombrotrophic peatlands seldom develop even when a precipitation surplus occurs. Once formed, a bog is remarkably resistant to conditions that alter the water balance and peat accumulation. The perched water table, the water-holding capacity of the peat, and its low pH create a microclimate that is stable under fairly wide environmental fluctuations.

Given the conditions of water surplus and peat accumulation, peatlands develop through *terrestrialization* (the infilling of shallow lakes) or *paludification* (the blanketing of terrestrial ecosystems by overgrowth of peatland vegetation). Three major bog formation processes are commonly seen: (1) quaking bog succession, (2) paludification, and (3) flowthrough succession.

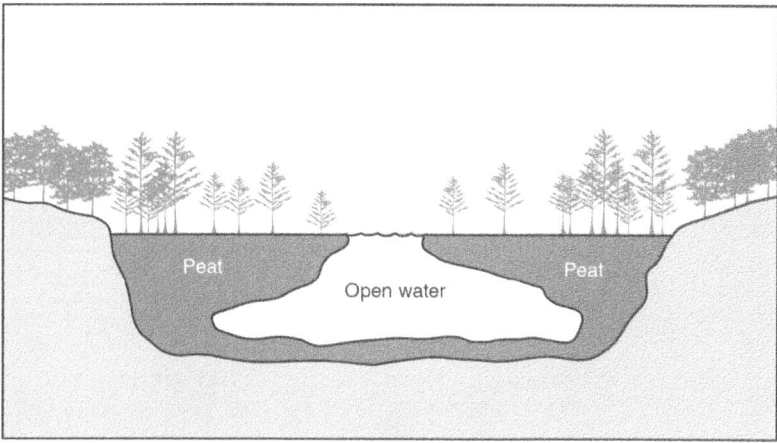

Figure 12.2 Typical profile of a quaking bog.

Quaking Bog Succession

Quaking bog succession is the classical process of terrestrialization, as described in most introductory botany or limnology courses. Bog development in some lake basins involves the filling in of the basin from the surface, creating a *quaking bog* (or *Schwingmoor* in German; Fig. 12.2). Plant cover, only partially rooted in the basin bottom or floating like a raft, gradually develops from the edges toward the middle of the lake. A mat of reeds, sedges, grasses, and other herbaceous plants develops along the leading edge of a floating mat of peat that is soon consolidated and dominated by *Sphagnum* and other bog flora. The mat has all of the characteristics of a *raised bog* except hydrologic isolation. The older peat is often colonized by shrubs and then forest trees such as pine, tamarack, and spruce, which form uniform concentric rings around the advancing floating mat.

These peatlands develop only in small lakes that have little wave action; they receive their name from the quaking of the entire surface that can be caused by walking on the floating mat. After peat accumulates above the water table, isolating the *Sphagnum*-dominated flora from their nutrient supply, the bog becomes increasingly nutrient poor. The development of a perched water table also isolates the peatland from groundwater and nutrient renewal. The result is a classic concentric, or excentric, raised ombrotrophic raised bog.

The hydrology of raised bogs has been investigated and found to be more complicated than originally thought, particularly for bogs that are on the edge of the boreal zone. Studies in the Lake Agassiz region of Minnesota showed that bogs and fens are part of a regional hydrology, with fens receiving groundwater and raised bogs generally recharging groundwater (Fig. 12.3a). Raised bogs are normally assumed to be disconnected from groundwater and fed only by precipitation. In a normal wet climate, this pattern of bog hydrology is true as a downward flow of excess precipitation deflects upwardly moving groundwater from mineral soil well below the surface

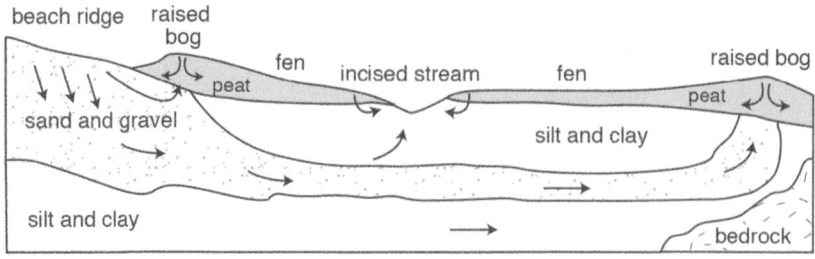

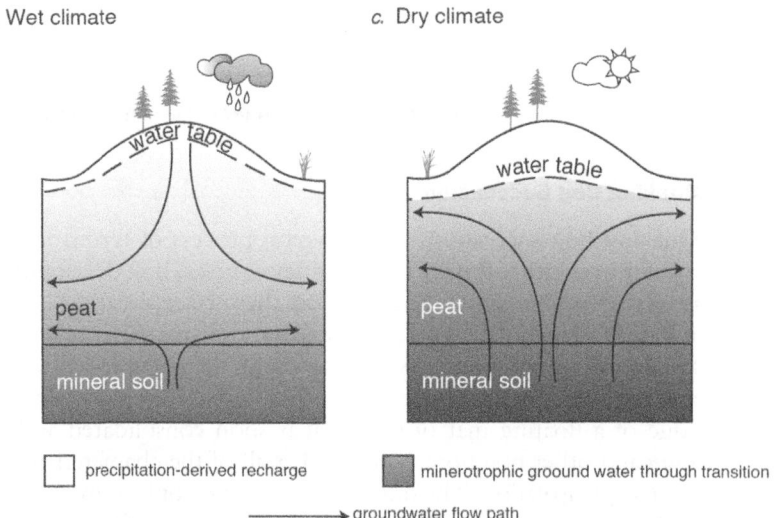

Figure 12.3 (a) Regional linkages between groundwater and raised bogs in the Lake Agassiz region of Minnesota (area is approximately 10 km long and 30 m thick). Detailed patterns of subsurface hydrology in the raised bogs are illustrated for (b) wet climate and (c) dry climate. During wet periods, precipitation-derived recharge maintains a head that flushes mineral-rich groundwater from the peat. During droughts, the water mound drops and mineral-rich groundwater can move upward into the raised-bog peat. (After Siegel et al., 1995; Glaser et al., 1997a)

(Fig. 12.3b). This accelerates peat accumulation, which, in turn, maintains the peat and, hence, hydrologic mound in the landscape. During droughts, which can be frequent events in peatlands on the edge of the boreal region, groundwater can move upward to within 1 to 2 m of the peat surface (Fig. 12.3c) and dramatically influence the peatland chemistry.

Paludification

A second pattern of bog evolution occurs when blanket bogs exceed basin boundaries and encroach on formerly dry land. This process of *paludification* can be brought about by climatic change, geomorphological change, beaver dams, logging of forests, or the natural advancement of a peatland. Often the lower layers of peat compress and become impermeable, causing a perched water table near the surface of what was formerly mineral soil. This causes wet and acid conditions that kill or stunt trees and allow only ombrotrophic bog species to exist. In some situations, the progression from forest to bog can take place in only a few generations of trees (Heilman, 1968).

Flowthrough Succession

Intermediate between terrestrialization and paludification is flowthrough succession (also termed *topogenous development*), in which the development of peatland modifies the pattern of surface water flow. It involves the development of a bog from a lake basin that originally had continuous inflow and outflow of surface water and groundwater. As the peat continues to build, the major inflow of water may be diverted and areas may develop that become inundated only during high rainfall. In the final stage, the bog remains above the groundwater level and becomes a true ombrotrophic bog.

Classification of Peatlands

Peatlands develop within a complex interaction of climate, hydrology, topography, chemistry, and vegetation development (succession). Because the physical and biotic processes that form peatlands are complex and differ somewhat from region to region, many different classification systems have been proposed over the past century (Table 12.1). Classification schemes have been based on at least seven features:

1. Floristics
2. Vegetation structure
3. Geomorphology (succession or development)
4. Hydrology
5. Chemistry
6. Stratigraphy
7. Peat characteristics

The last is used primarily for economic exploitation purposes. The other six are closely interrelated, leading to classification schemes that combine several natural features.

Landscape Classification

The developmental processes described above determine large-scale patterns of peatland development that have been divided into the following four landscape classifications.

1. *Raised bogs.* These are peat deposits that fill entire basins, are raised above groundwater levels, and receive their major inputs of nutrients from precipitation.

Table 12.1 Historical classification schemes for peatlands

Principal Basis for Classification	Mineral-Influenced Peatlands	Transition Peatlands	Precipitation-Dominated Peatlands	Reference
Topography	Fen		Bog or rasied bog	General use
	Niedermoore (low moor)	Ubergangsmoore	Hochmoore (high moor)	Weber (1907)
Hydrology	Geogenous Limnogenous Topogenous Soligenous		Ombrogenous	von Post and Granlund (1926), Sjörs (1948), Du Rietz (1949), Damman (1986)
	Rheophilous	Transition	Ombrophilous	Kulczynski (1949)
	Soligenous		Ombrogenous	Walter (1973)
	Minerogenous		Ombrogenous	Warner and Rubec (1997)
Water chemistry	Rich fen	Poor fen	Bog	General use; Sjörs (1948)
	Minerotrophic	Mesotrophic	Ombrotrophic	Moore and Bellamy (1974)
	Rheotrophic		Ombrotrophic	Moore and Bellamy (1974)
Nutrition	Nährstoffreichere	Mittelreiche	Nährstoffearme	Weber (1907)
	Eutrophic	Mesotrophic	Oligotrophic	Weber (1907), Pjavchenko (1982)
Vegetation	Emergent or forested fen	Transitional	Moss–lichen or forested bog	Cowardin et al. (1979), Gorham and Janssens (1992)

Source: Revised from Bridgham et al. (1996)

These bogs are found primarily in the boreal and northern deciduous biomes. When a concentric pattern of pools and peat communities forms around the most elevated part of the bog, the bog is called a *concentric domed bog*. Bogs that form from previously separate basins on sloping land and form elongated hummocks and pools aligned perpendicular to the slope are called *excentric raised bogs*. In Europe, the former are found near the Baltic Sea, and the latter are found primarily in the North Karelian region of Finland.

2. *Aapa peatlands*. These wetlands, which also are called *string bogs* and *patterned fens* (Figs. 12.4), are found throughout the boreal region, often north of the raised bog region. The dominant feature of these wetlands is the long, narrow alignment of the higher peat hummocks (*strings*) that form ridges perpendicular to the slope of the peatland and are separated by deep pools (*flarks* in Swedish). In appearance, they resemble a hillside of terraced rice fields. The strings and flarks develop perpendicular to the direction of the water flow. The pattern begins as a series of scattered pools on the down slope, wetter edge of the water track. These pools gradually coalesce into linear flarks. Peat accumulation in the adjacent strings and the increasing impermeability of decomposing peat in the flarks accentuate the pattern. Within the large water tracks, tree islands appear to be remnants of continuous swamp forests that were replaced by sedge lawns in the expanding water tracks.

Figure 12.4 Two oblique aerial images of string bogs in North America: (a) aerial photo of Cedarburg Bog in southwestern Wisconsin, showing a pattern of parallel peat ridges (strings) alternating with water-filled depressions (flarks) running diagonally across the lower half of the photograph; (b) a string fen in Labrador, Canada. The strings stand out because they are vegetated with ericaceous shrubs and scrub trees over sphagnum moss, whereas flarks are dominated by mosses and herbs or, in the case of the Canadian site, extensive standing water. (Photograph (a) by G. Guntenspergen, reprinted with permission (b) by D. Wells, reprinted by permission of C. Rubec and reprinted from Mitsch et al., 1994, p. 30, Fig. 30, with permission from Elsevier Science)

3. *Paalsa bogs.* These bogs, found in the southern limit of the tundra biome, are large plateaus of peat (20–100 m in breadth and length and 3 m high) generally underlain by frozen peat and silt. The peat acts like an insulating blanket, actually keeping the ground ice from thawing and allowing the southernmost appearance of the discontinuous permafrost. In Canada, as much as 40 percent of the land area is influenced by cyrogenic factors. When peat overlies frozen sediments, it influences the pattern of the landscape. Many distinctive forms are similar to European aapa and paalsa peatlands but are embedded in a continuous peat-covered landscape.

4. *Blanket bogs.* These wetlands along the northwestern coast of Europe and throughout the British Isles and are a result of paludification described above. The favorable humid Atlantic climate allows the peat literally to "blanket" very large areas far from the site of the original peat accumulation. Peat in these areas generally can advance on slopes of up to 18 percent; extremes of 25 percent have been noted on slopes covered by blanket bogs in western Ireland.

Chemistry-based Classification

The developmental processes described previously lead to increasing isolation of bogs from surface and subsurface flows of both water and mineral nutrients. The degree of hydrologic isolation of mires leads to a simple classification that is probably the most frequently used today and is based on the degree to which the peatland receives groundwater inflow as compared to only precipitation.

1. *Minerotrophic peatlands.* These are true fens that receive water that has passed through mineral soil. These peatlands generally have a high groundwater level and occupy a low point of relief in a basin. They are also referred to as *rheotrophic peatlands* and *rich fens* in general use.

2. *Mesotrophic peatlands.* These peatlands are intermediate between mineral-nourished (minerotrophic) and precipitation-dominated (ombrotrophic) peatlands. Another term used frequently for this class is *transitional peatlands* or *poor fens*.

3. *Ombrotrophic peatlands.* These are the true raised bogs that have developed peat layers higher than their surroundings and that receive nutrients and other minerals exclusively by precipitation.

Another "trophic" classification of peatlands, found in older European literature (Weber, 1907) and originally developed to classify peatlands and not lakes (Hutchinson, 1973), is the three-level trophic classification familiar to limnologists:

1. *Eutrophic peatlands.* Nutrient-rich peatlands; described by Weber (1907) as *Nährstoffreichere* (eutrophe).

2. *Mesotrophic peatlands.* Same as before; described by Weber (1907) as *Mittelreiche* (mesotrophe).

3. *Oligotrophic peatlands.* Nutrient-poor peatlands; described by Weber (1907) as *Nährstoffearme* (oligotrophe).

Hutchinson (1973) suggested that the process of peatland development could be called *oligotrophication*. The terms *eutrophic* and *oligotrophic* were applied to lakes and their current limnological use by Naumann (1919) twelve years after Weber (1907) applied the terms to peatlands Russian scientists such as Pjavchenko (1982) and Bazilevich and Tishkov (1982) continued to use this nomenclature for peatlands well into the 1980s.

Bridgham et al. (1996) argued for caution in the use of the "-trophic" suffix for classifying peatlands because the classic peatland gradient from minerotrophic to ombrotrophic, characterized by surface water chemistry such as pH, conductivity, and alkalinity, does not necessarily correlate with the eutrophic to oligotrophic gradient, which is defined in terms of nutrient (e.g., nitrogen, phosphorus, and potassium) availability. Bridgham et al. (1998) found evidence to suggest that there was higher phosphorus availability in bogs and higher nitrogen availability in fens. In other words, a strict correlation between measures of dissolved minerals and available nutrients has never been established. They suggested resurrecting the terms *eutrophic* and *oligotrophic*, which are rarely used in peatland literature today, because they clearly refer to nutrients and not to other minerals. Such a resurrection did not occur.

Hydrology-based Classification

Terms such as *soligenous* and *ombrogenous* actually refer to the hydrological and topographic origins of the peatlands, not to the mineral conditions of the inflowing water. A true hydrologic classification of peatlands based on the following two categories is illustrated in Figure 12.5:

1. *Ombrogenous peatlands.* Open only to precipitation
2. *Geogenous peatlands.* Open to outside hydrologic flows other than precipitation:
   a. *Limnogenous peatlands.* Develop along slow-flowing streams or lakes
   b. *Topogenous peatlands.* Develop in topographic depressions with at least some regional groundwater flow
   c. *Soligenous peatlands.* Develop with regional interflow and surface runoff

Canadian Classification

One of the more complete classifications developed for wetlands in general and peatlands in particular is the Canadian Wetland Classification System (Warner and Rubec, 1997). This classification uses the terms *minerotrophic* and *ombrotrophic* in its water chemistry classification and *minerogenous* and *ombrogenous* in its hydrological classification. Their simple classification of peatlands is:

1. *Bog.* Peatland receiving water exclusively from precipitation and not influenced by groundwater; sphagnum-dominated vegetation
2. *Fen.* Peatland receiving water rich in dissolved minerals; vegetation cover composed dominantly of graminoid species and brown mosses

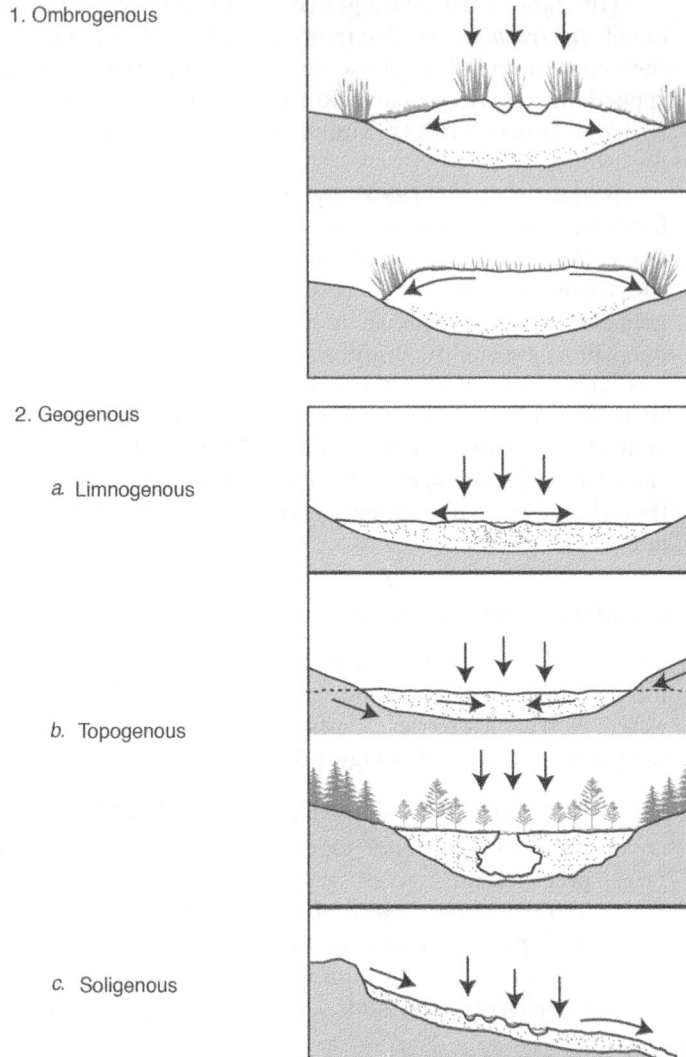

Figure 12.5 Classification of peatlands based on hydrology. Two major categories are geogenous peatlands, which are open to surface and groundwater flow, and ombrogenous peatlands, which only receive precipitation. (After Damman, 1986)

3. *Swamp*. Peatland dominated by trees, shrubs, and forbs; waters rich in dissolved minerals

Biogeochemistry

Soil and water chemistry are among the most important factors in the development and structure of the peatland ecosystems. Factors such as pH, mineral concentration, available nutrients, and cation exchange capacity influence the vegetation types and their productivity. Conversely, the plant communities influence the chemical properties of the soil water. In few wetland types is this interdependence so apparent as in northern peatlands. The major features of peatland biogeochemistry are discussed here.

Acidity and Exchangeable Cations

The pH of peatlands generally decreases as the organic content increases with the development from a minerotrophic fen to an ombrotrophic bog (Fig. 12.6). Fens are dominated by minerals from surrounding soils whereas bogs rely on a sparse supply of minerals from precipitation. Therefore, as a fen develops into a bog, the supply of metallic cations (Ca^{2+}, Mg^{2+}, Na^+, K^+) drops sharply. At the same time, as the organic content of the peat increases because of the slowing of the decomposition rate, the capacity of the soil to adsorb and exchange cations increases. These changes lead to the domination by hydrogen ions, and the pH falls sharply. Fens, in contrast, can range from slightly acidic (poor fens) to strongly alkaline (rich fens) depending on groundwater flow rate and chemistry (Bedford and Godwin, 2003). Gorham (1967) found that bogs in the English Lake District had a pH range of 3.8 to 4.4 compared to noncalcareous fens, which had a pH range of 4.8 to 6.0. The Russian scientist Pjavchenko (1982) assigned a pH range of 2.6 to 3.3 to oligotrophic bogs and a

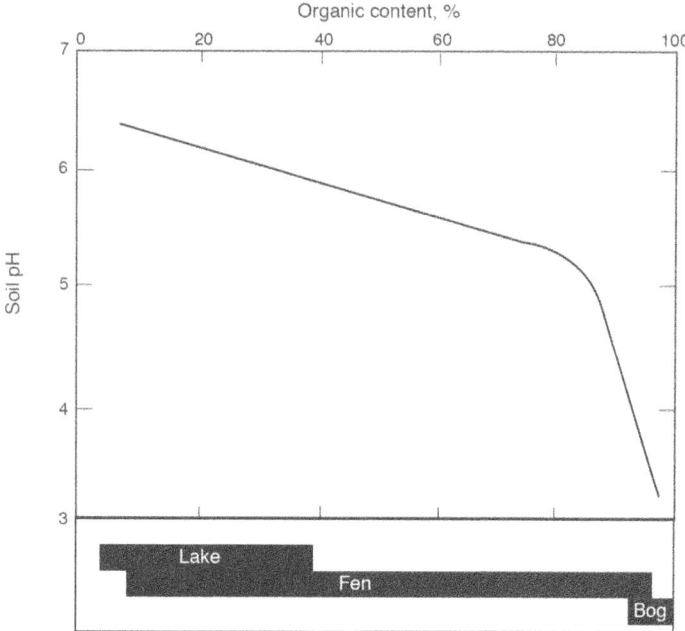

Figure 12.6 **Soil pH as a function of organic content of peat soil. (After Gorham, 1967)**

range of 4.1 to 4.8 to mesotrophic bogs; a pH greater than 4.8 defined a eutrophic (minerotrophic) fen.

As little as 10 percent of the water supply from groundwater may change the pH of a bog from 3.6 to 6.8, that is, from an ombrotrophic bog to a minerotrophic-rich fen. In an upper peat of a Minnesota raised bog, pH and conductivity, both indicators of mineral groundwater, increased dramatically in a drought year compared to a wet year.

The causes of bog acidity are not entirely clear, but five causes usually are cited for the low pH:

1. *Cation exchange by Sphagnum.* Cation exchange may be the most important mechanism for the generation of acidity in peatlands. There is a direct relationship between pH and the exchangeable hydrogen in peat, presumably the result of the metabolic activity of the plants. *Sphagnum* peats have a high exchangeable hydrogen and, consequently, a lower pH than sedge-dominated peats.

2. *Oxidation of sulfur compounds to sulfuric acid.* Organic sulfur reserves in peat may be oxidized to acidic compounds.

3. *Atmospheric deposition.* Sulfur deposition is a significant source of acidity, depending on the oxidation state of the sulfur and the location of the bog. Acid sources in precipitation and dry deposition are usually small except close to sources of atmospheric pollution.

4. *Biological uptake of cations by plants.* Ions in the peat water are concentrated by evaporation and are differentially absorbed by the mosses. This affects acidity, for example, by the uptake of cations that are exchanged with plant hydrogen ions to maintain the charge balance.

5. *Buildup of organic acids by decomposition.* Gorham et al. (1984) presented evidence supporting this source of bog acidity. Organic acids help buffer the system against the alkalinity of metallic cations brought in by rainfall and local runoff.

A detailed hydrogen budget constructed for a Minnesota bog complex implicated nutrient uptake as a major source of acidity (Table 12.2). About 15 percent of this hydrogen budget represents ion exchange on the cell walls of *Sphagnum*. Most of this acidity is neutralized by the release of cations during decomposition. Most of the rest of the acidity is generated by organic acid production from fulvic and other acids that result from the incomplete oxidation of organic matter and that buffer the pH of bogs throughout the world at a value of about 4. In addition to decomposition, the major source of alkalinity to neutralize the acids, the weathering of iron and aluminum and runoff are major processes.

Limiting Nutrients

Bogs are exceedingly deficient in available plant nutrients; fens that contain groundwater and surface water sources generally have considerably more nutrients. The paucity

Table 12.2 Acidity balance for a Minnesota bog complex

Sources	Acidity (meq m^{-2} yr^{-1})
Wet and dry deposition	−0.20 ± 10.7
Upland runoff	−44.3 ± 18.6
Nutrient uptake	827 ± 248
Organic acid production	263 ± 50
Total	1,044
Sinks	
Denitrification	12.2
Decomposition	784
Weathering	76
Outflow	142 ± 50
Total	1,044

meq = milliequivalents
Source: Urban et al. (1985)

of nutrients in bogs leads to two significant results, which are discussed in more detail later in this chapter: (1) The productivity of nutrient-poor bogs is lower than that of nutrient-rich fens; and (2) the characteristic plants, animals, and microbes have many special adaptations to the low-nutrient conditions. Many studies have attempted to find the ultimate limiting factor for bog primary productivity; this may be a complex and academic question because all available nutrients are in short supply, and the growing season is short and cool. Although calcium and potassium have been shown to be limiting, nitrogen and phosphorus are the major limiting chemicals in bog and fen productivity. When these nutrients are added in significant amounts to peatlands, major vegetation shifts occur; with management such as mowing, the limiting factor can change from nitrogen to phosphorus. Bog formation in its latter stages is essentially limited to nutrients brought in by precipitation. The effects on peatlands of increased atmospheric sources of nitrogen throughout the developed world due to fossil fuel burning has yet to be assessed adequately.

Vegetation

Bogs can be simple sphagnum moss peatlands, sphagnum–sedge peatlands, sphagnum–shrub peatlands, bog forests, or any number or combination of acidophilic plants. Mosses, primarily those of the genus *Sphagnum*, are the most important peat-building plants in bogs throughout their geographical range. Mosses grow in cushionlike, spongy mats; water content is high, with water sometimes held higher than it normally would be held by capillary action. *Sphagnum* grows shoots actively only in the surface layers (at a rate of about 1–10 cm annually); the lower layers die off and convert to peat.

In North American peatlands, *Sphagnum* often grows in association with cotton grass (*Eriophorum vaginatum*), various sedges (*Carex* spp.), and certain ericaceous shrubs, such as heather (*Calluna vulgaris*), leatherleaf (*Chamaedaphne calyculata*), cranberry and blueberry (*Vaccinium* spp.), and Labrador tea (*Ledum palustre*). Trees such as pine (*Pinus sylvestris*), crowberry (*Empetrum* spp.), spruce (*Picea* spp.), and tamarack (*Larix* spp.) are often found in bogs as stunted individuals that may be scarcely 1 m high yet several hundred years old. Fens in the United States tend to be dominated by a diverse community of plants that are distinct from boreal peatlands and typically include bryophytes, sedges (*Carex* and other genera of Cyperaceae), dicotyledonous herbs, and grasses (Amon et al., 2002; Bedford and Godwin, 2003).

Vegetation Patterns in a Minnesota Peatland

Heinselman (1970) described seven vegetation associations in the Lake Agassiz peatlands of northern Minnesota that are typical of many of those in North America. These occur in an intricate mosaic across the landscape, reflecting the topography, chemistry, and previous history of the site. The vegetation zones correspond closely to the underlying peat and to the present nutrient status of the site. The seven major zones are:

1. *Rich swamp forest*. These forested wetlands form narrow bands in very wet sites around the perimeter of peatlands. The canopy is dominated by northern red cedar (*Thuja occidentalis*); there are also some species of ash (*Fraxinus* spp.), tamarack, and spruce. A shrub layer of alder, *Alnus rugosa*, is often present, as are hummocks of *Sphagnum* moss.
2. *Poor swamp forest*: These swamps, occurring downslope of the rich swamp forests, are nutrient-poor ecosystems and are the most common peatland type in the Lake Agassiz region. Tamarack is usually the dominant canopy tree, with bog birch (*Betula pumila*) and leatherleaf in the understory and *Sphagnum* forming 0.3- to 0.6-m-high hummocks.
3. *Cedar string bog and fen complex*. This is similar to zone 2 except that trees'-edge fens alternate with cedar (*Thuja occidentalis*) on the bog ridges (strings) and treeless sedge (mostly *Carex*) in hollows (flarks) between the ridges.
4. *Larch string bog and fen*. In this type of string bog, similar to zones 2 and 3, tamarack (*Larix*) dominates the bog ridges.
5. *Black spruce–feathermoss forest*. This type is a mature black spruce (*Picea mariana*) forest that also contains a carpet of feathermoss (*Pleurozium*) and other mosses. The trees are tall, dense, and of similar ages. This peatland occurs near the margins of ombrotrophic bogs and generally does not have standing water.

> 6. *Sphagnum–black spruce–beatherleaf bog forest*. This is a widespread wetland type in northern North America. Stunted black spruce is the only tree, and there is a heavy shrub layer of leatherleaf, laurel (*Kalmia* spp.), and Labrador tea growing in large "pillows" of *Sphagnum* moss between spruce patches. This association is found in convex relief and is isolated from mineral-bearing water.
> 7. *Sphagnum–leatherleaf–Kalmia–spruce heath*. A continuous blanket of *Sphagnum* moss is the most conspicuous feature; a low shrub layer and stunted trees (usually black spruce) are present in 5 to 10 percent of the area. Zones 6 and 7 occur on a raised bog.
>
> In the water chemistry classification presented earlier in the chapter, zones 1 through 4 would be classified as minerotrophic, zone 5 as transitional, zone 6 as semiombrotrophic, and zone 7 as ombrotrophic.

Although *Sphagnum* species are the characteristic peat-forming ground cover of bogs, as sedges are of poor fens, there is a considerable overlap of species along the chemical gradient from mineral poor to mineral rich and from low pH to high pH. In a direct gradient analysis of vascular plants found in both bogs and fens in northern Minnesota, the sedges *Carex oligosperma* and *Eriophorum spissum* decrease in cover abundance with mineral enrichment of the peat, whereas tamarack (*Larix laricina*) increases in abundance. Black spruce and the ericaceous shrubs Labrador tea and leatherleaf, however, show dual peaks, indicating that their distribution is not controlled by mineral water chemistry but by another gradient, such as water level or possibly nitrogen or phosphorus availability.

Nicholson et al. (1996) investigated climatic and ecological gradients and how they affected bryophyte distribution in the Mackenzie River basin in northwestern Canada. They found that the most important variables that explained bryophyte species distributions were water chemistry (Mg^{2+}, Ca^{2+}, H^+), height above the water table, precipitation, and annual temperature. As a result of examining these gradients, seven peatland groups were clustered from the original 82 sites in the basin: (1) poor fens, (2) peat plateaus with thermokarst pools, (3) low-boreal bogs, (4) bogs and peat plateaus without thermokarst pools, (5) low-boreal dry poor fens, (6) wet moderate-rich fens, and (7) wet extremely rich fens. Thermokarst pools are features of a permafrost landscape where permafrost thawing and subsequent ice melting creates an uneven topography of mounds, sinkholes, caverns, and lake basins.

Locky et al. (2005) investigated black spruce (*Picea mariana*) swamps, fens, and bogs in the southern boreal region of Manitoba, Canada. They emphasized the distinction between black spruce swamps and other peatlands in the region pointing out their tendency to occur on gradual slopes, adjacent to water bodies, to contain larger trees with significant cover, and to occur on shallower peat than other peatlands. Chemically, these swamps are similar to moderate-rich fens.

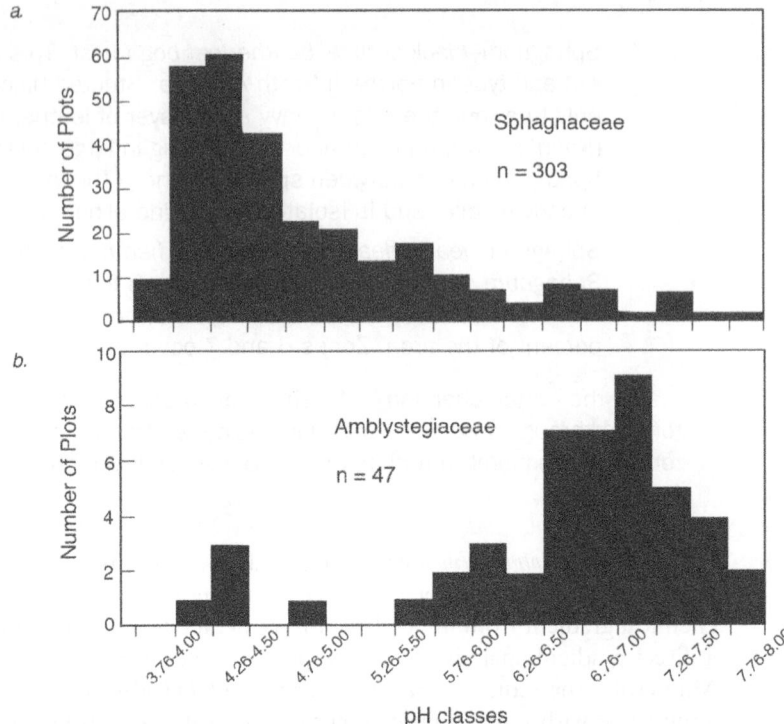

Figure 12.7 Distribution of two bryophyte families (Sphagnaceae and Amblysteglaceae) versus surface water pH for 440 peatland plots across North America. Plots were counted if they had at least one species of the family covering more than 25 percent of the total area. The bimodal pattern suggests a classification of peatlands based on moss vegetation. (After Gorham and Janssens, 1992)

In another study that attempted to relate vegetation directly to water chemistry, Gorham and Janssens (1992) investigated two families of mosses (Sphagnaceae and Amblystegiaceae) at 440 sites across northern North America (Fig. 12.7). They found a clear bimodal split in their occurrence, with Sphagnaceae most often in low-pH peatlands (mode pH, 4.0–4.25) and Amblystegiaceae in high-pH peatlands (mode pH, 6.76–7.0).

Black Spruce Peatlands

One of the dominant forested wetlands in the world is the black spruce peatland of the taiga of Canada and Alaska. Black spruce (*Picea mariana*), often growing in association with tamarack (*Larix laricina*), is the tree species most associated with forested peatlands in the boreal regions of North America. These wetlands are estimated to encompass about half of the palustrine shrub–scrub wetlands in Alaska and cover an estimated 14 million ha in the state. Black spruce is mostly associated with ombrotrophic (bog) rather than minerotrophic (fen) communities. In bogs, it is found in associations with leatherleaf (*Chamaedaphne calyculata*), Labrador tea (*Ledum* spp.), laurel (*Kalmia latifolia*), blueberry (*Vaccinium* spp.), and bog rosemary *(Andromeda polifolia)*. *Sphagnum* spp., of course, is found as ground cover in these bogs. In Alaska, common associations include *P. mariana* with *Vaccinium uliginonsum, Ledum groenlandicum,* and feathermoss (*Pleurozium schreberi*) and *P. mariana* with *Sphagnum* spp. and *Cladina* spp. (Post, 1996). In regions where permafrost is prevalent, black spruce wetlands often occur in paalsa hummocks.

Carolina Pocosins

In contrast to the more northern peatlands, the woody vegetation of pocosins found mostly near the southeastern United States coastal region of North and South Carolina is dominated by evergreen trees and shrubs. Two broad community classes have been identified, and their presence was related to fire frequency, soil type, and hydroperiod. A *Pinus–Ericalean* (pine and heath shrub) community develops on deep organic soils with long hydroperiods and frequent fire. Three associations within this community are (1) pond pine (*Pinus serotina*) canopy with titi (*Cyrilla racemiflora*) and zenobia (*Zenobia pulverulenta*) shrubs, (2) pond pine and loblolly bay (*Gordonia lasianthus*) canopy with fetterbush (*Lyonia lucida*), and (3) pond pine canopy with titi and fetterbush shrubs. A *conifer–hardwood* community type is found on shallow organic soils with slightly shorter hydroperiods. Two associations in this group are (1) pond pine canopy with titi, fetterbush, red maple (*Acer rubrum*), and black gum (*Nyssa sylvatica*) shrubs; and (2) pond pine and pond cypress (*Taxodium distictium* var. *nutans*) canopy with red maple, titi, fetterbush, and black gum shrubs.

Peatland Adaptations

The vegetation in bogs and peatlands both controls and is controlled by its physical and chemical environment. Some of the conditions for which adaptations are necessary in peatlands are discussed here.

Waterlogging

Many bog plants, in common with wetland vegetation in general, have anatomical and morphological adaptation to waterlogged anaerobic environments. These include (1) the development of large intercellular spaces (aerenchyma or lacunae) for oxygen supply, (2) reduced oxygen consumption, and (3) oxygen leakage from the roots to produce a locally aerobic root environment. *Sphagnum*, conversely, is morphologically adapted to maintain waterlogging. The compact growth habit, overlapping leaves, and rolled branch leaves form a wick that draws up water and holds it by capillarity. These adaptations enable *Sphagnum* to hold water up to 15 to 23 times its dry weight.

Acidification of the External Interstitial Water

Sphagnum has the unique ability to acidify its environment, probably through the production of organic acids, especially polygalacturonic acids located on the cell walls. The galacturonic acid residues in the cell walls increase the cation exchange capacity to double that of other bryophytes. The adaptive significance of this peculiarity of *Sphagnum* is unclear. The acid environment retards bacterial action and hence decomposition, enabling peat accumulation despite low primary production rates. It has been suggested that the high cation exchange capacity also enables the plant to maintain a higher and more stable pH and cation concentration in the living cells than in the surrounding water.

Adaptations to Nutrient Deficiency

Many bog plants have adaptations to the low nutrient supply that enable them to conserve and accumulate nutrients. Adaptations seen in bog plants include evergreenness; sclerophylly, or the thickening of the plant epidermis to minimize grazing; uptake of amino acids; and high root biomass. Some bog plants, notably cotton grass (*Eriophorum* spp.), translocate nutrients back to perennating organs prior to litterfall in the autumn. These nutrient reserves are available for the following year's growth and seedling establishment. The roots of other bog plants penetrate deep into peat zones to bring nutrients to the surface. Bog litter has been demonstrated to release potassium and phosphorus, often the most limiting nutrients, more rapidly than other nutrients, an adaptation that keeps these nutrients in the upper layers of peat. Many ericaceous plants have adapted to low concentrations of nitrogen by effectively utilizing ammonium nitrogen in place of limited nitrate nitrogen under low-pH conditions, by efficiently using nitrogen and even by utilizing organic nitrogen sources). Some bog plants also carry out symbiotic nitrogen fixation. The bog myrtle (*Myrica gale*) and the alder develop root nodules characteristic of nitrogen fixers and have been shown to fix atmospheric nitrogen in bog environments.

Carnivorous Plants

Another well-known adaptation to nutrient deficiency in bogs is the ability of carnivorous plants to trap and digest insects. This special feature is seen in several unique insectivorous bog plants, including the pitcher plant (*Sarracenia purpurea*; Fig. 12.8) and sundew (*Drosera* spp.). A nutrient limitation study developed for *Sarracenia* in Minnesota showed that although nutrient and insect additions did not increase biomass, there were respective nutrient increases in the leaves of the plant. It was estimated that insect capture accounts for approximately 10 percent of the plant's nitrogen and phosphorus needs (Chapin and Pastor, 1995). Pitcher plants are obligate host to more invertebrate species than any other bog plant (Rymal and Folkerts, 1982). In the water-filled pitcher in the plant sketch shown in Figure 12.8, a mosquito, a midge, two sarcophagid flies, and a mite are suggested as captives. An aphid and three moths feed exclusively on the tissue. Other insects are associated with other parts of the plant.

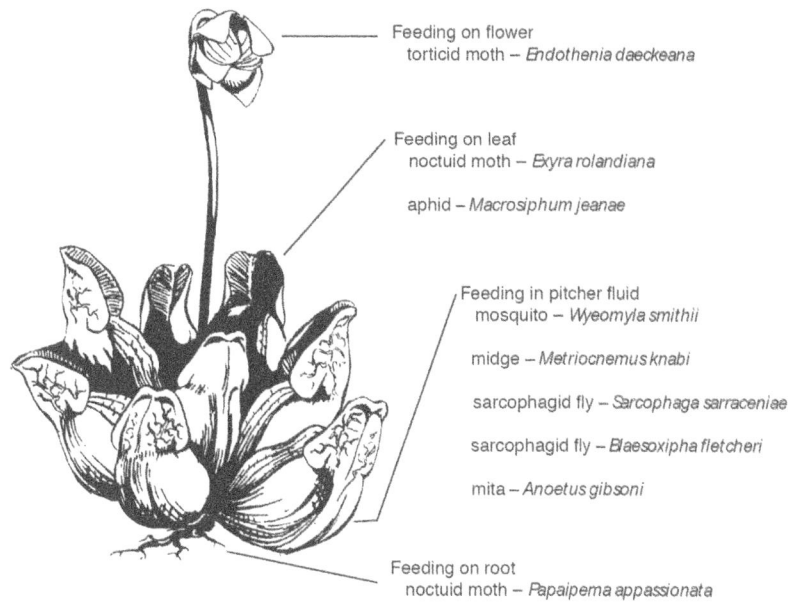

Figure 12.8 The pitcher plant (*Sarracenia purpurea*) including invertebrates that associate with the plant. (After Damman and French, 1987)

Overgrowth by Peat Mosses

Many flowering plants are faced with the additional problem of being overgrown by peat mosses as the mosses grow in depth and in area covered. Adapting plants must raise their shoot bases by elongating their rhizomes or by developing adventitious roots. Trees such as pine, birch, and spruce are often severely stunted because of the moss growth and poor substrate; they grow better on bogs where the vertical growth of moss has stopped.

Consumers

Mammals

The populations of animals in bogs are generally low because of the low productivity and the unpalatability of bog vegetation. Animal density is closely related to the structural diversity of the peatland vegetation. For example, forested peatlands tend to support the greatest number of small-mammal species, especially close to upland habitats. Large mammals tend to roam over larger landscapes and are thus not specific to individual peatland types. In northern Minnesota and New England, moose (*Alces alces*) are frequently found in small peatlands. White-tailed deer (*Odocoileus virginianus*) browse heavily in white cedar bogs in winter. Black bear (*Ursus americanus*) use peatlands for escape cover and for food. The woodland caribou (*Rangifer tarandus*) was the largest mammal that was largely restricted to peatlands, but it disappeared from Minnesota in 1936, probably as a result of hunting pressure. Several large predatory mammals have been reported to use or inhabit peatlands in North America, including the gray wolf (*Canis lupus*), red wolf (*Canis rufus*), puma (*Felis concolor cougar*), and grizzly bear (*Ursus arctos horriblis*) (Bedford and Godwin, 2003). Smaller mammals closely associated with peatlands are beaver (*Castor canadensis*), lynx (*Lynx canadensis*), fishers (*Martes pennant*), and snowshoe hares (*Lepus americanus*). The beaver is a fairly recent import into Minnesota peatlands. It moved in along drainage ditches, seldom penetrating deep into large peatlands, but it has had a significant effect on peatland flooding in northern Minnesota (Naiman et al., 1991). Wet forests are becoming the only habitats where wide-ranging mammals such as the black bear, otter, and mink are found (Sharitz and Gibbons, 1982; L. D. Harris, 1989). This is not so much because peatlands are obligate habitats but because the clearing of upland forests has forced the remaining population into the remaining large tracts of forested wetlands.

Amphibians and Reptiles

Glaser (1987) reported only seven species of amphibians and four species of reptiles in northern Minnesota peatlands. Acid waters below pH 5 appear to be the major limiting factor in their ability to colonize bogs. Fens may have a more diverse array of faunal species including several that are rare. In their review of fens of the United States, Bedford and Godwin (2003) listed the bog turtle (*Clemmys muhlenbergii*) and eastern massasauga (*Sisturus catenatus*) as federally listed (threatened, endangered, or considered for listing) reptiles that use fens with high frequency. Other rare or uncommon species associated with fens include mole salamanders (*Ambystomia talpoideum*) and four-toed salamanders (*Hemidactylium scutatum*) that frequent small mountain fens in the Appalachians (Murdock, 1994).

Birds

Many bird species are seen in peatlands during different times of the year (Fig. 12.9). For example, Warner and Wells (1980) reported 70 species during the breeding season. Most of these are also common on upland sites, but a few depend on peatlands for survival. These include the sandhill crane (*Grus candensis*), great gray owl (*Strix nebulosa*), short-eared owl (*Asio flammeus*), sora (*Porzana carolina*), and sharp-tailed sparrow (*Ammospiza caudacuta*). In New England, as one moves from the Canadian border south, the species change, but the new species have analogous positions along the gradient (Fig. 12.9). Pocosins in the Southeast United States that support mature pond pine (*Pinus serotina*) can be inhabited by the endangered red-cockaded woodpecker (*Picoides borealis*) (Richardson 2003).

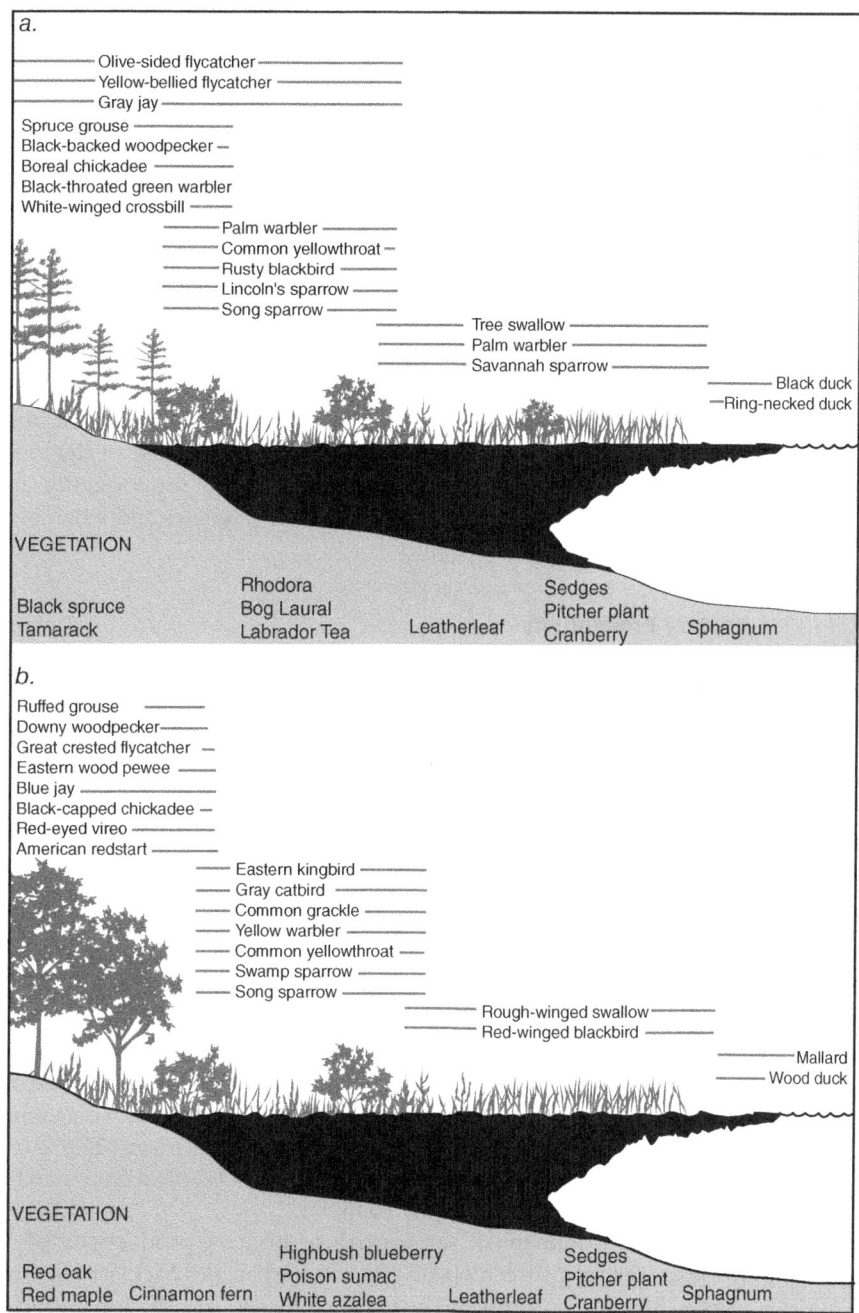

Figure 12.9 Comparison of bird distribution, typical of a lake-border bog in the northern and southern parts of the northeastern United States. (After Damman and French, 1987)

Ecosystem Function

The dynamics of peatlands reflect the realities of the harsh physical environment and the scarcity of mineral nutrients. These conditions result in three major features:

1. Bogs are systems of low primary productivity; fens are generally more productive; *Sphagnum* mosses often dominate bogs, and other vegetation is stunted in growth.
2. Bogs and fens are peat producers whose rates of accumulation are controlled by a combination of complex hydrologic, chemical, and topographic factors. This peat contains a great store of nutrients, most of it below the rooting zone and thus unavailable to plants.
3. Low-nutrient peatlands in cold climates have developed several unique pathways to obtain, conserve, and recycle nutrients. The amount of nutrients in living biomass is small. Cycling is slow because of the low temperatures, the nutrient deficiency of the litter, and the waterlogging of the substrate. It is more active when peat production stagnates and when bogs receive increased nutrient inputs.

Primary Productivity

Major organic inputs to bog systems come from the primary production of the vascular plants, liverworts, mosses, and lichens. Among vascular plants, ericaceous shrubs and sedges are the most important primary producers, and much of this production is below ground. Mosses, especially sphagnum mosses, account for one-third to one-half of the total production. Bogs and fens are usually less productive than most other wetland types and are generally less productive than the climatic terrestrial ecosystems in their region, about half that of a coniferous forest and a little more than a third that of a deciduous forest (Table 12.3). According to Pjavchenko (1982), forested peatlands produce a range of 260 to 400 g organic matter m^{-2} yr^{-1}, with the low value that of an om brotrophic bog and the high value that of a minerotrophic fen. Malmer (1975) cited a typical range of 400 to 500 g m^{-2} yr^{-1} for nonforested, raised (ombrotrophic) bogs in western Europe. In contrast, Lieth (1975) estimated the net primary productivity in the boreal forest to average 500 g m^{-2} yr^{-1} and in the temperate forest to average 1,000 g m^{-2} yr^{-1}. The estimate for boreal forests probably includes bog forests as well as upland forests. Annual above-ground productivity for control plots in a northern Minnesota fen ranged from 87 ± 2 to 459 ± 34 g m^{-2} yr^{-1} over a four-year period with below-ground biomass (estimated at the last year) at 470 ± 79 g m^{-2} (Weltzin et al., 2005).

The measurement of the growth or primary productivity of *Sphagnum* mosses presents special problems not encountered in productivity measurements of other plants. The upper stems of the plant elongate, and the lower portions gradually die off, become litter, and eventually form peat. It is difficult to measure the sloughing off of dead material to litter. It is equally hard to measure the biomass of the plant at any one time because it is difficult to separate the living and dead material of the

Table 12.3 Net primary productivity of peatlands in Europe and North America

Location	Type of Peatland	Living Biomass (g dry wt m^{-2})	Net Primary Productivity (g dry wt m-2 yr^{-1})	Reference
Europe				
Western Europe	general nonwooded raised bog	1,200	400–500	Malmer (1975)
Western Europe	forested raised bog	3,700	340	Moore and Bellamy (1974)
Russia	eutrophic forested bog	9,700–11,000	400	Pjavchenko (1982)
	mesotrophic forested bog	4,500–8,900	350	
	oligotrophic forested bog	2,200–3,600	260	
Russia	mesotrophic *Pinus-Sphagnum* Bog	8,500	393	Bazilevich and Tishkov (1982)
England	blanket bog		659 ± 53[a]	Forrest and Smith (1975)
England	blanket bog		635	Heal et al. (1975)
Ireland	blanket bog		316	Doyle (1973)
North America				
Michigan	rich fen		341[b]	Richardson et al. (1976)
Minnesota	forested peatland	15,941	1,014	Reiners (1972)
	fen forest	9,808	651[b]	
Manitoba	peatland bog		1,943	Reader and Stewart (1972)
Alberta	bog		280[b]	Szumigalski & Bayley (1996a)
	poor fen		310[b]	
	moderate-rich fen		360[b]	
	lacustrine sedge fen		214[b]	
	extreme-rich fen		245[b]	
Alberta	bog		390[b]	Thormann & Bayley (1997)
	floating sedge fen		356[b]	
	lacustrine sedge fen		277[b]	
	riverine sedge fen		409[b]	
Quebec	poor fen		114[b]	Bartsch & Moore (1985)
	rich fen		335[b]	
	transitional fen		176[b]	

[a] Mean ± standard deviation for seven sites.
[b] Above ground only.

peat. The following two methods for measuring *Sphagnum* growth give comparable results: (1) the use of "innate" time markers, such as certain anatomical or morphological features of the moss; and (2) the direct measurement of changes in weight. Growth rates for *Sphagnum* determined by these two techniques generally fall in the range of 300 to 800 g m^{-2} yr^{-1} (Table 12.4). Although Damman (1979) and Wieder and Lang (1983) suggested that annual production should increase with decreasing latitude, only *S. magellanicum* shows such a trend in Table 12.4. Evidently, local and regional factors are more important than latitude.

It is generally expected that peatlands are nutrient limited. However, this can vary by plant species and community. Chapin et al. (2004) experimentally loaded

Table 12.4 Comparison of selected data on production of *Sphagnum* species in order of decreasing latitude

Species[a]	Growth (mm/yr)	Production (g m^{-2} yr^{-1})	Latitude (N)	Location	Mean Annual Precipitation (mm)	Mean Annual Temperature (°C)	Source
fus	1.4–3.2	70	68°22'	N Sweden	600	2.9	Rosswall and Heal (1975)
fus	—	250	63°09'	S Finland	532	3.5	Silvola and Hanski (1979)
fus	—	220–290	63°09'	S Finland	532	3.5	K. Tolonen, in Rochefort et al. (1990)
fus	7–16	195	60°62'	S Finland	632	4–4.8	Pakarinen (1978)
mag	9.5	70	59°50'	S Norway	1,250	5.9	Pedersen (1975)
ang	14.7	500	—	—	—	—	
fus	9.8	90	56°05'	S Sweden	800	7.9	Damman (1978)
mag	7.8	100	—	—	—	—	
mag	10–18	50–100	55°09'	England	1,270	9.3	S. B. Chapman (1965)
ang	28–34	110–240	54°46'	England	1,980	7.4	Clymo and Reddaway (1971)
mag	14–15	230	54°46'	England	1,980	7.4	Forrest and Smith (1975)
ang	—	240–330	—	—	—	—	
ang	38–43	110–440	54°46'	England	1,980	7.4	Clymo (1970)
fus	6–7	75–83	54°43'	Quebec	791	4.9	Bartsch and Moore (1985)
ang	4–17	19–127	—	—	—	—	T. R. Moore (1989)
fus	—	270	54°28'	England	1,375	7.4	Bellamy and Rieley (1967)
fus	30	424–801	54°20'	N Germany	714	8.4	Overbeck and Happach (1957)
mag	35–51	252–794	—	—	—	—	
ang	120–160	488–1,656	—	—	—	—	
fus	—	50	49°53'	S Manitoba	517	2.5	Reader and Stewart (1971)
fus	17–24	240	49°52'	NE Ontario	858	0.8	Pakarinen and Gorham (1983)
fus	7–31	69–303	49°40'	NW Ontario	714	2.6	Rochefort et al. (1990)
mag	11–34	52–240	—	—	—	—	
ang	20–39	97–198	—	—	—	—	
mag	62	540	39°07'	West Virginia	1,330	7.9	Wieder and Lang (1983)

[a]fus = *Sphagnum fuscum*; mag = *S. magellanicum*; ang = *S. angustifolium*.
Source: Rochefort et al. (1990).

nitrogen, phosphorus,, and calcium carbonate (to raise pH) into a bog and fen in northern Minnesota and examined plant community and species productivity in response. In the bog, calcium carbonate and low additions of N ($2\,g\,N\,m^{-2}\,yr^{-1}$) both increased above-ground net primary productivity (ANPP) while higher loads of N ($6\,g\,N\,m^{-2}\,yr^{-1}$) actually inhibited growth. Fen graminoid growth responded to increased P additions. Within both wetlands, there were variable responses to experimental conditions among plant types and species. The authors surmised that although nutrient availability is low in peatlands, this does not necessarily mean that peatlands are nutrient limited.

Decomposition

The accumulation of peat in bogs is determined by the production of litter (from primary production) and the destruction of organic matter (decomposition). As with primary production, the rate of decomposition in peat bogs is generally low because of (1) waterlogged conditions, (2) low temperatures, and (3) acid conditions. In fact, the accumulation of peat in peatlands is due more to slow decomposition processes than to net community productivity. Besides leading to peat accumulation, slow decomposition leads to slower nutrient recycling in an already nutrient-limited system.

The pattern of *Sphagnum* decomposition is highest near the surface, where aerobic conditions exist. By 20 cm depth, the rate is about one-fifth of that at the surface. This pattern is caused by anaerobic conditions. The bulk of the organic decomposition that does occur in peat bogs is by microorganisms, although the total numbers of bacteria in these wetland soils are much fewer than in aerated soils. As pH decreases, the fungal component of the decomposer food web becomes more important relative to bacterial populations. Verhoeven et al. (1994) used a cotton-strip decomposition method and found substantially lower decay rates in ombrogenous bogs compared to other peatlands and mineral-soil wetlands. Total phosphorus (positive correlation) and soil organic matter (negative correlation) explained 75 percent of the decay rates. Thus, low nutrients and high organic matter (which keeps the soils reduced) contribute significantly to the low decay rates in the bogs. Szumigalski and Bayley (1996b) found the following progression of rates of decay of litter from peatlands in central Alberta: *Carex* > *Betula* > mosses. The highest decomposition rates were with plant material with the highest nitrogen content. Using a standard litter material of *Carex lasiocarpa*, litter losses were in the following order:

poor fen > wooded rich fen > bog > open rich fen > sedge fen

In the same region, Bayley and Mewhort (2004) compared peat-accumulating marshes and moderate-rich fens. Although these wetlands can appear similar, the decomposition rates for fens were notably slower than marshes, which were attributed to the higher water levels in the marshes.

There has been considerable speculation about factors that give rise to patterned peatlands. The pattern of strings and flarks or hummocks and hollows, for example, appears to be related to differential rates of peat accumulation. Rochefort et al. (1990) determined that differential accumulation in a poor-fen system in northwestern Ontario, Canada, was caused more by differences in peat decomposition rates than by differences in primary production rates. They found that, even though the production rates of *Sphagnum* in hummocks were generally about equal to the rates in hollows or even lower than the rates in minerotrophic hollows, hummock species had slower decomposition rates than those of hollow species. As a result, peat accumulated faster on hummocks than in hollows, and hummocks may be expanding at the expense of hollows.

Peat Accumulation

The vertical accumulation rate of peat in bogs and fens is generally thought to be between 20 and 80 cm/1,000 yr in European bogs (Moore and Bellamy, 1974), although Cameron (1970) gave a range of 100 to 200 cm/1,000 yr for North American bogs and Nichols (1983) reported an accumulation rate for peat of 150 to 200 cm/1,000 yr in warm, highly productive sites. Malmer (1975) described a vertical growth rate of 50 to 100 cm/1,000 yr as typical for western Europe. Assuming an average density of peat of 50 mg/mL, this rate is equivalent to a peat accumulation rate of 25 to 50 g m^{-2} yr^{-1}. Hemond (1980) estimated a rapid accumulation rate of 430 cm/1,000 yr, eqivalent to 180 g m^{-2} yr^{-1}, for Thoreau's Bog, Massachusetts.

Comparing Bog Energy Flow Estimates on Two Continents

One of the earliest energy budgets for any ecosystem was determined in the classic study by Lindeman (1942) of Cedar Bog Lake, a small bog in northern Minnesota (Fig. 12.10a). Although this energy budget is crude, the main features have stood the test of time. Very little of the incoming radiation (<0.1 percent) is captured in photosynthesis. The two largest flows of organic energy are to respiration (26 percent) and to storage as peat (70 percent). Energy flow in the simplified food web is primarily to herbivores (13 percent), and about 3.5 percent goes to decomposers. As the following two more recent budget measurements show, the peat storage term is exceedingly high, and decomposition losses are probably underestimated.

Bazilevich and Tishkov (1982) and Alexandrov et al. (1994) presented a detailed energy flow through a mesotrophic (transition) bog in the European region of Russia (Fig. 12.10b). The bog is a sphagnum–pine (*Sphagnum girgenoshnii–Pinus sylvestris*) community containing shrubs such as bilberry (*Vaccinium myrtillus*). The total energy stored in the bog was estimated to be in excess of 137 kg dry organic matter/m^2, with dead organic matter (to a depth of 0.6 m of peat) accounting for 94 percent of the storage. Living biomass was 8.5 kg/m^2, or about 6 percent of the organic storage. Gross primary productivity was 987 g m^{-2} yr^{-1}, or about 4,400 kcal m^{-2} yr^{-1} (assuming 1 g organic matter = 4.5 kcal), with about 60 percent consumed by plant respiration. The distribution of the net primary production came from trees (39 percent), algae (28 percent), shrubs (21 percent), mosses and lichens (9 percent), and grasses (3 percent). The net primary production was primarily consumed by decomposers; much less was consumed in grazing food webs. Net accumulation of peat was 100 g m^{-2} yr^{-1} (or about 450 kcal m^{-2} yr^{-1}). Losses other than biotic decomposition, which accounted for most of the loss of organic matter, were chemical oxidation and surface and subsurface flows.

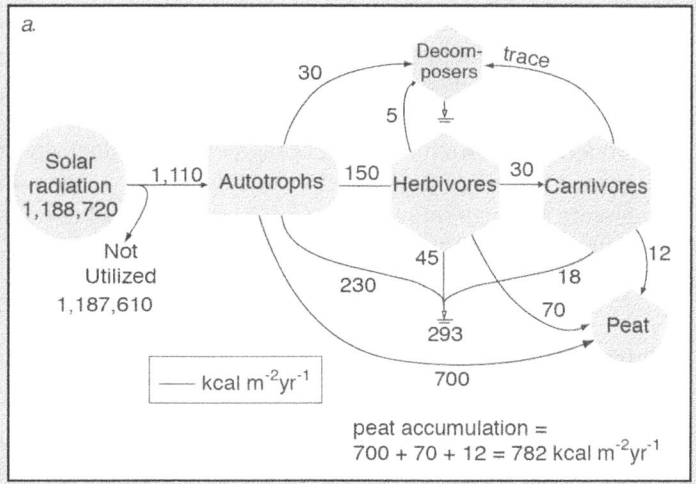

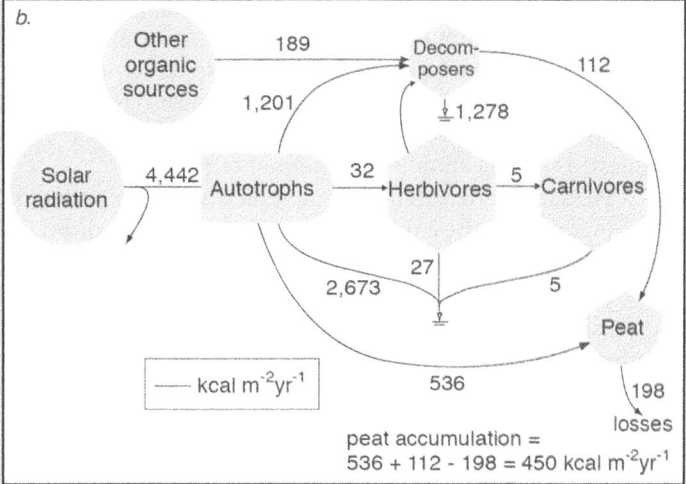

Figure 12.10 Diagrams of the energy flow in peatlands: (a) Cedar Bog Lake, Minnesota, and (b) a Russian transition peatland. Flow in kcal m^{-2} yr^{-1}. Flows in (a) were originally published in calories whereas flows in (b) were published in gram dry weight and converted to energy as 4.5 kcal g^{-1}. ((a) After Lindeman, 1942; (b) after Bazilevich and Tishkov, 1982; Alexandrov et. al., 1994)

Comparison of these two energy budgets from Russia and the United States, carried out several decades apart, illustrates several points. First, Lindeman's Cedar Bog was approximately one-fourth as productive as Bazilevich and Tishkov's Russian peatland, a possibility, given that the Russian site

> was described as transitional between a bog and a fen. Second, the American bog accumulated more peat than did the Russian peatland, results that could reflect either the sophistication of the measuring techniques for the time or what happens to peatlands as they transition from fens to true bogs. Lindeman shows a high percentage of the productivity stored permanently as peat. Assuming 50 g/L as the density of peat, Lindeman's peat accumulation results in a high rate 350 cm/1,000 yr, whereas the Russian study has a more reasonable 200 cm/1,000 yr.

Nutrient Budgets

Nitrogen

Nitrogen budgets for two peatlands—Thoreau's Bog in Massachusetts and a perched raised-bog complex in Minnesota—make an interesting comparison (Fig. 12.11). Although total nitrogen input is comparable in both systems, the Minnesota perched-bog system catches some runoff from surrounding uplands, whereas nitrogen fixation is the largest source of biologically active nitrogen in Thoreau's Bog. Otherwise, the budgets are remarkably similar despite the differences in bog type and location. Both accumulate nitrogen in peat and lose a significant portion through runoff. Denitrification is an uncertain term.

Peatlands have often been identified as diverse wetlands. Increased nutrient loading appears to reduce that diversity. Drexler and Bedford (2002) traced various nutrients in a small peatland adjacent to a farm field in central New York. They found that the farm field was a nutrient source with phosphorus and potassium loading coming primarily from overland flow while nitrogen loading was delivered primarily via groundwater. Peat concentrations of phosphorus and potassium, along with groundwater flux of nitrate-nitrogen and NH_4–N, were all negatively correlated with plant diversity. Nutrient loading promoted the growth of taller, monotypic stands of vascular plants (*Calamagrostis canadensis*, *Carex lacustris*, *Epilobium hirsutum*, and *Typha latifolia*) at the expense of plant diversity.

Peatlands appear to have the capacity for denitrification (Hemond, 1983), but the magnitude *in vivo* is uncertain. Some recent studies have investigated nitrogen dynamics in peatlands. Wray and Bayley (2007) estimated annual denitrification rates for boreal marshes and fens in Alberta at 11 and 24 g N m^{-2} y^{-1}, respectively, with N_2 being the dominant product.

Bridgham et al. (2001) examined nutrient availability along an ombrotrophic-minerotrophic gradient for 16 peatlands in northern Minnesota. They found that N availability generally increased along the gradient. They detected seasonal patterns associated with N availability with NO_3–N being more available in the summer and NH_4–N more available in the winter. Highest P availability was found in minerotrophic swamps and beaver meadows, with the least availability found in fens and bogs.

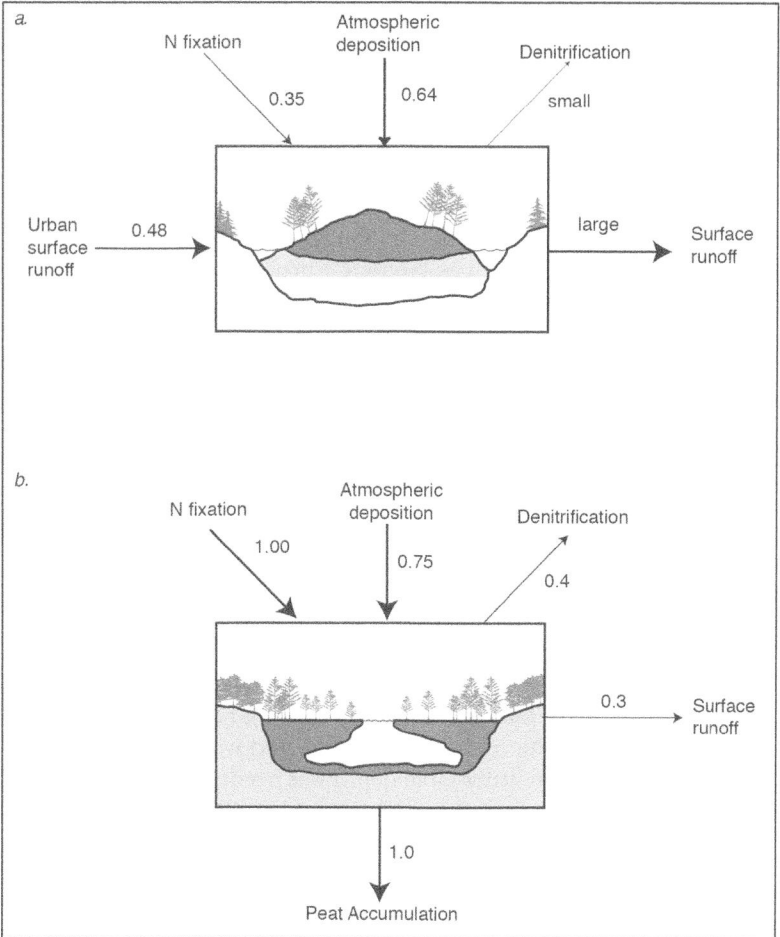

Figure 12.11 Nitrogen budgets for two northern ombrotrophic bogs: (a) perched raised-bog complex in northern Minnesota and (b) small floating-mat sphagnum bog (Thoreau's Bog) in Massachusetts. Values are in g N m^{-2} yr^{-1}. (Data from Hemond, 1983; Urban and Eisenreich, 1988)

Carbon

Carbon budgets for peatlands have drawn a great deal of interest, given the importance of these ecosystems in global carbon dynamics (see Chapter 17: "Wetlands and Climate Change" for more discussion on this topic). High-latitude peatlands are known to store tremendous amounts of carbon. In the Western Siberian lowlands, Kremenetski et al. (2003) estimated that peatlands have an average peat depth of 2.6 m and store greater than 53.8 million metric tons of carbon. It is generally accepted that boreal peatlands were once carbon sinks, but there is little consensus that they are contemporary sinks. Carbon budgets have been developed for small peatlands and for large peatland-dominated watersheds (Rivers et al., 1998). The latter, a 1,500-km^2 watershed in the Lake Agassiz peatlands in Minnesota, illustrated that the peat watershed had a net carbon storage of 12.7 g C m^{-2} yr^{-1} but that there was a tenuous balance between the watershed being a source and a sink of carbon. Inflows of carbon are groundwater, precipitation, and net community productivity, whereas outtflows are groundwater and surface flow and outgassing of methane. It was estimated from a companion study (Glaser et al., 1997b) that peat is accumulating at a rate of 1 mm/yr (100 cm/1,000 yr). This budget illustrates the importance of accurate hydrologic measurements as well as biological productivity measurements in determining accurate nutrient budgets. Mitsch et al. (2013) report a net carbon retention of 29 ± 8 g C m^{-2} yr^{-1} for eight boreal peatlands where both carbon sequestration and methane emissions were measured, an average rate considerably lower than net carbon retention rates for temperate or tropical wetlands.

Methane emissions from peatlands have also been studied closely because of their potential influence as greenhouse gases and the expansive store of carbon in peatlands. Annual precipitation and water table position also seems to be a primary controller of methane flux in peatlands (Heikkinen et al., 2002; Huttunen et al., 2003; Smemo and Yavitt, 2006). Methane emission from a Finnish minerotrophic peatlands ranged between 8 and 330 mg m^{-2} d^{-1}, and there was a positive correlation between methane emission and water table level (Huttunen et al., 2003). Bubier et al. (2005) found a similar seasonal average range of CH$_4$ emission (10–350 mg m^{-2} d^{-1}) but emphasized there was considerable spatial variability. They too found a significant relationship between water table position and mean CH$_4$ flux (Fig. 12.12). They noted the log linear relationship between CH$_4$ flux and water table position suggesting that only a small increase in water table depth was needed to increase CH$_4$ emission substantially. In their review of the forested peatland literature, Trettin et al. (2006) concluded that

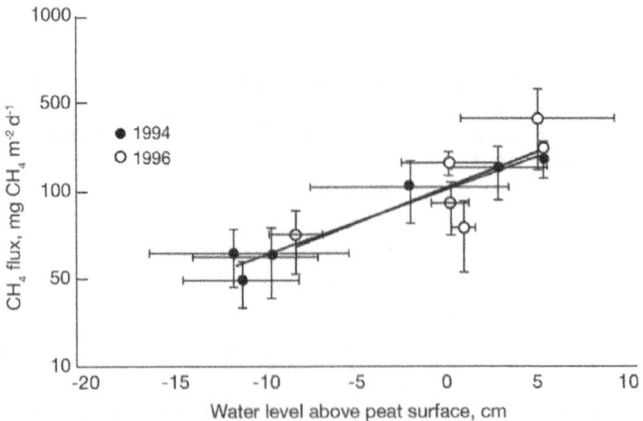

Figure 12.12 **Methane flux as a function of mean water level in Canadian peatlands. Water table indicates depth below (negative) or above (positive) the peat surface. Error bars are standard deviations.** (After Bubier et al., 2005)

decreased water table levels will result in decreased CH_4 emission and increased CO_2 emission from the peat surface. They emphasized this does not mean that these peatlands will necessarily decrease in their soil carbonpool as the difference may be made up by changes in plant succession and increased productivity.

Recommended Readings

Rochefort, L., M. Strack, M. Poulin, J. S. Price, and C. Lavoie. 2012. Northern Peatlands. In: D. R. Batzer and A. H. Baldwin, eds., *Wetland Habitats of North America: Ecology and Conservation Concerns*, pp. 119–134. Los Angeles: University of California Press.

Wieder, R. K., and D. H. Vitt, eds., 2006. *Boreal Peatlands Ecosystems*. Berlin: Springer-Verlag.

References

Alexandrov, G. A., N. I. Bazilevich, D. A. Logofet, A. A. Tishkov, and T. E. Shytikova. 1994. Conceptual and mathematical modeling of mater cycling in Tajozhny Log Bob ecosystem (Russia). In B. C. Patten, ed. *Wetlands and Shallow Continental Water Bodies*, Vol. 2. SPB Academic Publishing, The Hague, the Netherlands, pp. 45–93.

Amon, J. P., C. A. Thompson, Q. J. Carpenter, and J. Miner. 2002. Temperate zone fens of the glaciated midwestern USA. *Wetlands* 22: 301–317.

Banaszuk, P., and A. Kamocki. 2008. Effects of climatic fluctuations and land-use changes on the hydrology of temperate fluviogenous mire. *Ecological Engineering* 32: 133–146.

Bartsch, I., and T. R. Moore. 1985. A preliminary investigation of primary production and decomposition in four peatlands near Schefferville, Quebec. *Canadian Journal of Botany* 63: 1241–1248.

Bayley, S. E., and R. L. Mewhort. 2004. Plant community structure and functional differences between marshes and fens in the southern boreal region of Alberta, Canada. *Wetlands* 24: 277–294.

Bazilevich, N. I., and A. A. Tishkov. 1982. Conceptual balance model of chemical element cycles in a mesotrophic bog ecosystem. In *Ecosystem Dynamics in Freshwater Wetlands and Shallow Water Bodies*, Vol. 2. SCOPE and UNEP, Centre of International Projects, Moscow, USSR, pp. 236–272.

Bedford, B. L., and K. S. Godwin. 2003. Fens of the United States: Distribution, characteristics, and scientific connection versus legal isolation. *Wetlands* 23: 608–629.

Bellamy, D. J., and J. Rieley. 1967. Some ecological statistics of a "miniature bog." *Oikos* 18: 33–40.

Botch, M. S., K. I. Kobak, T. S. Vinson, and T. P. Kolchugina. 1995. Carbon pools and accumulation in peatlands of the former Soviet Union. *Global Biogeochemical Cycles* 9: 37–46.

Bridgham, S. D., J. Pastor, J. A. Janssens, C. Chapin, and T. J. Malterer. 1996. Multiple limiting gradients in peatlands: A call for a new paradigm. *Wetlands* 16: 45–65.

Bridgham, S. D., K. Updegraff, and J. Pastor. 1998. Carbon, nitrogen, and phosphorus mineralization in northern wetlands. *Ecology* 79: 1545–1561.

Bridgham, S. D., K. Updegraff, and J. Pastor. 2001. A comparison of nutrient availability indices along an ombrotrophic-minerotrophic gradient in Minnesota wetlands. *Soil Science Society of America Journal* 65: 259–269.

Bubier, J., T. Moore, K. Savage, and P. Crill. 2005. A comparison of methane flux in a boreal landscape between a dry and a wet year. *Global Biogeochemical Cycles* 19: GB1023.

Buxton, R. P., P. N. Johnson, and P. R. Espie. 1996. Sphagnum Research Programme: The Ecological Effects of Commercial Harvesting. Science for Conservation: 25. New Zealand Department of Conservation, Wellington, 33 pp.

Cameron, C. C. 1970. Peat deposits of northeastern Pennsylvania U.S. Geological Survey Bull. 1317-A, 90 pp.

Chapin, C. T., and J. Pastor. 1995. Nutrient limitations in the northern pitcher plant *Sarracenia purpurea*. *Canadian Journal of Botany* 73: 728–734.

Chapin, C. T., S. D. Bridgham, and J. Pastor. 2004. pH and nutrient effects on above-ground net primary production in a Minnesota, USA bog and fen. *Wetlands* 24: 186–201.

Chapman, S. B. 1965. The ecology of Coom Rigg Moss, Northumberland: Some water relations of the bog system. *Journal of Ecology* 53: 371–384.

Clymo, R. S. 1970. The growth of *Sphagnum*: Methods of measurement. *Journal of Ecology* 58: 13–49.

Clymo, R. S., and E. J. F. Reddaway. 1971. Productivity of *Sphagnum* (bog-moss) and peat accumulation. *Hydrobiologia (Bucharest)* 12: 181–192.

Coles, B., and J. Coles. 1989. *People of the Wetlands, Bogs, Bodies and Lake-Dwellers*. Thames & Hudson, New York. 215 pp.

Cowardin, L. M., V. Carter, F. C. Golet, and E. T. LaRoe. 1979. *Classification of Wetlands and Deepwater Habitats of the United States*. FWS/OBS-79/31. U.S. Fish and Wildlife Service, Washington, DC. 103 pp.

Cromarty, D., and D. Scott. 1996. *A Directory of Wetlands in New Zealand*. Department of Conservation, Wellington, New Zealand.

Damman, A. W. H. 1978. Distribution and movement of elements in ombrotrophic peat bogs. *Oikos* 30: 480–495.

Damman, A. W. H. 1979. Geographic patterns in peatland development in eastern North America. In *Proceedings of the International Symposium of Classification of Peat and Peatlands*. Hyytiala, Finland, International Peat Society, pp. 42–57.

Damman, A. W. H. 1986. Hydrology, development, and biogeochemistry of ombrogenous peat bogs with special reference to nutrient relocation in a western Newfoundland bog. *Canadian Journal of Botany* 64: 384–394.

Damman, A. W. H., and T. W. French. 1987. *The ecology of peat bogs of the glaciated northeastern United States: A community profile*. U.S. Fish and Wildlife Service, Washington, DC. 100 pp.

Doyle, G. J. 1973. Primary production estimates of native blanket bog and meadow vegetation growing on reclaimed peat at Glenamoy, Ireland. In L. C. Bass and F. E. Wielgolaski, eds., *Primary Production and Production Processes, Tundra Biome*. Tundra Biome Steering Committee, Edmonton, Canada, pp. 141–151.

Drexler, J. Z., and B. L. Bedford. 2002. Pathways of nutrient loading and impacts on plant diversity in a New York peatland. *Wetlands* 22: 263–281.

Du Rietz, G. E. 1949. Huvudenheter och huvudgränser i Svensk myrvegetation. *Svensk Botanisk Tidkrift* 43: 274–309.

Forrest, G. I., and R. A. H. Smith. 1975. The productivity of a range of blanket bog types in the Northern Pennines. *Journal of Ecology* 63: 173–202.

Glaser, P. H. 1987. *The Ecology of Patterned Boreal Peatlands of Northern Minnesota: A Community Profile*. U.S. Fish and Wildlife Service Biological Report 85 (7.14), Washington, DC, 98 pp.

Glaser, P. H., D. I. Siegel, E. A. Romanowicz, and Y. P. Shen. 1997a. Regional linkages between raised bogs and the climate, groundwater, and landscape of northwestern Minnesota. *Journal of Ecology* 85: 3–16.

Glaser, P. H., P. C. Bennett, D. I. Siegel, and E. A. Romanowicz. 1997b. Palaeo-reversals in groundwater flow and peatland development at Lost River, Minnesota, USA. *Holocene* 6: 413–421.

Glob, P. V. 1969. *The Bog People: Iron Age Man Preserved*, trans. by R. Bruce-Mitford. Cornell University Press, Ithaca, NY. 200 pp.

Gorham, E. 1967. Some chemical aspects of wetland ecology. Technical Memorandum 90, Committee on Geotechnical Research, National Research Council of Canada, pp. 2–38.

Gorham, E. 1991. Northern peatlands: Role in the carbon cycle and probable responses to climatic warming. *Ecological Applications* 1: 182–195.

Gorham, E., and J. A. Janssens. 1992. Concepts of fen and bog reexamined in relation to bryophyte cover and the acidity of surface waters. *Acta Societatis Botanicorum Poloniae* 61: 7–20.

Gorham, E., S. J. Eisenreich, J. Ford, and M. V. Santelmann. 1984. The chemistry of bog waters. In W. Stumm, ed. *Chemical Processes in Lakes*. John Wiley & Sons, New York, pp. 339–363.

Harris, L. D. 1989. The faunal significance of fragmentation of southeastern bottomland forests. In D. D. Hook and R. Lea, eds., *Proceedings of the Symposium the Forested Wetlands of the Southern United States*, U.S. Department of Agriculture, Forest Service, Southeastern Forest Experiment Station, Asheville, NC, pp. 126–134.

Heal, O. W., H. E. Jones, and J. B. Whittaker. 1975. Moore House, U.K. In T. Rosswall and O. W. Heal, eds., *Structure and function of tundra ecosystems. Ecological Bulletin 20*. Swedish Natural Science Research Council, Stockholm, pp. 295–320.

Heikkinen J. E. P., V. Elsakov, and P. J. Martikainen. 2002. Carbon dioxide and methane dynamics and annual carbon budget in tundra wetland in NE Europe, Russia. *Global Biogeochemical Cycles* 16. doi:10.1029/2002GB0001930

Heilman, P. E. 1968. Relationship of availability of phosphorus and cations to forest succession and bog formation in interior Alaska. *Ecology* 49: 331–336.

Heinselman, M. L. 1970. Landscape evolution and peatland types, and the Lake Agassiz Peatlands Natural Area, Minnesota. *Ecological Monographs* 40: 235–261.

Hemond, H. F. 1980. Biogeochemistry of Thoreau's Bog, Concord, Mass. *Ecological Monographs* 50: 507–526.

Hemond, H. F. 1983. The nitrogen budget of Thoreau's Bog. *Ecology* 64: 99–109.

Hutchinson, G. E. 1973. Eutrophication: The scientific background of a contemporary practical problem. *American Scientist* 61: 269–279.

Huttunen, J. T., H. Nykänen, J. Turunen, and P. J. Martikainen. 2003. Methane emissions from natural peatlands in the northern boreal zone in Finland, Fennoscandia. *Atmospheric Environment* 37: 147–151.

Kulczynski, S. 1949. Peat bogs of Polesie. *Acad. Pol. Sci. Mem.*, Ser. B, No. 15. 356 pp.

Kremenetski, K. V., A. A. Velichko, O. K. Borisova, G. M. MacDonald, L. C. Smith, K. E. Frey, and L. A. Orlova. 2003. Peatlands of Western Siberian lowlands: Current knowledge on zonation, carbon content and Late Quaternary history. *Quaternary Science Review* 22: 703–723.

Lieth, H. 1975. Primary production of the major units of the world. In H. Lieth and R. H. Whitaker, eds., *Primary Productivity of the Biosphere*. Springer-Verlag, New York, pp. 203–215.

Lindeman, R. L. 1942. The trophic-dynamic aspect of ecology. *Ecology* 23: 399–418.

Locky, D. L., S. E. Bayley, and D. H. Vitt. 2005. The vegetational ecology of black spruce swamps, fens, and bogs in southern boreal Manitoba, Canada. *Wetlands* 25: 564–582.

Malmer, N. 1975. Development of bog mires. In A. D. Hasler, ed. *Coupling of Land and Water Systems*. Ecology Studies 10. Springer-Verlag, New York, pp. 85–92.

Mitsch, W. J., R. H. Mitsch, and R. E. Turner. 1994. Wetlands of the Old and New Worlds—Ecology and Management. In W. J. Mitsch, ed. *Global Wetlands: Old and New*. Elsevier, Amsterdam, the Netherlands, pp. 3–56.

Mitsch, W. J., B. Bernal, A. M. Nahlik, U. Mander, L. Zhang, C. J. Anderson, S. E. Jørgensen, and H. Brix. 2013. Wetlands, carbon, and climate change. *Landscape Ecology* 28: 583–597.

Moore, P. D., and D. J. Bellamy. 1974. *Peatlands*. Springer-Verlag, New York. 221 pp.

Moore, T. R. 1989. Growth and net production of *Sphagnum* at five fen sites subarctic eastern Canada. *Canadian Journal of Botany* 67: 1203–1207.

Murdock, N. A. 1994. Rare and endangered plants and animals of southern Appalachian wetlands. *Water, Air, and Soil Pollution* 77: 385–405.

Naiman, R. J., T. Manning, and C. A. Johnston. 1991. Beaver population fluctuations and tropospheric methane emissions in boreal wetlands. *Biogeochemistry* 12: 1–15.

Naumann, E. 1919. Nagra sypunkte angaende planktons ökilogi. Med. Sarskild hänsyn till fytoplankton. *Svensk Botanisk Tidkrift* 13: 129–158.

Nichols, D. S. 1983. Capacity of natural wetlands to remove nutrients from wastewater. *Journal of the Water Pollution Control Federation* 55: 495–505.

Nicholson, B. J., L. D. Gignac, and S. E. Bayley. 1996. Peatland distribution along a north-south transect in the Mackenzie River Basin in relation to climatic and environmental gradients. *Vegetatio* 126: 119–133.

Overbeck, F., and H. Happach. 1957. Über das Wachstum und den Wasserhaushalt einiger Hochmoor Sphagnum, *Flora* 144: 335–402.

Pakarinen, P. 1978. Production and nutrient ecology of three Sphagnum species in southern Finnish raised bogs, *Annales Botanici Fennici* 15: 15–26.

Pakarinen, P., and E. Gorham. 1983. Mineral Element Composition of *Sphagnum fuscum* peats collected from Minnesota, Manitoba, and Ontario. In *Proceedings of the International Symposium on Peat Utilization*. Bemidji State University Center for Environmental Studies, Bemidji, MN, pp. 417–429.

Pedersen, A. 1975. Growth measurements of five *Sphagnum* species in South Norway. *Norwegian Journal of Botany* 22: 277–284.

Pjavchenko, N. J. 1982. Bog ecosystems and their importance in nature. In D. O. Logofet and N. K. Luckyanov, eds., *Proceedings of International Workshop on Ecosystems Dynamics in Wetlands and Shallow Water Bodies, Vol. 1, SCOPE and UNEP Workshop*. Center for International Projects, Moscow, USSR, pp. 7–21.

Post, R.A. 1996. Functional Profile of Black Spruce Wetlands in Alaska. Report EPA 910/R-96-006, U.S. Environmental Protection Agency, Seattle, WA, 170 pp.

Reader, R J., and J. M. Stewart. 1971. Net primary productivity of the bog vegetation in southeastern Manitoba. *Can. J. Bot.* 49:1471–1477.

Reader, R. J., and J. M. Stewart. 1972. The relationship between net primary production and accumulation for a peatland in southeastern Manitoba. *Ecology* 53: 1024–1037.

Reiners, W. A. 1972. Structure and energetics of three Minnesota forests. *Ecological Monographs* 42: 71–94.

Richardson, C. J. 2003. Pocosins: Hydrologically isolated or integrated wetlands on the landscape? *Wetlands* 23: 563–576..

Richardson, C. J., W. A. Wentz, J. P. M. Chamie, J. A. Kadlec, and D. L. Tilton. 1976. Plant growth, nutrient accumulation and decomposition in a central Michigan peatland used for effluent treatment. In D. L. Tilton, R. H. Kadlec, and C. J. Richardson, eds., *Freshwater Wetlands and Sewage Effluent Disposal*. University of Michigan, Ann Arbor, pp. 77–117.

Rivers, J. S., D. I. Siegel, L. S. Chasar, J. P. Chanton, P. H. Glaser, N. T. Roulet, and J. M. McKenzie. 1998. A stochastic appraisal of the annual carbon budget of a large circumboreal peatland, Rapid River Watershed, northern Minnesota. *Global Biogeochemical Cycles* 12: 715–727.

Rochefort, L., D. H. Vitt, and S. E. Bayley. 1990. Growth, production, and decomposition dynamics of Sphagnum under natural and experimentally acidified conditions. *Ecology* 71: 1986–2000.

Rosswall, T., and O. W. Heal. 1975. Structure and function of tundra ecosystems. *Ecological Bulletin (Sweden)* 20: 265–294.

Rymal, D. E., and G. W. Folkerts. 1982. Insects associated with pitcher plants (Sarracenia, salraceniaceae), and their relationship to pitcher plans conservation: A review. *Journal of Alabama Academy of Sciences* 53: 131–151.

Sharitz, R. R., and J. W. Gibbons, eds. 1982. *The Ecology of Southeastern Shrub Bogs (Pocosins) and Carolina Bays: A Community Profile*, U.S. Fish Wildlife Service, Division of Biological Services, Washington, DC, FWS/OBS-82/04.

Siegel, D. I., A. S. Reeve, P. H. Glaser, and E. A. Romanowicz. 1995. Climate-driven flushing of pore water in peatlands. *Nature* 374: 531–533.

Silvola, J., and I. Hanski. 1979. Carbon accumulation in a raised bog. *Oecologia (Berlin)* 37: 285–295.

Sjörs, H. 1948. Myrvegetation i bergslagen. *Acta Phytogeographica Suecica* 21: 1–299.

Smemo, K. A., and J. B. Yavitt. 2006. A multi-year perspective on methane cycling in a shallow peat fen in central New York State, USA. *Wetlands* 26: 20–29.

Szumigalski, A. R., and S. E. Bayley. 1996a. Net above-ground primary production along a bog-rich fen gradient in central Alberta, Canada. *Wetlands* 16: 467–476.

Szumigalski, A. R., and S. E. Bayley. 1996b. Decomposition along a bog to rich fen gradient in central Alberta, Canada. *Canadian Journal of Botany* 74: 573–581.

Thormann, M. N., and S. E. Bayley. 1997. Aboveground net primary productivity along a bog-fen-marsh gradient in southern boreal Alberta, Canada. *Ecoscience* 4: 374–384.

Trettin, C. C., R. Laiho, K. Minkkinen, and J. Laine. 2006. Influence of climate change factors on carbon dynamics in northern forested peatlands. *Canadian Journal of Soil Science* 86: 269–280.

Urban, N. R., S. J. Eisenreich, and E. Gorham. 1985. Proton cycling in bogs: geographic variation in northeastern North America. In T. C. Hutchinson and K. Meema, eds., *Proceedings of the NATO Advanced Research Workshop on the Effects of Acid Deposition on Forest, Wetland, and Agricultural Ecosystems*. Springer-Verlag, New York, pp. 577–598.

Urban, N. R., and S. J. Eisenreich. 1988. Nitrogen cycling in a forested Minnesota bog, *Canadian Journal of Botany* 66: 435–449.

Verhoeven, J. T. A., D. F. Whigham, M. van Kerkhoven, J. O'Neill, and E. Maltby. 1994. Comparative study of nutrient-related processes in geographically separated wetlands: Toward a science base for functional assessment procedures. In W. J. Mitsch, ed., *Global Wetlands: Old World and New*. Elsevier, Amsterdam. pp. 91–106 von Post, L., and E. Granlund. 1926. Södra Sveriges Torvtillgångar. Sveriges Geologiska Undersökning Ser. C Avhandlingar och uppsater, No. 355. *Arsbok* 19 (2): 1–127.

Warner, B. G., and C. D. A. Rubec, eds. 1997. *The Canadian Wetland Classification System*. National Wetlands Working Group, Wetlands Research Centre, University of Waterloo, Waterloo, Ontario, Canada.

Warner, D., and D. Wells. 1980. Bird population structure and seasonal habitat use as indicators of environment quality of peatlands. Minnesota Department of Natural Resources, St. Paul, MN, 84 pp.

Weber, C. A. 1907. Aufbau und Vegetation der Moore Norddutschlands. *Beibl. Bot. Jahrbüchern*. 90: 19–34.

Weltzin, J. F., J. K. Keller, S. D. Bridgham, J. Pastor, P. B. Allen, and J. Chen. 2005. Litter controls plant community composition in a northern fen. *Oikos* 110: 537–546.

Wieder, R. K., and G. E. Lang. 1983. Net primary production of the dominant bryophytes in a Sphagnum-dominated wetland in West Virginia. *Bryologist* 86: 280–286.

Wieder, R. K., D. H. Vitt, and B. W. Benscoter. 2006. Peatlands and the boreal forest. In R. K. Wieder and D. H. Vitt, eds., *Boreal Peatland Ecosystems*. Springer Verlag, Berlin, pp. 1–8.

Wray, H. E., and S. E. Bayley. 2007. Denitrification rates in marsh fringes and fens in two boreal peatlands in Alberta, Canada. *Wetlands* 27: 1036–1045.

Part IV

Traditional Wetland Management

Chapter 13

Wetland Classification

Wetlands have been classified since the early 1900s, beginning with the peatland classifications of Europe and North America. The U.S. Fish and Wildlife Service has developed two major wetland classifications as the bases for wetland inventories. The early (1956) classification described 20 wetland types based on flooding depth, dominant forms of vegetation, and salinity regimes.
Classification of Wetlands and Deepwater Habitats of the United States, *published in 1979, uses a hierarchical approach based on systems, subsystems, classes, subclasses, dominance types, and special modifiers to define wetlands and deepwater habitats precisely. Canadian and international wetland classification systems provide alternative systems that recognize 49 and 32 different wetland types, respectively. More recently, classifications based on wetland function have been developed, including a functionally based approach called the hydrogeomorphic classification. Wetland inventories are carried out at many different scales with several different imageries and with both aircraft and satellite platforms.*

To deal realistically with wetlands on a regional scale, wetland scientists and managers have found it necessary both to categorize the different types of wetlands that exist and to determine their extent and distribution. The first of these activities is called *wetland classification*, and the second is called a *wetland inventory*. Some of the earliest efforts were undertaken to find wetlands that could be drained for human use; later classifications and inventories centered on the desire to compare different types of wetlands in a given region, often for their value to waterfowl. The protection of multiple ecological values of wetlands came later; now it is the most common reason for wetland classification and inventory. Recognition of wetland "value" has led some to now seek wetland classifications based on priorities for protection, with highest protection afforded to those wetlands with the greatest value. As with other techniques, classifications and inventories are valuable only when the user is familiar with their scope and limitations.

Our textbook uses a simple five-chapter classification of wetland ecosystems (Chapters 8–12), divided into two major groups: (1) *coastal*—tidal marshes (salt marshes and tidal freshwater marshes) and mangrove swamps, and (2) *inland*—freshwater marshes, freshwater swamps, and peatlands (Tables 8.1 and 10.1). Other types of wetlands, such as inland saline marshes, may fall between the cracks in this simple wetland classification, but these five categories cover most wetlands currently found in the world.

Why Do We Classify Wetlands?

Several attempts have been made to classify wetlands into categories that follow their structural and functional characteristics. These classifications depend on a well-understood general definition of wetlands (see Chapter 2: "Wetland Definitions"), although a classification contains definitions of individual wetland types. A primary goal of wetland classifications, according to Cowardin et al. (1979), "is to impose boundaries on natural ecosystems for the purposes of inventory, evaluation, and management." These authors identified four major objectives of a classification system:

1. To describe ecological units that have certain homogeneous natural attributes;
2. To arrange these units in a unified framework for the characterization and description of wetlands, that will aid decisions about resource management;
3. To identify classification units for inventory and mapping; and
4. To provide uniformity in concepts and terminology.

The first objective deals with the important task of grouping ecosystems that have similar characteristics in much the same way that taxonomists categorize species in taxonomic groupings. The wetland attributes that are frequently used to group and compare wetlands include the geomorphic and hydrologic regime, vegetation physiognomic type, and plant and/or animal species.

The second objective, to aid wetland managers, can be met in several ways when wetlands are classified. Classifications (which are definitions of different types of wetlands) enable wetland managers to deal with wetland regulation and protection consistently from region to region and from one time to the next. Classifications also enable wetland managers to pay selectively more attention to those types of wetlands that are most threatened or functionally the most valuable to a given region.

The third and fourth objectives, to provide consistency in the formulation and use of inventories, mapping, concepts, and terminology, are also important in wetland management. The use of consistent terms to define particular types of wetlands is needed in the field of wetland science (see Chapter 2: "Wetland Definitions"). These terms should then be applied uniformly to wetland inventories and mapping so that different regions can be compared and so that there will be a common understanding of wetland types among wetland scientists, wetland managers, and wetland owners.

Wetland Classifications

Peatland Classifications

Many of the earliest wetland classifications were undertaken for the northern peatlands of Europe and North America. An early peatland classification in the United States, developed by Davis (1907), described Michigan bogs according to three criteria: (1) the landform on which the bog was established, such as shallow lake basins or deltas of streams; (2) the method by which the bog was developed, such as from the bottom up or from the shores inward; and (3) the surface vegetation, such as tamarack or mosses. Based on the work of Weber (1907), Potonie (1908), Kulczynski (1949), and others in Europe, Moore and Bellamy (1974) described seven types of peatlands based on flowthrough conditions. Three general categories, called rheophilous, transition, and ombrophilous, describe the degree to which peatlands are influenced by outside drainage. The more modern terminology is minerotrophic, transition, and ombrotrophic peatlands (see Chapter 12: "Peatlands"). Most peatlands are limited to northern temperate climes and do not include all or even most types of wetlands in North America. These classifications, however, served as models for more inclusive classifications. They are significant because they combined the chemical and physical conditions of the wetland with the vegetation description to present a balanced approach to wetland classification.

Circular 39 Classification

In the early 1950s, the U.S. Fish and Wildlife Service recognized the need for a national wetlands inventory to determine "the distribution, extent, and quality of the remaining wetlands in relation to their value as wildlife habitat" (Shaw and Fredine, 1956). A classification was developed for that inventory (Martin et al., 1953), and the results of both the inventory and the classification scheme were published in U.S. Fish and Wildlife Circular 39 (Shaw and Fredine, 1956). Twenty types of wetlands were described under four major categories (Table 13.1).

Table 13.1 Early "Circular 39" wetland classification by U.S. Fish and Wildlife Service

Type Number	Wetland Type	Site Characteristics
Inland Fresh Areas		
1	Seasonally flooded basins or flats	Soil covered with water or waterlogged during variable periods, but well drained during much of the growing season; in upland depressions and bottomlands
2	Fresh meadows	Without standing water during growing season; waterlogged to within a few centimeters of surface
3	Shallow fresh marshes	Soil waterlogged during growing season; often covered with 15 cm or more of water
4	Deep fresh marshes	Soil covered with 15 cm to 1 m of water
5	Open fresh water	Water less than 2 m deep
6	Shrub swamps	Soil waterlogged; often covered with 15 cm or more of water
7	Wooded swamps	Soil waterlogged; often covered with 30 cm of water; along sluggish streams, flat uplands, shallow lake basins
8	Bogs	Soil waterlogged; spongy covering of mosses
Inland Saline Areas		
9	Saline flats	Flooded after periods of heavy precipitation; waterlogged within few centimeters of surface during the growing season
10	Saline marshes	Soil waterlogged during growing season; often covered with 0.7 to 1 m of water; shallow lake basins
11	Open saline water	Permanent areas of shallow saline water; depth variable
Coastal Fresh Areas		
12	Shallow fresh marshes	Soil waterlogged during growing season; at high tide, as much as 15 cm of water; on landward side, deep marshes along tidal rivers, sounds, deltas
13	Deep fresh marshes	At high tide, covered with 15 cm to 1 m water; along tidal rivers and bays
14	Open fresh water	Shallow portions of open water along fresh tidal rivers and sounds
Coastal Saline Areas		
15	Salt flats	Soil waterlogged during growing season; sites occasionally to fairly regularly covered by high tide; landward sides or islands within salt meadows and marshes
16	Salt meadows	Soil waterlogged during growing season; rarely covered with tide water; landward side of salt marshes
17	Irregularly flooded salt marshes	Covered by wind tides at irregular intervals during the growing season; along shores of nearly enclosed bays, sounds, etc.
18	Regularly flooded salt marshes	Covered at average high tide with 15 cm or more of water; along open ocean and along sounds
19	Sounds and bays	Portions of saltwater sounds and bays shallow enough to be diked and filled; all water landward from average low-tide line
20	Mangrove swamps	Soil covered at average high tide with 15 cm to 1 m of water; along coast of southern Florida

Source: Shaw and Fredine (1956)

Types 1 through 8 are freshwater wetlands that include bottomland hardwood forests (type 1), infrequently flooded meadows (type 2), freshwater nontidal marshes (types 3 and 4), open water less than 2 m deep (type 5), shrub-scrub swamps (type 6), forested swamps (type 7), and bogs (type 8). Types 9 through 11 are inland wetlands that have saline soils. They are defined according to the degree of flooding. Types 12 through 14 are wetlands that, although freshwater, are close enough to the coast to be influenced by tides. Types 15 through 20 are coastal wetlands that are influenced by both saltwater and tidal action. These include salt flats and meadows (types 15 and 16), true salt marshes (types 17 and 18), open bays (type 19), and mangrove swamps (type 20).

This wetland classification was the most widely used in the United States until 1979, when the current National Wetlands Inventory classification was adopted. The earlier system is still referred to today by some wetland managers and is regarded by many as elegantly simple compared with its successor. It primarily used the physiognomy (life-forms) of vegetation and the depth of flooding to identify the wetland type. Salinity was the only chemical parameter used, and although wetland soils were addressed in the Circular 39 publication, they were not used to define wetland types.

Coastal Wetland Classification

H. T. Odum et al. (1974) described coastal ecosystems by their major forcing functions (e.g., seasonal programming of sunlight and temperature) and stresses (e.g., ice) (Fig. 13.1). Coastal wetland types in this classification include salt marshes and

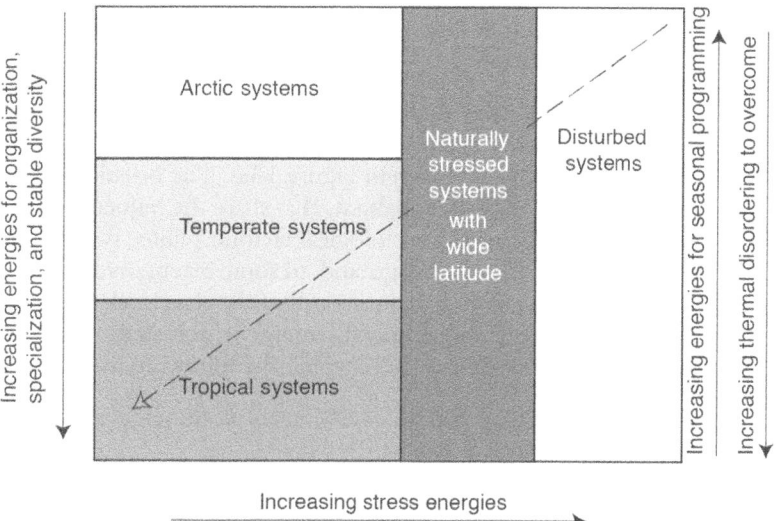

Figure 13.1 **Coastal ecosystem classification system based on latitude (and, hence, solar energy) and major stresses. (After H. T. Odum et al., 1974)**

mangrove swamps. Salt marshes, found in the type C category of natural temperate ecosystems with seasonal programming, have "light tidal regimes" and "winter cold" as forcing function and stress, respectively. Mangrove swamps are classified as type B (natural tropical ecosystems) because they have abundant light, show little stress, and reflect little seasonal programming. Three additional classes, type A (naturally stressed systems of wide latitudinal range), type D (natural arctic ecosystems with ice stress), and type E (emerging new systems associated with human activity), were included in this classification. The last class, which includes new systems formed by pollution, such as pesticides and oil spills, is an interesting concept that still could be applied to other wetland classifications.

The United States *Classification of Wetlands and Deepwater Habitats*

The U.S. Fish and Wildlife Service began an inventory of the nation's wetlands in 1974. Because this inventory was designed to fulfill several scientific and management objectives, a new classification scheme, broader than the Circular 39 classification, was developed and finally published in 1979 as a *Classification of Wetlands and Deepwater Habitats of the United States* (Cowardin et al., 1979). Because wetlands were found to be continuous with deepwater ecosystems, both categories were addressed in this classification. It is thus a comprehensive classification of all continental aquatic and semiaquatic ecosystems. As described in that publication in 1979:

> This classification, to be used in a new inventory of wetlands and deepwater habitats of the United States, is intended to describe ecological taxa, arrange them in a system useful to resource managers, furnish units for mapping, and provide uniformity of concepts and terms. Wetlands are defined by plants (hydrophytes), soils (hydric soils), and frequency of flooding. Ecologically related areas of deep water, traditionally not considered wetlands, are included in the classification as deepwater habitats.

This classification is based on a hierarchical approach analogous to taxonomic classifications used to identify plant and animal species. The first three levels of the classification hierarchy are given in Figure 13.2. The broadest level is *systems*: "a complex of wetlands and deepwater habitats that share the influence of similar hydrologic, geomorphologic, chemical, or biological factors." Thus, systems, subsystems, and classes are based primarily on geologic and, to some extent, hydrologic considerations. Broad vegetation types are included primarily at the class level, and, even here, the vegetation types are generic (i.e., perennial, emergent, forested, scrub-shrub, or moss-lichen). Systems shown in Figure 13.2 include the following five:

1. *Marine*. Open ocean overlying the continental shelf and its associated high-energy coastline.
2. *Estuarine*. Deepwater tidal habitats and adjacent tidal wetlands that are usually semienclosed by land but have open, partially obstructed, or sporadic access to the ocean and in which ocean water is at least occasionally diluted by freshwater runoff from the land.

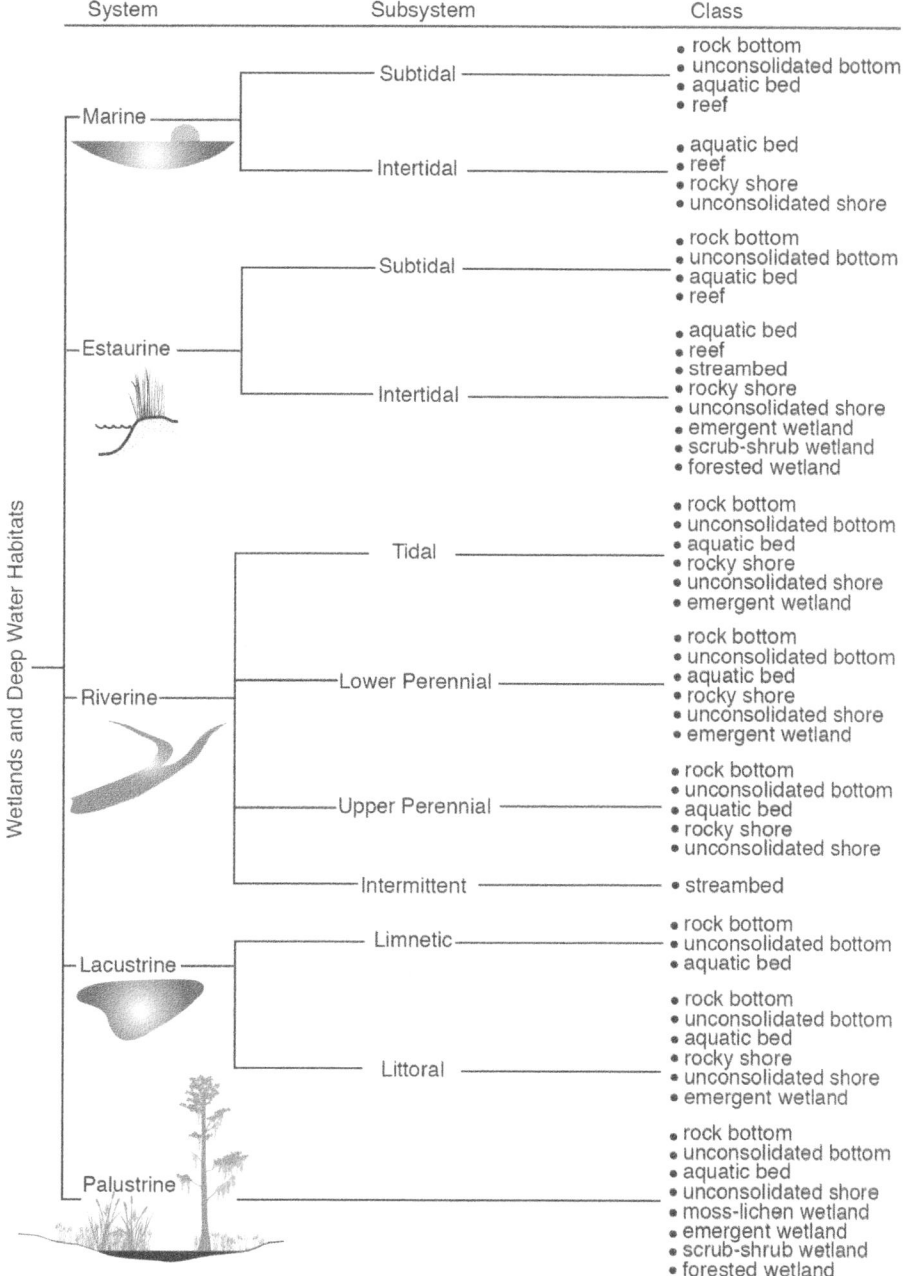

Figure 13.2 Current U.S. Fish and Wildlife Service wetland and deepwater habitat classification hierarchy showing 5 major systems, 10 subsystems, and numerous classes. (After Cowardin et al., 1979)

3. *Riverine.* Wetlands and deepwater habitats contained within a channel with two exceptions: (1) wetlands dominated by trees, shrubs, persistent emergents, emergent mosses, or lichens; and (2) deepwater habitats with water containing ocean-derived salts in excess of 0.5 ppt.
4. *Lacustrine.* Wetlands and deepwater habitats with all of the following characteristics: (1) situated in a topographic depression or a dammed river channel; (2) lacking trees, shrubs, persistent emergents, emergent mosses, or lichens with greater than 30 percent areal coverage; and (3) total area in excess of 8 ha. Similar wetland and deepwater habitats totaling less than 8 ha are also included in the lacustrine system when an active wave-formed or bedrock shoreline feature makes up all or part of the boundary or when the depth in the deepest part of the basin exceeds 2 m at low water.
5. *Palustrine.* All nontidal wetlands dominated by trees, shrubs, persistent emergents, emergent mosses, or lichens, and all such wetlands that occur in tidal areas where salinity stemming from ocean-derived salts is below 0.5 ppt. It also includes wetlands lacking such vegetation but with all of the following characteristics: (1) area less than 8 ha; (2) lack of active wave-formed or bedrock shoreline features; (3) water depth in the deepest part of the basin of less than 2 m at low water; and (4) salinity stemming from ocean-derived salts of less than 0.5 ppt.

Subsystems, as shown in Figure 13.2, give further definition to the systems. These include the following eight:

1. *Subtidal.* Substrate continuously submerged
2. *Intertidal.* Substrate exposed and flooded by tides, including the splash zone
3. *Tidal.* For riverine systems, gradient low and water velocity fluctuates under tidal influence
4. *Lower perennial.* Riverine systems with continuous flow, low gradient, and no tidal influence
5. *Upper perennial.* Riverine systems with continuous flow, high gradient, and no tidal influence
6. *Intermittent.* Riverine systems in which water does not flow for part of the year
7. *Limnetic.* All deepwater habitats in lakes
8. *Littoral.* Wetland habitats of a lacustrine system that extends from shore to a depth of 2 m below low water or to the maximum extent of nonpersistent emergent plants

The *class* of a particular wetland or deepwater habitat describes the general appearance of the ecosystem in terms of either the dominant vegetation life form or the

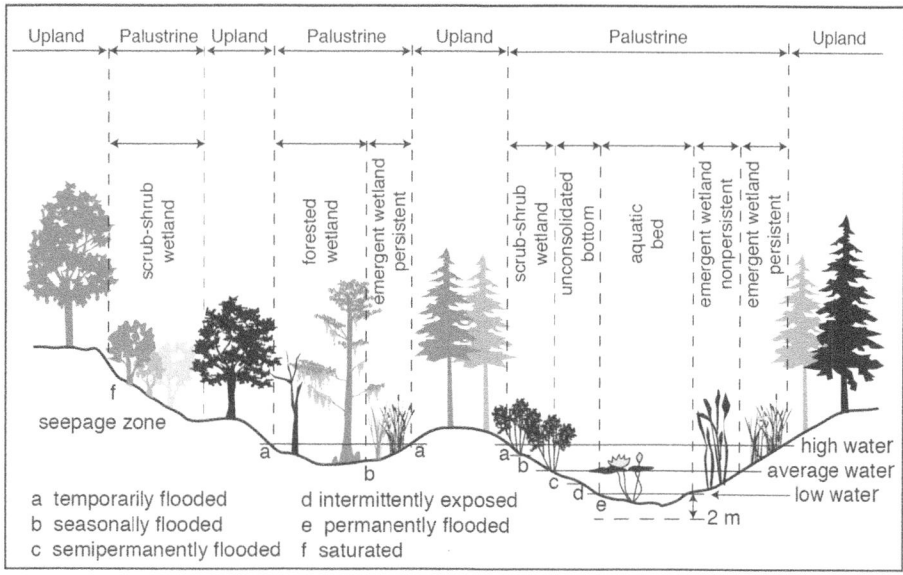

Figure 13.3 Features and examples of wetland classes and hydrologic modifiers in the palustrine system. (After Cowardin et al., 1979)

substrate type. When more than 30 percent cover by vegetation is present, a vegetation class is used (e.g., shrub-scrub wetland). When less than 30 percent of the substrate is covered by vegetation, then a substrate class is used (e.g., unconsolidated bottom). The typical demarcation of many of the classes of the palustrine system is shown in Figure 13.3.

Most inland wetlands fall into the palustrine system, in the classes moss-lichen, emergent, scrub-shrub, or forested wetland. Coastal wetlands are classified in the same classes within the estuarine system and intertidal subsystem. Only nonpersistent emergent wetlands are classified into other systems.

Further descriptions of the wetlands and deepwater habitats are possible through the use of *subclasses, dominance types*, and *modifiers*. Subclasses such as "persistent" and "nonpersistent" give further definition to a class such as emergent vegetation. Type refers to a particular dominant plant species (e.g., bald cypress, *Taxodium distichum*, for a needle-leaved deciduous forested wetland) or a dominant sedentary or sessile animal species (e.g., eastern oyster, *Crassostrea virginica*, for a mollusk reef). Modifiers (Table 13.2) are used after classes and subclasses to describe more precisely the water regime, the salinity, the pH, and the soil. For many wetlands, the description of the environmental modifiers adds a great deal of information about their physical and chemical characteristics. Unfortunately, those parameters are difficult to measure consistently in large-scale surveys such as inventories.

Table 13.2 Modifiers used in current wetland and deepwater habitat classification by U.S. Fish and Wildlife Service

Water Regime Modifiers (Tidal)
Subtidal—substrate permanently flooded with tidal water
Irregularly exposed—land suface exposed by tides less often than daily
Regularly flooded—alternately floods and exposes land surfaces at least daily
Irregularly flooded—land surface flooded less often than daily

Water Regime Modifiers (Nontidal)
Permanently flooded—water covers land surface throughout year in all years
Intermittently exposed—surface water present throughout year except in years of extreme drought
Semipermanently flooded—surface water persists throughout growing season in most years; when surface water is absent, water table is at or near surface
Seasonally flooded—surface water is present for extended periods, especially in early growing season but is absent by the end of the season
Saturated—substrate is saturated for extended periods during growing season but surface water is seldom present
Temporarily flooded—surface water is present for brief periods during growing season but water table is otherwise well below the soil surface
Intermittently flooded—substrate is usually exposed but surface water is present for variable periods with no seasonal periodicity

Salinity Modifiers

Marine and Estuarine	Riverine, Lacustrine, and Palustrine	Salinity (ppt)
Hyperhaline	Hypersaline	>40
Euhaline	Eusaline	30–40
Mixohaline (brackish)	Mixosaline	0.5–30
Polyhaline	Polysaline	18.0–30
Mesohaline	Mesosaline	5.0–18
Oligohaline	Oligosaline	0.5–5
Fresh	Fresh	<0.5

pH Modifiers

	Acid	pH <5.5
	Circumneutral	pH 5.5–7.4
	Alkaline	pH >7.4

Soil Material Modifiers

Mineral	1. <20% organic carbon and never saturated with water for more than a few days, or
	2. Saturated or artificially drained and has
	a. <18% organic carbon if 60% or more is clay
	b. <12% organic carbon if no clay
	c. a proportional content of organic carbon between 12 and 18% if clay content is between 0 and 60%
Organic	Other than mineral as described above

Source: Cowardin et al. (1979).

Canadian Wetlands Classification System

The Canadian Wetland Classification System (Warner and Rubec, 1997) is designed to be practical as well as hierarchical. Its three major features include:

1. *Classes.* Based on natural features of the wetlands rather than on interpretation for various uses. They have direct application to large wetland regions. Wetland classes are recognized on the basis of properties that reflect the overall "genetic origin" of the wetland system and the nature of the wetland environment. Division into classes allows ready identification in the field and delineation on maps. Classes are also convenient groupings for data storage, retrieval, and interpretation.
2. *Forms.* Subdivisions of wetland classes based on surface morphology, water type, and morphology characteristics of the underlying mineral soil. Some forms are further subdivided into subforms. Forms are easily recognized features of the landscape and are the basic wetland-mapping unit.
3. *Types.* Subdivisions of wetland forms and subforms based on physiognomic characteristics of the vegetation communities. They are comparable to the modifiers used in the U.S. Fish and Wildlife Service classification system. Types are most useful for evaluation of wetland values and benefits, management for wetland hydrology and wildlife habitat, and conservation and protection of rare and endangered species.

Currently, the system recognizes five wetland classes (bog, fen, swamp, marsh, shallow-water marsh), 49 wetland forms, and 75 subforms, although geomorphological, hydrologic, and chemical characteristics do not appear in the classification.

International Wetland Classification System

Table 13.3 compares the international Ramsar Convention classification system with the U.S. and Canadian classification systems. The Ramsar system has 32 classes, divided into a marine/coastal group and an inland group. Because it attempts to be global, it has categories that neither the U.S. nor the Canadian systems have—for example, underground karst systems and oases. The U.S. system, because it is hierarchical, has fewer classes at the system and subsystem level but uses modifiers to identify specific wetland types. The Canadian system also has only five classes but with forms and subforms reaches a total of over 70 different categories by which to classify wetlands. The Ramsar Convention itself is described in more detail in Chapter 15: "Wetland Laws and Protection."

Hydrogeomorphic Wetland Classification

Mark Brinson (1993) developed a wetland classification system modeled after the hydrogeomorphic (HGM) classification systems for mangroves (see Chapter 9: "Mangrove Swamps") and cypress swamps (see Chapter 11: "Freshwater Swamps and Riparian Ecosystems") that came from the H. T. Odum program at University of

Table 13.3 The Ramsar Convention International Wetland classification system compared to the U.S. Fish and Wildlife wetland and deepwater habitat classification and the Canadian wetland classification system

Ramsar Convention Code Name		U.S. Fish and Wildlife System[a]	Canadian System[b]
Marine/Coastal Wetlands			
A	Marine water <6 m	Marine subtidal	Shallow (<2 m) water marsh
B	Marine subtidal aquatic beds	Marine subtidal aquatic bed	
C	Coral reefs	Marine subtidal reef	—
D	Rocky marine shores	Marine intertidal rock bottom	—
E	Sandy shore or dune	Marine intertidal unconsolidated	—
F	Estuarine waters	Estuarine subtidal	Estuarine marsh, water
G	Intertidal flats	Estuarine intertidal unconsolidated bottom	Estuarine water, tidal water
H	Intertidal marshes	Estuarine intertidal emergent wetland	Tidal marsh
I	Intertidal forested wetland	Estuarine intertidal forested wetland	Tidal swamp
J	Coastal saline lagoon	Estuarine subtidal unconsolidated, saline	Estuarine water
K	Coastal fresh lagoon	Estuarine subtidal unconsolidated, fresh	Estuarine water
Zk(a)	Marine/coastal karst	Estuarine subtidal rocky shore	—
Inland Wetlands			
L	Permanent inland deltas	Riverine perennial	Riparian delta marsh
		Estuarine delta marsh	
		Shallow riparian delta water	
M	Permanent rivers/streams	Riverine perennial swamp, marsh	Shallow riparian water
N	Intermittent rivers/streams	Riverine intermittent	—
O	Permanent fresh lakes	Lacustrine littoral or limnetic	Shallow lacustrine water
	Riparian water (oxbows)		
P	Intermittent fresh lakes	Lacustrine or riparian littoral	—
Q	Permanent saline lakes	Lacustrine littoral unconsolidated, saline	—
R	Intermittent saline lakes	Lacustrine littoral intermittent, saline	—
Sp	Permanent saline marshes/pools	Palustrine emergent wetland or unconsolidated bottom, saline	Estuarine marsh, inland salt swamp
Ss	Intermittent saline marshes/pools	Palustrine emergent wetland or unconsolidated bottom, intermittent	Spring, slope, or basin marsh
Tp	Permanent fresh marsh/pools (<8 ha)	Palustrine emergent wetland or unconsolidated bottom, fresh	Shallow basin water, lacustrine marsh
Ts	Intermittent fresh marsh/pools, inorganic soils	Palustrine emergent wetland, intermittently flooded	Shallow basin water
U	Nonforested peatlands	Palustrine emergent wetland, persistent	Bogs, fens
Va	Alpine wetlands	Palustrine emergent wetland, persistent	—
Vt	Tundra wetlands	Palustrine emergent wetland, persistent	Bogs, fens, shallow basin water
W	Shrub-dominated wetlands	Palustrine scrub-shrub wetland	Riparian, flat, slope, discharge or mineral-rise swamp
Xf	Fresh forested wetlands on inorganic soils	Palustrine forested wetland	Riparian swamp
Xp	Forested peatlands	Palustrine forested or scrub-shrub wetland	Flat bog, flat or raised peatland swamp
Y	Freshwater springs, oases	—	Hummock marsh
Zg	Geothermal wetlands	—	—
Zg(b)	Inland karst systems, underground	—	—

[a] Ramsar class can be further approximated with additional modifiers.
[b] Class and form indicated only; subforms can approximate Ramsar more closely. The term *shallow* in the wetland type indicates the shallow marsh class.

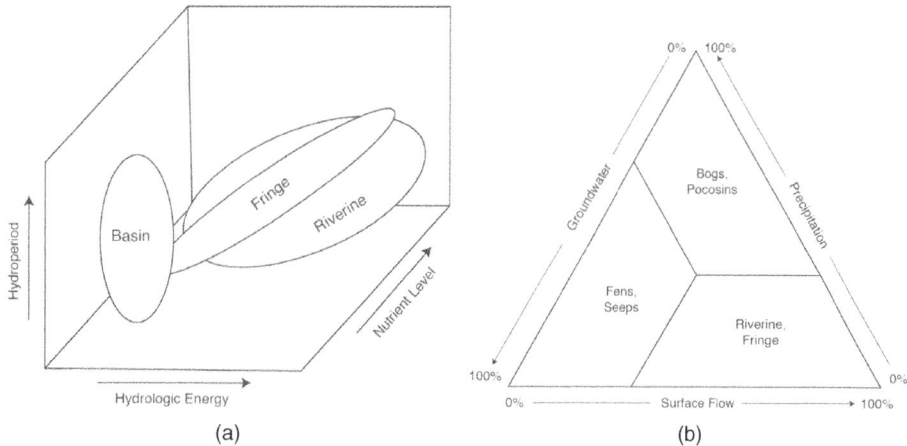

Figure 13.4 Basis of the hydrogeomorphic (HGM) classification system: (a) geomorphic settings (basin, fringe, and riverine) arranged around three core factors of hydroperiod, hydrologic energy, and nutrient levels; (b) the relative contribution of a combination of three water sources—precipitation, groundwater discharge, and surface inflow—to determining the type of wetlands. (After Brinson, 1993)

Florida in the 1970s. It was designed to be used for evaluation of wetland functions and is currently being used as a means of assessing the physical, chemical, and biological functions of wetlands. It is useful for comparing the level of functional integrity of wetlands within a functional class or for evaluating the impact of proposed human activities on wetlands and mitigation alternatives (Fig. 13.4; Table 13.4).

The classification is based primarily on hydrodynamic differences as they function within four geomorphic settings. Thus, the three core components of the classification system are geomorphology, water source, and hydrodynamics (Table 13.4). Geomorphic setting is the topographic location of a wetland in the surrounding landscape. Four geomorphic settings are identified: depressional, riverine, fringe, and extensive peatlands. The first three are clearly related to the hydrologic setting. Extensive peatlands are different because the dominant influence on hydrology is biogenic accretion. Water sources are precipitation, surface or near-surface flow, and groundwater discharge (into a wetland). The term *hydrodynamics* refers to the direction and strength of water movement within a wetland. The three core features are heavily interdependent, so it is difficult to describe any one without the other two. Taken as a group, the three core features may be pooled in 36 combinations, but because of the interdependence, not all combinations are found in nature. It becomes clear when examining Table 13.4 how interrelated the three core features are. The water entering a wetland is seldom from only one of the three sources—precipitation, groundwater discharge, and surface inflow—although ombrotrophic peat wetlands typically are dominated by precipitation; mineral fens and seep wetlands, by groundwater discharge; and riverine and fringe wetlands, by surface flows.

The core component hydrodynamics is an expression of the fluvial energy that drives the system. This ranges from low-energy water table fluctuations typical

Table 13.4 Functional classification of wetlands by geomorphology, water source, and hydrodynamics

Core Component	Description	Example
Geomorphic Setting	Topographic location of a wetland in the surrounding landscape	
Depressional	Wetlands in depressions that typically receive most moisture from precipitation, hence often ombrotrophic; found in dry and moist climates	Kettles, potholes, vernal pools, Carolina bays, groundwater slope wetlands
Extensive peatlands	Peat substrate isolates wetland from mineral substrate; peat dominates movement and storage of water and chemicals	Blanket bogs, tussock tundra
Riverine	Linear strips in landscape; subject predominantly to unidirectional surface flow	Riparian wetlands along rivers, streams
Fringe	Estuarine and lacustrine wetlands with bidirectional surface flow	Estuarine tidal wetlands, lacustrine fringes subject to winds, waves, and seiches
Water Source	Relative importance of three main sources of water to a wetland	
Precipitation	Wetlands dominated by precipitation as the primary water source; water level may be variable because of evapotranspiration	Ombrotrophic bogs, pocosins
Groundwater discharge	Primary water source from regional or perched mineral groundwater sources	Fens, groundwater slope wetlands
Surface inflow	Water source dominated by surface inflow	Alluvial swamps, tidal wetlands, montane streamside wetlands
Hydrodynamics	Motion of water and its capacity to do work	
Vertical fluctuation	Vertical fluctuation of the water table resulting from evapotranspiration and replacement by precipitation or groundwater discharge	Usually depressional wetlands, bogs (annual), prairie potholes (multiyear)
Unidirectional flow	Unidirectional surface or near-surface flow; velocity corresponds to gradient	Usually riverine wetlands
Bidirectional flow	Occurrence in wetlands dominated by tidal and wind-generated water-level fluctuations	Usually fringe wetlands

Source: Brinson (1993)

of depressional wetlands to unidirectional flows found in riverine wetlands to bidirectional flows of tidal and high-energy lacustrine wetland systems. Uni- and bidirectional surface flows range widely in energy from hardly perceptible movement to strong erosive currents. Combined with the geographic setting, hydrodynamics can result in a range of different wetland types.

This classification system was designed to be independent of plant communities, because it depends on the geomorphic and hydrologic properties of the wetlands. In practice, however, vegetation often provides important clues to the hydrogeomorphic forces at work. Because most modern classification systems developed for inventory purposes have some basis in hydrogeomorphology, their classes often give important clues as to function.

Rating Wetlands

Since the introduction of legislation and regulations aimed at wetland conservation, especially in the United States, there has been considerable interest in classifying wetlands for their value to society in order to simplify the issuance of permits for wetlands activities; that is, high-value wetlands presumably would receive more protection than low-value wetlands. Also, when wetland loss is being mitigated, it is important to compare the functions are being lost and those that are gained with the mitigation strategy. Almost every state in the United States has procedures for rating wetlands. Three examples are described here.

1. Washington

The state of Washington was one of the first places where such a wetland "rating" was developed. Its system, with the state divided into two hydrogeomorphic areas, western and eastern, established a method for including the state's wetlands in one of four categories based on their sensitivity to disturbance, their significance, their rarity, our ability to replace them, and the functions they provide (Hruby, 2004).

Category I. Unique, rare, relatively irreplaceable, or high level of valuable functions

Category II. Wetlands that are difficult though not impossible to replace

Category III. Moderate level of functions

Category IV. Lowest level of valuable functions and often heavily disturbed

The rating scores were based on water quality and hydrologic and habitat functions (Hruby, 2004). Effective 2015, the state of Washington revised its 2004 rating system by changing some of the rating questions. See http://www.ecy.wa.gov/programs/sea/wetlands/ratingsystems/2014updates.html

2. Ohio

Ohio developed a similar wetland assessment technique based on earlier versions of the Washington model. The Ohio Rapid Assessment Method (ORAM) for wetlands attempts to determine the function of wetlands in three categories, put in reverse order to the Washington system (Mack, 2001):

Category 1. Wetlands that provide minimal functions or habitat

Category 2. Wetlands that provide moderate functions and are dominated by native species

Category 3. Wetlands that provide superior functions, have high levels of diversity, and are habitat for rare and endangered species

A quantitative ranking system on a 0-to-100 scale, based on field and remote-sensed data, is at the heart of this system. Six metrics to determine the wetland score include:

1. Wetland area
2. Upland buffers and surrounding land use
3. Hydrology
4. Habitat alteration and development
5. Special wetlands
6. Plant communities

Details are provided in Mack (2001), which is available at www.epa.state.oh.us/dsw/401/ecology.aspx.

3. Florida

The state of Florida, which has more remaining wetlands than any other state except Alaska and has one of the highest rates of human population increases of any state, has major challenges of balancing human development, wetland protection, and mitigation of wetland loss (see Chapter 18: "Wetland Creation and Restoration" for discussion on mitigating wetland losses). In an attempt to quantify and compare wetland losses with gains resulting from mitigation of those losses, the state developed a state-wide wetland assessment method that is called a Uniform Mitigation Assessment Method (UMAM). It became effective in 2004. UNAM "is designed to assess any type of impact and mitigation, including the preservation, enhancement, restoration, and creation of wetlands, as well as the evaluation and use of mitigation banks" (http://sfrc.ufl.edu/ecohydrology/UMAM_Training_Manual_ppt.pdf). For the quantitative section, wetlands (to compare current conditions and expected mitigation conditions, for example) are evaluated in the following three categories on a scale of 0 to 10:

1. Location and landscape support—the ecological context in which the wetland system functions;
2. Water environment—the hydrologic condition including degree to which it is altered and impaired; and
3. Community structure (vegetation and/or benthic/sessile communities)—vegetation measurements include plant cover, species lists, invasive exotics, plant conditions, and topographic features; benthic and sessile community structure is evaluated where submerged benthic communities are present such as oyster reefs, corals, and soft-bottom systems such as riverine systems.

> When mitigation of wetland loss is being proposed, the comparison of the estimated indices from the above three factors are adjusted using a preservation factor, a factor for time lag, and a factor that incorporates the anticipated risk. Details of this wetland assessment method are at www.dep.state.fl.us/water/wetlands/mitigation/umam. A useful training manual for UMAM can be found at http://sfrc.ufl.edu/ecohydrology/UMAM_Training_Manual_ppt.pdf.

Wetland Remote Sensing and Inventory

One of the major objectives of wetland classification is to be able to inventory the location, extent, and type of wetlands in a region of concern. An inventory can be made of a small watershed, a political unit such as a county or parish, an entire state or province, or an entire nation. Whatever the size of the area to be surveyed, the inventory must be based on some previously defined classification and should be constructed to meet the needs of specific users of information on wetlands. Generally, inventories require not only information about the types and extent of wetlands but also documentation of their geographic locations and boundaries. To accomplish this goal, *remote platforms*—aircraft and/or satellites—produce *imagery*—photographs or digital information that can be used to make images. This imagery must be *interpreted* to identify the locations, boundaries, and types of wetlands on the image.

Remote-Sensing Platform

In the early days of wetland classification, wetlands were mapped from surveyors' records and from boats. Later, interpretation of low-altitude aerial photographs combined with verification in the field made the process both faster and more accurate. High-altitude imagery from aircraft such as that from the U-2 were available in the past, but satellite imagery is rapidly becoming the norm. The commonly used satellite systems—Landsat Multispectral Scanner (MSS), Landsat Thematic Mapper (TM), and Systéme Pour l'Observation de la Terre (SPOT)—now have been joined by a number of high-resolution environmental satellites (Table 13.5). These remote platforms are an effective way of gathering data for large-scale wetland surveys. Satellites offer repeated coverage that allow seasonal monitoring of wetlands as well as providing data on the surrounding landscape readily translatable to a geographic information system (GIS) format (Ozesmi and Bauer, 2002). The choice of which platform to use depends on the resolution required, the area to be covered, and the cost of the data collection. Low-altitude aircraft surveys offer relatively inexpensive and fairly effective ways to survey small areas. High-altitude aircraft offer much greater coverage in each image (photograph) and may be less expensive per unit area than low-altitude aircraft when costs of photo interpretation are included. The limitations of satellite remote sensing include the lack of ability to separate different wetland types or to even separate wetlands from upland forests or agricultural land.

Table 13.5 Examples of spatial resolution, revisiting time, and altitude of recent satellites available for wetland inventories and studies

	IKONOS	QuickBird	OrbView-3	WorldView-1	GeoEye-1	WorldView-2
Sponsor	Space Imaging	Digital Globe	Orbimage	Digital Globe	GeoEye	Digital Globe
Launch Date	Sept 1999	Oct 2001	June 2003	Sept 2007	Sept 2008	Oct 2009
Spatial resolution, m (panchromatic)	1.0	0.61	1.0	0.5	0.41	0.5
Spatial resolution, m (multispectral)	4.0	2.44	4.0	n/a	1.65	2
Swath width, km	11.3	16.5	8	17.6	15.2	16.4
Revisit time, days	2.3–3.4	1–3.5	1.5–3	1.7–3.8	2.1–8.3	1.1–2.7
Orbital altitude, km	681	450	470	496	681	770

Source: Klimas, 2013

Orbiting satellites have been providing data for Earth resources classification since the launching of the first of the Landsat satellites in 1972. Today, several highly effective satellites appropriate for wetland inventories and management orbit the Earth (Klimas, 2013). One problem with early satellites was the poor resolution (Landsat has a resolution of 30 m). Today's satellites (Table 13.5) can resolve features of Earth's surface to as little as 1 m, but high resolution is not an unmixed blessing. It requires the ability to transmit and process enormous amounts of data, because the data generated for a given surface area quadruples every time the resolution doubles.

Remote-Sensing Imagery

In addition to choosing the remote-sensing platform, the wetland scientist or manager has the choice of several types of imagery from different types of sensors. Color photography and color–infrared photography were popular for many years for wetland inventories from aircraft (see Shuman and Ambrose, 2003), although black-and-white photography has been used with some success. Color–infrared film and now digital imagery provide good definition of plant communities and is the film of choice. Satellites, and some aircraft, gather digital data in one or more electromagnetic spectral bands. For example, most of the satellites listed in Table 13.5 have panchromatic and multispectral capability in several color bands, including infrared. The U.S. National Wetlands Inventory (see box) relies to some extent on imagery interpretation, but computerized interpretation of satellite imagery is used more and more for wetland mapping, especially on agricultural landscapes.

> **The U.S. National Wetlands Inventory**
>
> The U.S. National Wetlands Inventory (NWI) is a good example of a major wetlands mapping project and illustrates some of the problems encountered in any mapping enterprise. The Cowardin et al. (1979) classification

scheme has provided the basic mapping units for the NWI being carried out by the U.S. Fish and Wildlife Service for several decades. The Service announced as of May 2014 that all of the wetlands of the entire lower 48 states of the United States, Hawaii, and dependent territories as well as 35 percent of Alaska are now digitally mapped (www.fws.gov/wetlands/Documents/Completion-of-National-Wetlands-Database-News-Release.pdf), and users can now access wetlands data and maps through an online wetland mapping service.

For the NWI, aerial photography at scales ranging from 1:60,000 to 1:130,000 was the primary source of data, with color–infrared photography providing the best delineation of wetlands (Wilen and Pywell, 1981; Tiner and Wilen, 1983). In the 1970s, maps were created mostly from 1:80,000-scale black-and-white photography. It is now supplemented with Environmental Systems Research Institute satellite imagery, which provides resolution of 1 m or better for the United States. Photointerpretation and field reconnaissance are then used to define wetland boundaries according to the wetland

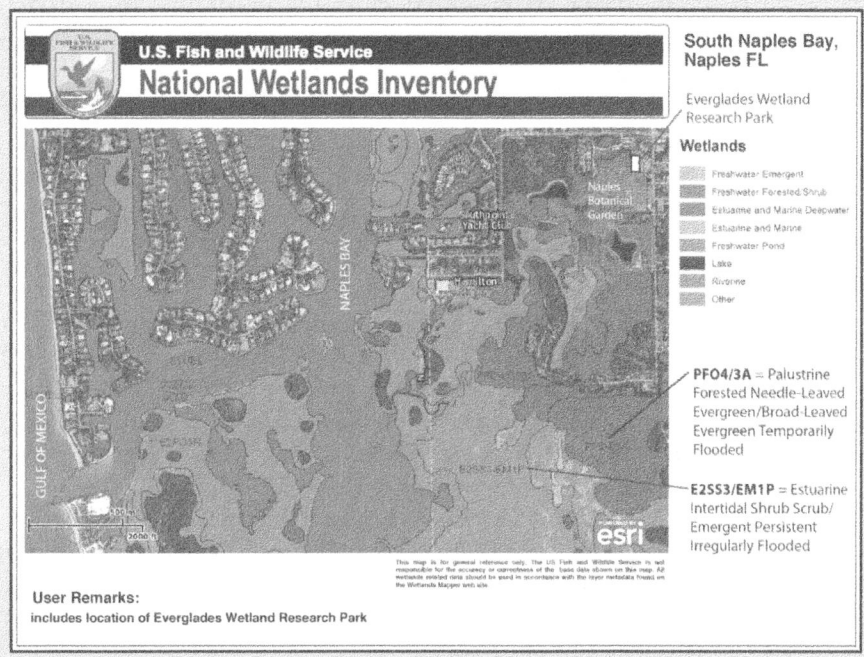

Figure 13.5 Sample of map of wetlands in Naples, Florida, created from the U.S. National Wetlands Inventory at www.fws.gov/wetlands/index.html, showing two examples of a classification notation, one for an estuarine intertidal shrub scrub (= mangrove swamp) and the other for a palustrine forested needle-leaved (= cypress swamp). Map also shows location of Everglades Wetland Research Park (http://fgcu.edu/swamp) at the Naples Botanical Garden.

classification system. The information is summarized on base maps using an alphanumeric system based on the U.S. Fish and Wildlife Classification system (Cowardin et al., 1979).

Nowadays wetland maps can be crafted at any scale by users at the NWI Inventory site at www.fws.gov/wetlands/index.html using their "Wetland Mapper" system (see Fig. 13.5). In Figure 13.5, two wetland indicators are highlighted; one indicates an expansive mangrove swamp in an estuarine system and the other shows a cypress swamp/hardwood swamp combination in the palustrine system.

Recommended Readings

Brinson, M. M. 1993. *A Hydrogeomorphic Classification for Wetlands*. Wetlands Research Program Technical Report WRP-DE-4. Vicksburg, MS: U.S. Army Corps of Engineers Waterways Experiment Station.

Cowardin, L. M., V. Carter, F. C. Golet, and E. T. LaRoe. 1979. *Classification of Wetlands and Deepwater Habitats of the United States*. Washington, DC: U.S. Fish and Wildlife Service, FWS/OBS-79/31.

Tiner, R. W. 1999. *Wetland Indicators: A Guide to Wetland Identification, Delineation, Classification, and Mapping*. Boca Raton, FL: CRC Press.

U.S. National Wetland Inventory web page: www.fws.gov/wetlands/index.html.

References

Brinson, M. M. 1993. *A Hydrogeomorphic Classification for Wetlands*. Wetlands Research Program Technical Report WRP-DE-4, U.S. Army Corps of Engineers Waterways Experiment Station, Vicksburg, MS.

Cowardin, L. M., V. Carter, F. C. Golet, and E. T. LaRoe. 1979. *Classification of Wetlands and Deepwater Habitats of the United States*. Washington, DC: U.S. Fish and Wildlife Service, FWS/OBS-79/31.

Davis, C. A. 1907. Peat: Essays on its origin, uses, and distribution in Michigan. In *Report of the State Board Geological Survey Michigan for 1906*, pp. 95–395.

Hruby, T. 2004. Washington state wetland rating system for western Washington—Revised. Publication # 04–06–025. Washington State Department of Ecology, Olympia, 113 pp. + appen.

Klimas, V. 2013. Using remote sensing to select and monitor wetland restoration sites: An overview. *Journal of Coastal Research* 29: 958–970.

Kulczynski, S. 1949. Peat bogs of Polesie. *Acad. Pol. Sci. Mem.*, Ser. B, No. 15. 356 pp.

Mack, J. J. 2001. Ohio Rapid Assessment Method for Wetlands, v. 5. User's Manual and Scoring Forms, Ohio EPA Technical Report WET/2001-1. Ohio Environmental Protection Agency, Division of Surface Water/Wetland Ecology Unit, Columbus, Ohio, 66 pp. + forms.

Martin, A. C., N. Hutchkiss, F. M. Uhler, and W. S. Bourn. 1953. *Classification of Wetlands of the United States*. Special Science Report—Wildlife 20, U.S. Fish and Wildlife Service, Washington, DC. 14 pp.

Moore, P. D., and D. J. Bellamy. 1974. *Peatlands*. Springer-Verlag, New York. 221 pp.

Odum, H. T., B. J. Copeland, and E. A. McMahan, eds. 1974. *Coastal Ecological Systems of the United States*. Conservation Foundation, Washington, DC. 4 vols.

Ozesmi, S. L., and M. E. Bauer. 2002. Satellite remote sensing of wetlands. *Wetlands Ecology and Management* 10: 381–402.

Potonie, R. 1908. Aufbau und Vegetation der Moore Norddeutschlands. *Englers botanische jahrbücher 90*. Leipzig, Germany.

Shaw, S. P., and C. G. Fredine. 1956. Wetlands of the United States, their extent, and their value for waterfowl and other wildlife. Circular 39, U.S. Fish and Wildlife Service, U.S. Department of Interior, Washington, DC. 67 pp.

Shuman, C. S., and R. F. Ambrose. 2003. A comparison of remote sensing and ground-based methods for monitoring wetland restoration success. *Restoration Ecology* 11: 325–333.

Tiner, R. W., and B. O. Wilen. 1983. U.S. Fish and Wildlife Service National Wetlands inventory project. Unpublished report, U.S. Fish and Wildlife Service, Washington, DC. 19 pp.

Warner, B. G., and C. D. A. Rubec, eds. 1997. The Canadian Wetland Classification System. National Wetlands Working Group, Wetlands Research Centre, University of Waterloo, Ontario.

Weber, C. A. 1907. Aufbau und Vegetation der Moore Norddutschlands. *Beibl. Bot. Jahrbüchern*. 90: 19–34.

Wilen, B. O., and H. R. Pywell. 1981. *The National Wetlands Inventory*. Paper presented at In-Place Resource Inventories: Principles and Practices—A National Workshop, Orono, ME, August 9–14. 10 pp.

Chapter 14

Human Impacts and Management of Wetlands

Wetland impacts have included both wetland alteration and wetland destruction. In earlier times, wetland drainage was considered the only policy for managing wetlands. The most common alterations of wetlands have been draining, dredging, and filling of wetlands; modification of the hydrologic regime; highway construction; mining and mineral extraction; and water pollution. Peat resources, estimated to be 1.9 trillion tons in the world, are harvested in many countries as a source of fuel and horticultural materials. Wetlands can also be managed close to their natural state for certain objectives, such as fish and wildlife enhancement, agricultural and aquaculture production, water quality improvement, and flood control. Management of wetlands for coastal protection has now taken on more significance with potential sea-level increases.

The concept of wetland management has had different meanings at different times to different disciplines and in different parts of the world. Until the middle of the twentieth century, the term *wetland management* usually meant wetland drainage to many policy makers, except for a few resource managers who maintained wetlands for hunting, fishing, and waterfowl/wildlife protection. Landowners were encouraged through government programs to tile and drain wetlands to make the land suitable for agriculture and other uses. Dredging for navigation and filling for land development destroyed countless coastal and inland wetlands.

Until the last quarter of the twentieth century, there was little understanding of and concern for the inherent values of wetlands except by those who recognized wetlands as wildlife habitats, particularly for waterfowl. A whole science of "marsh management" developed in the middle part of the twentieth century around the idea of maintaining specific hydrologic conditions to optimize fish or waterfowl populations. Only since the mid-1970s have other values, such as flood control, coastal protection, and water quality enhancement, been recognized. It has taken disasters such as the 1993 Upper Mississippi River Basin flooding, the 2004 Indian Ocean tsunami, and the 2005 Hurricane Katrina disaster in New Orleans to cause societies to focus on the potential lives that could be saved and property damage minimized if wetland buffer systems were provided at our land–water margins.

Today, the management of wetlands usually means setting several objectives, depending on the priorities of the wetland managers, current environmental regulations, and wishes of a myriad of stakeholders who are usually involved. In some cases, objectives such as preventing pollution from reaching wetlands and using wetlands as sites of water quality improvement can be conflicting. Many floodplain wetlands are now managed and zoned to minimize human encroachment and maximize floodwater retention. Coastal wetlands are now included in coastal zone protection programs for storm protection and as sanctuaries and subsidies for estuarine fauna. In the meantime, wetlands continue to be altered or destroyed throughout the world by drainage, filling, conversion to agriculture, water pollution, and mineral extraction.

We are thankful to have witnessed a slowing of the destruction rate of wetlands, even since we wrote the first edition of this *Wetland* textbook (Mitsch and Gosselink, 1986), at least in the United States. We are not as certain that destruction of the world's wetlands is being slowed, but we are aware that there is a much greater international appreciation of wetlands than before. Vigilance is required, however, to make sure that wetland values continue to be protected. Wetland conservation and even wetland restoration and creation (see Chapter 18: "Wetland Creation and Restoration," for details on this type of wetland management) have accelerated, particularly in the developed world over the past 40 years. But there are few if any regulations or restrictions on wetland destruction or pollution in developing parts of the world. This may be the next frontier of wetland protection.

Early History of Wetland Management

The early history of wetland management, a history that still influences many people today, was driven by the misconception that wetlands were wastelands that should be avoided or, if possible, drained and filled. Throughout the world, as long as there have been humans, there has been hydrologic alteration of the landscape. As summarized by Joe Larson and Jon Kusler (1979): "For most of recorded history, wetlands were regarded as wastelands if not bogs of treachery, mires of despair, homes of pests, and refuges for outlaw and rebel. A good wetland was a drained wetland free of this mixture of dubious social factors."

In the United States, this opinion of wetlands and shallow-water environments led to the destruction of more than half of the total wetlands in the lower 48 states over a 200-year period. In New Zealand, settlement by Europeans that began in earnest in the mid-1800s contributed significantly to a 90 percent loss of wetlands in a relatively short time. Preliminary estimates suggest that, over human history, about half of the world's wetlands have been lost (see Chapter 3: "Wetlands of the World").

Table 14.1 Human actions that cause direct wetland losses and degradation[a]

Cause	Estuaries	Floodplains	Freshwater Marshes	Lakes/Littoral Zone	Peatlands	Swamp Forest
Agriculture, forestry, mosquito control drainage	xx	xx	xx	x	xx	xx
Stream channelization and dredging; flood control	x		x			
Filling—solid-waste disposal; roads; development	xx	xx	xx	x		
Conversion to aquaculture/mariculture	xx					
Dikes, dams, seawall, levee construction	xx	x	x	x		
Water pollution—urban and agricultural	xx	xx	xx	xx		
Mining of wetlands of peat and other materials	x	x		xx	xx	xx
Groundwater withdrawal		x	xx			

[a] xx = common and important cause of wetland loss and degradation.
x = present but not a major cause of wetland loss and degradation.
Blank indicates that effect is generally not present except in exceptional situations.
Source: Dugan (1993)

With over 70 percent of the world's population living on or near coastlines, coastal wetlands have long been destroyed through a combination of excessive harvesting, hydrologic modification and seawall construction, coastal development, pollution, and other human activities. Likewise, inland wetlands have been continually affected, particularly through hydrologic modification and agricultural and urban development. Human activities, such as agriculture, forestry, stream channelization, aquaculture, dam, dike, and seawall construction, mining, water pollution, and groundwater withdrawal, all had impacts, some severe, on wetlands (Table 14.1). Wetlands are degraded and destroyed indirectly as well through alternation of sediment patterns in rivers, hydrologic alteration, highway construction, and land subsidence (Table 14.2). A third possibility is the loss of wetlands from natural causes (Table 14.3), although wetlands are normally resilient and can recover from natural events. For example, many coastal wetlands that were devastated by the 2004 Indian Ocean tsunami or the 2005 hurricane that destroyed much of New Orleans have long since recovered.

The propensity in the East was not to drain valuable wetlands entirely, as has been done in the West, but to work within the aquatic landscape, albeit in a heavily managed way. Dugan's (1993) interesting comparison between *hydraulic civilizations* (European in origin), which controlled water flow through the use of dikes, dams, pumps, and drainage tile, and *aquatic civilizations* (Asian in origin), which better adapted to their surroundings of water-abundant floodplains and deltas, is an interesting way to view humans' use of wetlands. The former approach of controlling nature rather than working it is becoming more dominant around the world today; that is why we continue to find such high losses of wetlands worldwide.

Table 14.2 Human activities that indirectly cause wetland losses and degradation[a]

Cause	Estuaries	Floodplains	Freshwater Marshes	Lakes/Littoral Zone	Peatlands	Swamp Forest
Sediment retention by dams and other structures	xx	xx	xx			
Hydrologic alteration by roads, canals, etc.	xx	xx	xx	xx		
Land subsidence due to groundwater, resource extraction, and river alternations	xx	xx	xx			

[a] xx = common and important cause of wetland loss and degradation.
x = present but not a major cause of wetland loss and degradation.
Blank indicates that effect is generally not present except in exceptional situations
Source: Dugan (1993)

Table 14.3 Natural events that cause wetland losses and degradation[a]

Cause	Estuaries	Floodplains	Freshwater Marshes	Lakes/Littoral Zone	Peatlands	Swamp Forest
Subsidence	x			x	x	x
Sea-level rise	xx					xx
Drought	xx	xx	xx	x	x	x
Hurricanes, tsunamis, and other storms	xx			x	x	
Erosion	xx	x			x	
Biotic effects		xx	xx	xx		

[a] xx = common and important cause of wetland loss and degradation.
x = present but not a major cause of wetland loss and degradation.
Blank indicates that effect is generally not present except in exceptional situations
Source: Dugan (1993)

Wetland Drainage History in the United States

Had not politics intervened, George Washington may have succeeded in draining the Great Dismal Swamp in Virginia in the mid-eighteenth century (see Chapter 3) instead of leading a new nation. Draining swamps and other wetlands was an acceptable and even desired practice from the time Europeans first settled in North America. In the United States, public laws actually encouraged wetland drainage. Congress passed the Swamp Land Act of 1849, which granted to Louisiana the control of all swamplands and overflow lands in the state for the general purpose of controlling floods in the Mississippi River basin. In the following year, the act was extended to the states of Alabama, Arkansas, California, Florida, Illinois, Indiana, Iowa, Michigan, Mississippi, Missouri, Ohio, and Wisconsin. Minnesota and Oregon were added in 1860. The act was designed to decrease federal involvement in flood control and drainage by transferring federally owned wetlands to the states, leaving to them the initiative of "reclaiming" wetlands through activities such as levee construction and drainage.

By 1954, an estimated 26 million ha of land had been ceded to those 15 states for reclamation. Ironically, although the federal government passed the Swamp Land Act to get out of the flood control business, the states sold those lands to individuals for pennies per acre, and the private owners subsequently successfully lobbied both national and state governments to protect these lands from floods. Further, governments are now paying enormous sums to buy the same lands back for conservation purposes. Although current government policies are generally in direct opposition to the Swamp Land Act and it is now disregarded, the act cast the initial wetland policy of the U.S. government in the direction of wetland elimination.

Other actions led to the rapid decline of the nation's wetlands. An estimated 23 million ha of wet farmland, including some wetlands, were drained under the U.S. Department of Agriculture's Agricultural Conservation Program between 1940 and 1977. An estimated 18.6 million ha of land, much of it wetlands, was drained in seven states in the upper Mississippi River basin alone. Some of the wetland drainage activity was hastened by projects of groups such as the Depression-era Works Progress Administration, the Soil Conservation Service, and other federal agencies. Coastal marshes were eliminated or drained and ditched for intercoastal transportation, residential developments, mosquito control, and even for salt marsh hay production. Interior wetlands were converted primarily to provide land for urban development, road construction, and agriculture.

Go South, Young Man?

Typical of the prevalent attitude toward wetlands in the mid-twentieth century is the following quote by Norgress (1947) discussing the "value" of Louisiana cypress swamps:

> With 1,628,915 acres of cutover cypress swamp lands in Louisiana at the present time, what use to make of these lands so that the ideal cypress areas will make a return on the investment for the landowner is a serious problem of the future....
>
> The lumbermen are rapidly awakening to the fact that in cutting the timber from their land they have taken the first step toward putting it in position to perform its true function—agriculture....
>
> It requires only a visit into this swamp territory to overcome such prejudices that reclamation is impracticable. Millions of dollars are being put into good roads. Everywhere one sees dredge boats eating their way through the soil, making channels for drainage.
>
> After harvesting the cypress timber crop, the Louisiana lumbermen are at last realizing that in reaping the crop sown by Nature ages ago, they have left a heritage to posterity of an asset of permanent value and service—land, the true basis for wealth.

> The day of the pioneer cypress lumberman is gone, but we need today in Louisiana another type of pioneer—the pioneer who can help bring under cultivation the enormous areas of cypress cutover lands suitable for agriculture. It is important to Louisiana, to the South, and the Nation as a whole, that this be done. Would that there were some latter-day Horace Greeleys to cry, in clarion tones, to the young farmers of today, "Go South, young man; go South!"

As an example of state action leading to wetland drainage, Illinois passed the Illinois Drainage Levee Act and the Farm Drainage Act in 1879, which allowed counties to organize into drainage districts to consolidate financial resources. This action accelerated draining to the point that 30 percent of Illinois and Indiana and 20 percent of Iowa and Ohio are now under some form of drainage, and almost all of the original wetlands in these states (80 to 90 percent) have been destroyed. Chapter 3 described two very large wetlands in this region of the United States—the Great Kankakee Marsh in Indiana and the Black Swamp in Ohio—that essentially no longer exist. Drainage was absolute there.

Wetland Alteration

In a sense, wetland alteration or destruction is an extreme form of wetland management. One model of wetland alteration (Fig. 14.1) assumes that three main factors influence wetland ecosystem health: water level, nutrient status, and natural disturbances. Through human activity, the modification of any one of these factors can lead to wetland alteration, either directly or indirectly. For example, a wetland can be disturbed through decreased water levels, as in draining and filling, or through increased water levels, as in downstream drainage impediments. Nutrient status can be affected through upstream flood control that decreases the frequency of nutrient inputs or through increased nutrient loading from agricultural areas.

The most common alterations of wetlands have been (1) draining, dredging, and filling of wetlands; (2) modification of the hydrologic regime; (3) highway construction; (4) mining and mineral extraction; and (5) water pollution. These wetland modifications are described in more detail next.

Wetland Conversion: Draining, Dredging, and Filling

The major cause of wetland loss around the world continues to be conversion to agricultural use. Drainage for farms in the United States progressed at an average rate of 490,000 ha/yr over much of the twentieth century (slope of line in Fig. 14.2a). Less drainage occurred during the Great Depression of the 1930s and World War II years. This conversion was particularly significant in the vast midwestern United States

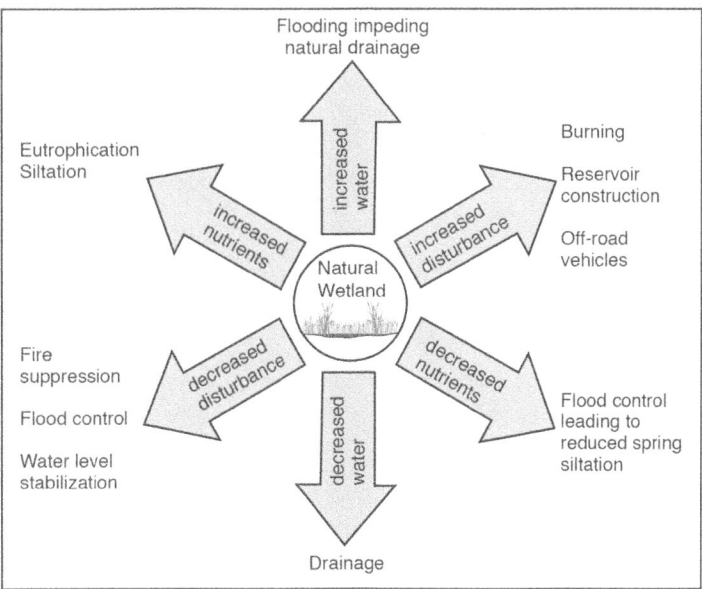

Figure 14.1 Model of human-induced impacts on wetlands, including effects on water level, nutrient status, and natural disturbance. By either increasing or decreasing any one of these factors, wetlands can be altered. (After Keddy, 1983)

"breadbasket," which has provided the bulk of the grain produced on the continent (Fig. 14.2b). Some of the world's richest farming is in the former wetlands of Ohio, Indiana, Illinois, Iowa, and southern Minnesota. When drained and cultivated, the fertile soils of the prairie pothole marshes and east Texas *playas* also produce excellent crops. With ditching and modern farm equipment, it has been possible to farm former wetlands routinely (Fig. 14.3). The modern farm equipment of today and mass-produced reels of plastic drainage pipe also make it possible to drain much more area per day than was ever possible with earlier equipment and the use of clay tiles.

Some of the most rapid wetland losses have occurred in the bottomland hardwood forests of the lower Mississippi River alluvial floodplain (Fig. 14.4). As populations increased along the river, the floodplain was channeled and leveed so that it could be drained and inhabited. Since colonial times, the floodplain has provided excellent cropland, especially for cotton and sugarcane. Cultivation, however, was restricted to the relatively high elevation of the natural river levees, which flooded regularly after spring rains and upstream snowmelts but drained rapidly enough to enable farmers to plant their crops. Because spring floods naturally fertilized river levees, they required no additional fertilizers to grow productive crops. One of the results of drainage and flood protection is the additional cost of fertilization. The lower parts of the floodplain, which are too wet to cultivate, were left as forests but harvested for timber. As pressure for additional cropland increased, these agriculturally marginal forests were clear-cut at an unprecedented rate. This was feasible, in part, because of the

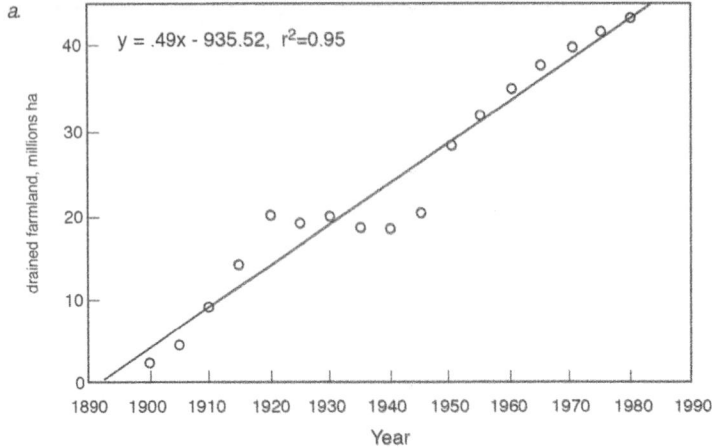

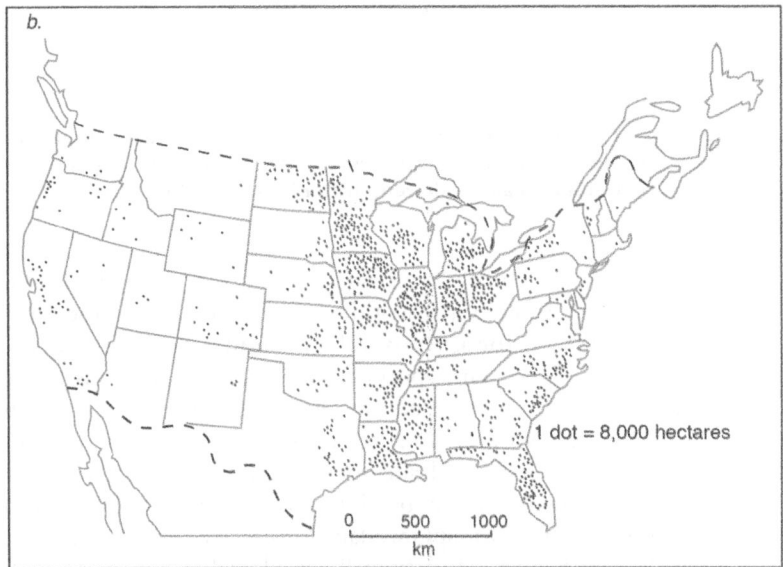

Figure 14.2 Artificially drained land in the United States: (a) Trend from 1900 to 1980; (b) extent and location of drainage through mid-1980s. Each dot represents 8,000 ha (20,000 acres), and the total area drained is 43 million ha. ((a) After Gosselink and Maltby, 1990; (b) after Dahl, 1990)

development of soybean varieties that mature rapidly enough to be planted in June or even early July, after severe flooding has passed. Often the land thus reclaimed was subsequently incorporated behind flood control levees, where it was kept dry by pumps. Clear-cutting of bottomland forests is still proceeding. Most of the available

Figure 14.3 Modern drainage machinery such as that illustrated in these photos is able to drain dozens of hectares per day: (a) Detail of the drainage machinery; (b) results of about 1 minute of drainage, showing new ditch and plastic pipe installed. (Photos by W. J. Mitsch)

wetland has been converted in Arkansas and Tennessee; Mississippi and Louisiana are experiencing large losses.

Along the nation's coasts, especially the East and West Coasts, the major cause of wetland loss is draining and filling for urban and industrial development or wetland loss due to subsidence. Compared to land converted to agricultural use, the area involved is rather small. Nevertheless, in some coastal states, notably California, almost all coastal wetlands have been lost. The rate of coastal wetland loss from 1954 to 1974 was closely tied to population density. This finding underscores two facts: (1) Two-thirds of the world's population lives along coasts; and (2) population density puts great pressure on coastal wetlands as sites for expansion. The most rapid development of coastal wetlands occurred after World War II. In particular, several large airports were built in coastal marshes. Since the passage of federal legislation controlling wetland development, the rate of conversion has slowed.

486 Chapter 14 Human Impacts and Management of Wetlands

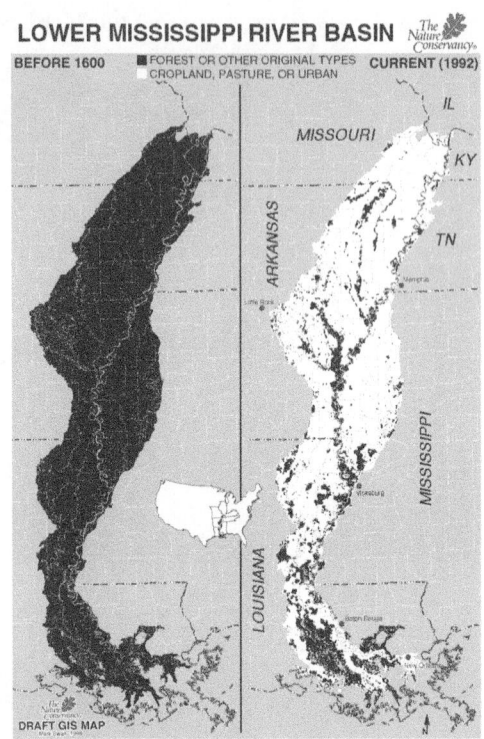

Figure 14.4 Historical and current distribution of bottomland wetland forests in the Mississippi River floodplain. (After The Nature Conservancy, 1992)

Hydrologic Modifications

Ditching, draining, and levee building are hydrologic modifications of wetlands specifically designed to dry them out. Other hydrologic modifications destroy or change the character of thousands of hectares of wetlands annually. Usually these hydrologic changes were made for some purpose that had nothing to do with wetlands; wetland destruction is an inadvertent result. Canals, ditches, and levees are created for three primary purposes:

1. *Flood control.* Most of the canals and levees associated with wetlands are for flood control. The canals have been designed to carry floodwaters off the adjacent uplands as rapidly as possible. Normal drainage through wetlands is slow surface sheet flow; straight, deep canals are more efficient. Ditching marshes and swamps to drain them for mosquito control or biomass harvesting is a special case designed to lower water levels in the wetlands. Along most of the nation's major rivers are systems of levees constructed to prevent overbank flooding of the adjacent floodplain. The U.S. Army Corps of Engineers built most of these levees after Congress passed flood control

legislation following the disastrous floods of the 1920s and 1930s. (For a fascinating account of the great flood of 1927, the disruption it caused, and the social and political reverberations that led to flood control legislation, see Barry, 1997.) These levees, by separating the river from its floodplain, isolated wetlands so that they could be drained expeditiously. For example, along the lower Mississippi River, the construction of levees created a demand from farmers for additional floodplain drainage. The sequence of response and demand was so predictable that farmers bought and cleared floodplain forests in anticipation of the next round of flood control projects.

2. *Navigation and transportation.* Navigation canals tend to be larger than drainage canals. They traverse wetlands primarily to provide water transportation access to ports and to improve transport among ports. For example, the Intracoastal Waterway was dredged through hundreds of miles of wetlands in the northern Gulf Coast. In addition, when highways were built across wetlands, fill material for the roadbed was often obtained by dredging soil from along the right-of-way, thus forming a canal parallel to the highway.

3. *Industrial activity.* Many canals are dredged to obtain access to sites within a wetland to sink an oil well, build a surface mine, or other kinds of development. Usually pipelines that traverse wetlands are laid in canals that are not backfilled.

The result of all of these activities can be a wetland crisscrossed with canals, such as in the immense coastal wetlands of the northern Gulf Coast. These canals modify wetlands in many ecological ways by changing normal hydrologic patterns. Straight, deep canals in shallow bays, lakes, and marshes capture flow, depriving the natural channels of water. Canals are hydrologically efficient, allowing the more rapid runoff of freshwater than the normal shallow, sinuous channels do. As a result, water levels fluctuate more rapidly than they do in unmodified marshes, and minimum levels are lowered, drying the marshes. In addition, when deep, straight channels connect low-salinity areas to high-salinity zones, as with many large navigation channels, tidal water, with its salt, intrudes farther upstream, changing freshwater wetlands to brackish. In extreme cases, salt-intolerant vegetation is killed and is not replaced before the marsh erodes into a shallow lake. On the Louisiana coast, the natural subsidence rate is high; wetlands go through a natural cycle of growth followed by decay to open bodies of water. There, canals accelerate the subsidence rate by depriving wetlands of natural sediment and nutrient subsidies.

Highway Construction

Highway construction can have a major effect on the hydrologic conditions of wetlands. Although few definitive studies have been able to document the extent of wetland damage caused by highways, the major effects of highways are alteration of the hydrologic regime, sediment loading, and direct wetland removal. In general, wetlands are more sensitive to highway construction than uplands are, particularly through the disruption of hydrologic conditions. Many early studies (Clewell et al., 1976; Evink, 1980; and Adamus, 1983) found that highway construction led to negative effects on wetlands through hydrologic isolation. Other than solar energy, the most important driving forces for wetlands are hydrologic, including tides, gradient currents (e.g., streamflow), runoff, and groundwater flow. The importance of protecting the hydrologic regime during highway construction is based on the contention presented in Chapter 4, "Wetland Hydrology," that the hydrology of wetlands is the most important determinant of a wetland's structure and function.

Peat Mining

World resources of peat, principally in peatlands in the Northern Hemisphere, are estimated to be 1.9×10^{12} t (trillion metric tons), of which countries that comprise the former Soviet Union have about 770×10^9 t (billion tons) and Canada about 510×10^9 t. In the United States, deposits of peat occur in most states, with estimated resources of about 310×10^9 t, or about 16 percent of the world total. Surface peat mining has been a common activity in several European countries, particularly Ireland and countries in eastern and northeastern Europe, since the eighteenth century (Fig. 14.5). These countries account for almost 75 percent of peat mining in the world; some of this peat is still used as a fuel for electric power production. For centuries but no longer, turf (dried-out peat) was used for home heating in Ireland (see Chapter 1: "Wetlands: Human Use and Science"). Peat projection in the world was estimated to be 25.5×10^6 t/yr in 2012, down from previous years and about the same amount of peat production as 14 years prior (Table 14.4).

Figure 14.5 Peat mining near Tartu, Estonia. Peat is burned in power plant shown with smokestack in background. (From J. S. Aber; printed with permission)

Table 14.4 World peat production by country for 1998 and 2012 in metric tons (t) per year

	1998 Peat Production ($\times 10^3$ t/yr)			2012 Peat Production ($\times 10^3$ t/yr)		
Country[a]	Fuel	Horticulture	Total	Fuel	Horticulture	Total
Finland	7,000	400	7,400	4,000	760	4,760
Ireland	4,500	300	4,800	1,452	500	1,952
Russia	3,000		3,000			1,300
Germany	180	2,800	2,980		3,048	3,048
Canada		1,127	1,127		973	973
Sweden	800	250	1,050	1,880	1,420	3,300
Ukraine	1,000		1,000			735
Estonia			1,000	360	567	927
United States		676	676		488	488
United Kingdom		500	500			1
Latvia			450			13,800
Belarus	300		300	250	3,000	3,250
Netherlands		300	300			
Moldova						475
Denmark		205	205	15	130	130
France		200	200		200	200
Poland			200			736
Lithuania			195	15	371	386
Spain			60		60	60
Hungary		45	45		25	25
Norway		30	31		440	440
Australia		15	15			n/a
Argentina		5	5		6	6
Burundi		5	5		8	8
Turkey						150
Rwanda					19	19
Grand total	16,800	6,900	25,500	11,200	9,200	25,500

[a] In addition to the countries listed, Austria, Chile, Iceland, Italy, and Romania produced negligible amounts of peat.
Source: Jasinski (1999) and United States Geological Survey (2013)

Peat produced in the United States was about 448,000 t (metric tons) per year in 2012, ranking the country twelfth in total production in the world. Since its inception, peat mining in North America has been primarily for horticultural and agricultural applications. The fibrous structure and porosity of peat promote a combination of water retention and drainage, which makes it useful for applications such as potting soils, lawn and garden soil amendments, and turf maintenance on golf courses. Peat is also used as a filtering medium to remove toxic materials and pathogens from wastewater, sewage effluent, and stormwater. It is generally classified as reed-sedge peat, whereas the imports from Canada typically are a weakly decomposed *Sphagnum* peat, which has a higher market value per ton. Approximately 95 percent of domestic peat is sold for horticulture/agriculture usage, including, in order of importance, general soil improvement, potting soils, earthworm culture, the nursery business, and golf course maintenance and construction.

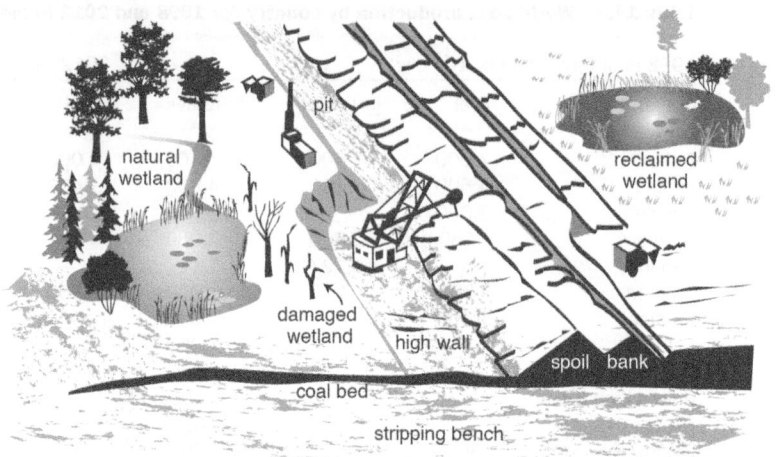

Figure 14.6 The impact of coal surface mining on wetlands and the possible use of wetlands in reclamation of coal surface mines for wildlife enhancement and control of mine drainage.

Mineral and Water Extraction

Surface mining activity for materials other than peat often affects major wetlands regions. Phosphate mining in central Florida is carried out over 120,000 ha and has had a significant impact on wetlands in the region (Brown, 2005). Thousands of hectares of wetlands may have been lost in central Florida because of this activity alone, although the reclamation of phosphate-mined sites for wetlands is now a common practice. H. T. Odum et al. (1981) argued that "managed ecological succession" on mined sites could be an economical alternative to current expensive reclamation techniques involving massive earth moving and reclamation planting.

Surface mining of coal has also affected wetlands in some parts of the United States (Brooks et al., 1985). Forty-six thousand hectares of wetlands in western Kentucky in the early 1980s, mostly bottomland hardwood forests, were or could have been affected by surface coal mining. The recognition of the potential benefits of including wetlands as part of the reclamation of coal mines has not been as widespread as one would have expected (Fig. 14.6), because of strict interpretation of measures regulating the return of the land to its original contours and because of liability questions. This is in contrast to the widespread acceptance of the reclamation of wetlands on phosphorus mine sites in Florida.

In some parts of the country, the withdrawal of water from aquifers or minerals from deep mines has resulted in accelerated subsidence rates that are lowering the elevations of marshes and built-up areas alike, sometimes dramatically. Land subsidence, which can also result in the creation of lakes and wetlands, is a geologically common phenomenon in Florida. Often, when excessive amounts of water are removed from karst deposits, underground cave-ins occur, causing surface slumpage. Some believe that the cypress domes in north-central Florida are an indirect result of a similar natural process, whereby fissure and dissolutions of underground limestone cause slight surface slumpage and subsequent wetland development.

Water Pollution

Wetlands are altered by pollutants from upstream or local runoff and, in turn, change the quality of the water flowing out of them. The ability of wetlands to cleanse water has received much attention in research and development and is discussed elsewhere in this book. The effects of polluted water on wetlands have received less attention, although water quality standards for wetlands have now been established in several regions of United States.

Species composition may also change with eutrophication of wetlands. For example, increased agricultural runoff, laden with phosphorus, is believed to have caused a spread of *Typha domingensis* in conservation areas that are part of the original Everglades in Florida (Fig. 14.7). This, in turn, has increased fears that the phosphorus will eventually lead to invasion of *Typha* in the Everglades National Park, replacing the natural sawgrass (*Cladium jamaicense*) (see Case Study 1, Chapter 18: "Wetland Creation and Restoration," and Case Study 3, Chapter 19: "Wetlands and Water Quality").

When metals, oils or other toxic organic compounds are the pollutants, effects on the wetland can be dramatic such as in the 2010 Deepwater Horizon Gulf of Mexico oil spill, where about 1800 km of coastal wetlands were affected (NRC, 2013). Another case of pollution in wetlands occurred two decades before when sulfates were discharged into a forested wetland in Florida (J. Richardson et al., 1983). Acid drainage from active and abandoned coal mines has been shown to affect wetlands seriously. In a study of wetlands adjacent to coal surface mining in western Kentucky, Mitsch et al. (1983a, 1983b, 1983c) described the extensive ecological damage that could occur where waters with low pH and high iron and sulfur were discharged from the mines into or through wetlands.

In one of the most publicized and dramatic cases of water pollution of a wetland, selenium from farm runoff contaminated marshes in Kesterson National Wildlife Refuge in California's San Joaquin Valley (Ohlendorf et al., 1986, 1990; Presser and Ohlendorf, 1987; T. Harris, 1991). The selenium contamination led to excessive death and deformities of wildlife and to eventual "closing" of the contaminated marsh in the mid-1980s, amid much controversy.

Wetland Management by Objective

Wetlands are managed for environmental protection, for recreation and aesthetics, and for the production of renewable resources. Twelve specific goals of wetland management are applicable today:

1. Maintain water quality.
2. Reduce erosion.

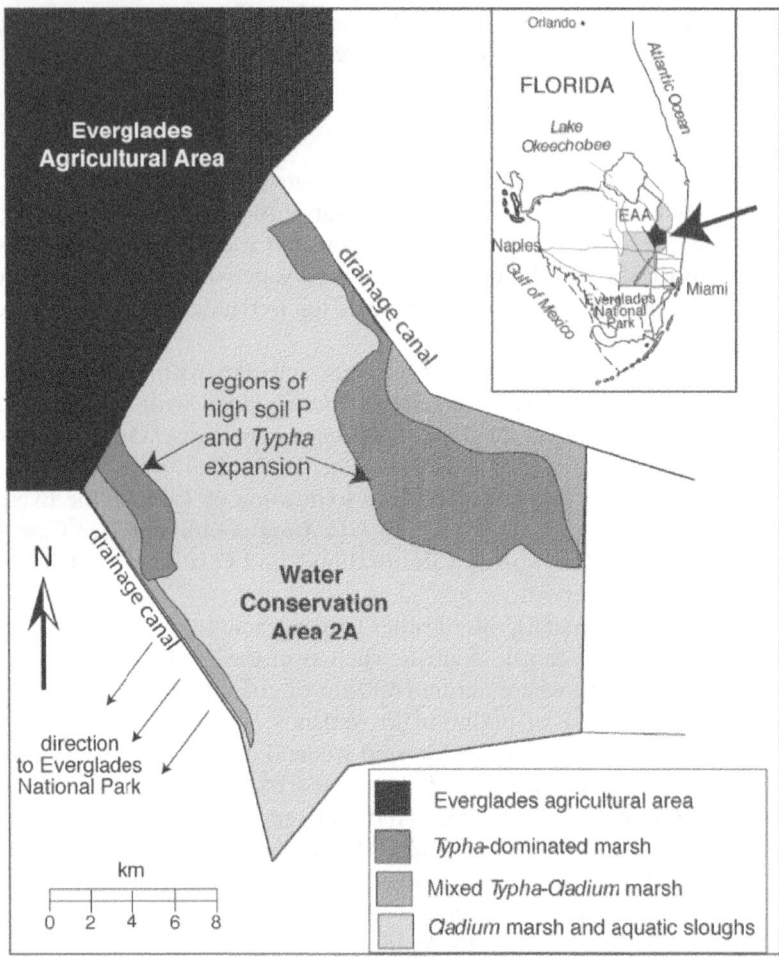

Figure 14.7 Water Conservation Area 2A (44,700 ha) in the south Florida Everglades, showing the area that has received high-nutrient surface overflow from agricultural land drainage since the 1960s. Excess nutrients from the Everglades Agricultural Area to the northwest have caused the spread of *Typha domingensis* and the loss of *Cladium jamaicense* over the 8,000-ha area shaded. (After Koch and Reddy, 1992)

3. Protect from floods and storm damage.
4. Provide a natural system to process airborne pollutants.
5. Provide a buffer between urban residential and industrial segments to ameliorate climate and physical impact such as noise.
6. Maintain a gene pool of marsh plants and provide examples of complete natural communities.
7. Provide aesthetic and psychological support for human beings.

8. Produce wildlife.
9. Control insect populations.
10. Provide habitats for fish spawning and other food organisms.
11. Produce food, fiber, and fodder (e.g., timber, cranberries, cattails for fiber).
12. Expedite scientific inquiry.

One management approach is to fence in a wetland to preserve it. Although simple, this is an act of conservation of a valuable natural ecosystem involving no substantive changes in management practices. Often, however, management has one or more specific objectives that require positive manipulation of the environment. Efforts to maximize one objective may be incompatible with the attainment of others, although, in recent years, most management objectives have been broadly stated to enhance multiple objectives. Multipurpose management generally focuses on system-level support rather than individual species. This has often been achieved indirectly through plant species manipulation, because plants provide food and cover for the animals. In the management of many small wetland areas in proximity, the use of different practices or staggered management cycles, so that the different areas are not all treated the same way at the same time, not only increases the diversity of the larger landscape but also attracts wildlife.

Waterfowl and Wildlife Management

The best wetland management practices are those that enhance the natural processes of the wetland ecosystem involved. One way to accomplish this is to maintain conditions as close as possible to the natural hydrology of the wetland, including hydrologic connections with adjacent rivers, lakes, and estuaries. Unfortunately, this cannot easily be accomplished in wetlands managed for wildlife; the vagaries of nature, especially in hydrologic conditions, make planning difficult. Hence, marsh management for wildlife, particularly waterfowl, has often meant water-level manipulation. Dikes (impoundments), weirs (solid structures in marsh outflows that maintain a minimum water level), control gates, and pumps control water level. In general, the results of the management activity depend on how well the water-level control is maintained, and control depends on the local rainfall and on the sophistication of the control structures. For example, weirs provide the poorest control; all they do is maintain a minimum water level. Pumps provide positive control of drainage or flooding depth at the desired time; and the management objectives usually can be met, although the cost is much higher than fixed weirs.

Baldassarre and Bolen (2006) summarize several of the wetland management techniques used for waterfowl and water birds. They conclude that the practices fall into two general categories: (1) *natural management* that takes advantage of natural attributes of wetlands, such as seed banks, plant succession, water-level fluctuations, and herbivory; and (2) *artificial management* that includes practices such as planting, ditching, and island building. One of the most frequently used management techniques—perhaps a combination of the two above categories—is a water-level

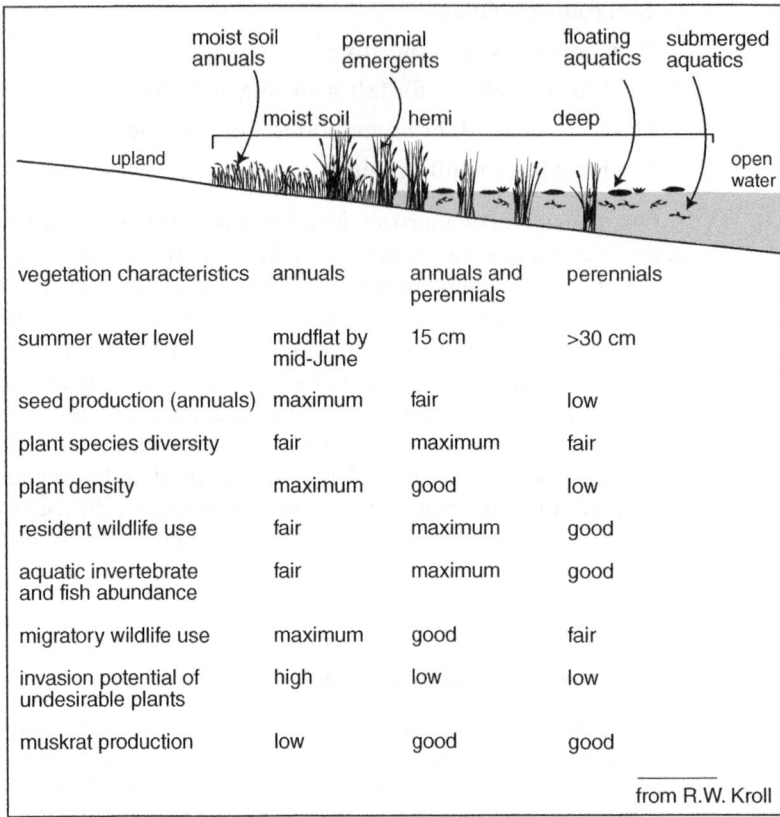

Figure 14.8 Generalizations of water-level management for vegetation, wildlife use, and other characteristics as practiced on impounded marshes near Lake Erie in northern Ohio. (From Roy Kroll, unpublished illustration, provided with permission)

drawdown. Drawdowns are carried out to recycle nutrients from otherwise undecomposed organic matter, to allow for "moist-soil management" to enhance vegetation regeneration from the wetland seed bank, and sometimes to manage for a diversity of macroinvertebrate (important source of protein for ducks) communities. Quite often, trade-offs occur during water-level manipulations.

To illustrate the trade-offs in wetland management for wildlife enhancement, some generalizations about water-level manipulation of Lake Erie (Ohio) coastal marshes are shown in Figure 14.8. Maximum migratory wildlife use of the marshes occurs in moist-soil conditions, but these conditions are also the best for the invasion of potentially undesirable plants and are generally least favorable for the overall abundance and diversity of resident plant and animal populations. Shallow-water (called *hemi* conditions by marsh managers; around 15 cm depth in summer) usually result in the highest plant species diversity and greatest fish and resident wildlife use but less migratory wildlife. Deepwater conditions (>30 cm) offer the least potential for both annual emergent plants and invading, undesirable plants, and desirable migratory waterfowl use is only fair in deep water. Kroll et al. (1997) and Gottgens et al. (1998) point out that because the landward advance of marshes during high lake water times is restricted by human development, and because carp (*Cyprinus carpio*) are present in the lake, long-term above-average water levels probably mean that removal of dikes along the Great Lakes would lead to an irreversible loss of wetland vegetation fringing the lakes. Mitsch et al. (2001) found that only 25 percent of the existing marshes encompassed by dikes would have had the right conditions to be emergent marshes more than 50 percent of the time during the twentieth century.

The set of management recommendations by Weller (1978) for prairie pothole marshes in the north-central United States and south-central Canada is another example of multipurpose wildlife enhancement. Those recommendations mimic the natural cycle of marshes in the middle of North America. Although they may seem drastic, they are entirely natural in their consequences. In sequence, the six practices are:

1. When a pothole is in the open stage and there is little emergent vegetation, the cycle should be initiated by a spring drawdown. This stimulates the germination of seedlings on the exposed mud surfaces.
2. A slow increase in water level after the drawdown maintains the growth of flood-tolerant seedlings without shading them out in turbid water. Shallowly flooded areas attract dabbling ducks during the winter.
3. The drawdown cycle should be repeated for a second year to establish a good stand of emergent plants.
4. Low water levels should be maintained for several more seasons to encourage the growth of perennial emergent plants such as *Typha*.
5. Maintaining stable, moderate water depths for several years promotes the growth of rooted submerged perennial aquatic plants and associated benthic fauna that make excellent food for waterfowl. During that period, the emergent vegetation will gradually die out and will be replaced by shallow ponds. When that occurs, the cycle can be initiated again, as described in step 1.
6. Different wetland areas maintained in staggered cycles provide all stages of the marsh cycle at once, maximizing habitat diversity.

Weller (1994) made the distinction between a complete drawdown of water levels, as described before—a management option when vegetation is completely lost because of high water levels, herbivory, winter kill, or plant disease—and a partial drawdown, which can be implemented when vegetation is reduced but not eliminated or when wildlife use has declined but not disappeared.

Wildlife management in coastal salt marshes such as those found in Louisiana uses a similar strategy, although the short-term cycle is not as pronounced there. Drawdowns to encourage the growth of seedlings and perennials preferred by ducks are common practices, as is fall and winter flooding to attract dabbling ducks. As it happens, there is general agreement that stabilizing water levels is not good management, even though our society seems to feel intuitively that stability is a good thing. Wetlands thrive on cycles, especially flooding cycles, and practices that dampen these cycles also reduce wildlife productivity. Although the management practices described previously enhance waterfowl production, they are generally deleterious for wetland-dependent fisheries in coastal wetlands because free access between the wetlands and the adjacent estuary is restricted; the wetlands' role in regulating water quality is also often underutilized.

There is a tendency to want to control all external variables when we manage wetlands by objectives. This management tendency, although understandable, is particularly strong when herbivores, such as geese, nutria, beavers, or muskrats, "invade" a managed wetland. These animals can be discouraged and/or trapped to keep their influence on vegetation at a minimum, but one has to remember that these animals are not invaders at all but are simply coming to a habitat that is generally well suited to their needs. In the ecosystem context, these animals are often nature's "ecosystem engineers" and provide many functions that, in the long term, may enhance marshes. Beavers cause water-level manipulations just as humans do. Muskrats and geese remove large areas of vegetation but open up the system to allow for other vegetation to come into the wetland.

Whether management of ecosystem managers is a wise strategy is a complex issue. In coastal Louisiana, for example, muskrats and especially nutria (a South American immigrant) can "eat-out" extensive marsh areas, which do not recover because of rising sea level and high marsh subsidence rates induced, in part, by human activities. Trapping used to keep the rodent populations in check, but the worldwide slump in fur sales no longer makes trapping profitable, and rodent populations are rapidly escalating.

Baldassarre and Bolen (2006) presented seven general principles that provide a useful set of rules for wetland managers. Many of the principles on wetland restoration described in Chapter 18 mirror these wetland management principles:

1. Protect wetland complexes that include a wide variety of wetland hydroperiods and wetland sizes.
2. Protect small wetlands, as these wetlands are most vulnerable to being lost along with their unique biota.
3. Consider all wetland-dependent wildlife when managing wetlands, not just one or two species.
4. Protect large wetlands too for species that require large areas.
5. Recognize the importance of wetland complexes for species with complex life-history requirements.
6. Recognize that protected sites often require direct management intervention to protect wildlife values.
7. Protect and restore upland habitats that are contiguous with wetlands.

Agriculture and Aquaculture

When wetlands are drained for agricultural use, they no longer function as wetlands. They are, as local farmers say, "fast lands" removed from the effects of periodic flooding, and they grow terrestrial, flood-intolerant crops. Some use is made of more or less undisturbed wetlands for agriculture, but it is minor. In New England, high salt marshes were harvested for salt marsh hay (*Spartina patens*), which was considered an excellent bedding and fodder for cattle. In fact, the proximity of fresh and salt hay marshes was a major factor in selecting the sites for many towns in New England before 1650. Subsequently, marshes were ditched to allow the intrusion of tides to promote the growth of salt marsh hay, but the extent of this practice has not been well documented. On parts of the coast of the Gulf of Mexico where marshes are firm underfoot, they are still used extensively for cattle grazing. To improve access, small embankments or raised earthen paths are constructed in these marshes.

The ancient Mexican practice of *marceno* is unique. In the freshwater wetlands of the northern coast of Mexico, small areas were cleared and planted in corn during the dry season. These native varieties were tolerant enough to withstand considerable flooding. After harvest (or apparently sometimes before harvest), the marshes were naturally reflooded, and native grasses were reestablished until the next dry season. This practice is no longer followed, but there has been some interest in reviving it.

On a global scale, the production of rice in managed wetlands contributes a major proportion of the world's food supply. There are approximately 1.3 million km^2 of rice paddies in the world (Chapter 3), of which almost 90 percent are in Asia. In North America, especially in Minnesota, there are several commercial wild rice (*Zizania aquatica*) operations in wetlands and several other locations where Native American tribes have harvested wild rice in natural marshes for centuries.

Aquaculture, the farming of fish and shellfish, which produced less than 1 million tons per year in the early 1950s, now produces 70 million metric tons, or almost half of the annual worldwide total fish and shellfish harvest of 160×10^6 t (million metric tons)/yr (Fig. 14.9). The Food and Agriculture Organization of the United Nations (http://www.fao.org/fishery/topic/13540/en) estimates that "to maintain the current level of per capita consumption, global aquaculture production will need to reach 80 million ton by 2050." Most of this aquaculture production occurs in Asia, with China by far the largest producer. The United States is the major consumer of aquaculture products but accounts for only a small percentage of worldwide production, mostly salmon and crayfish.

Fish farming practices vary. The most environmentally benign approach, similar to the Mexican *marceno* described previously, intercrops shellfish with a grain crop, usually rice. Typical is crayfish farming in the United States and Indian shrimp aquaculture in rotation with rice. The practice is described for crayfish in the southern United States. Crayfish are an edible delicacy in the southern United States and in many foreign countries. They live in burrows in shallow flooded areas, such as swamp forests and rice fields, emerging with their young early in the year to forage for food. The young grow to edible size within a few weeks and are harvested in the spring. When floodwaters retreat, the crayfish construct burrows, where they remain until the

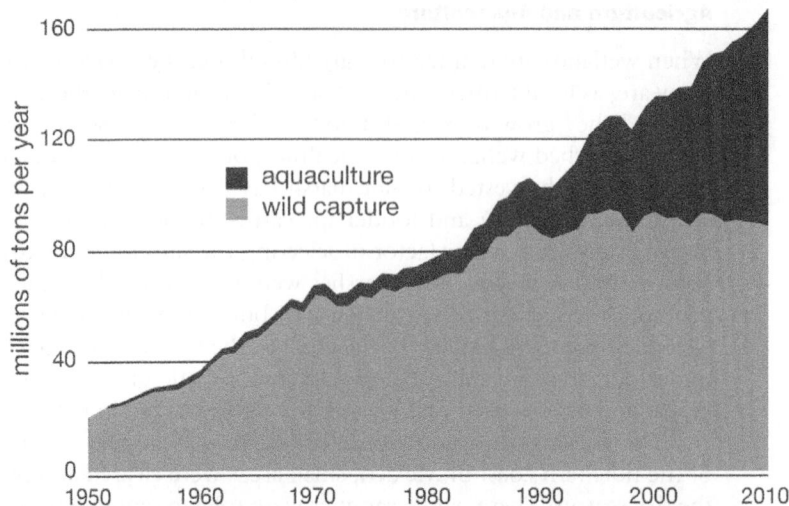

Figure 14.9 World's fisheries harvest, 1950 to 2010, showing increased contribution of aquaculture to overall production. (From http://en.wikipedia.org/wiki/Seafood)

next winter flood. In crayfish farms, this natural cycle is enhanced by controlling water levels. An area of swamp forest is impounded; it is flooded deep during the winter and spring and drained during the summer. This cycle is ideal for crayfish, which thrive. Fish predators are controlled within the impoundments to improve the harvest. The hydrologic cycle is also favorable for forest trees. It simulates the hydrologic cycle of a bottomland hardwood forest; forest tree productivity is high, and seedling recruitment is good because of the summer drawdown. Species composition tends toward species typical of bottomland hardwoods.

Some rice farmers have also found that they can take advantage of the annual flooding cycle typically used to grow rice to combine rice and crayfish production. Rice fields are drained during the summer and fall, when the rice crop matures and is harvested. Then the fields are reflooded, allowing crayfish to emerge from their burrows in the rice field embankments and forage on the vegetation remaining after the rice harvest. The crayfish harvest ends when the fields are replanted with rice. When this rotation is practiced, extreme care has to be exercised in the use of pesticides.

The most intensive aquaculture techniques control all aspects of production. Wetlands, salt flats, mangrove forests, and even high-quality farmland are dredged to form ponds in which water levels are controlled by pumps. "Seed" organisms, the young postlarvae, are raised in separate hatcheries. The young organisms are fed in the ponds on synthetic diets, often composed of fish from commercial fisheries. Water quality is monitored, the ponds are aerated, and, in the most sophisticated operations, wastes are treated. Yields from this kind of operation can be several metric tons per hectare per crop, and in tropical areas two crops per year are expected.

Whereas aquaculture farms in Asia have historically been small operations managed by local farmers, the worldwide boom in aquaculture, fueled by the high demand for fishery products, has led many countries to offer large incentives to initiate new fish farms and draw in large corporations to invest in the industry. Worldwide, 50,000 shrimp farms cover more than 10,000 km^2 of coastal lands. This has resulted in a serious loss of wetlands (coastal wetlands are required habitats for most commercial marine fish), especially mangrove forests. These fish farms not only disrupt natural ecosystems but also bring in diseases, create enormous waste problems, deplete oxygen in shallow coastal waters, and reduce water quality. These disruptions have been cited as one reason for the decline in commercial fisheries in the areas where shrimp culture is concentrated.

Water Quality Enhancement

Several studies have shown natural wetlands to be sinks for certain chemicals, particularly sediments and nutrients. It is now common to cite the water quality role of natural wetlands in the landscape as one of the most important reasons for their protection. The idea of applying domestic, industrial, and agricultural wastewaters, sludges, and even urban and rural runoff to wetlands to take advantage of this nutrient sink capacity has also been explored in countless studies. The basic principles and practices of these so-called treatment wetlands are covered in detail in Chapter 19: "Wetlands and Water Quality."

Flood Control and Stormwater Protection

Wetlands can be managed, often passively, for their role in the hydrologic cycle. Hydrologic values of wetlands include streamflow augmentation, groundwater recharge, water supply potential, and flood protection. It is not altogether clear how well wetlands carry out these functions, nor do all wetlands perform these functions equally well. It is known, for example, that wetlands do not necessarily always contribute to low flows or recharge groundwater. Some wetlands, however, should be, and often are, protected for their ability to hold water and slowly return it to surface-water and groundwater systems during periods of low water. If wetlands are impounded to retain even more water from flooding downstream areas, considerable changes in vegetation will result as the systems adapt to the new hydrologic conditions. The values of wetlands for coastal protection and flood mitigation are discussed in more detail in Chapter 16: "Wetland Ecosystem Services."

Recommended Reading

Baldassarre, G. A., and E. G. Bolen. 2006. *Waterfowl Ecology and Management*, 2nd ed. Malabar, FL: Krieger.

References

Adamus, P. R. 1983. *A Method for Wetland Functional Assessment, Vol. 1: Critical Review and Evaluation Concepts, and Vol. 2: FHWA Assessment Method.* Federal Highway Reports FHWA-IP-82-83 and FHWA-IP-82-84, U.S. Department of Transportation, Washington, DC. 16 pp. and 134 pp.

Baldassarre, G. A., and E. G. Bolen. 2006. *Waterfowl Ecology and Management*, 2nd ed. Krieger Publishing, Malabar, Florida, 567 pp.

Barry, J. M. 1997. *Rising Tide: The Great Mississippi Flood of 1927 and How It Changed America.* Simon & Schuster, New York.

Brooks, R. P., D. E. Samuel, and J. B. Hill, eds. 1985. *Wetlands and Water Management on Mined Lands.* Proceedings of a conference, October 23–24, 1985. Pennsylvania State University Press, University Park. 393 pp.

Brown, M. T. 2005. Landscape restoration following phosphate mining: 30 years of co-evolution of science, industry, and regulation. *Ecological Engineering* 24: 309–329.

Clewell, A. F., L. F. Ganey, Jr., D. P. Harlos, and E. R. Tobi. 1976. *Biological Effects of Fill Roads across Salt Marshes.* Report FL-E.R-1-76. Florida Department of Transportation, Tallahassee.

Dahl, T. E. 1990. Wetlands losses in the United States, 1780s to 1980s. U.S. Department of Interior, Fish and Wildlife Service, Washington, DC. 21 pp.

Dugan, P. 1993. *Wetlands in Danger.* Michael Beasley, Reed International Books, London. 192 pp.

Evink, G. L. 1980. *Studies of Causeways in the Indian River, Florida.* Report FL-ER-7-80. Florida Department of Transportation, Tallahassee. 140 pp.

Gosselink, J. G., and E. Maltby. 1990. Wetland losses and gains. In M. Williams, ed., *Wetlands: A Threatened Landscape.* Basil Blackwell Ltd., Oxford, pp. 296–322.

Gottgens, J. F., B. P. Swartz, R. W. Kroll, and M. Eboch. 1998. Long-term GIS-based records of habitat changes in a Lake Erie coastal marsh. *Wetlands Ecology and Management* 6: 5–17.

Harris, T. 1991. *Death in the Marsh.* Island Press, Washington, DC. 245 pp.

Jasinski, S. M. 1999. Peat. In *Minerals Yearbook 1999: Volume 1—Metals and Minerals.* Minerals and Information, U.S. Geological Survey, Reston, VA.

Keddy, P. A. 1983. Freshwater wetland human-induced changes: Indirect effects must also be considered. *Environmental Management* 7: 299–302.

Koch, M. S., and K. R. Reddy. 1992. Distribution of soil and plant nutrients along a trophic gradient in the Florida Everglades. *Soil Science Society of America Journal* 56: 1492–1499.

Kroll, R. W., J. F. Gottgens, and B. P. Swartz. 1997. Wild rice to rip-rap: 120 years of habitat changes and management of a Lake Erie coastal marsh. *Transactions of the 62nd North American Wildlife and Natural Resources Conference* 62: 490–500.

Larson, J. S., and J. A. Kusler. 1979. Preface. In P. E. Greeson, J. R. Clark, and J. E. Clark, eds., *Wetland Functions and Values: The State of Our Understanding.* American Water Resources Association, Minneapolis, MN.

Mitsch, W. J., J. R. Taylor, and K. B. Benson. 1983a. Classification, modelling and management of wetlands—A case study in western Kentucky. In W. K. Lauenroth, G. V. Skogerboe, and M. Flug, eds., *Analysis of Ecological Systems: State-of-the-Art in Ecological Modelling*. Elsevier, Amsterdam, the Netherlands, pp. 761–769.

Mitsch, W. J., J. R. Taylor, K. B. Benson, and P. L. Hill, Jr. 1983b. *Atlas of Wetlands in the Principal Coal Surface Mine Region of Western Kentucky*. FWS/OBS-82/72, U.S. Fish and Wildlife Service, Washington, DC. 135 pp.

Mitsch, W. J., J. R. Taylor, K. B. Benson, and P. L. Hill, Jr. 1983c. Wetlands and coal surface mining in western Kentucky—A regional impact assessment. *Wetlands* 3: 161–179.

Mitsch, W. J., and J. G. Gosselink. 1986. *Wetlands*. Van Nostrand Reinhold, New York. 539 pp.

Mitsch, W. J., N. Wang, and V. Bouchard. 2001. Fringe wetlands of the Laurentian Great lakes: Effects of dikes, water level fluctuations, and climate change. *Verh. Internat. Verein. Limnol.* 27: 3430–3437.

National Research Council (NRC). 2013. *An Ecosystem Services Approach to Assessing the Impacts of the Deepwater Horizon Oil Spill in the Gulf of Mexico*. The National Academies Press, Washington DC.

Norgress, R. E. 1947. The history of the cypress lumber industry in Louisiana. *Louisiana Historical Quarterly* 30: 979–1059.

Odum, H. T., P. Kangas, G. R. Best, B. T. Rushton, S. Leibowitz, J. R. Butner, and T. Oxford. 1981. *Studies on Phosphate Mining, Reclamation, and Energy*. Center for Wetlands, University of Florida, Gainesville. 142 pp.

Ohlendorf, H. M., D. J. Hoffman, M. K. Saiki, and T. W. Aldrich. 1986. Embryonic mortality and abnormalities of aquatic birds: Apparent impacts of selenium from irrigation drainwater. *Science of the Total Environment* 52: 49–63.

Ohlendorf, H. M., R. L. Hothem, C. M. Bunck, and K. C. Marois. 1990. Bioaccumulation of selenium in birds at Kesterson Reservoir, California. *Archives of Environmental Contamination and Toxicology* 19: 495–507.

Presser, T. S., and H. M. Ohlendorf. 1987. Biogeochemical cycling of selenium in the San Joaquin Valley. *Environmental Management* 11: 805–821.

Richardson, J., P. A. Straub, K. C. Ewel, and H. T. Odum. 1983. Sulfate-enriched water effects on a floodplain forest in Florida. *Environmental Management* 7: 321–326.

The Nature Conservancy. 1992. *The Forested Wetlands of the Mississippi River: An Ecosystem in Crisis*. The Nature Conservancy, Baton Rouge, LA, 25 pp.

U.S. Geological Survey. 2013. *2012 Minerals Yearbook: Peat* (advanced release). Edited by L. E. Apodaca. http://minerals.usgs.gov/minerals/pubs/commodity/peat/myb1-2012-peat.pdf.

Weller, M. W. 1978. Management of freshwater marshes for wildlife. In R. E. Good, D. F. Whigham, and R. L. Simpson, eds., *Freshwater Wetlands: Ecological Processes and Management Potential*. Academic Press, New York, pp. 267–284.

Weller, M. W. 1994. *Freshwater Marshes*, 3rd ed. University of Minnesota Press, Minneapolis. 192 pp.

Chapter **15**

Wetland Laws and Protection

Wetlands are now protected by a myriad of laws and regulations in the United States and by some treaties internationally. The United States has relied on federal executive orders and court decisions, a "no net loss" policy, and sections of the Clean Water Act for wetland protection, augmented by some wetland conservation programs and by the development of wetland delineation as a formal technique for identifying wetlands. Three Supreme Court decisions in the twenty-first century have limited the jurisdiction of federal protection of some wetlands in the United States but illustrate how important these ecosystems are viewed by the legal world in the country. International cooperation in wetland protection, particularly through the Ramsar Convention and the North American Waterfowl Management Plan, has been emphasized in recent years as policy makers realize that the functions of local wetlands cross international boundaries.

Wetlands are now the focus of institutional and legal protection efforts throughout the world, but because of this focus, they are beginning to be defined by legal fiat as much as by the application of ecological principles. Chapter 2: "Wetland Definitions," reviewed the major definitions of wetlands that have developed in the United States and internationally. Some definitions are scientific, whereas others are principally to allow legal protection of wetlands. Protection has been implemented through a variety of policies, laws, and regulations, ranging from animal and plant protection to land use and zoning restrictions, to enforcement of dredge-and-fill laws. In the United States, wetland protection has historically been a national initiative, often with assistance and implementation provided by individual states. In the international arena, agreements to protect ecologically important wetlands throughout the world have been negotiated and ratified and are becoming more important every year.

Legal Protection of Wetlands in the United States

The policy of the United States for more than 120 years was to drain wetlands. The Swamp Land Acts of 1849, 1850, and 1860, described in Chapter 14: "Human Impacts and Management of Wetlands," were precursors to one of the most rapid and dramatic changes in the landscape that has ever occurred in history, even though the acts were deemed to be largely ineffective in their intended purpose (National Research Council, 1995). By the mid-1970s, about half of the wetlands in the lower 48 states were drained (see Chapter 3: "Wetlands of the World"). In the early 1970s, interest in wetland protection started as scientists began to identify and quantify the many values (now referred to as ecosystem services) of wetlands. This interest in wetland protection began to be translated at the federal level in the United States into interpretation of existing laws, regulations, and public policies. Prior to this time, federal policy on wetlands was vague and often contradictory. Policies in agencies such as the U.S. Army Corps of Engineers, the Soil Conservation Service (now the Natural Resources Conservation Service), and the Bureau of Reclamation had encouraged the destruction of wetlands, whereas policies in the Department of the Interior, particularly in the U.S. Fish and Wildlife Service, had long encouraged their protection. Some states also developed inland and coastal wetland laws and policies during the 1970s.

The primary wetland protection mechanisms used by the U.S. federal government are summarized in Table 15.1. Some of the more significant activities of the federal government that led to a more consistent wetland protection policy have included presidential orders on wetland protection and floodplain management, implementation of a dredge-and-fill permit system to protect wetlands, coastal zone management policies, and initiatives and regulations issued by various agencies. Despite all of this activity related to federal wetland management, two major points should be emphasized:

1. *There is no specific national wetland law in the United States.* Wetland management and protection result from the application of many laws intended for other purposes. Jurisdiction over wetlands has also been spread over several agencies, and, overall, federal policy continually changes and requires considerable interagency coordination.

2. *Wetlands have been managed under regulations related to both land use and water quality.* Neither of these approaches, taken separately, can lead to a comprehensive wetland policy. This regulatory split mirrors the scientific split noted by many wetland ecologists, who must develop expertise in both aquatic and terrestrial systems. Rarely do individuals possess expertise in both areas.

Early Presidential Orders

President Jimmy Carter issued two executive orders in May 1977 that established the protection of wetlands and riparian systems as the official policy of the federal

Table 15.1 Major federal laws, directives, and regulations in the United States used for the management and protection of wetlands

	Date	Responsible Federal Agency
Directive or Statute		
Rivers and Harbors Act, Section 10	1899	U.S. Army Corps of Engineers
Fish and Wildlife Coordination Act	1967	U.S. Fish and Wildlife Service
Land and Water Conservation Fund Act	1968	U.S. Fish and Wildlife Service, Bureau of Land Management, Forest Service, National Park Service
National Environmental Policy Act	1969	Council on Environmental Quality
Federal Water Pollution Control Act (PL 92-500) as amended (Clean Water Act)	1972, 1977, 1982	
Section 404—Dredge-and-Fill Permit Program		U.S. Army Corps of Engineers with assistance from Environmental Protection Agency and U.S. Fish and Wildlife Service
Section 208—Areawide Water Quality Planning		U.S. Environmental Protection Agency
Section 303—Water Quality Standards		U.S. Environmental Protection Agency
Section 401—Water Quality Certification		U.S. Environmental Protection Agency (with state agencies)
Section 402—National Pollutant Discharge Elimination System		U.S. Environmental Protection Agency (or state agencies)
Coastal Zone Management Act	1972	Office of Coastal Zone Management, Department of Commerce
Flood Disaster Protection Act	1973, 1977	Federal Emergency Management Agency
Federal Aid to Wildlife Restoration Act	1974	U.S. Fish and Wildlife Service
Water Resources Development Act	1976, 1990	U.S. Army Corps of Engineers
Executive Order 11990—Protection of Wetlands	May 1977	All agencies
Executive Order 11988—Floodplain Management	May 1977	All agencies
Food Security Act, swampbuster provisions	1985	Department of Agriculture, Natural Resources Conservation Service
Emergency Wetland Resources Act	1986	U.S. Fish and Wildlife Service
Executive Order 12630—Constitutionally Protected Property Rights	1988	All agencies
National list of Plant Species that Occur in Wetlands (original and update)	1988, 2012	U.S. Fish and Wildlife Service, U.S. Army Corps of Engineers
Wetlands Delineation Manual (various revisions)	1987, 1989, 1991	All agencies
"No Net Loss" Policy	1988	All agencies
North American Wetlands Conservation Act	1989	U.S. Fish and Wildlife Service
Coastal Wetlands Planning, Protection and Restoration Act	1990	U.S. Army Corps of Engineers
Wetlands Reserve Program (WRP)	1991	Department of Agriculture, Natural Resources Conservation Service
Executive Order 12962—Conservation of Aquatic Systems for Recreational Fisheries	1995	All agencies
Federal Agriculture improvement and Reform Act	1996	Department of Agriculture, Natural Resources Conservation Service
Agricultural Conservation Easement Program (consolidates WRP with other programs)	2014	Department of Agriculture, Natural Resources Conservation Service

(continued)

Table 15.1   (Continued)

	Date	Responsible Federal Agency
Policy and Technical Guidance		
Water Quality Standards Guidance	1990	U.S. Environmental Protection Agency
Non-Point Source Guidance	1990	U.S. Environmental Protection Agency
Mitigation/Mitigation Banking	1990, 1995	U.S. Army Corps of Engineers
Wetlands on Agricultural Lands, memo of agreement	1990, 1994	U.S. Army Corps of Engineers, Department of Agriculture
Wetlands and Forestry Guidance	1995	U.S. Army Corps of Engineers, Department of Agriculture
Regulatory Guidance Letter on Wetland Mitigation	2001, 2002	U.S. Army Corps of Engineers
Final Rules for Compensatory Mitigation	2008	U.S. Army Corps of Engineers, U.S. Environmental Protection Agency
Regional Supplements to Wetland Delineation Manual	2009 - 2014	U.S. Army Corps of Engineers
Proposed rule to define "Waters of the United States" under the Clean Water Act	2014	U.S. Army Corps of Engineers, U.S. Environmental Protection Agency

government. Executive Order 11990, Protection of Wetlands, required all federal agencies to consider wetland protection as an important part of their policies:

> Each agency shall provide leadership and shall take action to minimize the destruction, loss or degradation of wetlands, and to preserve and enhance the natural and beneficial values of wetlands in carrying out the agency's responsibilities for (1) acquiring, managing, and disposing of Federal lands and facilities; and (2) providing federally undertaken, financed, or assisted construction and improvement; and (3) conducting Federal activities and programs affecting land use, including but not limited to water and related land resources planning, regulating, and licensing activities.

Executive Order 11988, Floodplain Management, established a similar federal policy for the protection of floodplains, requiring agencies to avoid activity in the floodplain wherever practicable. Furthermore, agencies were directed to revise their procedures to consider the impact that their activities might have on flooding and to avoid direct or indirect support of floodplain development when other alternatives are available.

Both of these executive orders were significant because they set in motion a review of wetland and floodplain policies by almost every federal agency. Agencies, such as the U.S. Environmental Protection Agency (U.S. EPA) and the Soil Conservation Service, established policies of wetland protection prior to the issuance of these executive orders, and many other agencies, such as the Bureau of Land Management, were compelled to review or establish wetland and floodplain policies.

No Net Loss

A significant initiative in developing a national wetlands policy was undertaken in 1987, when a National Wetlands Policy Forum was convened by the Conservation Foundation at the request of the U.S. EPA to investigate the issue of wetland management in the United States (National Wetlands Policy Forum, 1988). The 20 distinguished members of this forum (which included three governors, a state legislator, state and local agency heads, chief executive officers of environmental groups and businesses, farmers, ranchers, and one of the coauthors of this book [James G. Gosselink]) published a report that set significant goals for the nation's remaining wetlands. The forum formulated one overall objective: "To achieve no overall net loss of the nation's remaining wetlands base and to create and restore wetlands, where feasible, to increase the quantity and quality of the nation's wetland resource base" (National Wetlands Policy Forum, 1988).

The group recommended as an interim goal that the holdings of wetlands in the United States should decrease no further—*no net loss*—and as a long-term goal that the number and quality of the wetlands should increase—*net gain*. In his 1988 presidential campaign and in his 1990 budget address to Congress, President George Bush presented the "no net loss" concept as a national goal, shifting the activities of many agencies such as the Department of the Interior, the U.S. EPA, the U.S. Army Corps of Engineers, and the Department of Agriculture toward achieving a unified and seemingly simple goal. It was not anticipated that there would be a complete halt of wetland loss in the United States when economic or political reasons dictated otherwise. Consequently, implied in the concept is wetland creation and restoration to replace destroyed wetlands. The "no net loss" concept became a cornerstone of wetland conservation in the United States and remains so to this day.

Clean Water Act

The primary vehicle for wetland protection and regulation in the United States for 40 years has been Section 404 of the Federal Water Pollution Control Act (FWPCA) amendments of 1972 (PL 92-500) (also known as the Clean Water Act). Section 404 required that anyone dredging or filling in "waters of the United States" must request a permit from the U.S. Army Corps of Engineers. This requirement was an extension of the 1899 Rivers and Harbors Act, in which the Corps had responsibility for regulating the dredging and filling of navigable waters.

The use of Section 404 for wetland protection has been controversial and the subject of continued lower and Supreme Court actions and revisions of regulations. The surprising point about the importance of the Clean Water Act in wetland protection is that wetlands are not directly mentioned in Section 404, and at first this directive was interpreted narrowly by the Corps to apply only to navigable waters. The definition of waters of the United States was expanded to include wetlands in two 1974–1975 court decisions, *United States v. Holland* and *Natural Resources Defense Council v. Calloway*. These decisions, along with Executive Order 11990, Protection of Wetlands, put the

Army Corps of Engineers squarely in the center of wetland protection in the United States. On July 25, 1975, the Corps issued revised regulations for the Section 404 program that enunciated the policy of the United States on wetlands:

> As environmentally vital areas, [wetlands] constitute a productive and valuable public resource, the unnecessary alteration or destruction of which should be discouraged as contrary to the public interest.
> —*Federal Register*, July 25, 1975

Wetlands were defined in these regulations to encompass coastal wetlands ("marshes and shallows and ... those areas periodically inundated by saline or brackish waters and that are normally characterized by the prevalence of salt or brackish water vegetation capable of growth and reproduction") and freshwater wetlands ("areas that are periodically inundated and that are normally characterized by the prevalence of vegetation that requires saturated soil conditions for growth and reproduction") (*Federal Register*, July 25, 1975). By these actions, the jurisdiction of the Corps was extended to include 60 million ha of wetlands, 45 percent of which are in Alaska. Several times since 1975, the Corps has issued revised regulations for the dredge-and-fill permit program, and in 1985, the U.S. Supreme Court, in *United States v. Riverside Bayview Homes*, rejected the contention that Congress did not intend to include wetland protection as part of the Clean Water Act.

The procedure for obtaining a "404 permit" for dredge-and-fill activity in wetlands is complex. As a starting point, no discharge of dredged or fill material can be permitted in wetlands if a practicable alternative exists. So in the initial screening of a project that involves potential effects on wetlands, the following three approaches are evaluated in sequence:

1. *Avoidance*. Taking steps to avoid wetland impacts where practicable
2. *Minimization*. Minimizing potential impacts to wetlands
3. *Mitigation*. Providing compensation for any remaining, unavoidable impacts through the restoration or creation of wetlands (see Chapter 18: "Wetland Creation and Restoration")

An individual Section 404 permit is usually required for potentially significant impacts, but for many activities that have minimal adverse effects, the Army Corps of Engineers used to issue general permits. The decision to issue a permit rests with the Corps' district engineer, and it must be based on several considerations, including conservation, economics, aesthetics, and other factors. Assistance to the Corps on the dredge-and-fill permit process in wetland cases is provided by the U.S. EPA, the U.S. Fish and Wildlife Service, the National Marine Fisheries Service, and state agencies. The U.S. EPA has statutory authority to designate wetlands subject to permits and also has veto power on the Corps' decisions. Some states require state permits as well as Corps permits for wetland development. The district engineer, according to

Corps regulations, should not grant a permit if a wetland is identified as performing important functions for the public, such as biological support, wildlife sanctuary, storm protection, flood storage, groundwater recharge, or water purification. An exception is allowed when the district engineer determines "that the benefits of the proposed alteration outweigh the damage to the wetlands resource and the proposed alteration is necessary to realize those benefits" (*Federal Register*, July 19, 1977). The effectiveness of the Section 404 program has varied since the program began and has also varied from district to district.

Swampbuster

Normal agricultural and silvicultural activities were exempted from the Section 404 permit requirements for the first decade of the permit program, thereby still allowing wetland drainage on farms and in commercial forests. Allowing such exemptions created conflict within the federal government: The U.S. Army Corps of Engineers and the U.S. EPA were encouraging wetland conservation through the Clean Water Act, and the Department of Agriculture was encouraging wetland drainage by providing federal subsidies for drainage projects. The conflict ended when Congress passed, as part of the 1985 Food Security Act, "swampbuster" provisions that denied federal subsidies to any farm owner who knowingly converted wetlands to farmland after the act became effective. The swampbuster provisions of the act drew the U.S. Soil Conservation Service (now the Natural Resources Conservation Service, or NRCS) into federal wetland management, primarily as an advisory agency helping farmers identify wetlands on their farms. The NRCS also administers the Wetlands Reserve Program (WRP) that was set up in 1990 to acquire federal easements.

In August 1993, U.S. President Bill Clinton's administration released a document entitled "Protecting America's Wetlands: A Fair, Flexible, and Effective Approach." The document reaffirmed no net loss, established that 21.5 million ha (53 million acres) of previously converted wetlands would not be subject to regulations, and established the NRCS as the lead agency for identifying wetlands on agricultural land under both the Clean Water Act and the Food Security Act swampbuster provisions. The policy was agreed to in a January 6, 1994, memorandum of agreement among the four principal federal agencies involved in wetland policy in the United States (U.S. Fish and Wildlife Service, Natural Resources Conservation Service, U.S. Army Corps of Engineers, and U.S. EPA). Since that time, some of that collaboration has diminished, and the agencies' programs diverged again.

Wetland Delineation

To determine whether a particular piece of land was a wetland and, therefore, if it was necessary to obtain a Section 404 permit to dredge or fill that wetland, federal agencies, beginning with the Army Corps of Engineers, began to develop guidelines for the demarcation of wetland boundaries in a process that came to be called *wetland delineation*. In 1987, the U.S. Army Corps of Engineers (1987) published a technical manual for wetland delineation (*1987 Wetlands Delineation Manual*). This manual specified three mandatory technical criteria—hydrology, soils, and vegetation—for a parcel of land to be declared a wetland [see the next box for details]. Subsequently, the U.S. EPA, the Soil Conservation Service, and the U.S. Fish and Wildlife Service developed separate documents for their respective roles in wetland protection.

After months of political and scientific debate and negotiation among the agencies, a single draft *Federal Manual for Identifying and Delineating Jurisdictional Wetlands* was published by the four federal agencies in August 1989 to unify the government's approach to wetlands. This 1989 manual, while also requiring the three mandatory technical criteria for a parcel of land to be declared a wetland, allowed one criterion to infer the presence of another (e.g., the presence of hydric soils to infer hydrology). The manual also provided some guidance about how to use field indicators such as watermarks on trees or stains on leaves to determine recent flooding, wetland vegetation (from published lists), and hydric soil indicators such as mottling.

The development of a wetland delineation manual that everyone could agree on led to a contentious and quite heady period in U.S. wetland history between 1989 and 1992 (see selected political cartoons in Fig. 15.1), when the 1989 manual (liberal in defining wetlands) and a proposed 1991 manual (conservative in defining wetlands) were introduced in quick succession. The 1991 manual was pushed from the White House through the executive branch U.S. EPA in response to heavy lobbying by developers, agriculturalists, and industrialists for a relaxing of the wetland definitions, in order to lessen the regulatory burden on the private sector. That manual was published for public comment in August 1991 (referred to here as the *1991 Wetlands Delineation Manual*) but was quickly and heavily criticized for its lack of scientific credibility and unworkability (Environmental Defense Fund and World Wildlife Fund, 1992); it was eventually abandoned in 1992, but not before many recommended that the question of a definition of a wetland be turned over to the apolitical National Academy of Sciences (NAS). Results from the eventual NAS study are described later in the section called "National Academy of Sciences Studies."

At present, the 1987 Corps technical manual (U.S. Army Corps of Engineers, 1987), which was generally agreed to be a version ecologically and politically between the "liberal" 1989 manual and the "conservative" 1991 manual, continues to be used as the official way in which wetlands are determined, and there is no reason to believe that this practice will change in the near future. But one major change in the delineation process began in 2009–2010 when, based on recommendations given by the National Research Council (NRC, 1995), regional "addendums" for the wetland delineation manual were developed to better fit to the diverse biomes and ecosystems found throughout the United States. Ten such supplements based on 10 ecological regions (Fig. 15.2) are now all published in their second revisions and are available at www.usace.army.mil/Missions/CivilWorks/RegulatoryProgramandPermits/reg_supp.aspx

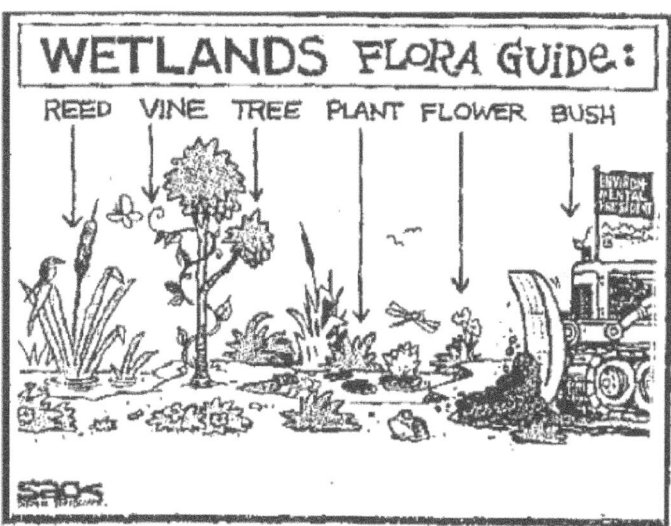

Figure 15.1 Wetland political cartoons were frequent in the early 1990s in the United States when wetland protection and regulation were front-page stories. (Top, by Henry Payne, copyright 1991 by United Media. Bottom, by Steve Sack, copyright by *Minneapolis Star Tribune*)

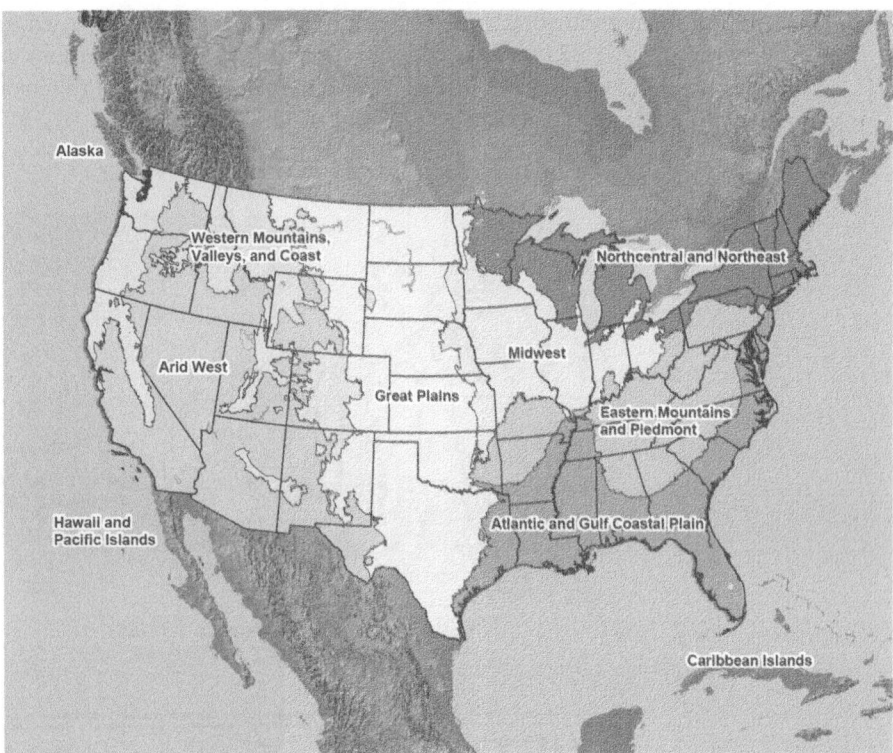

Figure 15.2 Map of the United States listing or showing 10 regions for which addendums to the 1987 wetland delineation manual have been written. (Map and addendums from www.usace.army.mil/Missions/CivilWorks/RegulatoryProgramandPermits/reg_supp.aspx)

Delineating Wetlands in the United States

Guidelines follow the U.S. Army Corps of Engineer's definition of wetlands (*Federal Register*, 1980; *Federal Register*, 1982; See Chapter 2: "Wetland Definitions"): "those areas that are inundated or saturated by surface or ground water [*hydrology*] at a frequency and duration sufficient to support, and that under normal circumstances do support, a prevalence of vegetation [*vegetation*] typically adapted for life in saturated soil conditions [*soil*]" (bracketed words added for emphasis). The definition refers to (1) wetland-adapted vegetation, (2) soil, and (3) flooding or saturating hydrology. Wetland delineation according to the 1987 manual (U.S. Army Corps of Engineers, 1987) depends on determining the boundaries of the area for which these three parameters are met.

Vegetation

Wetland vegetation is defined as macrophytes typically adapted to inundated or saturated conditions. Plants are grouped into five categories (Table 15.2): (1) obligate wetland plants (OBL); (2) facultative wetland plants (FACW); (3) facultative plants (FAC); (4) facultative upland plants (FACU); and (5) obligate upland plants (UPL). To meet the wetland vegetation requirement, more than 50 percent of the dominant species must be OBL, FACW, or FAC. Species lists of plants in these categories are available from several sources (U.S. Army Corps of Engineers, 1987). Other indicators of wetland plants may also be used, including morphological, physiological, and reproductive adaptations, such as buttressed tree trunks, pneumatophores, adventitious roots, and enlarged lenticels. Furthermore, the technical literature may provide additional information about the ability of plants to endure saturated soils.

Table 15.2 Plant indicator status categories used in wetland delineation

Indicator Category	Indicator Symbol	Definition
Obligate wetland plants	OBL	Plants that occur almost always (estimated probability >99%) in wetlands under natural conditions, but that may also occur rarely (estimated probability <1%) in nonwetlands. Examples: *Spartina alterniflora, Taxodium distichum*.
Facultative wetland plants	FACW	Plants that occur usually (estimated probability >67–99%) in wetlands, but also occur (estimated probability 1–33%) in nonwetlands. Examples: *Fraxinus pennsylvanica, Cornus stolonifera*.
Facultative plants	FAC	Plants with a similar likelihood (estimated probability 33–67%) of occurring in both wetlands and nonwetlands. Examples: *Gleditsia triacanthos, Smilax rotundifolia*.
Facultative upland plants	FACU	Plants that occur sometimes (estimated probability 1–<33%) in wetlands, but occur more often (estimated probability >67–99%) in nonwetlands. Examples: *Quercus rubra, Potentilla arguta*.
Obligate upland plants	UPL	Plants that occur rarely (estimated probability <1%) in wetlands, but occur almost always (estimated probability >99%) in nonwetlands under natural conditions. Examples: *Pinus echinata, Bromus mollis*.

Source: U.S. Army Corps of Engineers (1987)

Hydric Soils

A *hydric soil* is a soil that is saturated, flooded, or ponded long enough during the growing season to develop anaerobic conditions that favor the growth and regeneration of hydrophytic vegetation (see Chapter 5: "Wetland Soils"). All *histosols* (organic soils) except folists are hydric. Soils in a few other groups are hydric, particularly *aquic soils* that are poorly drained, are saturated, or have shallow (typically less than 15 cm) water tables for a significant period (usually more than one week) during the growing season In general, the hydric condition of mineral soils is determined by using a Munsell® Soil Color Chart as described in Chapter 5. When a hydric soil is drained, it may not be referred to as hydric, unless the vegetation is hydrophytic and indicators of hydrology support the designation as a hydric soil. Hydric soil designation can be supported by additional indicators (defined in detail in the manual), such as low permeability, appropriate soil chroma, development of mottles, and iron or manganese concretions.

Wetland Hydrology

Areas with evident characteristics of wetland hydrology are those in which the presence of water has an overriding influence on characteristics of vegetation and soils caused by anaerobic and reducing conditions, respectively. Generally, determination of wetland hydrology depends on the frequency, timing, and duration of inundation, or soil saturation, as presented in Table 15.3 for nontidal areas. Zone I is aquatic, and Zone VI is upland. Zones II through IV are wetlands. Zone V may or may not be considered wetland, depending on other indicators. Additional indicators of wetland hydrology use recorded data from stream, lake, or tidal gauges, flood predictions, and historical data on flooding. Visual observations are also indicators, such as soil saturation, watermarks on trees or other structures, drift lines, sediment deposits, and drainage patterns.

Table 15.3 Hydrologic zones for nontidal areas used in hydrology determinations for wetland delineation

Zone	Name	Duration[a]	Comments
I	Permanently inundated	100%	Inundation >2 m mean water depth. Aquatic, not wetlands
II	Semipermanently to nearly permanently inundated or saturated	>75%–<100%	Inundation defined as <2 m mean water depth
III	Regularly inundated or saturated	>25%–75%	
IV	Seasonally inundated or saturated	>12.5%–25%	
V	Irregularly inundated or saturated	≥5%–12.5%	Many areas having these hydrologic characteristics are not wetlands
VI	Intermittently or never inundated or saturated	<5%	Areas with these hydrologic characteristics are not wetlands

[a] Refers to duration of inundation and/or soil saturation during the growing season.
Source: U.S. Army Corps of Engineers, 1987

Delineation Procedure

Routine delineation methods require a combination of office gathering and synthesis of available data on the site, combined with on-site inspection and additional data generation. A flowchart (Fig. 15.3) shows the steps to determine, first, if on-site inspection is necessary and, second, if unnecessary, to determine whether the area is a jurisdictional wetland. Comprehensive delineation methods are reserved for particularly sensitive cases and usually require significant time and effort to obtain the needed quantitative data.

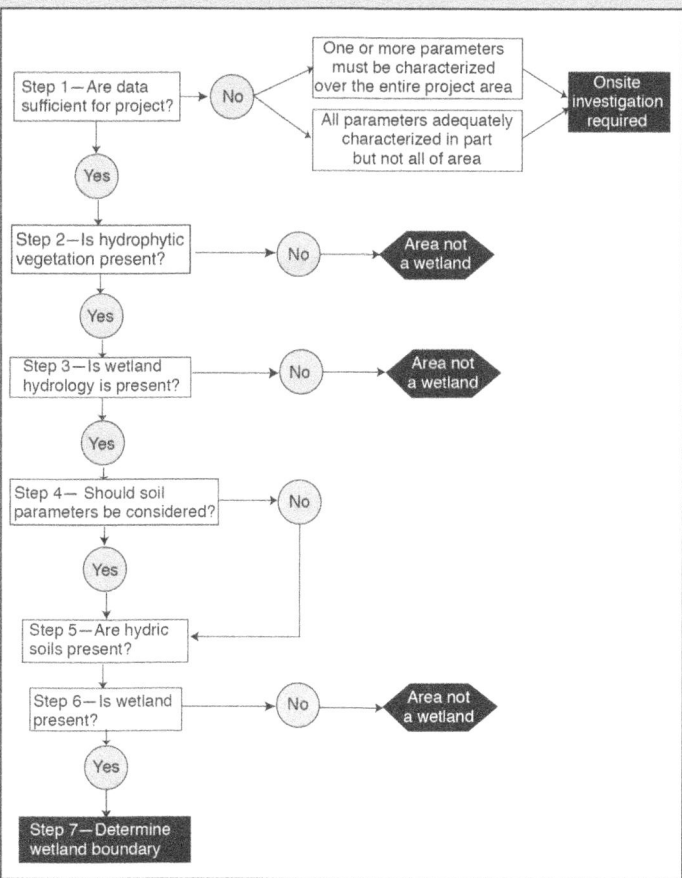

Figure 15.3 Flowchart of steps involved in making a wetland determination when an on-site inspection is unnecessary (U.S. Army Corps of Engineers, 1987)

All methods begin with accumulation of available data on the site to be delineated, data such as U.S. Geological Survey (USGS) quadrangle

> maps, National Wetlands Inventory (NWI) wetland maps, plant surveys, soil surveys, gauge data, environmental assessments or impact statements, remotely sensed data, local expertise, and the applicant's survey plans and engineering designs (often with topographic surveys). These data are synthesized into a preliminary determination of whether the information is adequate to make a wetland delineation of the entire tract in question.
>
> The 1987 manual also details methods, depending on the size of the project area, for on-site evaluation when available data are inadequate. These methods may include, for example, the use of transects when the area is too large to survey in its entirety. The intensity of the on-site investigation is determined by the available information on the site, the type of project anticipated, the ecological sensitivity of the area, and other factors. The objective of on-site investigations is to obtain adequate data to determine whether all or part of the project area fits the criteria for wetlands and, if so, where the wetland boundaries lie.

National Academy of Science Studies

Two notable studies related to wetlands were carried out, at the request of the federal government, by the National Academy of Science's (NAS) operating arm, the National Research Council (NRC), during the 1990s. The NAS is a nongovernmental agency set up in the nineteenth century by Abraham Lincoln to provide scientific reviews of subjects chosen and paid for by the federal government. It has its strengths in its independence from the government and in its ability to recruit scientists and engineers from anywhere in the country for its committees.

The first NRC study dealt with the proper procedures for delineating wetlands, which had become a hot political issue during the three wetland delineation manual period of 1989 to 1992. About that time, many scientists began to call for the National Academy of Science to answer the question: What is a wetland? In April 1993, the U.S. EPA, at the request of the U.S. Congress, asked the NRC to appoint a committee to undertake a scientific review of scientific aspects of wetland characterization. The 17-member committee was selected in the summer of 1993 and met over a two-year period. The committee was charged with considering (1) definition of wetlands; (2) adequacy of science for evaluating hydrologic, biological, and other ways that wetlands function; and (3) regional variation. The report from that committee (NRC, 1995) presented a new definition of wetlands (see Chapter 2) and gave 80 recommendations on topics such as fine-tuning the delineation procedure, dealing with especially controversial wetlands, regionalization, mapping, modeling, administrative issues, and functional assessment of wetlands. The report, in essence, suggested that use of the 1987 manual was appropriate with a few minor modifications. The report was released in early 1995, just as the U.S. Congress was considering two bills on wetlands (House Bill 961 and Senate Bill 851) that would have drastically changed the definitions and

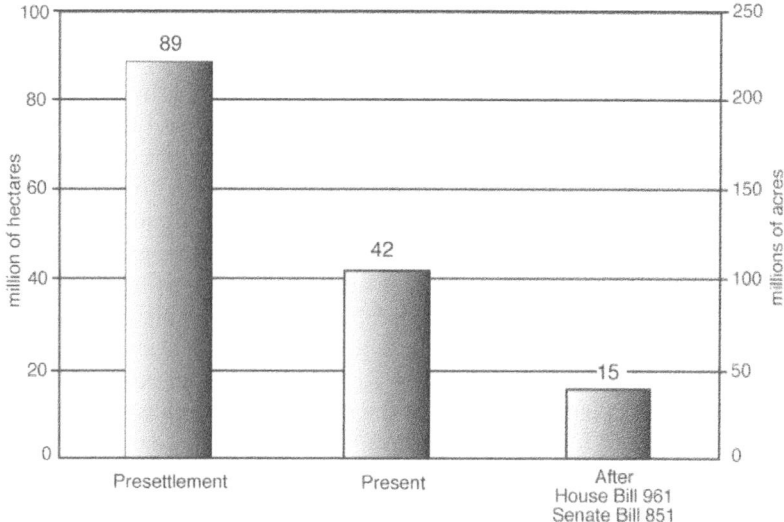

Figure 15.4 Estimated extent of wetlands in the lower 48 states of the United States for presettlement times (1780s) and present day. The numbers in the first two bars, already presented in Chapter 3, are compared with an estimate of the extent of wetlands that would have remained legally protected if House Bill 961 or Senate Bill 851 in the U.S. Congress had been passed in 1995. Each proposed law contained formal definitions of wetlands. These proposed laws would have protected only 11 million to 15 million ha of "legal" wetlands in the United States. Neither law passed, but this potential "loss" of wetlands by redefinition by the U.S. Congress illustrates that wetlands can be lost either by drainage or by legal fiat that redefines wetlands.

management of wetlands in the United States (Fig. 15.4). It may have been a result of the release of the NRC report or just by coincidence, but neither bill became law.

The second NRC study in the late 1990s was in response to questions about whether ecological function was being replaced in wetlands created and restored to mitigate wetland loss in compliance with the no net loss policy described earlier in this chapter. That study report (NRC, 2001) concluded the following:

- The goal of no net loss of wetlands was not being met for wetland functions by the mitigation program, despite progress in the last 20 years;
- A watershed approach would improve permit decision making; and
- Performance expectations in Section 404 permits have often been unclear, and compliance has often not been assured or attained.

The U.S. Army Corps of Engineers, as the lead agency with interest in both wetland delineation and mitigation of wetland loss, responded to both NRC reports by tightening up both delineation procedures and replacement wetland standards. But soon afterward, the Corps' hands would be tied again, at least on defining wetlands, by decisions coming from the U.S. Supreme Court.

Other Federal Activity

Several other federal laws and activities have led to wetland protection since the 1970s. The Coastal Zone Management Program, established by the Coastal Zone Management Act of 1972, has provided up to 80 percent of matching-funds grants to states to develop plans for coastal management based on establishing a high priority to protecting wetlands. The National Flood Insurance Program offers some protection to riparian and coastal wetlands by offering federally subsidized flood insurance to state and local governments that enact local regulations against development in flood-prone areas. The Clean Water Act, in addition to supporting the Section 404 program, supported the U.S. Fish and Wildlife Service to complete its inventory of wetlands of the United States (see Chapter 13: "Wetland Classification"). The Emergency Wetlands Resource Act passed by Congress in 1986 required the U.S. Fish and Wildlife Service to update its report on the status of and trends in wetlands every 10 years. (See Chapter 3 for the conclusions of these reports to date.)

The purpose of the North American Wetlands Conservation Act was to encourage voluntary, public-private partnerships to conserve North American wetland ecosystems. This law, passed in 1989, provides grants, primarily to state agencies and private and public organizations, to manage, restore, or enhance wetland ecosystems to benefit wildlife. From 1991 through mid-1999, almost 650 projects in Canada, Mexico, and the United States were approved for funding. Approximately 3.5 million ha (8.6 million acres) of wetlands and associated uplands were acquired, restored, or enhanced in the United States and Canada. The act also paid for a significant amount of wetland conservation education and management plan projects in Mexico.

The "Takings" Issue

One of the dilemmas of valuing and protecting wetlands is that the values accrue to the public at large but rarely to individual landowners who happen to have a wetland on their property. If government laws that protect wetlands or other natural resources lead to a loss of the use of that land by the private landowner, the restriction on that use has been referred to as a "taking" (denial of an individual's right to use his or her property). Many legal scholars believed that wetland and other land-use laws could result in takings and thus be against the Fifth Amendment of the U.S. Constitution. In a major ruling in June 1992 (*Lucas v. South Carolina Coastal Council*), the U.S. Supreme Court ruled that regulations denying "economically viable use of land" require compensation to the landowner, no matter how great the public interest served by the regulations (Runyon, 1993). This case was referred back to the state of South Carolina to determine if the developer, David Lucas, was denied all economically viable use of his land (beachfront property that was rezoned by South Carolina in response to the 1980 Coastal Zone Management Act). The ultimate result of this Supreme Court decision on wetland legal protection was originally thought to be important but mostly turned out to be inconclusive. The major days for wetlands in the U.S. Supreme Court were yet to come.

U.S. Supreme Court Decisions in the Twenty-first Century

The U.S. Supreme Court has ruled on cases regarding wetland regulations no fewer than three times in the twenty-first century. No other ecosystem has such a distinction, whether this is a dubious honor or otherwise. But it does suggest that wetlands may be becoming as much a legal entity as an ecological entity in the United States.

2001: *SWANCC v. Army Corps of Engineers* (Cook County, Illinois)

In January 2001, the U.S. Supreme Court, in a 5–4 decision in the case *Solid Waste Agency of Northern Cook County (SWANCC) v. U.S. Army Corps of Engineers*, limited the scope of the Corps' Section 404 authority applied to "isolated wetlands." The case was brought forward by SWANCC, a consortium of Chicago suburban municipalities, when it was prohibited from using a 216-ha landfill site that had become a wooded wetland complex with more than 200 permanent and seasonal ponds and wetlands and substantial wildlife, including 121 species of birds. The basic issue brought up by SWANCC was that the wetlands were not specifically connected to interstate streams and should not fall under the authority of the federal government but rather should be the responsibility of the state of Illinois. The Corps of Engineers had denied a permit request from SWANCC for a landfill, partly because the wetland had become the second-largest heron rookery in northeastern Illinois and because the landfill could have an impact on a drinking water aquifer below the site (Downing et al., 2003).

In that case, the Supreme Court also held that the Corps' "migratory bird rule" exceeded its authority. In 1996, the U.S. Army Corps of Engineers adopted a migratory bird rule, which stated that areas that fell under Section 404 jurisdiction as interstate waters included those areas (a) that are or would be used as habitat by birds protected by migratory bird treaties; or (b) that are or would be used as habitat by other migratory birds which cross state lines. Before the Supreme Court disallowed this rule, the Corps was using both water and birds to show that wetlands were related to interstate commerce. The real issue of this court decision was that it reintroduced the connection of wetlands to "navigable waters of the United States" that was the original basis of Section 404 of the Clean Water Act (Downing et al., 2003).

A new term called "significant nexus" to navigable bodies of water entered the general wetland vocabulary as a result of this case. It came into more prominent use with the Supreme Court decision described next.

2006: *Rapanos* and *Carabell* cases (Michigan)

In a second U.S. Supreme Court decision on wetlands in the twenty-first century, the Supreme Court agreed to hear two "waters of the United States" cases from Michigan—*Rapanos v. United States* and *Carabell v. U.S. Army Corps of Engineers*—and ruled on these cases in June 2006. By a 5–4 vote, the Supreme Court continued to question the Corps' regulation of isolated wetlands under the Clean Water Act. The 5–4 vote remanded the case back to the lower courts in Michigan. The ruling has caused more confusion than clarity because it had three dominant opinions. Four justices took a narrow view of interstate wetlands in the Clean Water Act and believed that the act should consider "only those wetlands with a continuous surface connection to [other regulated waters]" (Justice Scalia opinion, *Rapanos v. United States*, 126 S. Ct. 2208, 2006). Four other justices "took a broad view of the Act's jurisdiction, deferring to the Corps' current categorical regulation of all tributaries and their adjacent wetlands" (Murphy, 2006).

The ninth judge, Justice Kennedy, took the middle road, rejecting both of these positions and finding that waters need to have a significant nexus to navigable waters and that this nexus needs to be determined on a case-by-case basis. Justice Kennedy gave the definition of nexus:

> Wetlands possess the requisite nexus, and thus come within the statutory phrase "navigable waters" if the wetlands, either alone or in combination with similarly situated lands in the region, significantly affect the chemical, physical, and biological integrity of other covered waters more readily understood as "navigable." When, in contrast, wetlands' effects on water quality are speculative or insubstantial, they fall outside the one fairly encompassed by the statutory term "navigable waters."
> —**Justice Kennedy opinion, *Rapanos v. United States*, 126 S. Ct. 2208**

Because his was a middle opinion, Justice Kennedy's opinion got the most attention. The overall effect of this decision remains unclear, although "significant nexus" will be the test for many decisions in the future on specific wetland cases. As pointed out in a review of this decision by Murphy (2006), "the Court's decision was, to use a phrase only water attorneys could love, quite turbid."

2013: *Koontz v. St. Johns River Water Management District* (Florida)

The third time in the twenty-first century that the Supreme Court ruled on wetlands was on June 25, 2013, in the case *Koontz v. St. Johns River Water Management District*. Developer Coy Koontz was denied a permit to develop a 6-ha (14.9 acre) site east of Orlando, Florida, in 1972 because of an inadequate mitigation plan. Florida had, at the same time (1972), enacted its Water Resources Act that divided the state into five water management districts. The act required the petitioner to obtain a management and storage of surface water (MSSW) permit. Florida also passed the Henderson Protection Act in 1984, which made it illegal to dredge and fill surface waters without a wetland resource management (WRM) permit. Koontz applied for both the MSSW and WRM permits in 1984 to develop 1.5 ha of the same land while deeding 4.4 ha to the state as a conservation easement. The St. Johns River Water Management District considered the easement to be inadequate and proposed several additional requirements. Koontz disagreed with these additional requirements and filed suit. The Florida District Court agreed with Koontz and reversed the decision, which was then reversed by State Supreme Court in 2011. The case went to U.S. Supreme Court, which took on the case because it touched on federal laws.

The Supreme Court ruled that the St. Johns River Water Management District interfered with the landowner's constitutional rights in its mitigation demands. According to *Florida Times Union* reporter Steve Patterson (June 26, 2013 http://jacksonville.com/news/metro/2013-06-26/story/supreme-court-ruling-unsettles-water-management-districts-wetlands-rule), "the ruling could shift standards nationally about how governments can regulate development, and it was cheered by property-rights advocates." Others have argued that it will make land use planning more difficult and more probable that agencies will just say no to petitions to avoid legal entanglements.

The *Koontz* Supreme Court decision on wetlands was the subject of workshop held at Stetson University College of Law in November 2013. Some of those presentations are published in the *National Wetlands Newsletter* (36, No. 2, March/April 2014). Gardner (2014) summarized 10 takeaways from the decision, including his belief that this may result in less rigorous mitigation requirements required by federal and state agencies, and we should expect more similar wetland litigation. Goldman-Carter (2014) summarized five key results from this Supreme Court case from her perspective:

- The Court acknowledged the state's interest in wetland and floodplain conservation and mitigation.
- The majority now puts the burden on state and local resource managers to prove the "essential nexus" and "rough proportionality" between the impacts of developing in these waters and the permit conditions required to mitigate those impacts.
- Water resource managers must be very careful what they ask for. They must now prove this nexus and proportionality even—as in the *Koontz* case—for possible mitigation conditions they might discuss with developers in trying to negotiate an environmentally responsible development permit.
- After *Koontz*, the prudent course of action for water resource managers may be to *just say no*. Proposing to permittees innovative and flexible mitigation conditions can be a trap, ensnaring state and local governments in costly and wasteful litigation.
- What mustn't happen is for water resource agencies to simply approve development projects in wetlands and floodplains—abandoning their duty to protect the public interest and putting communities and wildlife in harm's way.

International Wetland Conservation

The Ramsar Convention

Intergovernmental cooperation on wetland conservation has been spearheaded by the Convention on Wetlands of International Importance, more commonly referred to as the Ramsar Convention because it was initially adopted at an international conference held in Ramsar, Iran, in 1971. The global treaty provides the framework for the international protection of wetlands as habitats for migratory fauna that do not observe international borders and for the benefit of human populations dependent on wetlands. The convention's mission is "the conservation and wise use of all wetlands through local, regional, and national actions and international cooperation, as a contribution toward achieving sustainable development throughout the world" (www.ramsar.org, 2014). A permanent secretariat headquartered at the International

Union of Conservation of Nature and Natural Resources (IUCN) in Switzerland was established in 1987 to administer the convention, and a budget based on the United Nations scale of contributions was adopted.

The specific obligations of countries that have ratified the Ramsar Convention are the following "three pillars":

1. Member countries shall formulate and implement their planning so as to promote the "wise use" of all wetlands in their territory and develop national wetland policies.
2. Member countries shall designate at least one wetland in their territory for the "List of Wetlands of International Importance." The so-called Ramsar sites should be developed based on their international significance in terms of ecology, botany, zoology, limnology, or hydrology.
3. Member countries shall cooperate over shared species and development assistance affecting wetlands.

Early in the Ramsar process, the emphasis was on the protection of migratory fauna, particularly waterfowl. The importance of wetlands for many other biological functions has been recognized more recently, and currently eight criteria are used to evaluate potential wetland sites for formal designation as "wetlands of international importance" (Table 15.4). Group A sites must meet criterion 1 that they contain representative, rare, or unique wetland types. Group B sites, internationally important for conserving biological diversity, are judged on seven criteria involving questions of rare and endangered communities, biodiversity, habitat for waterfowl, or habitat or food source for indigenous fish species.

As of early 2015, 168 contracting parties have joined the Ramsar Convention, and they have registered 2,186 wetland sites totally almost 209 million ha (2.1×10^6 km^2). (See Appendix B for the Ramsar web address for an update of these numbers). The program is advancing rapidly in international interest (Fig. 15.5). For example, in 1993 there were 582 Ramsar wetland sites, comprising almost 37 million ha in the world, 18 percent of the current total area 22 years later. In 2000, there were 117 member countries with half the current number of sites and area: 1,021 Ramsar wetland sites, totaling 74.8 million ha. In 2006, there were 150 million ha of Ramsar sites in 154 member countries. Overall, the Ramsar program has done a credible job of bringing needed attention to wetland conservation and protection around the world.

North American Waterfowl Management Plan

The United States and Canada, partially as a result of collaboration begun by the Ramsar Convention, established the *North American Waterfowl Management Plan* in 1986 to conserve and restore about 2.4 million ha of waterfowl wetland habitat in Canada and the United States. This treaty was formulated as a partial response to the steep decline in waterfowl in Canada and the United States that had become apparent in the early 1980s (see Chapter 16: "Wetland Ecosystem Services"). This bilateral treaty is jointly administered by the U.S. Fish and Wildlife Service and the Canadian Wildlife Service, but also involves public and private participation by groups such as

Table 15.4 Ramsar Convention criteria for identifying wetlands of international importance

Group A. Sites Containing Representative, Rare, or Unique Wetland Types
Criterion 1   A wetland should be considered internationally important if it contains a representative, rare, or unique example of a natural or near-natural wetland type found within the appropriate biogeographic region.

Group B. Sites of International Importance for Conserving Biological Diversity
Criteria based on species and ecological communities
Criterion 2   A wetland should be considered internationally important if it supports vulnerable, endangered, or critically endangered species or threatened ecological communities.
Criterion 3   A wetland should be considered internationally important if it supports populations of plant and/or animal species important for maintaining the biological diversity of a particular biogeographic region.
Criterion 4   A wetland should be considered internationally important if it supports plant and/or animal species at a critical stage in their life cycles, or provides refuge during adverse conditions.

Specific Criteria Based on Waterbirds
Criterion 5   A wetland should be considered internationally important if it regularly supports 20,000 or more waterbirds.
Criterion 6   A wetland should be considered internationally important if it regularly supports 1 percent of the individuals in a population of one species or subspecies of waterbird.

Specific Criteria Based on Fish
Criterion 7   A wetland should be considered internationally important if it supports a significant proportion of indigenous fish subspecies, species or families, life-history stages, species interactions, and/or populations that are representative of wetland benefits and/or values and thereby contributes to global biological diversity.
Criterion 8   A wetland should be considered internationally important if it is an important source of food for fishes, spawning ground, nursery, and/or migration path on which fish stock, either within the wetland or elsewhere.

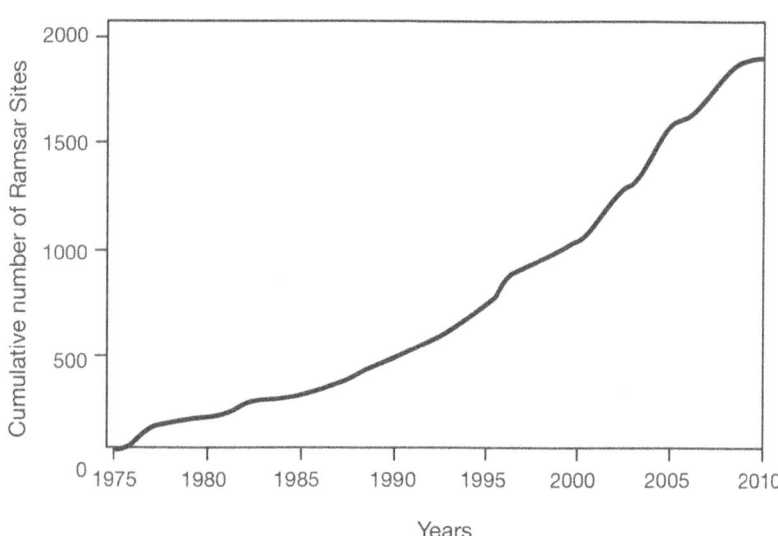

Figure 15.5 Cumulative number of wetlands designated by Ramsar as Wetlands of International Importance from, 1975 to 2010 (From Ramsar Convention, 2011)

Ducks Unlimited. Mexico became a participant in the plan in 1994. To date, the plan has developed joint private-public ventures that have invested $7.5 billion to protect, restore, or enhance about 8.9 million ha of waterfowl habitat, mostly wetlands. Major emphasis has been placed on sites that cross international borders, including the prairie pothole region, the lower Great Lakes–St. Lawrence River basin, and the Middle–Upper Atlantic Coastline.

Recommended Readings

Connolly, K. D., S. M. Johnson, and D. R. Williams. 2005. *Wetlands Law and Policy*. Chicago: American Bar Association.

National Research Council. 1995. *Wetlands: Characteristics and Boundaries*. Washington, DC: National Academy Press.

National Research Council. 2001. *Compensating for Wetland Losses under the Clean Water Act*. Washington, DC: National Academy Press.

Ramsar Convention on Wetlands Web site: www.ramsar.org

References

Downing, D. M., C. Winer, and L. D. Wood, 2003. Navigating through Clean Water Act Jurisdiction: A Legal Review. *Wetlands* 23: 475–493.

Environmental Defense Fund and World Wildlife Fund. 1992. *How Wet Is a Wetland? The Impact of the Proposed Revisions to the Federal Wetlands Delineation Manual*. Environmental Defense Fund and World Wildlife Fund, Washington, DC. 175 pp.

Gardner, R. C. 2014. Contemplating Koontz: Ten takeaways. *National Wetlands Newsletter* 36(2): 9–11.

Goldman-Carter, J. 2014. U.S. Supreme Court to water resource managers: Be careful what you ask for. *National Wetlands Newsletter* 36(2): 12–14.

Murphy, J. E. 2006. *Rapanos v. United States*: Wading through murky waters. *National Wetlands Newsletter* 28(5): 1.

National Research Council. 1995. *Wetlands: Characteristics and Boundaries*. National Academy Press, Washington, DC. 306 pp.

National Research Council. 2001. *Compensating for Wetland Losses under the Clean Water Act*. National Academy Press, Washington, DC. 158 pp.

National Wetlands Policy Forum. 1988. *Protecting America's Wetlands: An Action Agenda*. Conservation Foundation, Washington, DC. 69 pp.

Ramsar Convention. 2011. Ramsar's Liquid Assets, 40 years of the Convention on Wetlands. Ramsar Wetland Convention, Gland, Switzerland, 36 pp.

Runyon, L. C. 1993. *The Lucas Court Case and Land-Use Planning*. National Conference of State Legislators, Denver, CO, Supplement to State Legislatures, vol. 1, no. 10 (March).

U.S. Army Corps of Engineers. 1987. Corps of Engineers Wetlands Delineation manual. Technical Report Y-87-1. U.S. Army Corps of Engineers Waterways Experiment Station, Vicksburg, MS. 100 pp. and appendices.

Part V

Ecosystem Services

Chapter 16

Wetland Ecosystem Services

Wetlands provide many services and commodities to humanity. Using the Millennium Ecosystem Assessment terminology, provisioning services include harvesting of wetland-dependent fish, shellfish, fur-bearing animals, waterfowl, timber, and peat. Regulating ecosystem services from wetlands include moderating the effects of floods, improve water quality, protect coastlines from storms, hurricanes, and tsunamis, climate regulation, and aquifer recharge. Cultural ecosystem services have aesthetic and heritage values and subsistence of ancient and sustainable cultures. Valuation techniques include nonmonetary scaling and weighting approaches for comparing different wetlands or different management options for the same wetland and common-denominator approaches that reduce the various values to some common term, such as dollars, embodied energy, or emergy. These common-denominator methodologies can include willingness to pay, replacement value, energy analysis, and emergy analysis. None of these approaches is without problems, and no universal agreement about their use has been reached. But when compared to other ecosystems or uses of the landscape, sustainable ecosystem services of wetlands are often among the highest of any ecosystems.

The terms *value* and *services* impose an anthropocentric orientation on a discussion of wetlands. In ordinary parlance, the words connote something worthy, desirable, or useful to humans. The reasons that wetlands are often legally protected have to do with their value to society, not with the abstruse ecological processes that occur in wetlands; this is the sense in which the words *value* and *services* are used in this chapter. Perceived values arise from the functional ecological processes described in previous chapters but are determined also by human perceptions, the location of a particular wetland, the human population pressures on it, and the extent of the resource.

Regional wetlands are integral parts of larger landscapes—drainage basins and/or estuaries. Their functions and their values to people in these landscapes depend on both their extent and their location. Thus, the value of a forested wetland varies. If it lies along a river, it probably plays a greater functional role in stream water quality and downstream flooding than if it was isolated from the stream. If situated at the headwaters of a stream, a wetland functions differently from a wetland located near the stream's mouth. The fauna it supports depend on the size of the wetland relative to the home range of the animal. Thus, to some extent, each wetland is ecologically unique. This complicates the measurement of its value.

Wetland Ecosystem Services

A series of publications in the mid-2000s referred to as the Millennium Ecosystem Assessment (2005) brought focus to the importance of natural ecosystems by pointing out four main findings with regard to humans and the planet's ecosystems:

1. Humans have changed the planets ecosystems over the last 50 years of the twentieth century more than any comparable period in human history.
2. The changes that have been made to ecosystems have contributed to substantial gains in well being and economic development, but at the cost of losing many ecosystem services.
3. The degradation is expected to grow significantly worse in the first half of the twenty-first century.
4. Reversing this degradation of ecosystems will involve significant changes in policies, practices, and institutions that are not yet in place.

The term *ecosystem services* replaced *ecosystem values*, a concept used in earlier editions of this textbook since its first edition in 1986. In those previous editions, we divided the values of wetlands into three hierarchical levels familiar to ecologists: population, ecosystem, and global. Population values include those related to providing habitats for animals harvested for pelts, waterfowl and other hunted and watched birds, fish and shellfish production, timber and peat harvesting, and support of endangered and threatened species. Ecosystem values included water quality improvement, storm and flood mitigation, aquifer recharge, and even sustenance of human cultures. Global values included maintaining water and air quality influences on a much broader scale than that of the ecosystem level, especially in regional and global cycles of nitrogen, sulfur, and carbon.

The current paradigm of ecosystem services as organized by the Millennium Ecosystem Assessment (2005) also divided ecosystem services into three categories but with a division related to human well-being (Fig. 16.1).

1. *Provisioning ecosystem services* include products obtained from ecosystems, such as food, water, timber, fiber, or genetic resources.

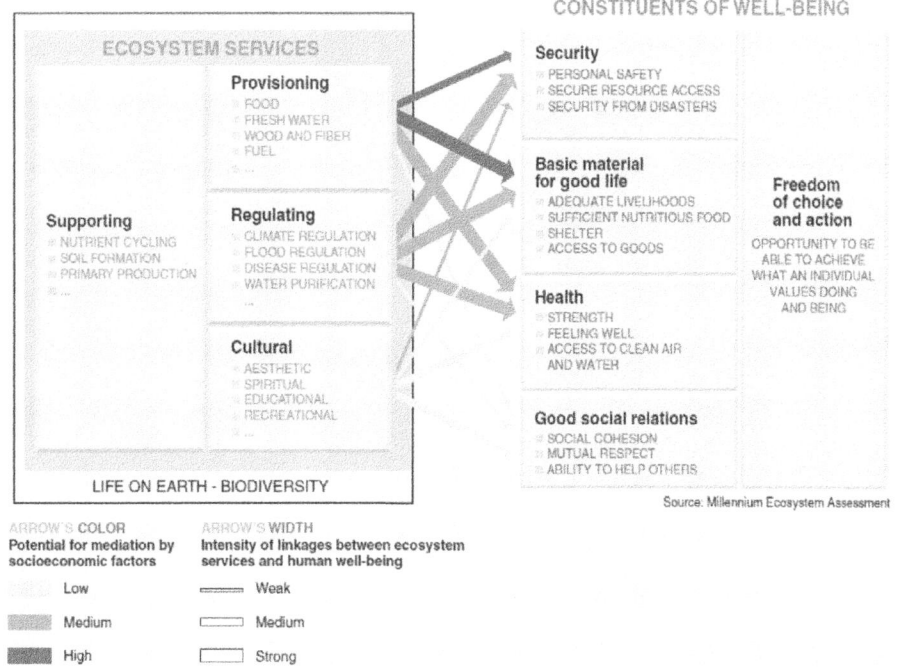

Figure 16.1 Illustration of provisioning, regulating, and cultural ecosystem services and their connection to human well-being. (From Millennium Ecosystem Assessment, 2005, copyright World Resources Institute, reprinted with permission)

2. *Regulating ecosystem services* include air quality regulation, climate regulation, water purification, disease regulation, pest regulation, pollination, and natural hazard regulation.
3. *Cultural ecosystem services* include benefits that people obtain from ecosystems related to spiritual enrichment, recreation, ecotourism, aesthetics, formal and informal education, inspiration, and cultural heritage.

This is the system we use in this current edition of *Wetlands* to describe wetland ecosystem services.

Provisioning Ecosystem Services

Animals Harvested for Pelts

Fur-bearing mammals, and even alligators and crocodiles, are harvested for their pelts throughout the world. In contrast to most other commercially important wetland species, these animals typically have a limited range and spend their lives within a short distance of their birthplaces. The most abundant fur-bearer historically

Figure 16.2 Three fur-bearing animals found in wetlands that have been historically harvested for their pelts: (a) muskrat (*Ondatra zibethicus*), (b) nutria (*Myocastor coypus*), (c) beaver (*Castor canadensis*).

harvested in wetlands in the United States is the muskrat (*Ondatra zibethicus*). Muskrats (Fig. 16.2a) are found in wetlands throughout the United States except, strangely, the south Atlantic Coast. They prefer fresh inland marshes but along the northern Gulf Coast are more abundant in brackish marshes. About 50 percent of the nation's harvest is from the Midwest and 25 percent from along the northern Gulf of Mexico, mostly Louisiana. The nutria (*Myocastor coypus*), an ecological analog of the muskrat, is the next most abundant species. It is very much like a muskrat but is larger and more vigorous (Fig. 16.2b). This species was imported from South America to Louisiana and escaped from captivity in 1938, spreading rapidly through the state's coastal marshes. In the 1940s, the animal was promoted by state agents for controlling aquatic weeds, particularly water hyacinth (*Eichhornia crassipes*). It is now abundant in freshwater swamps and in coastal freshwater marshes, from which it may have displaced muskrats to more brackish locations, and is spreading up the coastal Atlantic states well beyond Louisiana. In order of decreasing abundance in the United States, other harvested fur animals are beaver, mink, and otter. Beavers (*Castor canadensis*; Fig. 16.2c) once had a population of 60 million in North America, and were associated with a major colonial beaver-trapping industry, especially for

European women's fur and men's and women's hats and maintained largely by the French in Canada and what is now the midwestern United States. Nowadays beavers are associated with forested wetlands, especially in the Midwest. Minnesota harvests a high percentage of the nation's beaver catch in the United States. Beavers are also now harvested for castoreum—a fungus exudate from the castor sacs of mature beavers that is used for perfume, medicine, and a food additive.

Waterfowl and Other Birds

Birds, as our only remaining evolutionary link to the dinosaurs that once roamed Earth, may have survived the dinosaur die-off precisely because of wetlands (Gibbons, 1997; Weller, 1999). Although not all current bird species require wetlands as their primary habitat, a great many do, and several are synonymous with wetlands around the world (Figs. 16.3 and 16.4). Eighty percent of America's breeding bird population and more than 50 percent of the 800 species of protected migratory birds rely on wetlands. Wetlands, which are probably known best for their waterfowl abundance, also support a large and valuable recreational hunting industry. We use the term *industry* because hunters spend large sums of money in the local economy for guns, ammunition, hunting clothes, travel to hunting spots, food, and lodging.

Most of the birds hunted are hatched in marshes in the far North, sometimes above the Arctic Circle, but are shot during their winter migrations to the southern United States and Central America. There are exceptions—the wood duck (*Aix sponsa*) breeds locally throughout the continent—but the generalization holds for most species. Different groups of geese and ducks have different habitat preferences, and these preferences change with the maturity of the duck and the season.

A broad diversity of wetland habitat types is important for waterfowl success. The freshwater prairie potholes of North America are the primary breeding place for waterfowl in North America. There, an estimated 50 to 80 percent of the continent's main game species are produced. Wood ducks prefer forested wetlands. During the winter, diving ducks (*Aythya* spp. and *Oxyura* spp.) are found in brackish marshes, preferably adjacent to fairly deep ponds and lakes. Dabbling ducks (*Anas* spp.) prefer freshwater marshes and often graze heavily in adjacent rice fields and in very shallow marsh ponds. Gadwalls (*Anas strepera*) like shallow ponds with submerged vegetation.

The waterfowl value of wetlands such as the prairie pothole region of North America (see Chapter 3: "Wetlands of the World") is unmistakable. When waterfowl census data for the prairie pothole region over the 30-year period were compared to the number of potholes flooded in May of each year, there was a clear positive correlation, indicating the importance of wetland hydrology in the breeding success of waterfowl. On average, there are almost 22 million waterfowl (dabbling and diving ducks) in the region, dominated by the mallard. Generally, the duck population of North America has shown a 10- to 20-year cycle of increase and decline, with low points in the early 1960s and 1990s and highs in the mid-1950s, mid-1970s, and late 1990s (Table 16.1). Populations of 9 of the 10 duck species listed in Table 16.1 were lower than historical averages after the dry years 1987 to 1991, while populations of 7 of the same 10 duck species were higher than historical averages after the

(a)

(b)

Figure 16.3 Two wetland waterfowl known around the world: (a) Mallard (*Anas platyrhynchos*) and (b) Canada goose (*Branta canadensis*). (Photos courtesy of Alan and Elaine Wilson)

wet years 1995 to 1998. Over that period, from dry period to wet period, the total number of ducks increased by 60 percent. The trends of below-average populations during dry periods and above-average populations during wet periods are particularly apparent for mallards, green-winged and blue-winged teals, northern shovelers, and canvasback. Climatic changes that influence the number of ponds from year to year in the breeding grounds appear to be the major cause of year-to-year fluctuations.

Figure 16.4 Herons are consummate symbols of wetlands throughout the world. Different species that dominate this wading niche in parts of the world include: (a) great blue heron (*Ardea herodias*) from North America; (b) white-necked heron (*Ardea cocoi*) from South America; (c) black-headed heron (*Ardea melanocephala*) from eastern Africa; (d) white-faced heron (*Ardea novaehollandiae*) from Australia/New Zealand; (e) gray heron (*Ardea cinerea*) from Europe and Africa. (Photograph (a) by T. Daniel, Ohio Department of Natural Resources; (b), (c) by W. J. Mitsch; (d) by B. Harcourt, courtesy of New Zealand Department of Conservation; (e) by P. Marion; reprinted by permission)

Table 16.1 Population estimates of the 10 most common species of breeding ducks and 4 species of goose in North America for a dry year (1991) and a wet year (1998) in the prairie pothole region, with percentage change in 1991 and 1998 compared to 1995–1990 and 1955–1997 averages, respectively

Species	Population (×1,000)		Percentage Change	
	1991 (Dry Year)	1998 (Wet Year)	1991[a]	1998[b]
All species	24,200	39,100		20
Mallard (*Anas platyrhynchos*)	5,353 ± 188	9,640 ± 302	−27	+32
Gadwall (*Anas strepera*)	1,573 ± 94	3,742 ± 206	+22	+149
American wigeon (*Anas americana*)	2,328 ± 135	2,858 ± 145	−14	5
Green-winged teal (*Anas crecca*)	1,601 ± 88	2,087 ± 139	−4	+16
Blue-winged teal (*Anas discors*)	3,779 ± 245	6,399 ± 332	−10	+36
Northern shoveler (*Anas clypeata*)	1,663 ± 84	4,120 ± 194	−8	+106
Northern pintail (*Anas acuta*)	1,794 ± 199	3,558 ± 194	−62	−36
Redhead (*Aythya americana*)	437 ± 37	918 ± 77	−26	+48
Canvasback (*Aythya valisneria*)	463 ± 57	689 ± 57	−16	+28
Scaup (*Aythya* spp.)	5,247 ± 333	4,122 ± 234	−7	−35
Average of 10 duck species			−15	+34
Canada goose (*Branta canadensis*)	3,750	4,683		
Snow goose (*Chen caerulsecens*)	2,440	3,776		
White-fronted goose (*Anser albifrons*)	492	941		
Brant (*Branta bernicla*)	275	276		

[a] Compared to average for 1955 to 1990.
[b] Compared to average for 1955 to 1997.
Source: U.S. Fish and Wildlife Service. Duck surveys on summer breeding grounds; goose surveys during summer, fall, and winter.

Fish and Shellfish

A direct relationship between shrimp and fish harvests and wetland area has been illustrated for many fisheries around the world, including marine, freshwater, and pond raised (Fig. 16.5). Over 95 percent of the fish and shellfish species that are harvested commercially in the United States are wetland dependent (Feierabend and Zelazny, 1987). The degree of dependence on wetlands varies widely with species and with the type of wetland. Some important species are permanent residents; others are merely transients that feed in wetlands when the opportunity arises. Some shallow wetlands, which may exhibit several other wetland values, may be virtually devoid of fish, whereas other types of deepwater and coastal wetlands may serve as important nursery and feeding areas.

Virtually all of the freshwater species are dependent, to some degree, on wetlands, often spawning in marshes bordering lakes or in riparian forests during spring flooding. These species are primarily recreational, although some small local commercial fisheries exploit them. The saltwater species tend to spawn offshore, move into the coastal marsh "nursery" during their juvenile stages, and then emigrate offshore as they mature. They are often important for both commercial and recreational fisheries. The menhaden is caught only commercially, but competition between commercial and

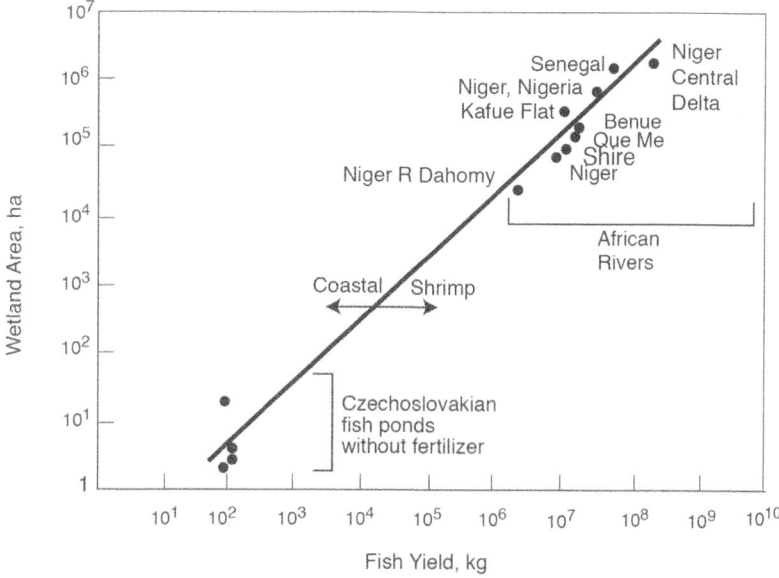

Figure 16.5 Relationship between wetland area and fish harvests. The linear slope describes the line of about 60 kg/ha yield. (After Turner, 1982)

sport fishermen for shrimp, blue crab, oyster, catfish, sea trout, and striped bass can be intensive and acrimonious. Anadromous fish probably use wetlands less than the other two groups. However, young anadromous fish fry sometimes linger in estuaries and adjacent marshes on their migrations to the ocean from the freshwater streams in which they were spawned.

Analyses of fishery harvests from wetlands show the importance of recreational fishing. Although the commercial harvest is usually much better documented, several studies have shown that the recreational catch far outweighs the commercial catch for certain species. Furthermore, the value to the economy of recreational fishing is usually far greater than the value of the commercial catch, because sports fishermen spend more money per fish caught (they are less efficient) than their commercial counterparts.

Timber and Other Vegetation Harvest

Wetlands often provide an abundance of building materials and foodstuffs for local economies. Timber from forested wetlands was one of the staples of the economy of southeastern United States. The antebellum homes of the South were often supported by giant trusses of cypress trees harvested from nearby swamps. The Mississippi River alluvial floodplain and the floodplains of rivers entering the South Atlantic are mostly deciduous wetlands, whereas the forested wetlands along the northern tier of states are primarily evergreen. The former are more extensive and potentially more valuable commercially because of the much faster growth rates in the South.

In addition to the timber harvest, the production of herbaceous vegetation in marshes is a potential source of energy, fiber, and other commodities. These prospects have not been explored widely in North America but are viable options elsewhere. For example, many commercial products are harvested from restored and natural salt marshes and freshwater marshes in China. The productivity of many wetland species (e.g., *Spartina alterniflora* (salt marsh cordgrass), *Phragmites australis* (reed grass), *Typha* spp. (cattail), *Eichhornia crassipes* (water hyacinth), *Cyperus papyrus* (papyrus) is as great as our most vigorous agricultural crops.

Peat Harvesting

In addition to the annual production of living vegetation in wetlands, great reservoirs of buried peat exist around the world. Peat harvesting was described in detail in Chapter 14: "Human Impacts and Management of Wetlands." This buried peat is a nonrenewable energy source that destroys the wetland habitat when it is mined. In the United States and Canada, peat is mined primarily for horticultural peat production, but in other parts of the world—for example, several republics of the former Soviet Union and in Finland—it has been used as a fuel source for hundreds of years. It is used to generate electricity, formed into briquettes for home use, and gasified or liquefied to produce methanol and industrial fuels.

Endangered and Threatened Species

Wetland habitats are necessary for the survival of a disproportionately high percentage of endangered and threatened species. Table 16.2 summarizes the statistics but imparts no information about the particular species involved, their location, wetland habitat requirements, degree of wetland dependence, and factors contributing to their demise. Although wetlands occupy only about 3.5 percent of the land area of the United States, of the more than 200 animal species listed as endangered, about 50 percent depend on wetlands for survival and viability. Almost one-third of native North American

Table 16.2 Threatened and endangered species associated with wetlands

Taxon	Number of Species Endangered	Number of Species Threatened	Percentage of U.S. Total Threatened or Endangered
Plants	17	12	28
Mammals	7	—	20
Birds	16	1	68
Reptiles	6	1	63
Amphibians	5	1	75
Mussels	20	—	66
Fish	26	6	48
Insects	1	4	38
Total	98	25	

Source: Niering (1988)

freshwater fish species are endangered, threatened, or of special concern. Almost all of these were adversely affected by habitat loss. Sixty-three species of plants and 34 species of animals that are considered endangered, threatened, or candidates for listing occupy southern U.S. forested wetlands. Of these, amphibians and many reptiles are especially linked to wetlands. In Florida, where the number of amphibian and reptile species is about equal to the number of mammal and breeding bird species, 18 percent of all amphibians and 35 percent of all reptiles are considered threatened or endangered or their status is unknown (Harris and Gosselink, 1990).

The fate of one wetland-dependent endangered species is discussed here to illustrate the ecological complexity of species endangerment and also hope of recovery of endangered species. Whooping cranes (*Grus americana*) nest in wetlands in the Northwest Territories of Canada, in water 0.3 to 0.6 m deep, during the spring and summer. In the fall, they migrate to the Aransas National Wildlife Refuge, Texas, stopping off in riverine marshes along the migration route. In Texas, they winter in tidal marshes. All three types of wetlands are important for their survival. The decline in the once-abundant species has been attributed both to hunting and to habitat loss. The last whooping crane nest in the United States was seen in 1889. In 1941, the flock consisted of 13 adults and 2 young. Since then, the flock has been gradually built up to about 600 birds wild and captive.

The American Alligator: From Endangered to Plentiful

The American alligator (*Alligator mississippiensis*; Fig. 16.6) represents a dramatic success story of the return from the edge of extinction to a healthy U.S. population. Alligators are abundant in fresh and slightly brackish lakes and streams and build nests in adjacent marshes and swamps in the southwestern United States, especially in Florida and Louisiana. Alligators have an interesting role in wetlands—they depend on them, and, in return, the character of the wetland is shaped by the alligator, at least in the south Florida Everglades. They are another example of an ecosystem engineer (See Chapter 7: "Wetland Vegetation and Succession." As the annual dry season approaches, alligators dig "gator holes." The material thrown out around the holes forms a berm high enough to support trees and shrubs in an otherwise treeless prairie. The trees provide cover and breeding grounds for insects, birds, turtles, and snakes. The hole is a place where the alligator can wait out the dry period until the winter rains. It also provides a refuge for dense populations of fish and shellfish (up to $1,600/m^2$). These organisms, in turn, attract top carnivores, and so the gator holes are sites of concentrated biological activity that may be important for the survival of many species.

American alligator populations were reduced by hunters and poachers to such low levels that the species was declared endangered in the 1970s.

Figure 16.6 The American alligator (*Alligator mississippiensis*) in Corkscrew Sanctuary, Naples, Florida. (Photograph by W. J. Mitsch)

The species was threatened by severe hunting pressure, not by habitat loss. When that pressure was removed, its numbers increased rapidly. The animal is now harvested under close regulation and grown commercially in both Louisiana and Florida. About 250,000 alligators are harvested in the wild and in farms annually in Louisiana, yet the population remains constant or is slightly increasing. Alligator hunting and farming in Louisiana has increased dramatically; it was worth $16 million in 1992 and $26 million in 2004 for both wild and farm-raised animals.

In Florida, where limited hunting is permitted, the harvest in the wild and on farms is considerably less than that in Louisiana, but the compatibility of alligators and a rapidly increasing human population is constantly being challenged. It is probably extraordinary that there have been fewer than 20 confirmed fatal alligator attacks on humans per recent 50 years in Florida, given the high number of both alligators and people in the state.

In addition to the harvest of alligators for their meat, alligator skins from both Florida and Louisiana are sold worldwide, particularly for high-end luxury handbags, wallets, belts, and boots. Apparently with the increased interest in wetlands and wildlife, the fashion world has gone reptile chic.

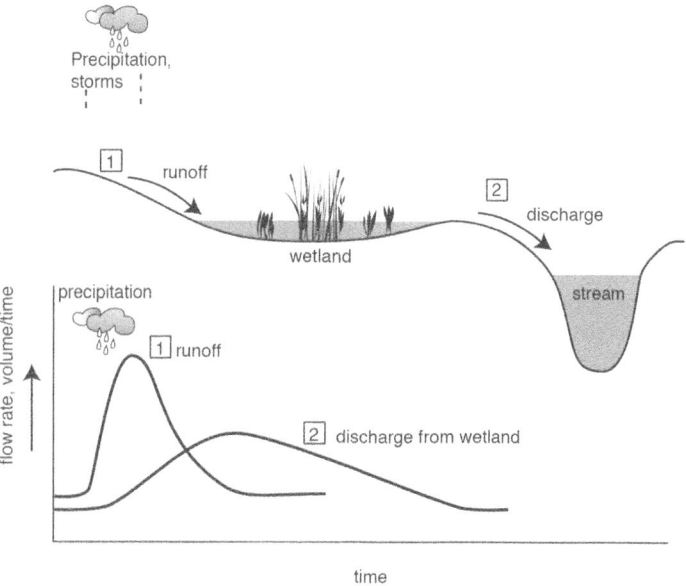

Figure 16.7 The general effect of wetlands on streamflow and stormwater runoff.

Regulating Ecosystem Services

Flood Mitigation

Chapter 4 dealt with the importance of hydrology in determining the character of wetlands. In addition, wetlands influence regional hydrology. One way they do this is to intercept storm runoff and to store storm waters, thereby changing sharp runoff peaks to slower discharges over longer periods of time (Fig. 16.7). Because it is usually the peak flows that produce flood damage, the effect of the wetland area is to reduce the danger of flooding. Riverine wetlands are especially valuable in this regard. In a classic study on the Charles River in Massachusetts, the floodplain wetlands were deemed so effective for flood control by the U.S. Army Corps of Engineers that it purchased them rather than build expensive flood control structures to protect Boston (U.S. Army Corps of Engineers, 1972). The study on which the Corps' decision was based demonstrated that if the 3,400 ha of wetlands in the Charles River basin were drained and leveed off from the river, flood damages would increase by $17 million per year.

Bottomland hardwood forests along the Mississippi River before European settlement stored floodwater equivalent to about 60 days of river discharge. Storage capacity has been reduced to only about 12 days as a result of leveeing the river and draining the floodplain. The consequences—the confinement of the river to a narrow channel and the loss of storage capacity—are major reasons that flooding is increasing along the lower Mississippi River.

Novitzki (1985) analyzed the relationship between flood peaks and the percentage of basin area in lakes and wetlands. In the Chesapeake Bay drainage basin, where the wetland area was 4 percent, flood flow was only about 50 percent of that in basins containing no wetland storage. However, in Wisconsin river basins that contained 40 percent lakes and wetlands, spring streamflow was as much as 140 percent of that in basins that do not contain storage. This apparent anomaly is probably related to a reduction in the proportion of precipitation that can infiltrate the soil and to a lack of additional storage capacity in lakes and wetlands that are already at full capacity during spring floods. Thus, the location of wetlands in the river basin can complicate the response downstream. For example, detained water in a downstream wetland of one tributary can combine with flows from another tributary to increase the flood peak rather than to desynchronize flows.

Ogawa and Male (1983, 1986) used a hydrologic simulation model to investigate the relationship between upstream wetland removal and downstream flooding. Their study found that for rare floods—that is, those predicted to occur only once in 100 or more years—the increase in peak stream flow was significant for all sizes of streams when wetlands were removed. The authors concluded that the usefulness of wetlands in reducing downstream flooding increases with (1) an increase in wetland area, (2) the distance that the wetland is downstream, (3) the size of the flood, (4) the closeness to an upstream wetland, and (5) the lack of other upstream storage areas such as reservoirs.

Storm Abatement and Coastal Protection

Coastal wetlands absorb the first fury of ocean storms as they come ashore (Fig. 16.8). Salt marshes and mangrove wetlands act as giant storm buffers (Barbier et al., 2013; Das and Crépin, 2013; Marois and Mitsch, 2015). This value can be seen in the context of wetland conservation versus development. Natural marshes and mangrove

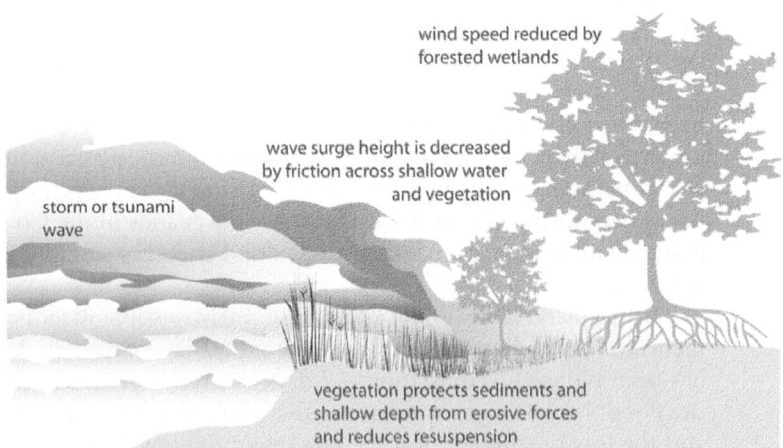

Figure 16.8 The general protection that coastal wetlands provide to buffer coastlines from tidal surges caused by hurricanes, typhoons, and tsunamis.

forests, which sustain little permanent damage from these storms, can shelter inland developed areas. Buildings and other structures on the coast are vulnerable to storms, and hurricane and typhoon damage in the world is increasing almost every year. Inevitably, the public pays much of the cost of this damage through taxes for public assistance, rebuilding public services such as roads and utilities, and federally guaranteed insurance. Two coastal disasters in the first decade of the twenty-first century poignantly illustrate in hindsight the value of coastal wetlands for coastal protection (see boxes). In both of these cases, as the memory of the disaster fades, there will be the tendency to go back to the ways things were done in the past.

Mangrove Swamps and the Indian Ocean Tsunami of December 2004

On December 26, 2004, an earthquake-caused tsunami produced unprecedented damage and loss of life (estimated that 230,000 were killed or missing) around the entire Indian Ocean. The earthquake activity center was off the west coast of Sumatra, Indonesia, and so the greatest devastation occurred in that region. This Boxing Day Tsunami has been called "one of the deadliest natural disasters in recorded history" (http://en.wikipedia.org/wiki/2004 Indian Ocean earthquake).

While no coastal defense system is capable of buffering areas that were hit with a 10-m-high wall of water, it is clear that the destruction of mangrove wetlands for shrimp farms and tourist meccas and the habitation of these areas by humans are at least partially responsible for the carnage. The mangrove swamps suffered significant temporary destruction as well, but they have evolved to survive a violent seascape and certainly restored themselves. The same cannot be said for human settlements that were built in areas of former mangrove swamps.

One year prior to the Indian Ocean tsunami event, simulation models had illustrated that a wide (100 m) belt of dense mangrove trees (referred to as a "greenbelt") could reduce a tsunami pressure flow by more than 90 percent (Hiraishi and Harada, 2003). That information was not made public quickly, and the Indian Ocean tsunami happened with little to no warning. In the five countries hit hardest by the tsunami, at least 1.5 million ha of mangrove wetlands, or 26 percent of the mangrove cover, were destroyed between 1980 and 2000 (FAO, 2003; Check, 2005).

The protective role that mangrove wetlands provided during the Indian Ocean tsunami was illustrated in hindsight for a region along the southeast coastline in Tamil Nadu, India (Danielsen et al., 2005). In an area without mangroves and coastal *Casuarina* plantations, a sand spit was totally removed and parts of the local village was destroyed; there were "significantly less damaged" areas where mangroves and plantations were present. Danielsen et al. (2005) concluded that "conserving or replanting coastal mangroves and greenbelts should buffer communities from future tsunami events." There is hope that such a tsunami disaster will never occur again, but conserving and restoring mangrove swamps for coastal protection now has the attention of all tropical and subtropical countries that face open oceans (See also Case Study 6 in Chapter 18: "Wetland Creation and Restoration.")

Hurricane Katrina of 2005 and New Orleans, Wetland Wet Suit

Hurricane Katrina struck the Louisiana coastland and the city of New Orleans, Louisiana, in late August 2005 with devastating results to lives and property (Fig. 16.9). One of the reasons for the extensive destruction is the fact

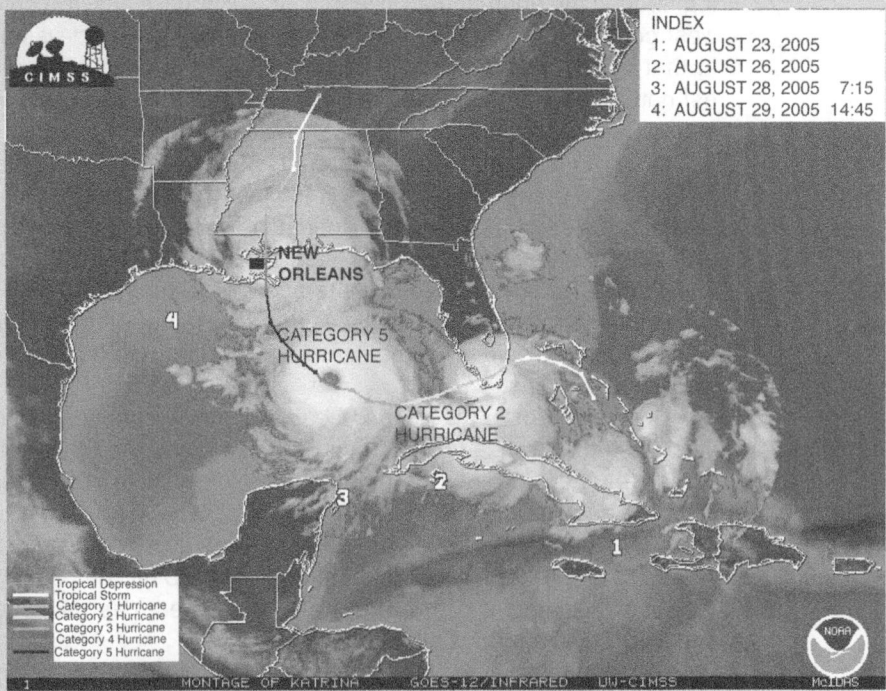

Figure 16.9 Path of Hurricane Katrina across Florida, the Gulf of Mexico, Louisiana, and Mississippi in August 2005. The hurricane crossed to the east of New Orleans on August 29, 2005, and a tidal wave caused by the hurricane caused extensive damage and loss of life in New Orleans and surrounding parishes. A more robust system of coastal wetlands and barrier beaches, many of which have been lost in the last century, would have provided more protection for the city. (From National Oceanic and Atmospheric Administration)

that New Orleans and Louisiana are losing their deltaic wetlands due to land subsidence caused by natural and human effects. Studies over 50 years in Louisiana led to the conclusion that "New Orleans was becoming a more vulnerable city with each passing year" (Costanza et al., 2006). The formerly extensive salt marshes and other wetlands that used to surround New Orleans could have provided some coastal protection from the 6-m storm surge that overwhelmed the city's levee system during Hurricane Katrina. But the wetlands have been lost at a rate of 65 km^2 per year since the beginning of the 20th century, after 6,000 years of gradual land building. Almost 4,800 km^2 of coastal wetlands have been lost since the 1930s alone (Day et al., 2005).

Since marsh plants hold and accrete sediments (Cahoon et al., 1995), often reduce sediment resuspension (Harter and Mitsch, 2003), and consequently maintain shallow water depths, the presence of vegetation contributes in two ways: (1) by actually decreasing surges and waves, and (2) by maintaining the shallow depths that also accomplish the same. Because wetlands indicate shallow water, the presence of wetland vegetation is also an "indicator" of the degree to which New Orleans and other human settlements are protected. While few experimental studies or modeling efforts have specifically addressed the effect of coastal marshes on storm surges, anecdotal data accumulated after Hurricane Andrew in 1992 in Louisiana suggested that the storm surge from that hurricane was reduced about 4.7 cm per km of marsh that it traveled over (Louisiana Coastal Wetlands Conservation Task Force and Wetlands Conservation and Restoration Authority, 1998). Extrapolating from this number, a storm tracking from the south of New Orleans through existing coastal marshes could have its surge reduced by 3.7 m if it crossed 80 km of marsh before reaching the city. Barbier et al. (2013) found that a 1 percent increase in the wetland/open water ratio in the Louisiana delta could decrease a storm surge by 8 to 11 percent, equivalent to a 1 m reduction in storm surge per 9 to 13 km of additional wetlands. They translated this storm surge reduction in to substantial reduction in property damage in southeastern Louisiana. It is not inappropriate to refer to the disappearing marshes around New Orleans as that city's wet suit.

Climate Regulation

Wetlands may be significant factors in the global cycles of nitrogen, sulfur, and carbon. The natural supply of ecologically useful nitrogen comes from the fixation of atmospheric nitrogen gas (N_2) by a small group of plants and microorganisms that can convert it into organic form. Currently, ammonia is manufactured from N_2 for fertilizers, at more than double the rate of all natural fixation. Wetlands may be important in returning a part of this "excess" nitrogen to the atmosphere through denitrification. Denitrification requires the proximity of an aerobic and a reducing environment, such as the surface of a marsh, as well as a source of organic carbon, something abundant in most wetlands. Because most temperate wetlands are the receivers of fertilizer-enriched agricultural runoff and are ideal environments for denitrification, it is likely that they are important to the world's available nitrogen balance. The phenomenon of nitrogen enrichment of coastal waters causing "dead zones," or hypoxia (dissolved oxygen <2.0 mg/L in the hypolimnion now occurs worldwide. (See "The Nitrogen Cycle, Wetlands, and Hypoxia" in Chapter 6: "Wetland Biogeochemistry.") Wetland restoration and creation in the watershed have been recommended as a fundamental approach to solving this eutrophication (Mitsch et al., 2001; Mitsch and Day, 2006).

The global carbon cycle and wetlands are tightly linked. Wetlands, particularly northern peatlands, have stored enormous quantities of carbon in the peat. When these peatlands are protected and their water table is not affected, this carbon remains essentially in storage forever. When this peat is oxidized, whether by burning directly as a fuel or indirectly by altering the hydrology and causing drying and oxidation of the peat, the peatlands could become important sources of carbon dioxide to the atmosphere. Wetlands can be significant sinks of carbon if they are still building peat or accumulating carbon in their soil. This could be a significant advantage for tropical wetlands and for created and restored wetlands that are still building carbon storage in their soils compared to terrestrial systems that accumulate organic carbon in the soil slowly. Wetland carbon sequestration and greenhouse gas emissions are discussed in more detail in Chapter 17: "Wetlands and Climate Change."

Aquifer Recharge

Another value of wetlands related to hydrology is groundwater recharge. This function has received too little attention, and the magnitude of the phenomenon has not been well documented. Some hydrologists believe that, although some wetlands recharge groundwater systems, most wetlands do not. The reason for the absence of recharge is that soils under most wetlands are impermeable. In the few studies available, recharge occurred primarily around the edges of wetlands and was related to the edge : volume ratio of the wetland. Thus, recharge appears to be relatively more important in small wetlands, such as prairie potholes, than in large ones. These small wetlands can contribute significantly to recharge of regional groundwater.

Water Quality

Under favorable conditions, wetlands have been shown to remove organic and inorganic nutrients and toxic materials from water that flows across them. The concept of wetlands as sinks for chemicals was discussed in Chapter 6: "Wetland Biogeochemistry," and the practice of using wetlands for wastewater treatment and water quality improvement is discussed in detail in Chapter 19: "Wetlands and Water Quality." Wetlands have six attributes that influence the chemicals that flow through them, whether the chemicals are naturally added or artificially applied:

1. Wetlands cause a reduction in water velocity as streams enter wetlands, causing sediments and chemicals sorbed to sediments to drop out of the water column;
2. Many anaerobic and aerobic processes occur in close proximity in wetlands, promoting denitrification, chemical precipitation, and other chemical reactions that remove certain chemicals from the water;
3. High productivity in many wetlands can lead to high rates of mineral uptake by vegetation and subsequent burial in sediments when the plants die;
4. A diversity of decomposers and decomposition processes occur in wetland sediments;
5. There is a large contact surface of water with sediments because of the shallow water, leading to significant sediment–water exchanges; and
6. Organic peat accumulates in many wetlands, causing the permanent burial of chemicals.

Cultural Ecosystem Services

Aesthetics

A real but difficult aspect of a wetland to capture is its aesthetic value, often hidden under the dry term *nonconsumptive use values*, which simply means that people enjoy being out in wetlands. There are many aspects of this kind of wetland use. Wetlands are excellent "biological laboratories," where students in elementary, secondary, and higher education can learn natural history firsthand. They are visually and educationally rich environments because of their ecological diversity. Their complexity makes them excellent sites for research. Many visitors to wetlands use hunting and fishing as excuses to experience wildness and solitude, expressing that frontier pioneering instinct that may lurk in all of us. In addition, wetlands are a rich source of information about our cultural heritage. The remains of prehistoric Native American villages and mounds of shells or middens have contributed to our understanding of Native American cultures and of the history of the use of our wetlands.

Many artists—the Georgia poet Sidney Lanier, the painters John Constable and John Singer Sargent, and others who paint and photograph wetlands—have been drawn to them. Two artists—one a photographer and the other a painter—took a one-year excursion through the wetlands of the Louisiana delta in 2004 and 2005 (Lockwood and Gary, 2005). Their works, shown as exquisite photographs and paintings, have been shown in several museums throughout the United States.

Subsistence Use

In many regions of the world, the subsistence use of wetlands is extensive. There, wetlands provide the primary resources on which village economies are based. These societies have adapted to the local ecosystems over many generations and are integrated into them. Some of these cultures, including the Camarguais in France, the

Louisiana Cajuns in the United States, and the Marsh Arabs in Iraq, are described in Chapter 1: "Wetlands: Human Use and Science."

Quantifying Ecosystem Services

Efforts have been made to quantify the "free services" and amenities that wetlands provide to society for more than 40 years. Starting with the economics via energy flux approaches of H. T. Odum of the 1960s and 1970s that influenced a generation of new scientists, publication of *The Southern River Swamp—A Multiple-Use Environment* (Wharton, 1970) and *The Value of the Tidal Marsh* (Gosselink et al., 1974), a significant literature now exists in the general field of ecological economics on ascribing values to wetlands for the services they provide. Costanza et al. (1997, 2014) took these types of calculations one step further by estimating the public service functions of all Earth's ecosystems, including wetlands. These studies and others have generated a new vocabulary on ecosystem values with terms such as *public service function*, *natural capital*, *environmental services*, and *ecosystem goods and services*. All of these terms mean essentially the same thing. Nature, including wetlands, provides value to humans, and the value needs to be recognized whenever wetlands are either threatened or conserved (Söderquist et al., 2000; Mitsch and Gosselink, 2000).

Several approaches to the valuation of wetlands have been advanced. Because of the complexities described previously, there is no universal agreement about which approach is preferable. In part, the choice depends on the circumstances. Valuations fall broadly into two classes: ecological (or functional) evaluation and economic (or monetary) valuation. The former evaluation generally is necessary before attempting the latter valuations; ecological functions are the causes of monetary values.

Ecological Valuation

Habitat Evaluation Procedures

Table 16.3 shows an example of the application of the Habitat Evaluation Procedure (HEP) of the U.S. Fish and Wildlife Service to different development plans for a cypress–gum swamp ecosystem. The present value of the swamp for a representative group of terrestrial and aquatic animals was evaluated (baseline condition) using a habitat suitability index (HSI) based on a range of 0 to 1 for the optimum habitat for the species in question. The evaluation resulted in a mean terrestrial HSI of 0.8 and a mean aquatic HSI of 0.4. This baseline condition was compared with the projected habitat condition in 50 and 100 years under three projected scenarios: Plan A, Plan B, and a no-project projection. The results suggest that Plan A would be detrimental to the environment, whereas Plan B would have no effect on terrestrial habitat values and would improve aquatic ones. Whether to proceed with either of these plans is a decision that requires weighing the projected environmental effects against the projected economic benefits of the project.

One often-neglected feature of the analysis is the effect of aggregating HSIs for different species. Although, overall, Plan B appears to be about equivalent environmentally to the no-project option, scrutiny of Table 16.3 shows that Plan B is expected to improve the habitat for swamp rabbits and large-mouthed bass but decrease its value

Table 16.3 Habitat Evaluation Procedure of the impact of two management plans and a no-management control in a cypress–gum swamp in southeastern USA[a]

Species	Baseline Condition	Future with Project Plan A[b]		Future with Project Plan B[c]		Future without Project	
		50 Years	100 Years	50 Years	100 Years	50 Years	100 Years
Terrestrial							
Raccoon	0.7	0.5	0.6	0.8	0.8	0.7	0.9
Beaver	0.7	0.2	0.2	0.4	0.3	0.6	0.4
Swamp rabbit	0.7	0.2	0.2	0.8	0.8	0.7	0.4
Green heron	0.9	0.2	0.1	0.8	0.9	0.9	1.0
Mallard	0.8	0.3	0.2	1.0	0.9	0.9	1.0
Wood duck	0.8	0.3	0.2	0.9	1.0	1.0	1.0
Prothonotary warbler	0.8	0.3	0.1	0.6	0.7	0.8	0.9
Snapping turtle	0.8	0.4	0.3	0.8	0.7	0.8	0.9
Bullfrog	0.9	0.3	0.2	0.8	0.9	1.0	1.0
Total terrestrial HSI	7.1	2.7	2.1	6.9	7.0	7.4	7.5
Mean terrestrial HSI	0.8	0.3	0.2	0.8	0.8	0.8	0.8
Aquatic							
Channel catfish	0.3	0.3	0.4	0.4	0.4	0.4	0.4
Largemouth bass	0.4	0.2	0.3	0.7	0.8	0.4	0.4
Total aquatic HSI	0.7	0.5	0.7	1.1	1.2	0.8	0.8
Mean aquatic HSI	0.4	0.3	0.4	0.6	0.6	0.4	0.4

[a] Numbers in the tables are habitat suitability index (HSI) values, which have a maximum value of 1 for an optimal habitat.
[b] Channelization of water and clearing of swamp for agricultural development with a loss of 324 ha of wetland.
[c] Construction of levees around swamp for flood control with no loss of wetland area.
Source: Schamberger et al. (1979)

for warblers and turtles. This kind of detailed scrutiny may be important because it indicates a change in the quality of the environment, but it is often neglected when the "apples and oranges" are combined into "fruit."

Hydrogeomorphic Analysis

The hydrogeomorphic (HGM) classification described in Chapter 13: "Wetland Classification" also allows a quantification of the functions of wetlands. Its uniqueness lies in its quantification of natural wetland functions without regard to their significance to society. This is done by comparing the wetland of interest to a reference site that is characteristic of the same HGM class. Brinson et al. (1994) summarized the assessment procedure:

1. *Group wetlands into HGM classes with shared properties.* (The classification is discussed in Chapter 13.)
2. *Define the relationship between HGM properties and the functions of wetlands.* The goal is to select functions that are linked clearly and logically to wetland HGM properties and that have hydrologic, geomorphic, and ecological significance. This step represents the scientific basis for the presence of the function.

3. *Develop functional profiles for each wetland class.* These can range from descriptive narratives to multivariate data sets covering numerous sites.
4. *Develop a scale for expressing functions within each wetland class, by using indicators and profiles from the reference wetlands of that class.* These scales serve as benchmarks for each wetland class. Reference wetlands should include the full range of natural and human-induced variations due to stress and disturbance.
5. *Develop the assessment methodology.* The assessment relies on indicators to reveal the likelihood that the functions being evaluated are present in the wetland and depends on reference populations to scale the assessment. The reference wetlands are also used to set goals for compensatory mitigation.

Evaluating Alternatives with the HGM Technique—Illustration from North Carolina

In an illustration of the method to estimate the impact of a project or restoration on wetland functions, Rheinhardt et al. (1997) apply the HGM method to evaluate mitigation strategies in mineral soil forested pine (*Pinus palustris*)

Table 16.4 Field parameters used to estimate ecosystem function in a hydrogeomorphic assessment of forested wetlands in southeastern North Carolina

Variable	Description
Hydrology/Topography	
V_{DITC}	Lack of ditches nearby (<50 m)
V_{MICR}	Microtopographic complexity
Herbaceous Vegetation	
V_{GRAM}	Percentage cover of graminoids
V_{FORB}	Percentage cover of forbs
Canopy Vegetation	
V_{TREE}	Total basal area for trees (m^2/ha; >10 cm DBH)
V_{TDEN}	Density of canopy trees (stems/ha; >10 cm DBH)
V_{TDIA}	Average tree diameter (m)
V_{CVEG}	Sorensen similarity index of canopy importance value
Subcanopy Vegetation	
V_{SUBC}	Density of subcanopy (stems/ha)
V_{SDLG}	Percentage cover of trees and shrubs <1 m tall
V_{SVEG}	Sorensen similarity index of subcanopy importance value
Litter/Standing Dead	
V_{LTR}	Litter depth (cm)
V_{SNAG}	Density of standing dead stems (stems/ha)
V_{CWD}	Volume of coarse woody debris (cm^3/ha)

Source: Rheinhardt et al. (1997)

Table 16.5 Predicted changes in hydrologic regime function resulting from a hypothetical airport construction on one wetland site and the comparison of the mitigation required for two different wetland restoration alternatives (variables are defined in Table 16.4)

	Reference Wetland		Wetland Being Destroyed				Restoration Alternataive 1[a]				Restoration Alternative 2[b]			
			Now		After Airport		Now		After Restoration		Now		After Restoration	
Variable	Raw	Index	Raw	Index	Raw	Index	Raw	Index	Raw	Index	Raw	Index	Raw	Index
V_{TREE}	14.7	1.0	14.6	1.0	—	0.0	0.0	0.0	0.0	0.0	15.3	1.0	10.0	0.7
V_{SUBC}	12,550	1.0	13,314	1.0	—	0.0	0.0	0.0	6,963	0.5	18,402	0.5	9,800	0.8
V_{MICR}	2.5	1.0	2.5	1.0	—	0.0	0.0	0.0	2	1.0	4.2	1.0	4.2	1.0
V_{DITC}	1.0	1.0	0.5	0.5	—	0.0	0.0	0.0	1.0	1.0	0.5	0.5	1.0	1.0
Functional index[c]		1.0		0.71		0.0		0.0		0.71		0.64		0.91
Relative impact						−0.71				+0.71				+0.27
Mitigation ratio[d]										0.71/0.71 = 1:1				0.71/0.27 = 2.6:1

[a] Restoration of an agricultural field (former wetland) to a forested wetland.
[b] Restoration of a pine plantation to a forested wetland.
[c] Hydrologic functional index = $[((V_{TREE} + V_{SUBC} + V_{MICR})/3) \pm V_{DITC}]^{1/2}$.
[d] Ratio of wetland must be restored to area of wetland destroyed to achieve functional equivalent hydrologic regime.
Source: Rheinhardt et al. (1997).

flats in North Carolina. Fourteen variables were used to estimate the function of both study and reference wetlands (Table 16.4). Absolute values of some of the variables (e.g., tree density) are then translated into indices on a scale of 0.0 to 1.0 by comparing those functions to a reference wetland site. Such indices, in turn, are applied to model functions, such as "maintain hydrologic regime" as in Table 16.5, and comparisons of human impact on wetlands can be assessed. Table 16.5 shows a hypothetical case in which an airport is destroying a wetland (with an overall loss index of 0.71 when compared to a nearby reference, which, by definition, has an index of 1.0), and two restoration alternatives are being considered. The analysis shows that restoration of a cropland back to a wetland (Restoration Alternative 1) would be a good alternative because the cropland currently has 0.0 value in maintaining hydrologic regime. Thus, the restoration is estimated to require only 1 ha of that cropland (gain = +0.71) for every hectare of wetland lost due to the airport (loss = −0.71). This is a 1:1 mitigation ratio (the ratio of area of wetland restored to wetland lost).

> Restoration of an existing pine plantation to a natural pine wetland, however, would probably be easier but, functionally, the plantation already has some of the desired values of wetlands. (It rates a functional index of 0.64 before any restoration takes place and would rate an index of 0.91 after restoration, a net change of +0.27.) So that restoration strategy would require 2.6 ha (0.71/0.27) of pine plantation to be restored for every hectare of wetland lost for the airport (mitigation ratio = 2.6:1).

Economic Evaluation

Evaluation systems that seek to compare natural wetlands to human economic systems usually reduce all values to monetary terms (thus losing sight of the apples and oranges). Conventional economic theory assumes that in a free economy, the economic benefit of a commodity is the dollar amount that the public is willing to pay for the good or service rather than be without it.

Although this characterization of value is reasonable under most conventional economic conditions, it leads to real problems in monetizing nonmarket commodities, such as pure water and air, and in pricing wetlands whose value in the marketplace is determined by their value as real estate, not by their "free services" to society. Consequently, attempts to monetize wetland values have generally emphasized the commercial crops from wetlands: fish, shellfish, furs, and recreational fishing and hunting, for which pricing methodologies are available. This kind of pricing ignores ecosystem- and global-level ecosystem services related to clean air and water and other life-support functions. Even in the cases of market commodities from wetlands, available data are seldom adequate to develop reliable demand curves.

Economists recognize four more or less independent aspects of "value" that contribute to the total. These aspects are:

1. *Use value*. The most tangible portion of total value derived from identifiable direct benefits to the individual; hunting, harvesting fish, and nature study are examples.
2. *Social value*. Those amenities that accrue to a societal group rather than an individual; examples are improved water quality, flood protection, and the maintenance of the global sulfur balance.
3. *Option value*. The value that exists for the conservation of perceived benefits for future use.
4. *Existence value*. The benefits deriving from the simple knowledge that the valued resource exists, irrespective of whether it is ever used. For example, the capacity of an extant wetland to conserve biological diversity is an existence value.

As we have seen, use value is the easiest to estimate. The other three values, which are more difficult to quantify and also generally reflect longer-term viewpoints, have been addressed by economists using alternative methods, as illustrated next.

Willingness-to-Pay Methods

In the absence of a well-developed free-market alternative, pricing methodologies have been applied. One of these, *willingness to pay*, establishes a more or less hypothetical (contingency) market for nonmarket goods or services. Willingness to pay or, more accurately, *net* willingness to pay, is "the amount society would be willing to pay to produce and/or use a good beyond that which it actually does pay" (Scodari, 1990). The principle is illustrated as follows: Suppose a fisherman were willing to pay $30 a day to use a particular fishing site but had to spend only $20 per day in travel and associated costs. The net benefit, or economic value, to the fisherman of a fishing day at the site is not the $20 expenditure but the $10 difference between what he was willing to spend and what he had to spend. If the fishing opportunity at the site was eliminated, the fisherman would lose $10 worth of satisfaction fishing; the $20 cost that he would have incurred would be available to spend elsewhere. In the case of commercial goods, such as harvested fish, the total value of a wetland is the sum of the net benefit to the consumer plus the net benefit to the producer (the fisherman).

Opportunity Costs

A second approach to resource evaluation in the absence of a free-market model is the *opportunity cost* approach. In general terms, the opportunity cost associated with a resource is the net worth of that resource in its best alternative use. For example, "the opportunity cost of conserving a wetland area is the net benefit which might have been derived from the best alternate use of the area which must be foregone in order to preserve it in its natural state" (Bardecki, 1987). Because determining the opportunity cost associated with wetland conservation would require the evaluation of each wetland service as well as the identification and valuation of the best alternative use, in practice, a comprehensive evaluation of the opportunity cost of wetland conservation is far from possible. Nevertheless, it may represent a useful approach to the valuation of specific wetland functions.

Replacement Value

If one could calculate the cheapest way of replacing various services performed by a wetland and could make the case that those services would have to be replaced if the wetland was destroyed, then the figure arrived at would be the replacement value. Some of the replacement technologies that might be necessary to replace services provided by wetland processes are listed in Table 16.6. A sample calculation of the replacement cost method is shown in Table 16.7. In this example, a fish hatchery is used to calculate fishery production, a flood reservoir to calculate flood and drought control, sediment dredging to estimate sediment retention, and wastewater treatment to estimate water quality enhancement.

Table 16.6 Some replacement technologies for societal support values provided by wetlands

Societal Support	Replacement Technologies
Peat Accumulation	
Accumulating and storing organic matter (peat)	Artificial fertilizers
	Artificial flooding
Hydrologic Functions	
Maintaining drinking water quality	Water transport
	Pipeline to distant source
Maintaining groundwater level	Well-drilling
	Saltwater filtering
Maintaining surface water level	Dams for irrigation
	Pumping water to dam
	Irrigation pipes and machines
	Water transport for domestic animals
Moderation of water flows	Regulating gate
	Pumping water to stream
Biogeochemical Functions	
Processing sewage; cleansing nutrients and chemicals	Mechanical sewage treatment
	Sewage transport
	Sewage treatment plant
	Clear-cutting ditches and stream
Maintaining drinking water quality	Water quality inspections
	Water purification plant
	Silos for manure from domestic animals
	Nitrogen filtering
	Water transport
Filter to coastal waters	Nitrogen reduction in sewage treatment plants
Food Chain Functions	
Providing food for humans and domestic animals	Agriculture production
	Import of food
Providing cover	Roofing materials
Sustaining anadromous trout populations	Releases of hatchery-raised trout
	Farmed salmon
Sustaining other fish species and wetland-dependent flora and fauna	Work by nonprofit organizations
Species diversity; storehouse for genetic material	Replacement not possible
Bird watching, sport fishing, boating, and other recreational values	Replacement not possible
Aesthetic and spiritual values	Replacement not possible

Source: Folke (1991)

This approach has the merit of being accepted by some conventional economists. For certain functions, it gives very high values compared with those of other valuation approaches discussed in this section. For example, the tertiary treatment of wastewater is extremely expensive, as is the cost of replacing the nursery function of marshes for juvenile fish and shellfish. Serious questions, however, have been raised about whether these functions would be replaced by treatment plants and fish nurseries if the wetlands

Table 16.7 Estimated value of 770-ha riparian wetlands along the Kankakee River, northeastern Illinois, estimated by replacement value approach and by energy analysis

Replacement Cost Approach	$/Year	Total Value
Ecosystem Function (Replacement Technology)		
Fish productivity (fish hatchery)	$91,000	
Flood control/drought prevention (flood control reservoir)	$691,000	
Sediment control (sediment dredging)	$100,000	
Water quality enhancement (wastewater treatment)	$57,000	
Total replacement cost	$939,000	
Value/area $939,000 yr^{-1}/770 ha =		U.S.$1,219 ha^{-1} yr^{-1}
Energy Flow Approach		
Energy Flow Parameter	Number	Total Value
Ecosystem gross primary productivity (kcal m^{-2} yr^{-1})	20,000	
Energy quality conversion, (kcal GPP/kcal fossil fuel)	20	
Energy conversion in U.S. economy (kcal fossil fuel /U.S.$)	14,000	
Value/area =		U.S.$714 ha^{-1} yr^{-1}

kcal = kilocalorie; GPP = gross primary productivity.
Source: Mitsch et al. (1979)

were destroyed. Some ecologists and economists argue that, in the long run, either the services of wetlands would have to be replaced or the quality of human life would deteriorate. Other individuals argue that this assertion cannot be supported in any convincing manner.

Energy Analysis

A completely different approach uses the idea of energy flow through an ecosystem or the similar concept of embodied energy. The concepts of embodied energy (Costanza, 1980), and *emergy* (= energy memory; H. T. Odum, 1988, 1989, 1996) both attempt to estimate the total energy required to produce something and then translate the energy analysis into economic terms. It is assumed to be a valid index of the totality of ecosystem functions and is applicable to human systems as well. In this way, both natural and human systems can be evaluated on the basis of one common currency: energy. Because there is a clear relationship between energy and money in our society, energy flow can be translated to the more familiar currency of dollars at the end of the evaluation.

A simple calculation using the annual energy flow of a bottomland forested wetland in Illinois is illustrated in Table 16.7. Here, an estimated ecosystem energy flow (gross primary productivity [GPP]) of 20,000 kcal m^{-2}yr^{-1} yielded an estimated value of $714 ha^{-1}yr^{-1}. The energy analysis method gave a number about 60 percent of the replacement value. The concept of energy "quality" was used in this calculation to differentiate between energy flow in the ecosystem (based on gross primary productivity) and energy flow in the human-based fossil fuel economy. This is a precursor to the current approach of emergy discussed below.

Louisiana Coastal Wetlands: Comparing Energy and Economic Analyses

Costanza et al. (1989) showed that the economist's willingness-to-pay approach and energy analysis converge to a surprising degree for coastal marshes in Louisiana, although both methods result in a great deal of uncertainty (Table 16.8). The energy analysis approach yielded higher wetland values, but the ranges overlap. The sensitivity of both conventional and energy analysis methods to the choice of a discount rate, which has been vital for decades in the outcome of cost–benefit studies, is also demonstrated in this comparison. The energy analysis method is based on using the total amount of energy captured by natural ecosystems as a measure of their ability to do useful work (for nature and hence for society). The gross primary productivity (GPP) of representative coastal marsh systems, which ranges from 48,000 to 70,000 kcal m^{-2} yr^{-1}, is converted to monetary units by multiplying by a conversion factor of 0.05 units fossil fuel energy/unit GPP energy and dividing by the energy/money ratio for the economy (15,000 kcal fossil fuel/1983 $). These calculations resulted in an estimate of annual coastal wetland value of about \$1,560 ha^{-1} yr^{-1}, which, when converted to present value for an infinite series of payments, yields the range of capitalized values of \$16,000 to \$70,000 ha^{-1} for the discount rates used in Table 16.8.

Table 16.8 Estimates of wetland values in \$/ha of Louisiana coastal marshes based on willingness-to-pay and energy analysis at two discount rates

Method	Discount Rate	
	3%	8%
Willingness to pay		
Commercial fishery	\$2,090	\$783
Fur trapping (muskrat and nutria)	991	373
Recreation	447	114
Storm protection	18,653	4,732
Total willingness-to-pay value	\$22,181	\$6,002
Energy analysis	\$42,000–\$70,000	\$16,000–\$26,000
Best estimate	\$22,000–\$42,000	\$6,000–\$16,000

Source: Costanza et al. (1989)

In comparison, the willingness-to-pay estimates reflect the assessment that a reasonable range of wetland value for coastal Louisiana is between \$6,000 and \$22,000 ha^{-1}, depending on the discount rate applied to determine the present value. Costanza et al. (1989) used this range from the willingness-to-pay and energy analysis approaches to suggest that the annual loss of Louisiana coastal wetlands is costing society from $77 million to $544 million per year.

Emergy Analysis

Emergy analysis is a variation on the energy analysis (both terms were pioneered by H. T. Odum at the University of Florida in the 1970s and 1980s). The key to emergy analysis is the determination of transformities, or ratios that allow the conversion of one form of energy to another, as was done previously for gross primary productivity and fossil fuel energy described in the example above. These ratios are usually expressed in terms of solar emjoules (sej) per joule (or similar unit) of base energy or ecosystem flow. An example of an emergy flow analysis used for wetlands is illustrated in the next box.

Emergy Analysis of Wetlands in Florida

A comparison was made among three types of wetlands in Florida—a forested wetland, a shrub-scrub wetland, and a marsh (Bardi and Brown, 2001)—to compare their ecosystem services. The services considered were not only gross primary productivity but also infiltration of water to the groundwater (groundwater recharge) and transpiration. In addition, the storages of natural capital (stored water, biomass, and basin structure) were added. When all of the environmental services and natural capital are first converted to solar emjoules (sej) and then to dollars (Table 16.9), the data suggest that a 1-ha

Table 16.9 Results of emergy analysis comparing the economic value of three types of wetlands in Florida for their environmental services and natural capital (Values are U.S.$/ha)

Ecosystem Type	Environmental Services[a]	Natural Capital[b]	Total Value
Forested wetland	$231,880	$1,322,723	$1,554,603
Shrub/Scrub wetland	$31,831	$1,075,536	$1,107,366
Freshwater marsh	$13,173	$626,645	$639,817

[a]Environmental services include gross primary productivity, infiltration, and transpiration.
[b]Natural capital includes live biomass, peat, water, and basin structure (formed by geological processes).
Source: Bardi and Brown (2001).

forested wetland is approximately 2.4 times more valuable than a similar-size marsh. Furthermore, the analysis points out that the wetland values range from $640,000 to $1.5 million per ha. At the time, the going rate for buying wetland mitigation credit in Florida was $187,000/ha. Thus, the rate being paid for mitigation credit was one-third to one-eighth that of the values calculated for these wetlands. According to this estimate, wetlands were being sold to destruction at too low a price.

Energy and emergy analyses, although imprecise because of the many conversion factors needed, are more satisfying to many scientists than conventional cost-accounting methods, because they are based on the inherent function of the ecosystem, not on perceived values that may change from generation to generation and from location to location.

Valuing Ecosystem Goods and Services

Costanza et al. (1997) wrote a highly cited paper on the value of the goods and services from ecosystems and suggested that the world's ecosystems were worth $33 trillion per year (1995 $). Updated, this is equivalent to U.S.$46 trillion per year (in 2007 U.S.$). That study used ecosystem unit estimators that showed that wetlands, especially inland swamps and floodplains, were considerably more valuable than lakes and rivers, forests, and grasslands (Table 16.10). Only coastal estuaries had higher unit values than inland and coastal wetlands from the 1997 study.

Balmford et al. (2002) argued that the *net marginal benefits* of ecosystems should be estimated rather than the aggregated numbers developed by Costanza et al. (1997), which often were simple replacement values. The net marginal benefit is the difference between values of relatively intact ecosystems and the values to humans of the same ecosystems converted to human use. After investigating over 300 case studies,

Table 16.10 Estimated unit values of ecosystems (all numbers normalized to 2007 U.S.$)

Ecosystem	1997 Estimate Unit Value (U.S.$ ha^{-1} yr^{-1})	2011 Estimate Unit Value (U.S.$ ha^{-1} yr^{-1})
Estuaries	31,509	28,916
Inland swamps/floodplains	27,021	25,681
Tidal marshes/mangroves	13,786	193,843
Lakes/rivers	11,727	12,512
Forests	1,338	3,800
Grasslands	321	4,166

Source: Costanza et al. (2014); 1997 estimates from Costanza et al. (1997) but revised to 2007 U.S.$

Balmford et al. (2002) came up with only five studies worldwide where economic estimates were available for both conditions—intact ecosystems and the same landscape heavily managed. Two of those five case studies were of wetlands (Fig. 16.10). An economic analysis of a mangrove swamp in Thailand showed that conversion of a swamp to aquaculture made economic sense in the short term, but in the long term, the total economic value of an intact mangrove swamp was $60,400, about 3.6 times that of the value of converting the swamp to shrimp aquaculture. The values provided by the natural mangrove swamp included timber, charcoal, nontimber forest products, offshore fisheries, and storm protection. In a similar comparison, a freshwater marsh in Canada was found to have a total economic value of $8,800/ha, about 2.4 times the value realized by converting the wetland to intensive agriculture. Here, the major values of the natural marsh were for sustainable hunting, fishing, and trapping. The Balmford et al. (2002) estimates were used extensively by the Millennium Ecosystem

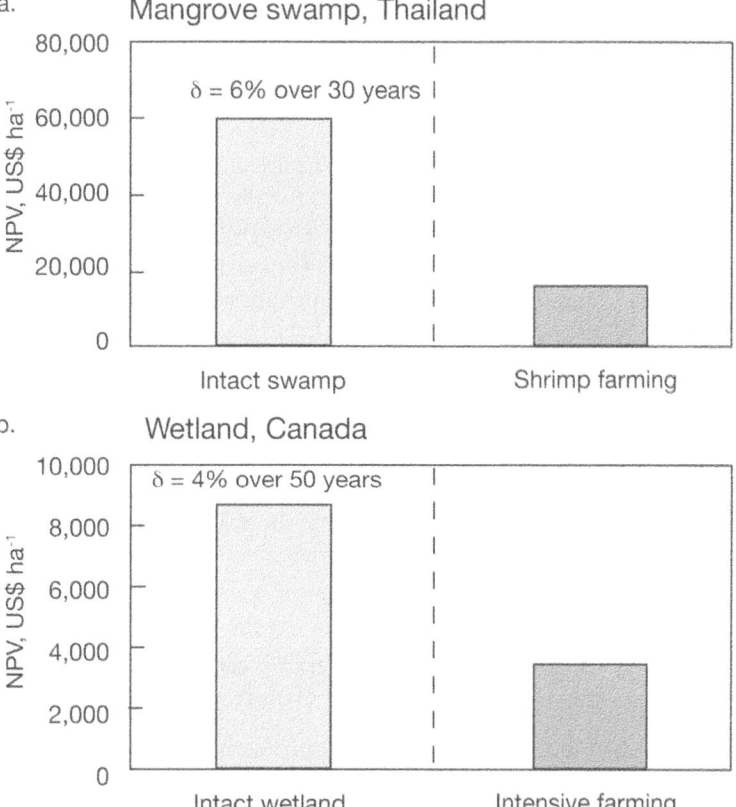

Figure 16.10 Two case studies of the marginal benefits of natural wetlands versus conversion of the wetland to intensive human industry: (a) mangrove system in Surat Thani, southern Thailand, and (b) freshwater marshes in Canada. ∂ indicates discount rates; NPV indicates net present value in year 2000 U.S.$/ha. (From Balmford et al., 2002)

Assessment (2005) that was published a few years later, two of only four case studies worldwide featured to show that sustainably managed ecosystems provide more economic benefit than do conversion to agricultural and aquaculture uses.

Costanza et al. (2014) revisited the calculations of his 1997 paper, using some revised unit values determined by deGroot et al. (2012) and others from a new United Nations–sponsored post–Millennium Ecosystem Assessment. The unit values for ecosystems are given in the last column in Table 16.10. The inland swamps/floodplain number stayed approximately the same while the tidal marsh/mangroves unit value increased 14-fold, "largely due to new studies of the storm protection, erosion protection, and waste treatment values" of these tidal wetlands. The overall value of ecosystems of the world, updated from the Costanza et al. (1997) paper, is U.S.$125 trillion to $145 trillion per year (Costanza et al., 2014). This reinvestigation of "what nature is worth" has received significant press coverage (e.g., Zimmer, 2014) in the *New York Times*[1] and Rosen (2014) in *The Atlantic*.[2]

Problems and Paradoxes of Quantifying Wetland Values

Regardless of which kind of ecosystem evaluation is used, eight generic problems and paradoxes to quantifying wetland values should be appreciated:

1. The terms *value* and *service* are anthropocentric; hence, assigning values to different natural processes usually reflects human perceptions and needs rather than intrinsic ecological processes.
2. The most valuable products of wetlands are public amenities that have no commercial value for the private wetland owner.
3. The ecological value, but not necessarily the economic value, of a wetland depends on its context in the landscape.
4. The relationships among wetland area, surrounding human population, and marginal value are complex.
5. Commercial values are finite, whereas wetlands provide values in perpetuity.
6. A comparison of economic short-term gains with wetland value in the long term is often not appropriate.
7. Estimates of values and services, by their nature, are colored by the biases of individuals and society and by the economic system.
8. A landscape view of wetlands is required to make intelligent decisions about the values of created and managed wetlands.

If one ignores the technical problems of functional ecosystem substitution, the idea attracts many people because of the common perception among economists that any commodity can be replaced. As scarcity of one product drives the price up, the

---

[1] www.nytimes.com/2014/06/05/science/earth/putting-a-price-tag-on-natures-defenses.html?_r=2.
[2] www.theatlantic.com/business/archive/2014/06/how-much-are-the-worlds-ecosystems-worth/372862/.

creativity of the free market will surely result in the development of a cheaper substitute. This is not true of ecosystems, however. Much of the value of an ecosystem, especially an open system such as a wetland, depends on its landscape context and on strong interactions among the parts of the landscape. Thus, the value of a riparian forest depends on its ecological links to the adjacent stream on one side and the upland fields or forest on the other.

Faustian Bargain

Because of the many problems documented in this chapter relate to valuation of natural ecosystem services, many ecologists oppose economic valuation of ecosystems. It implies that natural systems can be equated in the marketplace to other market products. Attempts to place dollar values on natural ecosystems, however, such as those cited in this chapter, have raised public awareness (see, e.g., Zimmer, 2014; Rosen, 2014) of the high value of the goods and services of nature, and in this way helped in efforts to conserve natural resources. Thus, ecologists are caught in a Faustian bargain with the devil, trying to make the case for ecosystem services in the common currency of our civilization while clearly documenting the reasons why natural ecosystem conservation should not depend on the operation of free-market forces. There is no easy answer to this dilemma.

Recommended Readings

Costanza, R., R. de Groot, P. Sutton, S. van der Ploeg, S. J. Anderson, I. Kubiszewski, S. Farber, and R. K. Turner. 2014. Changes in the Global Value of Ecosystem Services. *Global Environmental Change* 26: 152–158.

Millennium Ecosystem Assessment. 2005. *Ecosystems and Human Well-Being: Synthesis*. Washington, DC: Island Press.

References

Balmford, A., A. Bruner, P. Cooper, R. Costanza, S. Farber, R. E. Green, M. Jenkins, P. Jefferiss, V. Jessamy, J. Madden, K. Munro, N. Myers, S. Naeem, J. Paavola, M. Rayment, S. Rosendo, J. Roughgarten, K. Trumper, and R. K. Turner. 2002. Economic reasons for conserving wild nature. *Science* 297: 950–953.

Barbier, E. B., I. Y. Georglou, B. Enchelmeyer, and D. J. Reed. 2013. The value of wetlands in protecting Southeast Louisiana from hurricane storm surges. *PLoS ONE* 8 (3): e58715.

Bardecki, M. J. 1987. *Wetland Evaluation: Methodology Development and Pilot Area Selection*. Report 1, Canadian Wildlife Service and Wildlife Habitat Canada, Toronto.

Bardi, E., and M. T. Brown, 2001. Emergy evaluation of ecosystems: A basis for environmental decision making. In M. T. Brown, ed., *Emergy Synthesis: Theory and Applications of the Emergy Methodology*. Proceedings of a conference held at Gainesville, FL, September 1999. Center for Environmental Policy, University of Florida, Gainesville, pp. 81–98.

Brinson, M. M., W. Kruczynski, L. C. Lee, W. L. Nutter, R. D. Smith, and D. F. Whigham. 1994. Developing an approach for assessing the functions of wetlands. In W. J. Mitsch, ed., *Global Wetlands: Old World and New*. Elsevier, Amsterdam, the Netherlands, pp. 615–624.

Cahoon, D. R., D. J. Reed, and J. W. Day, Jr. 1995. Estimating shallow subsidence in microtidal salt marshes of the southeastern United States—Kaye and Barghoorn revisted. *Marine Geology* 128: 1–9.

Check, E. 2005. Roots of recovery. *Nature* 438: 910–911.

Costanza, R. 1980. Embodied energy and economic evaluation. *Science* 210: 1219–1224.

Costanza, R., S. C. Farber, and J. Maxwell. 1989. Valuation and management of wetland ecosystems. *Ecological Economics* 1: 335–361.

Costanza, R., R. d'Arge, R. de Groot, S. Farber, M. Grasso, B. Hannon, K. Limburg, S. Naeem, R. V. O'Neill, J. Paruelo, R. G. Raskin, P. Sutton, and M. van den Belt. 1997. The value of the world's ecosystem services and natural capital. *Nature* 387: 253–260.

Costanza, R., W. J. Mitsch, and J. W. Day. 2006. A new vision for New Orleans and the Mississippi delta: Applying ecological economics and ecological engineering. *Frontiers in Ecology and the Environment* 4: 465–472.

Costanza, R., R. de Groot, P. Sutton, S. van der Ploeg, S. J. Anderson, I. Kubiszewski, S. Farber, and R. K. Turner. 2014. Changes in the global value of ecosystem services. *Global Environmental Change* 26: 152–158.

Danielsen, F., M. K. Sørensen, M. F. Olwig, V. Selvam, F. Parish, N. D. Burgess, T. Hiraishi, V. M. Karunagaran, M. S. Rasmussen, L. B. Hansen, A. Quarto, and N. Suryadiputra. 2005. The Asian tsunami: A protective role for coastal vegetation. *Science* 310: 643.

Das, S, and A-S. Crépin. 2013. Mangroves can provide protection against wind during storms. *Estuarine, Coastal and Shelf Science* 134: 98–107.

Day, J. W., Jr., J. Barras, E. Clairain, J. Johnston, D. Justic, G. P. Kemp, J.-Y. Ko, R. Lane, W. J. Mitsch, G. Steyer, P. Templet, and A. Yañez-Arancibia. 2005. Implications of global climatic change and energy cost and availability for the restoration of the Mississippi Delta. *Ecological Engineering* 24: 253–265.

deGroot, R., Brander, L., van der Ploeg, S., Costanza, R., Bernard, F., Braat, L., Christie, M., Crossman, N., Ghermandi, A., Hein, L., Hussain, S., Kumar, P., McVittie, A., Portela, R., Rodriguez, L.C., ten Brink, P., and van Beukering, P. 2012. Global estimates of the value of ecosystems and their services in monetary units. *Ecosystem Services* 1: 50–61.

Feierabend, S. J., and J. M. Zelazny. 1987. Status report on our nation's wetlands. National Wildlife Federation, Washington, DC. 50 pp.

Folke, C. 1991. The societal value of wetland life-support. In C. Folke and T. Kaberger, eds., *Linking the Natural Environment and the Economy*. Kluwer Academic Publishers, Dordrecht, the Netherlands, pp. 141–171.

Food and Agriculture Organization of the United Nations (FAO). 2003. *State of the World's Forests*. FAO, Rome, Italy.

Gibbons, A. 1997. Did birds fly through the K–T extinction with flying colors? *Science* 275: 1068.

Gosselink, J. G., E. P. Odum, and R. M. Pope. 1974. The value of the tidal marsh. Publication LSU-SG-74-03. Center for Wetland Resources, Louisiana State University, Baton Rouge. 30 pp.

Harris, L. D., and J. G. Gosselink. 1990. Cumulative impacts of bottomland hardwood forest conversion on hydrology, water quality, and terrestrial wildlife. In J. G. Gosselink, L. C. Lee, and T. A. Muir, eds., *Ecological Processes and Cumulative Impacts*. Illustrated by Bottomland Hardwood Wetland Ecosystems. Lewis Publishers, Chelsea, MI, pp. 259–322.

Harter, S. K., and W. J. Mitsch. 2003. Patterns of short-term sedimentation in a freshwater created marsh. *Journal of Environmental Quality* 32: 325–334.

Hiraishi, T., and K. Harada. 2003. Greenbelt tsunami prevention in south-Pacific region. Report of the Port and Airport Research Institute, vol. 24, no. 2. EqTAP Project, Japan. www.drs.dpri.kyoto-u.ac.jp/eqtap/report/papua/tsunam_prevention/tsunami_pari_report3.pdf.

Lockwood, C. C., and R. Gary. 2005. *Marsh Mission: Capturing the Vanishing Wetlands*. Louisiana State University Press, Baton Rouge. 106 pp.

Louisiana Coastal Wetlands Conservation and Restoration Task Force and the Wetlands Conservation and Restoration Authority. 1998. *Coast 2050: Toward a Sustainable Coastal Louisiana, an Executive Summary*. Louisiana Department of Natural Resources, Baton Rouge. 12 pp.

Marois, D. and W. J. Mitsch. 2015. Coastal protection from tsunamis and cyclones provided by mangrove wetlands—a review. *International Journal for Biodiversity Science, Ecosystems Services and Management*, doi.org/10.1080/21513732.2014.997292

Millennium Ecosystem Assessment, 2005. *Ecosystems and Human Well-Being: Synthesis*. Island Press and World Resources Institute, Washington, DC. 137 pp.

Mitsch, W. J., M. D. Hutchison, and G. A. Paulson. 1979. The Momence Wetlands of the Kankakee River in Illinois—an assessment of their value. Document 79/17, Illinois Institute of Natural Resources, Chicago. 55 pp.

Mitsch, W. J., and J. G. Gosselink. 2000. The value of wetlands: Importance of scale and landscape setting. *Ecological Economics* 35: 25–33.

Mitsch, W. J., J. W. Day, Jr., J. W. Gilliam, P. M. Groffman, D. L. Hey, G. W. Randall, and N. Wang. 2001.Reducing nitrogen loading to the Gulf of Mexico from the Mississippi River Basin: Strategies to counter a persistent ecological problem. *BioScience* 51: 373–388.

Mitsch, W. J., and J. W. Day, Jr. 2006. Restoration of wetlands in the Mississippi-Ohio-Missouri (MOM) River Basin: Experience and needed research. *Ecological Engineering* 26: 55–69.

Niering, W. A. 1988. Endangered, threatened and rare wetland plants and animals of the continental United States. In D. D. Hook et al., eds., *The Ecology and Management of Wetlands, Vol. 1: Ecology of Wetlands*. Timber Press, Portland, OR, pp. 227–238.

Novitzki, R. P. 1985. The effects of lakes and wetlands on flood flows and base flows in selected northern and eastern states. In H. A. Groman, D. Burke, T. Henderson, and H. Groman, eds., *Proceedings of a Conference—Wetlands of the Chesapeake*. Environmental Law Institute, Washington, DC, pp. 143–154.

Odum, H. T. 1988. Self-organization, transformity, and information. *Science* 242: 1132–1139.

Odum, H. T. 1989. Ecological engineering and self-organization. In W. J. Mitsch and S. E. Jørgensen, eds., *Ecological Engineering*. John Wiley & Sons, New York, pp. 79–101.

Odum, H. T. 1996. *Environmental Accounting: Energy and Environmental Decision Making*. John Wiley & Sons, New York. 370 pp.

Ogawa, H., and J. W. Male. 1983. *The Flood Mitigation Potential of Inland Wetlands*. Publication 138, Water Resources Research Center, University of Massachusetts, Amherst. 164 pp.

Ogawa, H., and J. W. Male. 1986. Simulating the flood mitigation role of wetlands. *Journal of Water Resource Planning and Management* 112: 114–128.

Rheinhardt, R. D., M. M. Brinson, and P. M. Farley. 1997. Applying wetland reference data to functional assessment, mitigation, and restoration. *Wetlands* 17: 195–215.

Rosen, R. J. 2014. How much are the world's ecosystems worth? *The Atlantic*, June 16, 2 pp. www.theatlantic.com/business/archive/2014/06/how-much-are-the-worlds-ecosystems-worth/372862/.

Schamberger, M. L., C. Short, and A. Farmer. 1979. Evaluation wetlands as a wildlife habitat. In P. E. Greeson, J. R. Clark, and J. E. Clark, eds., *Wetland Functions and Values: The State of Our Understanding*. American Water Resources Association, Minneapolis, MN, pp. 74–83.

Scodari, P. F. 1990. *Wetlands protection: The role of economics*. Environmental Law Institute, Washington, DC. 89 pp.

Söderqvist, T., W. J. Mitsch, and R. K. Turner, eds. 2000. The Values of Wetlands: Landscape and Institutional Perspectives. *Special issue of Ecological Economics* 35: 1–132.

Turner, R. E. 1982. Protein yields from wetlands. In B. Gopal, R. E. Turner, R. G. Wetzel, and D. F. Whigham, eds., *Wetlands: Ecology and Management*. National Institute of Ecology and International Scientific Publications, Jaipur, India, pp. 405–415.

U.S. Army Corps of Engineers. 1972. Charles River Watershed, Massachusetts. New England Division, Waltham, MA. 65 pp.

Weller, M. W. 1999. *Wetland Birds*. Cambridge University Press, Cambridge, UK.

Wharton, C. H. 1970. *The Southern River swamp—A multiple-use environment*. Bureau of Business and Economic Research, Georgia State University, Atlanta. 48 pp.

Zimmer, C. 2014, June 5. Putting a price tag on nature's defenses. *New York Times*. http://www.nytimes.com/2014/06/05/science/earth/putting-a-price-tag-on-natures-defenses.html?_r=1.

Chapter 17

Wetlands and Climate Change

Earth's climate is changing, as witnessed by higher atmospheric temperatures, decreased snow and ice cover, and increasing sea levels, especially for the last 30 or 40 years. Wetlands emit 20 to 25 percent of global methane emissions to Earth's atmosphere, yet they also have the best capacity of any ecosystem to retain carbon through permanent burial (sequestration). Both processes have implications for climate change. Of the total storage of organic carbon in Earth's soils, 20 to 30 percent or more is stored in wetlands, and that storage is vulnerable to loss back to the atmosphere if the climate warms or becomes drier. Our estimates show that the world's wetlands are climate-change positive; that is, the negative effects of methane emissions on climate are more than compensated for by carbon sequestration into peat or wetland soils.

The effects of climate change on coastal wetlands could be significant if sea level rises, particularly in large river deltas where land subsidence is already occurring and where inland migration of wetlands is prevented by human development in a process called coastal squeeze. For inland wetlands, change in precipitation patterns and warmer temperatures likewise can have detrimental effects on wetland function.

Wetlands have significant yet still underappreciated roles in the global carbon cycle. They are also positioned in the landscape where climate change could affect them more than most other ecosystems. So their roles both as players in and recipients of climate change are the subject of this chapter.

Climate Change

There is little doubt that something significant is happening to our climate. According to the consensus of hundreds of scientists who have been involved in the

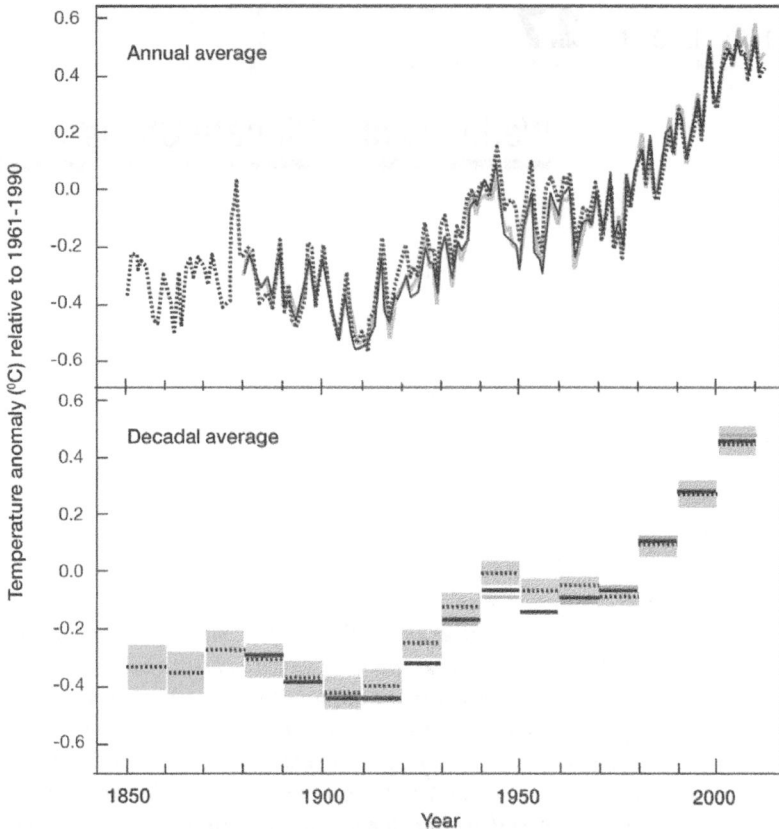

Figure 17.1 Observed globally averaged land and ocean surface temperature anomaly, 1850 to 2012 relative to the period 1961 to 1990, for two data sets: (a) annual averages and (b) decadal averages with estimated uncertainty for black line data set. (From IPCC, 2013)

Intergovernmental Panel on Climate Change (IPCC), some major findings should concern anyone interested in our planet and its future. The IPCC was established by the World Meteorological Organization and the United Nations Environmental Programme to assess scientific, technical, and socioeconomic information relevant for understanding climate change, its potential impacts, and options for adaptation and mitigation. Some of the dominant conclusions of the panel, drafted in its most recent multivolume reports and summaries (IPCC, 2013, 2014a,b; The Royal Society and The National Academy of Sciences, 2014) are summarized here:

- *The global average surface temperature has increased over the period 1880 to 2012 by about 0.85°C.* This trend is illustrated in Figure 17.1. The temperate increase was about 0.25°C more than that estimated by the IPCC (2001) for the twentieth century (0.6°C). This temperature increase in the twentieth century had also been determined to be the largest increase in the last 1,000 years.

- *Each of the last three decades has been successively warmer at Earth's surface than any preceding decade since 1850.* In the Northern Hemisphere, 1983 to 2012 was likely the warmest 30-year period of the last 1,400 years.
- *The surface 75 m of oceans warmed by 0.11° C per decade over the period 1992 to 2005.* It is also likely that regions of high salinity, where evapotranspiration (ET)≫precipitation (P), have become more saline while humid regions, where P>ET, have become fresher since the 1950s.
- *The average rate of ice loss from glaciers around the world, excluding glaciers on the periphery of the ice sheets, was very likely 226×10^9 t (Gt) yr^{-1} over the period 1971 to 2009 and very likely 275 Gt yr^{-1} over the period 1993 to 2009.* Over the last two decades, the Greenland and Antarctic ice sheets have been losing mass, glaciers continue to shrink worldwide, and Arctic sea ice and Northern Hemisphere spring snow cover have continued to decrease.
- *Sea level has risen globally about 1.7 mm yr^{-1} (19 cm total) between 1901 and 2010 and at a much greater rate of 3.2 mm yr^{-1} from 1993 to 2010* (Fig. 17.2). Glacial mass loss and ocean thermal expansion explains 75 percent of this observed global sea level rise.

Causes of Climate Change

The cause of climate change is the increasing concentration of the so-called greenhouse gases in the atmosphere, mostly caused by anthropocentric emissions. These gases adsorb several wavelengths of long-wave radiation, causing Earth to be a little warmer if the gas concentrations increase. The primary greenhouse gas is carbon dioxide (CO_2), which is released through the burning of fossil fuels and also by cement production. Atmospheric CO_2 is estimated to have increased by over 30 percent since

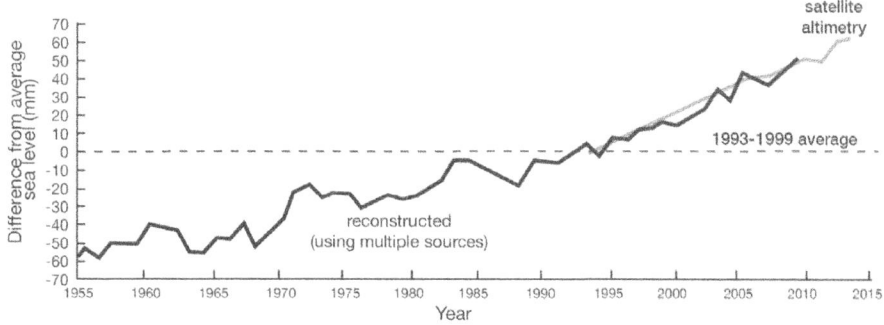

Figure 17.2 Relative global mean sea level for 1955 to 2014 (Grey line reconstructed from several data sets aligned to have the same value in 1993, the first year of satellite altimetry data). (From the Royal Society and the National Academy of Sciences, 2014)

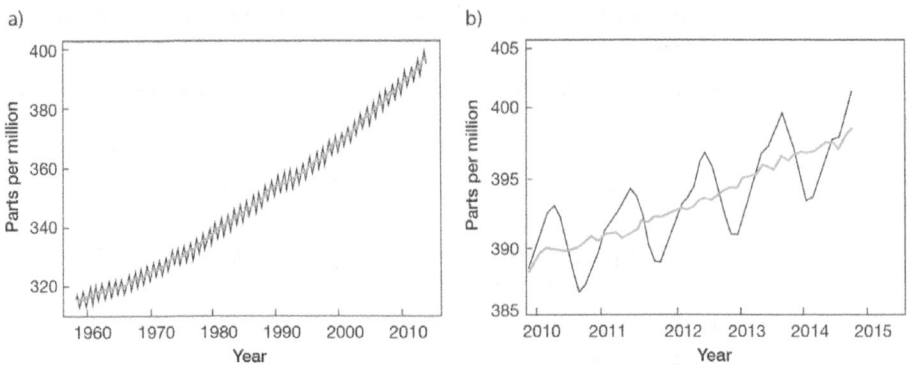

Figure 17.3 (a) Concentration of CO_2 in atmosphere at Mauna Loa Observatory in Hawaii for 1958 through mid-2014. (b) Details of last four years of seasonal CO_2 fluctuations. Peaks are at the beginning of the Northern Hemisphere growing season, after which photosynthesis reduces the CO_2 until the end of the growing season. (Data from Scripps Institution of Oceanography and NOAA Earth System Research Laboratory)

the mid-eighteenth century. The longest record of continuous monitoring of CO_2 in the atmosphere is at Mauna Loa, Hawaii, started by C. David Keeling of the Scripps Institution of Oceanography in March 1958 at a National Oceanic and Atmospheric Administration (NOAA) facility (Fig. 17.3a). CO_2 increased at a rate of 2.2 ppm yr^{-1} over 2009 to 2013, more than double the rate at which it was increasing in the 1960s. Monthly average concentrations reached 400 ppm during the spring of 2014 (Fig. 17.3b).

There has been much discussion about sources of CO_2 besides fossil fuel burning, such as tropical forest deforestation and burning. IPCC (2013) estimates that fossil fuel combustion and cement production combined have released 375 Pg (= petagram = Gt = 10^{15} g) of CO_2 as carbon to the atmosphere while deforestation and other land-use changes are estimated to have released 180 Pg to the atmosphere. Fossil fuel consumption continues to rise, from 6.7 Pg/yr in the mid-2000s to 10 Pg/yr in 2013. The second most important greenhouse gas is actually water vapor, but it is not known to have any trend or change. It is one of the most abundant gases in the troposphere. When water vapor and other aerosols condense, they have a net negative radiative forcing on the atmosphere, offsetting a major portion of the global mean radiative forcing from other greenhouse gases (IPCC, 2013).

The third most important greenhouse gas is methane (CH_4), which has been estimated to have more than doubled in concentration, from about 720 parts per billion (ppb) in preindustrial times to about 1,803 ppb in 2011. Before about 1980, CH_4 was assumed to be a stable concentration in the atmosphere, but it increased by 13 percent between 1978 and 1999 alone (Whalen, 2005). Wetlands were described in Chapter 6: "Wetland Biogeochemistry" as being sources of CH_4 gas, and that will be put in context with other sources and sinks later in this chapter. What should be clear is that if one argues that Earth has lost half of the world's wetlands as a result of human activity over the last 100 years when CH_4 concentrations are increasing, there is a disconnect. If wetlands were the major source of CH_4, we would have seen a decrease in CH_4 in the atmosphere over the last 100 years.

A fourth important greenhouse gas, nitrous oxide (N_2O), also comes from wetlands as a result of nitrification and especially denitrification (see Chapter 6). While N_2O is a normal product of denitrification, it is usually a small percentage of denitrification products, with most of nitrates converted to dinitrogen (N_2) gas. N_2O has increased by about 20 percent in the atmosphere since preindustrial times.

Wetlands in the Global Carbon Cycle

Although soil carbon in wetland soils is recognized as an important component of global carbon budgets and future climate change scenarios, very little work has been done to consider the role of wetlands, particularly those in temperate and tropical regions of the world, in the global carbon cycle. A carbon budget for the world, with wetlands included to show their relative contributions, is shown in Figure 17.4. Following, we discuss the role of wetlands in this carbon budget in terms of carbon storage in peat, carbon sequestration through peat and organic soil development, and CH_4 emissions. This budget is a major revision from the budget published in the fourth edition of *Wetlands* (Mitsch and Gosselink, 2007). The major changes are a significantly higher carbon sequestration of 1 Pg yr^{-1} (=1000 Tg (teragram) yr^{-1} = 10^{15} g yr^{-1}) estimated for the world's wetlands, based on new data from several wetlands around

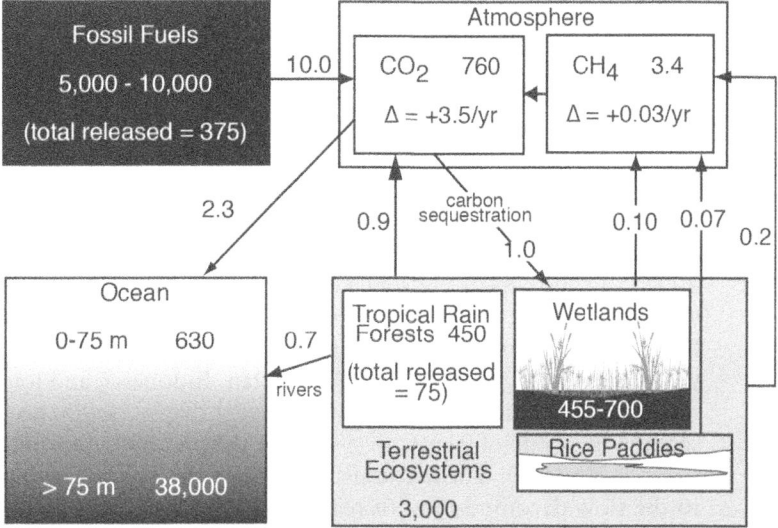

Figure 17.4 Global carbon budget with estimated role of wetlands in the carbon cycle. Fluxes are in Pg/yr; storages are in Pg. Pg = 10^{15} g. (CH_4 emissions from wetlands and rice paddies from Bloom et al., 2010; terrestrial ecosystem and fossil fuel inputs to CO_2 from IPCC, 2013; carbon sequestration by wetlands from Mitsch et al., 2013)

the world (Mitsch et al., 2013), and a continual increase in carbon emissions from fossil fuel combustion from a mid-2000s estimate of 6.3 Pg/yr to the current rate of 10 Pg/yr, a 60 percent increase in emissions between two editions of this textbook.

Peat Storage and a Global Carbon Budget

Peat deposits in the world's wetlands, particularly in boreal and tropical regions, are substantial storages of carbon (C) in the lithosphere. Of the total storage of C in Earth soils of 1,400 to 2,500 Pg-C (Pg = 10^{15} g), anywhere from 20 to 30 percent is stored in wetlands (Mitsch and Wu, 1995; Roulet, 2000; Hadi et al., 2005; Lal, 2008). These peat deposits, if disturbed, however, could contribute significantly to worldwide atmospheric CO_2 levels, depending on the balance between draining and oxidation of the peat deposits and their formation in active wetlands.

Carbon Sequestration

Many studies have now estimated carbon sequestration in wetlands in a variety of temperate and tropical settings to augment the frequent estimates that already existed for northern peatlands (Table 17.1).

For northern peatlands, the vertical accumulation rate of between 20 and 200 cm/1,000 yr (see Chapter 12: "Peatlands") usually results in carbon accumulation in the range of 10 to 50 g-C m^{-2} yr^{-1}. This is a typical range of carbon accumulation in peatlands (Table 17.1). A reasonable average of 29 g-C m^{-2} yr^{-1} was found in a review of the literature for eight recent peatlands around the world where carbon sequestration was measured (Table 17.1).

Most of the rates for carbon sequestration in tropical/subtropical wetlands and for coastal mangroves and salt marshes are in the range of 150 to 250 g-C m^{-2} yr^{-1} (Table 17.1). Carbon sequestration by coastal wetlands (salt marshes, mangroves, sea grasses) now has enormous international support and recognition, partially because it is referred to in the literature and popular press as *blue carbon* (Mcleod et al., 2011; Vaidyanathan, 2011; World Wildlife Fund, 2012; also see http://thebluecarboninitiative.org/). The tropical wetlands included some high rates of carbon sequestration such as seen for *Cyperus* wetlands in Uganda (Saunders et al., 2007) but also relatively low rates of carbon sequestration in seasonally flooded wetlands in Costa Rica and Botswana (Bernal and Mitsch, 2013b). In a study of long-term accumulation in the tropics, Page et al. (2004) investigated a 9.5-m core of peat from a tropical peatland in Kalimantan, Indonesia, and found an average carbon sequestration of the core of 56 g-C m^{-2} yr^{-1} over a 24,000-year period and a carbon sequestration rate of 94 g-C m^{-2} yr^{-1} for the past 500 years in the upper meter of the core (Table 17.1). The accumulation of peat in tropical wetlands may be due more to the slow decomposition of recalcitrant lignin in roots and woody material under constant high water rather than to high productivity of these systems (Chimner and Ewel, 2005). The lower rates of carbon sequestration in seasonally flooded tropical wetlands are probably due to the high temperatures year-round, especially in the dry season, when some of the carbon is oxidized, or in some cases the presence of fire.

Table 17.1 Carbon Sequestration in wetlands (updated from Mitsch et al., 2013) m^{-2}

Wetland Type	Carbon Sequestration, g-C m^{-2} $year^{-1}$	Reference
Northern Peatlands		
Boreal peatlands	29 ± 13 (n = 8)	Mitsch et al. (2013)
Boreal peatlands	15 – 26	Turunen et al. (2002)
Temperate peatlands	10 – 46	Turunen et al. (2002)
Russian tundra peatlands	−8 – 38	Heikkinen et al. (2002)
Coastal Wetlands		
Mangroves, North America	180	Chmura et al. (2003)
Salt marshes, North America	220	Chmura et al. (2003)
Tidal freshwater wetlands, North America	140 ± 20	Craft (2007); Craft et al. (2009)
Brackish marshes, North America	240 ± 30	Craft (2007); Craft et al. (2009)
Salt marshes, North America	190 ± 40	Craft (2007); Craft et al. (2009)
Mangrove swamps, S.E. Asia	90 – 230	Suratman (2008)
Coastal wetlands, S.E. Australia, Undisturbed sites	105 – 137	Howe et al. (2009)
Coastal wetlands, S.E. Australia, Disturbed sites	64 – 89	Howe et al. (2009)
Mangroves (global)	160 ± 40	Breithaupt et al. (2012)
Mangroves (global)	226 ± 39	Mcleod et al. (2011)
Tropical/Subtropical Freshwater Wetlands		
Tropical/subtropical wetlands	194 ± 56 (n = 6)	Mitsch et al. (2013)
Florida Everglades, general	86–387	Reddy et al. (1993)
Tropical freshwater wetland, Indonesia	56 (for 24,000-year core)	Page et al. (2004)
Tropical freshwater wetland, Indonesia	94 (for last 500-year core)	Page et al. (2004)
Cyperus wetland in Uganda	480	Saunders et al. (2007)
Cypress (*Taxodium*) swamp, Florida	122	Craft et al. (2008)
Cypress (*Taxodium*) swamp, Georgia	36	Craft et al. (2008)
Everglades (*Cladium*) marsh, Florida	19–46	Craft et al. (2008)
Tropical flowthrough swamp, Costa Rica	222–465 (ave = 306 for 3 sites)	Bernal and Mitsch (2013b)
Tropical forest basin wetland, Costa Rica	61–131 (ave = 84 for 3 sites)	Bernal and Mitsch (2013b)
Seasonally dry tropical floodplain wetland, Cost Rica	80–89 (ave = 84 for 3 sites)	Bernal and Mitsch (2013b)
Seasonally flooded tropical floodplain wetland, Botswana	33–53 (ave = 42 for 3 sites)	Bernal and Mitsch (2013b)
Florida Everglades—cypress strand/swamp	98	Villa and Mitsch (2015)
Florida Everglades—pond cypress	64	Villa and Mitsch (2015)
Florida Everglades—wet prairie	39	Villa and Mitsch (2015)
Florida Everglades—upland pine flatwood	22	Villa and Mitsch (2015)
Temperate Freshwater Wetlands		
Temperate wetlands	278 ± 42 (n = 7)	Mitsch et al. (2013)
Temperate flowthrough wetlands, northern Ohio	140 ± 16 (n = 3)	Bernal and Mitsch (2012)
Depressional wetlands, Ohio	317 ± 93 (n = 3)	Bernal and Mitsch (2012)
Reed (*Phragmites*) marsh, Denmark	504	Brix et al. (2001)

(continued)

Table 17.1 (*Continued*)

Wetland Type	Carbon Sequestration, g-C m^{-2} year^{-1}	Reference
Created and Restored Wetlands		
Prairie pothole wetlands, North America		Euliss et al. (2006)
Restored (semipermanently flooded)	305	
Reference wetland	83	
Abandoned peat meadow, Netherlands	280	Hendriks et al. (2007)
Created temperate riverine flowthrough marshes, Ohio		
10 years old	181–193	Anderson and Mitsch (2006)
15 years old	219–267	Bernal and Mitsch (2013a)
Reference wetland	140	Bernal and Mitsch (2013a)

Temperate freshwater wetlands showed some of the highest rates of carbon sequestration of any of the three climates investigated by Mitsch et al. (2013). Carbon sequestration in temperate-zone wetlands range from 230 to 320 to g-C m^{-2} yr^{-1} (Table 17.1). Brix et al. (2001) estimated a high rate of more than 500 g-C m^{-2} yr^{-1} in a productive *Phragmites* marsh in Denmark.

Created and restored wetlands might be the best opportunity for carbon sequestration. A carbon sequestration rate of 180 to 190 g-C m^{-2} yr^{-1} for two created wetland basins in Ohio (Anderson and Mitsch, 2006) 10 years after the wetlands were created increased to 220 to 270 g-C m^{-2} yr^{-1} by the time the wetlands were 15 years old (Bernal and Mitsch, 2013a; Figure 17.5). About one-fourth of that carbon sequestration was as inorganic carbon, precipitated as calcite/calcium carbonate ($CaCO_3$) due to high productivities in the water column. Euliss et al. (2006) compared the carbon sequestration in several wetlands that had been restored for more than a decade in the prairie pothole wetlands of North America and found 305 g-C m^{-2} yr^{-1}, one of the highest numbers in Table 17.1. This is not surprising, because restoration in these cases meant reflooding agricultural land, allowing organic carbon to once again build up in the soil. For comparison, Euliss et al. (2006) estimated an accumulation rate in reference (natural) marshes in the region of 83 g-C m^{-2} yr^{-1} based on average sedimentation rates of 2 mm/yr.

Methane Emissions

Wetlands are estimated to emit about 20 to 25 percent of current global CH_4 emissions or about 115 to 170 Tg-CH_4 yr^{-1} (Tg = 10^{12} g; Table 17.2). Thus, in climate change discussions concerning wetlands, these "natural emissions" often receive the most attention. Rice paddies, which are essentially domestic wetlands, account for another 60 to 80 Tg-CH_4 yr^{-1}. Other anthropogenic sources account for most of the rest. CH_4 emissions are a concern because CH_4 is estimated to be 25 times more effective as a greenhouse gas on a molecular basis than is CO_2 after 100 years.

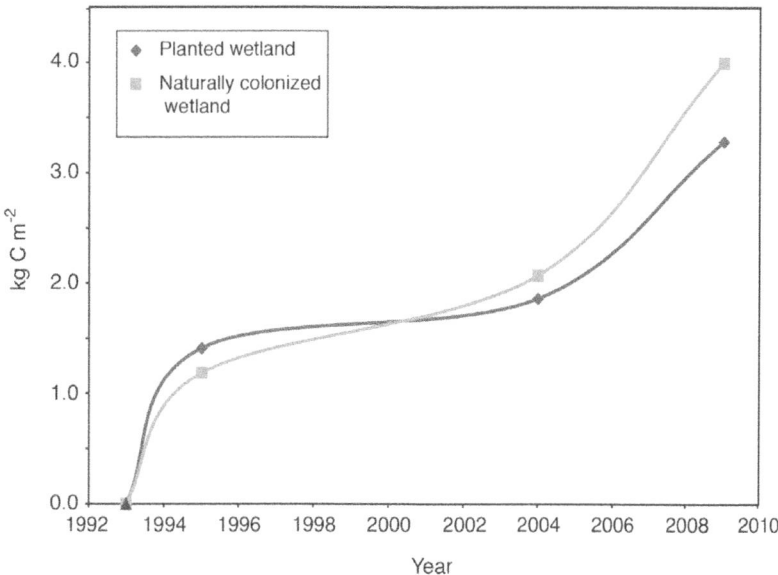

Figure 17.5 Total soil carbon accumulation in two primary-succession, flowthrough, 1-ha created wetlands in central Ohio over 15 years (1994 to 2009). 1995 carbon data are from Nairn (1996); 2004 data are from Anderson et al. (2005) and Anderson and Mitsch (2006); and 2009 data are from Bernal and Mitsch, 2013a). "Planted wetland" was planted with 2,500 individual plants representing 13 native plant species in May 1994; the "Naturally colonizing wetland" remained as an unplanted control. Each wetland had identical hydrology for the 15 years (Mitsch et al., 2012). (From Bernal and Mitsch, 2013a)

Table 17.2 Estimates of annual fluxes of methane from wetlands and other sources, Tg-CH_4/yr[a]

Sources	Megonigal et al. (2004)	Whalen (2005)	Bloom et al. (2010)
Natural wetlands	115	145	170
Tropics	65		
Northern latitude	40		
Others	10		
Other Natural Sources[b]	45	45	
Anthropogenic			
Rice Paddies	60	80	57
Other[c]	315	330	
TOTAL SOURCES	535	600	

[a] $Tg = 10^{12}$ g
[b] Other natural sources include termites, ocean, freshwater, and geological sources.
[c] Other anthropogenic sources include fossil fuels, landfills, domestic wastewater treatment, animal waste, enteric fermentation (ruminants), and biomass burning.

Tropical wetlands have been described recently as more important than originally thought for methane emissions (IPCC, 2013). Bloom (2010) suggests that 58 percent (132 Tg-CH_4 yr^{-1}) of the total methane emissions from wetlands and rice paddies (227 Tg-CH_4 yr^{-1}; see Table 17.2) comes from the tropics. Sjögersten et al. (2014) used a web analysis of current literature to estimate 90 ± 77 Tg-CH_4 yr^{-1} of methane emissions from tropical wetlands. They suggest that the methane emissions in the tropics are greater from mineral soil wetlands than organic soil wetlands.

CH_4 emissions are actually the result of two competing processes going on at the same time by microbial communities (see p. 197, "Methanogenesis," and "Methane Oxidation," in Chapter 6: "Wetland Biogeochemistry") (Fig. 17.6). The degradation of organic matter by aerobic respiration is fairly efficient in terms of energy transfer. Because of the anoxic nature of wetland soils, anaerobic processes, which are less

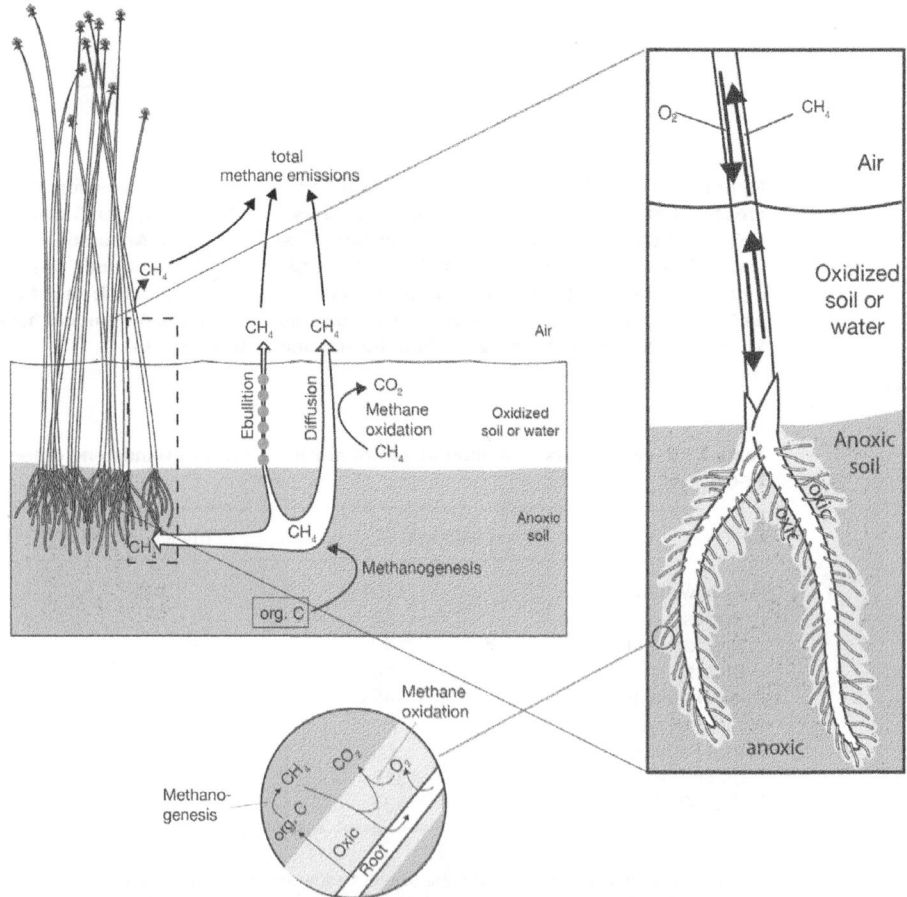

Figure 17.6 Conceptual model of CH_4 emissions, ebullition, and CH_4 oxidation in wetland soils and plant vascular system. (From Conrad, 1993 and Whalen, 2005)

efficient in terms of energy transfer, occur in close proximity to aerobic processes. *Methanogenesis* occurs when microbes called methanogens use CO_2 as an electron acceptor for the production of gaseous CH_4 or, alternatively, use a low-weight organic compound, such as one from a methyl group. CH_4 production requires extremely reduced conditions, with a redox potential of less than −200 mv, after other terminal electron acceptors oxygen (O_2), nitrates (NO_3^-), and sulfates ($SO_4^=$) have been reduced.

Conversely, nonflooded upland soils (e.g., forests, grasslands, arable land) are regarded as the major biological sink of atmospheric CH_4 (the major sink overall is tropospheric photochemistry). Obligate aerobic methanotrophic bacteria use molecular oxygen to oxidize CH_4 to CO_2 and cellular carbon. The consumption of atmospheric CH_4 is the result of two physiologically distinct microbial groups: (1) the methanotrophs, which have a membrane-bound enzyme system, and (2) an autotrophic nitrifier community. Methanotrophs are estimated to consume about 30 Tg CH_4 yr^{-1} (Whalen, 2005).

CH_4 production is much higher in the freshwater wetlands than from saltwater wetlands. A major reason for low CH_4 emissions from saltwater wetlands is the high concentration of sulfates in seawater relative to freshwater that competes with carbon for oxidizable substrate (see "Carbon–Sulfur Interactions" in Chapter 6). CH_4 emissions from studies of various freshwater wetlands around the world show have a considerable range (Table 17.3) and measurements at a given wetland are rarely normally distributed. Ebullition (see Fig. 17.6) is frequent yet hard to measure with enough frequency. In a word, it is extraordinarily difficult to obtain accurate and repeatable CH_4 emission measurements from wetlands.

Most early CH_4 emission studies were done in northern peatlands (bogs and fens) in cold climates. Moore and Roulet (1995) suggested that most annual CH_4 emission flux measurements in Canada are less than 10 g CH_4 m^{-2} yr^{-1} with the primary controlling mechanisms being soil temperature, water table position, or a combination of both. We estimate from recent studies using modern field and laboratory methods that the general range of CH_4 emissions from boreal wetlands is from 15 to 25 g-C m^{-2} yr^{-1} (Table 17.3). An early estimate of CH_4 emissions by Gorham (1991) that has been used for determining the global contributions of northern peatlands is 28 g-C m^{-2} yr^{-1}. In general, CH_4 emissions from bogs are much lower than those from the more mineral-rich fens. Aselmann and Crutzen (1989) assumed rates of CH_4 emissions in the order of increasing emissions is bogs<fens<swamps<marshes<rice paddies. Temperate wetlands emit CH_4 generally in the range of 40 to 75 g-C m^{-2} yr^{-1} (Table 17.3) although numbers are often quite variable.

In an interesting comparison of created versus natural wetlands in temperate climates, Nahlik and Mitsch (2010) found that CH_4 emissions in a reference natural flowthrough wetland in Ohio were almost twice the emission rates found in 15-year-old created flowthough marshes in Ohio (57 vs. 30 g-C m^{-2} $year^{-1}$; Table 17.3). This suggests that created and restored wetland CH_4 emissions, even 15 years after the wetlands are created, may not be nearly at rates comparable to natural wetlands yet.

In those same created wetlands in Ohio a few years before, CH_4 emissions were compared during a year when both wetlands were pulsed with six hydrologic pulses with pumped water floods, one each in months January through June 2004, the normal wet season with the following year (2005) when pumped water was steady flow all year long. CH_4 emissions during the seasonally pulsed year were considerably lower in continuously flooded zones in the flood pulsing year than in the steady flowing year (Altor and Mitsch, 2008) (Fig. 17.3). There was also a considerable difference in CH_4 emissions in both years between the continuously flooded zones and the intermittently flooded edge zones of the wetlands (Table 17.3). These results have significant

Table 17.3 Methane emissions from freshwater wetlands (Updated from Mitsch et al., 2013)

Climate and Wetland Type	Methane Emissions, $g\text{-}C\ m^{-2}\ y^{-1}$	Reference
Boreal Wetlands		
Peatlands, general	19 ± 7 (n =8)	Mitsch et al. (2013)
Canadian peatlands	<7.5	Moore and Roulet (1995)
Russian peatlands	−1.2 – 12	Heikkinen et al. (2002)
Tropical/Subtropical Freshwater Wetlands		
Tropical/subtropical wetlands	119 ± 40 (n=6)	Mitsch et al. (2013)
Amazon basin, Brazil	40 – 215	Devol et al. (1988)
Amazon basin, Brazil	30	Melack et al. (2004)
Orinoco floodplain, Veneuzela	9	Smith et al. (2000)
Tropical flowthrough wetland, Costa Rica	33 ± 5	Nahlik and Mitsch (2011)
Disturbed tropical floodplain wetland, Costa Rica	263 ± 64	Nahlik and Mitsch (2011)
Tropical rain forest basin wetland, Costa Rica	220 ± 64	Nahlik and Mitsch (2011)
Tropical seasonally flooded marsh, Botswana	72 ± 8	Mitsch et al. (2013)
Subtropical cypress strand, southwest Florida (4 wetland communities)	1 – 49	Villa and Mitsch (2014)
Temperate Freshwater Wetlands		
Temperate wetlands	58 ± 15 (n=7)	Mitsch et al. (2013)
Australian billabong	12 – 22	Sorrell and Boon (1992)
Temperate forested wetlands	35	Bartlett and Harriss (1993)
Freshwater marsh, Virginia	62	Whiting and Chanton (2001)
Louisiana freshwater marshes	3 – 225	Delaune and Pezeshki (2003)
Louisiana bottomland hardwood forest	10	Yu et al. (2008)
Spring-fed wetlands, Mississippi	51	Koh et al. (2009)
Flowthrough wetlands, Ohio (experimental pulsing and steady flow years in edge zones)	19 ± 6	Altor and Mitsch (2008)
Flowthrough wetlands, Ohio (experimental pulsing year in continuously flooded zones)	49 ± 9	Altor and Mitsch (2008)
Flowthrough wetlands, Ohio (experimental steady flow year in continuously flooded zones)	97 ± 19	Altor and Mitsch (2008)
Created temperate marshes, Ohio	30 ± 14	Nahlik and Mitsch (2010)
Reference flowthrough wetlands, Ohio	57 ± 18	Nahlik and Mitsch (2010)

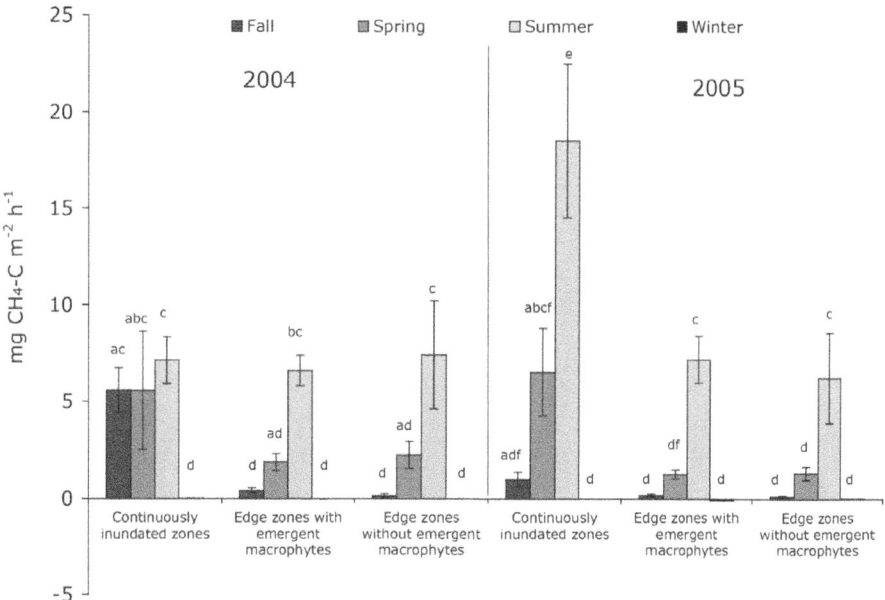

Figure 17.7 Mean methane flux rates from experimental wetlands in created riverine wetlands in central Ohio during a flood pulsed year (2004) and a steady flow year (2005). Different letters represent a significant difference (p <0.05) between wetland zones, seasons, or years. Bars represent standard error. (From Altor and Mitsch, 2008)

implications for keeping rivers and riverine wetlands free flowing with periodic floods, as opposed to flow regulation where water levels and flow rates are maintained at constant rates. Floods are good for minimizing CH_4 emissions.

Comparing Apples and Oranges: The Net Balance of Methane Production and Carbon Sequestration of Wetlands

There is a lot of confusion on the part of wetland conservationists, ecological engineers who are creating and restoring wetlands, and climatologists as to where wetlands fit into climate change. On one hand, wetlands are creating a greenhouse gas, CH_4 (and have been doing so for the ages), but on the other hand, wetlands of the world are sequestering carbon, some at significant rates. In fact, some of the fossil fuels that are now running our economy come from the organic carbon sequestered by swamps. So are wetlands good or bad for climate change?

Mitsch et al. (2013) developed a dynamic carbon model (Fig. 17.8) that included both soil carbon sequestration and CH_4 emissions to investigate this question. The model featured two carbon exchanges with the

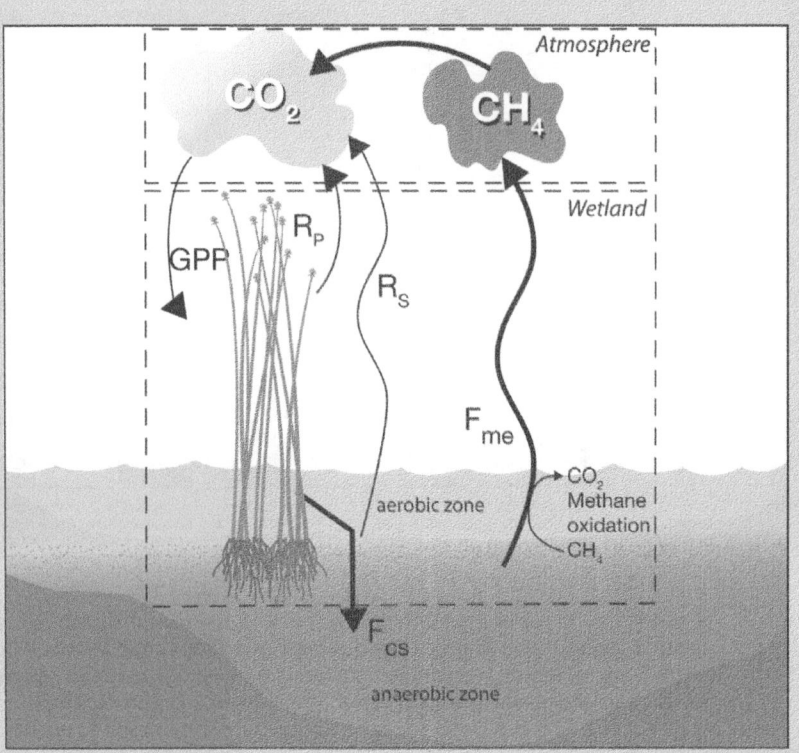

Figure 17.8 Wetland carbon simulation model designed to estimate the net effects of carbon sequestration and CH_4 emissions over time, assuming a global warming potential (GWP) of methane relative to CO_2 and linear atmospheric decay of methane. (GPP = gross primary productivity; R_p = plant respiration; R_s = soil respiration; F_{cs} = net carbon sequestration; F_{me} = methane emissions). (From Mitsch et al., 2013)

atmosphere—CH_4 emissions from the wetland to the atmosphere and CO_2 exchange to the wetland from the atmosphere. Model parameters include a half-life of seven years for CH_4 and a global warming potential (GWP) for CH_4. CH_4 emission and carbon sequestration data from 16 natural wetlands from around the world were used as inputs for the model simulations. The CO_2 equivalent is determined as:

$$CO_2 eq = CO_2 + (GWP_M \times M_{CH4}) \qquad (17.1)$$

where CO_2 = atmospheric carbon dioxide, g-CO_2 m^{-2}
M_{CH4} = atmospheric methane, g-CH_4 m^{-2}
GWP_M = global warming potential for methane = 25 for 100 years

> Model simulations showed that most of the 16 wetlands become net sinks of radiative forcing well within the 100 years. This is because the impact of CH_4 emissions is temporary in the atmosphere; CH_4 eventually decays to CO_2 and is "trumped" by the permanent burial of carbon in the wetland soil. In this set of simulations, only 2 of the 16 wetlands remain radiative sources; both were Russian peatlands that were already CO_2 sources because they had been drained. If a wetland is a CO_2 source, then it will always be a source of radiative forcing. This model results suggests that if the natural hydrology of a wetland is intact and the wetland is sequestering some CO_2 from the atmosphere, it will, with little question, be a net sink of radiative forcing and thus good for the climate.

Climate Change Feedbacks

One of the interesting questions about the vast storages of peat in northern climes related to the potential positive feedback to climate change that could occur. Because there is significantly more carbon stored in the world's soils than in the atmosphere (see Fig. 17.4), there is the potential that if the climate were to warm and accelerate decomposition of peatlands, then these peatlands would become an additional major source of carbon, through aerobic respiration and possibly fires, to the atmosphere. Davidson and Janssens (2006) summarize the comparison of uplands, which have good drainage and aeration and are therefore less prone to having large releases of CO_2 in the event of warming, to peatlands, where drainage is poor and soils are anaerobic. They describe peatland soils as enormously vulnerable to climate change compared to upland soils (Table 17.4), even though peatland soils make up a relatively small percentage of Earth's landscape. The release of 100 petagrams of carbon (Pg-C) from peatlands by the year 2100 would mean that for several years, carbon would be released at rates comparable to those currently caused by fossil fuels. If peatland productivity were to increase with the increase in temperature, it could offset this positive feedback and even lead to a negative feedback, where more carbon is sequestered than released.

Table 17.4 Below-ground carbon stocks in the world and their vulnerabilities to loss by 2100 due to global warming

Carbon Pool	Carbon Size, Pg-C	Potential Loss by 2100 from Global Warming
Upland soil inventory (3 m depth)	2,300	0–40
Peatlands (3 m depth)	450	100
Permafrost	400	100

Source: Davidson and Janssens (2006)

Christensen (1991) predicted that, as a result of a 5 percent global warming, the tundra would change from being a net sink of CO_2 to a net source of up to 1.25 Pg/yr carbon because of a combination of thermokarst erosion, deepening of the active layer in permafrost areas, lowering of the water table, and higher temperatures. Tarnocai (2006) was more direct and predicted severe degradation of the frozen peatlands in the subarctic and northern boreal Canada and severe drying in the southern boreal regions as well, but a scenario of 3° to 5°C increase in air temperature and 5° to 7°C increase over the oceans by the end of the twenty-first century. The affected area represents about 50 percent of all the organic carbon mass occurring in all Canadian wetlands.

In general, both the increase in temperature and the changes in water levels are important variables in the production of CH_4 and CO_2 from wetlands, but their relative importance for CH_4 generation is poorly understood. Using a model with inputs of a 3°C rise in temperature and a decrease in the water table between 14 and 22 cm for a subarctic fen, Roulet et al. (1992) estimated that the increased temperature raised the CH_4 flux between 5 and 40 percent, but the lowered water table decreased the CH_4 flux by 74 to 81 percent. This decrease in CH_4 flux in drier conditions was caused by a decrease in the zone of active methanogenesis and by an increase in CH_4 oxidation in the aerobic layer. Thus, the influence of global temperature rise would depend locally on the temperature increase relative to the induced change in the moisture regime.

Carbon Budgets

Carbon budgets for peatlands have drawn a great deal of interest, given the importance of these ecosystems in global carbon dynamics. A carbon budget for individual created wetland basins was already presented in Chapter 6. It is accepted that boreal peatlands were once carbon sinks, but there is little consensus that they are contemporary sinks. Carbon budgets have been developed for small peatlands (Carroll and Crill, 1997; Waddington and Roulet, 1997) and for substantial-size peatland-dominated watersheds (Rivers et al., 1998). In the latter, a 1,500-km^2 watershed in the Lake Agassiz peatlands in Minnesota illustrated that the peat watershed had a net carbon storage of 12.7 g-C m^{-2} yr^{-1} but that there was a tenuous balance between the watershed being a source and a sink of carbon (Fig. 17.9). Inflows of carbon are groundwater, precipitation, and net community productivity, while outflows are groundwater and surface flow and outgassing of CH_4. It was estimated from a companion study (Glaser et al., 1997) that peat is accumulating at a rate of 1 mm/yr (100 cm/1000 yr). This budget illustrates the importance of accurate hydrologic measurements as well as biological productivity measurements in determining accurate carbon budgets for wetlands and wetland landscapes.

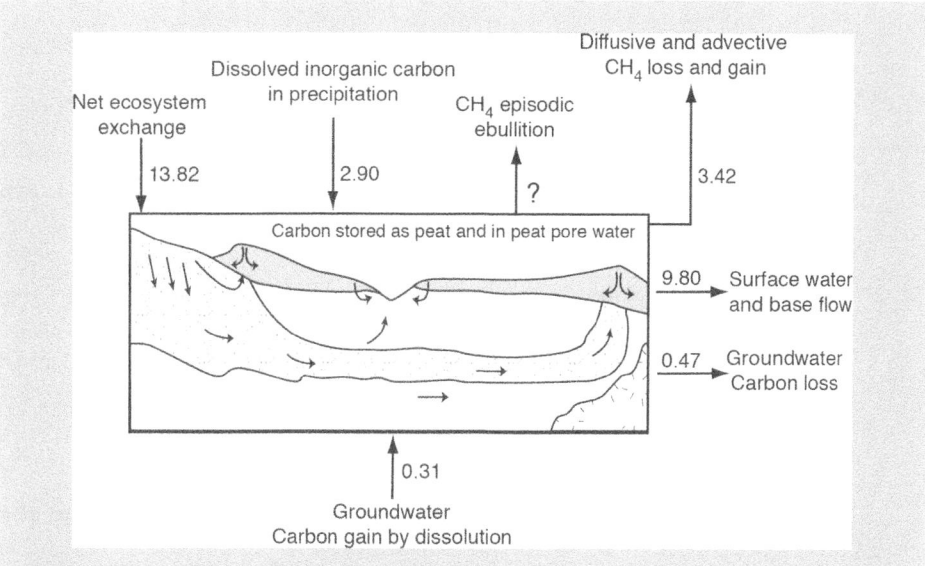

Figure 17.9 Carbon budget of the 1,500-km² Rapid River watershed in the Lake Agassiz peatland basin of northern Minnesota. Fluxes are in g-C m⁻² yr⁻¹. (After Rivers et al., 1998)

Effects of Climate Change on Wetlands

Wetlands may be key ecosystems for mitigating the effects of fossil fuel emissions on climate. Conversely, sea-level and temperature changes may have significant impacts on coastal and inland wetlands.

Coastal Wetlands

One of the major impacts of possible climate changes on wetlands is the effect that sea-level rise will have on coastal wetlands. Estimates of sea-level rise over the next century range from 50 to 200 cm. (Fig. 17.2 shows the current rate of sea level rise of 32 cm/century.) It has been estimated that if sea level were to rise by 100 cm, half of the wetlands designated by the Ramsar Convention as wetlands of international importance would be threatened (Nicholls, 2004). The regions where wetlands are most at risk, even for a 44-cm rise in sea level by 2080, are shown in Figure 17.10. If the rise in sea level is not accompanied by equivalent vertical accretion of marsh sediments, then coastal marshes will gradually disintegrate as a result of increased inundation, erosion, and saltwater intrusion. Because much of the coastline of the world is developed, efforts to protect dry upland from inundation by the construction of bulkheads or dikes will exacerbate the problem. In essence, the wetlands will be trapped between the rising sea and the protected dry land, a situation that has already occurred over

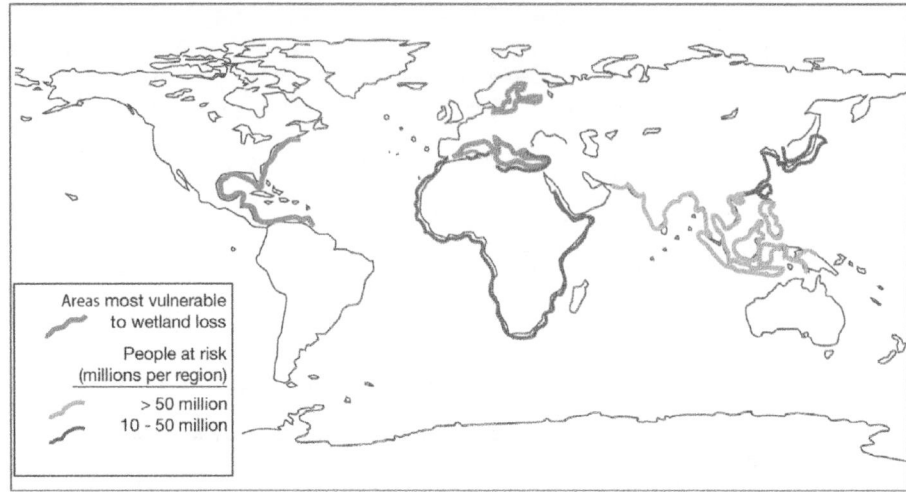

Figure 17.10 Coastal wetland areas most vulnerable to a sea-level rise of 44 cm by 2080. (From IPCC, 2001)

the centuries in the Netherlands and China. This effect has been termed the *coastal squeeze* of sea-level rise (Nicholls, 2004). Even in the absence of bulkheads in most of our regions where coastal wetlands exist, "the slope above the wetland is steeper than that of the wetlands; so a rise in sea level causes a net loss of wetland acreage" (Titus, 1991).

Estimates of the loss of coastal wetlands in the United States vary, with much of the variability dependent on the assumed sea-level rise and the degree to which dry land is protected at all cost (Table 17.5). If there is no shoreline protection, a sea-level rise of 1 m could reduce coastal wetlands by 26 to 66 percent. If the policy were to protect all dry land, then the estimated loss of wetlands increases dramatically to 50 to 82 percent. How well these figures can be extrapolated to the rest of the world is unclear. In long-developed coastlines, such as those of Europe and the Far East, the losses would probably be less.

Table 17.5 Estimated percentage coastal wetland loss in the United States with sea-level rise

	Sea-Level Rise		
	0.5 m	1 m	2 m
If no shores are protected	17–43%	26–66%	29–76%
If densely developed dry land is protected	20–45%	29–69%	33–80%
If all dry land is protected	38–61%	50–82%	66–90%

Source: Titus (1991)

The Mississippi River Delta in Louisiana may be a model for seeing the effects of global sea-level rise on coastal wetlands. Here, the "apparent" sea-level rise is already 1 m/100 yr (1 cm/yr), primarily because of sediment subsidence rather than actual sea-level rise. In this delta marsh, vertical accretion is not keeping up with subsidence, in part because the Mississippi River is carrying only about 20 percent of the sediment load it did in 1850 (Kesel and Reed, 1995) and its flow is contained within levees, so riverborne sediments no longer reach the wetlands during spring floods. As a result, this region has the highest rate of wetland loss in the United States. Day et al. (2005) describe the ramifications of global climate change on restoration efforts now under way in the delta. With a sea-level rise of 30 to 50 cm by 2100 possible, the relative sea-level rise will increase from 1 cm/yr (caused mostly by land subsidence) to 1.3 to 1.7 cm/yr, exacerbating an already difficult situation of wetland loss in the Louisiana Delta. In addition, Day et al. (2005) note that as a result of milder temperatures already, mangrove swamps were beginning to replace their temperate-zone analog, the salt marsh, in several locations in the delta. This mangrove expansion is another effect that would be expected in subtropical regions that were previously dominated by salt marshes. Mangroves are valuable coastal ecosystems, as are salt marshes, but the overall effects of this substitution of ecosystems is unclear.

Management of Coastal Wetlands

There are few management possibilities for managing coastal wetlands in the face of sea-level rise. Figure 17.11 shows two future conditions. In Future 1, the house is protected with a bulkhead in the face of rising sea level, and the salt marsh is lost

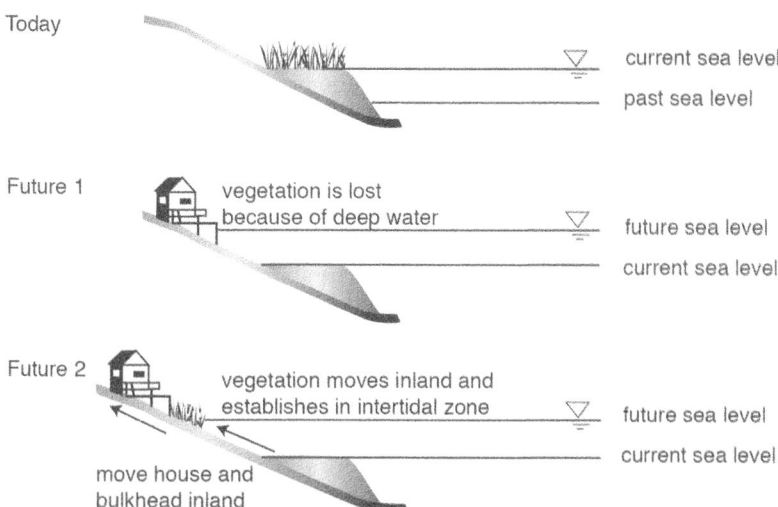

Figure 17.11 Coastal wetland management scenarios in the face of sea-level rise. Future 1 is without moving human habitation inland. Future 2 involves moving human activity inland to allow room for the wetland to move inland. (From Titus, 1991)

or "squeezed out." In Future 2, the house is moved upland to accommodate the wetland, which would begin to form if a gentle slope and adequate sediment sources were available.

Future 2 models the wetlands of the Laurentian Great Lakes, which, for centuries, were "wetlands on skateboards," moving inland and lakeward with frequent (over periods of decades) water-level changes in the lakes (Mitsch, 1992). With stabilization of the coastline in the past century, diking the remaining wetlands along the Great Lakes became necessary for their survival.

Day and Templet (1989) and Day et al. (2005) concluded, after extensive investigation of the apparent sea-level rise in coastal Louisiana, that we can manage coastal wetlands in periods of rising sea level through comprehensive, long-range planning and through the application of the principles of ecological engineering by using nature's energies, such as upstream riverine sediments and fresh water, vegetation productivity, winds, currents, and tides, as much as possible.

Inland Wetlands

In addition to the effects of climate change on coastal wetlands through sea-level rise, the change in climate, particularly temperature (Fig. 17.1), will probably affect the function and distribution of inland wetlands. In the tundra, any melting of the permafrost would result in the loss of wetlands. In boreal and temperate areas, climate change would result in changing rainfall patterns, thus affecting runoff and groundwater inflows to wetlands. In general, a decrease in precipitation or an increase in evapotranspiration will result in less-frequent flooding of existing wetlands, although the types of wetlands may not change. Greater precipitation patterns would increase the length and depth of flooding of inland wetlands. Most susceptible to these effects are depressional wetlands that have small watersheds and that are in regions between arid and mesic climates, such as the prairie potholes of North America.

The impact of climate change on the Prairie Pothole Region (PPR) of North America was investigated by Johnson et al. (2005). These wetlands provide 50 to 80 percent of the continent's duck population and are exactly on the edge between areas to the east with abundant precipitation and arid climates to the west. By using a wetland simulation model, Johnson et al. (2005) were able to predict areas in the pothole region that would have highly favorable water conditions for three climate scenarios: (1) a 3°C temperature increase with no change in precipitation; (2) a 3°C temperature increase with a 20 percent increase in precipitation; and (3) a 3°C temperature increase with a 20 percent decrease in precipitation (Fig. 17.12). Basically any temperature increase coupled with precipitation decrease shifted the area favorable for ducks to the east. Overall, the climate change would "diminish the benefits of wetland conservation in the central and western PPR. Simulations further indicate that restoration of wetlands along wetter fringes of the PPR may be necessary to ameliorate potential impacts of climate change on waterfowl populations" (Johnson et al., 2005).

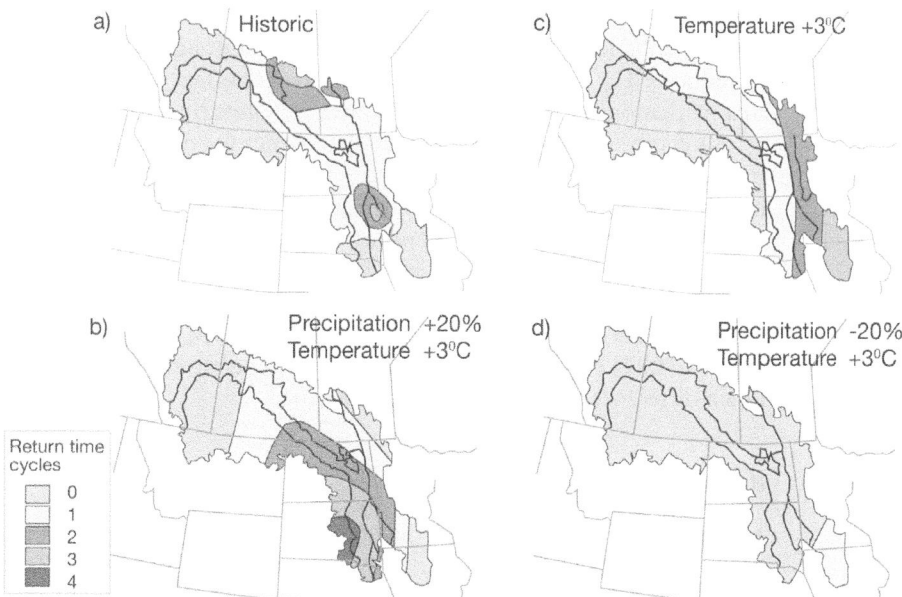

Figure 17.12 Simulation results for locations of highly favorable water and cover conditions in the Prairie Pothole Region of North America for waterfowl breeding under options (a) historic; (b) a 3°C temperature increase with no change in precipitation; (c) a 3°C temperature increase with a 20 percent increase in precipitation; and (d) a 3°C temperature increase with a 20 percent decrease in precipitation. (From Johnson et al., 2005, reprinted with permission)

Management of Inland Wetlands

Limited experimentation, especially in rice paddies, suggests some management alternatives that might be appropriate for inland wetlands, especially to reduce methane emissions. Sass et al. (1992) measured the effects on methane emissions of four different water management methods in some rice fields in Texas and found that temporary drainage (midseason drainage and multiple aeration) decreased CH_4 emission caused by both increased CH_4 consumption in the aerobic layer and decreased CH_4 production. Such management may be practical only in flat systems with sufficient control of water levels.

Nutrient and compost management may also offer opportunities for reducing methane emissions. There may be a relationship between the carbon:nitrogen (C:N) ratio of the organic matter in wetlands and CH_4 emissions, although the trends are not clear. Yagi and Minami (1990) found, in rice paddies in Japan, that compost with a low C:N ratio (enriched in nitrogen) causes lower emissions of methane than uncomposted rice straw with a high C:N ratio. Conversely, Schutz et al. (1989) found high emissions in fields applied with compost. Because of its competition with methanogenesis, enhancing sulfate reduction is often suggested as a management alternative to reduce CH_4 emissions. This has long been known as one of the primary reasons that methanogenesis is lower in saltwater wetlands than in freshwater wetlands.

One of the easiest management approaches for minimizing CH_4 emissions from freshwater wetlands is to allow the wetlands to have their natural fluctuating hydroperiods and, in some cases, a pulsing hydrology. Studies by Altor and Mitsch (2006, 2008) described above showed that a pulsing hydrology had CH_4 emissions that were much lower than those from permanently flooded sites.

We cannot estimate, at present, with much certainty whether wetlands are significant global carbon sources or sinks. Nevertheless, the opportunities for managing CO_2 and CH_4 emissions in wetlands are not generally on a scale large enough to make much difference to the global carbon balance.

Recommended Readings

Mitsch, W. J., B. Bernal, A. M. Nahlik, U. Mander, L. Zhang, C. J. Anderson, S. E. Jørgensen, and H. Brix. 2013. Wetlands, carbon, and climate change. *Landscape Ecology* 28: 583–597.

The Royal Society and the National Academy of Sciences. 2014. *Climate Change: Evidence and Causes.* An overview from the Royal Society and the US National Academy of Sciences. London and Washington DC: Royal Society and US NAS.

Whalen, S. C. 2005. Biogeochemistry of methane exchange between natural wetlands and the atmosphere. *Environmental Engineering Science* 22: 73–94.

References

Altor, A. E., and W. J. Mitsch. 2006. Methane flux from created riparian marshes: Relationship to intermittent versus continuous inundation and emergent macrophytes. *Ecological Engineering* 28: 224–234.

Altor, A. E., and W. J. Mitsch. 2008. Pulsing hydrology, methane emissions, and carbon dioxide fluxes in created marshes: A 2-year ecosystem study. *Wetlands* 28: 423–438.

Anderson, C. J., W. J. Mitsch, and R. W. Nairn. 2005. Temporal and spatial development of surface soil conditions in two created riverine marshes. *Journal of Environmental Quality* 34: 2072–2081.

Anderson, C. J., and W. J. Mitsch. 2006. Sediment, carbon, and nutrient accumulation at two 10-year-old created riverine marshes. *Wetlands* 26: 779–792.

Aselmann, I., and P. J. Crutzen. 1989. Global distribution of natural freshwater wetlands and rice paddies, their net primary productivity, seasonality and possible methane emissions. *Journal of Atmospheric Chemistry* 8: 307–358.

Bartlett, K. B., and R. C. Harriss. 1993. Review and assessment of methane emissions from wetlands. *Chemosphere* 26: 261–320.

Bernal, B., and W. J. Mitsch. 2012. Comparing carbon sequestration in temperate freshwater wetland communities. *Global Change Biology* 18: 1636–1647.

Bernal, B., and W. J. Mitsch. 2013a. Carbon sequestration in two created riverine wetlands in the Midwestern United States. *Journal of Environmental Quality* 42: 1236–1244.

Bernal. B., and W. J. Mitsch. 2013b. Carbon sequestration in freshwater wetlands of Costa Rica and Botswana. *Biogeochemistry.* 115: 77–93.

Bloom, A. A., P. I. Palmer, A. Fraser, D. S. Reay, and C. Frankenberg. 2010. Large-scale controls of methanogenesis inferred from methane and gravity spaceborne data. *Science* 327: 322–325.

Breithaupt, J. L., J. M. Smoak, T. J. Smith, C. J. Sanders, and A. Hoare 2012, Organic carbon burial rates in mangrove sediments: Strengthening the global budget. *Global Biogeochem.* Cycles 26: GB3011.

Brix, H., B. K. Sorrell, and B. Lorenzen. 2001. Are *Phragmites*-dominated wetlands a net source or net sink of greenhouse gases? *Aquatic Botany* 69: 313–324.

Carroll, P., and P. Crill. 1997. Carbon balance of a temperate poor fen. *Global Biogeochemical Cycles* 11: 349–356.

Chimner, R. A., and K. C. Ewel. 2005. A tropical freshwater wetland: II. Production, decomposition, and peat formation. *Wetlands Ecology and Management* 13: 671–684.

Chmura, G. L., S. C. Anisfeld, D. R. Cahoon, and J. C. Lynch. 2003. Global carbon sequestration in tidal, saline wetland soils. *Global Biogeochemical Cycles* 17:1111.

Christensen, T. 1991. Arctic and sub-Arctic soil emissions: Possible implications for global climate change. *Polar Record* 27: 205–210.

Conrad, R. 1993. Controls of methane production in terrestrial ecosystems. In M. O. Andreae and D. S. Schimel, eds. *Exchange of Trace Gases between Terrestrial Ecosystems and the Atmosphere.* John Wiley & Sons, New York, pp. 39–58.

Craft, C. 2007. Freshwater input structures soil properties, vertical accretion, and nutrient accumulation of Georgia and U.S. tidal marshes. *Limnology and Oceanography* 52:1220–1230.

Craft, C., C. Washburn, and A. Parker. 2008. Latitudinal trends in organic carbon accumulation in temperate freshwater peatlands. In J. Vymazal, ed. *Wastewater Treat Plant Dynamics and Management in Constructed Natural Wetlands.* Springer, Dordrecht, the Netherlands, p. 23–31.

Craft, C., J. Clough, J. Ehman, J. Samantha, R. Park, P. Pennings, H. Guo, and M. Machmuller. 2009. Forecasting the effects of sea-level rise on tidal marsh ecosystem services. *Frontiers in Ecology and the Environment* 7:73–78.

Davidson, E. A., and I. A. Janssens. 2006. Temperature sensitivity of soil carbon decomposition and feedbacks to climate change. *Nature* 440: 165–173.

Day, J. W., Jr., and P. H. Templet. 1989. Consequences of sea level rise: Implications from the Mississippi Delta. *Coastal Management* 17: 241–257.

Day, J. W., Jr., J. Barras, E. Clairain, J. Johnston, D. Justic, G. P. Kemp, J.-Y. Ko, R. Lane, W. J. Mitsch, G. Steyer, P. Templet, and A. Yañez-Arancibia. 2005.

Implications of global climatic change and energy cost and availability for the restoration of the Mississippi Delta. *Ecological Engineering* 24: 253–265.

Delaune, R. D., and S. Pezeshki. 2003. The role of soil organic carbon in maintaining surface elevation in rapidly subsiding U.S. Gulf of Mexico coastal marshes. *Water, Air and Soil Pollution* 3: 167–179.

Devol, A. H., J. E. Richey, W. A. Clark, S. L. King. 1988. Methane emissions to the troposphere from the Amazon floodplain. *Journal of Geophysical Research* 93: 1492–1583.

Euliss, N. H., R. A. Gleason, A. Olness, R. L. McDougal, H. R. Murkin, R. D. Robarts, R. A. Bourbonniere, and B. G. Warner. 2006. North American prairie wetlands are important nonforested land-based carbon storage sites. *Science of the Total Environment* 361: 179–188.

Glaser, P. H., P. C. Bennett, D. I. Siegel, and E. A. Romanowicz. 1997. Palaeo-reversals in groundwater flow and peatland development at Lost River, Minnesota, USA. *Holocene* 6: 413–421.

Gorham, E. 1991. Northern peatlands: Role in the carbon cycle and probable responses to climatic warming. *Ecological Applications* 1: 182–195.

Hadi, A., K. Inubushi, Y. Furukawa, E. Purnomo, M. Rasmadi, and H. Tsuruta. 2005. Greenhouse gas emissions from tropical peatlands of Kalimantan, Indonesia. *Nutrient Cycling in Agroecosystems* 71: 73–80.

Heikkinen J. E. P., V. Elsakov, and P. J. Martikainen. 2002. Carbon dioxide and methane dynamics and annual carbon budget in tundra wetland in NE Europe, Russia. *Global Biogeochemical Cycles* 16. doi:10.1029/2002GB0001930

Hendriks, D. M. D., J. van Huissteden, A. J. Dolman, and M. K. van der Molen. 2007. The full greenhouse gas balance of an abandoned peat meadow. *Biogeosciences* 4: 411–424.

Howe, A. J., J. F. Rodriguez, and P. M. Saco. 2009. Surface evolution and carbon sequestration in disturbed and undisturbed wetland soils of the Hunter Estuary, southeast Australia. *Estuarine, Coastal and Shelf Science* 84: 75–83.

Intergovernmental Panel on Climate Change (IPPC). 2001. *Climate Change 2001: The Scientific Basis*. Published for the Intergovernmental Panel on Climate Change. Cambridge University Press, UK.

Intergovernmental Panel on Climate Change (IPCC). 2013. *Climate Change 2013: The Physical Science Basis. Contribution of Working Group I to the Fifth Assessment Report of the Intergovernmental Panel on Climate Change*, ed. T. F. Stocker, D. Qin, G.-K. Plattner, M. Tignor, S. K. Allen, J. Boschung, A. Nauels, Y. Xia, V. Bex, and P. M. Midgley. Cambridge University Press, Cambridge, UK. 1535 pp.

Intergovernmental Panel on Climate Change. 2014a. *Climate Change 2014: Impacts, Adaptation, and Vulnerability. Parts A and B*. Cambridge University Press, Cambridge, UK. 1132 pp and 688 pp.

Intergovernmental Panel on Climate Change, 2014b. *Climate Change 2014: Mitigation of Climate Change, Working Group III Contribution to the IPCC 5th Assessment Report*. IPCC, http://www.ipcc.ch/report/ar5/wg3/. 99 pp.

Johnson, W. C., B. V. Millett, T. Gilmanov, R. A. Voldseth, G. R. Guntenspergen, and D. E. Naugle. 2005. Vulnerability of northern prairie wetlands to climate change. *BioScience* 55: 863–872.

Kesel, R., and D. J. Reed. 1995. Status and trends in Mississippi River sediment regime and its role in Louisiana wetland development. In D. J. Reed, ed. *Status and Historical Trends of Hydrologic Modification, Reduction in Sediment Availability, and Habitat Loss/Modification in the Barataria–Terrebonne Estuarine System*. BT/NEP Publication 20, Barataria–Terrebonne National Estuary Program, Thibodaux, LA, pp. 80–98.

Koh, H.-S., C. A. Ochs, and K. Yu. 2009. Hydrologic gradient and vegetation controls on CH_4 and CO_2 fluxes in a spring-fed forested wetland. *Hydrobiologia* 630: 271–286.

Lal, R. 2008. Carbon sequestration. *Philosophical Transactions of the Royal Society* B 363: 815–830.

Mcleod, E., G. L. Chmura, S. Bouillon, R. Salm, M. Björk, C. M. Duarte, C. E. Lovelock, W. H. Schlesinger, and B. R Silliman. 2011. A blueprint for blue carbon: toward an improved understanding of the role of vegetated coastal habitats in sequestering CO_2. *Frontiers of Ecology and the Environment* 9: 552–560.

Megonigal, J. P., M. E. Hines, and P. T. Visscher. 2004. Anaerobic metabolism: Linkages to trace gases and aerobic processes. In W. H. Schlesinger, ed. *Biogeochemistry*. Elsevier-Pergamon, Oxford, UK, pp. 317–424.

Melack, J. M., L. L. Hess, M. Gastil, B. R. Forsberg, S. K. Hamilton, I. B. T. Lima, and E. M. L. M. Novo. 2004. Regionalization of methane emissions in the Amazon Basin with microwave remote sensing. *Global Change Biology* 10:530–544.

Mitsch, W. J. 1992. Combining ecosystem and landscape approaches to Great Lakes wetlands. *Journal of Great Lakes Research* 18:552–570.

Mitsch, W. J., and X. Wu. 1995. Wetlands and global change. In R. Lal, J. Kimble, E. Levine, and B. A. Stewart, eds. *Advances in Soil Science, Soil Management, and Greenhouse Effect*. CRC Press/Lewis Publishers, Boca Raton, FL, pp. 205–230.

Mitsch, W. J., and J. G. Gosselink. 2007. *Wetlands*, 4th ed. John Wiley & Sons, Hoboken, New Jersey, 582 pp.

Mitsch, W. J., L. Zhang, K. C. Stefanik, A. M. Nahlik, C. J. Anderson, B. Bernal, M. Hernandez, and K. Song. 2012. Creating wetlands: Primary succession, water quality changes, and self-design over 15 years. *BioScience* 62: 237–250.

Mitsch, W. J., B. Bernal, A. M. Nahlik, U. Mander, L. Zhang, C. J. Anderson, S. E. Jørgensen, and H. Brix. 2013. Wetlands, carbon, and climate change. *Landscape Ecology* 28: 583–597.

Moore, T. R., and N. T. Roulet. 1995. Methane emissions from Canadian peatlands. In R. Lal, J. Kimble, E. Levine, and B. A. Stewart, eds. *Advances in Soil Science: Soils and Global Change*. CRC Press, Boca Raton, FL, pp. 153–164.

Nahlik, A. M., and W. J. Mitsch. 2010. Methane emissions from created riverine wetlands. *Wetlands* 30: 783–793. Erratum in *Wetlands* (2011) 31: 449–450.

Nahlik, A. M., and W. J. Mitsch. 2011. Methane emissions from tropical freshwater wetlands located in different climatic zones of Costa Rica. *Global Change Biology* 17: 1321–1334.

Nairn, R. W. 1996. Biogeochemistry of newly created riparian wetlands: Evaluation of water quality changes and soil development. Ph.D. dissertation, Graduate Program in Environmental Science, Ohio State University, Columbus.

Nicholls, R. J. 2004. Coastal flooding and wetland loss in the 21st century: Changes under the SRES climate and socio-economic scenarios. *Global Environmental Change* 14: 69–86.

Page, S. E., R. A. J. Wust, D. Weiss, J. O. Rieley, W. Shotyk, and S. H. Limin. 2004. A record of late Pleistocene and Holocene carbon accumulation and climate change from an equatorial peat bog (Kalimantan, Indonesia): Implications for past, present, and future carbon dynamics. *Journal of Quaternary Science* 19: 625–635.

Reddy, K. R., R. D. DeLaune, W. F. DeBusk, and M. S. Koch. 1993. Long-term nutrient accumulation rates in the Everglades. *Soil Science Society of America Journal* 57: 1147–1155.

Rivers, J. S., D. I. Siegel, L. S. Chasar, J. P. Chanton, P. H. Glaser, N. T. Roulet, and J. M. McKenzie. 1998. A stochastic appraisal of the annual carbon budget of a large circumboreal peatland, Rapid River Watershed, northern Minnesota. *Global Biogeochemical Cycles* 12: 715–727.

Roulet, N. T. 2000. Peatlands, carbon storage, greenhouse gases, and the Kyoto Protocol: Prospects and significance for Canada. *Wetlands* 20: 605–615.

Roulet, N. T., T. R. Moore, J. Bubier, and P. Lafleur. 1992. Northern fens: CH4 flux and climate change. *Tellus* 44B: 100–105.

Sass, R. L., F. M. Fisher, Y. B. Wang, F. T. Turner, and M. F. Jund. 1992. Methane emission from rice fields: The effect of floodwater management. *Global Biogeochemical Cycles* 6: 249–262.

Saunders M. J., M. B. Jones, and F. Kansiime. 2007. Carbon and water cycles in tropical papyrus wetlands. *Wetlands Ecology and Management* 15: 489–498.

Schutz, H., A. Holzapfel-Pschorn, R. Conrad, H. Rennenberg, and W. Seiler. 1989. A three year continuous record on the influence of daytime, season and fertilizer treatment on methane emission rates from an Italian rice paddy field. *Journal of Geophysical Research* 94: 16405–16416.

Sjögersten, S., C. R. Black, S. Evers, J. Hoyos-Santillan, E. L. Wright, and B. L. Turner. 2014. Tropical wetlands: A missing link in the global carbon cycle? *Global Biogeochem. Cycles* 28. doi:10.1002/2014GB004844.

Smith, K.A., K.E. Dobbie, B.C. Ball, et al. 2000. Oxidation of atmospheric methane in Northern European soils, comparison with other ecosystems, and uncertainties in the global terrestrial sink. *Global Change Biology* 6: 791-803.

Sorrell, B. K., and P. I. Boon. 1992. Biogeochemistry of billabong sediments. II. Seasonal variations in methane production. *Freshwater Biology* 27: 435–445.

Suratman, M. H. 2008. Carbon sequestration potential of mangroves in southeast Asia. In F. Bravo, V. LeMay, R. Jandl, and K. von Gadow, eds., *Managing Forest Ecosystems. The Challenge of Climate Change*. Springer Science, Dordrecht, Netherlands, pp. 297–315.

Tarnocai, C. 2006. The effect of climate change on carbon in Canadian peatlands. *Global and Planetary Change* 53: 222–232.

The Royal Society and the National Academy of Sciences. 2014. *Climate Change: Evidence and Causes*. An overview from the Royal Society and the US National Academy of Sciences, Royal Society and US NAS, London and Washington DC.

Titus, J. G. 1991. Greenhouse effect and coastal wetland policy: How Americans could abandon an area the size of Massachusetts at minimum cost. *Environmental Management* 15: 39–58.

Turunen, J., E. Tomppo, K. Tolonen, and E. Reinkainen. 2002. Estimating carbon accumulation rates of undrained mires in Finland: Application to boreal and subarctic regions. *The Holocene* 12: 79–90.

Vaidyanathan, G. 2011. "Blue carbon" plan takes shape. *Nature*. doi:10.1038/news.2011.112

Villa, J. A. and W. J. Mitsch. 2014. Methane emissions from five wetland plant communities with different hydroperiods in the Big Cypress Swamp region of Florida Everglades. *Ecohydrology & Hydrobiology* 14: 253–266.

Villa, J. A. and W. J. Mitsch. 2015. Carbon sequestration in different wetland plant communities in Southwest Florida. *International Journal for Biodiversity Science, Ecosystems Services and Management*. doi.org/10.1080/21513732.2014.973909.

Waddington, J. M., and N. T. Roulet. 1997. Groundwater flow and dissolved carbon movement in a boreal peatland. *Journal of Hydrology* 191: 122–138.

Whalen, S. C. 2005. Biogeochemistry of methane exchange between natural wetlands and the atmosphere. *Environmental Engineering Science* 22: 73–94.

Whiting, G. J., and J. P. Chanton. 2001. Greenhouse carbon balance of wetlands: Methane emission versus carbon sequestration. *Tellus* 53B: 521–528.

World Wildlife Fund. 2012. Blue carbon: A new concept for reducing the impacts of climate change by conserving coastal ecosystems in the Coral Triangle. WWF-Australia, Brisbane, Queensland, Australia. 20 pp.

Yagi, K., and K. Minami. 1990. Effect of organic matter application on methane emission from some Japanese paddy fields. *Soil Science and Plant Nutrition* 36: 599–610.

Yu, K., S. P. Faulkner, and M. J. Baldwin. 2008. Effect of hydrological conditions on nitrous oxide, methane, and carbon dioxide dynamics in a bottomland hardwood forest and its implication for soil carbon sequestration. *Global Change Biology* 14: 798–812.

Chapter **18**

Wetland Creation and Restoration

Loss rates of wetlands around the world and the subsequent recognition of wetland values have stimulated restoration and creation of these systems. Policies such as "no net loss" of wetlands in the United States have made wetland creation and restoration a veritable industry in that country. Wetland restoration involves returning a wetland to its original or previous wetland state, whereas wetland creation involves conversion of uplands or shallow open-water systems to vegetated wetlands. Wetland restoration and creation can occur for replacement of habitat, for coastal restoration, and for restoration of mined peatlands. Wetland mitigation banks may overcome many of the limitations of current approaches to replacing lost wetlands but they are controversial. Generally, wetland restoration and creation first involve establishment or reestablishment of appropriate natural hydrologic conditions, followed by establishment of appropriate vegetation communities. Although many of these created and restored wetlands have become functional, there have been some cases of failures of created or restored wetlands, generally caused by a lack of proper hydrology. Creating and restoring wetlands should be based on the concept of self-design, whereby any number of native propagules can be introduced, but the ecosystem adapts and changes itself according to its physical constraints, and success should not be determined solely by specific plant and animal presence. Giving these systems sufficient time to carry out their self-design is another factor that is generally overlooked.

There are two general starting points for anyone interested in getting involved in wetland restoration and creation:

1. Learn and understand wetland science and its principles first.
2. Broaden your horizons beyond the field that you were trained in so that you resist the ever-present temptation to overengineer, overbotanize, or overzoologize the wetlands that you create and restore.

The principles and practices of wetland creation and restoration are based on wetland science (hydrology, biogeochemistry, adaptations, and succession). Our advice if you are interested in creating and restoring wetlands is to first become an expert in wetland science. Know how the real wetlands work first. That was the intent of the first 17 chapters in this book. Only after you understand the function and structure of natural wetlands are you qualified to create and restore wetlands.

The second point is one that needs to be emphasized to all professions. Most of us have been taught in our lives and professions that we can improve on nature. Indeed, human civilization is based on that premise. But when we are attempting to create or re-create natural ecosystems, Mother Nature is in control. In all situations of wetland creation and restoration, human contribution to the design of wetlands should be kept simple and should strive to stay within the bounds established by the natural landscape.

The literature on wetland creation and restoration continues to explode, and it is impossible for us to include all of the possible principles, case studies, and techniques in this chapter. A critique of the policies and techniques of wetland creation and restoration in the United States was published as a National Academy of Sciences report (NRC 2001). Mitsch (2013) provided a summary of wetland creation and restoration and several case studies from around the world, some of which are updated here. Several of these case studies were later reviewed and received "mid-term grades" (Mitsch, 2014). Some notable papers discuss specific wetland creation and restoration projects for salt marshes (Alphin and Posey, 2000; Craft et al., 2002; Edwards and Proffitt, 2003; Callaway and Zedler, 2004; Peterson et al., 2005), mangrove swamps (Lewis, 2005; Lewis and Gilmore, 2007), freshwater marshes (Atkinson et al., 2005; Mitsch et al., 2012, 2014), peatlands (Gorham and Rochefort, 2003), and forested wetlands (Rodgers et al., 2004).

Definitions

Several terms are frequently used in connection with the creation and restoration of wetlands. Precise definitions are important, and confusion about the exact meaning of wetland creation, restoration, and related terms is common (Lewis, 1990a). Bradshaw (1996) concurred that "we must be clear in what is being discussed."

Wetland restoration refers to the return of a wetland from a disturbed or altered condition caused by human activity to a previously existing condition. The wetland may have been degraded or hydrologically altered, and restoration then may involve reestablishing hydrologic conditions to reestablish previous vegetation communities.

Wetland creation refers to the conversion of a persistent upland or shallow water area into a wetland by human activity.

Wetland enhancement refers to a human activity that increases one or more functions of an existing wetland.

> One type of created wetland, a *constructed wetland*, refers to a wetland that has been developed for the primary purpose of contaminant or pollution removal from wastewater or runoff.
>
> This last type of wetland is also referred to as a *treatment wetland* and is the main topic discussed in Chapter 19: "Wetlands and Water Quality."

Significant efforts now focus on the voluntary restoration and creation of wetlands. Part of the interest in wetland creation and restoration stems from the fact that we are losing or have lost so much of this valuable habitat (see Chapter 3: "Wetlands of the World"). Often interest is less voluntary and more in response to government policies, such as "no net loss" in the United States, that require the replacement of wetlands for those unavoidably lost. New Zealand, which has lost 90 percent of its wetlands, has major efforts under way to restore marshes and other wetlands in the Waikato River Basin on North Island and in the vicinity of Christchurch on South Island. In southeastern Australia, restoration of the Murray-Darling watersheds, particularly the riverine billabongs, has become a major undertaking, while coastal plain wetland restoration and creation are occurring in southwestern Australia.

There are concerted efforts to restore mangrove forests in the Mekong Delta of Vietnam, along South American coastlines where shrimp farming has destroyed thousands of hectares of mangroves, and around the Indian Ocean to provide tsunami and typhoon protection for coastal areas. Tidal marshes have been created along much of China's eastern coastline, and wetland creation and restoration are now occurring in the Yangtze Delta in Shanghai and upstream of the Three Gorges Dam on the upper Yangtze. Wetland restoration and creation are being proposed or implemented on very large scales to prevent more deterioration of existing wetlands (Everglades in Florida), to mitigate the loss of fisheries (Delaware Bay in eastern United States), to reduce land loss and provide protection from hurricanes (Mississippi Delta in Louisiana), to stabilize a watershed and provide water quality improvement (Skjern River, Denmark), and to solve serious cases of overenrichment of coastal waters (Baltic Sea in Scandinavia; Gulf of Mexico in United States; Laurentian Great Lakes).

Mitigating Wetland Habitat Loss

Wetland protection regulations in the United States and now elsewhere have led to the practice of requiring that wetlands be created, restored, or enhanced to replace wetlands lost in developments such as highway construction, coastal drainage and filling, or commercial development. This is referred to as the process of "mitigating" the original loss, and these "new" wetlands are often called *mitigation wetlands*. (Note: To *mitigate* means to "make less harsh or harmful." The term *mitigation wetland* or *wetland mitigation* is therefore poor use of English. We should rather refer to "mitigating the loss of a wetland.") Perhaps it might be more appropriate to refer to these wetlands as *replacement wetlands*.

594 Chapter 18 Wetland Creation and Restoration

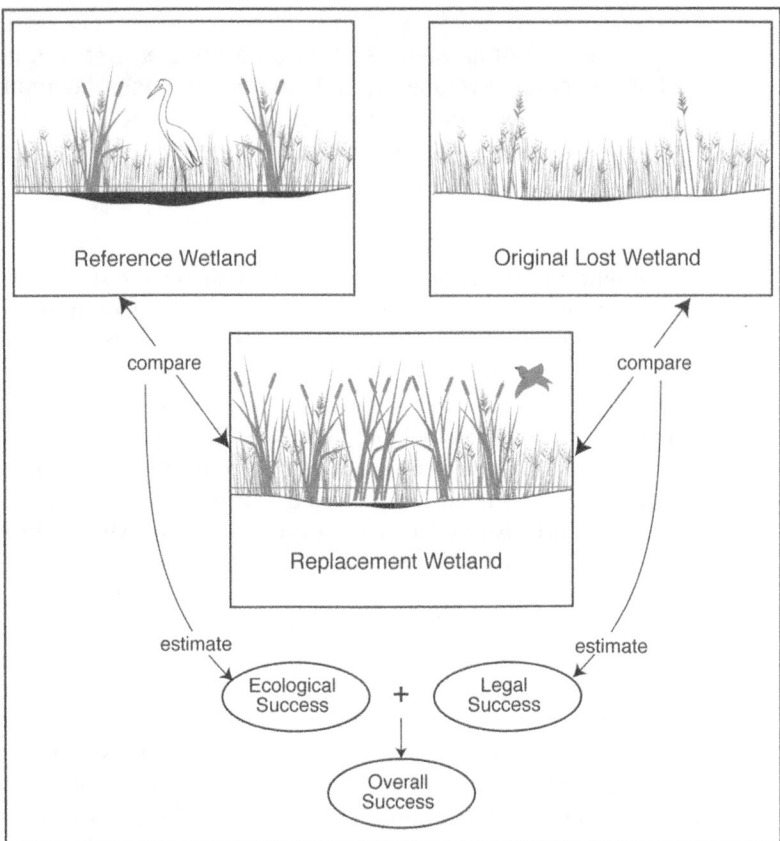

Figure 18.1 Proper wetland mitigation with comparisons with both what has been lost (legal success) and with regional reference natural wetlands (ecological success). (From Wilson and Mitsch, 1996)

Figure 18.1 illustrates conceptually how success should be measured for replacement wetlands. *Legal success* involves a comparison of the lost wetland function and area with that which is gained in the replacement wetland. *Ecological success* should involve a comparison of the replacement wetland with a reference wetland (natural wetlands of the same type that may occur in the same setting or generally accepted "standards" of regional wetland function). Overall success would then be gauged by a combination of the legal and ecological comparisons. While this model represents an ideal, the comparison involving both standards is rarely done.

In fact, the usual decision is based on the size of the wetland lost and little else. Replacement wetlands are designed to be at least the same size as the lost wetlands, but more often a *mitigation ratio* is applied so that more wetlands are created and/or restored than are lost. For example, a mitigation ratio of 2:1 means that 2 hectares (ha) of wetlands will be restored or created for every hectare of wetland lost to development. Considerable controversy exists, for example, in the United States, on the question as to whether wetland loss can be mitigated successfully or if it is essentially impossible (NRC, 2001). Robb (2002) reviewed several years' efforts on mitigating wetland loss in Indiana and suggested, based on failure rates of various wetland types, that there should be these mitigation ratios: 7.6:1 for wet meadows, 3.5:1 for forested wetlands, 1.2:1 for freshwater marshes, and 1:1 for open-water systems.

On paper, the U.S. Army Corps of Engineers' implementation of the U.S. policy of "no net loss" of wetlands over the past 20 years (see Chapter 15: "Wetland Laws and Protection") appears to be working. There was an estimated net gain of 8,000 ha/yr of wetlands and associated uplands in the United States over the 20-year period of 1993 through 2012 as a result of enforcement of the Clean Water Act through mitigation of wetland loss (Fig. 18.2). This number is the result of the issuing of permits for the destruction of 8,000 ha/yr of wetlands and the creation, restoration, enhancement, or preservation of approximately 16,000 ha/yr of wetlands and associated uplands. Over those 20 years of record, the United States has lost 161,000 ha of wetlands and "gained" 318,000 ha of mitigation credit.

There are two reasons why one should not be so euphoric about this accounting that shows a net gain of wetlands. First, it is impossible to tell from these general numbers just how successful this wetland trading has been, because few statistics exist on what functions were lost versus what functions were gained. There is some discussion of this at the end of this chapter. Second, the estimated gain of 157,000 ha over 20 years does not make much of an impact on the loss of 47,000,000 ha of wetlands that occurred from presettlement time up to the 1980s in the United States. There also appears to a distinct pattern of fewer wetlands permitted for drainage and therefore less mitigation in recent times (Fig. 18.2). About 11,000 ha/yr of wetlands were created or restored from 2007 to 2012 compared to 20,000 ha/yr from 1996 to

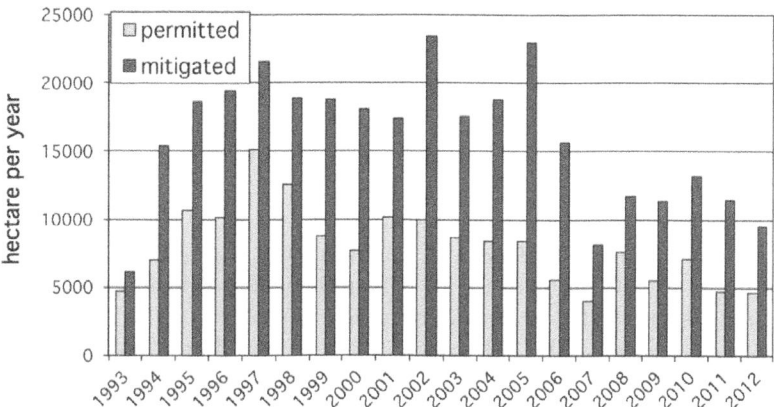

Figure 18.2 Patterns of mitigation for wetland loss in the United States for the 20-year period 1993 to 2012. "Permitted" refers to the area of wetlands that were permitted for drainage in a given year. "Mitigated" refers to the area of wetlands that were required as compensatory mitigation. (Data courtesy of Public Affairs Office, U.S. Army Corps of Engineers, Washington, DC)

2005. The lower numbers could be because of less wetland protection at the federal level as a result of the 2001 and 2006 U.S. Supreme Court decisions described in Chapter 15. Some of that decrease can also be explained by economic downturn in the USA and hence fewer housing and other developments in the late 2000s.

Mitigation Banks

One of the more interesting strategies that the private sector and government agencies have developed to deal with the piecemeal approach to mitigation of wetland loss is the concept of a *mitigation bank*. A mitigation bank is defined as "a wetland area that has been restored and protected to provide compensation for impacts to wetlands" (USEPA and US Army Corps of Engineers, 2008). In this approach, wetlands are usually built in advance of development activities that cause wetland loss, and credits of wetland area can be sold to those who are in need of mitigation for wetland loss. Banks are seen as a way of streamlining the process of mitigating wetland loss and, in many cases, providing a large, fully functional wetland rather than small, questionable wetlands near the site of wetland loss. The mitigation bank can be set up with bonds ensuring compliance. Arrangements are easier for wetland mitigation banks to be managed "in perpetuity" through conservation easements or transfer of titles to resource agencies. Financial resources can be arranged ahead of time for proper monitoring of the wetland bank. Mitigation banks can be publicly or privately owned, although there is a potential conflict of interest if public agencies run mitigation banks. Public agencies could be involved in enforcing regulations on mitigating wetland loss and then steer permittees toward their own banks rather than to private banks.

In 1992, there were only 46 wetland mitigation banks in the United States. By 2002, there were 219 mitigation banks, both private and public, covering 50,000 ha in 29 states in the country (Spieles, 2005). As of August 2013, there were over 1,800 mitigation bank sites listed in the United States. At one time, there were 62 formal mitigation banks (proposed and operating) and hundreds of quasi-mitigation banks in Florida alone (Ann Redmond, personal communication). It appears that, if mitigation of wetland loss continues to be the nation's policy and if regulation of mitigation banks can be developed that is fair and uncomplicated, the use of mitigation banks to solve this "wetland trading" issue will continue to increase well through the twenty-first century.

New regulations on mitigating the loss of wetlands were developed in 2008 by the U.S. Army Corps of Engineers and U.S. EPA (2008). These regulations were meant to increase the effectiveness of wetland mitigation banks and to strengthen requirements for the use of in-lieu fee mitigation (U.S. EPA and U.S. Army Corps of Engineers, 2008). The data in Figure 18.2 do not show any obvious improvement in the mitigation ratios after 2008 yet.

Agricultural Land Restoration

For many decades in the United States, farm pond creation was encouraged as a way of providing drinking water for domestic animals and other functions on the farm.

Although individually quite small (usually about 0.2 ha), the total number of constructed ponds is large. Several years ago, ponds were being constructed at a rate of about 50,000 per year. Marshes often develop around the perimeter of many of these ponds, while other ponds have converted to marshes. Many of these ponds were built with large, shallow areas to attract waterfowl, and these shallow zones have become typical pothole marshes.

Dahl (2006) estimated that between 1998 and 2004, wetland pond areas increased by 280,000 ha in the United States, a 12.6 percent increase. Most of this gain (141,000 ha) resulted on nonagricultural upland, while 29,000 ha were constructed on farmland. Many of the nonagricultural ponds are built in housing and commercial developments, especially in states like Florida, as stormwater runoff ponds. There are those who question the ecological value of these ponds; for example, some regulators do not like ponds because they have fish and therefore cannot support amphibians.

Conservation programs were set up in the 1990s to encourage individual farmers in the United States to restore wetlands on their land. Both the Conservation Reserve Program (CRP) and the Wetlands Reserve Program (WRP) under the U.S. Department of Agriculture (USDA) have led to significant areas of wetlands being restored or protected. CRP guidelines, announced in 1997, give increased emphasis to the enrollment and restoration of *cropped wetlands*—that is, wetlands that produce crops but serve wetland functions when crops are not being grown. The CRP also encourages wetland restoration, particularly through hydrologic restoration.

The WRP, a voluntary program established in 1992, and was specific for wetland restoration; it offered landowners the opportunity to protect, restore, and enhance wetlands on their property and provides funds for farmers to do so. The USDA Natural Resources Conservation Service provides technical and financial support to help landowners. The WRP options to protect, restore, and enhance wetlands and associated uplands include permanent easements, 30-year easements, or 10-year restoration cost-share agreements. As of 2012, approximately 900,000 ha of wetlands and adjacent uplands have been enrolled in the WRP in the United States, with the most intense activity in the lower Mississippi River basin and Florida. The U.S. Agricultural Act of 2014 established the Agricultural Conservation Easement Program (ACEP) in the U.S. Department of Agriculture. This act repealed the WRP but did not affect the validity or terms of any WRP agreement entered into prior to February 7, 2014. Wetland reserve easements to restore, protect, and enhance wetlands continue under the ACEP. The long-range affect that this administrative change will have on wetland conservation and restoration is unknown.

Forested Wetland Restoration

There is less experience with forested wetland restoration and creation compared to herbaceous marshes, although these wetlands have been lost at alarming rates, particularly in the southeastern United States. Forested wetland creation and restoration are different from marsh creation and restoration, because forest regeneration takes decades rather than years to complete, and there is more uncertainty about the results.

Much riparian forest restoration in the United States has centered on the 10-million-ha lower Mississippi River alluvial valley, where more than 182,000 to 220,000 ha have been reforested (Haynes, 2004), primarily with bottomland hardwood species and, to a lesser extent, deepwater swamp species. This is a small contribution to the restoration of this alluvial floodplain, where 7.2 million ha of bottomland hardwood forest were estimated to have been lost (Hefner and Brown, 1985).

Hydrologic Restoration of Wetlands

Lines often blur between wetlands created and restored for habitat restoration and those restored for water quality and hydrology improvement. In fact, most wetlands that are restored or created are done so for both reasons. One of the largest freshwater wetland restorations in the world is being carried out in the Florida Everglades to restore, at least to some degree, the natural hydrologic conditions, at least in the Everglades that are left (see Case Study 1). The restoration of the Mesopotamian Marshlands of Iraq (Case Study 2) is an example where hydrology was restored to a drained wetland of incredible cultural importance. There, the hydrology had been purposefully disrupted by the regime of Saddam Hussein and indirectly through upstream river management by Iraq's upstream neighbors. The Iraqi people, with some international assistance, are undertaking a hydrologic restoration of this historically and culturally important wetland (see also Chapter 1: "Wetlands: Human Use and Science").

CASE STUDY 1: Restoring the Florida Everglades

The restoration of the Florida Everglades, one of the largest wetland areas in the world, actually involves several separate initiatives being carried out in the 4.6-million-ha Kissimmee–Okeechobee–Everglades (KOE) region in the southern third of Florida (Fig. 18.3). The basic plan involves restoring something closer to the original hydrology of the KOE region (Fig 18.3a, b) by sending less of the water from the upper watershed to the Caloosahatchee River to the west and the St. Lucie Canal to the east than is currently the case and directing more of the water to the Everglades south of Lake Okeechobee.

The actual hydrologic flows for water years 2012 and 2013 (Fig. 18.3c and Table 18.1) were not close to either the historic or restoration goals of Figure 18.3a, b. Despite the fact that precipitation was slightly below normal in 2012 and almost exactly at normal amounts in 2013 (1,350 mm/yr), a significant portion of the Lake Okeechobee outflows were still shunted down the Caloosahatchee to the west and the St. Lucie Canal to the east in 2013:

746,000 m³/yr, or 58 percent of the 1,285,000 m³/yr that was diverted south into the Florida Everglades. There has been significant pollution and ecological problems in the estuaries of the Gulf of Mexico and Atlantic Ocean because of these excessive freshwater flows from Lake Okeechobee going east and west instead of south in the Florida Everglades as proposed by the original restoration goal shown in Figure 18.2c.

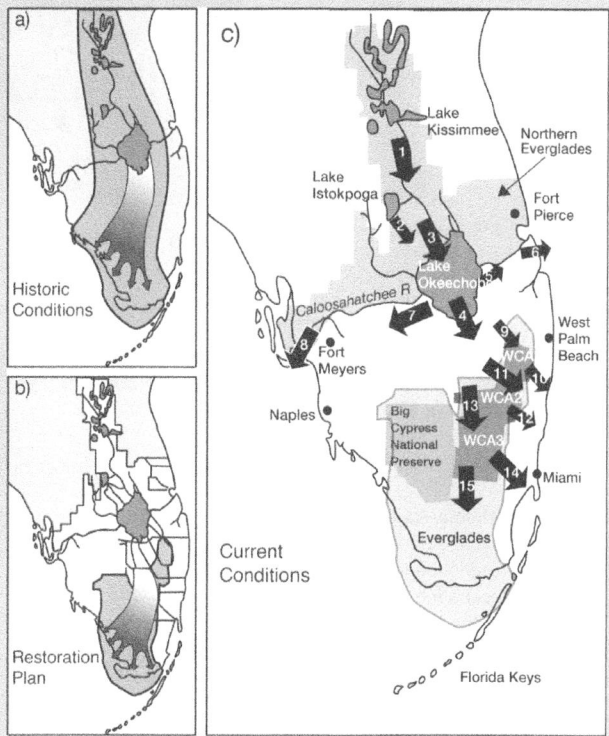

Figure 18.3 Illustrations of Florida Everglades wetland restoration: (a) historic conditions of the Florida Everglades hydrology in presettlement times; (b) the flow conditions desired when the Everglades is restored; (c) flow conditions in water years 2012 and 2013 in the Everglades where much of the water was still sent east and west to the sea rather than south to the Everglades (flows for numbered pathways are given in Table 18.1); (d) captured Burmese python (*Python molurus bivittatus*) in Florida Everglades as symbolic of major problems of invasive species that continue to occur. ((a), (b) from Mitsch and Jørgensen, 2004; (c) from South Florida Water Management District (SFWMD, 2014); (d) from Mike Rochford, reprinted with permission)

Figure 18.3 (*Continued*)

Table 18.1 Water flow through the Florida Everglades for water years 2012 and 2013 as illustrated in Figure 18.3c (Flows are × 1000 m³/yr.) Flow pathway number refers to pathways shown in Figure 18.3c

Flow pathway in Greater Florida Everglades	Water Year 2013	Water Year 2012
1. Lake Kissimmee outflows	543	1,004
2. Lake Istokpoga outflows	347	281
3. Lake Okeechobee inflows	2,590	2,246
4. Lake Okeechobee outflows	1,285	920
5. Flows into the St. Lucie Canal from Lake Okeechobee	128	58
6. Flows into the St. Lucie Estuary through the St. Lucie Canal	189	0
7. Flows into the Caloosahatchee Canal from Lake Okeechobee	618	222
8. Flows into the Caloosahatchee Estuary through the Caloosahatchee Canal	1,404	739
9. Water Conservation Area 1 inflows	449	210
10. Water Conservation Area 1 outflows	597	19
11. Water Conservation Area 2 inflows	1,325	476
12. Water Conservation Area 2 outflows	1,151	466
13. Water Conservation Area 3 inflows	1,631	1,110
14. Water Conservation Area 3 outflows	1,511	704
15. Everglades National Park inflows	1,847	918

Source: SFWMD (2014).

Specific problems in the Everglades have developed for four reasons:

1. Excessive nutrient loading to Lake Okeechobee and to the Everglades, primarily from agricultural runoff
2. Loss and fragmentation of habitat caused by urban and agricultural development
3. Spread of cattail (*Typha*), melaleuca (*Melaleuca quinquenervia*), and other invasives and exotics in the Everglades, replacing native vegetation
4. Hydrologic alteration due to an extensive canal and straightened river system built by the U.S. Army Corps of Engineers and others for flood protection, and maintained by water management districts

One major restoration project in the KOE region that initially received a lot of attention is the restoration of the Kissimmee River. As a result of the channelization of the river in the 1960s, a 166-km-long river was transformed into a 90-km-long, 100-m-wide ditch, and the extent of wetlands along the river decreased by 65 percent (Table 18.2). Waterfowl populations decreased by 90 percent as a result of this channelization (Blake, 1980). The restoration of the Kissimmee River is a major undertaking to reintroduce the sinuosity to the artificially straightened river. The river restoration work, expected to be completed in stages by 2019, will return 50 km^2 of lost wetland habitat to the riparian zone and reestablish 64 km of the river, at a total expected cost of almost US$1 billion (Koebel and Bousquin, 2014). It will also provide sinks for nutrients that are otherwise causing increased eutrophication in downstream Lake Okeechobee. Restoration to date is described in a special issue of *Restoration Ecology* (Bousquin, 2014). NRC (2014) concluded that the Kissimmee River restoration "is probably the most advanced in demonstrating substantial restoration of the natural system" in the

Table 18.2 Wetland changes due to channelization of the Kissimmee River in south Florida (Channelization took place between 1962 and 1971 and transformed a 166-km meandering river into a 90-km-long, 10-m-deep, 100-m-wide canal.)

Wetland Type	Prechannelization (ha)	Postchannelization (ha)	Percentage Change (%)
Marsh	8,892	1,238	−86
Wet prairie	4,126	2,128	−48
Scrub–shrub wetland	2,068	1,003	−51
Forested wetland	150	243	+62
Other	533	919	+72
Total	15,769	5,531	−65

Source: Toth et al. (1995).

Florida Everglades and that its long-term monitoring of restoration progress is a useful example for other restoration projects in the Florida Everglades.

Everglades restoration also involves halting the spread of high-nutrient cattail (*Typha domingensis*) through the low-nutrient sawgrass (*Cladium jamaicense*) communities that currently dominate the Everglades (see Chapter 14: "Human Impacts and Management of Wetlands" for general description of the water pollution problem in the Everglades and Chapter 19: "Wetlands and Water Quality" for a description of the created and restored wetlands being used to solve this problem).

Overall, the Everglades restoration, as now planned by the U.S. Army Corps of Engineers, has a budget of over $20 billion and may take decades more to complete.

A not-unrelated aspect of the restoration of the Florida Everglades is dealing with the vast number and often dominance of invasive species that have become entrenched there over the past century. There are about 250 nonnative plants in the ecosystems of the Florida Everglades, or about 16 percent of the total species count. Of these, 12 species are considered particularly important to the Everglades restoration (Rodgers et al., 2014). This short list includes plants that were purposefully introduced to south Florida and are now the targets of continual removal attempts in and around the Everglades, such as melaleuca, introduced in 1906 from Australia to enhance drainage of the Everglades, and Brazilian pepper (*Schinus terebinthifolius*), which was brought to Florida from South America as an ornamental in the mid-nineteenth century.

The invasive animal that has received more news media attention than any other is the Burmese python (*Python molurus bivittatus*) (Fig. 18.3d). The snakes, which can achieve lengths of up to 5.5 m, have been suggested to have a population as large as several thousand individuals in south Florida (M. Dorcas, personal communication, 2014). Dorcas et al. (2012) illustrated that over a 90 percent reduction in populations or mammals such as raccoons, opossums, bobcats, and rabbits occurred in the Everglades National Park coincident with the significant increase in python populations since 2000. Partly as a result of this study and the follow-up media attention given to this subject, a "Python Challenge" was sponsored in the winter 2013 in south Florida. Hundreds of hunters from all over the United States and Canada spent a month trying to capture and/or kill the reptile. In the end, only 68 snakes were killed or captured (http://phys.org/news/2013-02-python-everglades-nets.html).

NRC (2014) assessed the progress of the Florida Everglades restoration and discussed specific scientific and engineering issues that may impact further progress. They recommended a dedicated source of funding that could provide "ongoing long-term system-wide monitoring and assessment" of progress on restoration objectives. The report makes recommendations for restoration activities, project management strategies, management of invasive nonnative species, and high-priority research needs. Among the recommendations are "finding solutions to overcome current constraints related to authorization, funding, and water quality permitting" and dealing in a more programmatic way with climate change and sea-level rise that the report suggests "provide even more incentive for restoring the Everglades ecosystem." The report also brings attention to the importance of dealing with invasive species in the Everglades, a problem that did not exist prior to the drainage projects of the last century. The report concludes that "restoration progress made by CERP [Comprehensive Everglades Wetland Plan] projects to date remains fairly modest in scope." In fact, the authors note that some of the most important progress was with the Kissimmee River restoration and the C111 South Dade County spreader canal project, neither of which is part of the CERP.

CASE STUDY 2: Restoration of the Mesopotamian Marshlands

The Mesopotamian Marshlands of southern Iraq and Iran were described in Chapters 1 and 3. These wetlands, found at the confluence of the historic Tigris and Euphrates rivers, were 15,000 to 20,000 km^2 in area as recently as the early 1970s (Fig. 18.4a) but were drained and diked, especially in the 1990s, to less than 10 percent of that extent by 2000 (Fig. 18.4b). Among the main causes are upstream dams and drainage systems constructed in the 1980s and 1990s that altered the river flows and eliminated the flood pulses that sustained the wetlands.

Since the overthrow of Saddam Hussein's dictatorship in Iraq in 2003, there has been a concerted effort by the Iraqis and then the international community at restoring the marshlands (Richardson et al., 2005). The restoration has often occurred with local residents breaking dikes or removing impediments to flooding. Remote-sensing images showed that at least 37 percent of the wetlands were restored by 2005. *Frontiers in Ecology and the Environment* (3, No. 8, October 2005) reported in 2005 that "at least 74 species of migratory waterfowl and many endemic birds have been sighted in a survey of Iraq's marshland." It was also reported that as many as 90,000 Marsh Arabs have returned to the wetlands already (Azzam Alwash, personal communication).

Alwash, the director of the Eden Again effort, has suggested that perhaps as much as 75 percent of the marshlands can be restored (Fig. 18.4c, d). Several questions still remain unanswered about whether full restoration can occur, including whether adequate water supplies exist in the rivers, given the competition from Turkey, Syria, and Iran, and within Iraq itself, and whether landscape connectivity of the marshes can be reestablished (Richardson and Hussain, 2006).

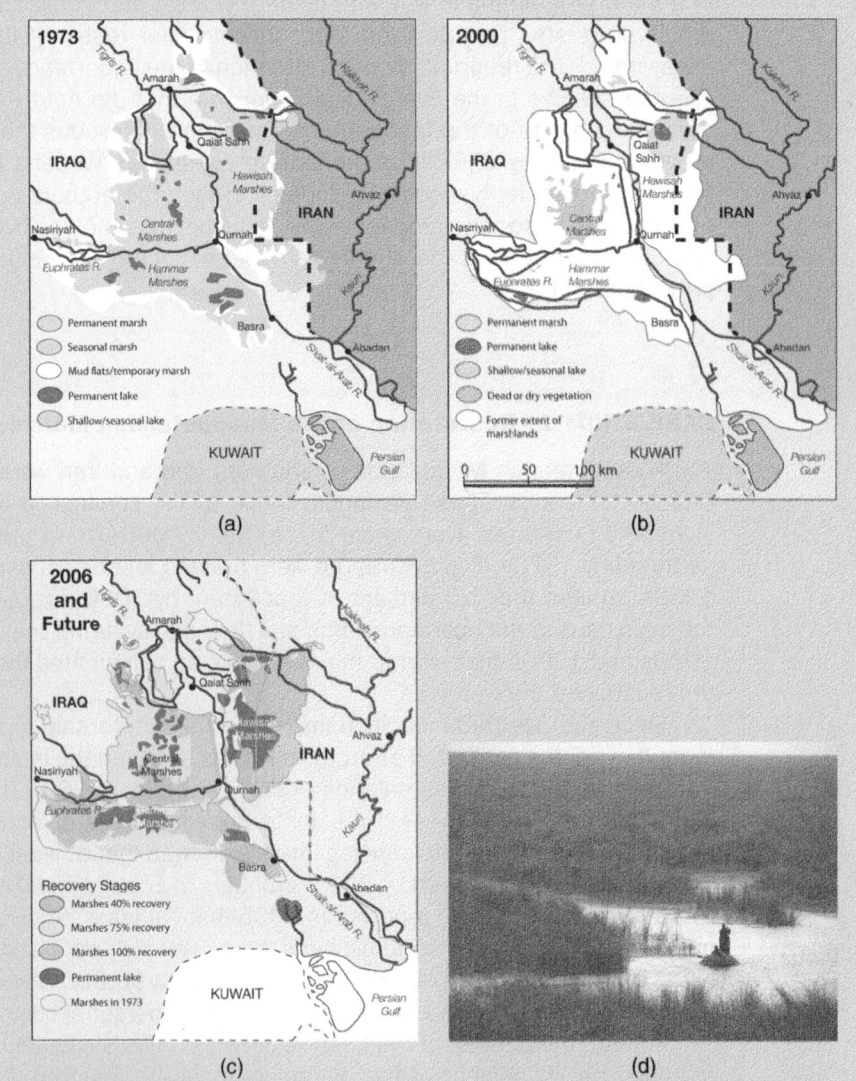

Figure 18.4 The Mesopotamian Marshlands of Iraq, with shading indicating extent of the marshlands: (a) in 1970 before extensive drainage; (b) in 2000 after extensive drainage; (c) as expected in the future with 75 percent restoration of the marshland; and (d) photo of the Mesopotamian Marshland after restoration. ((a) (b) from UNEP, 2001; (c), (d) from Azzam Alwash)

Peatland Restoration

Peatland restoration is a relatively new type of wetland restoration compared to other types and potentially could be the most difficult (Gorham and Rochefort, 2003). Early attempts with peatlands occurred in Europe, specifically in Finland, Germany, the United Kingdom, and the Netherlands. Increased peat mining in Canada and elsewhere has led to increased interest in understanding if and how mined peatlands can be restored. When peat surface mines are abandoned without restoration, the area rarely returns through secondary succession to the original moss-dominated system (Quinty and Rochefort, 1997). There is some promise that peatland restoration can be successful (Rochefort and Lode, 2006), but because surface mining causes major changes in local hydrology and peat accumulates at an exceedingly slow rate, restoration progress will be measured in decades rather than years.

In the 1960s and 1970s, block harvesting of peat was replaced by vacuum harvesting in southern Quebec and in New Brunswick, necessitating the development of different restoration techniques. While traditional block-cutting of peat left a variable landscape of high ground and trenches, vacuum harvesting leaves relatively flat surfaces bordered by drainage ditches. Abandoned block-cut sites appear to revegetate with peatland species more easily than do vacuum-harvested sites, and the latter can remain bare for a decade or more after mining (Rochefort and Campeau, 1997). Case Study 3 describes as peatland restoration research field lab in Quebec where the best ways to restore peatlands are being investigated.

CASE STUDY 3: Peatland Restoration Research at Bois-des-Bel, Quebec

Despite the vast expanses of peatlands in the world, whole-ecosystem experiments on restoration research on this type of wetland are rare. Bois-des-Bel peatland, located about 200 km northeast of Québec City, on the southern shore of the St. Lawrence River in Québec, Canada (Fig. 18.5), is a whole-ecosystem research site where scientists are evaluating the pace of peatland restoration after peat mining (Rochefort et al., 2003; Waddington et al., 2008). The entire peatland is about 210 ha; the research peatland area is about 11.5 ha that was drained in 1972 and mined by a vacuum extraction technique from 1973 to 1980. When mining stopped, a 2-m peat deposit remained. Restoration began in 1999 on 8.4 ha of the site, with the remaining as an unrestored control. The restored area was divided into four zones, each of which has two shallow pools (13 m x 5 m x 1.5 m max depth) for aquatic and amphibian habitat. Line Rochefort and her students and colleagues at Université Laval, Québec City, and at other Canadian universities have established the site as a long-term ecosystem research site to investigate the revegetation of mined peatlands (Price et al., 1998; Rochefort et al., 2003, 2012; Waddington et al., 2003, 2008;

Isselin-Nondedeu et al., 2007). The restoration involved terracing to produce better water distribution, reintroduction of *Sphagnum* diaspores harvested from a nearby natural wetland, and reflooding by blocking drainage ditches. The moss carpet increased by about 12 cm by 2007 and was three times the

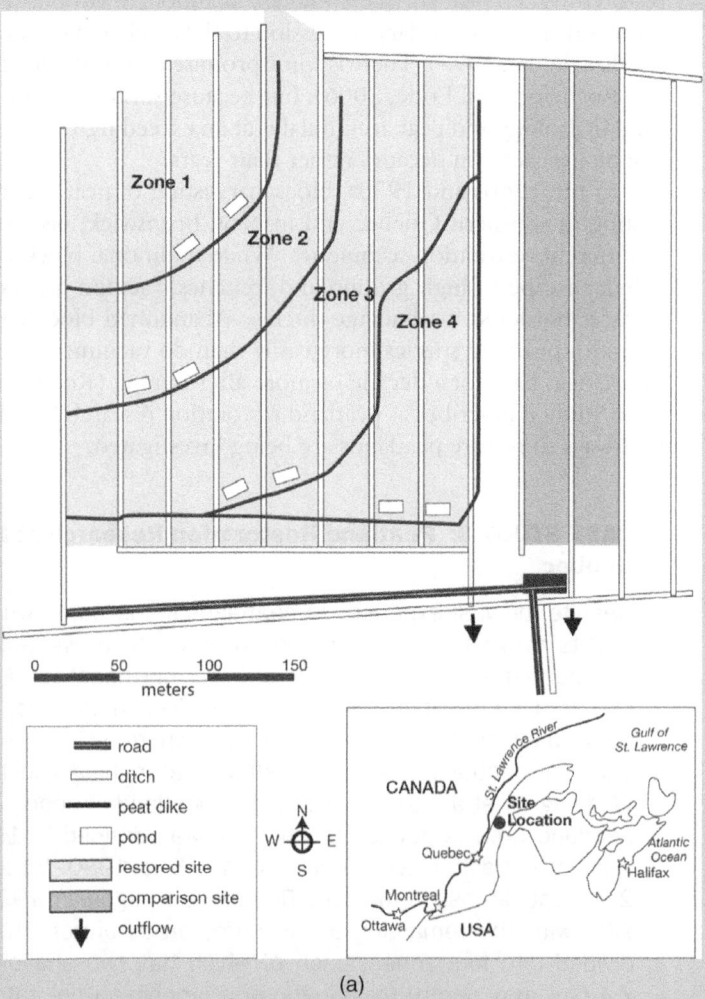

(a)

Figure 18.5 Bois-des-Bel experimental peatland on the southern shore of the St. Lawrence River in Québec, Canada. (a) map showing restored (west) and cutover (east) sites including four research zones in restored area for vegetation monitoring in created pool at Bois-des-Bel restored site; (b) shallow pools created in the restored area for aquatic and amphibian habitat. ((a) from Waddington et al., 2008; (b) photo courtesy of L. Rochefort, Peatland Ecology Research Group (PERG), Université Laval, Québec City, Canada, reprinted by permission)

(b)

Figure 18.5 (Continued)

thickness that it was in 2003. *Sphagnum* cover by 2005 was 60 percent of the area in the restored sites compared to only 0.25 percent in the nonrestored sites (Isselin-Nondedeu et al., 2007). The restored sites also exported less than half the dissolved organic carbon than did the cutover peatland sites (Waddington et al., 2008). This area has been monitored postrestoration for almost a decade (Rochefort et al., 2013).

Coastal Wetland Restoration

Salt Marsh Restoration

There is a great deal of interest in coastline restoration. Early pioneering work on salt marsh restoration was done in Europe (Lambert, 1964; Ranwell, 1967); China (Chung, 1982, 1989); and the United States on the North Carolina coastline (Woodhouse, 1979; Broome et al., 1988), in the Chesapeake Bay area (Garbisch et al., 1975; Garbisch, 1977, 2005), and along the coastlines of Florida, Puerto Rico (Lewis, 1990b, 1990c), and California (Zedler, 1988, 2000b; Josselyn et al., 1990). Some of this coastal wetland restoration has been undertaken for habitat development as mitigation for coastal development projects.

For coastal salt marshes in the eastern United States, the cordgrass *Spartina alterniflora* is the primary choice for coastal marsh restoration, but the same species is considered an invasive and unwanted plant on the West Coast of North America. *Spartina townsendii*, *S. anglica*, and *S. alterniflora* have been used to restore salt marshes in Europe, New Zealand, and China, although several of these species are now considered invasive in those regions. Salt marsh grasses tend to distribute easily through seed dispersal, and the spread of these grasses can be quite rapid once the reintroduction has begun, as long as the area being revegetated is intertidal—that is, the elevation is between ordinary high tide and low tide. The details of successful coastal wetland creation are site specific, but six generalizations seem to be valid in most situations:

1. Sediment elevation is the most critical factor determining the successful establishment of vegetation and the plant species that will survive. The site must be intertidal.
2. In general, the upper half of the intertidal zone is more rapidly vegetated than lower elevations.
3. Sediment composition does not seem to be a critical factor in colonization by plants unless the deposits are almost pure sand that is subject to rapid desiccation at the upper elevations.
4. The site needs to be protected from high wave energy. It is difficult or impossible to establish vegetation at high-energy sites.
5. Most sites revegetate naturally from seeds if the elevation is appropriate and the wave energy is moderate. Sprigging live plants has been accomplished successfully in some cases, and seeding also has been successful in the upper half of the intertidal zone.
6. Good stands can be established during the first season of growth, although sediment stabilization does not occur until after two seasons. Within four years, successfully planted sites are often indistinguishable superficially from natural marshes.

Several early studies emphasized the importance of restoring tidal conditions, including salinity, to marsh areas that had become more "freshwater" because of isolation from the sea. In cases such as this, the restoration is simple: Remove whatever impediment is blocking tidal exchange. Case Study 4 describes a salt marsh restoration where this has been done with some success: When the natural tidal hydrology was restored, the vegetation and aquatic species followed. Case Study 5 describes a many decade restoration of a coastal river/wetland complex in the New York City urban region.

CASE STUDY 4: Delaware Bay Salt Marsh Restoration

A large coastal wetland restoration project in the eastern United States involves the restoration, enhancement, and preservation of 5,000 ha of coastal salt marshes on Delaware Bay in New Jersey and Delaware (Figure 18.6a). This estuary enhancement, being carried out by New Jersey's electric utility (Public Service Enterprise Group [PSEG]), with advice from a team of scientists and consultants, was undertaken as mitigation for the potential impacts of once-through cooling from a nuclear power plant operated by PSEG on the bay. The reasoning was that the impact of once-through cooling on fin fish, through entrainment and impingement, could be offset by increased fisheries production from restored salt marshes. Because of uncertainties involved in this kind of ecological trading, the area of restoration was estimated as the salt marshes that would be necessary to compensate for the impacts of the power plant on fin fish times a safety factor of 4. Three distinct approaches are being utilized in this project to restore the Delaware Bay coastline:

1. *Reintroduce flooding*. The most important type of restoration involves the reintroduction of tidal inundation to about 1,800 ha of former diked salt-hay farms. Many marshes along Delaware Bay have been isolated by dikes from the bay, sometimes for centuries, and put into the commercial production of "salt hay" (*Spartina patens*). Hydrologic restoration was accomplished by excavating breaches in the dikes and, in most cases, connecting these new inlets to a system of re-created tidal creeks and existing canal systems.
2. *Reexcavate tidal marshes*. Additional restoration involves enhancing drainage by reexcavating higher-order tidal creeks in these newly flooded salt marshes, thereby increasing tidal circulation. This is particularly important in marshes that were formerly diked, because the isolation from the sea has led to the filling of former tidal creeks. After initial tidal creeks were established, it was expected that the system would self-design more tidal channels and increase the channel density.
3. *Reduce* Phragmites *domination*. In another set of restoration sites in Delaware and New Jersey, restoration involves the reduction in cover of the aggressive and invasive reed grass (*Phragmites australis*) in 2,100 ha of nonimpounded coastal wetlands. Alternatives that were investigated include hydrological modifications such as channel excavation, breaching remnant dikes, microtopographic changes, mowing, planting, and herbicide application.

Results of this study were reported in several presentations and reports and several early journal articles, including Teal and Weinstein (2002), and several papers in a special issue of *Ecological Engineering* (Peterson et al., 2005). From a hydrodynamic perspective, in those marshes where tidal exchange was restored, the development of an intricate tidal creek density from the originally constructed tidal creeks has been impressive. Figure 18.6b illustrates the development of a stream network at one of the newly restored

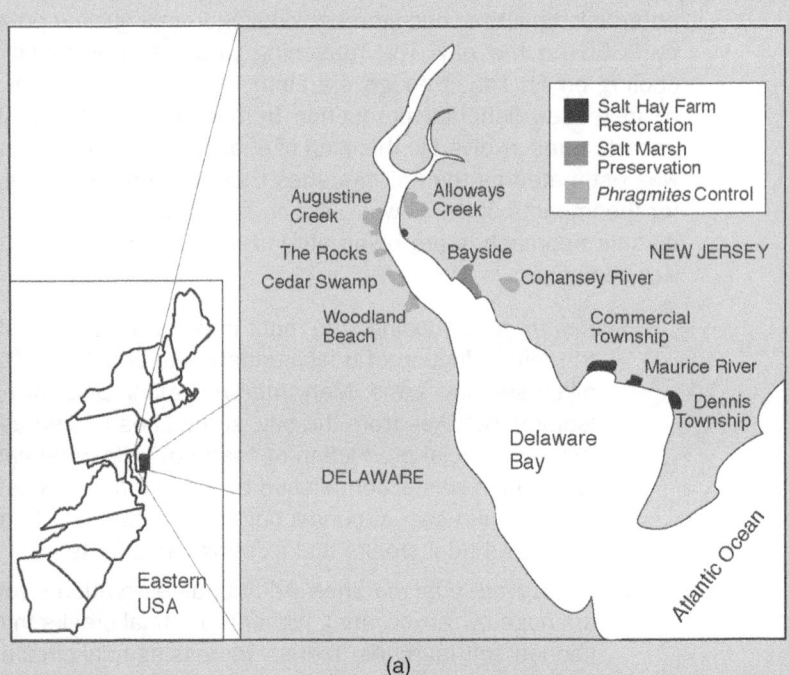

(a)

Figure 18.6 Delaware Bay salt marsh restoration from 1995 through present: (a) map of Delaware Bay between New Jersey and Delaware, showing locations of 5,800-ha salt marsh restoration that is being carried out to mitigate the loss of fin fish due to entrainment and impingement caused by once-through cooling at a nuclear power plant. Wetlands are being preserved, restored from salt-hay farms by reintroducing flooding, and enhanced by removal of *Phragmites australis*. (b) Total number of stream channels by channel class at Dennis Township (1996–2014) and Commercial Township (1997–2013) restored salt marshes on Delaware Bay. (c) Vegetative cover in 1995 prior to restoration at the Dennis Township restored salt marshes. White area indicates unvegetated region. (d) Vegetative cover in 2003, seven years after low-channel-class tidal creeks were restored and dikes were breached. Predominant gray areas are restored *Spartina alterniflora* and other desirable marsh vegetation. (e) Project consultant John Teal viewing a marsh area mostly restored to *Spartina alterniflora* of a portion of a 640-ha Alloway Creek, New Jersey, marsh on Delaware Bay that was restored from *Phragmites australis* with herbicide (mostly glyphosate) application over a period of more than 10 years. ((b) provided with permission, Kenneth A. Strait, PSEG Service Corporation, Salem, NJ; (c), (d) revised from Hinkle and Mitsch, 2005; (e) photo by W. J. Mitsch)

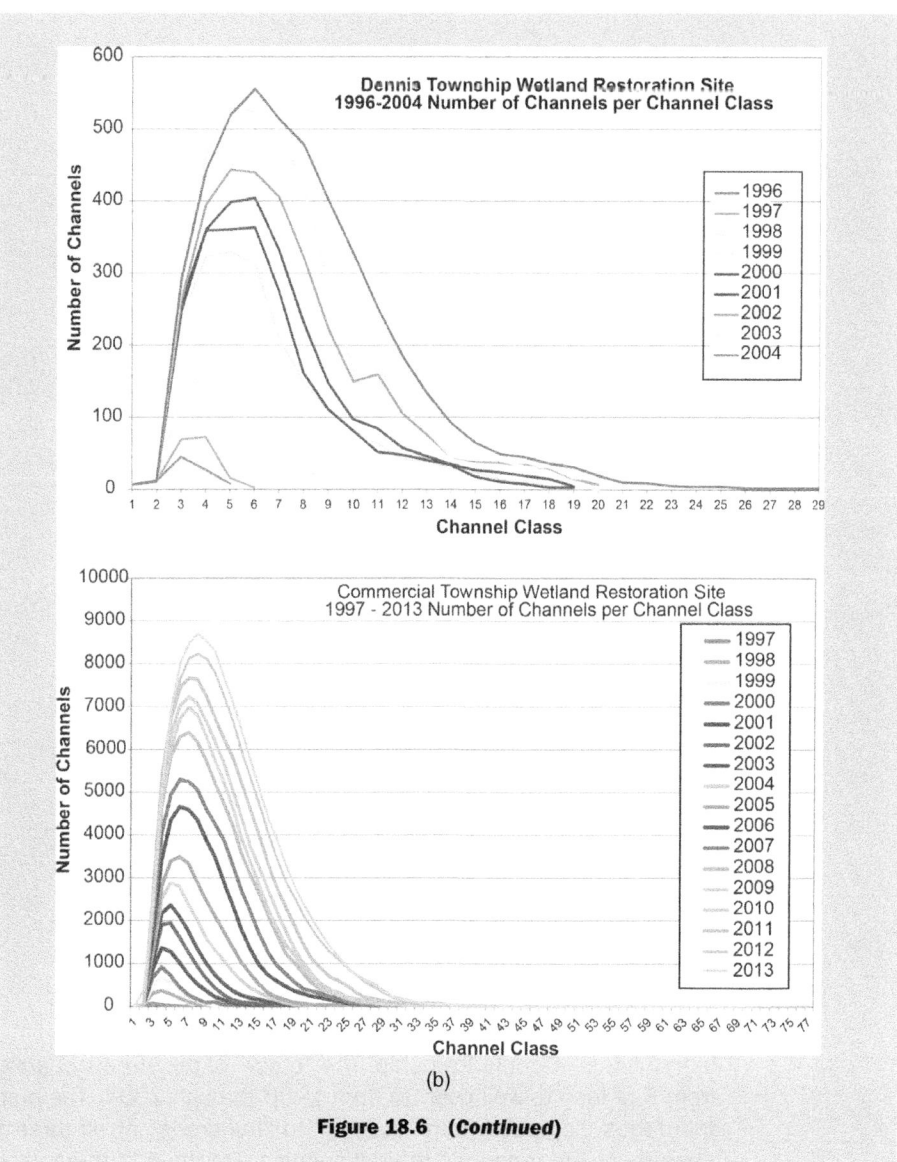

(b)

Figure 18.6 *(Continued)*

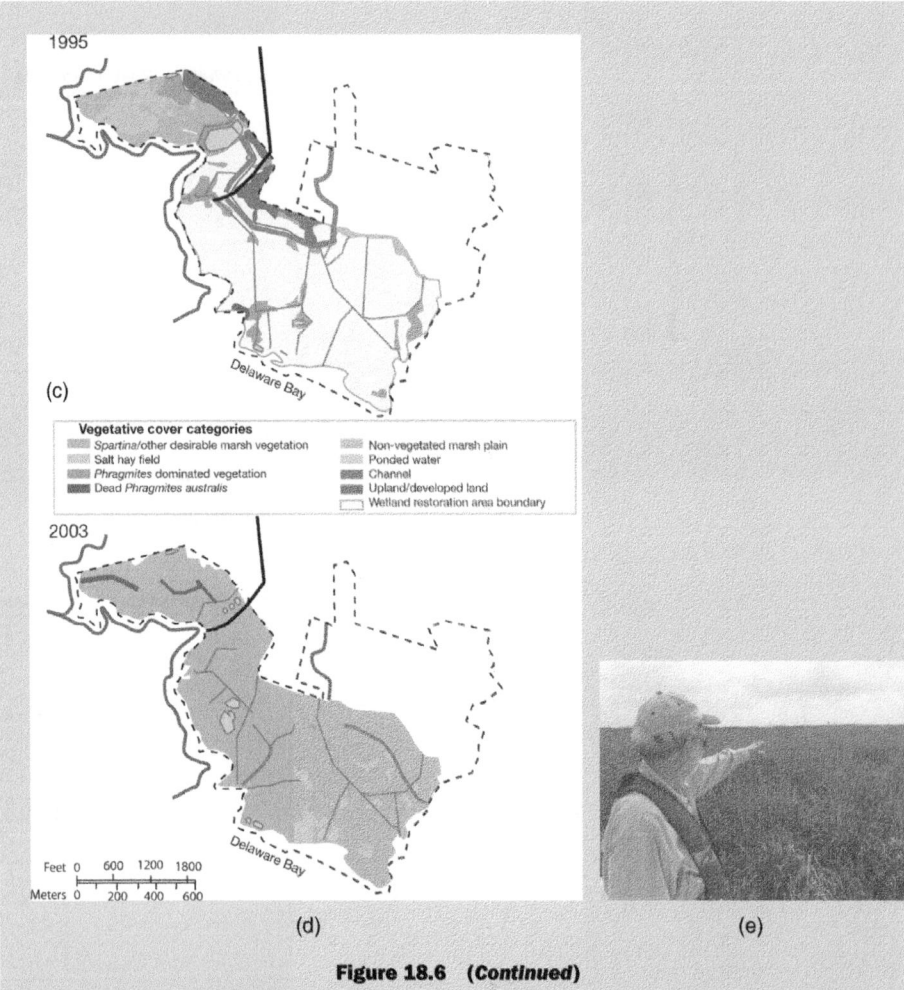

Figure 18.6 *(Continued)*

marsh sites—Dennis Township. The "order" of the stream channels increased from 5 or less to well over 20 from 1996 through 2004. The number of small tributaries increased from "dozens" to "hundreds" at all three salt-hay farm sites that were reopened to tidal flushing. For the first three years, there was a rapid increase in the growth of channel orders 3 through 9; in the next three years, there was a rapid increase in the channel orders 10 through 16. (Note: This definition uses channel order as opposite to the normal method on stream order; here the largest channels are designated as channel order 1.) Hydrologic design did occur in a self-design fashion after only initial cuts by construction of the first-order channels.

For the salt-hay farms that were flooded, typical goals include a high percentage cover of desirable vegetation such as *Spartina alterniflora*, a relatively low percentage of open water, and the absence of the invasive reed grass *Phragmites australis*. The success of this coastal restoration project, subject to a combination of legal, hydrologic, and ecological constraints, is also being estimated through comparison of restored sites to natural reference marshes. Results of this part of the project after almost two decades are encouraging. At the formerly diked salt-hay farms, reestablishment of *Spartina alterniflora* and other favorable vegetation has been rapid and extensive.

In Dennis Township, approximately 70 percent of the site was dominated by *Spartina alterniflora* after only two growing seasons and almost 80 percent by the fifth year after construction (Fig. 18.6c, d). Tidal restoration was completed at the Maurice River site, which is twice the size of the Dennis Township site in early 1998. Major revegetation by *Spartina alterniflora* and some *Salicornia* has already occurred, with 71 percent of the site showing desirable vegetation after four growing seasons. At the third and the largest salt-hay farm restoration site, at Commercial Township, which is five times larger than Dennis Township site, revegetation is occurring rapidly from the bayside. This study has shown that the speed with which salt marsh restoration takes place is dependent on three main factors:

1. The degree to which the tidal "circulatory system" works its way through the marsh
2. The size of the site being restored
3. The initial presence of *Spartina* and other desirable species

No planting was necessary on these sites, as *Spartina* seeds arrive by tidal fluxes, but the design of the sites to allow that tidal connectivity (and hence the importance of appropriate site elevations relative to tides) was critical. Self-design works when the proper conditions for propagule disbursement are provided. Extensive ponding in some areas of the marshes, especially at Commercial Township, which has the highest ratio of area to edge, has impeded the reestablishment of *Spartina* in some locations (Teal and Weinstein, 2002). Creating additional streams or waiting for the tidal forces to cause the same effect eventually allows these areas to develop tidal cycles and *Spartina* to establish itself.

Reducing *Phragmites* domination in another set of brackish marsh restoration sites in Delaware and New Jersey has required more years of effort, mainly through the use of herbicides, but there are now substantial areas of significant recovery of the marsh to desirable vegetation, including *Spartina alterniflora*. The *Phragmites* cover in the marsh shown in the photo at the Alloway Creek watershed in New Jersey (Fig. 18.6e) decreased from 60 percent cover in 1996 to less than 5 percent cover in 2013.

CASE STUDY 5: Urban Coastal Restoration in the New York City Region

The Hackensack Meadowlands (Fig. 18.7a) is adjacent to heavily industrialized and commercial Newark Bay, New Jersey, and is on one of the most polluted water courses in the United States. It is located a few km west of the

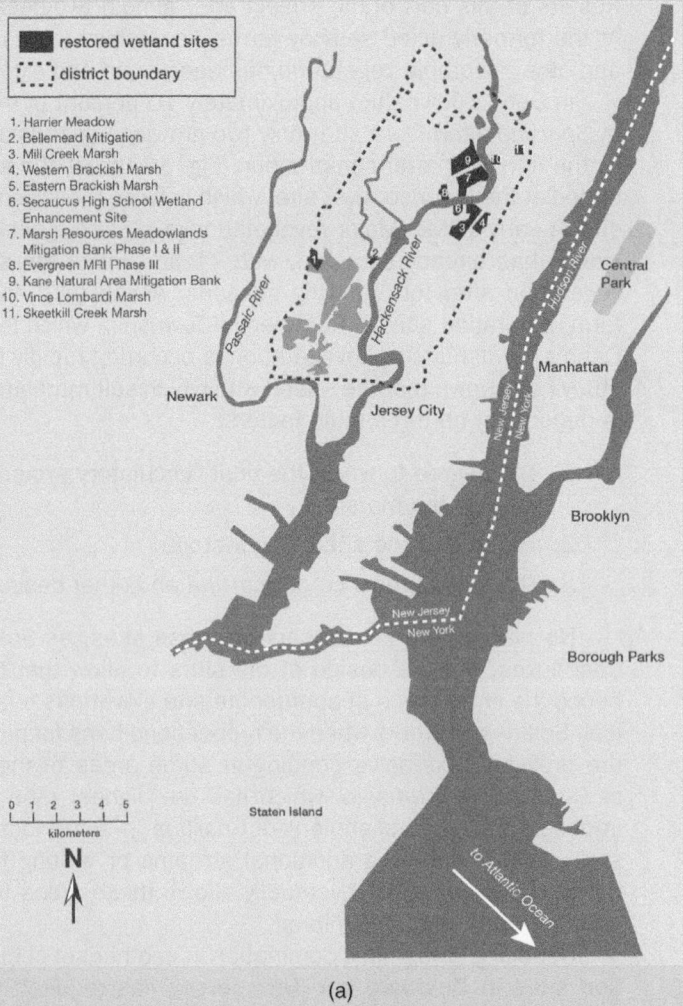

(a)

Figure 18.7 Hackensack Meadowlands, New Jersey, in New York City metropolitan area: (a) Map of Hackensack Meadowland District in New Jersey and location of wetland restoration projects in the district; (b) photo of Hackensack Meadowlands with New York City skyline in the distance. ((a) From maps provided by New Jersey Meadowlands Commission (NJMC)/Meadowlands Environmental Research Institute; (b) photo by W. J. Mitsch)

(b)

Figure 18.7 (Continued)

Hudson River and Manhattan Island in New York City (Fig. 18.7b). More than a century of ditching, filling, drainage, and diking have changed the lower reaches of the Hackensack River from a tidal salt and brackish marshland into highly urbanized mix of residential and industrial land uses interspersed with tidal creeks and marshes and mudflats. The Hackensack River watershed has four superfund sites, two power plants, three sewage treatment plants, and roughly 1,000 ha of landfills within 4 km of the river. The river itself is brackish with the salinity higher downstream. Tide gates prevent the free mixing of saline waters from the main stem of the river. Highest salinity (12–15 ppt is found in the main channel closest to the bay, and lowest salinity (<1 ppt) is found behind tide gates and up tidal creeks where freshwater from combined sewer overflows and sewage treatment plants are highest (Shin et al., 2013). The low marshes along the Hackensack are dominated with a tall vigorous form of *S. alterniflora*, mixed with bare mudflats; black grass marshes (*Juncus gerardii*) with stunted *Spartina alterniflora*. *Spartina patens* and *Distichlis spicata* dominate the high salt marshes. Extensive reed beds dominated by *Phragmites australis* are found throughout the Hackensack's brackish marshes (Artigas and Pechmann, 2010).

Eleven wetland restoration sites in the Hackensack Meadowlands District are identified in Figure 18.7a, probably covering less than 300 ha. At least

> 23 wetland mitigation sites are located in the district (R. M. Feltes, personal communication 2014), with the first constructed in 1983. Almost all of the mitigation conducted in the Meadowlands has been done in kind for impacts to tidal brackish waters. In-lieu fee agreements were used by the New Jersey Meadowlands Commission to acquire and enhance ecological functions at Harrier Meadow (constructed in 1998), Skeetkill Marsh (1998), Mill Creek Marsh (1999) and the Secaucus High School Wetland Enhancement Site (SHSWES; 2007). Although Mill Creek Marsh and SHSWES satisfied mitigation obligations for several different project permits, both public and private, these were not formal mitigation banks. Mitigation banks in the Meadowlands include Marsh Resources Meadowlands Mitigation Bank (MRI), Phases 1 and 2. Marsh Resources planting was completed in 2001 and continues to be monitored and managed for invasive plants. The 87-ha Richard P. Kane Natural Area Mitigation Bank, constructed in 2012, was designed largely as tidal *Spartina alterniflora* low marsh but has suffered from numerous design and financial challenges (R. M. Feltes, personal communication, 2014). Also a 20-ha Evergreen MRI Phase 3 Mitigation Bank was constructed and planted in 2012. The commission had an agreement with Rutgers University to conduct much more extensive monitoring of the Harrier Meadow, Skeetkill, and Mill Creek Marsh mitigation sites.

Mangrove Restoration

Restoring mangrove swamps in tropical regions of the world has some similar characteristics to restoring salt marshes in that the establishment of vegetation in its proper intertidal zone is the key to success. But that is generally where the similarities end. Mangrove restoration is more cosmopolitan in that it has been attempted throughout the tropical and subtropical world (Lewis, 2005); salt marsh restoration has been attempted primarily on the eastern North American and Chinese coastlines and to some extent in Europe and the West Coast of North America. Salt marsh restoration often can rely on waterborne seeds distributing through an intertidal zone; mangrove restoration often involves the physical planting of trees, although recent work has shown that these planting often fail (Samson and Rollon, 2008; Lewis, 2009). In countries such as Vietnam, mangrove declines have been attributed to: the spraying of herbicides during the Vietnam war; immigration of people to the coastal regions, leading to cutting of lumber for timber, fuel, wood, and charcoal; and extensive conversion of mangrove forests to shrimp aquaculture ponds.

Mangroves have being cleared for decades for construction of aquaculture ponds at unprecedented rates in Vietnam and many other tropical coastlines of the world (Benthem et al., 1999; Lewis and Brown, 2014). Most of the edible shrimp sold in the United States and Japan are produced in artificial ponds constructed in mangrove wetlands in Thailand, Indonesia, and Vietnam. Sold in the United States and Japan at very low prices, these products are the result of massive destruction of mangrove forests. More than 100,000 ha of abandoned ponds located in former mangrove swamps currently exist in these countries (R. Lewis, personal communication). In Vietnam, mangroves are being restored and protected to provide coastal protection and coastal fisheries support. In the Philippines, despite a presidential proclamation prohibiting the cutting of mangroves, it is estimated that the country still was losing 3,000 ha/yr in the late 1990s (2.4 percent/yr; deLeon and White, 1999). But the insatiable appetites in the United States, Japan, and several other developed countries for shrimp continue to cause mangroves to be destroyed. The shrimp ponds last only about five to six years before they develop toxic levels of sulfur; then they are abandoned and more mangroves are destroyed. These abandoned ponds present a challenge for mangrove restoration. Recent efforts to undertake restoration using Community Based Ecological Mangrove Rehabilitation in Indonesia have proved successful, and more large-scale restoration of these habitats is under way (Brown et al. 2014).

Lewis (2005), Lewis and Gilmore (2007), and Lewis and Brown (2014) argue that common ecological engineering approaches would work best in restoring mangrove swamps and that more of an analytic approach and less of a "gardening" approach should be taken. They recommend seven principles to correctly restore mangroves:

1. Get the hydrology right.
2. Do not initially build a nursery but carefully determine the reason for a lack of volunteer mangroves at a proposed restoration site.
3. See if the conditions that prevent natural colonization can be corrected. If they cannot, pick another site.
4. Examine normal hydrology and topography in reference mangrove swamps as your model for restored sites.
5. Remember that mangrove swamps do not have flat floors but have subtle topographic patterns.
6. Construct tidal creeks to facilitate flooding and drainage of tide waters and to improve fish and invertebrate access to mangroves for harvest by local fishers.
7. Evaluate the costs of restoration early in the project design to make the project as cost effective as possible.

Two manuals designed to assist with design and construction of successful mangrove restoration sites (Primavera et al., 2012; Lewis and Brown, 2014) describe what has been termed "ecological mangrove restoration" (EMR). EMR utilizes a step-wise approach to engineering large-scale mangrove restoration projects that generally avoid the wasted time and money of cultivating mangroves in nurseries but instead depend on natural recruitment of mangrove seeds and propagules to properly restored sites. Case Study 6 provides a description of the ambitions and failures of one of the largest mangrove restorations ever attempted in the world.

CASE STUDY 6: Restoring Mangroves after the 2004 Indian Ocean Tsunami

An estimated 230,000 people were killed in late December 2004 as a result of a massive tsunami around the Indian Ocean caused by an earthquake off the coast of Sumatra, Indonesia. (See "Mangrove Swamps and the Indian Ocean Tsunami of December 2004" in Chapter 16: "Wetland Ecosystem Services.") The 2004 Indian Ocean tsunami initially caused great interest in restoring mangrove and other coastal ecosystems to replace areas stricken by the tsunami as well as to provide coastal protection in the event of future tsunamis or other tidal surges. Immediately after the cleanup was under way, the whole world was made aware that the previous destruction of mangrove forests bears some of the blame for the high loss of human life and cultural impact. Soon afterward, mangrove and other coastal vegetation restoration was determined to be the best approach for local governments to ensure that a similar disaster would never happen again. Many countries and regions adopted this strategy, and efforts were undertaken all around the Indian Ocean. The governments of Malaysia, India, and Indonesia alone promised a total of U.S.$55 million to replant mangrove forests along their respective coastlands.

Mangrove restoration is really ecological engineering and many of the restored mangrove forests failed because the plantings occurred in regions where the tidal and hydrologic conditions are not appropriate (Lewis, 2005, 2010). Follow-up investigations of the success of the extensive plantings of mangroves after the tsunami have shown little real success, despite the planting of approximately 30 million mangrove seedlings (UNEP, 2007, 2008). At least half of all of the plantings failed within just a few years. In fact, Lewis (2010) declared that there is little evidence that mangroves have ever been restored on a large scale. He attributed this to two misguided assumptions regarding mangrove restoration: "(1) mangroves can only be restored by planting, and (2) sub-tidal mud flats are suitable for planting mangroves, when in fact they likely never supported a mangrove forest in the first place."

Millions of U.S. dollars were wasted worldwide partially because of the failure of land managers to recognize the overwhelming importance of self-design in mangrove recovery and to understand the basic ecology of mangrove vegetation. The other negative side of this project is that as we have passed the 10-year anniversary of the tsunami, interest in restoring mangroves has almost disappeared. Check (2005) reported that despite many local mangrove replantings and massive public assistance provided by international organizations, such as the United Nations, many tropical coastline regions returned to their old way of destroying mangrove forests for short-time profitable shrimp farms, making the regions even more susceptible than before to tropical storms and tsunamis.

Delta Restoration

As large rivers connect to the sea, multi-tributary deltas tend to develop, allowing the river to discharge to the sea in many channels. Many of these rich-soil deltas are among the most important ecological and economic regions of the world, from the ancient Nile delta in Egypt to the modern-day Mississippi River Delta in Louisiana. There should be two major ecological resource goals of delta areas: (1) protecting and restoring the functioning of the deltaic ecosystems in the context of a geologically dynamic framework; and (2) controlling pollution from entering the downstream lakes, oceans, gulfs, and bays. Delta restoration should have this dual emphasis where possible—ecosystem enhancement of the delta itself and improvement of coastal water quality downstream. The best strategy for delta restoration when "land building" is a necessary prerequisite is to restore the ability of the river to "spread out its sediments" in deltaic form through as wide an area as possible, particularly during flood events and by not discouraging (or encouraging and even creating) river distributaries. When river distributaries are not possible on a large scale because of navigation requirements or population locations, then restoring and creating riverine wetlands and constructing river diversions to divert river water to adjacent lands may be the best alternatives to maximize nutrient retention and sediment retention. In some cases, this involves the conversion of agricultural lands back to wetlands; in other cases, the dikes that "protect" wildlife protection ponds or retain rivers in their channels only need to be carefully breached to allow lateral flow of rivers during flood season. See further discussion of delta restoration for the dual purpose of wetland enhancement and water quality improvement in Case Study 4: Diverting the Mississippi River to the Louisiana Delta in Chapter 19: "Wetlands and Water Quality."

River Restoration

Rivers, well upstream of their deltas with the sea, are being restored around the world at a rapid rate (Bernhardt et al., 2005). In the past 10 years, there has been a paradigm shift in river restoration. Previously efforts centered on improving in-stream habitats by remeandering streams and adding physical structures, such as artificial riffles, in the stream itself (Hart et al., 2002). Current efforts involve restoration of the entire rivers and its corridor in what really is ecological engineering (Palmer et al., 2014). This restoration invariably involves a mix of creating and restoring a combination of fluvial- and shallow-water wetland systems, often with much of the emphasis on restoring the riparian edges to the river. Palmer et al. (2005) proposed five criteria for measuring success of river restoration:

1. The design of an ecological river restoration project should be based on a specified guiding image of a more dynamic, healthy river that could exist at the site.
2. The river's ecological condition must be measurably improved.
3. The river system must be more self-sustaining and resilient to external perturbations so that only minimal follow-up maintenance is needed.

4. During the construction phase, no lasting harm should be inflicted on the ecosystem.
5. Both pre- and post-assessment must be completed, and data must be made publicly available.

Now the emphasis is on restoring the entire river ecosystem that involves reconnecting the river to its floodplain and floodplain wetlands and restoration of the watershed itself (Mitsch et al., 2008; Kristensen et al., 2014). Case Study 7 describes one of the largest and most complete river restorations in Europe, where restoration involved both the stream itself and the floodplain wetlands. It is also a site where long-term monitoring has occurred and allowed significant opportunities for estimating if restoration success is possible in a decadal time frame.

CASE STUDY 7: Restoring the River Skjern in Denmark

The Skjern River in west-central Jutland drains a watershed of 2,490 km^2 and is the largest in Denmark. In Denmark's largest drainage project ever, 4,000 ha of wet meadow (Fig. 18.8a) was converted into arable land, and the lower Skjern was straightened to a fraction of its former meandering self. By the late 1980s, the river was essentially a straight line to the Ringkobing Fjord on the North Sea, eliminating thousands of hectares of marshland, meadows, and river habitat (Fig. 18.8b). The channelized river was diked, canals were built, and pumps were installed to hasten the downstream movement of water from the land. This public works project cost DDK 30 million (about U.S.$3.6 million) and was considered a success by the agricultural community at first as grains could now be grown in the formerly wet region. But the environment was paying a heavy price with this artificial river. The self-cleansing ability of the river was lessened, the downstream fjord was becoming polluted with nutrients and sediments, and the land that was draining began to subside due to peat oxidation and loss of water—up to 1 m or more in some locations. The human interference on this river has been described as "some of the most severe in northern Europe" by the Danish Ministry of Environment and Energy (DMEE, 1999). A few short years after the drainage, it appeared that another drainage project might be necessary, and public funding was requested. Instead, the Danish parliament (*Folketing*) passed a Public Works Act in 1998 by a huge majority that called for the restoration of the lower Skjern River and earmarked about US$40 million (DDK 254 million) for this project. The project was implemented in three phases for three reaches of the river. The river restoration in this case called for the following:

- Put back the meanders of the river wherever possible.
- Remove dikes along the river to allow adjacent meadows to be flooded once again.

- Define the project area by dikes far away from the river to prevent flooding of farmland outside of the project area.

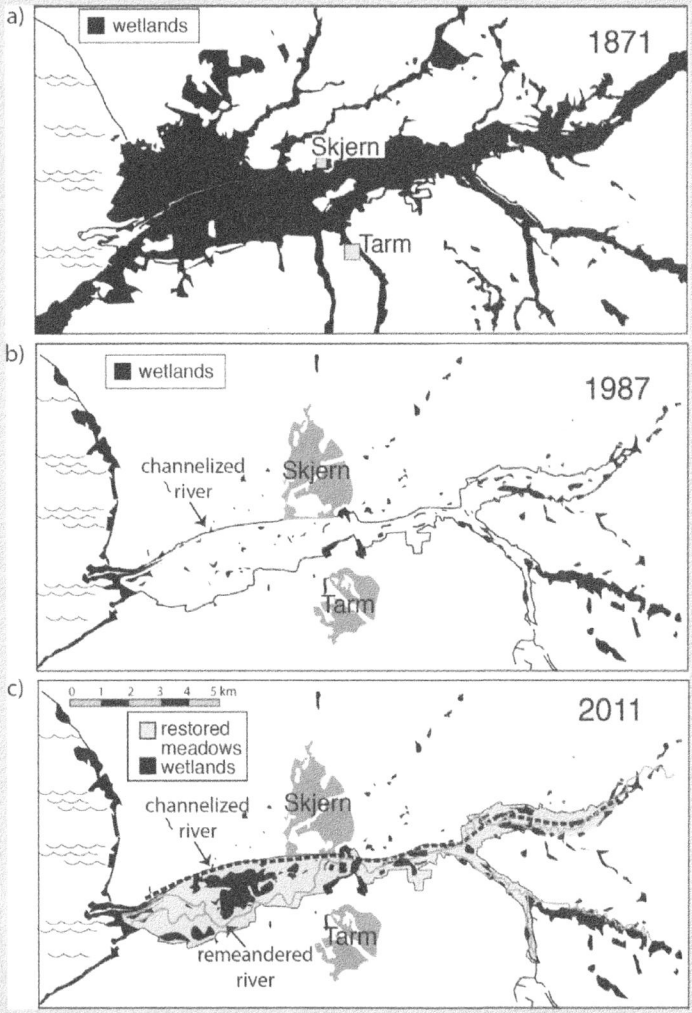

Figure 18.8 Restoration of Skjern River and its floodplain wetlands in western Denmark: (a) meadows and marshlands prior to channelization (1871); (b) channelized stream (1987) prior to restoration; and (c) stream, meadows, and marshlands almost a decade after restoration (2011); (d) photo of restored Skjern River and floodplain June 2012, 10 years after restoration. ((a), (b), (c) from DMEE, 1999, and Kristensen et al., 2014; (d) photo by W. J. Mitsch)

Figure 18.8 *(Continued)*

The restoration involved transforming 19 km of channelized river to 26 km of meandering river (Pedersen et al., 2007a,b). About 2,200 ha of the river valley wetlands were restored between 1998 and 2002 by removing dikes and levees that were adjacent to and moving them far away from the river to protect farmland outside of the project area. Early on, the project was successful in substantially increasing the biodiversity of aquatic macrophytes, invertebrates, amphibians, and mammals such as otters (Pedersen et al., 2007a). Kristensen et al. (2014) compared in-stream habitats and sedimentation/erosion as well as reconnections with the floodplain for three periods: (1) year 2000—before restoration; (2) year 2003—immediately after restoration; and (3) 2011—a decade after restoration (Fig. 18.8c, d). They found no long-term significant changes in in-stream habitats beyond what happened immediately after restoration in 2003. There was a net erosion on both banks of the restored stream but net sedimentation in the riverbed; overall, the restoration resulted in "a slightly wider and shallower river in 2011 than in 2001" (Kristensen et al., 2014). Processes that reshape the channels are slow and may take centuries. Reconnection between the river and its floodplain and the river channel was immediate, and 611 ha of riparian ecosystems were flooded 10 percent of the time, mostly during the winter; flooding of the riparian ecosystems was rare (<1 percent) during the growing season of May through August (Fig. 18.8d). It may take centuries for lost habitats on the floodplains, such as islands, backwaters, and oxbow lakes, to develop unless restoration engineering is employed to restore the lost habitats more quickly (Kristensen et al., 2014).

Wetland Creation and Restoration Techniques

Defining Goals

The design of an appropriate wetland or series of wetlands, whether for habitat recreation, the control of non–point source pollution, or wastewater treatment, should start with forming the overall objectives of the wetland. One view is that wetlands should be designed to maximize ecosystem longevity and efficiency and minimize cost. The goal, or a series of goals, should be determined before a specific site is chosen or a wetland is designed. If several goals are identified, one must be chosen as primary.

Placing Wetlands in the Landscape

In some cases, particularly when sites are being chosen for habitat replacement, many site choices are available in the landscape to locate a restored or created wetland. The natural design for a riparian wetland fed primarily by a flooding stream or river (Fig. 18.9a) allows for flood events of a river to deposit sediments and chemicals on a seasonal basis in the wetland and for excess water to drain back to the stream or river. Because there are natural and also often constructed levees along major sections of streams, it is often possible to create such a wetland with minimal construction. The wetland could be designed to capture flooding water and sediments and slowly release the water back to the river after the flood passes, or to receive flooding water and retain it through the use of flap gates.

Wetlands can be designed as in-stream systems by adding control structures to the streams or by impounding a distributary of the stream (Fig. 18.9b). Blocking an entire stream is a reasonable alternative only in headwater streams, and it is not generally cost-effective or ecologically advisable. This design is particularly vulnerable during flooding, and its stability might be unpredictable, but it has the advantage of potentially treating a significant portion of the water that passes that point in the stream. The maintenance of the control structure and the distributary might mean making significant management commitments to this design.

A riparian wetland fed by a pump (Fig. 18.9c) creates the most predictable hydrologic conditions for the wetland but at an obvious extensive cost for equipment and maintenance. If it is anticipated that the primary objective of a constructed wetland is the development of a research program to determine design parameters for future wetland construction in the basin, then wetlands fed by pumps is a good design. Two examples of wetlands of this type constructed primarily for research and education are the Des Plaines River wetlands in northeastern Illinois (Sanville and Mitsch, 1994) and the Olentangy River wetlands in central Ohio (Mitsch et al., 1998, 2012, 2014). If other objectives are more important, then the use of large pumps is usually not appropriate unless the wetland is constrained in an urban setting with no recourse.

Locating several small wetlands on small streams or intercepting ditches in the upper reaches of a watershed (but not in the streams themselves), rather than creating fewer larger wetlands in the lower reaches, should be considered (Fig. 18.9d). The usefulness of wetlands in decreasing flooding increases with the distance the wetland is downstream.

Figure 18.9e shows a design involving the creation of a wetland along a stream to intercept tile drains from agricultural fields. The stream is not diverted, but the

624 Chapter 18 Wetland Creation and Restoration

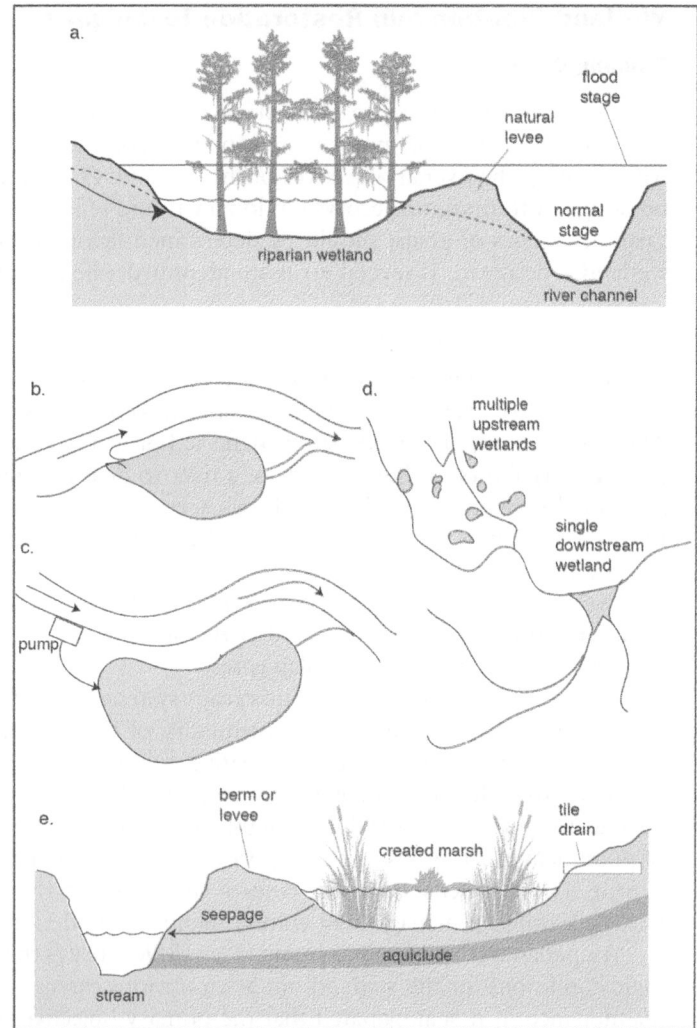

Figure 18.9 Landscape locations of created and restored wetlands in a riverine setting: (a) riparian wetland that both intercepts groundwater from uplands but also receives annual flood pulse from adjacent river; (b) riparian wetland with natural flooding; (c) riparian wetland with pump; (d) multiple upstream wetlands versus single downstream wetland; and (e) lateral wetland intercepting groundwater carried by tile drains.

wetlands receive their water, sediments, and nutrients from small tributaries, swales, and especially tile drains that otherwise would empty straight to the stream. If tile drains can be located and broken or blocked upstream to prevent their discharge into tributaries, they can be rerouted to make effective conduits to supply adequate water to constructed wetlands. Because tile drains are often the sources of the highest concentrations of chemicals, such as nitrates from agricultural fields, the lateral wetlands would be an efficient means of controlling certain types of non–point source pollution while creating a needed habitat in an agricultural setting.

Site Selection

Several important factors ultimately determine site selection. When the objective is defined, the appropriate site should allow for the maximum probability that the objective can be met, that construction can be done at a reasonable cost, that the system will perform in a generally predictable way, and that the long-term maintenance costs of the system are not excessive. These 12 factors are elaborated next.

1. Wetland restoration is generally more feasible than wetland creation.
2. Take into account the surrounding land use and the future plans for the land.
3. Undertake a detailed hydrologic study of the site, including a determination of the potential interaction of groundwater with the proposed wetland.
4. Find a site where natural inundation is frequent.
5. Inspect and characterize the soils in some detail to determine their permeability, texture, and stratigraphy.
6. Determine the chemistry of the soils, groundwater, surface flows, flooding streams and rivers, and tides that may influence the site water quality.
7. Evaluate on-site and nearby seed banks to ascertain their viability and response to hydrologic conditions.
8. Ascertain the availability of necessary fill material, seed, and plant stocks and access to infrastructure (e.g., roads, electricity).
9. Determine the ownership of the land and hence the price.
10. For wildlife and fisheries enhancement, determine if the wetland site is along ecological corridors, such as migratory flyways or spawning runs.
11. Assess site access.
12. Ensure that an adequate amount of land is available to meet the objectives.

Creating and Maintaining the Proper Hydrology

The key to restoring and creating wetlands is to develop appropriate hydrologic conditions. Groundwater inflow is often desired because it offers a more predictable and less seasonal water source. Surface flooding by rivers gives wetlands a seasonal pattern of flooding, but such wetlands can be dry for extended periods in flood-absent periods. Depending on surface runoff and flow from low-ordered streams can be the least predictable. Often wetlands developed in these conditions are isolated pools and potential mosquito havens for a good part of the growing season; their design should be carefully considered. It is generally considered to be optimum to build wetlands where they used to be and where hydrology is still in place for the wetland to survive. But tile drainage, ditches, and river downcutting have often changed local hydrology from prior conditions. Most biologists have difficulty estimating hydrologic conditions, while engineers often overengineer control structures that need substantial maintenance and are not sustainable.

Wetland basins are constructed either by establishing levees around a basin that may be partially excavated in the landscape or by excavating a depression without

626 Chapter 18 Wetland Creation and Restoration

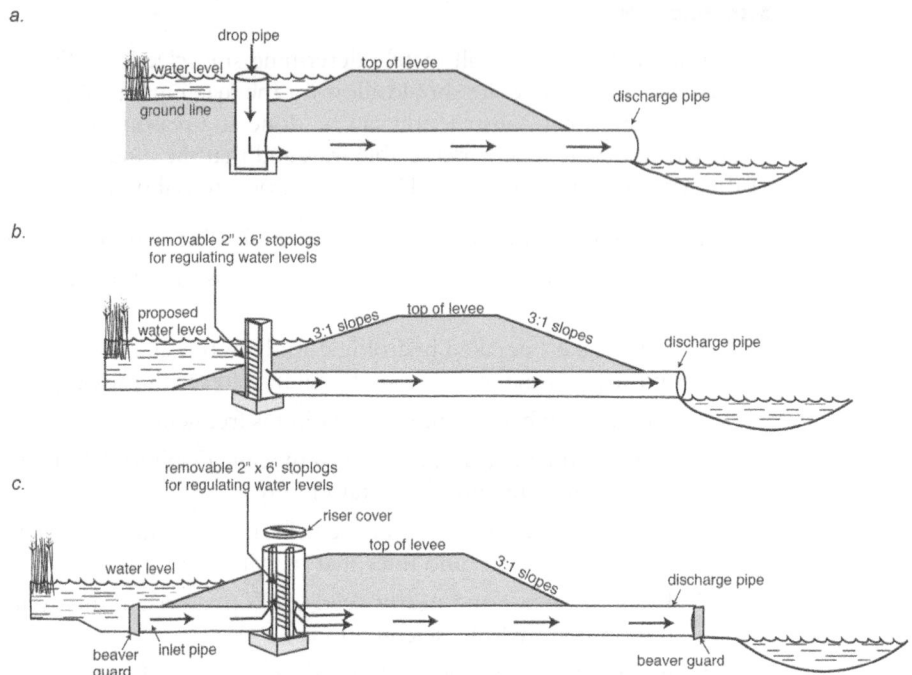

Figure 18.10 Designs for control systems for created and restored wetlands including (a) drop pipe, (b) flashboard riser, and (c) full-round riser. (From Massey, 2000)

constructing any levee. Construction engineers often note that if they use excavated soil for a levee, they can save large sums of money because excavation is often the largest cost of wetland restoration or creation. This levee construction is usually not a good idea, because levees are bound to have problems with leakage and, in many parts of the world, burrowing animals like muskrats (*Ondatra zibethicus*).

Some sort of control structure is often needed at the outflow of the wetland basin, whether there is a levee or not. The control devices are the outflow of the wetland. Three such control devices are shown in Figure 18.10: (1) drop pipes, (2) flashboard risers, and (3) full-round riser (combination of drop pipe and flashboard riser).

Each device has its own advantages and disadvantages. Drop pipes are the least flexible because they do not allow water-level manipulation. A flashboard riser is more flexible but can be easily vandalized. Full-round risers are a little more secure and can be designed for control of beavers, but they are a little more expensive. In the last two cases, the outflow risers include removable stoplogs that allow manual changes in water level. This option is desirable when the exact hydrology of the wetland basin is not known, because it allows flexibility.

But these types of control devices have several disadvantages. They require occasional maintenance, if only for removing accumulation of plant debris and resetting stoplogs. Also, stoplog removal is a favorite pastime of vandals. Control devices such as risers are also favorite locations for nature's ecological engineer—the beaver *Castor canadensis*—to provide its idea of water management, usually creating blockages that can raise the water level by a meter or more, changing the vegetation patterns dramatically.

The best design situation is when the local topography allows the wetland to be naturally flooded without control devices, but this opportunity is rarely available. For a reliable source of water, groundwater is generally less sensitive to seasonal highs and lows than is surface water. Also, a wetland fed by groundwater invariably has better water quality and generally fewer sediments that will eventually fill the wetlands.

Soils

Choice of the site for wetland creation and restoration is often limited by property ownership. If a choice exists, a wetland that is restored on former wetland (hydric) soils is much preferred over one constructed on upland soils. Hydric soils develop certain color and chemical patterns, because they have spent long periods flooded and thus under anaerobic conditions. The soil color is mostly black in mineral hydric soils, because iron and manganese minerals have been converted to reduced soluble forms and have leached out of the soil (see Chapter 5: "Wetland Soils"). In most cases, developing wetlands on hydric soils has three three advantages:

1. Hydric soils indicate that the site may still have or can be restored to appropriate hydrology.
2. Hydric soils may be a *seed bank* of wetland plants still established in the soil.
3. Hydric soils may have the appropriate soil chemistry for enhancing certain wetland processes. For example, mineral hydric soils generally have higher soil carbon than do mineral nonhydric soils. This soil carbon, in turn, stimulates wetland processes such as denitrification and methane production.

Otherwise, it is possible to create wetlands on upland soils. In the long run, those soils will develop characteristics typical of hydric soils, such as higher carbon content and seed banks (see Case Study 8).

> **CASE STUDY 8: Hydric Soil Development and Chemical Sequestration in Created Wetlands**
>
> It was not well known until recently how long it would take upland soils to develop wetland conditions; it was thought to be over a decade or even a century, depending on the soil types and the hydrology. In the study of the two wetlands at the Olentangy River Wetland Research Park in Ohio that were created in 1994, hydric soil conditions were shown to develop after only two years of continued flooding (Mitsch et al., 2005). Before the basins were first flooded, the most prevalent hue was 10YR, and the value/chroma soil color varied between 3/3 and 3/4. The chroma of 3 to 4 indicates nonhydric soils (see description of these soil color terms in Chapter 5). In 1995, about 18 months after flooding, chromas of 3 or less were common (median = 3/2).

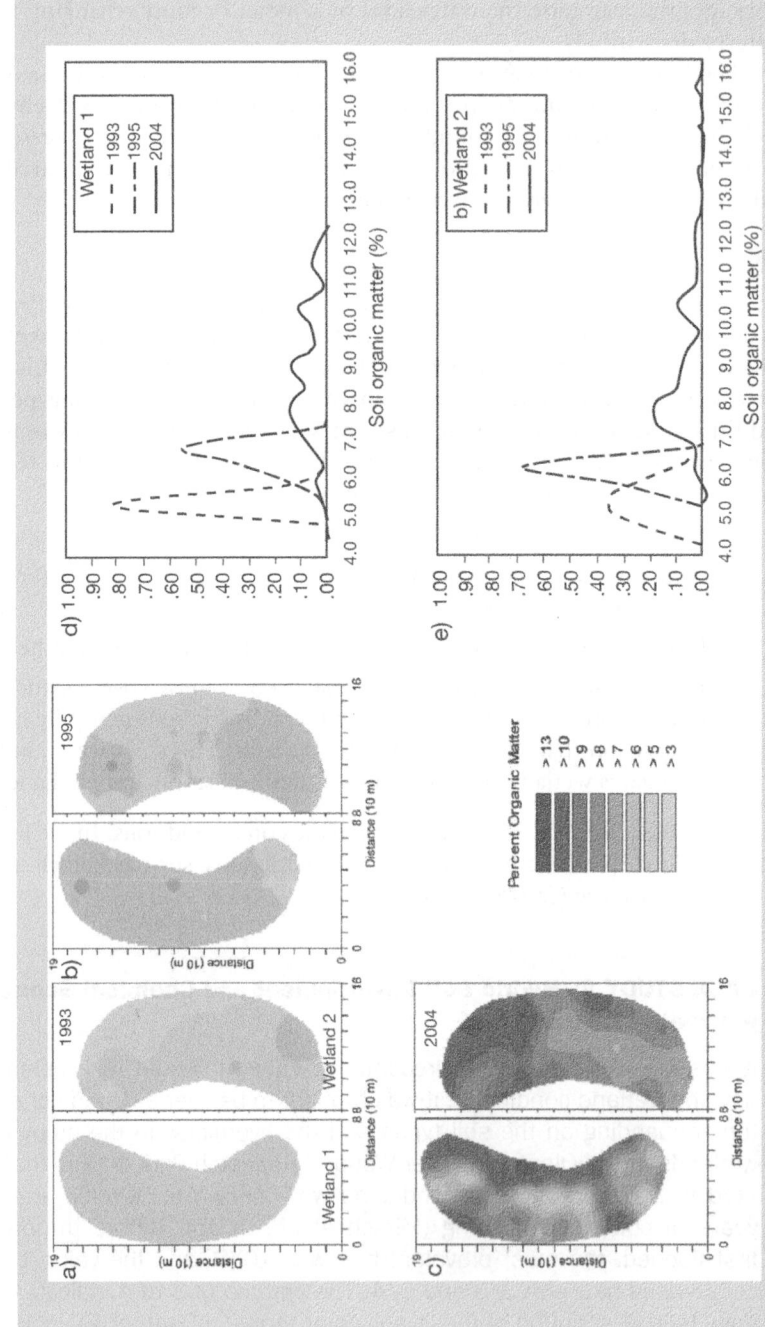

Figure 18.11 Soil organic matter development in two 1-ha created experimental wetlands over a 12-year period at the Olentangy River Wetland Research Park. Left: Spatial distribution maps of soil organic matter for Wetlands 1 and 2 shown for: (a) 1993, before water was added but after basins were excavated; (b) 1995, 16 months after pumping began; and (c) 2004, 10 years after flooding began. Right: Frequency distribution curves for soil organic matter in two experimental wetlands in 1993, 1995, and 2004 for (d) planted Wetland 1 and (e) unplanted Wetland 2. Wetlands were excavated in nonhydric alluvial floodplain soils in 1993, and water has been continuously pumped into both wetlands since March 4, 1994. (From Anderson et al., 2005)

The mean value in the surface samples was 3/2; subsurface sample median values were 4/2. Chromas started to consistently be of 2 or below in 1996, two years after flooding began. As of 2006, 13 years after the soils were flooded, almost all samples in these experimental wetlands show chromas of 2 or less in the surface sediments.

Lack of organic carbon in the soil is often described as a shortcoming of wetland creation projects. The organic content of the upper soils in these experimental wetlands increased steadily over the first decade, from 1994 when water was first added (Fig. 18.11; Table 18.3). The organic content of the surface (0–8 cm depth) soils increased from 5.3 ± 0.1 percent in 1993 (before water was added), to 6.1 ± 0.1 percent in 1995 (18 months after creation), to 8.9 ± 0.2 percent in 2004 (10 years after the wetland was created) (Anderson et al., 2005). Total carbon increased from 1.57 ± 0.04 percent in 1993, to 2.06 ± 0.12 percent in 1995, and to 3.76 ± 0.12 percent in 2004. In other words, the organic content and total carbon of these wetland surface sediments increased by 67 percent and 139 percent, respectively, over a decade. While most of the carbon increase is believed to be due to algal and macrophyte productivity, it is clear from other studies at these wetlands (Wu and Mitsch, 1998; Liptak, 2000; Tuttle et al., 2008) that a substantial amount of the carbon accumulation is probably due to inorganic calcium carbonate/ calcite ($CaCO_3$) (See inorganic carbon accumulation in Table 18.3) that precipitates at high rates in the growing season due to water column productivity in these wetlands. A rule of thumb from this study on newly created wetlands: Organic content in surface soils in newly created wetlands increases by about 1 percentage point every three years.

Table 18.3 Mean carbon and nutrient accumulation rates in two experimental wetlands at the Olentangy River Wetland Research Park, Ohio, 1994 to 2004

Parameter	Mean Annual Accumulation Rates g m^{-2} yr^{-1}
Total carbon	181–193
Organic carbon	152–166
Inorganic carbon	23–26
Total nitrogen	16.2–16.6
Total phosphorus	3.3–3.5
Total calcium	80.8–86.3

Source: Anderson and Mitsch (2006).

It has been argued that upland soils often do not allow the development of a major diversity of plant communities but often become *Typha* marshes instead because of the absence of seed banks. This domination by *Typha* is as much due to the fact that uplands converted to wetlands have often been used for agriculture for many years and are thus quite eutrophic. The high-nutrient conditions invariably lead to high-productivity, low-diversity systems. Again, the main advantage of using hydric soils in wetland restoration and creation is that they are indicators of appropriate hydrologic conditions.

Introducing Vegetation

The species of vegetation types to be introduced to created and restored wetlands depend on the type of wetland desired, the region, and the climate as well as the design characteristics described previously. Table 18.4 summarizes some of the plant species used for wetland creation and restoration projects, mostly in the United States. Vymazal (2013) found in a literature search of 643 surface-flow constructed wetlands described in 43 countries that 150 species have been used. The most commonly used genera are *Typha*, *Scirpus*, *Schoenoplectus*, *Phragmites*, *Juncus*, and *Eleocharis*. More details are given in Chapter 19 on plants used for water quality improvement.

Freshwater Marshes

Common plants used for freshwater marshes include bulrush (*Scirpus* spp. and *Schoenoplectus* spp.), cattails (*Typha* spp.), sedges (*Carex* spp.), and floating-leaved aquatic plants such as white water lilies (*Nymphaea* spp.) and spatterdock (*Nuphar* spp.). Submerged plants are not common in wetland design, and their propagation is often hampered by turbidity and algal growth in the early years of wetland development.

Coastal Marshes

For coastal salt marshes, *Spartina alterniflora* is the primary choice for coastal marsh restoration in the Eastern United States. Both *Spartina townsendii* and *S. anglica* have been used to restore salt marshes in Europe and in China. The details of successful coastal wetland creation are site specific, but several generalizations seem to be valid in most situations.

Forested Wetlands

Forested wetland restoration and creation usually involve the establishment of seedlings. In the southeastern United States, deciduous hardwood species typical of bottomland forests are planted. They include nuttall oak (*Quercus nuttallii*), cherrybark oak (*Q. falcata* var. *pagodifolia*), willow oak (*Q. phellos*), water oak (*Q. nigra*), cottonwood (*Populus deltoides*), sycamore (*Platanus occidentalis*), green ash (*Fraxinus pennsylvanica*), sweetgum (*Liquidambar styracifula*), and pecan (*Carya illinoensis*). There is less use of deepwater plants such as bald cypress (*Taxodium distichum*) and water tupelo (*Nyssa aquatica*), although *Taxodium* spp. was once the dominant genus of introduced species in many wetland restorations in Florida (Clewell, 1999). In Florida, a wide variety of wetland oaks, bays, gums, ashes, and pines are also used in forested wetland restoration.

Table 18.4 Selected plant species planted in created and restored wetlands

Scientific Name	Common Name	Scientific Name	Common Name
Freshwater Marsh—Emergent			
Acorus calamus	sweet flag	*Pontederia cordata*	pickerelweed
Cladium jamaicense	sawgrass	*Sagittaria rigida*	duck potato
Carex spp.	sedges	*Sagittaria latifolia*	duck potato; arrowhead
Eleocharis spp.	spike rush	*Saururus cernuus*	lizard's tail
Glyceria spp.	manna grass	*Schoenoplectus tabernaemontani**	soft-stem bulrush
Hibiscus spp.	rose mallow		
Iris pseudacorus	yellow iris	*Scirpus acutus*	hard-stem bulrush
Iris versicolor	blue iris	*Scirpus americanus*	three-square bulrush
Juncus effusus	soft rush	*Scirpus cyperinus**	Woolgrass
Leersia oryzoides	rice cutgrass	*Scirpus fluviatilis*	river bulrush
Panicum virgatum	switchgrass	*Sparganium eurycarpum*	giant bur reed
Peltandra virginica	arrow arum	*Spartina pectinata*	prairie cordgrass
Phalaris arundinacea	reed canary grass	*Typha angustifolia**	narrow-leaved cattail
*Phragmites australis**	giant reed	*Typha latifolia**	wide-leaved cattail
Polygonum spp.	smartweed	*Zizania aquatic*	wild rice
Freshwater Marsh—Submerged			
Ceratophyllum demersum	coontail	*Potamogeton pectinatus*	Sago pondweed
Elodea nuttallii	waterweed	*Vallisneria* spp.	wild celery; tape grass
Myriophyllum aquaticum	milfoil	*Najas guadalupensis**	Southern naiad
Freshwater Marsh—Floating			
Azolla caroliniana	water fern	*Nuphar luteum*	spatterdock
*Eichhornia crassipes**	water hyacinth	*Pistia stratiotes*	water lettuce
Hydrocotyle umbellata	water pennywort	*Salvinia rotundifolia*	floating moss
Lemna spp.	duckweed	*Wolffia sp.*	water meal
Nymphaea odorata	fragrant white water lily		
Bottomlands/Forested Wetland			
Acer rubrum	red maple	*Gordonia lasianthus*	loblolly bay
Acer floridanum	Florida maple	*Liquidambar styracifula*	sweetgum
Acer saccharinum	silver maple	*Platanus occidentalis*	sycamore
Alnus spp.	alder	*Populus deltoids*	cottonwood
Carya illinoensis	pecan	*Quercus falcata* var. *pagodifolia*	cherrybark oak
Celtis occidentalis	hackberry	*Ulmus Americana*	American elm
Cephalanthus occidentalis	buttonbush	*Quercus nigra*	water oak
Cornus stolonifera	red-osier dogwood	*Quercus nuttallii*	Nuttall oak
Fraxinus caroliniana	water ash	*Quercus phellos*	willow oak
Fraxinus pennsylvanica	green ash	*Salix* spp.	willow
Deepwater Swamp			
Nyssa aquatica	swamp tupelo	*Taxodium distichum*	bald cypress
Nyssa sylvatica var. *biflora*	black gum	*Taxodium distichum* var. *nutans* aka *Taxodium distichum* var. *imbricarium*	pond cypress

Table 18.4 (Continued)

Scientific Name	Common Name	Scientific Name	Common Name
Salt Marsh			
Distichlis spicata	spike grass	*Spartina foliosa*	cordgrass (Western U.S.)
Salicornia sp.	Saltwort	*Spartina patens*	salt meadow grass
Spartina alterniflora	cordgrass (Eastern U.S.)	*Spartina townsendii*	cordgrass (Europe)
Spartina anglica	cordgrass (Europe; China)		
Mangrove Swamp			
Rhizophora mangle	red mangrove	*Laguncularia racemosa*	white mangrove
Avicennia germinans	black mangrove		

*Commonly planted in treatment wetlands.

Wetland Planting Techniques

Plants can be introduced to a wetland by transplanting roots, rhizomes, tubers, seedlings, or mature plants; by broadcasting seeds obtained commercially or from other sites; by importing substrate and its seed bank from nearby wetlands; or by relying completely on the seed bank of the original and surrounding site. If planting stocks rather than site seed banks are used, it is most desirable to choose plants from wild stock rather than nurseries because the former are generally better adapted to the environmental conditions they will face in constructed wetlands. The plants should come from nearby if possible and should be planted within 36 hours of collection. If nursery plants are used, they should be from the same general climatic conditions and should be shipped by express service to minimize losses. Marshes should be planted at densities to ensure rapid colonization, adequate seed source, and effective competition with undesirable plants, such as *Typha* spp. Specifically, this could mean introducing from 2,000 to 5,000 plants/ha.

For emergent plants, the use of planting materials with stems of at least 20 to 30 cm is recommended, and whole plants, rhizomes, or tubers rather than seeds have been most successful. In temperate climates, both fall and spring planting times are possible for certain species, but spring plantings are generally more successful, because it is a better time to minimize destructive winter grazing of plants by migratory animals and the uprooting of the new plants by ice.

Transplanting plugs or cores (8–10 cm in diameter) from existing wetlands is another technique that has been used with success, for it brings seeds, shoots, and roots of a variety of wetland plants to the newly restored or created wetland.

If seeds and seed banks are used for wetland vegetation, several precautions must be taken. The seed bank should be evaluated for seed viability and species present. The use of seed banks from other nearby sites can be an effective way to develop wetland plants in a constructed wetland if the hydrologic conditions in the new wetland are similar. Seed bank transplants have been successful for many different species, including sedges (*Carex* spp.), *Sagittaria* sp., *Scirpus acutus*, *S. validus*, and *Typha* spp. The disruption of the wetland site where the seed bank is obtained must also be considered.

When seeds are used directly to vegetate a wetland, they must be collected when they are ripe and stratified if necessary. If commercial stocks are used, the purity of the seed stock should be determined. The seeds can be added with commercial drills or by broadcasting from the ground, watercraft, or aircraft. Seed broadcasting is most effective when there is little to no standing water in the wetland.

Natural Succession versus Horticulture

To develop a wetland that ultimately will be a low-maintenance one, natural successional processes need to be allowed to proceed. The best strategy is usually to introduce, by seeding and planting, as many native choices as possible to allow natural processes to sort out the species and communities in a timely fashion. Wetlands created or restored by this approach are called *self-design wetlands*. Providing some help to this selection process (e.g., selective weeding) may be necessary in the beginning, but ultimately the system needs to survive with its own successional patterns unless significant labor-intensive management is possible. A somewhat different approach, called *designer wetlands*, occurs when specified plant species are introduced, and the success or failure of those plants is used as indicators of success or failure of that wetland. This is akin to horticulture.

An important general consideration of wetland design is whether plant material is going to be allowed to develop naturally from some initial seeding and planting or whether continuous horticultural selection for desired plants will be imposed. W. E. Odum (1987) suggested, "In many freshwater wetland sites it may be an expensive waste of time to plant species which are of high value to wildlife. It may be wiser to simply accept the establishment of disturbance species as a cheaper although somewhat less attractive solution." As described above, Samson and Rollon (2008) and Lewis (2009) found that planting of mangrove seedlings was often a big waste of time and resources. The successful salt marsh restoration at Delaware Bay described in Case Study 4 (Teal and Weinstein, 2002; Hinkle and Mitsch, 2005) did not require any seeding or planting. Reinartz and Warne (1993) found that the way vegetation is established can affect the diversity and value of the mitigation wetland system. Their study showed that early introduction of a diversity of wetland plants may enhance the long-term diversity of vegetation in created wetlands. The study examined the natural colonization of plants in 11 created wetlands in southeastern Wisconsin. The wetlands under study were small, isolated, depressional wetlands. A two-year sampling program was conducted for the created wetlands, aged one to three years. Colonization was compared to five seeded wetlands where 22 species were introduced. The diversity and richness of plants in the colonized wetlands increased with age, size, and proximity to the nearest wetland source. In the colonized sites, *Typha* spp. comprised 15 percent of the vegetation for one-year wetlands and 55 percent for three-year wetlands, with the possibility of monocultures of *Typha* spp. developing over time in colonized wetlands. The seeded wetlands had a high species diversity and richness after two years. *Typha* cover in these sites was lower than in the colonized sites after two years.

Another study where the effects of planting versus not planting have been observed for several years was at two 1-ha experimental wetlands in central Ohio (Table 18.5; See also Case Study 8 above). One wetland was planted with 2,500 plants representing 13 macrophyte species; the other was left as a naturally colonizing control wetland. In essence, both wetlands were different degrees of self-design

Table 18.5 Summary of vegetation species richness in two 1-ha experimental wetlands in central Ohio through 17 years. Both wetlands were created in 1994. The "planted wetland" (W1) was planted with 2,500 individual plants representing 13 native wetland species. W2 is unplanted control

Wetland Age	Year	Number of Species			Number of Wetland Species			Number of Planted Species		Number of Woody Species		Number of Invasive Species	
		W1	W2	Total	W1	W2	Total	W1	W2	W1	W2	W1	W2
3	1996	67	56	72	43	31	44	9	1	5	7	1	1
5	1998	96	87	99	56	46	57	9	2	15	15	4	4
15	2008	101	97	116	55	52	61	9	2	18	21	7	9
17	2010	99	97	118	51	49	63	9	2	18	21	7	10

Source: Mitsch et al., 2012 updated.

because there were no expectations as to what the ultimate cover would be and there was no "gardening" to get to any endpoint. After three years, both wetlands were principally dominated by soft-stem bulrush *Schoenoplectus tabernaemontani* (= *Scirpus validus*) and were thought to be similar (Mitsch et al., 1998). After six years, however, several communities of vegetation continued to exist in the planted basin, but a highly productive monoculture of *Typha* dominated the unplanted basin where it did not have any competition from planted vegetation (Mitsch et al., 2005). By 2013, both wetlands had mostly converged on vegetation cover of *Typha* spp. The wetlands did have a few differences in wetland function that persisted a decade or more after planting that could be traced to effects of the initial planting (Mitsch et al., 2005, 2012, 2014). And 9 of the 13 planted species were still in the planted wetland 20 years after planting (only 2 "jumped over" to the unplanted basin), although most were not common (Mitsch et al., 2014). Both carbon sequestration and methane emissions were consistently higher in the naturally colonizing (unplanted) wetland due to its higher productivity (Nahlik and Mitsch, 2010; Sha et al., 2011; Mitsch et al., 2012, 2014; Bernal and Mitsch, 2013) whereas other ecological indicators, such as macrophyte community diversity, were almost always higher in the planted wetland. Planting at first appeared to have little to no effect on water quality but analyses of 15 years of data of nutrient fluxes showed that the planting enhanced overall retention of phosphorus but reduced the retention of total nitrogen (see also Case Study 5 in Chapter 19: "Wetlands and Water Quality").

If plant diversity is desired, planting makes sense. If productivity and carbon sequestration are desired, it may be a waste of effort to plant unless there are no sources of plant propagules (e.g., seed banks or inflowing rivers). In either case, a long-term effect on ecosystem function caused by introducing plants appeared to linger 20 years after a wetland is planted, but that effect is overshadowed by an overwhelming impact of natural propagule inputs and self-design.

Exotic or Undesirable Plant Species

In some cases, certain plants are viewed as desirable or undesirable because of their value to wildlife or their aesthetics. Reed grass (*Phragmites australis*) is often favored in constructed wetlands in Europe, and there is real concern for reed die-back around lakes and ponds in Europe. But reed grass is considered an invasive, undesirable plant in much of eastern North America, particularly in coastal freshwater and brackish marshes (Philipp and Field, 2005). *Spartina alterniflora* is the desired endpoint of salt marsh restoration on the East Coast of North America but is considered an invasive nuisance on the West Coast and now in China.

Some plants are considered undesirable in wetlands because they are aggressive competitors. In many parts of the tropics and subtropics, the floating aquatic plants water hyacinth (*Eichhornia crassipes*) and alligator weed (*Alternanthera philoxeroides*) are considered undesirable and, in eastern North America, particularly around the Great Lakes, the emergent purple loosestrife (*Lythrum salicaria*) is considered an undesirable alien plant in wetlands. Throughout the United States, cattail (*Typha* spp.) is championed by some and disdained by others, because it is a rapid colonizer but is of limited wildlife value. In other parts of the world, *Typha* is considered a perfectly acceptable plant in restored wetlands. In New Zealand, several species of willow (*Salix*) are invading marshes and other wetlands, and programs to eradicate them are common.

Estimating Success

There has not been a lot of positive analysis of created and restored wetlands in the literature, despite the fact that many good approaches have been developed. It is our belief that wetlands can (and should) be created and restored. The problem is more that those with the right understanding of wetland function are not the ones with the engineering consulting contracts creating and restoring them.

Few satisfactory methods are available to determine the success of a created or restored wetland or even a mitigation wetland created to replace the functions lost with the original wetland. Figure 18.1 illustrated conceptually how it should be done for replacement wetlands. It is clear from several studies of created and restored wetlands that some cases are successes, but there are still far too many examples of failures of created and restored wetlands to meet expectations. Examples of such studies done in the 1990s for southern Florida, northeastern Illinois, and Ohio are shown in Figure 18.12 and Table 18.6. In some cases, expectations were unreasonable; in other cases, the original wetland should not have been lost to begin with. Where expectations are ecologically reasonable, there is optimism that wetlands can be created and restored and that wetland function can be replaced.

A more recent analysis by Moreno-Mateos et al. (2012) using a meta-analysis of 621 wetland sites found that biological structure (mostly estimated from plant communities) and biogeochemical function (mostly estimated by carbon accumulation in the wetland soils) were 26 and 23 percent lower, respectively, in restored wetlands than in reference wetlands. They also found that large wetlands (>100 ha), tropical/temperate wetlands, and riverine wetlands fared better than did small wetlands, cold-climate wetlands, and depressional wetlands, respectively.

Hopple and Craft (2013) compared four restored and four natural wetlands in the glaciated northwest Indiana, including restored and natural wetland sites in the mostly drained Kankakee Marsh region (see Chapter 3) and found that, after 10 years, the

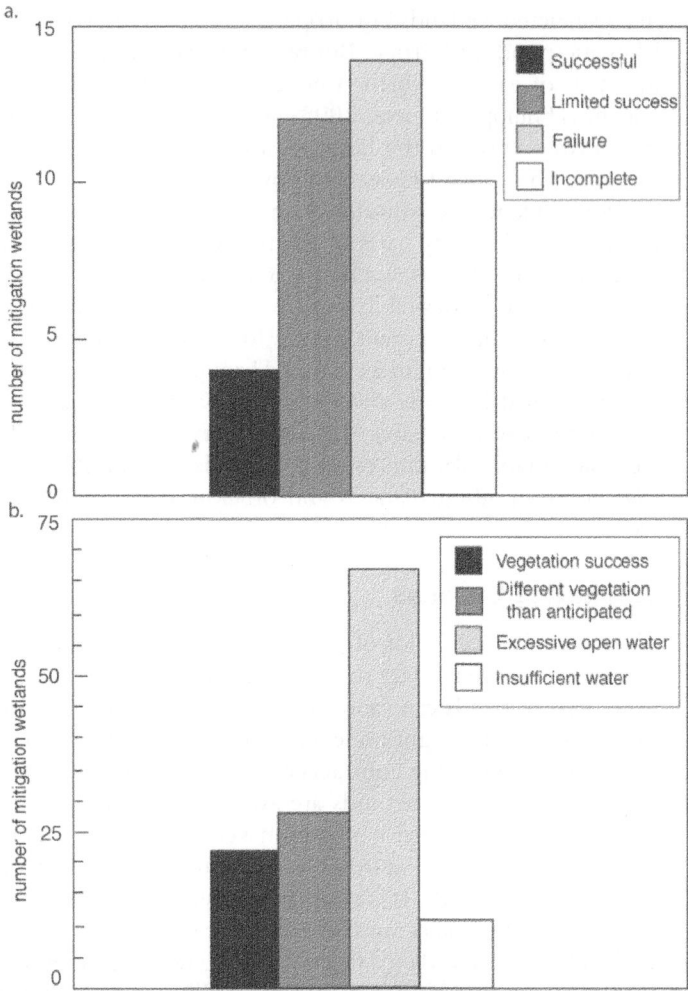

Figure 18.12 Evaluation of wetland mitigation projects in two regions of the United States: (a) 40 mitigation projects in south Florida involving wetland creation, mitigation, and preservation. The average age of the projects was less than three years. "Successful" meant that the project met all of its stated goals, whereas "failure" meant that few goals were met, or the created/restored wetland did not have functional equivalency to a reference wetland; (b) 128 wetland mitigation sites required by 61 permits in the six-county region around Chicago, Illinois. The permits were issued between 1990 and 1994, and this study began in 1996. ((a) From Erwin, 1991; (b) from Gallihugh and Rogner, 1998)

restored wetlands and the natural wetlands had similar plant richness (33.8 + 2.3 versus 27 + 6.4 species, respectively) and Floristic Quality Assessment Indices. They concluded that "the comparable plot and site diversity of restored wetlands is attributed to the use of management tools (such as seeding, prescribed burning, and herbicidal treatments) during restorations that enhance species richness and diversity and shorten the time required for the plant community of restored wetlands to converge with levels in natural wetlands."

Table 18.6 Permit requirements and compliance for five replacement wetlands investigated in Ohio

Location, County in Ohio	Wetland Area (ha)			Percentage of Required Area Replaced (%)
	Lost	Required	Happened	Location, County in Ohio
Delaware	3.7	5.4	~4.0	74
Franklin	15.0	28.0	3.2	11
Gallia	0.5	0.8	0.7	88
Jackson	4.8	7.2	7.5	105
Portage	0.4	0.6	0.6	100
Total	**24.4**	**42.0**	**~16.0**	**38**

Source: Wilson and Mitsch (1996).

It appears that wetland creation and restoration is being done in a more intelligent fashion now than it was 20 years ago. The earlier spotty record is due, in our opinion, to three factors:

1. Little understanding of wetland function by those creating and restoring the wetlands
2. Provision of insufficient time for the wetlands to develop
3. A complete lack of recognition or underestimation of the self-design capacity of nature by biologists and engineers alike

Understanding wetlands enough to be able to create and restore them requires substantial training in plants, soils, wildlife, hydrology, water quality, and engineering. Replacement projects and other restorations involving freshwater marshes need enough time, closer to 15 or 20 years than to 5 years, before success is apparent. Restoration and creation of forested wetlands, coastal wetlands, or peatlands may require even more time. Peatland restoration could take decades or more. Forested wetland restoration generally takes a lifetime. Finally, we should recognize that nature remains the chief agent of self-design, ecosystem development, and ecosystem maintenance; humans are not the only participants in these processes. Sometimes we refer to these self-design and time requirements for successful ecosystem restoration and creation as invoking "Mother Nature and Father Time" (Mitsch and Wilson, 1996; Mitsch et al., 1998, 2012).

Both wetland scientists and wetland engineers need to learn each other's trade for the field and the wetlands to be successful (Mitsch, 2014). Wetland science will continue to make significant contributions to the process of reducing our uncertainty about predicting wetland success. Wetland creation and restoration need to become part of an applied ecological science, not a technique without theoretical underpinnings. Scientists need to use quantitative and carefully designed experiments to make the connections between structure, such as vegetation density and diversity, and and function, such as productivity, wildlife use, organic sediment accretion, and nutrient retention. Engineers and managers need to recognize that systems that emphasize the role of self-design and sustainable structures are more ecologically viable in the long run than are heavily managed systems.

As Mitsch (2014) summarized:

> Ecosystem restoration, as currently practiced throughout the world, is done by practitioners who have little experience in design (scientists study systems, they do not design systems) and by engineers who do not appreciate the capabilities of ecosystems to self-design (engineering is a field devoted to removing uncertainty and controlling natural processes). The approach of many restorations is restoration by committee—that results in projects that are less successful than anticipated or are overdesigned by engineers with unsustainable technology.

Summary Principles

Seven general principles of ecological engineering that apply to the creation and restoration of wetlands are outlined next (Mitsch and Jørgensen, 2004):

1. Design the system for minimum maintenance and a general reliance on self-design.
2. Design a system that utilizes natural energies, such as the potential energy of streams, as natural subsidies to the system.
3. Design the system with the hydrologic and ecological landscape and climate in mind.
4. Design the system to fulfill multiple goals, but identify at least one major objective and several secondary objectives.
5. Give the system time.
6. Design the system for function, not form.
7. Do not overengineer wetland design with rectangular basins, rigid structures and channels, and regular morphology.

Many other principles can be invoked, but these are good starting points. Zedler (2000a) had suggested ten ecological principles that should be applied to wetland restoration, and they fit well with the preceding seven principles:

1. *Landscape context and position are crucial to wetland restoration.* See design principle 3 in the previous list. Wetlands are always a function of the watershed and ecological setting in which they are placed.
2. *Natural habitat types are the appropriate reference systems.* This suggests that while we may know how to build ponds, for example, are those the natural habitats of the area, even if they do increase waterfowl?
3. *The specific hydrologic regime is crucial to restoring biodiversity and function.* See design principles 2 and 3. In many cases, such as the Florida Everglades, the restoration is being done in the face of a massive change in the landscape's hydrologic character.
4. *Ecosystem attributes develop at different paces.* Give the system time; see design principle 5. Hydrology develops quickly, vegetation develops over several years, and soils develop over decades. Yet we are quick to review and criticize created and restored wetlands after a couple of years.

5. *Nutrient supply rates affect biodiversity recovery.* There are low-nutrient and high-nutrient wetland systems (see Chapter 7: "Wetland Vegetation and Succession"). Low-nutrient wetlands are often more difficult to create or restore. With very few exceptions, we live in a eutrophic landscape. (The Okavango Delta in Botswana might be one.) High-nutrient inflows cause wetlands to go for power, often instead of diversity.
6. *Specific disturbance regimes can increase species richness.* This can clearly be the case if we allow the word *disturbance* to include flood pulses, fire, and even tropical storms.
7. *Lack of seed banks and dispersal can limit recovery of plant species richness.* This is why restoring wetlands with seed banks can be so important. Another solution is to have a hydrologically or biologically "open" system with a multitude of inputs of propagules (plants, animals, microbes) more likely.
8. *Environmental conditions and life-history traits must be considered when restoring biodiversity.*
9. *Predicting wetland restoration begins with succession theory.* Again, design principle 5 says we need to give the system time. Ecological succession cannot be accelerated without other consequences. This also supports our contention that one must understand wetland science first before attempting to create and restore wetlands.
10. *Genotypes influence ecosystem structure and function.* This is an important but often overlooked principle about wetland restoration. Species are not the same everywhere. This has been shown in common garden experiments on *Spartina alterniflora* (Seliskar, 1995) and freshwater rush *Juncus effusus* (Weihe and Mitsch, 2000). A brackish/freshwater wetland plant with several genotypes that has invaded many natural and restored wetlands in the United States is *Phragmites australis*.

Recommended Readings

Carey, J. 2013. Architects of the Swamp. *Scientific American* 309 (6): 74–79.
Mitsch, W. J. 2013. Wetland Creation and Restoration. *Encyclopedia of Biodiversity*, 2nd ed. pp. 367–383. S. Levin, ed. Amsterdam: Elsevier.
Weinstein, M. P. and J. W. Day (eds.) 2014. Restoration Ecology in a Sustainable World. *Special Issue of Ecological Engineering* 65: 1–158.

References

Alphin, T. D., and M. H. Posey. 2000. Long-term trends in vegetation dominance and infaunal community composition in created marshes. *Wetlands Ecology and Management* 8: 317–325.
Anderson, C. J., W. J. Mitsch, and R. W. Nairn. 2005. Temporal and spatial development of surface soil conditions in two created riverine marshes. *Journal of Environmental Quality* 34: 2072–2081.

Anderson, C. J., and W. J. Mitsch. 2006. Sediment, carbon, and nutrient accumulation at two 10-year-old created riverine marshes. *Wetlands* 26: 779–792.

Artigas, F., and I. C. Pechmann. 2010. Balloon imagery verification of remotely sensed *Phragmites australis* expansion in an urban estuary of New Jersey, USA. *Landscape and Urban Planning* 95: 105–112.

Atkinson, R. B., J. E. Perry, and J. Cairns, Jr. 2005. Vegetation communities of 20-year-old created depressional wetlands. *Wetlands Ecology and Management* 13: 469–478.

Benthem, W., L. P. van Lavieren, and W. J. M. Verheugt. 1999. Mangrove rehabilitation in the coastal Mekong Delta, Vietnam. In W. Streever, ed., *An International Perspective on Wetland Rehabilitation*. Kluwer Academic Publishers, Dordrecht, the Netherlands, pp. 29–36.

Bernal, B., and W. J. Mitsch. 2013. Carbon sequestration in two created riverine wetlands in the Midwestern United States. *Journal of Environmental Quality* 42: 1236–1244.

Bernhardt, E. S., M. A. Palmer, J. D. Allen, G. Alexander, K. Barnas, S. Brooks, J. Carr, S. Clayton, C. Dahm, J. Follstad-Shah, D. Galat, S. Gloss, P. Goodwin, D. Hart, B. Hassett, R. Jenkinson, S. Katz, G. M. Kondolf, P. S. Lake, R. Lave, J. L. Meyer, T. K. O'Donnell, L. Pagano, B. Powell, and E. Sudduth. 2005. Synthesizing U.S. river restoration efforts. *Science* 308: 636–637.

Blake, N. M. 1980. *Land into Water—Water into Land: A History of Water Management in Florida*, University Press of Florida, Gainesville, 344 pp.

Bousquin, S. G.. ed. 2014. Kissimmee River restoration project. *Special section of Restoration Ecology* 22: 345–434.

Bradshaw, A. D. 1996. Underlying principles of restoration. *Canadian Journal of Fisheries and Aquatic Sciences* 53 (Suppl. 1): 3–9.

Broome, S. W., E. D. Seneca, and W. W. Woodhouse, Jr. 1988. Tidal salt marsh restoration. *Aquatic Botany* 32: 1–22.

Brown, B., R. Fadilla, Y. Nurdin, I. Soulsby and R. Ahmad. 2014. Community based ecological mangrove rehabilitation (CBEMR) in Indonesia. *SAPIENS* 7(2): 53–64.

Calloway, J. C., and J. B. Zedler. 2004. Restoration of urban salt marshes: Lessons from southern California. *Urban Ecosystems* 7: 107–124.

Check, E. 2005. Roots of recovery. *Nature* 438: 910–911.

Chung, C. H. 1982. Low marshes, China. In R. R. Lewis III, ed., *Creation and Restoration of Coastal Plant Communities*. CRC Press, Boca Raton, FL, pp. 131–145.

Chung, C. H. 1989. Ecological engineering of coastlines with salt marsh plantations. In W. J. Mitsch and S. E. Jörgensen, eds. *Ecological Engineering: An Introduction to Ecotechnology*. John Wiley & Sons, New York, pp. 255–289.

Clewell, A. F. 1999. Restoration of riverine forest at Hall Branch on phosphate-mined land, Florida. *Restoration Ecology* 7: 1–14.

Craft, C., S. Broome, and C. Campbell. 2002. Fifteen years of vegetation and soil development after brackish-water marsh creation. *Restoration Ecology* 10: 248–258.

Dahl, T. E. 2006. Status and trends of wetlands in the conterminous United States, 1998 to 2004. U.S. Department of the Interior, Fish and Wildlife Service, Washington, DC. 112 pp.

Danish Ministry of Environment and Energy (DMEE). 1999. The Skjern River Restoration Project, DMEE and National Forest and Nature Agency. Copenhagen, Denmark, 32 pp.

deLeon, R. O. D., and A. T. White. 1999. Mangrove rehabilitation in the Philippines. In W. Streever, ed., *An International Perspective on Wetland Rehabilitation*. Kluwer Academic Publishers, Dordrecht, the Netherlands, pp. 37–42.

Dorcas, M. E., J. D. Willson, R. N. Reed, R. W. Snow, M. R. Rochford, M. A. Miller, W. E. Meshaka, P. T. Andreadis, F. J. Mazzotti, C. M. Romagosa, and K. M. Hart. 2012. Severe mammal declines coincide with proliferation of invasive Burmese pythons in Everglades National Park. *Proceedings of the National Academy of Sciences* 109: 2418–2422.

Edwards, K. R., and C. D. Proffitt. 2003. Comparison of wetland structural characteristics between created and natural salt marshes in southwest Louisiana, USA. *Wetlands* 23: 344–356.

Erwin, K. L. 1991. *An Evaluation of Wetland Mitigation in the South Florida Water Management District*, Vol. 1. Final Report to South Florida Water Management District, West Palm Beach, FL. 124 pp.

Gallihugh, J. L., and J. D. Rogner. 1998. *Wetland Mitigation and 404 Permit Compliance Study*, Vol. 1. U.S. Fish and Wildlife Service, Region III, Burlington, IL, and U.S. Environmental Protection Agency, Region V, Chicago. 161 pp.

Garbisch, E. W. 1977. *Recent and Planned Marsh Establishment Work throughout the Contiguous United States: A Survey and Basic Guidelines.* CR D-77-3, U.S. Army Corps of Engineers Waterways Experiment Station, Vicksburg, MS.

Garbisch, E. W. 2005. Hambleton Island restoration: Environmental Concern's first wetland creation project. *Ecological Engineering* 24: 289–307.

Garbisch, E. W., P. B. Woller, and R. J. McCallum. 1975. *Salt Marsh Establishment and Development.* Technical Memo 52. U.S. Army Coastal Engineering Research Center, Fort Belvoir, VA.

Gorham, E., and L. Rochefort. 2003. Peatland restoration: A brief assessment with special reference to Sphagnum bogs. *Wetlands Ecology and Management* 11: 109–119.

Hart, D. D., T. E. Johnson, K. L. Bushaw-Newton, R. J. Horwitz, A. T. Bednarek, D. F. Charles, D. A. Kreeger, and D. J. Velinsky. 2002. Dam removal: Challenges and opportunities for ecological research and river restoration. *BioScience* 52: 669–681.

Haynes, R. J. 2004. The development of bottomland forest restoration in the Lower Mississippi River alluvial valley. *Ecological Restoration* 22: 170–182.

Hefner, J. M., and J. D. Brown. 1985. Wetland trends in the southeastern United States. *Wetlands* 4: 1–11.

Hinkle, R., and W. J. Mitsch. 2005. Salt marsh vegetation recovery at salt hay farm wetland restoration sites on Delaware Bay. *Ecological Engineering* 25: 240–251.

Hopple, A., and C. Craft. 2013. Managed disturbance enhances biodiversity of restored wetlands in the agricultural Midwest. *Ecological Engineering* 61P: 505–510.

Isselin-Nondedeu, F., L. Rochefort, and M. Poulin. 2007. Long-term vegetation monitoring to assess the restoration success of a vacuum-mined peatland *International Conference on Peat and Peatlands 2007*, 23: 153–166. Fédération des Conservatoires d'Espaces Naturels, Quebec, Canada.

Josselyn, M., J. Zedler, and T. Griswold. 1990. Wetland mitigation along the Pacific coast of the United States. In J. A. Kusler and M. E. Kentula, eds. *Wetland Creation and Restoration*. Island Press, Washington, DC, pp. 3–36.

Koebel, J. W., and S. G. Bousquin. 2014. The Kissimmee River restoration project and evaluation program, Florida, U.S.A. *Restoration Ecology* 22: 345–352.

Kristensen, E. A., B. Kronvang, P. Wiberg-Larsen, H. Thodsen, C. Nielsen, E. Amor, N. Friberg, M. L. Pedersen, and A. Baattrup-Pedersen. 2014. 10 years after the largest river restoration project in Northern Europe: Hydromorphological changes on multiple scales in River Skjern. *Ecological Engineering* 66: 141–149.

Lambert, J. M. 1964. The *Spartina* story. *Nature* 204: 1136–1138.

Lewis, R. R. 1990a. Wetland restoration/creation/enhancement terminology: Suggestions for standardization. In J. A. Kusler and M. E. Kentula, eds. *Wetland Creation and Restoration*. Island Press, Washington, DC, pp. 1–7.

Lewis, R. R. 1990b. Creation and restoration of coastal plain wetlands in Florida. In J. A. Kusler and M. E. Kentula, eds. *Wetland Creation and Restoration*. Island Press, Washington, DC, pp. 73–101.

Lewis, R. R. 1990c. Creation and restoration of coastal plain wetlands in Puerto Rico and the U.S. Virgin Islands. In J. A. Kusler and M. E. Kentula, eds. *Wetland Creation and Restoration*. Island Press, Washington, DC, pp. 103–123.

Lewis, R. R. 2005. Ecological engineering for successful management and restoration of mangrove forests. *Ecological Engineering* 24: 403–418.

Lewis, R. R. 2009. Methods and criteria for successful mangrove forest restoration. In G. M. E. Perillo, E. Wolanski, D. R. Cahoon, and M. M. Brinson, eds. *Coastal Wetlands: An Integrated Ecosystem Approach*. Elsevier, Amsterdam, the Netherlands, pp. 787–800.

Lewis, R. R. 2010. Mangrove field of dreams: If we build it, will they come? *Wetland Science and Practice* 27(1): 15–18.

Lewis, R. R., and R. G. Gilmore. 2007. Important considerations to achieve successful mangrove forest restoration with optimum fish habitat. *Bulletin of Marine Science* 80(3): 823–837.

Lewis, R. R., and B. Brown. 2014. *Ecological mangrove rehabilitation—a field manual for practitioners. Version 3*. Mangrove Action Project Indonesia, Blue Forests, Canadian International Development Agency, and OXFAM. 275 pp.

Liptak, M. A. 2000. Water column productivity, calcite precipitation, and phosphorus dynamics in freshwater marshes. Ph.D. dissertation, The Ohio State University, Columbus.

Massey, B. 2000. *Wetlands Engineering Manual*. Ducks Unlimited, Southern Regional Office, Hackson, MS, 16 pp.

Mitsch, W. J. 2013. Wetland creation and restoration. In S. Levin, ed., *Encyclopedia of Biodiversity*, 2nd ed. Elsevier, Amsterdam, pp. 367–383.

Mitsch, W. J. 2014. When will ecologists learn engineering and engineers learn ecology? *Ecological Engineering* 65: 9–14.

Mitsch, W. J., and R. F. Wilson. 1996. Improving the success of wetland creation and restoration with know-how, time, and self-design. *Ecological Applications* 6: 77–83.

Mitsch, W. J., X. Wu, R. W. Nairn, P. E. Weihe, N. Wang, R. Deal, and C. E. Boucher. 1998. Creating and restoring wetlands: A whole-ecosystem experiment in self-design. *BioScience* 48: 1019–1030.

Mitsch, W. J., and S. E. Jørgensen. 2004. *Ecological Engineering and Ecosystem Restoration*. John Wiley & Sons, Hoboken, NJ. 411 pp.

Mitsch, W. J., L. Zhang, C. J. Anderson, A. Altor, and M. Hernandez. 2005. Creating riverine wetlands: Ecological succession, nutrient retention, and pulsing effects. *Ecological Engineering* 25: 510–527.

Mitsch, W. J., L. Zhang, D. F. Fink, M. E. Hernandez, A. E. Altor, C. L. Tuttle and A. M. Nahlik. 2008. Ecological engineering of floodplains. *Ecohydrology & Hydrobiology* 8: 139–147.

Mitsch, W. J., L. Zhang, K. C. Stefanik, A. M. Nahlik, C. J. Anderson, B. Bernal, M. Hernandez, and K. Song. 2012. Creating wetlands: Primary succession, water quality changes, and self-design over 15 years. *BioScience* 62: 237–250.

Mitsch, W. J., L. Zhang, E. Waletzko, and B. Bernal. 2014. Validation of the ecosystem services of created wetlands: Two decades of plant succession, nutrient retention, and carbon sequestration in experimental riverine marshes. *Ecological Engineering* doi.org/10.1016/j.ecoleng.2014.09.108.

Moreno-Mateos, D., M. E. Power, F. A. Comin, and R. Yockteng 2012. Structural and functional loss in restored wetland ecosystems. *PLoS Biol* 10(1): e1001247. doi:10.1371/journal.pbio.1001247

Nahlik, A. M., and W. J. Mitsch. 2010. Methane emissions from created riverine wetlands. *Wetlands* 30: 783–793. *Erratum in Wetlands* (2011) 31: 449–450.

National Research Council. 2001. *Compensating for Wetland Losses under the Clean Water Act*. National Academy Press, Washington, DC. 158 pp.

National Research Council. 2014. *Progress toward Restoring the Everglades: The Fifth Biennial Review*. National Academies Press, Washington, DC. 240 pp.

Odum, W. E. 1987. Predicting ecosystem development following creation and restoration of wetlands. In J. Zelazny and J. S. Feierabend, eds. *Wetlands: Increasing Our Wetland Resources. Proceedings of the Conference Wetlands: Increasing Our Wetland Resources*. Corporate Conservation Council, National Wildlife Federation, Washington, DC, pp. 67–70.

Palmer, M. A., E. S. Bernhardt, J. D. Allan, P. S., Lake, G. Alexander, S. Brooks, J. Carr, S. Clayton, C. N. Dahm, J. Follestad Shan, D. L. Galat, S. G. Loss, P. Goodwin, D. D. Hart, B. Hassett, R. Jenkinson, G. M. Kondolf, R. Lave, L. J. Meyr, T. K. O'Donnell, L. Pagano, and E. Sudduth. 2005. Standards for ecologically successful river restoration. *Journal of Applied Ecology* 42: 208–217.

Palmer, M. A., S. Filoso, and R.M. Fanelli. 2014. From ecosystems to ecosystem services: Stream restoration as ecological engineering. *Ecol. Eng.* 65: 62–70.

Pedersen, M. L., J. M. Andersen, K. Nielsen, and M. Linnemann. 2007a. Restoration of Skjern River and its valley: Project description and general ecological changes in the project area. *Ecological Engineering* 30: 131–144.

Pedersen, M. L., Friberg, N., Skriver, J., Baattrup-Pedersen, A., 2007b. Restoration of Skjern River and its valley—short-term effects on river habitats, macrophytes and macroinvertebrates. *Ecol. Eng.* 30: 145–156.

Peterson, S. B., J. M. Teal, and W. J. Mitsch, eds. 2005. Delaware Bay Salt Marsh Restoration. *Special issue of Ecological Engineering* 25: 199–314.

Philipp, K. R., and R. T. Field. 2005. *Phragmites australis* expansion in Delaware Bay salt marshes. *Ecological Engineering* 25: 275–291.

Price, J., L. Rochefort, and F. Quinty. 1998. Energy and moisture considerations on cutover peatlands: Surface microtopography, mulch cover and Sphagnum regeneration. *Ecological Engineering* 10: 293–312.

Primavera, J. H., J. D. Savaris, B. Bajoyo, J. D. Coching, D. J. Curnick, R. Golbeque, A. Guzman, J. Q. Henderin, R. V. Joven, R. A. Loma, and H. J. Koldewey. 2012. *Manual on Community-Based Mangrove Rehabilitation.* Mangrove Manual Series No. 1. Zoological Society of London, London, UK. viii +240 pp.

Quinty, F., and L. Rochefort. 1997. Plant reintroduction on a harvested peat bog. In C. C. Trettin, M. F. Jurgensen, D. F. Grigal, M. R. Gale, and J. K. Jeglum, eds. *Northern Forested Wetlands: Ecology and Management.* CRC Press/Lewis Publishers, Boca Raton, FL, pp. 133–145.

Ranwell, D. S. 1967. World resources of *Spartina townsendii* and economic use of *Spartina* marshland. *Coastal Zone Management Journal* 1: 65–74.

Reinartz, J. A., and E. L. Warne. 1993. Development of vegetation in small created wetlands in southeast Wisconsin. *Wetlands* 13: 153–164.

Richardson C. J., P. Reiss, N. A. Hussain, A. J. Alwash, and D. J. Pool. 2005. The restoration potential of the Mesopotamian marshes of Iraq. *Science* 307: 1307–1311.

Richardson, C. J., and N. A. Hussain. 2006. Restoring the Garden of Eden: An ecological assessment of the marshes of Iraq. *BioScience* 56: 447–489.

Robb, J. T. 2002. Assessing wetland compensatory mitigation sites to aid in establishing mitigation ratios. *Wetlands* 22: 435–440.

Rochefort, L., and S. Campeau. 1997. Rehabilitation work on post-harvested bogs in south-eastern Canada. In L. Parkyn, R. E. Stoneman, and H. A. P. Ingram, eds. *Conserving Peatlands.* CAB International, Walingford, UK, pp. 287–284.

Rochefort, L., G. Quinty, S. Campeau, K. Johnson, and T. Malterer. 2003. North American approach to the restoration of *Sphagnum* dominated peatlands. *Wetlands Ecology and Management* 11: 3–20.

Rochefort, L., and E. Lode. 2006 Restoration of degraded boreal peatlands. In R. K. Wieder and D. H. Vitt, eds. *Boreal Peatland Ecosystems.* Springer, Berlin, Germany, pp. 381–424.

Rochefort, L., M. Strack, M. Poulin, J. S. Price, M. Graf, A. Desrochers, and C. Lavoie. 2012. Northern peatlands. In D. R. Batzer and A. H. Baldwin, eds. *Wetland Habitats of North America: Ecology and Conservation Concerns*. University of California Press, Los Angeles, pp. 119–134.

Rochefort, L., F. Isselin-Nondedeu, S. Boudreau, and M. Poulin. 2013. Comparing survey methods for monitoring vegetation change through time in a restored peatland. *Wetlands Ecology and Management* 21: 71–85.

Rodgers, H. L., F. P. Day, and R. Atkinson. 2004. Root dynamics in restored and naturally regenerated Atlantic white cedar wetlands. *Restoration Ecology* 16: 401–411.

Rodgers, L., M. Bodle, D. Black, and F. Laroche. 2014. Status of nonindigenous species. In *2014 South Florida Environmental Report, Volume 1: The South Florida Environment*. South Florida Water Management District, West Palm Beach, Chapter 7, pp. 7-1–7-55.

Samson, M. S., and R. N. Rollon. 2008. Growth performance of planted red mangroves in the Philippines: Revisiting forest management strategies. *Ambio* 37(4): 234–240.

Sanville, W., and W. J. Mitsch, eds. 1994. Creating Freshwater Marshes in a Riparian Landscape: Research at the Des Plaines River Wetland Demonstration Project. *Special Issue of Ecological Engineering* 3(4): 315–521.

Sartoris, J. J., J. S. Thullen, L. B. Barber, and D. E. Salas. 2000. Investigation of nitrogen transformations in a southern California constructed wastewater treatment wetland. *Ecological Engineering* 14: 49–65.

Seliskar, D. 1995. Exploiting plant genotypic diversity for coastal salt marsh creation and restoration. In M. A. Khan and I. A. Ungar, eds. *Biology of Salt-Tolerant Plants*. Department of Botany, University of Karachi, Pakistan, pp. 407–416.

Sha, C., W. J. Mitsch, Ü. Mander, J. Lu, J. Batson, L. Zhang and W. He. 2011. Methane emissions from freshwater riverine wetlands. *Ecological Engineering* 37: 16–24.

Shin, J. Y., F. Artigas, C. Hobble, and Y. S. Lee. 2013. Assessment of anthropogenic influences on surface water quality in urban estuary, northern New Jersey: Multivariate approach. *Environmental Monitoring and Assessment* 185: 2777–2794.

South Florida Water Management District. 2014. *South Florida Environmental Report 2014, Executive Summary*. SFWMD, West Palm Beach, FL. 28 pp.

Spieles, D. J. 2005. Vegetation development in created, restored, and enhanced mitigation wetland banks of the United States. *Wetlands* 25: 51–63.

Teal, J. M., and M. P. Weinstein. 2002. Ecological engineering, design, and construction considerations for marsh restorations in Delaware Bay, USA. *Ecological Engineering* 18: 607–618.

Toth, L. A., D. A. Arrington, M. A. Brady, and D. A. Muszick. 1995. Conceptual evaluation of factors potentially affecting restoration of habitat structure within the channelized Kissimmee River ecosystem. *Restoration Ecology* 3: 160–180.

Tuttle, C. L., L. Zhang, and W. J. Mitsch. 2008. Aquatic metabolism as an indicator of the ecological effects of hydrologic pulsing in flow-through wetlands. *Ecological Indicators* 8: 795–806.

United Nations Environmental Programme. 2001. *The Mesopotamian Marshlands: Demise of an Ecosystem.* UNEP/DEWA/TR.01–3 Revision 1. UNEP, Nairobi, Kenya.

United Nations Environmental Programme. 2007. *After the Tsunami—Coastal Ecosystems Restoration—Lessons Learnt.* UNEP, Nairobi, Kenya. iii +55 pp.

United Nations Environmental Programme. 2008. *UNEP Post-Tsunami Recovery Activities 2004–2007.* UNEP, Nairobi, Kenya. ii +68 pp.

U.S. Army Corps of Engineers and U.S. Environmental Protection Agency. 2008. *Compensatory Mitigation for Losses of Aquatic Resources; Final Rule.* Federal Register (April 10, 2008) 73 (70): 19594–19705.

U.S. Environmental Protection Agency and U.S. Army Corps of Engineers. 2008. *Wetlands Compensatory Mitigation Rule.* http://water.epa.gov/lawsregs/guidance/wetlands/upload/MitigationRule.pdf.

Vymazal, J. 2013. Emergent plants used in free water surface constructed wetlands: A review. *Ecological Engineering* 61P: 582–592.

Waddington, J. M., M. J. Greenwood, R. M. Petrone, and J. S. Price. 2003. Mulch decomposition impedes recovery of net carbon sink function in a restored peatland. *Ecological Engineering* 20: 199–210.

Waddington, J. M., K. Tóth, and R. Bourbonniere. 2008. Dissolved organic carbon export from a cutover and restored peatland. *Hydrologic Processes* 22: 2215–2224.

Weihe, P. E., and W. J. Mitsch. 2000. Garden wetland experiment demonstrates genetic differences in soft rush obtained from different regions (Ohio). *Ecological Restoration* 18: 258–259.

Wilson, R. F., and W. J. Mitsch. 1996. Functional assessment of five wetlands constructed to mitigate wetland loss in Ohio, USA. *Wetlands* 16: 436–451.

Woodhouse, W. W. Jr. 1979. *Building Salt Marshes along the Coasts of the Continental United States.* Special Report 4, U.S. Army Coastal Engineering Research Center, Fort Belvoir, VA.

Wu, X. and W. J. Mitsch. 1998. Spatial and temporal patterns of algae in newly constructed freshwater wetlands. *Wetlands* 18: 9–20.

Zedler, J. B. 1988. Salt marsh restoration: Lessons from California. In J. Cairns, ed., *Rehabilitating Damaged Ecosystems*, Vol. 1. CRC Press, Boca Raton, FL, pp. 123–138.

Zedler, J. B. 2000a. Progress in wetland restoration ecology. *TREE* 15: 402–407.

Zedler, J. B. 2000b. *Handbook for Restoring Tidal Wetlands.* CRC Press, Boca Raton, FL, 464 pp.

Chapter **19**

Wetlands and Water Quality

Wetlands created specifically to improve water quality are usually referred to as treatment wetlands. There are three types of wetlands used to treat wastewater or stormwater: natural wetlands, surface-flow constructed wetlands, and subsurface-flow constructed wetlands. Studies on using subsurface wetlands began in the 1960s in Europe. The use of natural wetlands to treat wastewater in Florida and Michigan in the 1970s pioneered the use of surface-flow–created wetlands. Wetlands have been used to treat a variety of threats to downstream water quality, including domestic wastewater, mine drainage, non–point source pollution, stormwater runoff, landfill leachate, and confined livestock operations. The design of treatment wetlands requires particular attention to hydrology, chemical loading, soil physics and chemistry, and wetland vegetation. Management issues include wildlife control and attraction, mosquito and pathogen control, and greenhouse gas and water-level management. Treatment wetlands are not inexpensive to build and operate, but they usually cost much less than chemical and physical treatment systems.

Wastewater and polluted water treatment by wetlands is an intriguing concept involving the forging of a partnership between humanity (our wastes) and an ecosystem (wetlands). Therefore, it is a good example of *ecological engineering* (see Mitsch and Jørgensen, 2004, and Mitsch, 2012, for definitions and applications of this field). In this chapter, we discuss the use of wetlands for removing unwanted chemicals from waters, be the waters municipal wastewater, non–point source runoff, or other forms of pollution.

As described in Chapter 6: "Wetland Biogeochemistry," wetlands can be sources, sinks, or transformers for a great number of chemicals. A wetland is a *sink* if it has a net retention of an element or a specific form of that element (e.g., organic or inorganic); that is, if the inputs are greater than the outputs. The desired situation for treatment wetlands is to optimize the wetlands' ability to serve as chemical (and sometimes biological) sinks.

German scientists investigated the use of constructed basins with macrophytes (*höhere Pflanzen*) for purification of wastewater. Later, researchers in Florida and Michigan in the United States investigated the role of natural wetlands to treat wastewater and thus recycle clean water back to groundwater and surface water. The two different approaches, one using artificial systems and the other utilizing natural wetlands, have converged into the general field of *treatment wetlands* (Kadlec and Wallace, 2009). The field now encompasses the construction and/or use of wetlands for a myriad of water quality applications. While water quality improvement is the primary goal of treatment wetlands, they also provide habitat for a wide diversity of plants and animals and can support many of the other wetland functions and services described in this book.

Classifications of Wastewater Treatment Wetlands

Three General Approaches

Three types of wetlands are used to treat wastewater. In the first approach, wastewater is purposefully introduced to existing natural wetlands rather than constructed wetlands (Fig. 19.1a). In the 1970s, studies involving application of wastewater to natural wetlands were carried out in locations of the United States such as Michigan (Kadlec, 2009b) and Florida (Odum et al., 1977; Ewel and Odum, 1984) where there were abundant wetlands. At that time, legal protection of wetlands had not been institutionalized. These pioneering studies elevated the importance of wetlands as "nature's kidneys" to the general public and governmental agencies. This importance was then translated, appropriately, into laws that protected wetlands. These same laws now often prohibit the addition of wastewater or polluted water to natural wetlands.

Two types of *constructed wetlands* are alternatives to using natural wetlands. *Surface-flow constructed wetlands* (Fig. 19.1b) mimic natural wetlands and can be a better habitat for certain wetland species because of standing water through most if not all of the year. *Subsurface-flow constructed wetlands* (Fig. 19.1c) more closely resemble wastewater treatment plants than wetlands. In these systems, the water flows horizontally or vertically through a porous medium, usually sand or gravel, supporting one or two of a relatively narrow list of macrophytes, such as *Phragmites australis*. There is rarely standing water in these subsurface systems as the wastewater passes laterally through the medium.

Subsurface treatment wetlands had their start in the Max-Planck Institute in Germany in the 1950s. Dr. Käthe Seidel performed many experiments with emergent macrophytes, *Schoenoplectus lacustris* in particular, and found that the plants contributed to the reduction of bacteria, and organic and inorganic chemicals (Seidel, 1964, 1966). This process was translated into a gravel-bed macrophyte system that became known as the *Max-Planck Institute process* or the *Krefeld system* (Seidel and Happl, 1981, as cited in Brix, 1994). The development of subsurface wetlands

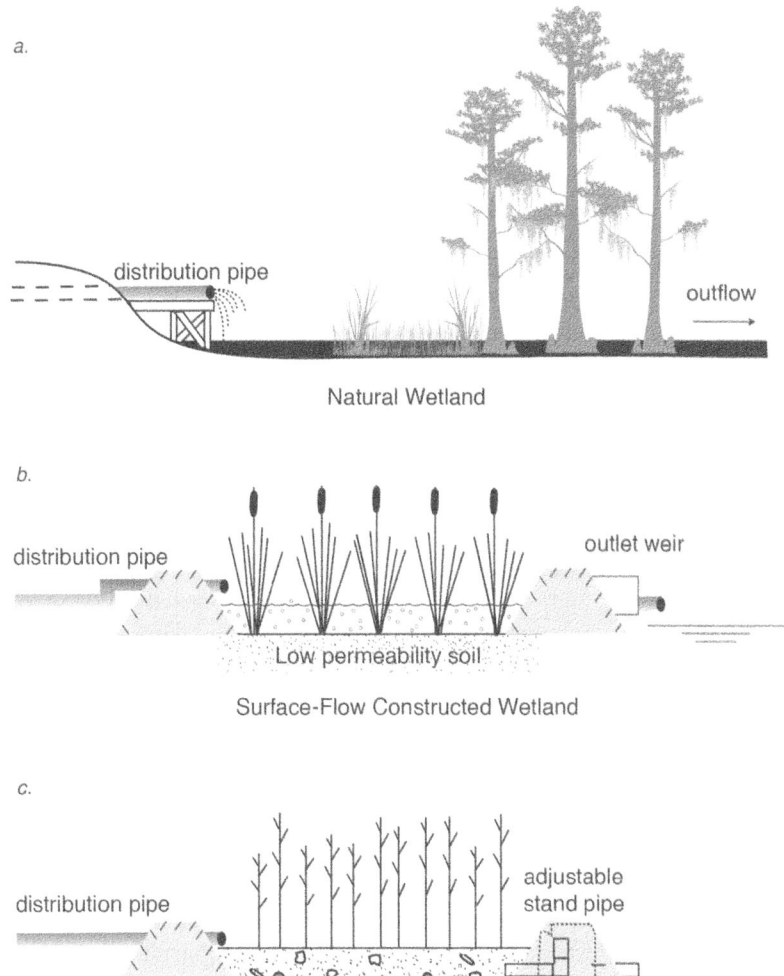

Figure 19.1 Three types of wetland treatment systems: (a) natural wetland, (b) surface-flow wetland, and (c) subsurface-flow wetland (After Kadlec and Knight, 1996 and Kadlec and Wallace, 2008)

continued in Europe using a system of subsurface-flow basins planted with *P. australis*. These systems were called the *root-zone method* (*Wurzelraumentsorgung*). Subsurface wetland systems continued to be studied and refined through the work of DeJong (1976) in Holland, Brix (1987) in Denmark, and many other scientists in Europe. The appeal of these more "artificial" types of wetlands in Europe (as opposed to free-water surface wetlands in North America) is due to two factors: (1) there are fewer natural wetlands remaining in Europe, and those that are left are protected for nature; and

(2) space is much more at a premium in Europe, and subsurface wetlands require less land area.

Classification According to Vegetation

Treatment wetlands can also be classified based on the life-form of their vegetation. In this case, there are five systems based on their vegetation:

1. Free-floating macrophyte systems—for example, water hyacinth (*Eichhornia crassipes*), duckweed (*Lemna* spp.)
2. Emergent macrophyte systems—for example, reed grass (*P. australis*), cattails (*Typha* spp.)
3. Submerged aquatic vegetation systems
4. Forested wetland systems
5. Multispecies algal systems, particularly algal-scrubber systems

Subsurface-flow constructed wetlands are limited to emergent macrophytes, whereas surface-flow constructed wetlands often utilize a combination of free-floating, emergent, and submerged macrophytes. Forested wetland treatment systems are generally not constructed wetlands at all but are natural wetlands to which wastewater is applied. They will often develop extensive communities of all of the other vegetation types described in this classification.

Treatment Wetland Types

The type of wastewater being treated can classify treatment wetlands. While many of these systems are used for municipal wastewater and that is often thought as the conventional system, there has been much interest in the use of wetlands to treat stormwater from urban areas, acid mine drainage from coal mines, non–point source pollution in rural landscapes, livestock and aquaculture wastewaters, and an array of industrial wastewaters.

Municipal Wastewater Wetlands

In Europe, most of the development of subsurface constructed wetlands was to replace both primary and secondary treatment to remove biochemical oxygen demand (BOD) and suspended solids as well as inorganic nutrients. Hundreds of subsurface wetland treatment systems for municipal wastewater have been constructed in Europe, particularly in the United Kingdom (Cooper and Findlater, 1990), Denmark (Brix and Schierup, 1989a, b; Brix, 1998; Brix and Arias, 2005), the Czech Republic (Vymazal, 1995, 1998, 2002, 2005; Vymazal and Kropfelova, 2005), Norway (Braskerud, 2002a, b), Spain (Solano et al., 2004), and Estonia (Teiter and Mander, 2005). There are also many applications of this technology in Australia (Mitchell et al., 1995; Greenway et al., 2003; Greenway, 2005; Headley et al., 2005; Davison et al., 2006), New Zealand (Cooke, 1992; Tanner, 1996; Nguyen et al., 1997; Nguyen,

2000), and Costa Rica (Nahlik and Mitsch, 2006). In North America, most but certainly not all of the wetlands built for treatment of municipal wastewater treatment are surface-water wetlands. Locations of wastewater wetlands that have been studied in some detail include Florida (Knight et al., 1987; J. Jackson, 1989), California (Gerheart et al., 1989; Gerheart, 1992; Sartoris et al., 2000; Thullen et al., 2005), Louisiana (Boustany et al., 1997; Day et al., 2004), Arizona (Wilhelm et al., 1989), Kentucky (Steiner et al., 1987; Steiner and Freeman, 1989), Pennsylvania (Conway and Murtha, 1989), Ohio (Spieles and Mitsch, 2000a, b), North Dakota (Litchfield and Schatz, 1989), and Alberta, Canada (White et al., 2000). Created wetlands for treating wastewater have been most effective for controlling organic matter (BOD), suspended sediments, and nutrients. Their value for controlling trace metals and other toxic materials is more controversial, not because these chemicals are not retained in the wetlands but because of concerns that they might concentrate in wetland substrate and fauna.

One of the longest running treatment wetland as far as data collection is the Houghton Lake treatment wetland in Michigan that had its beginnings in the early 1970s. This is described in detail in Case Study 1.

CASE STUDY 1: Long-Term Effectiveness of a Treatment Wetland at Houghton Lake, Michigan

Much of the interest in using surface-flow wetlands for water quality management was sparked by several studies begun in the early 1970s. In one of those studies, peatlands in Michigan were investigated by researchers from the University of Michigan for the wetlands' capacity to treat wastewater (Fig. 19.2). A pilot operation for disposing of up to 380 m^3/day (100,000 gallons per day) of secondarily treated wastewater in a 700-ha rich fen at Houghton Lake led to significant reductions in ammonia nitrogen and total dissolved phosphorus as the wastewater passed from the point of discharge through the wetlands. Inert materials, such as chloride, did not change as the wastewater passed through the wetland. In 1978, the flow was increased to approximately 5,000 m^3 day^{-1} over a much larger area, essentially using all the wastewater from the local treatment plant. Data after 30 years of operation show that the wetlands have remained effective over that long time in removing both total phosphorus (Fig. 19.2b,c) and inorganic nitrogen (Fig. 19.2d). Approximately 100 ha of the peatland was an irrigation zone (Fig. 19.2a) where most of the water quality improvement occurred (Kadlec, 2009b). Phosphorus decreased by 94 percent from 3.5 to 0.04 ppm (= mg-P/L) and dissolved inorganic nitrogen decreased by 95 percent. Nutrient retention on an annualized basis was 1.76 g-P m^{-2} yr^{-1} for phosphorus and 4.39 g-N m^{-2} yr^{-1} for dissolved inorganic nitrogen (Kadlec, 2009b). The peatland itself was not unaffected, with the vegetation in the irrigation area changing from a sedge-*Salix* community to a *Typha* spp. community, some of which has become floating mats (Kadlec, 2009b).

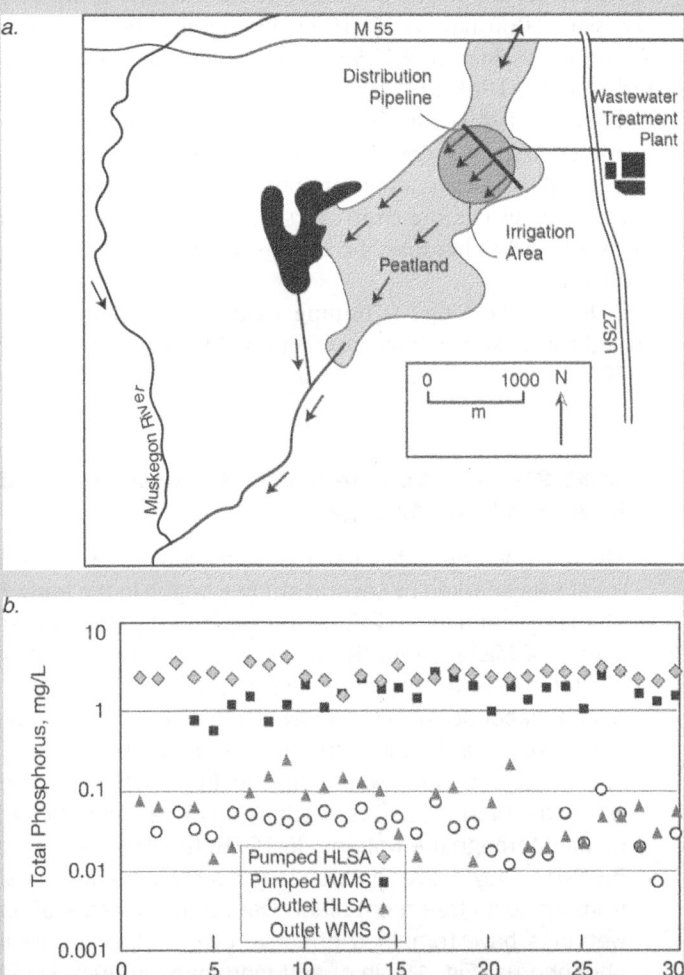

Figure 19.2 Houghton Lake treatment wetland in Michigan, where treated wastewater has been discharged into a peatland for 30 years: (a) map of site showing the irrigation area; (b) total phosphorus of influent and outlet as measured by Houghton Lake Sewer Authority (HLSA) and Wetland Management Services (WMS); (c) phosphorus inflow and outflow fluxes; (d) dissolved inorganic nitrogen inflow and outflow fluxes. (After Kadlec, 2009b)

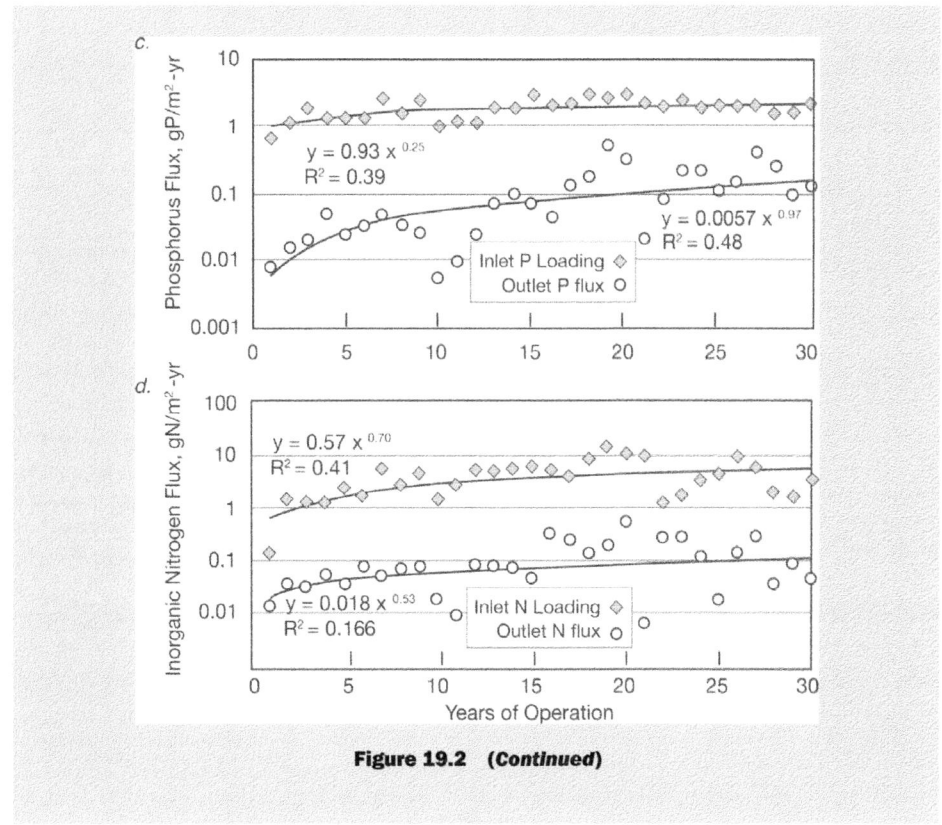

Figure 19.2 (Continued)

Mine Drainage Wetlands

Wetlands frequently have been used as downstream treatment systems for mineral mines. An example of such a system in southeast Ohio is described in Figure 19.3. Acid mine drainage water, with its low pH and high concentrations of iron, sulfate, aluminum, and trace metals, is a major water pollution problem in many coal mining regions of the world, and constructed wetlands are a viable treatment option. The use of wetlands for coal mine drainage control was probably first considered when volunteer *Typha* wetlands were observed near acid seeps in a harsh environment where no other vegetation could grow. By the 1980s, hundreds of wetlands had been constructed in the eastern United States alone to treat mine drainage water. The most common goal of these systems was usually the removal of iron from the water column to avoid its discharge downstream, but sulfate reduction and the alleviation of extremely acidic conditions were also appropriate goals (Wieder and Lang, 1984; Brodie et al., 1988; Fennessy and Mitsch, 1989; Mitsch and Wise, 1998; Tarutis et al., 1999).

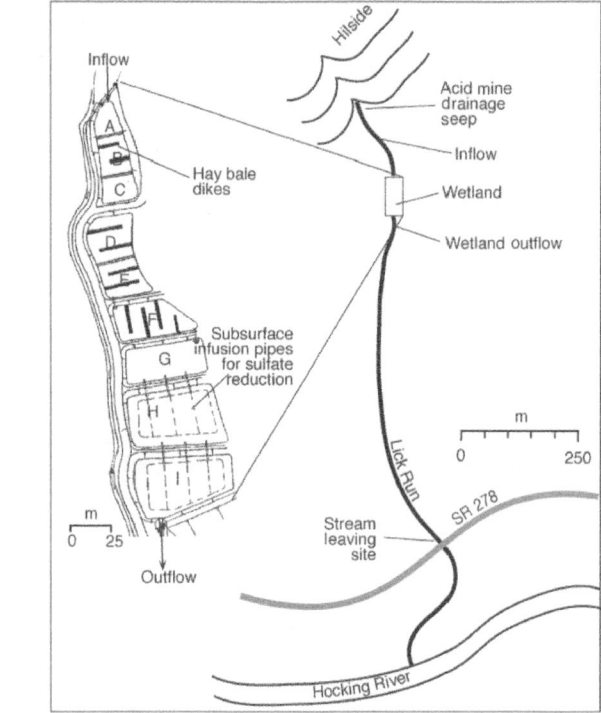

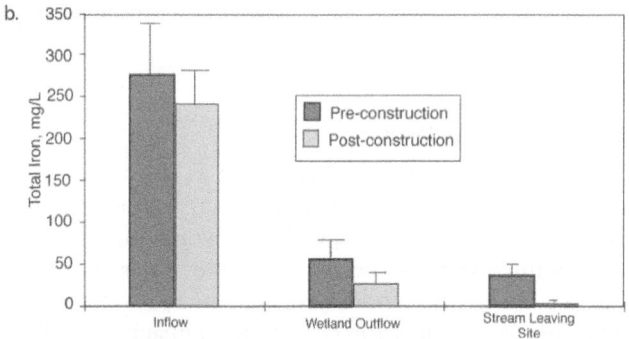

Figure 19.3 (a) A 0.4-ha acid mine drainage treatment wetland in southeastern Ohio; (b) total iron in wetland and downstream before wetland was built and after it was constructed. (After Mitsch and Wise, 1998)

Design criteria for these wetlands have been developed, but they are neither consistent from site to site nor generally accepted. Some are suggested in Table 19.1. Stark and Williams (1995) found design features that enhanced iron removal and decreased acidity included broad drainage basins, nonchannelized flow patterns, high plant diversity, southern exposure, low flow rates and loadings, and shallow depths. It is not

Table 19.1 Suggested design parameters for constructed wetlands used for controlling coal mine drainage

Parameter	Design	Reference
Hydrologic loading rate, cm/day	5	Fennessy and Mitsch, 1989
Retention time, days	>1	Fennessy and Mitsch, 1989
Iron loading, g-Fe m^{-2} day^{-1}		
pH <5.5	0.72	Brodie et al., 1988
pH >5.5	1.29	Brodie et al., 1988
For 90% removal, pH = 6	2–10	Fennessy and Mitsch, 1989
For 50% removal, pH = 6	20–40	Fennessy and Mitsch, 1989
pH = 3.5, ouflow <3.5 mg-Fe/L	2.5	Manyin et al., 1997
Basin characteristics		
Depth, m	<0.3	
Number of cells	>3	
Plant material	*Typha* spp.	
Substrate material	Organic peat over clay seal; spent mushroom material	

always cost effective to construct wetlands when extremely high (>85–90 percent) iron removal efficiencies are necessary or when the pH of the mine drainage water is less than 4. Hydraulic loading rates as high as 29 cm day^{-1} have been suggested for wetlands designed for acid mine drainage, although Fennessy and Mitsch (1989) recommended 5 cm/day as a conservative loading rate for this type of wetland and a minimum detention time of 1 day, with much longer periods for more effective iron removal (Table 19.1). The long-term suitability of wetland treatment systems is poorly understood, although it appears that *Typha*-dominated systems can survive decades in a mine drainage system. The accumulation of iron hydroxides can eventually cause mine drainage systems to begin to export materials, unless the design and management includes adequate storage capacity and/or material removal. Some researchers suggest that these wetlands, after several decades, can become mineral mines in their own sense, effectively recycling minerals that otherwise would be lost to downstream watersheds back to the economy. Where no other alternative is feasible, the use of wetlands to reduce this harsh water pollution should be viewed as a reasonable, low-cost alternative.

Urban Stormwater Treatment Wetlands

The control of stormwater pollution with wetlands is a valid and increasingly utilized application of wetland ecological engineering. Unlike municipal wastewater, stormwater and other non–point source pollution are seasonal, often quite sporadic, and variable in quality, depending on season and recent land use. Wetlands are one of several choices for systems to control urban runoff. More conventional approaches involve either dry detention ponds that fill only during storms or wet detention ponds that are usually deepwater systems, where the edge is usually stabilized with rocks and plant growth is actually discouraged.

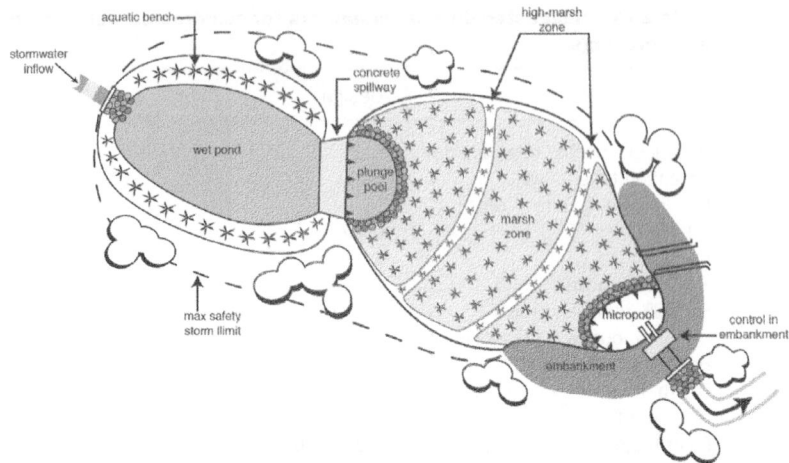

Figure 19.4 General design of a stormwater treatment wetland. (After Schueler, 1992)

Stormwater from urban areas is particularly rapid as it comes from impervious sources such as roofs, parking lots, and highways. One of the features of stormwater wetland systems is that severe storms have a dramatic effect on treatment efficiency. High flows resulting from high-intensity rainstorms usually result in lower nutrient and other chemical retention as a percentage of inflow, and sometimes the storms cause a net release of nutrients. The very nature of the sudden but short stormwater pulses makes management of these systems particularly difficult.

A layout of an ideal stormwater treatment wetland (Fig. 19.4) illustrates that a combination of deep ponds and marshes may be most appropriate. The first "wet pond" is a deep, usually unvegetated, basin designed to dampen the rapid stormwater pulse, allowing the downstream marshes to "treat" the runoff in a more effective manner. Multiple cells of marshes and a small outflow deepwater pond can contribute to the system's effectiveness. Sediment retention capability is the strong point of these wetlands, but if any significant construction projects occur upstream, even this capacity can be temporarily or permanently overwhelmed. Case Study 2 presents an application of this design.

CASE STUDY 2: Freedom Park: Intercepting Urban Runoff in an Urban Wetland Park in Florida

A 20-ha constructed wetland complex, called Freedom Park, was constructed in 2007–2008 in Naples, Florida, at an abandoned citrus grove to treat urban stormwater runoff (Fig. 19.5a). Its total construction cost was $10 million, and it was designed to treat 757,000 m^3 per year for an average hydraulic loading rate (HLR) of about 7.6 m/yr (2 cm/day). The system was designed

(a)

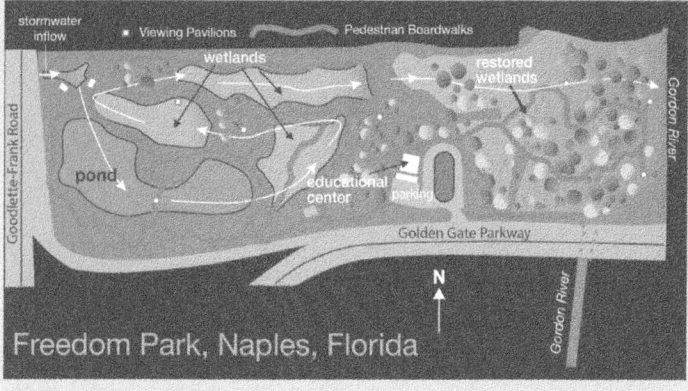

(b)

Figure 19.5 Freedom Park stormwater treatment wetlands in Naples Florida: (a) photo looking west with restored upland and wetland forests in the foreground and stormwater treatment wetland basins in the background; (b) map of Freedom Park, illustrating stormwater treatment ponds and wetlands on the left and restored forested wetlands on the right. Also shown are boardwalks through the system and visitors center; (c) total nitrogen concentrations in the inflow and outflow of the stormwater wetlands; (d) total phosphorus in the inflow and outflow of the stormwater wetlands. (Figures a, c, and d courtesy of Jim Bays, CH2M-Hill, Tampa, FL)

657

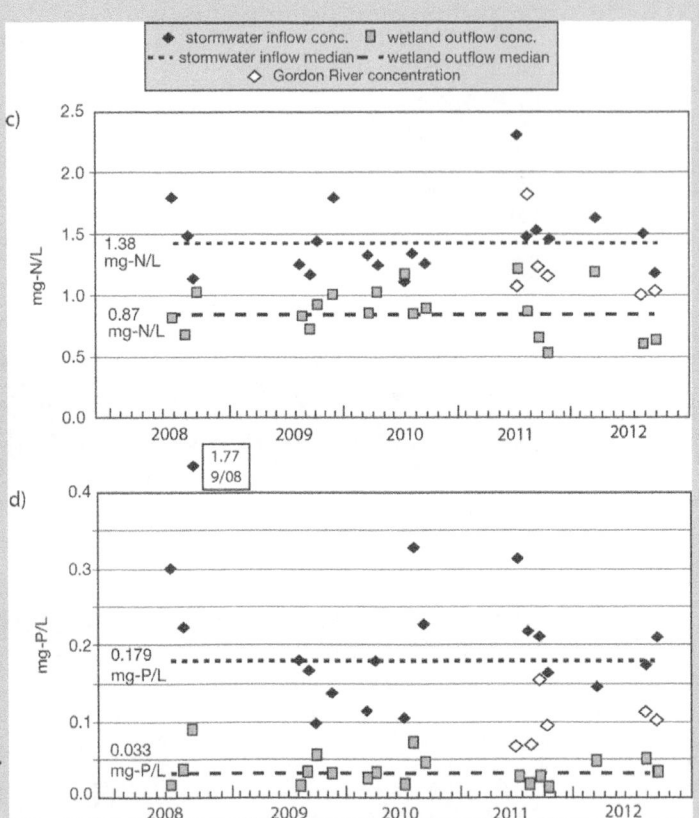

Figure 19.5 *(Continued)*

especially to treat urban stormwater runoff during the summer wet season when average flow is 10 to 100 times higher than in the dry season. During that wet season, the design calls for an average water detention time of 18 days. The wetland system includes an initial 1.9-ha deepwater pond that receives and temporarily stores the stormwater pulses, followed by 2.7 ha of shallower vegetated wetland ponds designed to sequentially improve water quality (Fig. 19.5b). The wetlands were planted with *Nuphar*, *Nymphaea*, *Thalia*, *Pontederia*, *Eleocharis*, *Sagittaria*, *Cladium*, and submerged aquatic vegetation. The last basin was designed as a shallow limestone bed vegetated with a periphyton-*Eleocharis* community for enhanced phosphorus removal. The water from the ponds/wetlands is then discharged as sheet flow through about 6 ha of restored forested hardwood hammock/wetland area before it flows into to the Gordon River. Water quality results from 2008 to 2012 show an average decrease of 37 percent of total nitrogen (Fig. 19.5c) and a decrease of total phosphorus of 81 percent (Fig. 19.d) through the ponds. Retention of both nutrients is important as the Gordon River flows first to Naples Bay in "old Naples" to the south and then to the coastal waters of the Gulf of Mexico. The treatment wetland site is also a city/county park with a visitors' center, several kilometers of 2-m-wide boardwalks, and many interpretative signs. An estimated 25,000 visitors came to Freedom Park in 2013 (J. Bays, personal communication).

Agricultural Stormwater Treatment Wetlands

One of the most important applications of wetland treatment systems—yet an application that is still behind municipal treatment wetlands in understanding design issues—is the use of non–point source wetlands for treating stormwater and runoff from agricultural fields. Research projects illustrating the effects and functioning of these types of wetlands in agricultural watersheds have been carried out in southeastern Australia (Raisin and Mitchell, 1995; Raisin et al., 1997), northeastern Spain (Comin et al., 1997), Illinois (Kadlec and Hey, 1994; Phipps and Crumpton, 1994; Mitsch et al., 1995; Kovacic et al., 2000; Larson et al., 2000; Hoagland et al., 2001), Florida (Moustafa, 1999; Reddy et al., 2006), Ohio (Fink and Mitsch, 2004), and Sweden (Leonardson et al., 1994; Jacks et al., 1994; Arheimer and Wittgren, 1994). Several wetland sites have received the equivalent of non–point source pollution but under somewhat controlled hydrologic conditions (e.g., river overflow to riparian basins) over several years of study. Bony Marsh, a constructed wetland located along the Kissimmee River in southern Florida, was investigated for nutrient retention of river water for nine years (1978–1986) by Moustafa et al. (1996), who found it to be a consistent sink of nitrogen and phosphorus but at relatively low levels.

As described in Chapters 14: "Human Impacts and Management of Wetlands," and 18: "Wetland Creation and Restoration," the water quality in the Florida Everglades is threatened by high nutrients coming from the upstream Everglades Agricultural Area (EAA). Case Study 3 describes what is probably the largest assemblage of treatment wetlands anywhere in the world. They have been created to intercept phosphorus coming from agricultural stormwater runoff before it reaches the Florida Everglades.

CASE STUDY 3: Creating Treatment Wetlands to Protect Downstream Wetlands in Florida's Everglades

An ambitious plan of agricultural stormwater treatment by wetlands is occurring in the Florida Everglades, where 23,000 ha of created wetlands, called stormwater treatment areas (STAs), have been created for phosphorus control from upstream agricultural areas. As described in Chapter 14, the main cause of the spread of cattail (*Typha domingensis*) in the otherwise nutrient-poor

Florida Everglades dominated by sawgrass (*Cladium jamaicense*) is nutrient enrichment—especially by phosphorus emanating from agricultural areas in the basin.

A prototype of the STAs, a 1,544-ha treatment wetland complex called the Everglades Nutrient Removal (ENR) project, was first designed and tested, and the results were published in a special issue of *Ecological Engineering* (Reddy et al., 2006). Water was pumped to the ENR wetland complex from adjacent drainage canals. Over its first six-year operating schedule (1994–1999), the wetland decreased total phosphorus and total nitrogen by 79 and 26 percent, respectively (Gu et al., 2006), with an average outflow concentration of 21 ppb (= µg-P/L over that period (Kadlec, 2006).

As a result of the success of the ENR project, six full-scale STAs treating agricultural runoff from the EAA south of Lake Okeechobee have since been created (Fig. 19.6a). Some of these systems have now been in operation for almost 20 years (Fig. 19.6b). Overall, from their start through 2012, these wetlands reduced phosphorus loads by 73 percent and lowered the average phosphorus concentrations from 140 to 37 ppb (Pietro, 2012; Fig. 19.6c). The dynamics of phosphorus in these STA wetlands has been investigated by a number of researchers (Newman and Pietro, 2001; Juston and DeBusk, 2006, 2011; Dierberg and DeBusk, 2008; Paudel et al., 2010; Paudel and Jawitz, 2012; Entry and Gottlieb, 2014). Juston and DeBusk (2006) suggested that mass loadings at or below $1.3\,g\,P\,m^{-2}\,yr^{-1}$ provided "a high likelihood of achieving outflow total P (TP) concentrations less than 30 µg/L." Submerged aquatic vegetation wetlands and emergent vegetation wetlands restored from historic wetlands rather than agriculture and with loading rates at or below $2\,g\,P\,m^{-2}\,yr^{-1}$ have resulted in outflow phosphorus concentrations consistently between 10 and 20 ppb and mass removal efficiencies consistently above 85 percent. For the period 2008 to 2012, the inflow concentration of phosphorus to the STAs was 191 ppb and an average outflow concentration was 35 ppb, resulting in an average retention rate, by concentration, of 82 percent. The average retention rate of $1.25\,g\text{-}P\,m^{-2}\,yr^{-1}$ is well within the range of phosphorus retention of 0.5 to $5\,g\text{-}P\,m^{-2}\,yr^{-1}$ for low-nutrient (nonmunicipal wastewater) "sustainable" treatment wetlands as summarized by Mitsch et al. (2000). These STA wetlands were originally designed to reduce phosphorus to 50 ppb, a 60 to 75 percent reduction in concentration (Chimney and Goforth, 2006).

That effluent goal has been reduced by authorities to 10 ppb, essentially the background concentrations of phosphorus in the oligotrophic Everglades. Reaching this mandated 10 ppb threshold of total phosphorus has not been achieved with any consistency from the STAs. A multiyear mesocosm study that investigated the effect that different plant communities had on reducing the outflow of the STAs to 10 ppb of phosphorus showed that 10 ppb may be possible when the inflow is the effluent coming from the STAs and the hydraulic loading rates are substantially lower than those in the current STAs (Mitsch et al., 2015).

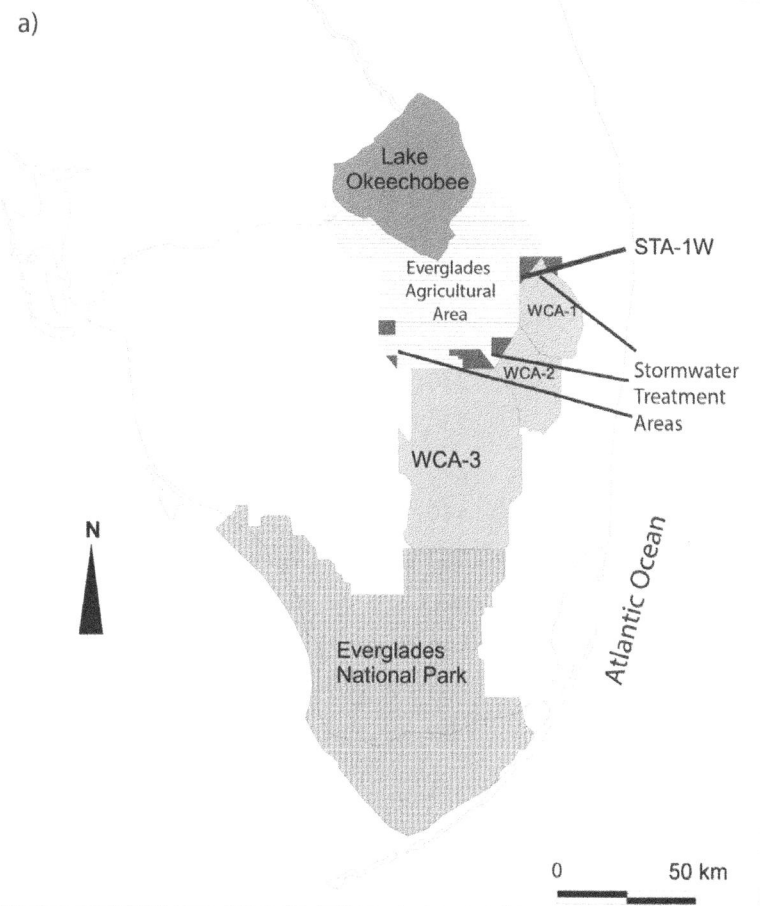

Figure 19.6 Stormwater treatment areas (STAs) downstream of the Everglades Agricultural Area created to reduce phosphorus inflow to the Florida Everglades: (a) location of the STAs; (b) inflow and outflow concentrations of total phosphorus in STA 1W, 1995–2012; (c) comparison of phosphorus concentrations in inflows and outflows of six STA complexes. (Data courtesy of South Florida Water Management District, West Palm Beach, FL)

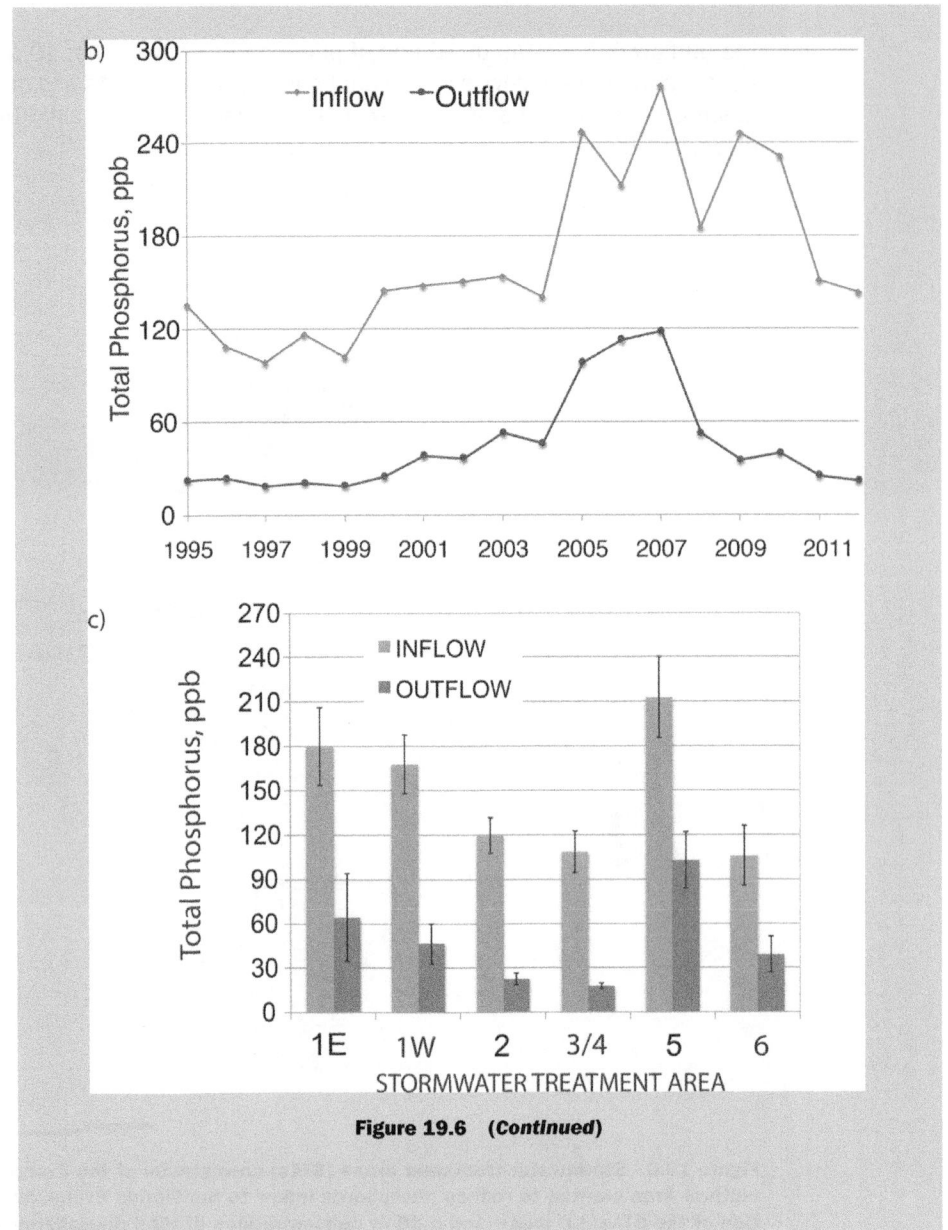

Figure 19.6 *(Continued)*

Agricultural Wastewater Wetlands

In addition to the nonpoint sources from agriculture that were discussed above, serious water pollution problems occur in many parts of the world resulting from runoff from confined animals, particularly dairy, cattle, and swine operations (Tanner et al., 1995;

Table 19.2 Hydrology and water quality of two wetlands constructed to deal with heavily polluted dairy milkhouse effluent

	Connecticut[1]		Maryland[2]	
Wetland area, m^2	400		1,160	
Flow, m^3/wk	18.8		—	
retention time, days	41		—	
	Inflow	Outflow	Inflow	Outflow
BOD, mg/L	2,680	611	1,914	59
Total N	103	74	170	13
Ammonia-N	8	52	72	32
Nitrate-N	0.3	0.1	5.5	10.0
Total P, mg/L	26	14	53	2.2
TSS, mg/L	1,284	130	1,645	65
coliform, #/100 mL	557,000	13,700	—	—

[1]Newman et al., 2000.
[2]Schaafsma et al., 2000.

Cronk, 1996; Knight et al., 2000). As more animals are concentrated per unit area to increase food production, the concentrations and volumes of effluents are becoming more noticeable, both by the public and by water pollution control authorities. Concentrations of organic matter, organic nitrogen, ammonia-nitrogen, phosphorus, and fecal coliforms from animal feedlots far exceed concentrations in most municipal sewer systems. Two examples from the eastern United States of the effectiveness of wetlands for treating wastewater from dairy milkhouses (Table 19.2) showed significant reductions in most pollutants in the treated water, although ammonia-nitrogen increased substantially in the Connecticut wetland and nitrate-nitrogen increased by 80 percent in the Maryland case. In addition to livestock waste from land-based agriculture, constructed wetlands have been used to treat effluent from several aquaculture operations, including shrimp ponds in Thailand and tilapia fish ponds in the United Kingdom.

River Diversion Wetlands

A somewhat different approach to cleaning up water is to pass river water through wetlands built on adjacent floodplains or backwaters. These are analogs of riverine oxbows or billabongs found throughout the world, and they have been shown to consistently improve water quality. These wetlands also are simulations of agriculture runoff wetlands, but with usually lower concentrations of nutrients. However, river sediment concentrations can be high, sometimes in excess of that found in agricultural runoff. River diversions have been done on a large scale in the Mississippi River Delta in Louisiana (see Case Study 4) and on a much smaller scale for research and water quality improvement in the midwestern United States (see Case Study 5). In both of these cases, significant improvement in diverted river water quality has been observed as the water is distributed to wetlands on the floodplains and deltas.

CASE STUDY 4: Diverting the Mississippi River to the Louisiana Delta

In the Louisiana delta, controlled diversions of the Mississippi River to the delta to renourish the delta and slow down wetland vegetation loss have been implemented at several sites along the river. It has also been recognized for some time that restoration of the delta could lead to reductions of the nutrient load, particularly from nitrogen, the primary cause of the Gulf of Mexico hypoxia (see Chapter 6: "Wetland Biogeochemistry"). It has been estimated that freshwater marshes at river diversions could denitrify up to 110 t-N km^{-2} yr^{-1} and overall remove up to 25 percent of the annual flux of nitrate-nitrogen to the Gulf, or 956,000 t-N yr^{-1} (Rivera-Monroy et al., 2013). The U.S. Army Corps of Engineers has been and will continue to invest hundreds of millions of U.S. dollars into river diversion systems in the Louisiana delta, now as part of a 2012 Coastal Management Plan. One of the largest diversions in operation on the river aimed at restoring deteriorating wetlands in the Mississippi delta is at Caernarvon (Fig. 19.7) on the east bank of the river south of New Orleans. It has a maximum flow of 226 m^3 sec^{-1} but an average discharge of only 21 m^3 sec^{-1} (Lane et al., 2006, 2007). River diversion began in August 1991, and peak discharge to date has been 140 m^3 sec^{-1}, which occurred in March 2007 (Day et al., 2013). Summer diversion flow rates are generally near the minimum, and winter flow rates are 50 to 80 percent of the maximum (Lane et al., 2004). The diversion delivers river water to the 260-km^2 Caernarvon freshwater wetlands, which eventually discharge into the larger brackish Breton Sound estuary on the Gulf of Mexico.

An intriguing issue is whether these downstream wetlands also retain nutrients, which is particularly important given the hypoxia in the Gulf of Mexico, discussed elsewhere in this book (see Chapter 6). The Caernarvon wetland and especially downstream Breton Sound were shown to have a loading rate of 3.5 g-N m^{-2} yr^{-1}, with overall retention efficiencies of that rate highest in the fall (98 percent) and lowest in the winter (74 percent) (Lundberg et al., 2014).

But as studies such as those by Deegan et al. (2012) in a whole-ecosystem, multiyear experiment in Massachusetts salt marshes illustrate, care must be taken so that the advantages of increasing the land elevation in the delta to restore the marshes and retaining nutrients to protect downstream deepwater coastal ecosystems is not outweighed by the negative effects that excess nutrients have on salt marsh stability. There continues to be discussion of the long-term effectiveness of these diversions. Some studies have suggested that the high-nutrient inputs to the delta weaken salt marsh plant structure and causes shallow rooting, which in turn has led to loss of vegetation cover during hurricanes (Kearney et al., 2011; Teal et al., 2012). This assumption of marsh plant demise was challenged by Day et al. (2013), who investigated wetland vegetation productivity downstream of the Caernarvon

the system. Ultimately, hydrologic conditions determine wetland function. Several parameters used to describe the hydrologic conditions of treatment wetlands include hydroperiod, depth, seasonal pulses, hydraulic loading rates, and retention time.

Hydroperiod and Depth

In wetlands, hydroperiod is the water depth or stage of a wetland over time (see Chapter 4: "Wetland Hydrology"). Wetlands that have a seasonal fluctuation of water depth have the most potential for developing a diversity of plants, animals, and biogeochemical processes. In a constructed wastewater wetland with a similar inflow of wastewater every day, water levels often vary little seasonally unless stormwater is part of the treatment inflow. During the start-up period of constructed wetlands, low water levels are needed to avoid flooding newly emerged plants. Start-up periods for the establishment of vegetation may take two to three years of careful attention to water levels.

While storms and seasonal patterns of floods rarely affect constructed wastewater wetlands built for municipal treatment (except when storm sewers are part of the inflow), they can significantly affect the performance of wetlands designed for the control of non–point source runoff. A variable hydroperiod, which exhibits dry periods interspersed with flooding, is a natural cycle in non–point source wetlands, and fluctuating water levels should be considered a natural feature. A fluctuating water level could provide needed oxidation of organic sediments and can, in some cases, rejuvenate a system to higher levels of chemical retention. There was a definitive seasonal cycle, plus sudden bursts of water levels during winter and spring storms at an agricultural wetland in Ohio described by Fink and Mitsch (2004). Furthermore, noting the importance of biology in wetland types, the water level in one basin in that wetland dropped almost 30 cm because of burrowing activity by muskrats (*Ondatra zibethicus*).

Hydraulic Loading Rate

The hydraulic loading rate, one of the most important variables in treatment wetlands, is defined as:

$$q = 100 \, Q/A \tag{19.1}$$

where

q = hydraulic loading rate (HLR), cm day^{-1}
Q = inflow rate, m^3 day^{-1}
A = wetland surface area, m^2

Table 19.3 summarizes several recommendations and measurements of HLR of surface-flow and subsurface-flow wastewater wetlands. Loading rates to surface-flow wetlands for small municipalities range from 1.4 to 22 cm day^{-1}, while rates to subsurface-flow constructed wetlands vary between 1.3 and 26 cm day^{-1}. Knight (1990) reviewed several dozen wetlands constructed for wastewater treatment and recommended an HLR of 2.5 to 5 cm/day for surface-flow constructed wetlands

Table 19.3 Recommended and actual hydrologic loading rates (HLR) for treatment wetlands

Type of Wetland	Recommended Loading Rate, cm/day [1]	Loading Rate, cm/day [2]	Median Loading Rate, cm/day [3]
Surface-flow treatment wetlands	2.5 – 5.0	5.4 ± 1.7 (n = 15)	3.0 (n = 205)
Subsurface-flow treatment wetlands	6.0 – 8.0	7.5 ± 1.0 (n = 23)	6.8 (n = 634)

and 6 to 8 cm day^{-1} for subsurface-flow wetlands. Kadlec (2009a) followed up with a review of more than 800 treatment wetlands 20 years later and found the median numbers for both surface-flow and subsurface-flow treatment wetlands to be almost exactly midpoint between those ranges.

Detention Time

Detention time of treatment wetlands is calculated as:

$$t = V\,p/Q \tag{19.2}$$

where

t = theoretical detention time, day
V = volume of wetland basin, m^3 (volume of water column for surface-flow wetlands; volume of medium for subsurface-flow wetlands
p = porosity of medium (e.g., sand or gravel for subsurface-flow wetlands)
  = 1.0 for surface-flow wetlands
Q = flow rate through wetland, m^3 day^{-1}

The optimum detention time (or nominal residence time) has been suggested to be from 5 to 14 days for treatment of municipal wastewater. Florida regulations on wetlands require that the volume in the permanent pools of the wetland must provide for a residence time of at least 14 days. Calculation of detention time or nominal residence time with Equation (19.2) is not always realistic because of short-circuiting and the ineffective spreading of the waters as they pass through the wetland. Tracer studies of flow through wetlands have illustrated the importance of not overrelying on the theoretical detention time to design treatment wetlands. Not all parcels of water that enter at a certain time leave the wetland at the same time. In some instances, water will short-circuit through the wetland, whereas other water will remain in backwater locations for considerably more time than the theoretical detention time.

Basin Morphology

Several aspects related to the morphology of constructed wetland basins need to be considered when designing wetlands. For example, Florida regulations for the Orlando area require, for littoral zones, a shelf with a gentle slope of 6:1 or flatter to a point of from 60 to 77 cm below the water surface. Slopes of 10:1 or flatter are even better. A flat littoral zone maximizes the area of appropriate water depth for emergent plants, thus allowing more wetland plants to develop more quickly and allowing wider bands of different plant communities. Plants will also have room to move "uphill" if water levels are raised in the basins because of flows being higher than predicted or to enhance treatment. Bottom slopes of less than 1 percent are recommended for wetlands built to control runoff, whereas a substrate slope, from inlet to outlet, of 0.5 percent or less has been recommended for surface-flow wetlands used to treat wastewater.

Flow conditions should be designed so that the entire wetland is effective in nutrient and sediment retention if these are desired objectives. This may necessitate several inflow locations and a wetland configuration to avoid channelization of flows. A length-to-width ratio (L/W) (called the *aspect ratio*) should be at least 10:1 if water is purposely introduced to the system. A minimum aspect ratio of 2:1 to 3:1 has been recommended for surface-flow wastewater wetlands.

Providing a variety of deep and shallow areas is optimum. Deep areas (>50 cm), while too deep for continuous emergent vegetation, offer habitat for fish, increase the capacity of the wetland to retain sediments, can enhance nitrification as a prelude to later denitrification if nitrogen removal is desired, and can provide low-velocity areas where water flow can be redistributed. Shallow depths (<50 cm) provide maximum soil–water contact for certain chemical reactions, such as denitrification, and can accommodate a greater variety of emergent vascular plants.

Individual wetland cells, placed in series or parallel, often offer an effective design to create different habitats or establish different functions. Cells can be parallel so that alternate drawdowns can be accomplished for mosquito control or redox enhancement, or they can be in a series to enhance biological processes.

Chemical Loadings

When water flows into a wetland, it brings chemicals that may be beneficial or possibly detrimental to the functioning of that wetland. In an agricultural watershed, this inflow will include nutrients such as nitrogen and phosphorus as well as sediments and possibly pesticides. Wetlands in urban areas can have all of these chemicals plus other contaminants such as oils and salts. Wastewater, when added to wetlands, has high concentrations of nutrients and, with incomplete primary treatment, high concentrations of organic matter (BOD) and suspended solids. At one time or another, wetlands have been subjected to all of these chemicals, and they often serve as effective sinks. Wetlands can be sized using design graphs, standard retention rates, or empirical models.

Design Graphs

The simplest model available to estimate the retention of nutrients or other chemicals by wetlands is to use design graphs that give some measure of chemical retention versus chemical loading, either areal (e.g., g m^{-2} yr^{-1}) or volumetric (e.g., g m^{-3} yr^{-1}). If a

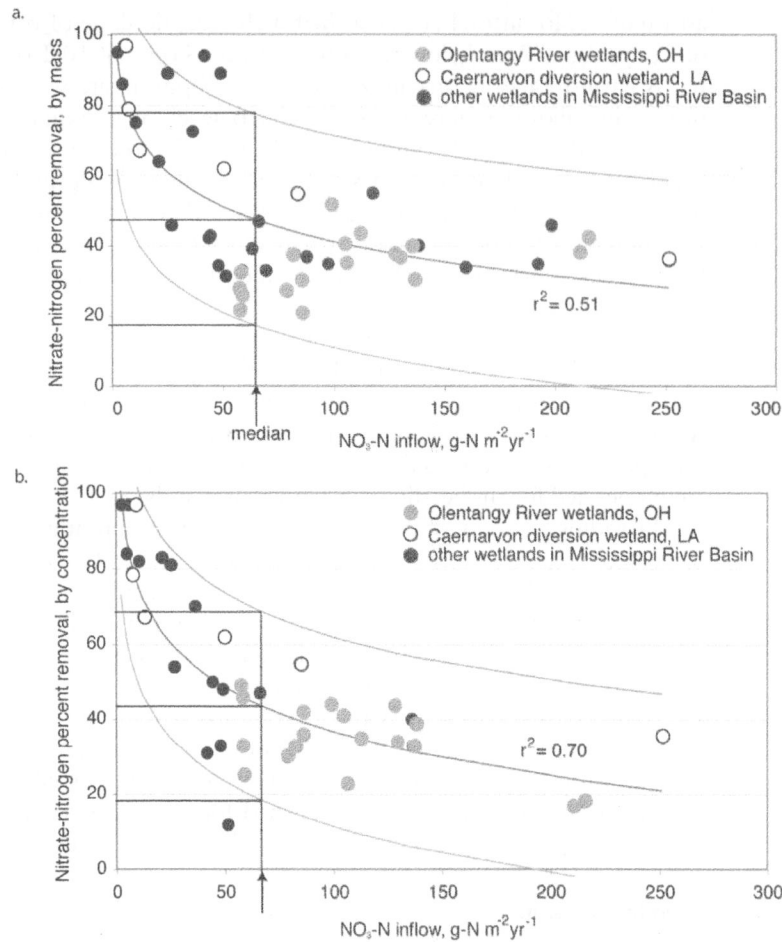

Figure 19.9 Decrease in nitrate-nitrogen by (a) mass, and (b) concentration for created and managed wetlands in the Mississippi River Basin. Each data point represents data for a complete year for a wetland. Outside lines are 95 percent confidence intervals. Vertical lines in graphs indicate median loading rate of 60 g-N m^{-2} yr^{-1}. (Mitsch et al., 2005b, copyright Elsevier, reprinted with permission)

wetland were designed to retain nutrients, for example, it would be desirable to know how well that retention would occur for various nutrient inflows. Data compiled from a large number of wetland sites in North America and Europe provide an indication of the nutrient retention of wetlands. For example, Figure 19.9, compiled from wetlands in the Mississippi River Basin, illustrates the percentage removal of nitrate-nitrogen versus loading for the midwestern United States in two ways: (1) mass retention per unit area, and (2) percentage retention by concentration. Each of the data points is based on one year's data at one wetland basin in either the Midwestern United States or the river delta in Louisiana.

Table 19.4 Nutrient and sediment removal rates and efficiency in constructed wastewater wetlands

Wetland Type Parameter	Loading (g m^{-2} yr^{-1})	Retention (g m^{-2} yr^{-1})	Percentage Retention
SURFACE-FLOW CONSTRUCTED WETLANDS			
Nitrate + nitrate nitrogen	29	13	44.4
Total nitrogen	277	126	45.6
Total phosphorus	4.7–56	2.1–45	46–80
Suspended solids	107–6,520	65–5,570	61–98
SUBSURFACE-FLOW CONSTRUCTED WETLANDS			
Nitrate + nitrate nitrogen	5,767	547	9.4
Total nitrogen	1,058	569	53.8
Total phosphorus	131–631	11–540	8–89
Suspended solids	1,500–5,880	1,100–4,930	49–89

Source: Kadlec and Knight, 1996.

Retention Rates

Another approach to estimating the retention of nutrients is to simply compare several studies and estimate the chemical retention that consistently happens in wetlands. Averages from data from many constructed wastewater wetlands are shown in Table 19.4. In general, as suggested by the HLR data in Table 19.3, subsurface wetlands receive more wastewater and thus receive greater loadings of chemicals and sediments. The high average mass retention of nitrate-nitrogen in subsurface wetlands is due more to these high loading rates in subsurface-flow wetlands than it is to any ability of these systems to sequester more nitrate-nitrogen. Note that the percent nitrate-nitrogen retention is much higher in the surface-flow than in the subsurface-flow wetlands. The retention of phosphorus tends to be more variable in subsurface wetlands than in surface wetlands.

Summaries of retention rates for several wetlands intercepting non–point source pollution are given in Table 19.5. Rules of thumb for this type of wetland are that wetlands can consistently retain phosphorus in amounts of 0.5 to 5 g-P m^{-2} yr^{-1} and nitrogen in amounts of about 10 to 40 g-N m^{-2} yr^{-1} (Mitsch et al., 2000). To maintain biological diversity in the plant community, the lower end of these loading rates should be used. The long-term rates for phosphorus and nitrogen for the 30-year-old Houghton Lake treatment wetland described in Case Study 1 in this chapter and the phosphorus retention rates for the stormwater urban and agricultural wetlands in south Florida described in Case Studies 2 and 3 fit in these ranges. Nitrate-nitrogen retention capabilities of freshwater marshes receiving non–point source pollution in seasonal to cold climates shows a range of nitrogen retention from 3 to 93 g-N m^{-2} yr^{-1} and a phosphorus retention rate of 0.1 to 6 g-P m^{-2} yr^{-1} (Table 19.5) Low retention numbers are generally from wetlands that are "underfed" nutrients. High numbers are usually only periodic and therefore would be inappropriate to use for design purposes.

Table 19.5 Nutrient retention in constructed and natural wetlands receiving low-concentration (i.e., non-wastewater, nutrient loading from rivers, overflows, or non–point source pollution)

Wetland Location and Type	Wetland Size, ha	Nitrogen g-N m^{-2} yr^{-1}	Phosphorus g-P m^{-2} yr^{-1}	Reference
WARM CLIMATE				
Everglades marsh, S. Florida	8000	—	[1]0.4–0.6	Richardson and Craft, 1993; Richardson et al., 1997
Florida Everglades stormwater treatment areas (STAs)	23,000	—	1.25	See Case 2, this chapter
Boney Marsh, S. Florida	49	4.9	0.36	Moustafa et al., 1996
Everglades Nutrient Removal Project, S. Florida	1545	10.8	0.94	Moustafa, 1999
Restored marshes, Mediterranean delta, Spain	3.5	69	—	Comin et al., 1997
Constructed rural wetland, Victoria, Australia	0.045	23	2.8	Raisin et al., 1997
Breton Sound Estuary, Louisiana Delta	110,000	3.5	—	Lundberg et al., 2014
COLD CLIMATE				
Houghton Lake, Michigan (30 years)	100	[2]4.39	1.76	Kadlec, 2009b
Constructed wetlands, NE Illinois				Phipps and Crumpton, 1994; Mitsch et al., 1995
river-fed and high-flow	2	[2]11–38	1.4–2.9	
river-fed and low-flow	2–3	[2]3–13	0.4–1.7	
Artificially flooded meadows, southern Sweden	180	43–46	—	Leonardson et al., 1994
Constructed wetland basins, Norway	0.035–0.09	50–285	26–71	Braskerud, 2002a,b
Palustrine freshwater wetlands, NW Washington				Reinhelt and Horner, 1995
urban area	2	—	0.44	
rural area	15	—	3.0	
Created instream wetland, OH	6	—	2.9	Niswander and Mitsch, 1995
Created riverine wetlands, OH	2	38.8	2.4	Mitsch et al., 1998, 2014; Spieles and Mitsch, 2000a; Nairn and Mitsch, 2000
Created river diversion wetland, OH	3	32	4.5	Fink and Mitsch, 2007; Mitsch et al., 2008
Agricultural wetlands, OH	1.2	[2]39	6.2	Fink and Mitsch, 2004
Agricultural wetlands, IL (3)	0.3–0.8	[2]33	0.1	Kovacic et al., 2000
Natural marsh, Alberta, Canada	360	—	[1]0.43	White et al., 2000

[1]estimated by phosphorus accumulation in soil
[2]nitrate-nitrogen only

Empirical Models

A third method for estimating the ability of wetlands to retain chemicals is to use equations that either have a theoretical base or are empirically determined from large databases of existing wastewater wetlands. One such general model, originally developed by Kadlec and Knight (1996) and others, is based on a mass-balance approach called the "k-C*" model" and is given as:

$$q(dC/dy) = k_A(C-C^*) \quad (19.3)$$

where

y = fractional distance from inlet to outlet, unitless
C = chemical concentration, g m^{-3}
k_A = areal removal rate constant, m yr^{-1}
C^* = residual or background chemical concentration, g m^{-3}

This equation is based on an assumption that processes can be described on an areal basis. Thus, the coefficient k_A has units of velocity and can be recognized as being similar to a settling velocity coefficient used in sedimentation models. C^* represents a background concentration of a chemical or constituent, below which it is generally agreed that treatment wetlands cannot go. Integrating this equation over the entire length of the wetland, the solution can be expressed as a first-order areal model:

$$[(C_o-C^*)/(C_i-C^*)] = e^{-kA/q} \quad (19.4)$$

where

C_o = outflow concentration, g m^{-3}
C_i = inflow concentration, g m^{-3}
q = hydraulic loading rate, m yr^{-1}

Estimates of the two parameters needed for this model, C^* and k_A, are listed in Table 19.6. This equation does not work equally well for all parameters, but it does provide a way of estimating the area of a wetland necessary for achieving a certain removal. Rearranging equations 19.4 and 19.1 gives the following calculation of wetland area for given results:

$$A = Q \ln[(C_o-C^*)/(C_i-C^*)]/k_A \quad (19.5)$$

where

Q = flow rate through wetland, m^3 yr^{-1}

Where this model is insufficiently backed with good data or does not work properly, strictly empirical relationships of the outflow concentration C_o as a function of the inflow concentration C_i and the hydraulic loading rate (q) have been developed (Table 19.7).

Table 19.6 Parameters for first-order areal model given in equations to for several constituents of wastewater wetlands (Subsurface-flow constructed wetlands and surface-flow constructed wetlands are given as wetland type where appropriate.)

Constituent and Wetland Type	k_A (m yr^{-1})	C^* (g m^{-3})
BOD, surface-flow	34	$3.5 + 0.053C_i$
BOD, subsurface-flow	180	$3.5 + 0.053C_i$
Suspended solids, surface-flow	1,000	$5.1 + 0.16C_i$
Total phosphorus, surface and subsurface-flow	12	0.02
Total nitrogen, surface-flow	22	1.5
Total nitrogen, subsurface-flow	27	1.5
Ammonia nitrogen, surface-flow	18	0
Ammonia nitrogen, subsurface-flow	34	0
Nitrate nitrogen, surface-flow	35	0
Nitrate nitrogen, subsurface-flow	50	0

Source: Kadlec and Knight, 1996.

Other Chemicals

Although most evaluations of the efficiency of wetlands have been concerned with this capacity to remove nutrients, sediments, and organic carbon (BOD), there is some literature on other chemicals, such as iron, cadmium, manganese, chromium, copper, lead, mercury, nickel, and zinc. Wetland soils or biota or both often easily sequester metals. That is the basic problem in using wetlands as sinks for such chemicals: They can accumulate in the food chain.

Soils

The topsoil is important to the overall function of a constructed wetland (Fig. 19.10). It is the primary medium supporting rooted vegetation, and, particularly for subsurface wetlands, it is part of the treatment system. The sediments retain certain chemicals and provide the habitat for micro- and macroflora and fauna that are involved in chemical transformations. Constructed wetland soil texture depends on whether surface flow over the substrate or subsurface flow through the substrate is being considered. Surface-flow wetland soils are generally less effective in removing pollutants per unit area but are closer in design to natural wetlands. Their ability to provide structure and nutrition to the wetland plants is important. Clay material, although favored as a subsurface liner, limits root and rhizome penetration and may prevent water from reaching plant roots. Silt clay or loam soils are preferable for the overlying soils in constructed wetlands. Sandy soil is less preferred for surface-flow wetlands. For subsurface-flow wetlands, high permeability is preferred. The material needs to be sand, gravel, or some other highly permeable media.

The subsoil of constructed wetlands (usually below the root zone and referred to as a *liner*) must have permeability low enough to cause standing water or saturated

Table 19.7 Empirical equations for the estimation of outflow concentrations or wetland area based on inflow concentrations and hydraulic retention time (Correlation coefficient (R^2) and number of wetlands used in analysis (n) are also given.). C_i, inflow concentration (g m^{-3}); C_o, outflow concentration (g m^{-3}); A, area of wetland (ha); Q, wetland inflow, (m^3/day); q, hydraulic loading rate, (cm/day).

Constituent	Equation[6]	r^2 (n)
BOD		
Surface-flow wetlands	$C_o = 4.7 + 0.173 C_i$	0.62 (440)
Subsurface-flow, soil	$C_o = 1.87 + 0.11 C_i$	0.74 (73)
Subsurface-flow, gravel	$C_o = 1.4 + 0.33 C_i$	0.48 (100)
Suspended solids		
Surface-flow wetlands	$C_o = 5.1 + 0.158 C_i$	0.23 (1,582)
Subsurface-flow wetlands	$C_o = 4.7 + 0.09 C_i$	0.67 (77)
Ammonia nitrogen		
Surface-flow wetlands	$A = 0.01 Q / \exp[1.527 \ln C_o - 1.05 \ln C_i + 1.69]$	
Surface-flow marshes	$C_o = 0.336 C_i^{0.728} q^{0.456}$	0.44 (542)
Subsurface-flow wetlands	$C_o = 3.3 + 0.46 C_i$	0.63 (92)
Nitrate nitrogen		
Surface-flow marshes	$C_o = 0.093 C_i^{0.474} q^{0.745}$	0.35 (553)
Subsurface-flow wetlands	$C_o = 0.62 C_i$	0.80 (95)
Total nitrogen		
Surface-flow marshes	$C_o = 0.409 C_i + 0.122 q$	0.48 (408)
Subsurface-flow wetlands	$C_o = 2.6 + 0.46 C_i + 0.124 q$	0.45 (135)
Total phosphorus		
Surface-flow marshes	$C_o = 0.195 C_i^{0.91} q^{0.53}$	0.77 (373)
Surface-flow swamps	$C_o = 0.37 C_i^{0.70} q^{0.53}$	0.33 (166)
Surface-flow wetlands	$C_o = 0.51 C_i^{1.10}$	0.64 (90)

soils. If clay is not available on site, it may be advisable to add a layer of clay to minimize percolation. Studies have also been undertaken to investigate other materials as liners for constructed wetlands. The most frequently used liners for constructed wetlands are clays, clay bentonite mixtures, or synthetic materials, such as polyvinyl chloride (PVC) and high-density polyethylene (HDPE). Experiments have been conducted in recycling materials such as coal combustion waste products. As it turns out, using calcium-rich sulfur-scrubber waste material was shown to actually increase the phosphorus-retention capability of the wetlands (Ahn et al., 2001; Ahn and Mitsch, 2001), but care must be taken that the material completely seals the wetland because leachate from this liner material is highly alkaline.

Subsurface flow through subsurface wetlands can be through soil media (*root-zone method*) or through rocks, gravel, or sand (*rock-reed filters*). Flow in both cases is 15 to 30 cm below the surface. Gravel is sometimes added to the substrate of subsurface-flow wetlands (*gravel-bed*) to provide a relatively high permeability that allows water to percolate into the root zone of the plants where microbial activity is high. Gravel can be silica based or limestone based; the former has less capacity for phosphorus retention. Another evaluation of the European-design subsurface wetlands indicated that they often decrease in hydrologic conductivity after several years and become clogged, essentially becoming partial-surface-flow wetlands.

Organic Content

The organic content of soils has some significance for the retention of chemicals in a wetland. Mineral soils generally have lower cation exchange capacity than organic soils do; the former is dominated by various metal cations, and the latter is dominated by the hydrogen ion. Organic soils can therefore remove some contaminants (e.g., certain metals) through ion exchange and can enhance nitrogen removal by providing an energy source and anaerobic conditions appropriate for denitrification. Organic matter in wetland soils varies between 5 and 75 percent, with higher concentrations in peat-building systems, such as bogs and fens, and lower concentrations in mineral-soil wetlands, such as riparian bottomland wetlands subject to mineral sedimentation or erosion. When wetlands are constructed, especially subsurface-flow wetlands, organic matter, such as composted mushrooms, peat, or detritus, is often added in one of the layers. For construction of many wetlands, however, organic soils are avoided because they are low in nutrients, can cause low pH, and often provide inadequate support for rooted aquatic plants.

Depth and Layering of Soil

The depth of substrate is an important design consideration for wastewater wetlands, particularly those that use subsurface flow. The depth of suitable topsoil or substrate should be adequate to support and hold vegetation roots. A common substrate depth for constructed wetlands is 60 to 100 cm. In some cases, layering more elaborate than that shown in Figure 19.10 is suggested.

Soil Chemistry

Although exact specifications of nutrient conditions in a wetland soil necessary to support aquatic plants are not well known, low nutrient levels characteristic of organic, clay, or sandy soils can cause problems for initial plant growth. Although fertilization may be necessary in some cases to establish plants and enhance growth, it should be avoided if possible in wetlands that eventually will be used as sinks for the same macronutrients. When fertilization is required to get plants started in constructed wetlands, slow-release granular and tablet fertilizers are often useful.

When soils are submerged and anoxic conditions result, iron is reduced from ferric (Fe^{+++}) to the ferrous (Fe^{++}) ions, releasing phosphorus that was previously held as insoluble ferric phosphate compounds. The Fe-P compound can be a significant source of phosphorus to overlying and interstitial waters after flooding and anaerobic conditions occur, particularly if the wetland was constructed on previously agricultural land. After an initial pulse of released phosphorus in such constructed wetlands, the iron and aluminum contents of a wetland soil exert significant influences on the ability of that wetland to retain phosphorus. All things being equal, soils with higher aluminum

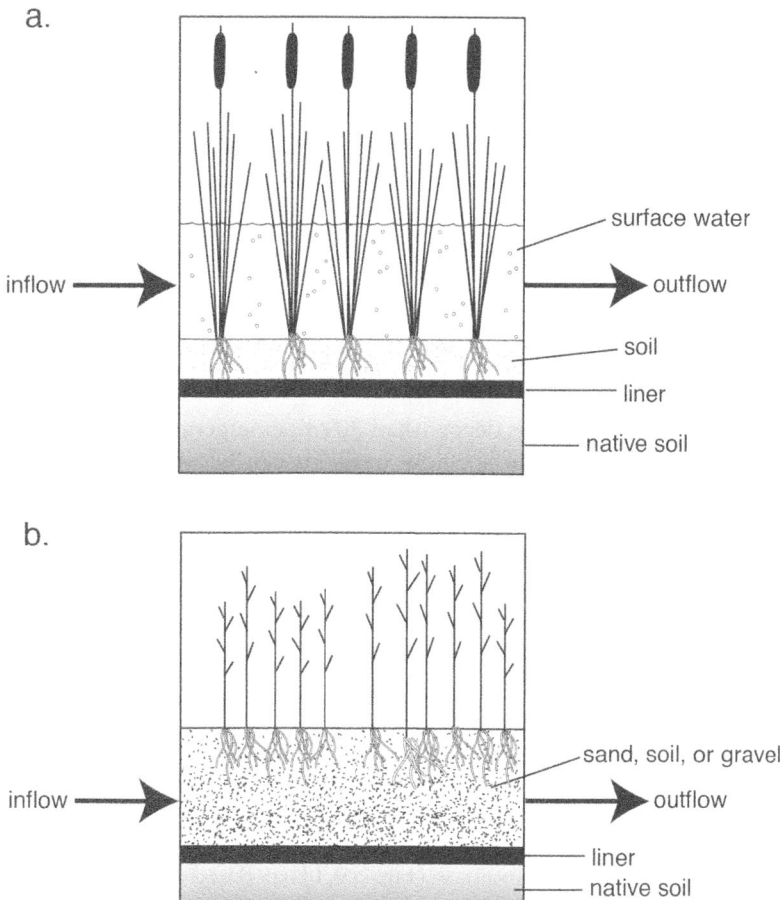

Figure 19.10 Soil cross-sections of (a) surface-flow wetland, and (b) subsurface-flow wetland. (After Knight, 1990)

and iron concentrations are more desirable because their affinity for phosphorus is higher.

Vegetation

Just as the question "What plants should be used?" arises for creating and restoring wetlands as discussed in Chapter 18, vegetation choice is also a consideration for treatment wetlands. But there is at least one significant difference for treatment wetlands: While creation and restoration of wetlands are done principally to develop a diverse vegetation cover and provide habitat, treatment wetlands are constructed with the main goal of improving water quality. The plants in created and restored wetlands are part of the solution; in treatment wetlands, they are the partial cause of the solution. Furthermore, treatment wetlands invariably have higher concentrations of chemicals in the water, which by its very nature limits the number of plant species that will survive in those wetlands. Experience has shown that relatively few plants thrive in the high-nutrient, high-BOD wastewaters that are applied to treatment wetlands. Vymazal (2013) found that a total of 150 different species of plants were used in 643 surface-water wetlands from 43 countries. Table 18.4 flagged some of the macrophyte species that are frequently used for treatment wetlands amid the hundreds that are used for creation of wetland habitat. Among those plants are cattails (*Typha* spp.), the bulrushes (*Schoenoplectus* spp., *Scirpus* spp.), and reed grass (*Phragmites australis*). The last is the preferred plant in subsurface-flow wetlands around the world but is not favored in many parts of North America because of its aggressive behavior in freshwater and brackish marshes. Other commonly used plants included *Juncus effusus*, *Eleocharis* spp., *Phalaris arundinacea*, and *Cyperus papyrus*, the last mostly in Africa.

When water is deeper than 30 cm, emergent plants often have difficulty growing. In these cases, surface-flow wetlands can become covered with duckweed (*Lemna* spp.) in temperate zones and water hyacinths (*Eichhornia crassipes*) and water lettuce (*Pistia* spp.) in the subtropics and tropics. While rooted floating aquatics, such as *Nymphaea*, *Nuphar*, and *Nelumbo*, are favored for their aesthetics, they thrive only in rare instances in treatment wetlands, where, due to high-nutrient conditions, they are easily overwhelmed by duckweed and filamentous algae.

Tanner (1996) compared relative nutrient uptake and pollutant removal of eight macrophytes in gravel-bed wetland mesocosms fed by dairy wastes in New Zealand (Fig. 19.11). Greatest above-ground biomass was seen in this highly polluted wastewater by *Glyceria maxima* and *Zizania latifolia*, while greatest below-ground biomass was seen with *Bolboschoenus fluviatilis* (*Scirpus fluviatilis* in the United States), which had below-ground biomass 3.3 times its above-ground biomass (Fig. 19.11a). Total nitrogen removed from these mesocosms was linearly correlated with total plant biomass (Fig. 19.11b). Based on key growth characteristics of the plants in this wastewater, three productive gramminoids (*Zizania latifolia*, *Glyceria maxima*, and *Phragmites australis*) had the highest overall scores. *Baumea articulata*, *Cyperus involucratus*, and *Schoenoplectus validus* had medium scores, while *Scirpus fluviatilis* and *Juncus effusus* had the lowest scores and are least likely to be effective plants in wastewater wetlands.

Establishing Vegetation

Vegetation can be established through the same general procedures outlined in Chapter 18—that is, by planting roots and rhizomes directly or by seeding. Because these wetlands are usually constructed on former upland with no connection to rivers or streams, reliance on nature bringing in plant propagules generally does not work. Field-harvested plants or nursery-grown stock can be used for plantings. The former have the advantage of establishing vegetation cover more quickly than would smaller nursery stock. Also, these plants, if harvested nearby, are adapted to the local climate and may be from the proper genotype for the region. Conversely, harvesting plants in large numbers from natural wetlands may threaten those wetlands. Nursery plants

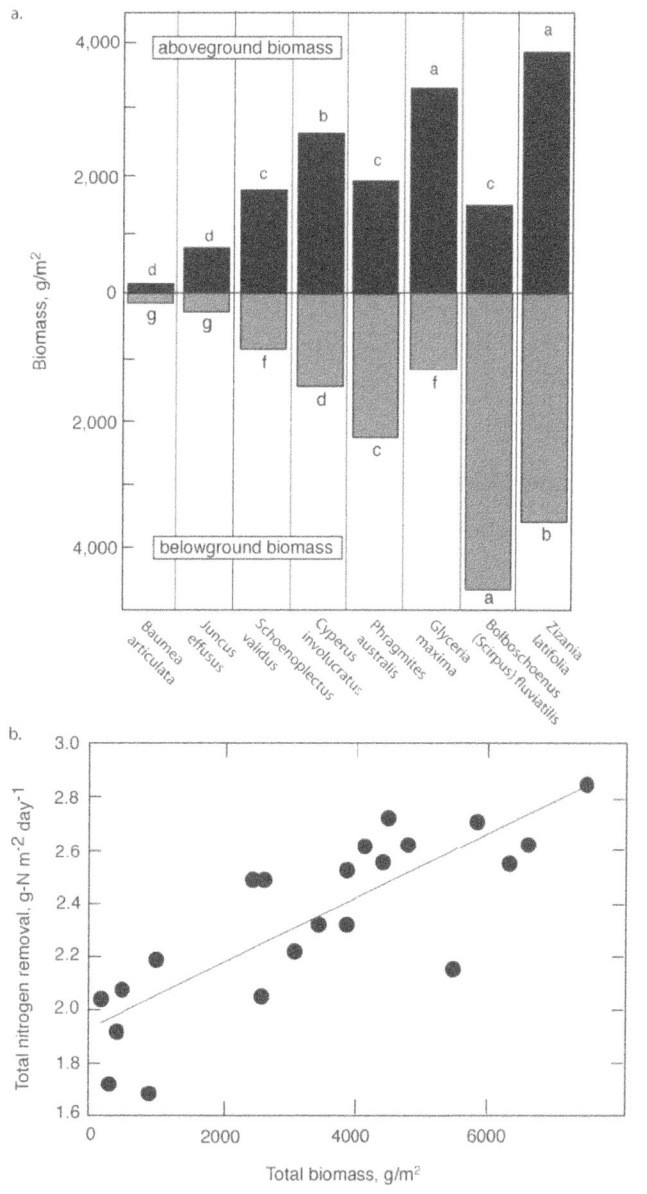

Figure 19.11 Results of a study comparing eight macrophytes commonly used in wastewater wetlands in New Zealand after 124 days of culture in dairy farm wastewater. (a) Mean above-ground and below-ground biomass accumulation of the eight macrophytes. Different letters indicate significant differences. (b) Total nitrogen removal from ammonium-rich dairy farm wastewater versus total plant biomass. Regression coefficient $r^2 = 0.66$. (After Tanner, 1996)

are easier to plant because of their generally small size, and a greater diversity and number of plants can be obtained from good nurseries. However, it is often unclear what genetic stock was used to start these plants, and they may not be from stock adapted to the local climate.

Water, either too much or too little, is the major reason why macrophytes do not become well established in wetlands constructed for wastewater treatment. When plants are first establishing themselves, the optimum conditions are moist soils or very shallow (<5 cm) water depth. If water is too deep, the new plants will be flooded out. If there is inadequate water and topsoil dries out, the plants will not survive. If the wastewater can be used in measured amounts to irrigate the plants, this is optimum. If not, artificial irrigation might be required to make sure that plants are successful.

Wetland Management after Construction

Wildlife Control

Although the development of wildlife is a welcomed and often desired aspect of treatment wetlands, managing plant and animal populations often becomes necessary maintenance of constructed wetlands. In North America, beavers (*Castor canadensis*) and muskrats (*Ondatra zibethicus*) create obstructions to inflows and outflows, destroy vegetation, or burrow into dikes. (This is one reason why dikes should not be built up around constructed wetlands if they can be avoided.) Major vegetation removal, particularly by herbivorous muskrats that use the plant material both for food and shelter, can turn a fully vegetated marsh into a plant-devoid pond in the matter of weeks or months. These events are referred to as eat-outs. There is very little that can be done to prevent these eat-outs except to trap and move the animals, which is a laborious task.

In other cases, animals such as beaver and muskrat and large birds such as Canada geese (*Branta canadensis*) and snow geese (*Chen* spp.) grazing on newly planted perennial herbs and seedlings are particularly destructive. The timing of planting is important, especially if migratory animals are involved in destructive grazing in the winter. Using gunshot devices and the extract of grape juice as a "hot foot" material on the adjacent landscape have all been suggested but without any permanent success. Probably the easiest approach we have noted in many years of observing geese is to have a wide band of emergent vegetation between where they land (on the water) and where they like to graze (upland lawns). But, of course, you will have to get the local muskrats to cooperate and not remove the vegetation.

Similarly, deeper wetlands often become havens for undesirable fish, such as carp (*Cyprinus carpio*), which can cause excessive turbidity and uproot vegetation. Carnivores such as northern pike (*Esox lucius*) have been discussed as a potential control of carp. Total removal of fish by drawdown is probably necessary if carp begin to degrade outflow water quality excessively. The problem is that this fish removal might affect mosquito control (see the following discussion).

Attracting Wildlife

Just as many animals can cause maintenance headaches, the attraction of wildlife to constructed wetlands is one of the reasons why public support for such projects can be high in the first place. So every attempt should be made to have a diverse ecosystem and not just a pond with water flowing through it. Weller (1994) recommends a 50:50 ratio of open water to vegetation cover in marshes to attract water birds, and this ratio, with proper development of the initial bathymetry of the ponds, is quite easy. Also, creating diverse habitats with live and dead vegetation, islands, and floating structures is desirable.

In many cases of wetland construction, wildlife enhancement begins soon after construction. At a constructed wetland at Pintail Lake in Arizona, the area's waterfowl population increased dramatically by the second year of use; duck nest density increased 97 percent over the first year (Wilhelm et al., 1989). A considerable increase in avian activity was also noted at the Des Plaines River Wetlands Demonstration Project in northeastern Illinois. Migrating waterfowl increased from 3 to 15 species and from 13 to 617 individuals between 1985 (preconstruction) and 1990 (one year after water was introduced to the wetlands). The number of wetland-dependent breeding birds increased from 8 to 17 species, and two state-designated endangered birds, the least bittern and the yellow-headed blackbird, nested at the site after wetland construction (Hickman and Mosca, 1991). At the Olentangy River Wetland Research Park in Ohio, a total of 174 bird species were listed for the site approximately 15 years after the initial wetlands were created. This count was high because part of the site included approximately 7 ha of bottomland forest on its eastern edge and an urban tree-dominated cemetery was on its western border.

One interesting question related to attracting birds to treatment wetlands is whether the birds might have an effect on the treatment capacity of the wetlands and, specifically, if birds are in high numbers, whether their excreta could undermine the effectiveness of the wetlands for nutrient and organic removals. Anderson et al. (2003) presented a several-year study done on a 10-ha treatment wetland in northern California on that possible effect. Bird use peaked at 12,000 individuals during the second year during a four-month period. Average daily inputs by birds of $2.5\,g\,N\,m^{-2}$ and $0.9\,g\,P\,m^{-2}$ were found, which represented less than 10 percent of the mean daily loading rates to the wetland. They concluded that bird use "does not lead to a significant reduction in treatment performance."

Mosquito Control

The subject of mosquito control will always be brought up when wetlands are being constructed, particularly when the wetlands receive runoff or wastewater. In general, it has been concluded that properly managed wastewater treatment wetlands pose no more mosquito threat than do natural wetlands (Knight et al., 2003). Mosquitoes can be controlled in constructed wetlands by changing the hydrologic conditions of the wetlands to inhibit mosquito larvae development (flowthrough conditions discourage mosquitoes) or by using chemical or biological control. Many researchers have proposed mosquito control by fish, especially the air-gulping mosquito fish (*Gambusia affinis*) or similar small fish. One reason to maintain some deeper areas in wetlands in temperate zones is to allow fish such as *Gambusia* and other top minnows and sunfish to survive the winter and feed on mosquito larvae. Little is known about the role that water quality has on encouraging or discouraging mosquitoes directly, but the effect of poor water quality by removing fish can have a dramatic effect in causing mosquito population increases. Bacterial insecticides (e.g., *Bacillus sphaericus* and *Bacillus thuringiensis* var. *israelensis*) and the fungus *Lagenidium giganteum* are known pathogens of mosquito larvae, but they have not been tested extensively, and there is always the possibility of resistance induction in mosquitoes (Knight et al. 2003). Constructing boxes to encourage nesting by swallows (Hirundinidae), swifts (Apodidae), and bats (Chiroptera) have also been used to control adult mosquito populations at constructed wetlands.

Some studies have evaluated the relative importance of different macrophyte species on the propensity of mosquito survival. In general, the denser the plant stands, the more difficult it is for both predators and mosquito control efforts to reach the mosquitoes. Thus, the highly productive plants (*Typha*, several *Schoenoplectus* = *Scirpus*, *Phragmites*, and *Eichhornia crassipes*) have the highest mosquito scores in Table 19.8. Knight et al. (2003) suggested the following strategy for treatment wetland design to minimize mosquitoes:

- Select plant species that optimize both wastewater treatment performance and mosquito control.
- Include deepwater zones that are free of emergent and aquatic plants to provide fish habitat and access to vegetated areas.
- Limit the width of emergent plant zones to facilitate access by predaceous fish and for application of chemical control agents.
- Design wetlands with steep embankments. (Although this is effective for mosquito control, it is not a good strategy to develop a diverse littoral zone around the wetland.)

Pathogens

Because many treatment wetlands are built specifically to deal with human and animal wastewater, proper sanitary engineering techniques should be used to minimize human exposure to pathogens. Treatment wetlands are meant to be biologically rich systems, and microbial activity is a major part of the treatment process. Measurements of indicator organisms, such as fecal and total coliforms, should be part of the monitoring of municipal wastewater treatment wetlands. Nearby wells should also be sampled, because water seeping from a wastewater wetland near potable water supplies should be monitored carefully. If a wetland is being used as tertiary treatment to a conventional treatment plant, the design of the disinfection system in the conventional treatment plant must be considered. Chlorine disinfection and the resulting

Table 19.8 Estimated mosquito production propensity of various wetland plant species. Scores less than 9 indicate minimal mosquito breeding problems; scores between 9 and 13 indicate a need to maintain a low coverage for this plant species; and scores of 14 and above indicate a need to minimize the occurrence of the plant species in the wetland to avoid mosquito issues.

Plant Group	Plant Species	Common Name	Mosquito Production Score[7]
Rooted emergent plants			
	Alisma geyeri	Water-plantain	7
	Alisma trivale	Water-plantain	7
	Alopercurus howellii	Foxtail	9
	Carex obnupta	Sedge	11
	Carex rostrata	Sedge	14
	Carex stipata	Sedge	13
	Cyperus aristatus	Flat sedge	9
	Cyperus difformis	Flat sedge	11
	Cyperus esculentus	Flat sedge	13
	Cyperus niger	Flat sedge	12
	Deschampsia danthonides	Grass	11
	Echinochloa crusgalli	Barnyard grass	11
	Echinodorus berteroi	Burhead	10
	Eleocharis palustris	Spikerush	10
	Equisetum arvense	Horsetail	14
	Frankenia grandifolia	Alkali heath	14
	Glyceria leptostachya	Mannagrass	12
	Juncus acutus	Softrush	13
	Juncus effusus	Softrush	10
	Jussiaea repens	Primrose	16
	Leersia oryzoides	Rice cutgrass	11
	Leptochloa fasicularis	Salt-meadow grass	10
	Ludwigia spp.	Primrose willow	9
	Lythrum californicum	Loosestrife	13
	Oryza sativa	Rice	9
	Phalaris arundinacea	Reed canary grass	14
	Phragmites australis	Common reed	17
	Plantago major	Common plantain	9
	Polygonum amphibium	Water smartweed	14
	Polygonum hydropiperoides	Smartweed	12
	Polygonum pennsylvanicum	Pinkweed	12
	Polygonum punctatum	Smartweed	12
	Polypogon elongatus	Rabbitfoot grass	11
	Potentilla palustris	Cinquefoil	11
	Pterididum aquilinum	Fern	13
	Sagittaria latifolia	Duck-potato	7
	Sagittaria longiloba	Arrowhead	7
	Sagittaria montevidensis	Giant arrowhead	8
	Scirpus acutus	Bulrush	15
	Scirpus americanus	Three-square bulrush	10

(continued)

Table 19.8 (Continued)

Plant Group	Plant Species	Common Name	Mosquito Production Score[7]
	Scirpus californicus	Giant bulrush	15
	Scirpus olneyi	Alkali bulrush	12
	Sparganium eurycarpum	Burreed	13
	Typha angustifolia	Narrowleaf cattail	16
	Typha glauca	Cattail	16
	Typha latifolia	Common cattail	17
	Zizania aquatica	Wildrice	13
Floating aquatic plants			
	Azolla filiculoides	Water fern	10
	Bacopa nobsiana	Water hyssop	13
	Brasenia schreberi	Water shield	12
	Eichhornia crassipes	Water hyacinth	18
	Hydrocotyle ranunculoides	Pennywort	15
	Hydrocotyle umbellata	Pennywort	15
	Lemna gibba	Duckweed	9
	Lemna minima	Duckweed	9
	Nasturtium officinale	Water cress	15
	Nuphar polysepalum	Spatterdock	11
	Pistia stratiotes	Water lettuce	18
	Potamogeton crispus	Curled pondweed	8
	Potamogeton diversifolius	Pondweed	8
	Ranunculus aquatilis	Buttercup	16
	Ranunculus flammula	Buttercup	15
	Spirodela polyhyiza	Duckmeat	9
	Wolffiella lingulata	Bog mat	9
Submerged aquatic plants			
	Callitriche longipedunculata	Water starwort	11
	Ceratophyllum demersum	Coontail	15
	Eleocharis acicularis	Spikerush	8
	Elodea canadensis	Waterweed	8
	Elodea densa	Waterweed	11
	Isoetes howellii	Quillwort	7
	Isoetes orcuttii	Quillwort	7
	Lilaeopsis occidentalis	Lilaeosis	7
	Myriophyllum spicatum	Water milfoil	14
	Najas flexilis	Naiad	11
	Najas graminea	Naiad	11
	Potamogeton filiformis	Pondweed	13
	Potamageton pectinatus	Sago pondweed	13
	Ruppia spiralis	Ditchgrass	11
	Utricularia gibba	Bladderwort	12
	Utricularia vulgaris	Bladderwort	13
	Zannichellia palustris	Horned pondweed	10

Source: Knight et al., 2003, and Collins and Resh, 1989.

chlorine residual would cause significant problems in treatment wetlands, so other means of disinfection (ozonation or ultraviolet radiation) should be used if disinfection is required before the wastewater enters the wetland.

Water-Level Management

The water level of surface-flow treatment wetlands is the key to both water quality enhancement and vegetation success. Most constructed municipal wastewater wetlands have little control on the overall inflow of wastewater. Flow and depth are first controlled by designing the basin large enough to create the proper HLR. Most constructed wetlands have a control structure, such as a flume or weir, to control outflow; these structures should be flexibly designed so they can be manipulated to control water depth. Too much water stresses macrophytes as much as too little water. Water depths of 30 cm or less are optimum for most herbaceous macrophytes used in treatment wetlands. Water depths greater than 30 cm can lead to vegetation reduction.

Compounding the effect that water level has on vegetation is the effect that it has on wastewater treatment. Deep water favors a high HLR and sediment and phosphorus retention associated with sedimentation and similar processes; it also leads to less resuspension, longer retention time, more organic matter accumulation, and lower redox conditions in the bottom waters. Shallow water leads to closer proximity of sediments and overlying water and more oxygen in the soil. Optimizing wastewater treatment and vegetation success is a continual balancing act.

Greenhouse Gas Emissions

Some concern has been expressed about treatment wetlands because of their emission of greenhouse gases, particularly methane (CH_4) In a survey of 158 published papers on constructed wetlands, median carbon dioxide (CO_2) emissions were significantly lower in surface-flow treatment wetlands (840 g-C m^{-2} yr^{-1}) than in subsurface flow wetlands (1200 g-C m^{-2} yr^{-1}) (Mander et al., 2014). Median methane emissions ranged from 35 g-C m^{-2} yr^{-1} from surface-water wetlands to 56 g-C m^{-2} yr^{-1} from horizontal subsurface wetlands. The CO_2 emissions from the treatment wetland soils are probably more than compensated for by vegetation productivity and soil carbon sequestration. The methane emissions are comparable to the rates presented in Chapter 17 (see Table 17.3), where an average methane emission rate of several temperate zone wetlands was reported to be 58 g-C m^{-2} yr^{-1}. Mander et al. (2014) pointed out that the methane emissions from treatment wetlands are one to two orders of magnitude lower than those coming from conventional wastewater treatment plants. Nitrous oxide (N_2O) emissions in this study had mediums of 0.8, 1.0, and 1.1 g-N m^{-2} yr^{-1}, respectively, for surface-water, vertical subsurface, and horizontal subsurface treatment wetlands. These low N_2O emissions were theorized by Mander et al. (2014) to be mostly residuals from denitrification, which mostly resulted in emissions of inert N_2 gas.

Economics and Values of Treatment Wetlands

It is generally believed that treatment wetlands are less expensive to build and maintain than conventional wastewater treatment, and that is the appeal of these systems to many people. However, cost comparisons should be made carefully before investing in these systems. Any estimate of the cost of a new wetland's development should include these four items:

1. Engineering plan
2. Preconstruction site preparation
3. Construction costs (e.g., labor, equipment, materials, supervision, indirect and overhead charges)
4. Cost of land

Capital Costs

An equation estimating the cost of constructing wetlands in general, including several wetlands that are not wastewater wetlands (not including the cost of land) is:

$$C_A = 196 A^{-0.511} \tag{19.6}$$

where

C_A = capital cost of wetland construction per unit area, U.S.$ × 1,000 ha^{-1}
A = area of wetland, ha

This relationship suggests that a 1-ha wetland would cost almost $200,000, a 10-ha wetland would cost $60,000 per ha, and a 100-ha wetland would cost $19,000 per ha. The data clearly suggest that there is an economy of scale involved in wetland construction. This equation included all types of created and constructed wetlands, not only treatment wetlands.

Kadlec (2009a) compared the costs of 92 surface water wetlands and 63 horizontal subsurface wetlands and came up with the relationships shown in Figure 19.12 and listed here:

$$\text{surface water wetlands } C = 194 A^{0.690} \quad 0.03 \text{ ha} < A < 10,000 \text{ ha} \tag{19.7}$$

and

$$\text{subsurface wetlands } C = 652 A^{0.704} \quad 0.005 \text{ ha} < A < 20 \text{ ha} \tag{19.8}$$

where

C = capital cost, U.S.$ × 1,000 (2006 $)
A = area of wetland, ha

Operating and Maintenance Costs

Operating and maintenance costs vary according to the wetland's use and to the amount and complexity of mechanical parts and plumbing that the wetland contains.

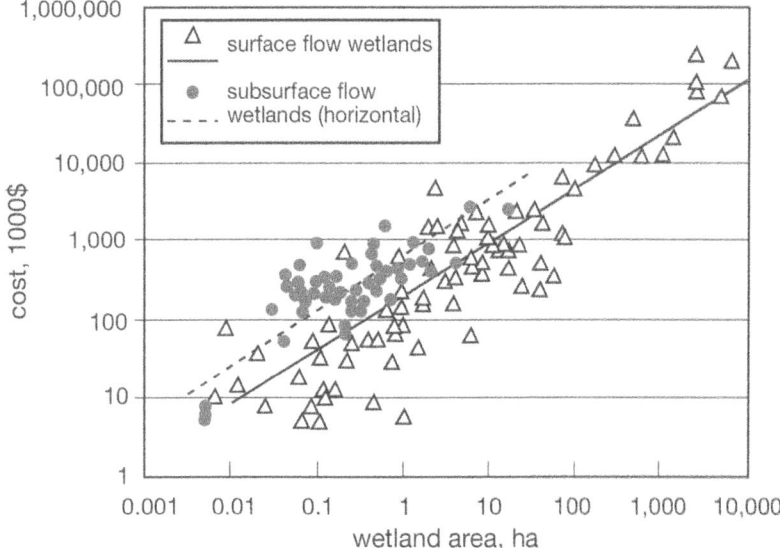

Figure 19.12 Costs of treatment wetlands versus wetland size for surface water wetlands and horizontal subsurface wetlands based on data from 92 surface water wetlands and 63 subsurface wetlands. (After Kadlec, 2009a)

Fewer data on operational costs are available. Kadlec and Knight (1996) estimated the operation and maintenance costs for one wastewater wetland to be about $85,500 per year. That estimate included $50,000 per year for personnel to be in charge of the 175-ha wetland. A wide range of $5,000 to $50,000 per year of operating and maintenance costs was estimated by those authors from smaller wetlands. Gravity-fed wetlands are far less expensive to maintain than highly mechanized wetlands that need significant plumbing and pumps. With current monitoring and legal reporting requirements for treatment wetlands, it is likely that current operating and maintenance costs would be at least $50,000 to $100,000 per year. This estimate would probably not include animal and vector control and unclogging (for subsurface wetlands) activities, which can be costly (Kadlec, 2009a).

Other Benefits of Treatment Wetlands

Subsurface treatment wetlands provide little additional benefit beyond the water quality improvement they were designed to provide, but surface-flow treatment wetlands have a variety of additional benefits. The watery habitat that is created can be a major ancillary benefit of these systems. In addition to providing habitat for mammals such as nutria, beavers, muskrats, amphibians, fish, and voles surface-flow treatment wetlands are often a haven for waterfowl and wading birds. Human uses, such as trapping and hunting, are not incompatible with some wastewater wetlands. If designed properly in an urban area, wetlands are locations where the public can visit and learn about their important water quality role. This message is a powerful one to the uninitiated, and they often become ardent wetland conservationists as a result of seeing wetlands at work.

Another benefit of using both natural and constructed wetlands for water quality improvement relates to areas where land-building is needed. In the subsiding environment of Louisiana's Gulf Coast, nutrients are permanently retained in the peat of wetlands receiving high-nutrient wastewater as the wetland aggrades to match subsidence. In this case, wastewater discharge into a wetland can occur without saturating the system and simultaneously helps counteract the deleterious effects of land subsidence.

Comparing Wetlands and Conventional Technology

A comparison of the construction and operating costs of a proposed large (>2,000 ha) wetland that was to be constructed in the Florida Everglades with conventional chemical treatment is illustrated in Table 19.9. In this example, if land costs are not considered, the wetland alternative has an 11 percent lower capital cost and a 56 percent lower operating cost than the chemical treatment alternative. Although land costs can be significant for treatment wetlands, particularly in urban areas (in essence, solar energy is being substituted for fossil fuel energy), it is generally not appropriate to use the cost of land in comparison with technological solutions that require little land. This is because the land being used by the wetland can be sold after the life of the wetland is completed, while the salvage value of the worn-out equipment used for conventional treatment alternatives is generally zero.

One of the more clever calculations of the difference between using wetlands versus conventional mechanical systems for wastewater treatment is an illustration of the

Table 19.9 Estimated cost comparison for phosphorus control in 760,000 m^3 day^{-1} agricultural runoff wetland in Florida

	Treatment Wetland	Chemical Treatment
Land cost	$34,434,000	$2,140,000
Capital costs (land free)	$95,836,000	$108,260,000
Total annual operation/maintenance	$1,094,000	$2,490,000
O&M, present worth	$33,443,000	$76,153,000
Total present worth, without land cost	$129,279,000	$185,637,000
Total present worth, with land cost	$163,713,000	$187,777,000

Source: Kadlec and Knight 1996.

Table 19.10 Net atmospheric generation of carbon for a 3,800 m^3 day^{-1} wastewater treatment facility using treatment wetlands or conventional mechanical treatment

Carbon Flow	Treatment Wetland (metric tons C day^{-1})	Conventional Treatment (metric tons C day^{-1})
Atmospheric carbon from power generation	53	1,350
Carbon sequestration	−3	0
Net atmospheric carbon	50	1,350

Source: Ogden 1999.

relative impact on the emission of the greenhouse gas CO_2. Normalizing estimates for a $3,800 \, m^3 \, day^{-1}$ (1 million gal/day) flow of wastewater, mechanical treatment leads to 27 times more emission of CO_2 to the atmosphere than does a treatment wetland (Table 19.10). The wetland system, in fact, has the additional benefit of sequestering a small amount of carbon. Conventional wastewater treatment uses 3.9 kg of fossil fuel carbon to remove 1 kg of carbon; a wetland treatment system uses 0.16 kg of fossil fuel carbon to remove 1 kg of carbon (Ogden, 1999).

Summary Considerations

Wastewater treatment wetlands are not the solution to all water quality problems and should not be viewed as such. Many pollution problems, such as excessive BOD or metal contamination, may require more conventional approaches. Yet the fact that thousands of wetlands have been constructed around the world for pollution control attests to their importance and value. Several considerations, both technical and institutional, must be considered as treatment wetlands are designed and built.

Technical Considerations

1. Values of the wetlands, such as wildlife habitat, should be considered in any treatment wetland development.
2. Acceptable pollutant and hydrologic loadings must be determined for the use of wetlands in wastewater management. Appropriate loadings, in turn, determine the size of the wetland to be constructed. Overloading a constructed wetland can be worse than not building it at all.
3. All existing characteristics of local natural wetlands, including vegetation, geomorphology, hydrology, and water quality, should be well understood so that natural wetlands can be "copied" in the construction of treatment wetlands.
4. Particular care should be taken in the wetland design to address public health, including mosquito control and protection of groundwater resources.

Institutional Considerations

1. Wastewater treatment by wetlands often can serve the dual purposes of both wetland habitat development and wastewater treatment and recycling. The creation of treatment wetlands as mitigation for lost wetlands is still generally unacceptable because of the lack of sustainability and the high level of pollution of treatment wetlands compared to restored wetlands, but opportunities for dual use of wetlands should continue to be explored.
2. Many permit processes in governments do not recognize treatment wetland systems as alternatives for wastewater treatment. In these cases, experimental systems should first be established for a given region. Modification of requirements for granting permits for pilot wetlands is needed to make effective progress in developing approaches.

It is useful to remember that wetland design is an inexact science and that perturbation and biological change are the only things we can be sure of in these created ecosystems. Traditional engineering approaches to wastewater wetlands, without an appreciation of self-design in ecosystems, are sure to cause disappointment. If a treatment wetland continues to function according to its main goal—that is, improving water quality—changes in plant species should not be viewed as that significant unless invasive nonnative plants become dominant.

Recommended Readings

IWA Specialists Group on Use of Macrophytes in Water Pollution Control. 2000. *Constructed Wetlands for Pollution Control.* Scientific and Technical Report No. 8. London: International Water Association (IWA).

Kadlec, R., and S. Wallace. 2009. *Treatment Wetlands*, 2nd ed. Boca Raton, FL: CRC Press.

References

Ahn, C., and W. J. Mitsch. 2001. Chemical analysis of soil and leachate from experimental wetland mesocosms lined with coal combustion products. *Journal of Environmental Quality* 30: 1457–1463.

Ahn, C., W. J. Mitsch, and W. E. Wolfe. 2001. Effects of recycled FGD liner material on water quality and macrophytes of constructed wetlands: A mesocosm experiment. *Water Research* 35: 633–642.

Anderson, D. C., J. J. Sartoris, J. S. Thullen, and P. G. Reusch. 2003. The effects of bird use on nutrient removal in a constructed wastewater-treatment wetland. *Wetlands* 23: 423–435.

Arheimer, B., and H. B. Wittgren. 1994. Modelling the effects of wetlands on regional nitrogen transport. *Ambio* 23: 378–386.

Boustany, R. G., C. R. Crozier, J. M. Rybczyk, and R. R. Twilley. 1997. Denitrification in a south Louisiana wetland forest receiving treated sewage effluent. *Wetlands Ecology and Management* 4: 273–283.

Braskerud, B. C. 2002a. Factors affecting nitrogen retention in small constructed wetlands treating agricultural non-point source pollution. *Ecological Engineering* 18: 351–370.

Braskerud, B. C. 2002b. Factors affecting phosphorus retention in small constructed wetlands treating agricultural non-point source pollution. *Ecological Engineering* 19: 41–61.

Brix, H. 1987. Treatment of wastewater in the rhizosphere of wetland plants: The root-zone method. *Water Science and Technology* 19: 107–118.

Brix, H. 1994. Use of constructed wetlands in water pollution control: Historical development, present status, and future perspectives. *Water Science and Technology* 30: 209–223.

Brix, H. 1998. Denmark. In J. Vymazal, H. Brix, P. F. Cooper, M. D. Green, and R. Haberl, eds. *Constructed Wetlands for Wastewater Treatment in Europe.* Backhuys Publishers, Leiden, the Netherlands, pp. 123–152.

Brix, H., and H.-H. Schierup. 1989a. The use of aquatic macrophytes in water- pollution control. *Ambio* 18: 100–107.

Brix, H., and H.-H. Schierup. 1989b. Sewage treatment in constructed reed beds—Danish experiences. *Water Science and Technology* 21: 1655–1668

Brix, H., and C. A. Arias. 2005. The use of vertical flow constructed wetlands for on-site treatment of domestic wastewater: New Danish guidelines. *Ecological Engineering* 25: 491–500.

Brodie, G. A., D. A. Hammer, and D. A. Tomljanovich. 1988. An evaluation of substrate types in constructed wetland drainage treatment systems. In *Mine Drainage and Surface Mine Reclamation, Vol. 1: Mine Water and Mine Waste*. U.S. Department of the Interior, Pittsburgh, PA, pp. 389–398.

Chimney, M. J., and G. Goforth. 2006. History and description of the Everglades Nutrient Removal Project, a subtropical constructed wetland in south Florida (USA). *Ecological Engineering* 27: 268–278.

Collins, J. N., and V. H. Resh. 1989. *Guidelines for the ecological control of mosquitoes in non-tidal wetlands of the San Francisco Bay area*. California Mosquito Vector Control Association and University of California Mosquito Research Program, Sacramento, CA.

Comin, F. A., J. A. Romero, V. Astorga, and C. Garcia. 1997. Nitrogen removal and cycling in restored wetlands used as filters of nutrients for agricultural runoff. *Water Science and Technology* 35: 255–261.

Conway, T. E., and J. M. Murtha. 1989. The Iselin Marsh Pond meadow. In D. A. Hammer, ed., *Constructed Wetlands for Wastewater Treatment*. Lewis Publishers, Chelsea, MI, pp. 139–144.

Cooke, J. G. 1992. Phosphorus removal processes in a wetland after a decade of receiving a sewage effluent. *Journal of Environmental Quality* 21: 733–739.

Cooper, P. F., and B. C. Findlater, eds. 1990. *Constructed Wetlands in Water Pollution Control*. Pergamon Press, Oxford, UK. 605 pp.

Cronk, J. K. 1996. Constructed wetlands to treat wastewater from dairy and swine operations: A review. *Agriculture, Ecosystems and Environment* 58: 97–114.

Davison, L., D. Pont, K. Bolton, and T. Headley. 2006. Dealing with nitrogen in subtropical Australia: Seven case studies in the diffusion of ecotechnological innovation. *Ecological Engineering* 28: 213–223.

Day, J. W., Jr., J. Ko, J., Rybczyk, D. Sabins, R. Bean, G. Berthelot, C. Brantley, L. Cardoch, W. Conner, J. N. Day, A. J. Englande, S. Feagley, E. Hyfield, R. Lane, J. Lindsey, J. Mistich, E. Reyes, and R. Twilley. 2004. The use of wetlands in the Mississippi delta for wastewater assimilation: A review. *Ocean and Coastal Management* 47: 671–691.

Day, J. W., Jr., R. Lane, M. Moerschbaecher, R. DeLaune, I. Mendelssohn, J. Baustian, and R. Twilley. 2013. Vegetation and soil dynamics of a Louisiana estuary receiving pulsed Mississippi River water following Hurricane Katrina. *Estuaries and Coasts* 36: 665–682.

Deegan, L. A., D. S. Johnson, R. S.Warren, B. J. Peterson, J. W. Fleeger, S. Fagherazzi, and W. M. Wollheim. 2012. Coastal eutrophication as a driver of salt marsh loss. *Nature* 490: 388–392

DeJong, J. 1976. The purification of wastewater with the aid of rush or reed ponds. In J. Tourbier and R. W. Pierson, Jr., eds., *Biological Control of Water Pollution*. University of Pennsylvania, Philadelphia, pp. 133–139.

Dierberg, F. E., and DeBusk, T. A. 2008. Particulate phosphorus transformations in south Florida stormwater treatment areas used for Everglades protection. *Ecological Engineering* 34, 100–115.

Entry, J. A., and A. Gottlieb. 2014. The impact of stormwater treatment areas and agricultural best management practices on water quality in the Everglades Protection Area. *Environmental Monitoring and Assessment* 186, 1023–1037.

Ewel, K. C., and H. T. Odum, eds. 1984. *Cypress Swamps*. University Presses of Florida, Gainesville. 472 pp.

Fennessy, M. S., and W. J. Mitsch. 1989. Treating coal mine drainage with an artificial wetland. *Research Journal of the Water Pollution Control Federation* 61: 1691–1701.

Fink, D. F., and W. J. Mitsch. 2004. Seasonal and storm event nutrient removal by a created wetland in an agricultural watershed. *Ecological Engineering* 23: 313–325.

Fink, D. F., and W. J. Mitsch. 2007. Hydrology and nutrient biogeochemistry in a created river diversion oxbow wetland. *Ecological Engineering.* 30: 93–162.

Gerheart, R. A. 1992. Use of constructed wetlands to treat domestic wastewater, city of Arcata, California. *Water Science and Technology* 26: 1625–1637.

Gerheart, R. A., F. Klopp, and G. Allen. 1989. Constructed free surface wetlands to treat and receive wastewater: Pilot project to full scale. In D. A. Hammer, ed., *Constructed Wetlands for Wastewater Treatment*. Lewis Publishers, Chelsea, MI, pp. 121–137.

Greenway, M. 2005. The role of constructed wetlands in secondary effluent treatment and water reuse in subtropical and arid Australia. *Ecological Engineering* 25: 501–509.

Greenway, M., P. Dale, and H. Chapman. 2003. An assessment of mosquito breeding and control in four surface flow wetlands in tropical-subtropical Australia. *Water Science & Technology* 48: 249–256.

Gu, B., M. J. Chimney, J. Newman, and M. K. Nungesser. 2006. Limnological characteristics of a subtropical constructed wetland in south Florida (USA). *Ecological Engineering* 27: 345–360.

Headley, T. R., E. Herity, and L. Davison. 2005. Treatment at different depths and vertical mixing within a 1-m deep horizontal subsurface-flow wetland. *Ecological Engineering* 25: 567–582.

Hickman, S. C., and V. J. Mosca. 1991. Improving habitat quality for migratory waterfowl and nesting birds: Assessing the effectiveness of the Des Plaines River wetlands demonstration project. Technical Paper 1, Wetlands Research, Chicago. 13 pp.

Hoagland, C. R., L. E. Gentry, M. B. David, and D. A. Kovacic. 2001. Plant nutrient uptake and biomass accumulation in a constructed wetland. *Journal of Freshwater Ecology* 16: 527–540

Jacks, G., A. Joelsson, and S. Fleischer. 1994. Nitrogen retention in forested wetlands. *Ambio* 23: 358–362.

Jackson, J. 1989. Man-made wetlands for wastewater treatment: Two case studies. In D. A. Hammer, ed., *Constructed Wetlands for Wastewater Treatment*. Lewis Publishers, Chelsea, MI, pp. 574–580.

Juston, J., and DeBusk, T. A. 2006. Phosphorus mass load and outflow concentration relationships in stormwater treatment areas for Everglades restoration. *Ecological Engineering* 26, 206–223.

Juston, J., and DeBusk, T. A. 2011. Evidence and implications of the background phosphorus concentration of submerged aquatic vegetation wetlands in stormwater treatment areas for Everglades restoration. *Water Resources Research* 41, 430–446.

Kadlec, R. H. 1999. Constructed wetlands for treating landfill leachate. In G. Mulamoottil, E. A. McBean, and F. Rovers, eds., *Constructed Wetlands for the Treatment of Landfill Leachates*. Lewis Publishers, Boca Raton, FL, pp. 17–31.

Kadlec, R. H. 2006. Free surface wetlands for phosphorus removal: The position of the Everglades Nutrient Removal Project. 2006. *Ecological Engineering* 27: 361–379.

Kadlec, R. H. 2009a. Comparison of free water and horizontal subsurface treatment wetlands. *Ecological Engineering* 35: 159–174.

Kadlec, R. H. 2009b. Wastewater treatment at the Houghton Lake wetland: Hydrology and water quality. *Ecological Engineering* 35: 1287–1311.

Kadlec, R. H., and D. L. Hey. 1994. Constructed wetlands for river water quality improvement. *Water Science and Technology* 29: 159–168.

Kadlec, R. H., and R. L. Knight. 1996. *Treatment Wetlands*. CRC Press/Lewis Publishers, Boca Raton, FL. 893 pp.

Kadlec, R. H., and S. Wallace. 2009. *Treatment Wetlands*, 2nd ed. CRC Press, Boca Raton, FL, 1016 pp.

Kearney, M. S., J. C. A. Riter, and R. E Turner. 2011. Freshwater river diversions for marsh restoration in Louisiana: Twenty-six years of changing vegetative cover and marsh area. *Geophysical Research Letters* 38, L16405. doi:10.1029/2011GL047847.

Knight, R. L. 1990. Wetland systems. In *Natural Systems for Wastewater Treatment, Manual of Practice FD-16*. Water Pollution Control Federation, Alexandria, VA, pp. 211–260.

Knight, R. L., T. W. McKim, and H. R. Kohl. 1987. Performance of a natural wetland treatment system for wastewater management. *Journal of the Water Pollution Control Federation* 59: 746–754.

Knight, R. L., V. W. E. Payne Jr., R. E. Borer, R. A. Clarke Jr., and J. H. Pries. 2000. Constructed wetlands for livestock wastewater management. *Ecological Engineering* 15: 41–55.

Knight, R. L., W. E. Walton, G. F. O'Meara, W. K. Reisen, and R. Wass. 2003. Strategies for effective mosquito control in constructed treatment wetlands. *Ecological Engineering* 21: 211–232.

Kovacic, D. A., M. B. David, L. E. Gentry, K. M. Starks, and R. A. Cooke. 2000. Effectiveness of constructed wetlands in reducing nitrogen and phosphorus export from agricultural tile drainage. *Journal of Environmental Quality* 29: 1262–1274.

Lane, R. R., J. W. Day, D. Justic, E. Reyes, J. N. Day, and E. Hyfield. 2004. Changes in stoichiometric Si, N and P ratios of Mississippi River water diverted through coastal wetlands to the Gulf of Mexico. *Estuarine Coastal & Shelf Science* 60: 1–10.

Lane, R. R., J. W. Day Jr., and J. N. Day. 2006. Wetland surface elevation, vertical accretion, and subsidence at three Louisiana estuaries receiving diverted Mississippi River water. *Wetlands* 26: 1130–1142.

Lane, R. R., J. W. Day Jr., B. D. Marx, E. Reyes, E. Hyfield, and J. N. Day. 2007. The effects of riverine discharge on temperature, salinity, suspended sediment and chlorophyll a in a Mississippi delta estuary measured using a flow-through system. *Estuarine Coastal & Shelf Science* 74: 145–154.

Larson, A. C., L. E. Gentry, M. B. David, R. A. Cooke, and D. A. Kovacic. 2000. The role of seepage in constructed wetlands receiving tile drainage. *Ecological Engineering* 15: 91–104.

Leonardson, L., L. Bengtsson, T. Davidsson, T. Persson, and U. Emanuelsson. 1994. Nitrogen retention in artificially flooded meadows. *Ambio* 23: 332–341.

Litchfield, D. K., and D. D. Schatz. 1989. Constructed wetlands for wastewater treatment at Amoco Oil Company's Mandan, North Dakota, refinery. In D. A. Hammer, ed., *Constructed Wetlands for Wastewater Treatment*. Lewis Publishers, Chelsea, MI, pp. 101–119.

Lundberg, C. J., R. R. Lane, and J. W. Day, Jr. 2014. Spatial and temporal variations in nutrients and water-quality parameters in the Mississippi River-influenced Breton Sound Estuary. *Journal of Coastal Research* 30: 328–336.

Mander, Ü, G. Dotro, Y. Ebie, S. Towprayoon, C. Chiemchaisri, S. F. Nogueira, B. Jamsranjav, K. Kasak, J. Truu, J. Tournebize, and W. J Mitsch. 2014. Greenhouse gas emission in constructed wetlands for wastewater treatment: a review. *Ecological Engineering* 66: 19–35.

Manyin, T., F. M. Williams, and L. R. Stark. 1997. Effects of iron concentration and flow rate on treatment of coal mine drainage in wetland mesocosms: An experimental approach to sizing of constructed wetlands. *Ecological Engineering* 9: 171–185.

Mitchell, D. S., A. J. Chick, and G. W. Rasin. 1995. The use of wetlands for water pollution control in Australia: An ecological perspective. *Water Science and Technology* 32: 365–373.

Mitsch, W. J. 2012. What is ecological engineering? *Ecological Engineering* 45: 5–12.

Mitsch, W. J., J. K. Cronk, X. Wu, R. W. Nairn, and D. L. Hey. 1995. Phosphorus retention in constructed freshwater riparian marshes. *Ecological Applications* 5: 830–845.

Mitsch, W. J., and K. M. Wise. 1998. Water quality, fate of metals, and predictive model validation of a constructed wetland treating acid mine drainage. *Water Research* 32: 1888–1900

Mitsch, W. J., X. Wu, R. W. Nairn, P. E. Weihe, N. Wang, R. Deal, and C. E. Boucher. 1998. Creating and restoring wetlands: A whole-ecosystem experiment in self-design. *BioScience* 48: 1019–1030.

Mitsch, W. J., A. J. Horne, and R. W. Nairn. 2000. Nitrogen and phosphorus retention in wetlands—Ecological approaches to solving excess nutrient problems. *Ecological Engineering* 14: 1–7.

Mitsch, W. J., and S. E. Jørgensen. 2004. *Ecological Engineering and Ecosystem Restoration*. John Wiley & Sons, Hoboken, NJ. 411 pp.

Mitsch, W. J., N. Wang, L. Zhang, R. Deal, X. Wu, and A. Zuwerink. 2005a. Using ecological indicators in a whole-ecosystem wetland experiment. In S. E. Jørgensen, F-L. Xu, and R. Costanza, eds., *Handbook of Ecological Indicators for Assessment of Ecosystem Health*. CRC Press, Boca Raton, FL, pp. 211–235.

Mitsch, W. J., J. W. Day Jr., L. Zhang, and R. Lane. 2005b. Nitrate-nitrogen retention by wetlands in the Mississippi River Basin. *Ecological Engineering* 24: 267–278.

Mitsch, W. J., L. Zhang, C. J. Anderson, A. Altor, and M. Hernandez. 2005c. Creating riverine wetlands: Ecological succession, nutrient retention, and pulsing effects. *Ecological Engineering* 25: 510–527.

Mitsch, W. J., L. Zhang, D. F. Fink, M. E. Hernandez, A. E. Altor, C. L. Tuttle and A. M. Nahlik. 2008. Ecological engineering of floodplains. *Ecohydrology & Hydrobiology* 8: 139–147.

Mitsch, W. J., L. Zhang, K. C. Stefanik, A. M. Nahlik, C. J. Anderson, B. Bernal, M. Hernandez, and K. Song. 2012. Creating wetlands: Primary succession, water quality changes, and self-design over 15 years. *BioScience* 62: 237–250.

Mitsch, W.J., L. Zhang, E. Waletzko, and B. Bernal. 2014. Validation of the ecosystem services of created wetlands: Two decades of plant succession, nutrient retention, and carbon sequestration in experimental riverine marshes. *Ecological Engineering* doi.org/10.1016/j.ecoleng.2014.09.108

Mitsch, W.J., L. Zhang, D. Marois, and K. Song. 2015. Protecting the Florida Everglades wetlands with wetlands: Can stormwater phosphorus be reduced to oligotrophic conditions? *Ecological Engineering*, in press.

Morris, J. T., G. P. Shaffer, and J. A. Nyman. 2013. Brinson Review: Perspectives on the influence of nutrients on the sustainability of coastal wetlands. *Wetlands* 33: 975–988.

Moustafa, M. Z. 1999. Nutrient retention dynamics of the Everglades nutrient removal project. *Wetlands* 19: 689–704.

Moustafa, M. Z., M. J. Chimney, T. D. Fontaine, G. Shih, and S. Davis. 1996. The response of a freshwater wetland to long-term "low level" nutrient loads—marsh efficiency. *Ecological Engineering* 7: 15–33.

Mulamoottil, G., E. A. McBean, and F. Rovers, eds. 1999. *Constructed Wetlands for the Treatment of Landfill Leachates*. Lewis Publishers, Boca Raton, FL. 281 pp.

Nahlik, A. M., and W. J. Mitsch. 2006. Tropical treatment wetlands dominated by free-floating macrophytes for water quality improvement in the Caribbean coastal plain of Costa Rica. *Ecological Engineering* 28: 246–257.

Nairn, R. W., and W. J. Mitsch. 2000. Phosphorus removal in created wetland ponds receiving river overflow. *Ecological Engineering* 14: 107–126.

Newman, J. M., J. C. Clausen, and J. A. Neafsey. 2000. Seasonal performance of a wetland constructed to process dairy milkhouse wastewater in Connecticut. *Ecological Engineering* 14: 181–198.

Newman, S., and K. Pietro. 2001. Phosphorus storage and release in response to flooding: Implications for Everglades stormwater treatment areas. *Ecological Engineering* 18: 23–38.

Nguyen, L. M. 2000. Phosphate incorporation and transformation in surface sediments of a sewage-impacted wetland as influenced by sediment sites, sediment pH and added phosphate concentration. *Ecological Engineering* 14: 139–155.

Nguyen, L. M., J. G. Cooke, and G. B. McBride. 1997. Phosphorus retention and release characteristics of sewage-impacted wetland sediments. *Water, Air, and Soil Pollution* 100: 163–179.

Niswander, S. F., and W. J. Mitsch. 1995. Functional analysis of a two-year-old created in-stream wetland: Hydrology, phosphorus retention, and vegetation survival and growth. *Wetlands* 15: 212–225.

Odum, H. T., K. C. Ewel, W. J. Mitsch, and J. W. Ordway. 1977. Recycling treated sewage through cypress wetlands in Florida. In F. M. D'Itri, ed., *Wastewater Renovation and Reuse*. Marcel Dekker, New York, pp. 35–67.

Ogden, M. H. 1999. *Constructed wetlands for small community wastewater treatment*. Paper presented at Wetlands for Wastewater Recycling Conference, November 3, 1999, Baltimore. Environmental Concern, St. Michaels, MD.

Paudel, R., J.-K. Min, and J. W. Jawitz. 2010. Management scenario evaluation for a large treatment wetland using a spatio-temporal phosphorus transport and cycling model. *Ecological Engineering* 36: 1627–1638.

Paudel, R., and J. W. Jawitz. 2012. Does increased model complexity improve description of phosphorus dynamics in a large treatment wetland? *Ecological Engineering* 42: 283–294.

Phipps, R. G., and W. G. Crumpton. 1994. Factors affecting nitrogen loss in experimental wetlands with different hydrologic loads. *Ecological Engineering* 3: 399–408.

Pietro, K. 2012. Synopsis of the Everglades stormwater treatment areas, water year 1996–2012. Technical Publication ASB-WQTT-12–001, South Florida Water Management District, West Palm Beach, FL. 56 pp.

Raisin, G. W., and D. S. Mitchell. 1995. The use of wetlands for the control of non-point source pollution. *Water Science and Technology* 32: 177–186.

Raisin, G. W., D. S. Mitchell, and R. L. Croome. 1997. The effectiveness of a small constructed wetland in ameliorating diffuse nutrient loadings from an Australian rural catchment. *Ecological Engineering* 9: 19–35.

Reddy, K. R., R. H. Kadlec, M. J. Chimney, and W. J. Mitsch, eds. 2006. The Everglades Nutrient Removal Project. *Special issue of Ecological Engineering* 27: 265–379.

Reinelt, L. E., and R. R. Horner. 1995. Pollutant removal from stormwater runoff by palustrine wetlands based on comprehensive budgets. *Ecological Engineering* 4: 77–97.

Richardson, C. J., and C. B. Craft. 1993. Effective phosphorus retention in wetlands—Fact or fiction? In G. A. Moshiri, ed., *Constructed Wetlands for Water Quality Improvement*. CRC Press, Boca Raton, FL, pp. 271–282.

Richardson, C. J., S. Qian, C. B. Craft, and R. G. Qualls. 1997. Predictive models for phosphorus retention in wetlands. *Wetlands Ecology and Management* 4: 159–175.

Rivera-Monroy, V. H., B. Branoff, E. Meselhe, A. McCorquodale, M. Dortch, G. D. Steyer, J. Visser, and H. Wang. 2013. Landscape-level estimation of nitrogen removal in coastal Louisiana wetlands: Potential sinks under different restoration scenarios. *Journal of Coastal Research* 67: 75–87.

Sartoris, J. J., J. S. Thullen, L. B. Barber, and D. E. Salas. 2000. Investigation of nitrogen transformations in a southern California constructed wastewater treatment wetland. *Ecological Engineering* 14: 49–65.

Schaafsma, J. A., A. H. Baldwin, and C. A. Streb. 2000. An evaluation of a constructed wetland to treat wastewater from a dairy farm in Maryland, USA. *Ecological Engineering* 14: 199–206.

Seidel, K. 1964. Abbau von Bacterium Coli durch höhere Wasserpflanzen. *Naturwissenschaften* 51: 395.

Seidel, K. 1966. Reinigung von Gewässern durch höhere Pflanzen. *Naturwissenschaften* 53: 289–297.

Seidel, K., and H. Happl. 1981. Pflanzenkläranlage "Krefelder system." *Sicherheit in Chemie und Umbelt* 1: 127–129.

Solano, M. L., P. Soriano, and M. P. Ciria. 2004. Constructed wetlands as a sustainable solution for wastewater treatment in small villages. *Biosystems Engineering* 87: 109–118.

Spieles, D. J., and W. J. Mitsch. 2000a. The effects of season and hydrologic and chemical loading on nitrate retention in constructed wetlands: A comparison of low and high nutrient riverine systems. *Ecological Engineering* 14: 77–91.

Spieles, D. J., and W. J. Mitsch. 2000b. Macroinvertebrate community structure in high- and low-nutrient constructed wetlands. *Wetlands* 20: 716–729.

Stark, L. R., and F. M. Williams. 1995. Assessing the performance indices and design parameters of treatment wetlands for H, Fe, and Mn retention. *Ecological Engineering* 5: 433–444.

Steiner, G. R., J. T. Watson, D. Hammer, and D. F. Harker, Jr. 1987. Municipal wastewater treatment with artificial wetlands—A TVA/Kentucky demonstration. In K. R. Reddy and W. H. Smith, eds., *Aquatic Plants for Wastewater Treatment and Resource Recovery*. Magnolia Publishing, Orlando, FL, p. 923.

Steiner, G. R., and R. J. Freeman, Jr. 1989. Configuration and substrate design considerations for constructed wetlands for wastewater treatment. In D. A. Hammer, ed., *Constructed Wetlands for Wastewater Treatment*. Lewis Publishers, Chelsea, MI, pp. 363–378.

Tanner, C. C. 1996. Plants for constructed wetland treatment systems—A comparison of the growth and nutrient uptake of eight emergent species. *Ecological Engineering* 7: 59–83.

Tanner, C. C., J. S. Clayton, and M. P. Upsdell. 1995. Effect of loading rate and planting on treatment of dairy farm wastewaters in constructed wetlands. II. Removal of nitrogen and phosphorus. *Water Research* 29: 27–34.

Tarutis, W. J., L. R. Stark, and F. M. Williams. 1999. Sizing and performance estimation of coal mine drainage wetlands. *Ecological Engineering* 12: 353–372.

Teal, J. M., R. Best, J. Caffrey, C. S. Hopkinson, K. L. McKee, J. T. Morris, S. Newman, and B. Orem. 2012. Mississippi River freshwater diversions in southern Louisiana: Effects on wetland vegetation, soils, and elevation. Final Report to the State of Louisiana and the U.S. Army Corps of Engineers through the Louisiana Coastal Area Science & Technology Program and National Oceanic and Atmospheric Administration. 49 pp.

Thullen, J. S., J. J. Sartoris, and S. M. Nelson. 2005. Managing vegetation in surface-flow wastewater-treatment wetlands for optimal treatment performance. *Ecological Engineering* 25: 583–593.

Teiter, S., and Ü. Mander. 2005. Emission of N_2O, N_2, CH_4, and CO_2 from constructed wetlands for wastewater treatment and from riparian buffer zones. *Ecological Engineering* 25: 528–541

Vymazal, J. 1995. Constructed wetlands for wastewater treatment in the Czech Republic—state of the art. *Water Science and Technology* 32: 357–364.

Vymazal, J. 1998. Czech Republic. In J. Vymazal, H. Brix, P. F. Cooper, M. B. Green, and R. Haberl, eds., *Constructed Wetlands for Wastewater Treatment in Europe*. Backhuys Publishers, Leiden, the Netherlands, pp. 95–121.

Vymazal, J. 2002. The use of sub-surface constructed wetlands for wastewater treatment in the Czech Republic: 10 years experience. *Ecological Engineering* 18: 633–646.

Vymazal, J., ed. 2005. Constructed wetlands for wastewater treatment. *Special Issue of Ecological Engineering* 25: 475–621.

Vymazal, J. 2013. Emergent plants used in free water surface constructed wetlands: A review. *Ecological Engineering* 61P: 582–592.

Vymazal, J., and L. Kropfelova. 2005. Growth of *Phragmites australis* and *Phalaris arundinacea* in constructed wetlands for wastewater treatment in the Czech Republic. *Ecological Engineering* 25: 606–621.

Weller, M. W. 1994. *Freshwater Marshes*, 3rd ed. University of Minnesota Press, Minneapolis. 192 pp.

White, J. S., S. E. Bayley, and P. J. Curtis. 2000. Sediment storage of phosphorus in a northern prairie wetland receiving municipal and agro-industrial wastewater. *Ecological Engineering* 14: 127–138.

Wieder, R. K., and G. E. Lang. 1984. Influence of wetlands and coal mining on stream water chemistry. *Water, Air, and Soil Pollution* 23: 381–396.

Wilhelm, M., S. R. Lawry, and D. D. Hardy. 1989. Creation and management of wetlands using municipal wastewater in northern Arizona: A status report. In D. A. Hammer, ed., *Constructed Wetlands for Wastewater Treatment*. Lewis Publishers, Chelsea, MI, pp. 179–185.

Appendix A

Wetland Losses by State in the United States, 1780s–1980s

State	Original Wetlands Circa 1780 (x1,000 ha)	National Wetlands Inventory, mid-1980s (x1,000 ha)	Change (percent)
Alabama	3,063	1,531	−50
Alaska	68,799	68,799	−0.1
Arizona	377	243	−36
Arkansas	3,986	1,119	−72
California	2,024	184	−91
Colorado	809	405	−50
Connecticut	271	70	−74
Delaware	194	90	−54
Florida	8,225	4,467	−46
Georgia	2,769	2,144	−23
Hawaii	24	21	−12
Idaho	355	156	−56
Illinois	3,323	508	−85
Indiana	2,266	304	−87
Iowa	1,620	171	−89
Kansas	340	176	−48
Kentucky	634	121	−81
Louisiana	6,554	3,555	46
Maine	2,614	2,104	−19
Maryland	668	178	−73
Massachusetts	331	238	−28
Michigan	4,533	2,259	50
Minnesota	6,100	3,521	−42
Mississippi	3,995	1,646	59
Missouri	1,960	260	−87

(continued)

Appendix A Wetland Losses by State in the United States, 1780s–1980s

State	Original Wetlands Circa 1780 (x1,000 ha)	National Wetlands Inventory, mid-1980s (x1,000 ha)	Change (percent)
Montana	464	340	−27
Nebraska	1,178	771	−35
Nevada	197	96	−52
New Hampshire	89	81	−9
New Jersey	607	370	−39
New Mexico	291	195	−33
New York	1,037	415	−60
North Carolina	4,488	2,300	−44
North Dakota	1,994	1,008	−49
Ohio	2,024	195	−90
Oklahoma	1,150	384	−67
Oregon	915	564	−38
Pennsylvania	456	202	−56
Rhode Island	42	26	−37
South Carolina	2,596	1,885	−27
South Dakota	1,107	720	−35
Tennessee	784	318	−59
Texas	6,475	3,080	−52
Utah	325	226	−30
Vermont	138	89	−35
Virginia	748	435	−42
Washington	546	380	−31
West Virginia	54	41	−24
Wisconsin	3,966	2,157	−46
Wyoming	809	506	−38
Total wetlands	158,395	111,060	−30
Total "lower 48"	89,491	42,240	−53

Source: Dahl, T. E. 1990. Wetlands losses in the United States, 1780s to 1980s. U.S. Department of Interior, Fish and Wildlife Service, Washington, DC. 21 pp.

Appendix B

Useful Wetland Web Pages

International

U.S. National Ramsar Committee, http://usnrc.net

The Ramsar Convention on Wetlands, www.ramsar.org

Wetlands International, www.wetlands.org

International Peat Society, www.peatsociety.org

International Association of Ecology (INTECOL), www.intecol.net/pages/index.php

United Nations Environment Programme, www.unep.org

IPCC, Intergovernmental Panel on Climate Change, www.ipcc.ch

The International Union for Conservation of Nature (IUCN), www.iucn.org

Wetland Restoration

The Marshlands of Mesopotamia, http://whc.unesco.org/en/tentativelists/1838/ The Bois-des-Bel experimental peatlands, Quebec, Canada www.gret-perg.ulaval.ca/recherche/themes-de-recherche/diversite-floristique/bois-des-bel

The Skjern River, Denmark www.globalrestorationnetwork.org/database/case-study/?id=115

The Comprehensive Everglades Restoration Plan (CREP) www.nps.gov/ever/naturescience/cerp.htm; www.evergladesplan.org

Florida Uniform Mitigation Assessment Method (UMAM) www.dep.state.fl.us/water/wetlands/mitigation/umam http://sfrc.ufl.edu/ecohydrology/UMAM_Training_Manual_ppt.pdf

Ohio Rapid Assessment Method (ORAM), www.epa.state.oh.us/dsw/401/ecology.aspx

State of Washington Wetland Rating System http://www.ecy.wa.gov/programs/sea/wetlands/ratingsystems/2014updates.html

University Wetland Programs

Everglades Wetland Research Park, FGCU, www.fgcu.edu/swamp

Duke University Wetland Center, http://nicholas.duke.edu/wetland

Howard T. Odum Center for Wetlands, University of Florida, http://cfw.essie.ufl.edu
LSU School of the Coast and Environment, www.sce.lsu.edu/about

U.S. Government
USGS National Wetlands Research Center, www.nwrc.usgs.gov
US EPA Wetlands, http://water.epa.gov/type/wetlands
USDA/NRCS Wetlands
   General: www.nrcs.usda.gov/wps/portal/nrcs/main/national/water/wetlands
   Plants: https://plants.usda.gov/core/wetlandSearch
   Soils: www.nrcs.usda.gov/wps/portal/nrcs/main/soils/use/hydric
U.S. Fish and Wildlife Service National Wetlands Inventory, www.fws.gov/wetlands
U.S. Army Corps of Engineers, www.usace.army.mil/Missions/CivilWorks/RegulatoryProgramandPermits.aspx

Societies/NGOs
Ducks Unlimited, www.ducks.org
Society of Wetland Scientists, www.sws.org
Association of State Wetland Managers, www.aswm.org
The Nature Conservancy, www.nature.org
Society for Ecological Restoration, www.ser.org
American Ecological Engineering Society, www.ecoeng.org
National Audubon Society, www.audubon.org

Scientific Journals
Ecological Engineering, www.journals.elsevier.com/ecological-engineering
Wetlands, www.sws.org/Publications/wetlands-journal.html
Wetlands Ecology and Management, http://link.springer.com/journal/11273
Estuaries and Coasts, http://link.springer.com/journal/12237

Adventure/Entertainment
"Bill and Ulo's Excellent Adventure" (in the Okavango Delta, Botswana), www.youtube.com/watch?v=ORg97C0zuRM

Mega Python vs. Gatoroid trailer, www.youtube.com/watch?v=S8pirKwkxb0

Swamp Thing trailer, YouTube video, Image Entertainment, Inc., uploaded on Apr 12, 2011, www.youtube.com/watch?v=kzbqK4nw3R8

"Olentangy River Wetland Research Park," YouTube video, WOSU Public Media (www.wosu.org), uploaded December 2, 2008, www.youtube.com/watch?v=KL-34AZPprE

Wetland and Carbon lecture—Bill Mitsch, Montpellier France, October 25, 2013, YouTube video, Everglades Wetland Research Park, uploaded December 4, 2013, www.youtube.com/watch?v=AaCUUSOg1xs

"Wetlands" by Bill Mitsch, Bonita Springs, Florida, April 4, 2013. YouTube Video Produced by Charlotte Harbor National Estuarine Program Office, www.youtube.com/watch?v=MYZGXXwlIOo&feature=youtu.be.

Appendix C

Useful Conversion Factors

Multiply	By	To Obtain
LENGTH		
centimeters (cm)	0.3937	inches
feet	0.3048	meters (m)
inches	2.54	centimeters (cm)
kilometers (km)	0.6214	miles
meters (m)	3.2808	feet
meters (m)	39.37	inches
meters (m)	1.0936	yards
miles	1.6093	kilometers (km)
yards	0.9144	meters (m)
AREA		
acres	0.4047	hectares (ha)
hectares (ha)	2.47	acres
hectares (ha)	10,000	square meters (m^2)
acres	4047	square meters (m^2)
hectares (ha)	0.01	square kilometers (km^2)
square kilometers (km^2)	100	hectares (ha)
square meters (m^2)	0.0001	hectares (ha)
VOLUME		
cubic feet	0.02834	cubic meters (m^3)
cubic meters (m^3)	35.31	cubic feet
cubic centimeters (cm^3)	10^{-3}	liters (L)
acre-feet	1223.5	cubic meters (m^3)
gallons	3.785	liters (L)
gallons	0.003785	cubic meters (m^3)
cubic meters (m^3)	264.2	gallons
liters (L)	0.2642	gallons

(continued)

Multiply	By	To Obtain
FLOW		
cubic feet per second (cfs)	0.002832	cubic meters per second ($m^3\ s^{-1}$)
cubic feet per second (cfs)	10.1952	cubic meters per hour ($m^3\ hr^{-1}$)
cubic feet per second (cfs)	448.86	gallons per minute (gpm)
cubic meters per second ($m^3\ s^{-1}$)	35.31	cubic feet per second (cfs)
cubic meters per second ($m^3\ s^{-1}$)	3600	cubic meters per hour ($m^3\ hr^{-1}$)
gallons per minute (gpm)	0.002228	cubic feet per second (cfs)
gallons per minute (gpm)	0.06308	liters per second ($L\ s^{-1}$)
gallons per minute (gpm)	0.00379	cubic meters per minute ($m^3\ min^{-1}$)
MASS		
grams (g)	0.002205	pounds
grams (g)	0.001	kilograms (kg)
kilograms (kg)	2.2046	pounds
kilograms (kg)	1000	grams (g)
pounds	453.6	grams (g)
pounds	0.4536	kilograms (kg)
metric tons (t)	2205	pounds
metric tons (t)	1000	kilograms (kg)
PRESSURE		
atmosphere (atm)	1.01325×10^5	pascal (Pa)
atmosphere (atm)	760	millimeters of mercury (mm Hg)
FLUX OF MASS		
grams per square meter per year ($g\ m^{-2}\ yr^{-1}$)	10	kilograms per hectare per year ($kg\ ha^{-1}\ yr^{-1}$)
grams per square meter per year ($g\ m^{-2}\ yr^{-1}$)	8.924	pounds per acre per year
kilograms per hectare per year ($kg\ ha^{-1}\ yr^{-1}$)	0.1	grams per square meter per year ($g\ m^{-2}\ yr^{-1}$)
pound per acre per year	1.12	kilograms per hectare per year ($kg\ ha^{-1}\ yr^{-1}$)
pounds per acre per year	0.112	grams per square meter per year ($g\ m^{-2}\ yr^{-1}$)
ENERGY		
British thermal units (BTU)	0.2530	kilocalories (kcal)
British thermal units (BTU)	1054	joules (J)
calories (cal)	4.1869	joules (J)
calories (cal)	0.001	kilocalories (kcal)
joules (j)	0.239	calories (cal)
joules (J)	2.390×10^{-4}	kilocalories (kcal)
kilocalories (kcal)	1000	calories (cal)
kilocalories (kcal)	3.968	British thermal units (BTU)
kilocalories (kcal)	4183	joules (J)
kilocalories (kcal)	4.183	kilojoules (kJ)
kilocalories (kcal)	0.001162	kilowatt-hours (kWhr)
kilojoule (kJ)	0.239	kilocalories (kcal)
kilowatt-hours (kWhr)	860.5	kilocalories (kcal)
kilowatt-hours (kWhr)	3.6×10^6	joules (J)

Multiply	By	To Obtain
langley (ly)[1]	1	calories per square centimeter (cal cm^{-2})
langley (ly)	10	kilocalories per square meter (kcal m^{-2})
POWER		
horsepower	0.7457	kilowatts (kW)
horsepower	10.70	kilocalories per minute (kcal/min)
kilocalories/day (kcal/day)	6.4937 × 10^{-5}	horsepower
kilocalories/day (kcal/day)	4.8417 × 10^{-5}	kilowatts (kW)
kilowatts (kW)	1.341	horsepower
kilowatts (kW)	14.34	kilocalories per minute (kcal/min)
kilowatts (kW)	1000	watts (W)
watt (W)	1	joule per second (J/sec)
PRIMARY PRODUCTIVITY/ENERGY FLOW		
grams dry weight (g-dw)	4.5	kilocalories (kcal)
grams dry weight (g-dw)	0.45	grams C (g-C)
grams O_2 (g-O_2)[2]	3.7	kilocalories (kcal)
grams O_2 (g-O_2)[2]	0.375	grams C (g-C)
grams C (g-C)[2]	10	kilocalories (kcal)
grams C (g-C)[2]	2.67	grams O_2 (g-O_2)
kilocalories (kcal)	4.18	kilojoules (kJ)
kilocalories (kcal)[2]	0.1	grams C (g-C)
STOICHIOMETRY OF ORGANIC MATTER[3]		
molar ratio	106C:16N:1P	
weight ratio	41C:7.2N:1P	
CONCENTRATIONS IN WATER		
part per thousand (ppt)	1	grams per liter (g L^{-1})
part per million (ppm)	1	milligrams per liter (mg L^{-1})
parts per million (ppm)	1	grams per cubic meter (g m^{-3})
parts per million (ppm)	1000	parts per billion (ppb)
parts per billion (ppb)	1	micrograms per liter (μg L^{-1})
milligrams per liter (mg L^{-1})	1000	micrograms per liter (μg L^{-1})
millimolarity (m mole/L)	molecular weight	milligrams per liter (mg L^{-1})
micromolarity (μ mole/L)	molecular weight	micrograms per liter (μg L^{-1})
microgram-atoms per liter (μg-atom L^{-1})	molecular weight	micrograms per liter (μg L^{-1})
milligrams per liter (mg L^{-1})	ionic charge/ molecular weight	milliequivatents per liter (meq L^{-1})
milliequivatents per liter (meq L^{-1})	molecular weight/ionic charge	milligrams per liter (mg L^{-1})
micromhos per cm (μmho cm^{-1})	1	microSiemens per centimeter (μS cm^{-1})

[1] Solar constant = radiant energy at outer limit of earth's atmosphere ~2.0 langleys per min (ly min^{-1}).
[2] Based on general photosynthetic equation showing the production of glucose:
$6CO_2 + 12H_2O + (118 \times 6 \text{kcal}) \rightarrow C_6H_{12}O_6 + 6O_2 + 6H_2O$.
[3] Based on Redfield (1958) of plankton organic matter $(CH_2O)_{106}(NH_3)_{16}(H_2PO_4)$.

Reference

Redfield, A. C. 1958. The biological control of chemical factors in the environment. *American Scientist* 46: 206–226.

Glossary

Aapa peatlands—Also called string bogs and patterned fens; peatlands identified by watertracks of long, narrow alignment of the high peat hummocks (strings) that form ridges perpendicular to the slope of the peatland and are separated by deep pools (flarks).

ADH—Alcohol dehydrogenase, the enzyme that catalyzes the reduction of acetaldehyde to ethanol in fermentation.

Adventitious roots—Roots that develop from some part of a vascular plant other than the seed. Usually they originate from the stem, and while common in most plants, they also develop as adaptations to anoxia in both flood-tolerant trees (e.g., *Salix* and *Alnus*) and herbaceous species, and flood-intolerant (e.g., tomato) plants just above the anaerobic zone when these plants are flooded.

Aerenchyma—Large air spaces in roots and stems of some wetland plants that allow the diffusion of oxygen from the aerial portions of the plant into the roots.

Alcohol dehydrogenase—*See* ADH.

Allochthonous—Pertains to material that is imported into an ecological system of interest from outside that system; usually refers to organic material and/or nutrients and minerals.

Allogenic succession—Ecosystem development whereby the distribution of species is governed by their individual responses to their environment with little or no feedback from organisms to their environment. Also called individualistic hypothesis, continuum concept, and Gleasonian succession.

Alluvial plain—The floodplain of a river, where the soils are alluvial deposits carried in by the overflowing river.

Ammonia volatilization—NH_3 released to the atmosphere as a gas.

Anadromous—Refers to marine species that spawn in freshwater streams.

Anammox—Abbreviation for anaerobic ammonium oxidation that leads to conversion of nitrite-nitrogen to dinitrogen gas.

Anaerobic—Refers to oxygenless conditions.

Anoxia—Waters or soils with no dissolved oxygen.

Artificial wetland—*See* Constructed wetland.

Aspect ratio—length to width ratio of a constructed wetland. Recommended to be 10:1 or higher for flow-through treatment wetlands.

Assimilatory nitrate reduction—Nitrate (NO_3) assimilated by plants or microbes and converted into biomass.

Assimilatory sulfate reduction—Process in the sulfur cycle whereby sulfur-reducing obligate anaerobes such as *Desulfovibrio* bacteria utilize sulfates as terminal electron acceptors in anaerobic respiration.

Autochthonous—Pertains to material that is produced within the ecological system of interest (e.g., organic material produced by photosynthesis). *See also* Allochthonous.

Autogenic succession—Clementian theory of succession of ecosystems whereby vegetation occurs in recognizable and characteristic communities; community change through time is brought about by the biota; changes are linear and directed toward a mature stable climax ecosystem.

Bankfull discharge—Streamflow at which a river begins to overflow onto its floodplain.

Basin wetland—A wetland that is hydrologically isolated with little or no flooding from streams, rivers, or tides.

Billabong—Australian term for a riparian wetland that is periodically flooded by the adjacent stream or river.

Biogeochemical cycling—The transport and transformation of chemicals in ecosystems.

Blanket bogs—In humid climates, peat that blankets large areas far from the site of the original peat accumulation, through the process of paludification.

BOD—Biochemical oxygen demand, a biological test for degradable organic matter in water.

Bog—A peat-accumulating wetland that has no significant inflows or outflows and supports acidophilic mosses, particularly *Sphagnum*.

Bottomland—Lowland along streams and rivers, usually on alluvial floodplains, that is periodically flooded.

Bottomland hardwood forest—Term used principally in southeastern and eastern United States to mean a mesic riparian forested ecosystem along a higher order stream or river that is subject to intermittent to frequent flooding from that stream or river; dominated by oaks and other deciduous hardwood tree species.

Bulk density—Dry weight of a known volume of soil, divided by that volume.

Buttress—Swollen bases of tree trunks growing in water.

Cajun—Term used for culture of former French-speaking immigrants who have lived for several centuries in the swamps of the Louisiana delta.

Carbon sequestration—The permanent retention of carbon in an ecosystem, usually in its soil. See also Sequestration.

Carr—Term used in Europe for forested wetlands characterized by alders (*Alnus*) and willows (*Salix*).

Cat clays—When coastal wetlands are drained, soil sulfides often oxidize to sulfuric acid, and the soils become too acidic to support plant growth.

Cation exchange capacity—The sum of exchangeable cations (positive ions) that a soil can hold.

Cheia—Annual period of flooding from March through May in the Pantanal region of South America that supports luxurious aquatic plant and animal life. *See also* Enchente, Seca, and Vazante.

Clay depletions—Clay is selectively removed along root channels after iron and manganese oxides have been depleted in wetland soils, only to redeposit as clay-coatings on soil particles below the clay depletions.

Coastal squeeze—A concept related to sea level change where coastal wetlands can be "squeezed out" during sea level rise if there are human-constructed impediments that restrict the wetlands from moving inland.

Coliforms—A quantitative biological test for the presence of colon bacteria or related forms; because of their ubiquitous presence, they are used as a presumptive index of general bacterial contamination of water.

Concentric domed bog—A concentric pattern of pools and peat communities formed around the most elevated part of a bog.

Constructed wetland—A wetland developed on a former uplands to create poorly drained soils and wetland flora and fauna for the primary purpose of contaminant or pollution removal from wastewater or runoff. *See also* Treatment wetland

Consumer surplus—In economics, the net benefit of a good to the consumer.

Continuum concept—*See* Allogenic succession.

Coprecipitation of phosphorus—Some calcium phosphate is precipitated along with the major precipitation of calcium carbonate in alkaline waters.

Created wetland—A wetland constructed where one did not exist before.

Cumbungi swamp—Cattail (*Typha*) marsh in Australia.

Cumulative loss—When ecosystems such as wetlands are lost, usually as a result of human development, one small piece at a time, with the cumulative loss being substantial.

Cypress domes—Also called cypress ponds or cypress heads; poorly drained to permanently wet depressions dominated by pond cypress (*Taxodium distichum* var. *nutans*). Called domes because the cypress grows more vigorously in the center than around the perimeter of the dome, giving it a domed appearance from a distance.

Cypress strand—A diffuse freshwater stream flowing through a shallow forested depression (dominated by *Taxodium*) on a gently sloping plain.

Dabbling duck—Waterfowl mainly in the family Anatidae (swans, geese and ducks) that feed mainly at the surface rather than by diving.

Dalton's law—Flux is proportional to a pressure gradient. An example of a process that follows Dalton's law is evaporation, which is proportional to the difference between the vapor pressure at the water surface and the vapor pressure in the overlying air.

Dambo—A seasonally waterlogged and grass-covered linear depressions in headwater zone of rivers with no marked stream channel or woodland vegetation. Term is ChiChewa (Central Africa) dialect meaning "meadow grazing."

Darcy's law—Groundwater equation that states that flow of groundwater is proportional to a hydraulic gradient and the hydraulic conductivity, or permeability, of the soil or substrate.

Delineation—Technique of determining an exact boundary of a wetland. Used for identifying jurisdictional wetlands in United States.

Delta—Location where rivers meet the sea and deposit sediments, often in a broad alluvial fan; there are also examples of inland deltas such as the Peace-Athabasca Delta in Canada and the Okavango Delta in Botswana where the water never reaches the sea.

Demand curve—Economist's estimate of consumer benefits.

Denitrification—Process in the nitrogen cycle carried out by microorganisms in anaerobic conditions, where nitrate acts as a terminal electron acceptor, resulting in the loss of nitrogen as it is converted to nitrous oxide (N_2O) and nitrogen gas (N_2).

Designer wetland—Created or restored wetland in which certain plant species or other organisms are introduced and the success or failure of those plants or organisms is used as the indicator of success or failure of that wetland.

Detention time—A measure of the length of time a parcel of water stays in a wetland; equivalent to the turnover time or retention time and the inverse of the turnover rate. Detention time is the term used most frequently for designing treatment wetlands. *See also* Retention time.

Discharge wetland—Wetland that has surface water (or groundwater) level lower hydrologically than the surrounding water table, leading to an inflow of groundwater.

Dissimilatory nitrate reduction to ammonia (DNRA)—Conversion of nitrate-nitrogen to ammonium-nitrogen.

Dissimilatory nitrogen reduction—Several pathways of nitrate reduction, particularly nitrate reduction to ammonia and denitrification. It is called dissimilatory as the nitrogen is not assimilated into a biological cell.

Diversion wetland—Wetland created or enhanced by diversion of an adjacent body of water, usually a river. Created diversion wetlands along rivers in upper watersheds are similar to oxbows or billabongs. River diversions in deltas are means to re-establish deltaic distributaries.

Diving duck—Ducks that feed by diving beneath the surface of the water. They are in a distinct subfamily, *Aythyinae*, of the large *Anatidae* family that includes ducks, geese, and swans. Also commonly called pochards or scaups.

DMS—Dimethyl sulfide, one of the gases given off by wetlands.

Drop roots—*See* Prop roots.

Duck stamps—Stamps sold in several countries to hunters to help pay for the protection of waterfowl habitat. The Duck Stamp program in the United States started in 1934.

Eat-out—A major wetland vegetation removal by herbivory, often by geese or muskrats.

Ebullitive flux—Flux of gases from wetland soils as bubbles or diffusion to the surface of the water and then to the atmosphere.

Ecological engineering—The design, creation, and restoration of ecosystems for the benefit of humans and nature.

Ecosystem engineers—Plants, animals, and microbes that carry out essential biological feedbacks in ecosystems. Examples in wetlands are beavers and muskrats.

Ecosystem services—values that ecosystems provide to humans; similar to ecosystem values. Has been divided into three categories related to human well-being: provisioning, regulating, and cultural.

Embodied energy—The total energy required to produce a commodity.

Emergy—Calculation of total energy requirement for any product in nature or humanity based on using transformities. Short for "energy memory." See H. T. Odum (1996).

Enchente—Period of rising waters from December through February in Pantanal region of South America. *See also* Cheia, Seca, and Vanzante.

Ericaceous plants—Flowering plants of the family Ericaceae, which, as a group, are acid-loving or acid-tolerant plants that often dominate bogs and other sites with acidic substrates.

Estuary—General location where rivers meet the sea and freshwater mixes with saltwater.

Eutrophic—Nutrient rich; generally used in lake classification, but is also applicable to peatlands.

Eutrophication—Process of aquatic ecosystem development whereby an ecosystem such as a lake, estuary, or wetland goes from an oligotrophic (nutrient poor) to eutrophic (nutrient rich) condition. If caused by humans, it is called cultural eutrophication.

Excentric raised bogs—Bogs that form from previously separate basins on sloping land and form elongated hummocks and pools aligned perpendicular to the slope.

Facultative—Adapted equally to either wet or dry condition. Usually used in the context of vegetation adapted to growing in saturated soils or upland soils.

Fen—A peat-accumulating wetland that receives some drainage from surrounding mineral soil and usually supports marshlike vegetation.

Fermentation—Partial oxidation of organic matter, when organic matter itself is the terminal electron acceptor in anaerobic respiration by microorganisms; forms various low-molecular-weight acids and alcohols and CO_2. Also called glycolysis.

Fibrists—*See* Peat.

Flarks—*See* Aapa peatlands.

Flood duration—The amount of time that a wetland is in standing water.

Flood frequency—The average number of times that a wetland is flooded during a given period.

Flood peak—Peak runoff into a wetland caused by a specific rainfall event.

Flood pulse concept (FPC)—Pulsing river discharge as the major force controlling biota in river floodplains, including the lateral exchange between floodplains and river channels.

Fluted trunk—Flared tree trunks at the ground surface that occurs on some trees growing in wet conditions.

Folists—Organic soils caused by excessive moisture (precipitation > evapotranspiration) that accumulate in tropical and boreal mountains; these soils are not classified as hydric soils as saturated conditions are the exception rather than the rule.

Functional guild—Categorization of plant communities into functional groups that can be defined by measurable traits.

Gardians—"Cowboys" who ride horses through the wetlands of southern France's Camargue.

Gator holes—Deep sloughs and solution holes that hold water during the dry season and that serve as wildlife refuges; term mostly used in the Florida Everglades.

Geogenous—Peatland subject to external flows.
Gleying—Development of black, gray, or sometimes greenish or blue-gray color in soils when flooded.
Glycolysis—*See* Fermentation.
Greenhouse gases (GHG)—Atmospheric gases that adsorb radiant energy at various wavelengths. The term is mostly used to refer to the gases CO_2, CH_4, and N_2O that are products of human activity and that could lead to atmospheric warming.
Guild—A group of functionally similar species in a community.
HAB—Harmful algal bloom.
Halophiles—"Salt-loving" organisms.
Halophytes—Salt-tolerant plants.
Hammock—Slightly raised tree islands, such as tree island freshwater hammocks or mangrove islands in the Florida Everglades.
Hatch–Slack–Kortschak pathway—Biochemical pathway of photosynthesis for C_4 plants.
Hemists—Mucky peat or peaty muck; conditions between saprist and fibrist soil.
HGM—*See* Hydrogeomorphic classification.
High marsh—Upper zone of a salt marsh that is flooded irregularly and generally is located between mean high water and extreme high water. Called inland salt marsh in Gulf of Mexico coastline.
Histosols—Organic soils that have organic soil material in more than half of the upper 80 cm, or that are of any thickness if they overlie rock or fragmental materials that have interstices filled with organic soil material.
HLR—*See* Hydraulic loading rate.
HSI—Habitat suitability index, a semi-quantitative measure of habitat value of an ecosystem for specific species.
Hydrarch succession—Development of a terrestrial forested climax community from a shallow lake with wetland as an intermediate sere. In this view, lakes gradually fill in as organic material from dying plants accumulates and minerals are carried in from upslope.
Hydraulic conductivity—*See* Permeability.
Hydraulic loading rate (HLR)—Amount of water added to a wetland, generally described as the depth of water (volume of flooding per wetland area) per unit time; generally used for treatment wetlands.
Hydric soils—Soils that formed under conditions of saturation, flooding, or ponding long enough during the growing season to develop anaerobic conditions in the upper part.
Hydrochory—Seed dispersal by water.
Hydrodynamics—An expression of the fluvial energy that drives a system.
Hydrogeomorphic classification (HGM)—Wetland classification system based on type and direction of hydrologic conditions, local geomorphology and climate.
Hydrogeomorphology—Combination of climate, basin geomorphology, and hydrology that collectively influences a wetland's function.
Hydroperiod—The seasonal pattern of the water level of a wetland. This approximates the hydrologic signature of each wetland type.
Hydrophyte—Plant adapted to the wet conditions.
Hydrophytic vegetation—Plant community dominated by hydrophytes.
Hypoxia—Waters with dissolved oxygen less than 2mg/L.
Interception—Precipitation that is retained in the overlying vegetation canopy.
Intermittently exposed—Refers to nontidal wetlands that are flooded throughout the year, except during periods of extreme drought.
Intermittently flooded—Refers to nontidal wetlands that are usually exposed, with surface water present for variable periods without detectable seasonal patterns.
Intertidal—Part of coastal wetland flooded periodically with tidal water.
Intrariparian continuum—The structure and function of riparian communities along a river system.
Irregularly exposed—Refers to coastal wetlands with surface exposed by tides less often than daily.

Irregularly flooded—Refers to coastal wetlands with surface flooded by tides less often than daily.
Isolated wetland—Legal term used in the United States to define wetlands that do not have an obvious surface-water connection to a navigable stream or river (*see also* Significant nexus).
Jurisdictional wetland—Term used in the United States to refer to wetlands that fall under the jurisdiction of federal laws for the purpose of permit issuance or other legal matters.
Kahikatea—Refers to both the tree (*Dacrycarpus dacrydiodes*) and the forested wetlands found throughout New Zealand. Referred to as "white pine" forests by locals.
Karst—A topography formed over limestone, dolomite, or gypsum.
Krefeld system or Max-Planck-Institute process—Gravel bed macrophyte subsurface flow treatment wetlands.
Lacustrine—Pertaining to lakes or lake shores.
Lagoon—Term frequently used in Europe to denote deepwater enclosed or partially opened aquatic system, especially in coastal delta regions.
Lentic—Related to slow-moving or standing water systems; usually refers to lake (lacustrine) and stagnant swamp systems.
Lenticels—Small pores found on mangrove tree prop roots and pneumatophores above low tide and presumed to be sites of oxygen influx for anaerobic roots survival.
Limnogenous peatland—Geogenous peatland that develops along a slow-flowing stream or a lake.
Littoral—Zone between high and low tide in coastal waters or the shoreline of a freshwater lake.
Loading rate—The amount of a material (e.g., a chemical) applied to a wetland, measured either per unit area (e.g., $g\ m^{-2}\ yr^{-1}$) or volumetrically (e.g., $g\ m^{-3}\ yr^{-1}$).
Lotic—Pertaining to running water (i.e., rivers and streams).
Low marsh—Intertidal or lower marsh in salt marsh that is located in the intertidal zone and is flooded daily. Called streamside salt marsh in coastal Gulf of Mexico.
Mangal—Same as mangrove.
Mangrove—Subtropical and tropical coastal ecosystem dominated by halophytic trees, shrubs, and other plants growing in brackish to saline tidal waters. The word "mangrove" also refers to the dozens of tree and shrub species that dominate mangrove wetlands.
Marginal value—The value of an additional increment of a commodity in a free market.
Marsh—A frequently or continually inundated wetland characterized by emergent herbaceous vegetation adapted to saturated soil conditions. In European terminology, a marsh has a mineral soil substrate and does not accumulate peat. *See also* Tidal freshwater marsh, Salt marsh.
Mesotrophic peatlands—Also called transition or poor fens. Peatlands intermediate between minerotrophic and ombrotrophic.
Methane emissions—Amount of methane released from a landscape as net result of methanogenesis minus methane oxidation.
Methane oxidation—Conversion by methane to methanol, formaldehyde, and carbon dioxide by obligate methanotrophic bacteria.
Methanogenesis—Carbon process under extremely reduced conditions when certain bacteria (methanogens) use CO_2 or low-molecular-weight organic compounds as electron acceptors for the production of gaseous methane (CH_4).
Methanogens—Bacteria that carry out methanogenesis.
Methanotrophs—Aerobic bacteria that oxidize methane.
Millennium Ecosystem Assessment—International study published in 2005 that focused on the changes that humans have caused to ecosystems and how those are affecting the services that they provide to humans.
Mineral soil—Soil that has less than 20 to 35 percent organic matter.
Minerotrophic peatlands—Also called rheotrophic peatlands or rich fens; peatlands that receive water that has passed through mineral soil.

Mire—Synonymous with any peat-accumulating wetland (European definition); from the Norse word "myrr." The Danish and Swedish word for peatland is now "mose."

Mitigate—To lessen or compensate for an impact. Used here in the context of mitigating wetland loss by restoring or creating wetlands.

Mitigation bank—A wetland area that has been restored and protected to provide compensation for impacts to wetlands.

Mitigation ratio—The ratio of restored or created wetland to wetland lost to development.

Mitigation wetland—*See* Replacement wetland.

Moor—Synonymous with peatland (European definition). A highmoor is a raised bog; a lowmoor is a peatland in a basin or depression that is not elevated above its perimeter. The primitive sense of the Old Norse root is "dead" or barren land.

Mottles (or redox concentrations)—Orange/reddish-brown (because of iron oxides) or dark reddish-brown/black (because of manganese oxides) accumulations in hydric soils throughout an otherwise gray (gleyed) soil matrix. Mottles suggest intermittently exposed soils and are relatively insoluble, enabling them to remain in soil long after it has been drained.

Muck—Sapric organic soil material with virtually all of the organic matter decomposed, not allowing for the identification of plant forms. Bulk density generally greater than $0.2 g/cm^3$ (more than peat).

Munsell soil color chart—Book of standard color chips for determining soil color value and chroma. Used to identify hydric soils.

Muskeg—Large expanse of peatlands or bogs; particularly used in Canada and Alaska.

NAD—Nicotinamide adenine dinucleotide, an enzyme that accumulates in anaerobic conditions.

NADP—NAD phosphate.

Nexus—Legal term emphasized by the U.S. Supreme Court to describe connections between wetlands and navigable waterways that are regulated. Wetlands must significantly affect the chemical, physical, and biological integrity of waters understood as "navigable" to have a significant nexus to those waters.

Nitrification—Ammonium nitrogen oxidized by microbes to nitrite nitrogen and nitrate nitrogen.

Nernst equation—Equation based on a hydrogen scale showing how redox potential is related to the concentrations of oxidants and reductants in a redox reaction.

Nitrogen fixation—Process in the nitrogen cycle whereby N_2 gas is converted to organic nitrogen through the activity of certain organisms in the presence of the enzyme nitrogenase.

No net loss—Wetland policy in the United States that began in the late 1980s and means that if wetlands are lost they must be replaced so that there is no "net loss" of wetlands overall.

Nutrient budget—Mass balance of a nutrient in an ecosystem.

Nutrient spiraling—The process whereby resources (organic carbon, nutrients, etc.) are temporarily stored, then released as they "spiral" downstream from organic to inorganic form and back again.

Obligate—Requiring a specific environment to grow, as in adapted only a wet environment. In the context of wetlands, obligate generally refers to plants requiring saturated soils.

Oligotrophic—Nutrient poor; generally used in lake classification, but is also applicable to peatlands.

Oligotrophication—Often the process of peatland development whereby a peatland eventually elevates itself above the surrounding landscape and goes from eutrophic (nutrient rich) to oligotrophic (nutrient poor).

Ombrogenous—Peatland with inflow from precipitation only; also called ombrotrophic.

Ombrotrophic—Literally rain fed, referring to wetlands that depend on precipitation as the sole source of water.

Opportunity cost—The net worth of a non–free market resource in its best alternative use; that is, the net benefit of the area in its best alternative use that has to be forgone in order to keep it in its natural state.

Organic soil—Soil that has more than 12 to 18 percent organic carbon, depending on clay content (*see* Fig. 06).

Osmoconformers—Marine animals in which the internal cell environment follows closely the osmotic concentration of the external medium.

Osmoregulators—Marine animals that control their internal cell environment despite a different osmotic concentration of the external medium.
Outwelling—Function of coastal wetlands as "primary production pumps" that feed large areas of adjacent waters with organic material and nutrients; analogous to upwelling of deep ocean water, which supplies nutrients to some coastal waters from deep water.
Overland flow—Nonchannelized sheet flow that usually occurs during and immediately following rainfall or a spring thaw, or as tides rise in coastal wetlands.
Oxbow—Abandoned river channel, often developing into a swamp or marsh, on a river floodplain.
Oxidation—Chemical process of giving up an electron (e.g., $Fe^{2+} \rightarrow Fe^{3+} + e^-$). Special cases involve uptake of oxygen or removal of hydrogen (e.g., $H_2S \rightarrow S^{2-} + 2H^+$).
Oxidized pore linings—*See* Oxidized rhizosphere.
Oxidized rhizosphere (also called oxidized pore linings)—Thin traces of oxidized soils through an otherwise dark matrix indicating where roots of hydrophytes were once found.
Paalsa peatlands—Peatlands found in the southern limit of the tundra biome; large plateaus of peat (20 to 100 m in breadth and length and 3 m high) generally underlain by frozen peat and silt.
Pakihi—Peatland in southwestern New Zealand dominated by sedges, rushes, ferns, and scattered shrubs. Most pakihi form on terraces or plains of glacial or fluvial outwash origin and are acid and exceedingly infertile.
Palmer Drought Severity Index (PDSI)—A relative measure of climatic "wetness." Used primarily to estimate the severity of droughts.
Paludification—The blanketing of terrestrial ecosystems by overgrowth of bog vegetation. *See also* Blanket bog.
Palustrine—Nontidal wetlands.
Panne—Bare, exposed, or water-filled depression in a salt marsh.
Patterned fens—*See* Aapa peatlands.
Peat—Fibric organic soil material with virtually all of the organic matter allowing for the identification of plant forms. Bulk density generally less than $0.1 g/cm^3$ (less than muck).
Peatland—A generic term of any wetland that accumulates partially decayed plant matter (peat).
Penman equation—Empirical equation for estimating evapotranspiration using an energy budget approach.
Perched wetland—Wetland that holds water well above the groundwater table.
Permanently flooded—Refers to nontidal wetlands that are flooded throughout the year in all years.
Permeability—The capacity of soil to conduct water flow. Also known as hydraulic conductivity. *See also* Darcy's law.
Petagram (Pg)—10^{15} grams.
Phreatophytes—Plants that obtain their water from phreatic sources (i.e., groundwater or the capillary fringe of the groundwater table).
Physiognomy—The appearance or life form of vegetation.
Piezometers—Groundwater wells that are only partially screened and thus measure the piezometric head of an isolated part of the groundwater.
Playa—An arid- to semiarid-region wetland that has distinct wet and dry seasons. Term used in the southwest United States for shallow depressional recharge wetlands occurring in the Great Plains region of North America that are formed through a combination of wind, wave, and dissolution processes.
Pneumatophores—"Air roots" that protrude out of the mud from the main roots of wetland plants such as black mangroves (*Avicennia*) and cypress (*Taxodium distichum*) and are thought to be organs for transport of oxygen and other gases to and from the roots of the plant. Called "knees" for cypress.
Pocosin—Peat-accumulating, nonriparian freshwater wetland, generally dominated by evergreen shrubs and trees and found on the southeastern Coastal Plain of the United States. The term comes from the Algonquin for "swamp on a hill."
Porosity—Total pore space in soil, generally expressed as a percentage.

Pothole—Shallow marshlike pond, particularly as found in the Dakotas and central Canadian provinces; the so-called prairie pothole region.

Prairie pothole—*See* Pothole.

Producer surplus or economic rent—The area over a good's supply curve bounded by price.

Prop roots—Above-ground arched roots that aid in support of some wetland trees such as the mangrove *Rhizophora*.

Pulse stability concept—Concept that pulses can be both a subsidy and a stress to an ecosystem, depending on their strength, with subsidies occurring with moderate pulses, while both weak and excessive pulses can result in stress responses.

Quaking bog—*Schwingmoor* in German. Bog in which the peat layer and plant cover is only partially attached in the basin bottom or is floating like a raft.

Quickflow—Direct runoff component of streamflow during a storm that causes an immediate increase in streamflow.

Raised bogs—Peat deposits that fill entire basins, are raised above groundwater levels, and receive their major inputs of nutrients from precipitation. *See* Ombrogenous and Ombrotrophic.

Ramsar Convention—International treaty originally started in Ramsar, Iran, in the early 1970s to protect wetland habitat around the world, especially for migratory waterfowl.

Raupo swamp—Cattail (*Typha*) marsh in New Zealand.

Recharge wetland—Wetland that has surfacewater (or groundwater) level higher hydrologically than the surrounding water table, leading to an outflow of groundwater.

Recurrence interval—The average interval between the recurrence of floods at a given or greater magnitude.

Redox concentrations—Bodies of accumulated iron and manganese oxides in wetland soils such as nodules and concretions, masses (formerly called "reddish mottles"), and pore linings (formerly called "oxidized rhizospheres").

Redox depletions—Bodies of low chroma (2 or less) where the natural (gray or black) color of the parent sand, silt, or clay results when soluble forms of iron, manganese, or clay are leached out of the soil. Generally have Munsell color values of 4 or greater. *See also* Clay depletions.

Redoximorphic features—Features formed by the reduction, translocation, and/or oxidation of iron and manganese oxides; used to identify hydric soils. Formerly called mottles and low-chroma colors.

Redox potential—Reduction-oxidation potential, a measure of the electron pressure (or availability) in a solution or measure of the tendency of soil solution to oxidize or reduce substances. Low redox potential indicates reduced conditions; high redox potential indicates oxidized conditions.

Reduced matrix—Soil that has low chroma and high value but whose color changes in hue or chroma when exposed to air.

Reduction—Chemical process of gaining an electron (e.g., $Fe^{3+} + e^- \rightarrow Fe^{2+}$). Special cases involve releasing oxygen or gaining hydrogen (hydrogenation) (e.g., $S^{2-} + 2H^+ \rightarrow H_2S$).

Reedmace swamp—Cattail (*Typha*) marsh in the United Kingdom.

Reedswamp—Marsh dominated by *Phragmites* (common reed); term used particularly in Europe.

Reference wetland—Natural wetland used as a reference or control site to judge the condition of another created, restored, or impacted wetland.

Regularly flooded—Refers to coastal wetlands with surface flooded and exposed by tides at least once daily.

Regulators (or avoiders)—In reference to biological adaptations to stress, organisms that actively avoid stress or modify it to minimize its effects.

Rehabilitation—Less than full restoration of an ecosystem to its predisturbance condition.

Renewal rate—*See* Turnover rate.

Replacement value—The sum of the cheapest way of replacing all the various services performed by a natural ecosystem area.

Replacement wetland—A wetland constructed to replace the functions lost by human development, usually in the same or an adjacent watershed.
Residence time—*See* Retention time.
Resource spiraling—*See* Nutrient spiraling.
Restoration—To return a site to an approximation of it condition before alteration. *See also* Wetland restoration.
Retention rate—The amount of a material retained in a wetland per unit time and area; usually refers to material more or less removed from water flowing over or through a wetland, as contrasted to *detention*, which is transitory.
Retention time—A measure of the average time that water remains in the wetland. Nominal residence time or retention time refers to the theoretical time that water stays in a wetland as calculated from the flowthrough and the water volume in the wetland. *See also* Detention time.
Rheotrophic peatlands—*See* Minerotrophic peatlands.
Riparian—Pertaining to the bank of a body of flowing water; the land adjacent to a river or stream that is, at least periodically, influenced by flooding.
Riparian ecosystem—Ecosystem with a high water table because of proximity to an aquatic ecosystem, usually a stream or river. Also called bottomland hardwood forest, floodplain forest, bosque, riparian buffer, and streamside vegetation strip.
River continuum concept (RCC)—Theory to describe the longitudinal patterns of biota found in streams and rivers.
Root zone method (*Wurzelraumentsorgung*)—Subsurface flow wetland basins, almost always found in Europe, and generally planted with *Phragmites australis*.
Runoff—Nonchannelized surfacewater flow.
Salt exclusion—A salinity adaptation by some wetland plants by which plants prevent salt from entering the plant at the roots.
Salt marsh—A halophytic grassland on alluvial sediments bordering saline water bodies where water level fluctuates either tidally or nontidally.
Salt secretion—A salinity adaptation by which some wetland plants excrete salt from specialized organs in the leaves.
Saprists—*See* Muck.
Saturated soils—Refers to nontidal wetlands where the soil or substrate is saturated for extended periods in the growing season, but standing water is rarely present.
Sclerophylly—Refers to the thickening of the plant epidermis.
Seasonally flooded—Refers to nontidal wetlands that are flooded for extended periods in the growing season, but with no surface water by the end of the growing season.
Seca—Dry period in Pantanal region of South America from September through November when the wetland reverts to vegetation typical of dry savannas. *See also* Cheia, Enchente, and Vazante.
Secondary treatment—Treatment of wastewater to remove organic material.
Sedge meadow—Very shallow wetland dominated by several species of sedges (e.g., *Carex, Scirpus, Cyperus*).
Seed bank—Seeds stored in soils, often for many years. In wetlands changing hydroperiod, as in wetland restoration or wetland drainage, can often lead to germination.
Self-design—The application of self-organization in the design of ecosystems. The process of ecosystem development whereby the continual or periodic introduction of species propagules (plants, animals, microbes) by humans or nature and their subsequent survival (or nonsurvival) provide the essence of the successional and functional development of an ecosystem.
Semipermanently flooded—Refers to nontidal wetlands that are flooded in the growing season in most years.
Sequestration—The permanent retention of a chemical or nutrient in an ecosystem. Often used to describe the permanent burial of carbon in wetland soils as carbon sequestration.

Serial discontinuity concept—Describes the effects that floodplains, dams, and the transverse dimension in general has on the functioning of a river system.

Shrub–scrub—Wetlands dominated by woody, low-stature vegetation such as freshwater buttonwood (*Cephalanthus*) or saltwater dwarf mangrove (*Rhizophora*) swamps.

Significant nexus—Legal term used in the United States to describe the connection of a wetland to an adjacent navigable water. The wetland should, by itself or in combination with other lands, significantly affect the chemical, physical, and biological integrity of the adjacent navigable water (*see also* Isolated wetland; Nexus).

Sink—Term used in the context of wetland nutrient budgets to define a wetland that imports more of a certain nutrient than it exports.

Slough—An elongated swamp or shallow lake system, often adjacent to a river or stream. A slowly flowing shallow swamp or marsh in the southeastern United States (e.g., cypress slough). From the Old English word "sloh" meaning a watercourse running in a hollow. *See* also Cypress strand.

Soligenous peatland—Geogenous peatland that develops with regional interflow and surface runoff.

Source—Term used in the context of wetland nutrient budgets to define a wetland that exports more of a certain nutrient than it imports.

Spit—A neck of land along a coastline behind which coastal wetlands sometimes develop.

SRP—Soluble reactive phosphorus, similar to orthophosphate; a measure of biologically available phosphorus.

Stem hypertrophy—Noticeable swelling of lower stem of vascular plant, usually caused by water or saturated soils. Includes tree buttresses and fluted trunks.

Stemflow—Precipitation that passes down the stems of vegetation. Used generally in connection with forests and forested wetlands.

Streamflow—Channelized surfacewater flow.

Stream order—Numerical system that classifies stream and river segments by size according to the order of its tributaries.

String bogs—*See* Aapa peatlands.

Strings—*See* Aapa peatlands.

Subsidence—Sinking of ground level, caused by natural and artificial settling of sediments over time.

Subsurface-flow constructed wetlands—Constructed wetlands through which water flows beneath the surface rather than over the surface. *See* also Root zone method.

Subtidal—Coastal wetland permanently flooded with tidal water.

Supply curve—Economist's estimate of producer benefits.

Surface-flow constructed wetlands—Constructed wetlands that mimic many natural wetlands with flow on surface rather than below the surface.

Swamp—Wetland dominated by trees or shrubs (U.S. definition). In Europe, forested fens and wetlands dominated by reed grass (*Phragmites*) are also called swamps (*see* Reedswamp).

Swampbuster—Provision of the U.S. Food Security Act that encourages farmers not to drain wetlands and thereby lose their farm subsidies.

Swamp gas (or marsh gas)—Methane.

Taking—The legal denial of an individual's right to use all or part of the area or structure (trees, wildlife, etc.) of his or her property.

Telmatology—A term originally coined to mean "bog science." From the Greek word "telma" for bog.

Temporarily flooded—Refers to nontidal wetlands that are flooded for brief periods in the growing season, but otherwise the water table is well below the surface.

Teragram (Tg)—10^{12} grams; a billion kilograms.

Terrestrialization—Generally in reference to succession of peatlands, the infilling of shallow lakes until they become, in appearance, a peat basin supporting terrestrial vegetation.

Tertiary treatment—Advanced treatment of wastewater after secondary treatment to remove inorganic nutrients and other trace materials. Wetlands are often used for this purpose.
Thornthwaite equation—Empirical equation for estimating potential evapotranspiration as a function of air temperature.
Throughfall—Precipitation that passes through vegetation cover to the water or substrate below. Used particularly in forests and forested wetlands.
Tidal creeks—Small streams that serve as important conduits for material and energy transfer between salt marshes or mangrove swamps and adjacent coastal bodies of water.
Tidal freshwater marsh—Marsh along rivers and estuaries close enough to the coastline to experience significant tides by nonsaline water. Vegetation is often similar to nontidal freshwater marshes.
Tolerators (also called resisters)—In reference to biological adaptations to stress, organisms that have functional modifications that enable them to survive and often to function efficiently in the presence of stress.
Topogenous—Refers to peatland development when the peatland modifies the pattern of surface water flow.
Total suspended solids—*See* TSS.
Transformer—Term used in the context of wetland nutrient budgets to define a wetland that imports and exports the same amount of a certain nutrient but changes it from one form to another.
Translocation—Movement of nutrients between below-ground and above-ground portions of plants.
Treatment wetland—Wetland constructed to treat wastewater or polluted runoff. *See also* Constructed wetland.
TSS—Total suspended solids, a measure of the sediments in a unit volume of water.
Turlough—Term is specific for these types of wetlands found mostly in western Ireland. Areas seasonally flooded by karst groundwater with sufficient frequency and duration to produce wetland characteristics. They generally flood in winter and are dry in summer and fill and empty through underground passages.
Turnover rate—Ratio of throughput of water to average volume of water within a wetland. This is the inverse of turnover time, residence time, or retention time of a wetland.
Turnover time—*See* Retention time.
Value—Something worthy, desirable, or useful to humanity; although the term is used often in ecology to refer to processes (e.g., primary production) or ecological structures (e.g., trees) as they are "valuable" to the way an ecosystem functions, the term generally should be limited to an anthropocentric connotation. Humans decide what is of "value" in an ecosystem.
Vazante—Period of declining water in Pantanal region of South America from June through August. *See also* Cheia, Enchente, and Seca.
Vernal pool—Shallow, intermittently flooded wet meadow, generally typical of Mediterranean-type climate with dry season for most of the summer and fall. Term is now used to indicate wetlands temporarily flooded during the spring throughout the United States.
Viviparity—The production of young in a living state.
Viviparous seedlings—Seedlings of trees germinate while still attached to the tree canopy, as with the mangrove genera *Rhizophora*. A specific case of viviparity.
Vleis—Seasonal wetland similar to a Dambo; term used in southern Africa.
Wad (pl. Wadden)—Unvegetated tidal flat originally referring to the northern Netherlands and northwestern German coastline. Now used throughout the world for coastal areas.
Watertracks—*See* Aapa peatlands.
Wetland—*See* various wetland definitions in Chapter 2. Generally, wetlands have the presence of shallow water or flooded soils for part of the growing season, have organisms adapted to this wet environment, and have soil indicators of this flooding such as hydric soils.
Wetland creation—The conversion of a persistent upland or shallow water area into a wetland by human activity.
Wetland delineation—The demarcation of wetland boundaries for legal purposes. *See* Jurisdictional wetlands.

Wetland Reserve Program (WRP)—a U.S. government policy for farmers to receive benefits to set aside wetlands on their property for conservation.

Wetlanders—People who live in proximity to wetlands and whose culture is linked to the wetlands.

Wetland restoration—The return of a wetland from a condition disturbed or altered by human activity to a previously existing condition.

Wetlands of international importance—Wetlands designated by the Ramsar Convention as important international wetlands because they contain rare wetland types, support biological diversity, waterfowl, and fish.

Wet meadow—Grassland with waterlogged soil near the surface but without standing water for most of the year.

Wet prairie—Similar to a marsh, but with water levels usually intermediate between a marsh and a wet meadow.

Willingness-to-pay, or net willingness-to-pay—A hypothetical market that establishes the amount society would be willing to pay to produce and/or use a good beyond that which it actually does pay.

Reference

Odum, H. T. 1996. *Environmental Accounting: Emergy and Environmental Decision Making*. John Wiley & Sons, New York. 370 pp.

Index

Aapa peatlands, 420–421
Acer rubrum (red maple), 59, 376
  in red maple swamps, 388
  root adaptations, 220
  root systems, 395
  swamp productivity/hydrology relationship, 147
Acidification, 432
Actophilornis africanus (African jacana), 85
Adaptations:
  of freshwater swamp vegetation, 391–395
  of mangrove vegetation, 322–326
  morphological, 216–220
  mutualism and commensalism, 226–227
  of peatland vegetation, 431–433
  physiological, 220–226
  of vascular plants, 216–227
  whole plant strategies, 226
Adenosine triphosphate (ATP), 226
ADH (alcohol dehydrogenase), 224
Adventitious roots, 218, 219, 227, 395
Aerenchyma, 216–218
Aerial habitat, 274–275
Aerobic bacteria, 281
Aerobic respiration, 195
Aesthetics of ecosystem services, 545
African jacana (*Actophilornis africanus*), 85
African reef heron (*Egretta gularis*), 88
African wetlands, 56, 83–89
  Congolian swamp forests, 86
  East Africa tropical marshes, 86–88
  Okavango Delta, 84–86
  West Africa mangrove swamps, 87–89
Agkistrodon piscivorus (cottonmouth moccasin), 396–397

Agricultural Conservation Program (U.S. Department of Agriculture), 481
Agricultural land restoration, 598
Agricultural stormwater treatment wetlands, 659–662
Agricultural wastewater wetlands, 662–663
Agriculture:
  in Nebraska sandhills, 62
  wetland management by objective for, 497–499
Alaska:
  black spruce peatlands, 430–431
  exclusion of wetlands from Food Security Act definition, 40, 41
  wetlands acreage, 49
Alder (*Alnus*), 82
  gas transport in, 223
  mutualism, 227
Algae:
  and decomposition process, 282–283
  in freshwater tidal marshes, 290
  and nitrogen fixation, 187, 188
  in salt marsh food web, 277–279
Alligator mississippiensis (American alligator), 114, 396, 537–538
Allogenic succession, 215, 228, 230
  autogenic succession vs., 227–235
  community concept and continuum idea, 232–234
  Lake Michigan dunes, 230–232
  linear directed change, 234
  seed banks, 234–235
Alluvial cypress swamp, 380
Alluvial river swamp, 379, 381
Alnus (alder), 82, 223, 227

Alnus barbata, 82
Alnus glutinosa, 82, 223
Alopochen aegyptiaca (Egyptian goose), 88
Alteration of wetlands, 482–492
  by drainage, dredging, and filling, 482–486
  for highway construction, 487–488
  hydrologic modifications, 486–487
  by mineral and water extraction, 490–491
  for peat mining, 488–489
  by water pollution, 491
Amazon basin, 77–78
Amazon River, 78, 119
American alligator (*Alligator mississippiensis*), 114, 396, 537–538
American *bulrush* (*Scirpus americanus*), 67, 68
American eel (*Anguilla rostrata*), 292
Ammonia, 198
Ammonia transformations, 183–184
Ammonia volatilization, 183, 208, 209
Ammonification, 181. *See also* Nitrogen mineralization
Ammonium nitrogen, 153, 183–184
Amphibians:
  freshwater marshes, 355–356
  freshwater swamps, 396–397
  peatlands, 434
  in tidal freshwater wetlands, 294–295
Anadromous species, 292
Anaerobic bacteria, 281
Anaerobic processes, 195–196
Anaerobic respiration, 224–226
Anaerobiosis, 279–280

Anammox (anaerobic ammonium oxidation), 188
Anas (dabbling ducks), 531
Anas erythrorhynchos (red-billed duck), 88
Anas platyrhynchos (mallard), 532
Anas undulata (yellow-billed duck), 88
Anguilla rostrata (American eel), 292
Anishinaabe, 7
Anoxia, 215, 220–221
Anoxygenic photosynthesis, 195
Aquaculture, 497–499
Aquatic civilizations, 5, 479
Aquatic habitat, 276
Aquic condition, 169–170
Aquifer recharge, 544
Archaea, 197
Arctic salt marshes, 262
Ardea cinerea (gray heron), 533
Ardea cocoi (white-necked heron), 533
Ardea herodias (great blue heron), 533
Ardea melanocephala (black-headed heron), 533
Ardea novaehollandiae (white-faced heron), 533
Arid riparian ecosystems:
  geographic extent, 377
  geomorphology and hydrology, 384
  vegetation, 390
Arid riparian forests, 390
Artificial management, 493, 494
Artificial wetland, 49
Asian wetlands, 56, 93–100
  China, 95–100
  Indian freshwater marshes, 93, 94
  Issyk Kul, 95
  southern Asia river deltas, 93–95
  Taiwan urban wetland park, 99, 100
  western Siberian lowlands, 93
Assimilatory nitrate reduction, 184
Assimilatory sulfate reduction, 194
Atchafalaya Basin, 65
Atchafalaya River, 65
Atchafalaya Swamp (Louisiana), 6
Athabasca River, 70
Atlantic coastal plain:
  swamp rivers of, 60–61
  tidal salt water marshes, 264
Atlantic white cedar (*Chamaecyparis thyoides*), 59, 375–376
ATP (adenosine triphosphate), 226
Australasia:
  loss of wetlands in, 50
  salt marshes, 265
  tidal salt marshes of, 265
  wetlands loss, 50
Australian wetlands, 56, 89–91
  eastern billabongs, 89–90
  losses of, 50

  tidal salt water marshes, 265
  western wetlands, 90–91
Autochthonous inputs, 344
Autogenic succession, 215, 227–228
  allogenic succession vs., 227–235
  community concept and continuum idea, 232–234
  Lake Michigan dunes, 230–232
  linear directed change, 234
  seed banks, 234–235
Avicennia (black mangrove), 314, 325
  adaptations, 324
  in basin mangroves, 317
  plant zonation, 321
  and pneumatophores, 220, 325
  rhizosphere oxygenation, 224
  salinity control, 323
  speciation, 312
Avicennia germinans, 314, 325

Bacillariophytes, 290
Backswamp, 382
Bacteria:
  and anoxygenic photosynthesis, 195
  as decomposers, 281
  and methanogenesis, 197
  and sulfate reduction, 194
  and sulfur cycle, 201
Bald cypress (*Taxodium distichum*), 220, 374, 375, 387
Bangladesh, 94–95
Bankfull discharge, 133
Barataria Bay estuary, Louisiana, 66
Barlett, Charles, 63
Basin mangroves, 317, 318
Basin morphology, 670–671
Basin wetland, 318, 320
Bay of Bengal Delta, 94–95
Bay of Fundy:
  salt marshes, 264
  tidal salt water marshes, 264
Beaver (*Castor canadensis*), 114, 295
  pelts from, 530–531
  as threat to reconstructed wetland, 682
Bengal tiger (*Panthera tigris tigris*), 95
Benthic habitat, 276
Beowulf, 9
Berezinski Bog, Byelorussia, 82–83
Big Cypress Swamp, Florida, 57, 119, 120
Billabongs, 33, 34, 89–90, 381, 382
Bi–Ob region (Russia), 93
Biogeochemical cycling, 179
Biogeochemistry, 179–212
  ammonia transformations and nitrification, 183–184
  anammox, 188
  carbon cycle, 195–202
  carbon–sulfur interactions, 201–202

  dissimilatory nitrate reduction to ammonia, 188
  ebullition and gaseous transport in plants, 200–201
  fermentation, 196–197
  freshwater marshes, 344–348
  freshwater swamps, 384–385
  and hypoxia, 189–192
  inflows and outflows, 180–181
  intrasystem cycling, 180
  iron and manganese transformations, 192
  methane emissions, 198–200
  methane oxidation, 197–198
  methanogenesis, 197
  nitrate transformations and denitrification, 184–187
  nitrogen cycle, 181, 183–192
  nitrogen fixation, 187–188
  nitrogen mineralization, 181
  nutrient budgets, 208–212
  oceans and estuaries, 205
  peatlands, 424–427
  phosphorus cycle, 202–204
  sinks, sources, and transformers of chemicals, 181, 182
  streams, rivers, and groundwater, 206–207
  sulfur cycle, 193–195
  water quality, 204–207
Biotic control of hydrology, 113–115
Birds:
  freshwater marshes, 357–358
  peatlands, 434–435
  as provisioning ecosystem service, 531–534
  in tidal freshwater wetlands, 293–294
Blackbird (*Icteridae*), 294, 358
Black-headed heron (*Ardea melanocephala*), 533
Black mangrove (*Avicennia*):
  adaptations, 324
  in basin mangroves, 317
  plant zonation, 321
  and pneumatophores, 220, 325
  rhizosphere oxygenation, 224
  speciation, 312
Black spruce (*Picea mariana*), 430
Black spruce-feathermoss forest, 428
Black spruce peatlands, 430–431
Blanket bogs, 422
Blue crab (*Callinectes sapidus*), 284
Blue-green algae, 188, 290
BOD (biochemical oxygen demand), 650
Bog, 33, 34, 413
Bog energy flow estimates, 440–442
Bog-iron, 192
Bottomland, 34
Bottomland hardwood forest, 118, 374, 377

Branta canadensis (Canada goose), 114, 532, 682
Building materials, 10–14
Bulk density, 163
Burmese python (*Python molurus bivittatus*), 601–602
Buttonwood (*Conocarpus erecta*), 321
Buttresses, 218, 219, 394
Byrd, William, III, 8

Cajuns, 5
California:
   algal productivity on coast, 279
   selenium contamination at Kesterson National Wildlife Refuge, 401
   wetland loss, 16, 66, 485
Callinectes sapidus (blue crab), 284
Calusa, 6
Camargue (France), 5, 6, 78, 79
Canada goose (*Branta canadensis*), 114, 532, 682
Canadian Wetland Classification System, 423–424, 465, 466
Canadian wetlands, 56
   definitions of, 37–38
   Great Lakes Wetlands/St. Lawrence Lowlands, 66–68
   Hudson–James Bay lowlands, 68–70
   losses of, 50, 54–55
   Peace–Athabasca Delta, 70–71
   peatland restoration, 605–607
   peatlands of Canada's central and eastern provinces, 68
   prairie potholes, 61–62
Capital costs of treatment wetlands, 688–689
Cap Tourmente freshwater marsh complex (Canada), 66–68
Carabell v. U.S. Army Corps of Engineers, 519
Carbon budget, peatlands, 443–445, 568, 578–579
Carbon cycle, 195–202. *See also* Global carbon cycle
   carbon–sulfur interactions, 201–202
   ebullition and gaseous transport in plants, 200–201
   fermentation, 196–197
   methane emissions, 198–200
   methane oxidation, 197–198
   methanogenesis, 197
Carbon export, 151–153
Carboniferous, 3
Carbon sequestration, 568–570
Carbon storage:
   in mangroves, 330–332
   in peatlands, 444, 567, 579
Carbon–sulfur interactions, 201–202

Carex (sedge):
   in peat composition, 164
   swamp productivity/hydrology relationship, 147
Carnivorous plants, 432–433
Carolina pocosins, 59, 431
Caspian sea, 82
Castor canadensis (beaver), 114, 295
   pelts from, 530–531
   as threat to reconstructed wetland, 682
Catadromous species, 292
Cat clays, 319
Cation exchange, 426
Cation exchange capacity, 165, 166
Cattail (*Typha*), 73, 74
   adaptations to waterlogging/flooding, 222
   in centrifugal organization model, 239
   in colonized wetlands, 635
   in Everglades, 601, 602
   and flow-through, 146
   freshwater marsh zonation, 349
   and nutrient enrichment, 659–660
   nutrient limitations, 365
   photosynthetic efficiency, 359–360
   in restored wetlands, 635
   and upland soils, 627–630
   and wetland primary succession, 246–249
Cedarburg Bog (Wisconsin), 421
Cedar string bog and fen complex, 428
Central American wetlands, 72–74
Centrifugal organization concept (succession), 238–240
Chamaecyparis thyoldes (white cedar), 59, 375–376, 394, 395
Channelized streamflow, 130–132
Cheia, 76
Chemical loadings:
   design graphs, 671–672
   empirical models, 675–676
   retention rates, 673–674
   in treatment wetlands design, 671–676
Chemosynthetic bacteria, 192
Chen (snow goose), 114, 682
China wetlands, 95–100
   loss of, 49, 50
   northeastern wetlands, 97
   Qinghai-Tibetan Plateau, 97, 98
   river deltas, 95, 96
   urban wetland parks, 98–99
   Yangtze River wetlands, 95, 97
Chlorophytes, 290
Chongming Island, China, 95, 96
CICOLMA (Coastal Research Center La Mancha, Veracruz, Mexico), 72
Circular 39 classification (U.S. Fish and Wildlife Service), 457–459
Cladium (sedge), 147
Cladium jamaicense (sawgrass), 57, 602

Classification of wetlands, *see* Wetland classification
Classification of Wetlands and Deepwater Habitats, 460–464, 466
Clay:
   depletions, 168, 170
   and phosphorus cycle, 204
   phosphorus sorption onto, 204
Clean Water Act, 39, 40, 507–509, 518
Clean Water Act amendments (1977), 39
Climate change, 563–584
   carbon budgets for peatlands, 578–579
   carbon sequestration, 568–570
   causes of, 565–567
   climate change feedbacks, 577–578
   coastal wetlands, 579–582
   effects on wetlands, 579–584
   inland wetlands, 582–584
   methane emissions, 570–577
   peat storage and global carbon budget, 568
   wetlands in global carbon cycle, 567–579
Climate regulation, 543–545
   aquifer recharge, 544
   water quality, 544–545
Climate zones, 47
Clinton, Bill, and administration, 509
Clonal dominant species, 236
Coastal marshes, 630. *See also* Tidal salt marshes
Coastal Plain (United States), 264
Coastal protection, 540–543
Coastal Research Center La Mancha (CICOLMA) (Veracruz, Mexico), 72
Coastal restoration, 607–622
   after Indian Ocean tsunami, 618
   Delaware Bay salt marsh, 608–613
   deltas, 619
   mangrove swamps, 616–618
   New York City region, 614–616
   rivers, 619–620
   River Skjern, Denmark, 620–622
   salt marshes, 607–616
Coastal salt marsh, 165. *See also* Tidal salt marshes
Coastal squeeze, 580
Coastal wetlands:
   classification of, 459–460
   effects of climate change on, 579–582
   hydrology/productivity relationship, 150–151
Coastal wetland ecosystem, 456
Coastal Zone Management Program, 518
Coliforms, 663
Colorless sulfur bacteria (CSB), 194
Columbus, Christopher, 74
Commensalism, 226–227

Common reed (*Phragmites australis*), 82, 89, 609, 613
  as invasive species, 353–354
  and reed swamps, 342
  in restored wetlands, 635
Community concept (succession), 232–234
Concentric domed bog, 420
Concretions, 170
Congolian swamp forests, 86
Conocarpus erecta (buttonwood), 321
Conservation of wetlands, 16–17
Conservation Reserve Program (CRP), 597
Constructed wetland:
  classification of, 648
  defined, 593
Consumers:
  freshwater marshes, 354–358
  freshwater swamps, 395–397
  mangrove swamps, 326–327
  peatlands, 433–435
  tidal freshwater wetlands, 290–295
  tidal salt marshes, 274–277
Consumption in tidal salt marshes, 281–283
Continuum idea (succession), 232–234
Convention of Wetlands, 39
Conversion factors, 709–712
Corkscrew Swamp, Greater Florida Everglades, 198, 200
Cotton grass (*Eriophorums*), 432
Cottonmouth moccasin (*Agkistrodon piscivorus*), 396–397
Cowles, H. C., 227, 230
Coypu (*Myocastor coypus*), 276–277
Crabs, 326
Cranberries, 9
Created wetland, 570, 571, 574. *See also* Restoration and creation of wetlands
Cropped wetlands, 597
Crow Wing County, Minnesota, 7
CRP (Conservation Reserve Program), 597
CSB (colorless sulfur bacteria), 194
Cultural ecosystem services, 545
Cumbungi swamp, 33, 34
Cumulative loss, 32
Cyanobacteria, 188, 290
Cypress (*Taxodium*), 57
  adaptations to waterlogging/flooding, 219
  and buttresses, 394
  longevity, 395
Cypress domes, 117, 378–380
Cypress strand, 379
Cypress swamps, 373
  geomorphology, 377–381
  hydrology/productivity relationship, 147
  vegetation, 386–388
Cypress-tupelo swamps, 374–375

Dabbling ducks (*Anas*), 293, 357, 531
DAI (days after inundation), 198, 200
Dalton's law, 137
Dambo, 34
Danube River delta, 82
Darcy's law, 137
Davis, John Henry, 318–321
Days after inundation (DAI), 198, 200
Day water lily (*Nymphaea nouchali* var. *caerulea*), 85
Decomposers, 281
Decomposition:
  freshwater marshes, 361–362
  of organic soils, 167
  peatlands, 439
  in tidal salt marshes, 281–283
Deepwater marsh, 33
Deepwater swamps, 598, 630
Defining wetlands, 27–42
  common terms for, 33–36
  by distinguishing features, 28–31
  formal definitions, 36
  legal definitions, 39–42
Deforestation, 77
Delaware Bay salt marsh, 608–613
Delineation, 28. *See also* Wetland delineation guidelines (U.S.)
Deltas, 34, 619. *See also specific deltas, e.g.*: Mississippi River Delta
Demand curve, 550
Denitrification, 184–187
  climate regulation, 543
  Gulf hypoxic zone, 189
  nutrient budgets, 301
  nutrient cycling, 153
  peatlands, 442
  as source of nitrous oxide, 567
  treatment wetlands, 671
Designer wetland, 246, 633
Design graphs, 671–672
Des Plaines River Wetlands, 666–668
Desulfovibrio, 194
Detention time, 670
Detritivores, 326
Diatoms, 290
Die-off of salt marshes, 284–285
Dimethyl sulfide, 194
Dinitrogen (N_2), 186
Discharge wetlands, 134–137
Dismal Swamp Land Company, 59
Dissimilatory nitrate reduction to ammonia (DNRA), 188
Dissimilatory nitrogenous oxide reduction, 184
Dissolved organic phosphorus (DOP), 203
Distinguishing features of wetlands, 28–31
Diversion wetlands, *see* River diversion wetlands

DMS (dimethyl sulfide), 194
DNRA (dissimilatory nitrate reduction to ammonia), 188
Dominance types, 463
DOP (dissolved organic phosphorus), 203
Drainage:
  alteration of wetlands by, 482–486
  U.S. wetland drainage history, 480–482
Drawdown, 494
Dredging, 482–486
Drop roots, 325
Ducks, 293–294, 531–534
Duck potato, 226
Duck stamps, 16–17
Dwarf cypress swamps, 378, 379
Dwarf mangroves, 317–318

East African tropical marshes, 86–88
Eastern Asia:
  salt marshes, 265
Eastern Australia billabongs, 89–90
Eastern North America:
  salt marshes, 264
Eat-outs, 114
Ebro Delta, 14, 78–79, 278
Ebullitive flux (ebullition), 201
Ecological engineering, 647
Ecological principles, 638–639
Ecological success, 594
Ecological valuation of ecosystem services, 546–550
  habitat evaluation procedures, 546
  hydrogeomorphic analysis, 547–550
Economic evaluation of ecosystem services, 550–558
  emergy analysis, 555–556
  energy analysis, 553–555
  opportunity costs, 551
  replacement value, 551–553
  valuing goods and services, 556–558
  willingness-to-pay methods, 551
Economics of treatment wetlands, 688–691
  additional benefits, 690
  capital costs, 688–689
  comparison of wetlands with conventional technology, 690–692
  operating and maintenance costs, 689
Ecosystem development, 240–251
  ecosystem engineers, 249
  landscape patterns, 250–251
  pulse stability, 244–245
  self-organization and self-design, 245–249
  strategy of, 242–249
  turnover rates and nutrient influxes, 242–244
  wetland insularity, 244
Ecosystem engineers, 114, 249

Ecosystem function:
  bog energy flow estimates, 440–442
  decomposition, 361–362, 439
  decomposition and consumption in salt marshes, 281–283
  die-off of salt marshes, 284–285
  energy flow, 402
  food webs, 362
  freshwater marshes, 358–366
  freshwater swamps, 397–403
  greenhouse gas emissions, 365–366
  herbivory, 361–362
  in mangrove swamps, 327–334
  nutrient budgets, 362–364, 402, 403, 442–445
  nutrient limitations, 364–365
  organic export in salt marshes, 283–284
  peat accumulation, 440
  peatlands, 436–445
  primary productivity, 277–281, 327–329, 358–361, 397–402, 436–438
  tidal freshwater wetlands, 295–301
  tidal salt marshes, 277–286
Ecosystem mass balance, 208
Ecosystem services, 527–559
  animals harvested for pelts, 529–531
  climate regulation, 543–545
  cultural, 545
  current paradigm of, 528–529
  ecological valuation, 546–550
  economic evaluation, 550–558
  endangered and threatened species, 536–538
  Faustian bargain in, 559
  fish and shellfish, 534–535
  flood mitigation, 539–540
  peat harvesting, 536
  problems and paradoxes of, 558–559
  provisioning services, 529–538
  quantifying, 545–559
  regulation of, 539–543
  storm abatement and coastal protection, 540–543
  timber and vegetation harvest, 535–536
  waterfowl and other birds, 531–534
Ecotones, 230
Ecotourism, 13–15
Egretta gularis (African reef heron), 88
Egyptian goose (*Alopochen aegyptiaca*), 88
Eleocharis sphaceiata, 199, 201
Elymus athericus, 273
Emergency Wetlands Resource Act, 518
Emergy, 553
Emergy analysis, 555–556
Empirical models, 675–676
Enchente, 76
Endangered species, 434, 536–538
Endla Bog, Estonia, 82–83
Energy analysis, 553–555

Energy flow:
  freshwater swamps, 402
  in tidal freshwater wetlands, 297–298
Ensenada del Pabellon (Mexico), 71
Environmental sieve model, 236–238
Epiphytic algae, 282
Ericaceous plants, 415, 432, 436
*Eriophorum*s (cotton grass), 432
Esox lucius (northern pike), 356
Estuaries:
  effects on mangrove swamps, 332–334
  water quality, 205
Estuarine system, 460, 461
Eucalyptus camaldulensis (river red gum), 90
Euprhates River, 89
European wetlands, 56, 78–83
  loss of, 49, 50
  Mediterranean Sea deltas, 78–79
  northern Europe coastal marshes, mud flats, and bays, 80–82
  peatlands, 82–83
  Rhine River delta, 79–80
  southeastern Europe inland deltas, 82, 83
Eutrophic (term), 423
Eutrophic peatlands, 422
Evapotranspiration, 137–142
  direct measurement of, 138–139
  and ecosystem development, 228
  effects of vegetation on, 141–142
  empirical estimates of, 139–141
  measuring, 154
  Thornthwaite equation, 139–141
Excentric raised bogs, 417, 420
Executive Orders, 504, 506–508
Exotic species, 634–635
Exports, hydrologic, 208

Fen, 33, 34, 146, 413
Fermentation:
  in anaerobic respiration, 225, 226
  in carbon cycle, 196–197
  carbon-sulfur interactions, 201
Fern, 288
Ferrous iron, 192
Ferrous sulfide, 195
Fibrists, 167. *See also* Peat
Fiddler crab (*Uca*), 326
Filling of wetlands, 482–486
Filter feeders, 326
Finch (Fringillidae), 294
Fish:
  freshwater marshes, 356–357
  freshwater swamps, 396
  as provisioning ecosystem service, 534–535
Flarks, 420, 421
Floating marshes, 288
Floods, 132–134

Flood control:
  hydrologic modifications for, 486–487
  wetland management by objective for, 499
Flood duration, 115
Flood frequency, 115
Flood mitigation, 539–540
Flood peak, 129
Flood pulse concept (FPC), 404–406
Flordia Everglades, 6–8
Florida:
  Big Cypress Swamp, 57, 119, 120
  energy analyses of wetlands, 555–556
  mangroves, 328
  mangrove swamps, 313–334
  rating wetlands in, 470–471
Florida Everglades, 9, 55–58, 492
  restoration of, 598–603
  treatment wetlands, 659–662
  water quality restoration, 598–602
Florida panther (*Puma concolor coryi*), 58
Flowthrough succession, 419
Fluted trunk, 219
Flycatcher (Tyrannidae), 294
Folists, 167
Food, 9–10
Food Security Act, 39
Food Security Act definition of wetlands, 40–41
Food webs, 362
Forested wetlands:
  introducing vegetation, 630
  restoration of, 597–598
Formal definitions of wetlands, 36
FPC (flood pulse concept), 404–406
France, 5, 6, 78, 79
Frankia alni, 227
Freedom Park, Florida, 656–659
Freshwater marshes, 341–366
  biogeochemistry, 344–348
  consumers, 354–358
  decomposition and herbivory, 361–362
  ecosystem function, 358–366
  food webs, 362
  greenhouse gas emissions, 365–366
  hydrology, 343–344
  introducing vegetation, 630
  invasive species, 353–354
  nutrient budgets, 362–364
  nutrient limitations, 364–365
  primary productivity, 358–361
  seed banks, 350–351
  species diversity, 350–353
  vegetation, 348–354
  zonation, 348–349
Freshwater swamps, 373–403. *See also* Riparian ecosystems
  biogeochemistry, 384–385
  consumers, 395–397

Freshwater swamps (*continued*)
  cypress swamps, 377–381, 386–388
  cypress-tupelo swamps, 374–375
  ecosystem function, 397–403
  energy flow, 402
  fish, 396
  geographic extent, 374–375
  geomorphology and hydrology, 377–381
  glaciated regions, 390–301
  invertebrates, 395–396
  nutrient budgets, 402, 403
  primary productivity, 397–402
  red maple swamps, 376, 381, 388–389
  reptiles and amphibians, 396–397
  tree adaptations, 391–395
  vegetation, 385–395
  white cedar swamps, 375–376, 381, 388
Freshwater tidal marshes, 259
Fringe mangroves, 316–318
Functional groups, 236
Functional guilds, 237
Fungi, 281

Gandau Nature Park (Taipei, Taiwan), 99–100
Ganges Delta, 94
Gardians, 78
Gaseous diffusion, 395
Gator holes, 114, 537
Geogenous peatlands, 423, 424
Geomorphology:
  arid ecosystems, 384
  cypress swamps, 377–381
  freshwater swamps, 377–381
  mesic ecosystems, 382–383
  red maple swamps, 381
  riparian ecosystems, 381–384
  white cedar swamps, 381
Geukensia demissa (ribbed mussel), 282
Glaciated regions, freshwater swamp vegetation in, 390–301
Gleasonian wetland, 238
Gleization, 168
Gleying, 168
Global carbon budget, 568
Global carbon cycle:
  carbon budgets for peatlands, 578–579
  carbon sequestration, 568–570
  climate change feedbacks, 577–578
  methane emissions, 570–577
  peat storage and global carbon budget, 568
  wetlands in, 567–579
Global Lakes and Wetlands Database (GLWD), 48
Glycolysis, 196. *See also* Fermentation
Goals for restoration, defining, 623–624
Gravel bed, 677–678
Gray heron (*Ardea cinerea*), 533

Great Black Swamp, 63–64
Great blue heron (*Ardea herodias*), 533
Great Dismal Swamp, 8, 59–60, 114, 480
Great Kankakee marsh, 63
Great Lakes Wetlands/St. Lawrence Lowlands, 66–68
Great Plains playas, 62–63
Green algae, 290
Greenhouse gases:
  carbon dioxide, 565, 566
  methane, 444, 566, 570, 572–577
  nitrous oxide, 186, 567
  water vapor, 566
Greenhouse gas emissions:
  freshwater marshes, 365–366
  from treatment wetlands, 687–688
Groundwater, 134–137
  Darcy's law, 137
  recharge and discharge wetlands, 134–137
  and runoff/streamflow composition, 206
  water quality, 206–207
Groundwater depression wetland, 136, 137
Groundwater flows, 154–155
Groundwater slope wetland, 136, 137
Grzimek, Bernhard, 87
Guilds, 236
Gulf of Mexico, 189–192
Gulf of Mexico oil spill (2010), 66

Habitat Evaluation Procedure (HEP), 546–547
Habitat loss, mitigating, 593–597
  measuring success of, 594–596
  mitigation banks, 596
Hackensack Meadowlands, New Jersey, 614–616
Haliclona implexiformis, 227
Halophytes, 323
Hammock, 57, 318
Hannah Bay Bird Sanctuary (Canada), 69
Hemists, 167
HEP (Habitat Evaluation Procedure), 546–547
Herbivory decomposition, 361–362
Heron, 533
HGM (hydrogeomorphic) analysis, 547–550
HGM (hydrogeomorphic) wetland classification, 465, 467–468
High marshes, 266
Highway construction, 487–488
Histosols, 167. *See also* Organic soils
HLR (hydraulic loading rate), 669–670
Hong Kong Wetland Park, 99, 100
Horticulture, natural succession vs., 633–635

Houghton Lake, Michigan, 651–653
Hudson–James Bay lowlands, 68–70
Human impact on wetlands, 4–8, 483.
  See also Wetland management
Humification, 167. *See also* Decomposition
Hurricanes:
  Andrew, 329–330
  effect on mangrove swamps, 329–330
  Katrina, 66, 542–543
  Rita, 66
Hussein, Saddam, 598
Hydrarch succession, 228, 229
Hydraulic civilizations, 5, 479
Hydraulic conductivity, 163–165
Hydraulic loading rate (HLR), 669–670
Hydric soils, 38, 161, 513–514, 627–630
Hydrochory, 394
Hydrodynamics, 467
Hydroelectric power dams, 77
Hydrogen sulfide, 194, 195
Hydrogeomorphic (HGM) analysis, 547–550
Hydrogeomorphic (HGM) wetland classification, 465, 467–468
Hydrogeomorphology, 112
  mangrove swamps, 314–318
  pannes, 267, 268
  tidal creeks, 267, 268
  tidal salt marshes, 265–268
Hydrologic exports, 208
Hydrologic modifications, 486–487
Hydrologic pathways, 208
Hydrology, 30, 111–155, 514
  accumulation of organic material, 151–153
  biotic control of, 113–115
  creation/maintenance of, 625–627
  cypress swamps, 377–381
  evapotranspiration, 137–142
  freshwater marshes, 343–344
  freshwater swamps, 377–381
  groundwater, 134–137
  hydroperiods, 115–121
  importance of, 112–115
  nutrient cycling and availability, 153
  and peatlands development, 416–419
  precipitation, 126–127
  and primary productivity, 146–151
  and productivity, 146–151
  recurrence interval, 133–134
  red maple swamps, 381
  renewal rate/turnover rate of water, 125
  riparian ecosystems, 381–384
  seiches, 142–144
  and species richness, 145–146
  study techniques for, 153–155
  surface flow, 127–133
  of tidal salt marshes, 266

tides, 142, 143
in treatment wetlands design,
    668–670
vegetation composition and species
    richness, 145–146
water budget, 119, 121–125
and wetland function, 143–153
white cedar swamps, 381
Hydrologic restoration, 598–604
    Florida Everglades, 598–602
    Mesopotamian Marshlands, 598,
        603–604
Hydroperiods, 115–121
    definitions of, 115
    and pulsing water levels, 119
    in treatment wetlands design, 669
    year-to-year fluctuations in, 119–121
Hydrophytes, 215, 216
Hydrophytic vegetation, 38
Hypoxia, 189–192

Igapó, 77
Illinois Drainage Levee Act, 482
Illinois Farm Drainage Act, 482
India, 94–95
Indian freshwater marshes, 93, 94
Indian Ocean tsunami (December 2004):
    coastal restoration following, 618
    mangrove swamps and, 541–542
Inflow, 128, 208
Inland marshes, 266
Inland wetlands:
    ecosystem, 456
    effects of climate change on, 582–584
Inputs, 208
Interception, 127
Intermittently exposed, 115
Intermittently flooded, 115
Intermittent subsystem, 461, 462
International legal protections, 521–524
    North American Waterfowl Management
        Plan, 522–524
    Ramsar Convention, 521–523
International Ramsar Convention
    classification system, 465, 466
International Union for the Conservation of
    Nature and Natural Resources
    (IUCN), 38–39
Interstitial perennials, 236
Intertidal subsystem, 461, 462
Intrasystem cycling, 180–181, 208
Introducing vegetation:
    forested wetlands, 630
    freshwater marshes, 630
    mangrove swamps, 633
    natural succession vs. horticulture,
        633–635
    planting techniques, 632–633
    salt marshes, 633

Invasive species:
    in Florida Everglades, 600, 602–603
    freshwater marshes, 353–354
    tidal salt marshes, 265
Inventory, see Wetland inventory
Invertebrates:
    freshwater marshes, 354–355
    freshwater swamps, 395–396
Iraq, 5, 89, 598, 603–604
Iron:
    and gleization, 168
    transformation of, 192
    transformations, 192
Iron bacteria, 192
Iron masses, 170
Irregularly exposed, 115
Irregularly flooded, 115
Irrigation, 62
Isolated wetlands, 28
Issyk Kul, 95
IUCN (International Union for the
    Conservation of Nature and Natural
    Resources), 38–39

Jacana spinosa (Northern jacana), 74
Jurisdictional wetlands, 41–42

Kahikatea Swamp, New Zealand, 91
Keeling, C. David, 565
Kennedy, Anthony, 520
Keoladeo National Park (Bharatpur, India),
    93, 94
Kesterson National Wildlife Refuge,
    California, 401
Kissimmee River, 58
Knees, 393–394
Konik horses, 79–81
Koontz, Coy, 520
*Koontz v. St. Johns River Water Management
    District*, 520–521
Krefeld system (Max-Planck-Institute
    process), 648
Kyrgystan, 95

Lacustrine system, 461, 462
Lacustrine wetlands, 116, 117
Lagoon, 34
Laguna de Términos (Campeche, Mexico),
    71
Laguncularia (white mangrove):
    in basin mangroves, 317
    salinity control, 323
Lake Drummond, 59
Lake-edge swamps, 378, 379
Lake Erie, 64
Lake Michigan dunes, 230–232
Landfill leachate wetlands, 668
Landscape patterns, 250–251
Lapland, 8

Larch string bog and fen, 428
Larix laricina (tamarack), 430
Legal definitions of wetlands, 39–42
Legal protection of wetlands, 503–524
    Clean Water Act, 507–509, 518
    Coastal Zone Management Program, 518
    early Executive Orders, 504, 506
    Emergency Wetlands Resource Act, 518
    international, 521–524
    major federal laws, directives, and
        regulations, 505–506
    National Academy of Science studies,
        516–517
    National Flood Insurance Program, 518
    no net loss concept, 507, 591, 593, 595
    North American Waterfowl Management
        Plan, 522–524
    North American Wetlands Conservation
        Act, 518
    Ramsar convention, 521–523
    "swampbuster" provisions of 1985 Food
        Security Act, 509
    "takings" issue, 518
    in United States, 504–521
    U.S./ Supreme Court decisions, 519–521
    wetland delineation guidelines, 509–516
Legal success, 594
Lenticels, 220, 325
Lignocellulose, 281
Limiting nutrients:
    freshwater marshes, 364–365
    peatlands, 426, 427
Limnetic subsystem, 461, 462
Limnogenous peatland, 423, 424
Linear directed change (succession), 234
Liner, 676–677
Linnaeus, Carl, 8
Literary references to wetlands, 79
Littoral subsystem, 461, 462
Littorina irrorata (snail), 284
Llanos, 75–76
Loading rate, 669–670
Losses (hydrologic exports), 208
Loss of wetlands:
    by state in United States, 701–703
    worldwide, 48–50
Lotus, 83
Louisiana:
    Cajuns, 5, 6
    crayfish harvesting in, 10
    cypress swamps, 482
    energy and economic analyses of
        wetlands, 554–555
    Hurricane Katrina, 66, 542–543
    landscape patterns in wetlands, 250
    tidal salt marsh, 258
Louisiana Delta, 64–66
Low chroma, 171
Lower perennial subsystem, 461, 462

Low marshes, 266
Lucas v. South Carolina Coastal Council, 518
Lythrum salicaria (purple loosestrife), 353

Maintenance costs of treatment wetlands, 689
Mallard (*Anas platyrhynchos*), 532
Mammals:
   freshwater marshes, 357
   peatlands, 433–434
   tidal freshwater wetlands, 295
   tidal salt marshes, 276–277
Management by objective, 491–499
   for agriculture and aquaculture, 497–499
   for flood control and stormwater protection, 499
   for waterfowl and wildlife, 493–496
   for water quality enhancement, 499
Managing wetlands, *see* Wetland management
Manchurian wild rice, 9
Mangal, 34
Manganese, 192
Manganese oxides, 168
Mangrove, 34
Mangrove swamps, 311–334
   adaptations of vegetation, 322–326
   consumers, 326–327
   ecosystem function, 327–334
   effects on estuaries, 332–334
   in Ganges Delta, 94–95
   geographical extent, 312–314
   hurricane effects, 329–330
   hydrodynamic classification, 315–318
   hydrogeomorphology, 314–318
   introducing vegetation, 630
   organic material storage and export, 330–332
   primary productivity, 327–329
   response to sea-level rise, 330
   restoration of, 616–618
   soil acidity, 319
   soils and salinity, 319
   vegetation, 319–326
   zonation, 320–322
Manning equation, 132
Marceno, 497
Marine system, 460, 461
Marshes, 33, 34. *See also specific types, e.g.:* Tidal marshes
Marsh Arabs, 5, 89
Maryland, 258
Masses, 170
Matrix perennials, 236
Matsalu State Nature Preserve, 81–82
Meander scroll, 382

Mediterranean Sea:
   salt marshes in, 264
   salt marsh primary productivity, 278
   tidal salt marshes of, 264
Mediterranean Sea deltas, 78–79
Mekong Delta, 94, 593
Mesic riparian ecosystems, 376, 382–383
Mesopotamian Marshlands, 89, 598, 603–604
Mesotrophic peatlands, 422
Methane emissions, 198–200, 298, 299
   in carbon cycle, 198–200
   and climate change, 570–577
   from freshwater marshes, 365
Methane oxidation, 197–198
Methanogenesis, 197, 572
Methanogens, 197–198
Methanotropic bacteria, 197
Mexico, 71–72
Miccosukee, 6, 7
Microbes, 113, 114
Middle East wetlands, 56, 89
Millennium Ecosystem Assessment, 528
Mills, Robert, 61
Mine drainage wetlands, 653–655
Mineral cycles, 241–242
Mineral extraction, alteration of wetlands in, 490–491
Mineral soils, 162, 167–171
   aquic condition in, 169–170
   cation exchange capacity, 166
   differences between organic soils and, 163–165
   hydraulic conductivity, 165
   as hydric soil, 170
   oxidized rhizosphere in, 168, 169
   physiochemical features, 163–165
   rate of formation, 168
   redox concentrations in, 168, 169
   reduced matrices and redox depletions, 168
Minerotrophic peatlands, 422
Mire, 34, 35
Mire ecologists, 17
Mississippi, 278
Mississippi-Ohio-Missouri (MOM) river basin, 189
Mississippi River, 64, 664–666
Mississippi River basin, 189
Mississippi River Delta, 65
   salt marshes in, 267
   and sea-level rise, 580, 581
Mississippi River deltaic marshes, 260
Mitigating habitat loss, 593–597
   measuring success of, 594–596
   mitigation banks, 596
Mitigation ratio, 594, 595
Mitigation wetland, 593
Modifiers (for classification), 463, 464

Momoge National Nature Reserve (Jilin Province, China), 97
MOM (Mississippi-Ohio-Missouri) river basin, 189
Montezuma cypress (*T. distichum* var. *mexicanum* Gordon), 374
Moor, 33, 34
Moose River Bird Sanctuary (Canada), 69
Morphological adaptations (vascular plants), 216–220
   adventitious roots, 218, 219
   aerenchyma, 216–218
   pneumatophores, 220
   stem elongation, root adaptations, and lenticels, 220
   stem hypertrophy, 218, 219
Mosquito control at treatment wetlands, 683–686
Mottles (redox concentrations), 168–170
Movies, wetland depictions in, 9, 10
Muck, 161, 163, 167
Mucky peat, 161, 167
Mud barrens, 268
Municipal wastewater wetlands, 650–653
Munsell soil color chart, 171
Murray–Darling watershed, 593
Muskeg, 33, 34
Muskrat (*Ondatra zibethicus*), 114, 276–277, 295, 496
   pelts from, 530
   as threat to reconstructed wetland, 682
Mussels, 81
Mutualism, 226–227
Myocastor coypus (coypu, nutria), 276–277, 295, 530

NAD (nicotinamide adenine dinucleotide), 226
National Academy of Science studies, 516–517
National Flood Insurance Program, 518
National Research Council (NRC), 38
National Wetlands Inventory (NWI), 472–474
National Wetlands Policy Forum, 507
National Wetlands Working Group (Canada), 37
Native Americans, 5–8
Natural management, 493
Natural Resources Conservation Service (NRCS), 40–41
Natural Resources Defense Council v. Callaway, 40, 507
Natural succession, horticulture vs., 633–635
Nature Conservancy, 62
Navigation, hydrologic modifications for, 487
Neap tides, 142

Nebraska sandhills, 62–63
Nekton, 291–293
Nelumbo nucifera (water lotus), 82
Nernst equation, 174
Nerodia (water snake), 295
Netherlands, 79–80
Net marginal benefit, 556
Net primary production, 296
New England:
　salt marshes, 264, 270–271
　tidal salt water marshes, 264
New Orleans, Louisiana, 66, 542–543
New York City region, 614–616
New Zealand, 56, 91–92
　sphagnum harvesting, 11
　wetlands loss, 49, 50, 478, 593
Niering, Bill, 240
Nitrate transformations, 184–187
Nitrification, 183–187
Nitrobacter, 184
Nitrogen:
　and decomposition process, 282
　in peatland nutrient budgets, 442–443
Nitrogenase, 187
Nitrogen cycle, 181, 183–192, 300, 301
　ammonia transformations and nitrification, 183–184
　anammox, 188
　dissimilatory nitrate reduction to ammonia, 188
　and hypoxia, 189–192
　hypoxia and, 189–192
　nitrate transformations and denitrification, 184–187
　nitrogen fixation, 187–188
　nitrogen mineralization, 181
Nitrogen fixation, 187–188
Nitrogen mineralization, 181
Nitrosomonas, 183
Nitrous oxide (N_2O), 186–187, 365
Nodules, 170
Nonconsumptive use values, 545
No net loss concept, 507, 591, 593, 595
Normandy (France), 278
North American Waterfowl Management Plan, 522–524
North American wetlands, 55–72. *See also* Canada; United States
　changes in, 49–55
　Florida Everglades, 55–58
　Great Black Swamp, 63–64
　Great Dismal Swamp, 59–60
　Great Kankakee marsh, 63
　Great Lakes Wetlands/St. Lawrence Lowlands, 66–68
　Hudson–James Bay lowlands, 68–70
　losses of, 50
　Louisiana Delta, 64–66
　Mexico, 71–72

Nebraska sandhills and Great Plains playas, 62–63
Okefenokee Swamp, 58–59
Peace–Athabasca Delta, 70–71
peatlands of Canada's central and eastern provinces, 68
Pocosins (Carolinas), 59
prairie potholes, 61–62
river deltas, 260
San Francisco Bay, 66
swamp rivers of South Atlantic coast, 60–61
wetlands loss, 50
North American Wetlands Conservation Act, 518
Northeastern China wetlands, 97
Northern Europe:
　coastal marshes, mud flats, and bays, 80–82
　salt marshes in, 262, 264
　tidal salt marshes of, 262, 264
Northern jacana (Jacana spinosa), 74
Northern jacana (*Jacana spinosa*), 74
Northern peatlands, 165
Northern pike (*Esox lucius*), 356
NRC (National Research Council), 38
NRCS (Natural Resources Conservation Service), 40–41
Nuphar lutea (water lily), 221
Nutria, 496
Nutria (*Myocastor coypus*), 276–277, 295, 530
Nutrients, limiting, 426, 427
Nutrient availability, 153, 164
Nutrient budgets, 208–212
　components, 180
　freshwater marshes, 362–364
　freshwater swamps, 402, 403
　peatlands, 442–445
　in tidal freshwater wetlands, 301
Nutrient cycling, 153
Nutrient deficiency, 432
Nutrient limitations, 364–365
Nutrient spiraling, 405–406
NWI (National Wetlands Inventory), 472–474
Nymphaea nouchali var. *caerulea* (day water lily), 85
Nyssa, 394
Nyssa aquatica (water tupelo), 374

Obligate annuals, 236
Ob River, 93
Oceans, water quality of, 205
Ogallala Aquifer, 62
Ohio, 469–470
Oil production, 65
Okavango Delta (Bostswana, southern Africa), 83–86

Okefenokee National Wildlife Refuge, 58
Okefenokee Swamp, 58–59
Olentangy River Wetland Research Park, Ohio State University, 666–668
Oligotrophication, 423
Oligotrophic peatlands, 422
Ombrogenous peatland, 423, 424
Ombrotrophic peatlands, 422
Ondatra zibethicus (muskrat), 114, 276–277, 295, 496
　pelts from, 530
　as threat to reconstructed wetland, 682
Ontario, Canada, 55
Operating costs of treatment wetlands, 689
Opportunity costs, 551
Organic export in tidal salt marshes, 283–284
Organic material:
　hydrology and accumulation of, 151–153
　in mangrove swamps, 330–332
　in tidal freshwater wetlands, 298–300
Organic soils, 161–162, 165–167
　botanic origin of, 166–167
　cation exchange capacity, 166
　characteristics of, 167
　classification and characteristics, 167
　decomposition of, 167
　differences between mineral soils and, 163–165
　groups of, 167
　physiochemical features, 163–165
Orinoco River delta, 74–75
Orthophosphates, 203
Outflow, 128, 208
Outwelling, 331
Outwelling hypothesis, 283–284
Overgrowth, 433
Overland flow, 128
Overwash mangrove islands, 317
Oxbow, 33, 34, 381, 382
Oxidation, 173
Oxidized rhizosphere (oxidized pore linings), 168–170, 224
Oxygen, 171–172, 192

Paalsa peatlands, 422
Pakihi, 34, 90, 415
Palmer Drought Severity Index (PDSI), 344
Palm swamps, 72
Palo Verde National Park (Costa Rica), 72–74
Paludification, 416–419
Palustrine system, 461, 462
Panicum hemitomon, 224
Pannes, 267–268
Pantanal, 76–77

Panthera tigris tigris (Bengal tiger), 95
Paraguay–Paraná River basin, 76
Particulate organic matter (POM), 331
Pathogens at treatment wetlands, 684, 687
Patterned fens, 420–421
Patuxent River, Maryland, 298
PDSI (Palmer Drought Severity Index), 344
Peace–Athabasca Delta, 70–71
Pearsall, W. H., 228
Peat, 161–163
  characteristics of, 167
  and ecosystem development, 228, 230
  from wetlands, 10–14
Peat accumulation, 440
Peat harvesting, 536
Peatlands, 33, 34, 90, 92, 413–445
  adaptations of vegetation, 431–433
  biogeochemistry, 424–427
  black spruce peatlands, 430–431
  bog energy flow estimates, 440–442
  of Canada's central and eastern provinces, 68
  Canadian Wetland Classification System, 423.424
  carbon budgets for, 578–579
  Carolina pocosins, 431
  chemistry-based classification, 422–423
  classification of, 419–424, 457
  and climate change, 578
  consumers, 433–435
  decomposition, 439
  ecosystem function, 436–445
  and ecosystem productivity, 146
  of Europe, 82–83
  geographic extent, 414–416
  hydrology and development of, 416–419
  hydrology-based classification, 423, 424
  hydroperiod of, 118–119
  landscape-based classification, 419–422
  nutrient budgets, 442–445
  peat accumulation, 440
  primary productivity, 436–438
  restoration of, 605–607
  soil permeability, 164
  vegetation, 427–433
Peat mining, 488–489
Peat mosses, 433
Peat storage, 568
Peaty muck, 167
Pelts, animals harvested for, 529–531
Penman equation, 140–141
Perched wetland, 135, 155
Permanently flooded, 115
Permeability, 137
PET (potential evapotranspiration), 141
Philippines, 10
Phosphate mining, 490

Phosphorus, 153, 203–204, 365
Phosphorus budget, 208, 210
Phosphorus cycle, 202–204
  and clay, 204
  co-precipitation of phosphorus, 203–204
  phosphorus release in anaerobic conditions, 204
Photosynthesis, 195
Phragmites (reed grass), 79, 82, 95, 416, 609
  and flow-through, 146
  in peat composition, 164
  photosynthetic efficiency, 359, 360
  and reed swamps, 342
Phragmites australis (common reed), 82, 89, 609, 613
  as invasive species, 353–354
  and reed swamps, 342
  in restored wetlands, 635
Phreatophytes, 390
Physiochemical environment, 30, 112, 113
Physiognomy, 459
Physiological adaptations (vascular plants), 220–226
  anaerobic respiration, 224–226
  lower water uptake, 224
  pressurized gas flow, 221–223
  rhizosphere oxygenation, 223–224
  sulfide avoidance, 224
Phytoplankton, 290
Picea mariana (black spruce), 430
Piezometers, 154–155
Pine (*Pinus serotina*), 59
Pitcher plant (*Sarracenia purpurea*), 432, 433
Plants, *see* Vegetation
Plant species functional groups (succession), 235–236
Playas, 34, 63
Pleistocene, 62
Pneumatophores, 219, 220, 325
  of freshwater swamps vegetation, 393–394
  in mangrove swamps, 325
Pocosins (Carolinas), 34, 59, 415–416, 431
Point bar, 382
Pokelogan, 34
Polar Bear Provincial Park (Ontario, Canada), 68–70
Pollution, *see* Water pollution
POM (particulate organic matter), 331
Pond cypress (*Taxodium distichum* var. *imbricarium*), 374, 375, 387
Pools, 208
Poor fens, 422
Poor swamp forest, 428
Pore linings, *see* Oxidized rhizosphere
Porosity, 163

Potential evapotranspiration (PET), 141
Pothole, 34
Prairie potholes, 61–62, 119, 120
Prairie Pothole Region (PPR), 582–583
Precipitation, 126–127
  measuring, 154
  and runoff, 206
Pressurized gas flow, 221–223
Primary productivity, 146–151
  freshwater marshes, 358–361
  freshwater swamps, 397–402
  hydrology and, 146–151
  in mangrove swamps, 327–329
  peatlands, 436–438
  in tidal freshwater wetlands, 295–297
  tidal salt marshes, 277–281
Prop roots, 219, 220, 325
Provisioning ecosystem services, 529–538
  animals harvested for pelts, 529–531
  endangered and threatened species, 536–538
  fish and shellfish, 534–535
  peat harvesting, 536
  timber and vegetation harvest, 535–536
  waterfowl and other birds, 531–534
Puerto Rico, 328
Pulse stability concept, 146, 148, 233, 244–245
Pulsing water levels, 119
Puma concolor coryi (Florida panther), 58
Purple loosestrife (*Lythrum salicaria*), 353
Python molurus bivittatus (Burmese python), 601–602

Qinghai-Tibetan Plateau, 97, 98
Quaking bog, 417
Quaking bog succession, 417–418
Quantifying ecosystem services, 545–559
  ecological valuation, 546–550
  economic evaluation, 550–558
  emergy analysis, 555–556
  energy analysis, 553–555
  Faustian bargain in, 559
  habitat evaluation procedures, 546
  hydrogeomorphic analysis, 547–550
  opportunity costs, 551
  problems and paradoxes of, 558–559
  replacement value, 551–553
  valuing goods and services, 556–558
  willingness-to-pay methods, 551
Quickflow, 128–129

Raised bogs, 417–420
Ramsar Convention, 38–39, 521–523
Ramsar Convention classification system, 465, 466
Rapanos v. United States, 519–520
Rapid stem elongation, 220
Rating curve, 130

Rating wetlands, 469–471
  in Florida, 470–471
  in Ohio, 469–470
  in Washington State, 469
Rational runoff method, 129–130
Raupo, 33
Raupo swamp, 34
RCC (river continuum concept), 404–406
Recharge wetlands, 134–137, 155
Recurrence interval, 133–134
Red-billed duck (*Anas erythrorhynchos*), 88
Red mangrove (*Rhizophora*):
  adaptations, 324
  inundation tolerance, 315
  mutualism, 227
  productivity, 331
  prop roots, 220
  speciation, 312
  viviparous seedlings, 226
Red maple (*Acer rubrum*), 59, 376
  in red maple swamps, 388
  root adaptations, 220
  root systems, 395
  swamp productivity/hydrology relationship, 147
Red maple swamps:
  geographic extent, 376
  geomorphology, 381
  hydrology/productivity relationship, 147
  vegetation, 388–389
Redox concentrations, 168–170
Redox depletions, 168, 170
Redoximorphic features, 167–168
Redox potential, 153, 173
Reduced matrices, 168, 170
Reduction, 173
Reduction/oxidation (in soils), 171–176
Reed grass (*Phragmites*), 33, 79, 82, 95, 416, 609
  and flow-through, 146
  in peat composition, 164
  photosynthetic efficiency, 359, 360
  and reed swamps, 342
Reedmace, 33
Reedmace swamp, 34
Reedswamp, 33, 34, 342
Reference wetland, 548–549, 594
Regularly flooded, 115
Regulation of ecosystem services, 539–543
  flood mitigation, 539–540
  storm abatement and coastal protection, 540–543
Remote platforms, 471
Remote-sensing platforms, 471–472
Renewal rate, 125, 242–244
Replacement value, 551–553
Replacement wetland, 593

Reptiles:
  freshwater swamps, 396–397
  peatlands, 434
  in tidal freshwater wetlands, 294–295
Residence time, 125
Resource spiraling, 405–406
Respiration, 195
Restoration and creation of wetlands, 591–638
  after Indian Ocean tsunami, 618
  agricultural land restoration, 596–597
  coastal marshes, 630
  coastal restoration, 607–622
  deepwater swamps, 630
  defining goals, 623–624
  Delaware Bay salt marsh, 608–613
  deltas, 619
  ecological principles for, 638–639
  estimating success of, 635–638
  exotic or undesirable species, 634–635
  Florida Everglades, 598–603
  forested wetland restoration, 598
  forested wetlands, 630
  freshwater marshes, 630
  hydrologic and water quality restoration, 598–604
  hydrology creation/maintenance, 625–627
  mangrove swamps, 616–618, 633
  Mesopotamian Marshlands, 598, 603–604
  mitigating habitat loss, 593–596
  natural succession vs. horticulture, 633–635
  New York City region, 614–616
  peatland restoration, 605–607
  planting techniques, 630–633
  rivers, 619–620
  River Skjern, Denmark, 620–622
  salt marshes, 607–616, 629
  site selection, 625
  soils, 627–630
  starting points for, 591–592
  terminology related to, 592–593
Retention rates in treatment wetlands design, 673–674
Reverse estuary, 89
Rheotrophic peatlands, 422
Rhine River delta, 79–80
Rhizobium, 188
Rhizophora (red mangrove):
  adaptations, 324
  adaptations to waterlogging/flooding, 219
  inundation tolerance, 315
  mutualism, 227
  plant zonation, 321
  productivity, 331
  prop roots, 220

  rhizosphere oxygenation, 224
  salinity control, 323
  speciation, 312
  viviparous seedlings, 226
Rhizophora mangle, 227, 321
Rhizosphere oxygenation, 223–224
Rhone River Delta, 5, 6, 78, 79, 278
Ribbed mussel (*Geukensia demissa*), 282
Rice Lake (Crow Wing County, Minnesota), 7
Rich fens, 422
Rich swamp forest, 428
Riparian ecosystems, 35
  arid and semiarid riparian forests, 390
  arid ecosystems, 377, 384
  flood pulse concept, 406
  geographic extent, 376–377
  geomorphology and hydrology, 381–384
  mesic ecosystems, 376, 382–383
  river continuum concept, 404–406
  and river exchanges, 404–406
  southeastern U.S. bottomland forests, 389–390
  vegetation, 389–390
Riparian wetlands, 132–134, 152
Rivers. *See also specific rivers*
  restoration of wetlands, 619–622
  water quality, 206–207
River continuum concept (RCC), 404–406
River diversion wetlands, 663–668
River exchanges:
  flood pulse concept, 406
  and riparian ecosystems, 404–406
  river continuum concept, 404–406
Riverine mangroves, 317, 318
Riverine system, 461, 462
Riverine wetlands, hydroperiod of, 117–118
River Murray, New South Wales, Australia, 199
River red gum (*Eucalyptus camaldulensis*), 90
River Skjern, Denmark, 620–622
River water, chemical concentrations of, 205
Rochefort, Line, 605
Rock-reed filter, 677
Roots:
  adaptations of, 220
  adventitious, 218, 219
  of freshwater swamps vegetation, 395
  morphological adaptations of, 220
  prop roots, 325
  shallow roots, 395
Root-zone method (*Wurzelraumentsorgung*), 649–650, 677
Ruderal annuals, 236
Runoff, 127–130, 206, 207
Ruppia (wigeon grass), 268

Sagittaria lancifolia, 365
Sagittaria latifolia, 226
St. Lawrence River, 55
St. Lawrence River Lowlands, 66–68
Salicornia, 79
Salinity:
  mangrove swamps, 319
  salt marshes, 269–270
  tidal salt marshes, 269–270
Salinity control, 323, 325
Salix (willow), 219
Salt, 9, 205
Salt exclusion, 323
Salt marshes, 35. *See also* Tidal salt marshes
  conditions for development of, 259
  die-off of, 284–285
  inland, 349–350
  introducing vegetation, 630
  and Louisiana delta, 64–65
  restoration of, 607–616
  vegetation zonation, 272
Salt marsh cordgrass (*Spartina*):
  and flow-through, 146
  in inland marshes, 280
Salt marsh hay (*Spartina patens*), 497
Salt secretion, 323, 325
Sand barrens, 268
Sandpiper (Scolopacidae), 294
San Francisco Bay, 66
Saprists, 167
Sarracenia purpurea (pitcher plant), 432, 433
Saturated, 115
Sawgrass (*Cladium jamaicense*), 57, 602
Scalia, Antonin, 520
Schoenoplectus tabernaemontani (soft-stemmed bulrush):
  in colonized wetlands, 634
  nutrient limitations, 365
Scirpus americanus (American bulrush), 67, 68
Sclerophylly, 432
Sea-level rise, mangrove swamp response to, 330
Seasonally flooded, 115
Seawater, chemical characteristics of, 205
Seca, 76
Secondary treatment, 650
Sedge (*Carex, Cladium*), 147
Sedge meadow, 33, 35
Seed banks, 234–235, 627
  freshwater marshes, 350–351
  tidal freshwater wetlands, 289–290
Seed germination and dispersal, 394–395
Seep wetland, 135
Seiches, 142–144
Seidel, Käthe, 648
Self-design, 245–249

Self-design wetlands, 591, 609, 612, 613, 618, 633, 634, 637–638
Self-organization, 245
Semianadromous species, 292
Semiarid riparian forests, 390
Seminole, 6–8
Semipermanently flooded, 115
Services, 527
Sesarma, 326
Shallow roots, 395
Shallow root systems, 220
Shelford, Victor, 230
Shelford curve, 198
Shellfish, 534–535
Shrub–scrub, 35
Significant nexus, 519, 520
Sinks, 181, 182
Site selection for wetland creation, 625
Skoagon Chippewa, 6
Slough, 35, 381, 382
Slow-flowing cypress strands, 378, 381
Snail (*Littorina irrorata*), 284
Snow goose, 67
Snow goose (*Chen*), 114, 682
Soft-stemmed bulrush (*Schoenoplectus tabernaemontani*):
  in colonized wetlands, 634
  nutrient limitations, 365
Soil(s), 161–176
  differences between organic and mineral soils, 163–165
  equic condition in, 169–170
  hydric, 170
  in mangrove swamps, 319
  mineral, 162–165, 167–171
  muck, 163
  organic, 161–167
  oxidized rhizosphere in, 168, 169
  peat, 162–163
  rate of formation, 168
  redox concentrations in, 168, 169
  reduced matrices and redox depletions, 168
  reduction/oxidation in, 171–176
  in restoring/creating wetland, 627–630
  salt marsh, 269
  tidal salt marshes, 269
  for treatment wetlands, 676–679
  in treatment wetlands design, 678–679
  types of, 161
Soil acidity, 319
Soil chemistry, 678–679
Soil color charts, 171. *See also* Munsell soil color chart
Soil profiles, 234
Solid Waste Agency of Northern Cook County (SWANCC) v. U.S. Army Corps of Engineers, 519
Soligenous peatland, 423, 424

Sources:
  of chemicals, 181, 182
  of nutrients, 181, 182
South American wetlands, 56, 73–78
  Amazon, 77–78
  Llanos, 75–76
  Orinoco River delta, 74–75
  Palo Verde National Park, 73–74
  Pantanal, 76–77
  tidal marshes, 265
South Atlantic coast, swamp rivers of, 60–61
Southeastern Europe inland deltas, 82, 83
Southeastern U.S. bottomland forests, 389–390
Southern Asia river deltas, 93–95
Sparrow, 294
Spartina (salt marsh cordgrass):
  and flow-through, 146
  in inland marshes, 280
Spartina alterniflora:
  anaerobic respiration, 225
  for coastal marsh restoration, 607
  in inland marshes, 280
  and marsh restoration, 630
  as pioneer species, 235
  in restored wetlands, 635, 639
  rhizosphere oxygenation, 223
  root photomicrograph, 218
  in salt marsh ecosystem, 270, 273
  and salt marsh restoration, 613, 615
  sulfur tolerance, 224
  tidal range and, 150–151
Spartina anglica, 630
Spartina patens (salt marsh hay), 497
Spartina townsendii, 630
Species diversity, 350–353
Species richness, 145–146
Sphagnum, 11
  and acidification, 432
  and bog acidity, 426
  cation capacity, 165
  decomposition, 439
  in northern peatlands, 166
  and organic soil density, 163
  as peat-building plant, 427–428
  and peatlands restoration, 606, 607
  primary productivity, 436–438
  and quaking bog succession, 417
  waterlogging tolerance, 431
Sphagnum–black spruce–beatherleaf bog forest, 429
Sphagnum bog, 29
Sphagnum–leatherleaf–*Kalmia*–spruce heath, 429
Sphagnum magellanicum, 437, 438
Spit, 266
Sponges, 227
Spring tides, 142

Spring wetland, 135
SRP (soluble reactive phosphorus), 203
Standing stocks, 208
Stem elongation, 220
Stemflow, 127
Stem hypertrophy, 218–219
Storm abatement, 540–543
Stormwater protection, 499
Strand, 35
Streams, 206–207
Streamflow, 128, 131–132, 206, 207
Streamside marshes, 266
String bogs, 420–421
String fen, 421
Study techniques (hydrology), 153–155
Subclass, 463
Subsistence as cultural ecosystem service, 545
Subsurface-flow constructed wetlands, 648, 649, 679
Subsystems, 461, 462
Subtidal, 115
Subtidal subsystem, 461, 462
Succession, 227–251
 allogenic vs. autogenic, 227–235
 centrifugal organization concept, 238–240
 community concept and continuum idea, 232–234
 ecosystem development, 240–251
 environmental sieve model, 236–238
 Lake Michigan dunes, 230–232
 landscape patterns, 250–251
 linear directed change, 234
 models of, 235–240
 plant species functional groups, 235–236
 seed banks, 234–235
 strategy of, 242–249
Sulfate reduction, 194
Sulfide avoidance, 224
Sulfide oxidation, 194–195
Sulfide toxicity, 195
Sulfur–carbon interactions, 201–202
Sulfur cycle, 193–195
 sulfate reduction, 194
 sulfide oxidation, 194–195
 sulfide toxicity, 195
Sundarbans (India and Bangladesh), 94–95
Supreme Court decisions, *see* U.S. Supreme Court decisions
Surface flow, 127–133
 channelized streamflow, 130–132
 floods and riparian wetlands, 132–134
 watersheds and runoff, 127–130
Surface-flow constructed wetlands, 648, 649, 679
Surface inflow, 128
Surface water depression wetland, 135, 136

Surface water slope wetland, 136–137
Sustainable cultures in wetlands, 5–8
Suwannee River, 58
Swallows (Hirundinidae), 294
Swamps, 33, 35, 289, 424. *See also specific types, e.g.*: Mangrove swamps
"Swampbuster" provisions (1985 Food Security Act), 509
Swamp gas (marsh gas), 197
Swamp Land Acts, 480–481, 504
Swamp rivers of South Atlantic coast, 60–61
SWANCC (Solid Waste Agency of Northern Cook County) v. U.S. Army Corps of Engineers, 519
Systems, 460

Taiwan urban wetland park, 99, 100
"Takings" issue, 518
Tamarack (*Larix laricina*), 430
Taxodium (cypress), 57
 adaptations to waterlogging/flooding, 219
 and buttresses, 394
 longevity, 395
Taxodium distichum (bald cypress), 220, 374, 375, 387
Taxodium distichum var. *imbricarium* (pond cypress), 374, 375, 387
Taxodium distichum var. *mexicanum* Gordon (Montezuma cypress), 374
Teal, John, 281, 283
Telmatology, 17
Temporarily flooded, 115
Terrace, 383
Terrestrialization, 416
Tertiary treatment, 668
Thermo-osmosis, 395
Thiobacillus, 194
Thoreau's Bog (Massachusetts), 440, 442, 443
Thornthwaite equation, 139–141
Threatened species, 536–538
Throughfall, 127, 154
Tidal creeks, 267, 268
Tidal freshwater marsh, 35
Tidal freshwater wetlands, 285–301
 amphibians and reptiles, 294–295
 birds, 293–294
 consumers in, 290–295
 ecosystem function, 295–301
 energy flow, 297–298
 floating marshes, 288
 mammals, 295
 marsh vegetation, 285, 287–288
 nekton, 291–293
 new marshes, 288
 nutrient budgets, 301

 organic import and export, 298–300
 primary productivity, 295–297
 seed banks, 289–290
 swamps, 289
 vegetation, 285–290
Tidal marshes, 259–301. *See also* Tidal salt marshes
 consumers in freshwater wetlands, 290–295
 ecosystem function of freshwater wetlands, 295–301
 estimated area of, 262
 freshwater wetlands, 285–301
 salinity, 269–270
 soil, 269
 vegetation, 285–290
Tidal salt marshes, 261–285
 aerial habitat, 274–275
 aquatic habitat, 276
 Arctic, 262
 Australasia, 265
 benthic habitat, 276
 consumers, 274–277
 decomposition and consumption, 281–283
 die-off of salt marshes, 284–285
 eastern North America, 264
 ecosystem function, 277–286
 geographic extent of, 262–265
 hydrogeomorphology of, 265–268
 hydrology, 266
 mammals, 276–277
 marsh development, 266–267
 Mediterranean, 264
 northern Europe, 262, 264
 organic export, 283–284
 pannes, 267, 268
 primary productivity, 277–281
 salinity, 269–270
 soil, 269
 South America, 265
 tidal creeks, 267, 268
 tropics, 265
 vegetation, 270–274
 western North America, 264–265
Tidal subsystem, 461, 462
Tides, 142, 143
Tigris–Euphrates Basin, 89
Tigris River, 89
Timber, 535–536
Timber industry, 59–60
Todania ignis, 227
Tomatoes, 218
Topogenous development, 419
Topogenous peatlands, 423, 424
Transformers (of chemicals), 181, 182
Transitional peatlands, 422
Translocation, 208
Transportation, 487

Treatment wetlands, 647–692
　agricultural stormwater treatment wetlands, 659–662
　agricultural wastewater wetlands, 662–663
　attracting wildlife after construction, 683
　basin morphology, 670–671
　capital costs, 688–689
　chemical loadings, 671–676
　classification of, 648–668
　comparison of wetlands with conventional technology, 690–692
　defined, 593
　design of, 668–682
　economics and values of, 688–691
　greenhouse gas emissions, 687–688
　hydrology, 668–670
　landfill leachate wetlands, 668
　management after construction, 682–691
　mine drainage wetlands, 653–655
　mosquito control, 683–686
　municipal wastewater wetlands, 650–653
　operating and maintenance costs, 689
　pathogens, 684, 687
　river diversion wetlands, 663–668
　soils, 676–679
　urban stormwater treatment wetlands, 655–659
　vegetation, 679–682
　water-level management, 687
　wildlife control, 682
Tropics, salt marshes of, 265
Tundra, 578
Tupelo/gum swamps, 373–374
Turlough, 35
Turnove rate of water, 125
Turnover time, 125
Typha (cattail), 73, 74
　adaptations to waterlogging/flooding, 221–222
　agricultural runoff and, 491
　in centrifugal organization model, 239
　in colonized wetlands, 635
　in Everglades, 601, 602
　and flow-through, 146
　freshwater marsh zonation, 349
　and nutrient enrichment, 659–660
　nutrient limitations, 365
　photosynthetic efficiency, 359–360
　in restored wetlands, 635
　and upland soils, 627–630
　and wetland primary succession, 246–249
Typha domingensis, 73, 74, 221–222, 491, 602, 659–660

Uca (fiddler crab), 326
Undesirable species, introducing, 634–635
U.S. Army Corps of Engineers:
　Louisiana Delta conservation efforts, 66
　wetland delineation, 509–516
　wetlands defined by, 39–40
　wetlands definition, 41, 512
U.S. Department of Agriculture, 40, 481
U.S. Fish and Wildlife Service:
　Circular 39 classification, 457–459
　Habitat Evaluation Procedure, 546–547
　peatland classifications, 457
　wetlands defined by, 36–37
　wetlands definition, 42
U.S. Fish and Wildlife Service Waterfowl Protection Area, 62
U.S. National Academy of Science, 38
U.S. Supreme Court decisions, 519–521
U.S. wetlands, 56
　Classification of Wetlands and Deepwater Habitats, 460–464, 466
　Clean Water Act, 507–509, 518
　coastal wetlands, 260–262
　Coastal Zone Management Program, 518
　drainage history, 480–482
　early Executive Orders, 504, 506
　Emergency Wetlands Resource Act, 518
　estimates of, at different times, 51
　estimates of changes in, 51
　Florida Everglades, 55–58
　Great Black Swamp, 63–64
　Great Dismal Swamp, 59–60
　Great Kankakee marsh, 63
　legal protections, 504–521
　loss of wetlands, 51, 52
　Louisiana Delta, 64–66
　major federal laws, directives, and regulations, 505–506
　mangrove swamps, 313–314
　National Academy of Science studies, 516–517
　National Flood Insurance Program, 518
　National Wetlands Inventory, 472–474
　Nebraska sandhills and Great Plains playas, 62–63
　no net loss concept, 507, 591, 593, 595
　North American Wetlands Conservation Act, 518
　Okefenokee Swamp, 58–59
　Pocosins (Carolinas), 59
　prairie potholes, 61–62
　San Francisco Bay, 66
　"swampbuster" provisions of 1985 Food Security Act, 509
　swamp rivers of South Atlantic coast, 60–61
　"takings" issue, 518
　tidal freshwater wetlands, 285
　U.S/ Supreme Court decisions, 519–521
　wetland delineation guidelines, 509–516

United States v. Holland, 507
United States v. Riverside Bayview Homes, Inc., 40, 508
Upper perennial subsystem, 461, 462
Urban stormwater treatment wetlands, 655–659
Urban wetland parks:
　in China, 98–99
　in Taiwan, 99, 100

Vallisneria gigantea, 199
Valuing ecosystem goods and services, 556–558
Várzea, 35, 77
Vascular plant adaptations, 216–227
　adventitious roots, 218, 219
　aerenchyma, 216–218
　anaerobic respiration, 224–226
　lower water uptake, 224
　morphological, 216–220
　mutualism and commensalism, 226–227
　physiological, 220–226
　pneumatophores, 220
　pressurized gas flow, 221–223
　rhizosphere oxygenation, 223–224
　stem elongation, root adaptations, and lenticels, 220
　stem hypertrophy, 218, 219
　sulfide avoidance, 224
　whole plant strategies, 226
Vazante, 76
Vegetation, 215–227
　allogenic vs. autogenic succession, 227–235
　cypress swamps, 386–388
　ebullition and gaseous transport in, 200–201
　ecosystem development, 240–251
　and evapotranspiration, 141–142
　freshwater marshes, 348–354
　freshwater swamps, 385–395
　in glaciated regions, 390–301
　harvesting as provisioning ecosystem service, 535–536
　hydrology and composition of, 145–146
　introducing, 630–635
　mangrove swamps, 319–326
　models of succession, 235–240
　morphological adaptations, 216–220
　mutualism and commensalism, 226–227
　peatlands, 427–433
　physiological adaptations, 220–226
　red maple swamps, 388–389
　riparian ecosystems, 389–390
　succession, 227–251
　tidal freshwater wetlands, 285–290
　tidal salt marshes, 270–274
　treatment wetlands classification by, 650
　in treatment wetlands design, 679–682

tree adaptations, 391–395
vascular plant adaptations, 216–227
white cedar swamps, 388
Vernal pool, 35, 117
Vietnam, 593, 616–617
Viviparous seedlings, 226, 325–326
Vivipary, 226
Vleis, 35
Volatilization, 183
Vulture (Cathartidae), 294

Wadden Sea, 81
Wad/wadden, 35
Washington State, 469
Wastewater treatment wetland classifications:
  agricultural stormwater treatment wetlands, 659–662
  agricultural wastewater wetlands, 662–663
  general approaches to, 648–650
  landfill leachate wetlands, 668
  mine drainage wetlands, 653–655
  municipal wastewater wetlands, 650–653
  river diversion wetlands, 663–668
  urban stormwater treatment wetlands, 655–659
  by vegetation, 650
Water budget, 119–125, 154, 380
  examples of, 122–125
  major components of, 122
Water extraction, 490–491
Waterfowl:
  as provisioning ecosystem service, 531–534
  wetland management by objective for, 493–496
Water-level management, 687
Water lily (*Nuphar lutea*), 221
Waterlogging, 431
Water lotus (*Nelumbo nucifera*), 82
Water pollution:
  alteration of wetlands by, 491, 599, 602
  and mine drainage wetlands, 653, 655
  and wastewater treatment wetlands, 655, 659, 662–663, 667, 673, 691, 692
Water quality, 204–207
  and climate regulation, 544–545
  oceans and estuaries, 205
  streams, rivers, and groundwater, 206–207
  wetland management by objective for enhancing, 499
  wetlands created to improve, *see* Treatment wetlands
Water quality restoration, 598–604
  Florida Everglades, 598–603
  Mesopotamian Marshlands, 603–604

Watersheds, 127–130
Water snake (*Nerodia*), 295
Water tupelo (*Nyssa aquatica*), 374
Water uptake, adapting, 224
Weirs, 131
West Africa mangrove swamps, 87–89
West Bengal, 94
Western Australia wetlands, 90–91
Western North America:
  salt marshes, 264–265
  tidal salt marshes, 264–265
Western Siberian lowlands, 93
Wetlands, 3–22. *See also* U.S. wetlands
  of Africa, 56, 83–89
  of Asia, 56, 93–100
  of Australia, 56, 89–91
  of Central America, 56, 72–74
  common terms used to describe, 33–36
  conservation of, 16–17
  defining, 28–32
  distinguishing features of, 28–31
  and ecotourism, 13–15
  of Europe, 56, 78–83
  food from, 9–10
  formal definitions of, 36
  global extent of, 45–48
  historical human impact on, 4–7
  legal definitions of, 39–42
  literary references to, 79
  management of, 21–22
  of Middle East, 56, 89
  of New Zealand, 56, 91–92
  of North America, 55–72. *See also* North American wetlands
  peat and building materials from, 10–14
  regional, 55, 56
  of South America, 56, 73–78
  specialization in study of, 17–21
  sustainable cultures in, 5–8
  as term, 27–28
  web pages related to, 705–707
  worldwide losses of, 48–50
Wetland classification, 455–474
  Canadian Wetland classification System, 465, 466
  Circular 39 classification, 457–459
  coastal wetland classification, 459–460
  hydrogeomorphic wetland classification, 465, 467–468
  international Ramsar Convention classification system, 465, 466
  peatland classifications, 457
  for rating wetlands, 469–471
  reasons for, 456–457
  U.S. *Classification of Wetlands and Deepwater Habitats*, 460–464, 466
  for wetland inventory, 471–474

Wetland creation, 592
Wetland delineation guidelines (U.S.), 509–516
Wetland ecologists, 17
Wetland ecology, 17
Wetland enhancement, 592
Wetlanders, 5
Wetland function:
  accumulation of organic material, 151–153
  hydrology and, 143–153
  nutrient cycling and availability, 153
  and primary productivity, 146–151
  vegetation composition and species richness, 145–146
Wetland inventory, 455, 471–474
  remote-sensing imagery, 472
  remote-sensing platform, 471–472
  U.S. National Wetlands Inventory, 472–474
Wetland management, 21–22, 477–499
  for agriculture and aquaculture, 497–499
  alteration of wetlands, 482–492
  drainage, dredging, and filling, 482–486
  early history of, 478–480
  for flood control and stormwater protection, 499
  and highway construction, 487–488
  hydrologic modifications, 486–487
  mineral and water extraction, 490–491
  by objective, 491–499
  peat mining, 488–489
  U.S. wetland drainage history, 480–482
  for waterfowl and wildlife, 493–496
  and water pollution, 491
  for water quality enhancement, 499
Wetland managers, 21–22
Wetland restoration, 592. *See also* Restoration and creation of wetlands
Wetland science, 17–21
Wetland scientists, 17
Wetlands of international importance, 17
Wetlands Reserve Program, 597
Wet meadow, 33, 35
Wet prairie, 33, 35
Whangamarino Wetland (North Island, New Zealand), 90
White cedar (*Chamaecyparis thyoldes*), 375
White cedar swamps:
  geographic extent, 375–376
  geomorphology, 381
  vegetation, 388
White-faced heron (*Ardea novaehollandiae*), 533
White mangrove (*Laguncularia*):
  in basin mangroves, 317
  salinity control, 323
White-necked heron (*Ardea cocoi*), 533

White pine, *see* Kahikatea
Wigeon grass (*Ruppia*), 268
Wildlife:
    at treatment wetlands, 682–683
    wetland management by objective for, 493–496
Wild rice, 7
Wild rice (*Zizania*), 68, 166, 497
Willingness-to-pay, 551
Willow (*Salix*), 219
Wilma H. Schiermeier Olentangy River Wetland Research Park (Ohio State University), 246

Wilson, L. R., 228
Wind tide, 143. *See also* Seiches
Wisconsin, 6
Wren (Troglodytidae), 294
Wurzelraumentsorgung (root-zone method), 649–650, 677

Xixi National Wetland Park (Hangzhou, China), 98, 99

Yangtze Delta, China, 95
Yangtze River wetlands, 95, 97

Yellow-billed duck (*Anas undulata*), 88
Yellow-billed stork (*Ibis ibis*), 87, 88

Zabel v. Tabb, 40
Zizania (wild rice), 7, 9, 68, 166, 497
Zizania aquatica, 7, 497
Zizania latifolia, 9
Zizania palustris, 68
Zonation:
    in freshwater marshes, 348–349
    in mangrove swamps, 320–322
Zooplankton, 298